Building
Engineering
and
Systems
Design

Building Engineering and Systems Design

Second Edition

Frederick S. Merritt
Consulting Engineer, West Palm Beach, Florida

and

James Ambrose
University of Southern California

 VAN NOSTRAND REINHOLD
New York

Library of Congress Catalog Card Number 89-14641

ISBN 0-442-20668-2

Printed in the United States of America

Van Nostrand Reinhold
115 Fifth Avenue
New York, New York 10003

Van Nostrand Reinhold International Company Limited
11 New Fetter Lane
London EC4P 4EE, England

Van Nostrand Reinhold
480 La Trobe Street
Melbourne, Victoria 3000, Australia

Nelson Canada
1120 Birchmount Road
Scarborough, Ontario M1K 5G4, Canada

16   15   14   13   12   11   10   9   8   7   6   5   4   3   2

Library of Congress Cataloging-in-Publication Data

Merritt, Frederick S.
  Building engineering and systems design / Frederick S. Merritt and
James Ambrose.—2nd ed.
     p.   cm.
  Includes bibliographies and index.
  ISBN 0-442-20668-2
  1. Building. 2. Systems engineering.   I. Ambrose, James E.
II. Title.
TH846.M47   1989
690—dc20   89-14641
                                                  CIP

Preface to Second Edition

This edition is based upon a firm conviction of the authors that the purpose of, and the need for the book, as described in the Preface to the First Edition, are as critical today as they were when the first edition was prepared. In fact, now, there is a greater need for applications of systems design to buildings. This need occurs because of rising construction costs, greater demand for more and improved building services, and better quality control of construction. In brief, this book explains what needs to be designed, and the issues to be addressed in the design process.

Revisions of the first edition have been aimed at refining the text and developing new topics which have emerged during the past decade. Increased attention is given to the involvement of architects in systems design, and to the inclusion of architectural goals and objectives in the value systems for optimized design. Traditionally, architects have been the only members of the building team whose formal training has included some work in all the major areas of building design. College courses in structures, plumbing, lighting, electrical power, mechanical systems, and building services in general, have, in the past, been included in most architectural education curricula. What is new is the tendency for architects to work directly and interactively with engineers, contractors, and other specialists during design development. This is facilitated by the use of shared computer-stored data and interactive computer-aided design processes.

While architects have traditionally been broadly educated for building design, engineers usually have not been so rounded in their education. One of the most valuable uses for this book is as a general education in the building design and construction process for those members of the building design team who did not experience comprehensive architectural training. This education is hard to obtain but of increasing importance as interactive design becomes more common.

To make the book more suitable for use in self study, the bibliographies and study materials have been arranged by chapter section, rather than by chapter, as in the first edition. Thus, study units are smaller and easier to handle for persons with limited study time. Chapter summaries have also been provided.

Learning any technology requires familiarity with a large new vocabulary. Many technical terms are defined and explained in this book, but a glossary would be too large for inclusion in the book. However this edition contains compilations, at the ends of most chapter sections, of terms used in those sections. The use of these lists will permit readers to develop a considerable technical vocabulary, by using the book index to find the definitions and explanations in the text. It would be advantageous, however, for readers to obtain at least one dictionary of building terminology.

While both authors of this edition have diverse backgrounds in education, writing, and management, our major focus in this work is on the needs of the building designer. That interest was the principal guide in the development of the text and in the general selection and emphasis of topics. In total, what we want to achieve are better buildings, and our major intention is to assist those persons who work in this field to accomplish that end.

FREDERICK S. MERRITT
JAMES AMBROSE

v

Preface to First Edition

As a consequence of technological, economic and sociological changes throughout the civilized world, new buildings are becoming ever more complex and costly; however, the public is demanding better buildings at less cost. To meet this challenge, building designers and constructors must improve their skills and develop better building methods. This book was written to help them.

Fundamentally, the book is a compendium of the best of the current building-engineering practices. It describes building materials, building components, types of construction, design procedures and construction methods that have been recommended by experts, and it covers nearly all disciplines. It presents the basics of building planning, structural engineering, fire safety, plumbing, air conditioning, lighting, acoustics, electrical engineering, escalator and elevator installation and many other technical skills needed in building design.

But if the challenge of constructing better buildings at less cost is to be met, future designers and builders will need more than just technical information. They will have to be more creative and ingenious in applying this information. In addition, they will have to organize more efficiently for design and construction and manage the design and construction processes in a more expert manner. The book also is intended to help meet these goals.

For the reasons cited above, a new concept of building design and construction is needed. Such a concept is the main theme of the book.

The concept requires that designers treat buildings as systems and apply techniques of operations research (more commonly known as systems design) to their design. Systems design employs the scientific method to obtain an optimum, or best, system and calls for an interdisciplinary approach to design. The techniques involved have been successfully used in machine design, but it was necessary, here, to adapt them to building design. However, the adaptation is accomplished in a way that will enable professionals accustomed to traditional procedures to convert easily to the new techniques and will also permit students who learn systems design from this book to fit readily into traditional organizations, if necessary for their employment.

The interdisciplinary approach to design advocated in the book requires that design be executed by a team, the building team. It consists of consultants specializing in various aspects of design and construction and also should include future users of the proposed building along with experienced building operators or managers.

For the team to function effectively, i.e. for intelligent participation in decision making, each member of the team, in addition to contributing his or her own special knowledge, skills and experience to the team effort, should also be acquainted with the duties, responsibilities and output of the other members of the team. In particular, the team leader should be more knowledgeable on all aspects of building design and construction, to lead, guide and coordinate the team. An important objective of this book, therefore, is to educate potential members of the building team for the roles they will have to play and to prepare professionals for leadership of the team.

For practical reasons, the book is restricted to presentation only of pertinent topics that the author considers basic and important. The treatment should be sufficient to provide a foundation on which the reader can build by additional reading and on-the-job experience. To assist toward this end, each chapter in the book concludes with a list of books for supplementary reading.

The book has been designed for use in either of two ways:

1. as a textbook in an introductory course for architecture, building engineering or construction management;
2. as a home-study book for professional building designers and builders who wish to learn how to use systems design in their work.

The book assumes that, at the outset, the reader has a knowledge of buildings, physics and mathematics comparable to that of a high-school graduate. Based on this assumption, the book describes building components, explains their functions and indicates how they are assembled to form a building. While these introductory discussions will be familiar to building professionals, they should find the review worthwhile as an introduction to the new design concept.

In preparation of this book, the author drew information and illustrative material from sources too numerous to list. He is indeed grateful to all who contributed and, where feasible, has given credit elsewhere in this book.

FREDERICK S. MERRITT

Contents

Chapter 1

New Directions in Building Design

Building construction is essential to the economy of nations. If building construction declines, the economy suffers. Buildings also are essential to the economic well-being of architects, engineers and contractors who engage in building design and construction. If potential clients do not wish to build, these professionals do not work. Thus, there are both personal and patriotic incentives for building designers and constructors to encourage building construction.

A potential client considers many things before deciding to proceed with a building project. But there are two conditions—cost and time—that when violated are almost certain to preclude construction of a project. If the proposed building will cost too much, it will not get built; if the proposed building will not be ready for occupancy when the owner wants it, the project will be canceled. Building designers and constructors know this and try to produce buildings that will meet the owners' budgets and schedules. (Sponsors of building projects are called *owners* in this book.)

Despite these efforts, many buildings that are needed do not get built because they would cost too much. The cost of construction, maintenance and operation exceeds what owners are willing to pay. As a result, some families that need housing have none. Some families have to live in substandard housing because they cannot afford decent accommodations. Schools may be inadequate and hospitals may be unavailable.

In addition to preventing construction of needed buildings, high building costs have other adverse effects. The costs of expensive buildings are passed along to users of the buildings or to purchasers of products manufactured in the buildings, and ultimately, as a consequence, the consumer pays higher prices. Despite this undesirable situation, building costs keep rising.

There are several reasons, beyond the control of building designers, for the continuous increase. One is inflation, a steady decrease in the purchasing power of money. Another consists of legal and social pressure for pleasant, healthy and safe living and working conditions in buildings. Still another is the result of technological changes that make it possible to do things in and with buildings that could not be done previously. Consider, for example, the change of status over time—from luxury, to occasional use, then to frequent use, and finally necessity—of items such as indoor plumbing, telephones, hot water, and air conditioning. All of these changes have made buildings more complex and more costly.

Consequently, the traditional efforts of building designers to control costs only for the purpose of meeting a construction cost within the owner's budget are no longer adequate. Designers must go further and bring down costs over the life of the building, including costs for construction, maintenance, and operation.

There is evidence, however, that traditional design methods have limited capability of decreasing costs, let alone any hope of halting their steady increase. Designers must find new ways of reducing the costs of constructing and using buildings.

One technique that shows great promise is systems design. It has been used successfully for other types of design, such as machine design, and can be adapted to building design. Systems design consists of a rational orderly series of steps that leads to the best decision for a given set of conditions. It is a general method and therefore is applicable to all sizes and types of buildings. When properly executed, systems design enables designers to obtain a clear understanding of the requirements for a proposed building and can help owners and designers evaluate proposed designs and select the best, or *optimum*, design. In addition, systems design provides a common basis of understanding and promotes cooperation between the specialists in various aspects of building design.

A major purpose of this book is to show how to apply systems design to buildings. In this book, systems design is treated as an integration of operations research and value analysis, or value engineering. In the adaptation of systems design to buildings, the author has tried to retain as much of traditional design and construction procedures as possible. Departures from the traditional methods of design, as described in this book, should not appear radical to experienced designers, because they are likely to have used some of the procedures before. Nevertheless, the modifications, incorporated in an orderly precise process, represent significant improvements over traditional methods, which rely heavily on intuitive conclusions.

Later in this chapter, the systems design approach to buildings is discussed. Also, this chapter examines the changing role of building designers with increasing complexity of buildings and indicates how they should organize for effective execution of systems design.

1.1. CHANGE FROM MASTER BUILDERS TO MANAGERS

The concepts of building design have changed with time, as have the roles of building designers and constructors along with the methods employed by them. These changes are still occurring, as building design moves in new directions.

Buildings that have survived through the ages testify to the ability of the ancients to construct beautiful and well-built structures. What they knew about building they learned from experience, which can be an excellent teacher.

Art and Empiricism

Until the 19th Century, buildings were simple structures. Nearly all of them might be considered to be merely shells compartmentalized into rooms, with decorations. Buildings primarily provided shelter from the weather and preferably were also required to be visually pleasing. Exterior walls were provided with openings or windows for light and ventilation. Candles or oil lamps were used for artificial illumination. Fireplaces for burning wood or coal were provided in rooms for heating. Generally, there was no indoor plumbing. Since stairs or ramps were the only available means of traveling from level to level, buildings generally did not exceed five stories in height. Floor and roof spans were short; that is, floors and roofs had to be supported at close intervals.

Design of such simple structures could be and was mastered by individuals. In fact, it was not unusual for designers also to be experts on construction and to do the building. These designers-builders came to be known as *master builders*.

To assist them, the master builders sought out and hired men skilled in handling wood and laying brick and stone in mortar. These craftsmen established the foundation on which the later subdivision of labor into trades was based.

Building design, as practiced by master builders, was principally an art. Wherever feasible, they duplicated parts of buildings they knew from experience would be strong enough. When they were required to go beyond their past experience, they used their judgment. If the advance succeeded, they would use the same dimensions under similar circumstances in the future. If a part failed, they would rebuild it with larger dimensions.

Early Specialization

By the 19th Century, however, buildings had become more sophisticated. Soaring costs of land in city centers brought about economic pressure for taller buildings. Factories and public buildings, such as railroad terminals, created a demand for large open spaces, which required longer floor and roof spans. More became known about building materials, and scientific methods could be applied in building design. Owners then found it expedient to separate the building process into two parts—design and construction—each executed by a specialist.

Building design was assigned to an *architect*. This professional was said to practice *architecture*, the art and science of building design. Construction was assigned to a *contractor*, who took full charge of transforming the architect's ideas into the desired building. The contractor hired craftsmen and supplied the necessary equipment and materials for constructing the whole building.

Basic Principles of Architecture

Basically, however, architecture has not changed greatly from ancient times. The Roman, Vitruvius, about 2,000 years earlier, had indicated that architecture was based on three factors: "convenience, strength and beauty." In the 17th Century the English writer, Sir Henry Wotton, referred to these as "commoditie, firmeness and delight." Thus:

1. A building must be constructed to serve a purpose.
2. The building must be capable of withstanding the elements and normal usage for a reasonable period of time.
3. The building, inside and out, must be visually pleasing.

Advent of the Skyscraper

In the middle of the 19th Century, a technological innovation marked the beginning of a radical change in architecture. Traveling from level to level in buildings by means of stairs had limited building heights, despite the economic pressures for taller buildings. Some buildings used hoists for moving goods from level to level, but they were not considered safe enough for people; if the hoisting ropes were to break, the platform carrying the people would fall to the

bottom of the hoistway. The fear of falling, however, was largely alleviated after E. G. Otis demonstrated in 1853 a safety brake he had invented. Within three years, a building with a passenger elevator equipped with the brake was constructed in New York City. Considerable improvements in elevator design followed; use of elevators spread. Under economic pressure to make more profitable use of central city land, buildings became taller and taller.

At this stage, however, building heights began to run up against structural limitations. In most buildings, floors and roof were supported on the walls, a type of construction known as *bearing-wall construction*. With this type, the taller a building, the thicker the walls had to be made (see Fig. 1.1). The walls of some highrise buildings became so thick at the base that

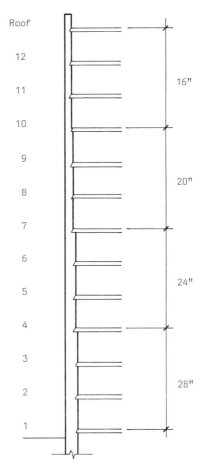

Fig. 1.1. Required thicknesses for brick bearing walls for a 12-story building. Building Code of the City of Chicago, 1928.

architects considered it impractical to make buildings any taller.

Then, another technological innovation eased the structural limitation on building height and permitted the radical change in architecture to continue. In 1885, architect W. L. Jenney took the first major step toward *skeleton framing* for high-rise buildings. (In skeleton framing, floors and roof are supported at relatively large intervals on strong, slender vertical members, called columns, rather than at short intervals on thick, wide masonry piers.) In the 10-story Home Insurance Building in Chicago, Jenney set cast iron columns, or posts, in the load-bearing masonry piers to support wrought-iron beams that carried the floors. (Also, in that year, another relevant event occurred. The first structural steel beams were rolled.) Two years later, architects Holabird & Roche took another step toward skeleton framing. By supporting floor-beams on cast-iron columns along the two street frontages of the 12-story Tacoma Building in Chicago, the architects eliminated masonry bearing walls on those two sides.

Cast-iron columns, however, have relatively low strength. Their continued use would have substantially limited building heights. Steel columns proved to be a stronger, more economical alternative. In 1889, the 10-story Rand McNally Building, designed by Burnham & Root, was constructed in Chicago with steel columns throughout. This set the stage for the final step to complete skeleton framing, with floors and roof carried on steel beams, in turn resting on steel columns (see Fig. 1.2). Thick walls were no longer necessary.

The full possibilities of skeleton framing was demonstrated in 1892 when it was used for the 21-story, 273-ft-high Masonic Temple in Chicago. Skeleton framing was then adopted in New York and other cities.

Meanwhile, development of reinforced concrete, a competitor of structural steel began. In 1893, construction of a concrete-framed museum building at Stanford University, Palo Alto, Calif., demonstrated the practicability of monolithic concrete construction. Ten years later, the first skyscraper with concrete framing, the

Fig. 1.2. Structural steel skeleton framing for a multistory building.

16-story Ingalls Building in Cincinnati, was completed.

Effects of Skyscrapers on Architecture

The trend to the skyscraper, which accelerated in the 20th Century, had several marked effects on architecture and its practice. For one thing, the external appearance of buildings underwent a radical change. Large expanses of masonry with small openings for windows (see Fig. 1.3) gave way to large glass windows with relatively small amounts of wall between them (see Fig. 1.4). Another effect was that use of skeleton framing developed a need for specialists capable of designing framing for safety and economy. Architects hired *structural engineers* for this purpose or retained consulting engineering firms. Still another effect was that indoor plumbing became essential. Pipes and fixtures had to be provided for water supply, waste disposal and gas for heating, cooking and illumination. In addition, central heating, with warm air, hot water or steam distributed throughout a building from a furnace in the basement, became a necessity. A need for specialists capable of designing plumbing and heating systems and elevators developed. To meet this need, architects hired *mechanical engineers* or retained consulting engineering firms. Thus, *building engineering* was incorporated in architecture.

At the same time, construction became more complex. In addition to masons, bricklayers and carpenters, contractors now needed to hire ironworkers, plumbers, window installers and

Fig. 1.3. Late 19th Century building, still expressing the basic forms of load-bearing wall construction, although its basic structure is steel framed. Auditorium Hotel, Chicago, by Adler and Sullivan. 80 years later, the worlds tallest steel frame structure, the Sears Tower, looms over it, clearly expressing the frame structure.

Fig. 1.4. Sears Tower (1974), Chicago, rises 110 stories, 1454 ft. Steel skeleton frame with bundled tube system for lateral load resistance.

elevator installers. Soon, companies were formed to offer such services to contractors. Thus, a building owner contracted construction of a building to a *general contractor*, who then subcontracted specialty work to *subcontractors*.

Humanization of Architecture

Advances in technology usually do not occur without mishaps. Floors, roofs and walls sometimes collapsed because of poor materials or workmanship, or sometimes because floor spans or wall heights were extended beyond the capabilities at the time. Also, many lives were lost in building fires. To prevent such mishaps, municipal authorities promulgated *building codes*, which established by law minimum design standards. Such codes contained provisions for minimum loads for structural design, minimum strength for materials, minimum thickness of walls, fire protection of structural components and emergency exits in case of fire. In the interests of health, regulations were incorporated governing plumbing installations and ventilation.

When electricity came into widespread use in buildings during the 20th Century, building codes incorporated provisions governing electrical installations. Specialists were needed to design such installations, so architects hired *electrical engineers* or retained consulting engineering firms. Similarly, general contractors subcontracted electrical work to electrical subcontractors.

During the 19th and 20th Centuries, industry developed rapidly. More and more factories were built, and more and more people were hired for manufacturing. Concern for the health and safety of these people led to establishment of government Labor Departments, which established regulations for employee conditions, many of which affected building design.

Concern for welfare, as well as health and safety, of people was demonstrated in the early part of the 20th Century, when municipal authorities promulgated *zoning codes*. These were intended to limit congestion in cities and prevent construction of buildings that would infringe unreasonably on the rights of occupants of neighboring buildings to light and air. Regulations in these codes had decided effects on architecture. Provisions indicated how much of a lot a building could occupy and, to some extent, where a building could be placed on a lot. Some codes placed specific limits on building heights, whereas others required the face of the building to be set back as it was made higher. In some cases, this requirement led architects to design buildings with facades sloping away from the adjoining street.

In addition, zoning codes generally indicated what type of building—residence, office building, shopping center, factory, etc.—and what type of construction—combustible or noncombustible—could be constructed in various city districts.

Concern for welfare of building occupants also was demonstrated by city Health Department regulations for heating of buildings in cold weather; however, by the middle of the 20th Century, commercial establishments began voluntarily to provide cooling in hot weather. To attract patrons, owners of theaters and retail stores installed cooling equipment, and so did owners of office buildings, to provide more efficient working conditions for employees. A convenient method of supplying the required cooling was by air conditioning, which also provided humidity control, and this method was widely adopted. Its use spread to residences, most public buildings and factories.

There was an effect on architecture but it was not very visible. Mechanical engineers took on the task of designing cooling installations, incorporating it in a general category HVAC (heating, ventilation and air conditioning). Heating subcontractors became HVAC subcontractors. Architects endeavored to make HVAC installations inconspicuous. They placed equipment in basements and other areas where it would not be noticeable, or they disguised equipment spaces with decorative treatment. The designers also hid ventilation ducts, when it was expedient, in enclosed shafts or between floors and ceilings.

During the last half of the 20th Century, concern for the effects of buildings on people became deeper. More stringent regulations for fire safety were promulgated. Other rules set minimum illumination and maximum sound levels in work areas. Requirements were established that prevented construction of a building until its full environmental impact could be assessed. And the need for energy conservation in building operation to conserve natural resources became apparent. These requirements placed additional constraints on building design. Both design and construction became even more complex.

New Twist in Construction Management

While complex buildings demanded by owners made design more difficult than before, owners now encountered problems even more difficult

than in the past, from the start of a project to its completion. Few owners were sophisticated enough to cope successfully with these problems. Consequently, projects often were completed late and construction costs exceeded expectations. Some owners consequently sought new ways to control costs.

With respect to cost control, the subdivision of the building process into design and construction by separate specialists was proving to be counterproductive. By specializing in design, architects and their design consultants gave up control of construction methods and equipment, exerted little influence on construction scheduling and lost intimate contact with actual construction costs. Hence, orthodox building designers could provide little help to owners in controlling construction costs and time.

There was one alternative. Master builders had not become extinct. Often, though, they had become transformed from an individual designer-builder to a corporation consisting of architects, engineers and construction management personnel. Under a turnkey contract, such companies would design and build a project for a stipulated sum of money. Some owners liked this arrangement because they knew what their maximum cost would be almost from the start of the project. Others disliked it because they were uncertain that they were getting the best possible design or the lowest possible cost.

Seeking a better alternative, some owners continued to engage architects and engineers for design only, in the hope of getting the best possible design for their money, but sought different means of controlling construction costs and time. Public agencies, for example, awarded prime contracts to former major subcontractors, such as HVAC, plumbing and electrical, as well as to a general contractor. This was done in the expectation that open competitive bidding on major cost items would result in lower total cost. However, there never was any certainty that the expectation would be realized.

Experienced owners often found that awarding a construction contract to the lowest bidder gave undesirable results—shoddy materials and workmanship, construction delays and cost overruns. Some owners therefore found it worthwhile to select a reputable contractor and pay a fee over actual costs for construction. Owners were uncertain, though, as to actual costs and especially as to whether costs could have been lowered.

To meet the challenge, a new breed of contractor evolved in the second half of the 20th Century. Called a *construction manager*, this contractor usually did not do any building. Instead, for a fee, the manager engaged a general contractor, supervised selection of subcontractors and controlled construction costs and time. Engagement of a construction manager also offered the advantage that his knowledge of costs could be tapped by the building designers during the design phase. Many large and complex projects have been successfully built under the control of construction managers.

Nevertheless, whether construction managers, reputable general contractors or multiple prime contractors are used, good construction management has demonstrated capability for keeping costs within estimates; however, such management is generally restricted primarily to the task of transforming the concepts of building designers into a structure. With the design function in the hands of others, constructors are limited in opportunities for lowering construction costs. If costs are to be lowered, designers probably will have to show the way. For that, they will need new methods.

References

S. Gideon, *Space, Time, and Architecture*, Harvard Univ. Press, 1954.

H. Gardner, *Art Through the Ages*, Eighth Ed. Harcourt, Brace, New York, 1986.

W. Jordy and W. Pierson, *American Buildings and Their Architects*, Doubleday, New York, 1970.

S. Timoshenko, *History of Strength of Materials*, McGraw-Hill, New York, 1953.

Words and Terms

Architect
Building code
Building engineering
Construction manager
Electrical engineer
HVAC
Master builder
Mechanical engineer
Structural engineer
Zoning codes

Significant Relations, Functions, and Issues

Change in building design and construction processes over time.

Roles of the architect, contractor, subcontractors, consulting engineers, construction manager.

Effects of the emergence of building codes and zoning codes.

1.2. BASIC TRADITIONAL BUILDING PROCEDURE

Before any new approaches to building design can be explored, a knowledge of current design practices is essential. Furthermore, the systems design method proposed in this book is a modification of current practices. Therefore, current practices are reviewed in this section. For this purpose, a commonly followed procedure is described. It is called the basic traditional building procedure. While other procedures are often used, they can readily be adapted to systems design in much the same way as the basic traditional procedure.

What Designers Do

Generally, an owner starts the design process by engaging an architect. In selecting the architect, owners do not always act in their own best interest. They should choose an architect who has established a reputation for both good design and low construction costs. Instead, some owners shop around for the architect with the lowest fee. Yet, a good designer can provide a high-quality building and, at the same time, save the owner several times the design fee in lower construction costs.

The architect usually selects the consulting engineers and other consultants who will assist in the design. A good architect selects engineers who have established a reputation for both good design and low construction costs.

Building design may be considered divided into two steps, planning and engineering, which necessarily overlap.

Planning consists generally of determining:

1. What internal and external spaces the owner needs
2. The sizes of these spaces
3. Their relative location

4. Their interconnection
5. Internal and external flow, or circulation, of people and supplies
6. Degree of internal environmental control
7. Other facilities required
8. Enhancement of appearance inside and outside (aesthetics)
9. How to maximize beneficial environmental impact and minimize adverse environmental impact of project.

In some cases, planning also includes locating, or layout, of machinery and other equipment to meet an owner's objectives.

Engineering consists generally of the following processes:

1. Determining the enclosures for the desired spaces
2. Determining the means of supporting and bracing these enclosures
3. Providing the enclosures and their supports and bracing with suitable characteristics, such as high strength, stiffness, durability, water resistance, fire resistance, heat-flow resistance and low sound transmission.
4. Determining the means of attaining the desired environmental control (HVAC, lighting, noise)
5. Determining the means of attaining the desired horizontal and vertical circulation of people and supplies
6. Providing for water supply and waste removal
7. Determining the power supply needed for the building and the means of distributing the required power to the places where it is needed in the building
8. Providing for safety of occupants in emergency conditions, such as fire.

Legally, the architect acts as an agent of the owner. Thus, at the completion of design, the architect awards a construction contract to a general contractor and later inspects construction on behalf of the owner, who is obligated to pay the contractor for work done.

What Contractors Do

In effect, the owner selects the general contractor. The architect provides advice and assists the owner in reaching a decision. The owner may pick a contractor on the basis of price alone (bidding) or may negotiate a price with a contractor chosen on the basis of reputation.

The general contractor selects the various subcontractors who will be needed. Selection is generally based on the lowest price obtained (bidding) from reputable companies with whom the contractor believes it will be easy to work. The contractor compensates the subcontractors for the work performed.

Construction consists of the processes of assembling desired enclosures and their supports and bracing to form the building specified by the architect. Construction also includes related activities, such as obtaining legal permission to proceed with the work, securing legal certification that the completed building complies with the law and may be occupied, supplying needed materials, installing specified equipment, providing for the safety of construction employees and the general public during construction, and furnishing power, excavation and erection equipment, hoists, scaffolding and other things essential to the work.

Programming

The basic traditional building design procedure is a multistep process. It starts with the collection of data indicating the owner's needs and desires and terminates with award of the construction contract (see Fig. 1.5).

The procedure starts with preparation of a *building program*. The program consists mainly of a compilation of the owner's requirements. The program also contains descriptions of conditions that will affect the building process and that will exist at the start of construction, such as conditions at the building site. It is the duty of the architect to convert the program into spaces, which then are combined to form a building. Hence, before planning of a building can start, a program is needed. The architect prepares the program from information supplied by the owner, owner representatives, or a building committee.

In collecting data for the program, it is important for the architect to learn as soon as possible how much the owner is willing to pay for the building (the budget) and if there is a specific date on which the building must be ready for

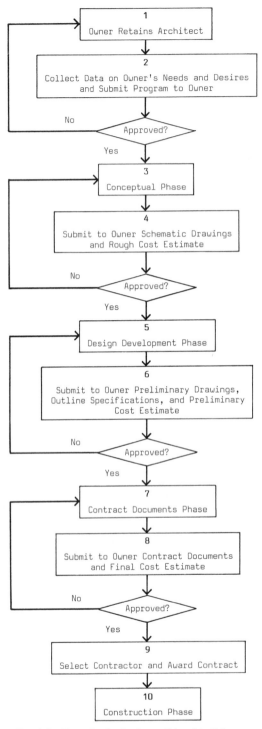

Fig. 1.5. Steps in the basic traditional building procedure.

occupancy. If either the budget or construction time are unrealistic, the owner should be informed immediately, in writing. If realistic figures cannot be negotiated, the architect

should not proceed. If he does and the owner suffers economic injury, the architect may not receive compensation for work performed on the project.

The data supplied by the owner should indicate clearly what his objectives are, so that the functions, or purposes, of the building are evident. The architect should also ascertain how the owner expects to attain those objectives—the activities to be performed in the building, approximate space needed for each activity, number of employees per activity, relationship between activities or work flow, equipment that will be installed for the activities, desired environmental conditions (HVAC, lighting and sound control) and other requirements that will be needed for design of the building.

Information also will be needed on the site on which the building will be erected. This information should cover subsurface conditions as well as surface conditions. If the owner has already purchased a building lot before the program has been prepared, the architect will have to adapt the building to the site. A much more desirable situation is one in which a site has not yet been bought, because the architect will then have greater planning flexibility; the architect can assist the owner in deciding on a site.

The owner is responsible for providing information on the site necessary for design and construction of the building. The architect, however, acting as the owner's agent, generally engages a land surveyor to make a site survey, and foundation consultants for subsurface investigations.

The architect should then submit the completed program to the owner for approval. If there are any omissions or misconceptions of the requirements, they should be rectified before planning starts, to save time and money. Approval of the owner should be obtained in writing.

Conceptual Phase

During data collection, the architect may have formulated some concepts of the building, but on completion of the program, he formalizes the concepts—translates requirements into spaces, relates the spaces and makes sketches illustrating his ideas. To see how other designers have met similar requirements for building design, the architect may visit other buildings. Then, by a combination of intuition, judgment based on past experience and skill, he decides on a promising solution to the requirements of the program.

Cost estimators then prepare an estimate of the construction cost for the selected solution. Since at this stage, practically no details of the building design have been decided, the result is called a rough cost estimate. If the estimate is within the owner's budget, the solution can be prepared for submission to the owner for approval. Otherwise, the scheme must be modified, usually by making the layout more efficient or by reducing allotted floor areas or building volume.

Efficiency of layout is sometimes measured by the *tare*, or ratio of useful floor area to the gross floor area (total floor area enclosed within the outer faces of the exterior walls). Efficiency for some types of buildings also may be measured by the floor area per occupant or unit of production.

The proposed solution is submitted to the owner mainly as sketches, known as schematic drawings, along with the rough cost estimate (see Fig. 1.5). Though lacking in detail, the schematics show the owner what the building will be like. They include a site plan indicating the orientation of the building and its location on the site, as well as the access to be provided to the site and the building. The schematics should also include major floor plans, showing the location of rooms and corridors and floor areas allotted. In addition, exterior views, or elevations, should be provided to illustrate the proposed finished appearance of the building exterior. The plans and elevations should indicate the basic materials that have been selected. Besides the schematics, the architect may submit to the owner perspective drawings or a model to give a better indication of how the building will look.

The owner may suggest modifications of the plans or may reject the entire scheme. In the latter case, a new concept must be developed. Because of this possibility, time and money are saved in the conceptual stage by developing no more detail than necessary to present a possible solution to the program requirements.

The conceptual phase is further discussed in Sec. 4.3.

Design Development

After the architect receives, in writing, the owner's approval of the schematic drawings and rough cost estimate, the design is developed in detail (see Fig. 1.5). In this phase, the designers concentrate on technology. The objective of this phase is to bring the building into clearer focus and to a higher level of resolution. The phase culminates in completion of preliminary construction drawings, outline specifications and preliminary cost estimate.

In the conceptual phase, the architect's aesthetic concerns were mainly with function, mass and space. During design development, the architect pays more attention to surface and detail.

The structural engineer prepares drawings showing the framing and sizes of components. The mechanical engineer shows the layout of pipes, air ducts, fixtures and HVAC equipment and provides data on escalators and elevators. The electrical engineer indicates in drawings the location and type of lighting fixtures and layout of electric wiring and control equipment, such as switches and circuit breakers.

The designers also prepare outline specifications to record, for review, the basic decisions on materials and methods that will later be incorporated in the contract documents. These specifications need not be as precisely worded as the final specifications; they may be brief, in the form of notes.

When the preliminary drawings and outline specifications have been completed, cost estimators can prepare a more accurate estimate of the construction cost for the building. If the refined estimate is not within the owner's budget, changes are made to reduce costs. It should not, however, be necessary to revise the basic concepts approved in the conceptual phase, but it may be necessary to modify the structural framing, switch window types, change the exterior facing, specify less expensive heating or cooling equipment, pick different lighting fixtures, or even omit some features desired by the owner but not really essential.

When construction cost estimates are brought to the desired level, the preliminary drawings, outline specifications and estimated cost are submitted to the owner for approval. Revisions are made, as necessary, to obtain the owner's written approval.

Design development is further discussed in Sec. 4.4.

Contract Documents Phase

The ultimate objective of the design effort is production of information and instructions to constructors to insure that a building will be produced in complete accordance with the design agreed on by the owner and the architect. The information and instructions are provided to the builder in the form of working, or construction, drawings and specifications (see Fig. 1.5). These are incorporated in the construction contract between the owner and the builder and therefore become legal documents. As such, they must be prepared with extreme care to be certain that they are precise and their intent is clear.

In the contract documents phase of design, the designers' efforts are concentrated mostly on details and refinements, inasmuch as the main features of the building were worked out in design development and approved by the owner. If changes have to be made in the design at this stage, they are likely to be much more costly than if they had been made in earlier phases. The architect, with the advice of legal counsel, also prepares the construction contract.

With final details of the design worked out, a more accurate estimate of construction cost can now be made. If this estimate exceeds the owner's budget, the designers have to revise drawings or specifications to bring costs down. When they have done this, the contract documents are submitted to the owner for approval. Again, revisions are made, as necessary, to obtain written approval; but with the high cost of changes at this stage, a sophisticated owner would restrict requests for modifications only to corrections of mistakes.

Contract documents are further discussed in Chap. 5.

Contract Award

After the contract documents have been approved, the architect assists the owner in obtaining bids from contractors or in negotiating a contract with a qualified contractor (see Fig. 1.5). The architect also aids in evaluating proposals submitted by contractors and in awarding the contract.

For private work, for example construction not performed for a public agency, the owner usually awards a single contract to a general contractor. The contractor then awards subcontracts to specialists who perform most or all of the work. The owner may negotiate a contract with a general contractor with whom the owner has had previous experience or who has been recommended by the architect or other advisers. Or the owner may select a contractor on the basis of bids for the work.

For public work, such as a city or state project, there may be a legal requirement that bids be taken and separate construction contracts be awarded for the major specialties, such as the mechanical and electrical trades. In addition, a separate contract must be awarded to a general contractor, who is assigned responsibility for coordinating the trades and execution of all of the work. Usually, bidding is open to anyone wishing to bid, and the contracts must be awarded to the lowest responsible bidders.

Bidding requirements and contract awards are further discussed in Sec. 5.5.

General Critique

The basic traditional building process described in this chapter and extended to the construction phase in Chap. 5 evolved into its present form over many years, and is widely used. Clients, designers and contractors are familiar with it and generally produce good buildings with it.

The basic traditional building procedure usually yields buildings that meet functional requirements well, are aesthetic, with safe structure, good lighting, adequate heating and cooling, and good horizontal and vertical circulation. In addition, the procedure is geared to submission of bids for construction that are within the owner's budget. The architect submits cost estimates to the owner for approval at the start of the conceptual phase, at the conclusion of the conceptual phase, at the end of design development and with the contract documents. At any stage, if the estimate is too high, the design is revised to reduce estimated costs. Also, if contractors' bids or negotiated prices are too high, changes in the design are made to bring prices down.

If the procedure produces good buildings at prices owners are willing to pay, why then should the procedure be changed?

Should it be changed because the charge can be made that the owner may be paying too much for the building provided, though he is willing to pay the price? This may be true, but it also is probably true of almost every conceivable design procedure. Enough research and study can always produce a better design. But the cost of such research and study may not warrant these efforts. Furthermore, the time available for design and construction may not be sufficient. Consequently, changes in the basic traditional procedure must be justified by more specific defects.

One drawback is the frequent occurrence of construction costs that exceed bid or negotiated prices. Such situations generally occur because the owner orders design changes while the building is under construction. Such changes almost never reduce construction costs and almost always are costly.

These situations may occur partly because of the type of construction contract used. For example, when a contractor takes a job for a fixed price, there is a profit incentive to encourage change orders. Design changes during construction usually yield higher profits to the contractor. To low bidders, change orders often mean the difference between profit and loss for a project. Nevertheless, as reputable contractors can point out, change orders often are necessary because of design mistakes or omissions. (Occasionally, changed conditions affecting the owner's requirements for the building may compel issuance of change orders.) Modifications of the design procedure therefore could have the objective of reducing the number of mistakes and omissions in design.

Another defect arises because of the separation of design and construction into different specialties. If designers do not build, they do not have firsthand knowledge of construction costs and consequently often cannot prepare cost estimates with needed accuracy.

In addition, construction costs usually depend on the construction methods used by the contractor. Since the contractor generally is free to choose construction methods, designers can only base their cost estimates on the probable choice of methods. This can introduce further inaccuracies in the estimates.

Also, knowledge of the construction market at the time when and the place where the building will be constructed is necessary. This requires familiarity with availability of subcontractors, construction workers, construction equipment, building materials, and equipment to be installed when needed for construction. Contractors take such conditions into account in establishing construction costs; designers rarely do. Thus, further inaccuracies may be introduced into their cost estimates.

The result often is that the owner pays too much for the building provided, though the price may be within his budget. Modifications of the design procedure therefore could have the objective of bringing construction experts into the design process.

Still another drawback is that construction costs are kept within the budget by permitting maintenance and operating costs to rise. Cheap building materials and equipment are specified to cut initial costs, but they prove expensive in the long run. Sometimes this condition is made necessary because an owner could otherwise not afford to build and is willing to risk high maintenance and replacement costs. He is willing to pay the higher maintenance and operating costs until he becomes affluent enough to replace the costly materials and equipment. Often, however, owners are not aware of excessive life-cycle costs until after they occupy the building. (*Life-cycle costs* are the sum of initial installation costs and maintenance and operating costs over a long period of time, usually at least ten years for buildings.) Changes in the design procedure consequently could have the objective of placing relevant emphasis on construction and life-cycle costs.

Another common defect is the lack of coordination of the work of the various design specialists. The architect develops building forms and room layouts with little advice from engineering consultants. The latter, in turn, practice their specialties with little concern for each other's products, except when the architect discovers that two different building components are scheduled to occupy the same space. Usually, then, one of the consultants is compelled to move an overlapping component.

Often, there is no effort to integrate components designed by different specialists into a single multipurpose component, with consequent reduction in construction costs. Hence, the objective of revamping the design procedure could be production and installation in buildings of more multipurpose building components.

Furthermore, the whole philosophy of design with respect to the basic traditional procedure may be questioned. Under existing economic pressures and time schedules, each designer proposes one scheme for his specialty, based on intuition, judgment or experience. This design may or may not be the *optimum* for the cost or the least costly; but the decision may not be questioned, especially when bid or negotiated prices fall within the owner's budget. Thus, there is no pressure for further reduction of construction costs. Consequently, an important reason for changing the design procedure is the need for reducing construction costs without increasing life-cycle costs.

Other variations of the basic traditional design procedure often used include engagement of a consulting engineer or an architect-engineer firm instead of an architect. These variations generally have about the same disadvantages as the traditional procedure. All need to be changed to reduce construction costs while maintaining high-quality design.

References

Architect's Handbook of Professional Practice (Volumes 1, 2, and 3), American Institute of Architects.

Guide for Supplementary Conditions (Publ. No. A511), AIA.

General Conditions of the Contract for Construction (Publ. No. A201), AIA.

Standard Form of Agreement Between Owner and Architect (Publ. No. B141), AIA.

J. Sweet, *Legal Aspects of Architecture, Engineering, and the Construction Process*, West Publishing Co., 1970.

C. Dunham, et al., *Contracts, Specifications, and Law for Engineers*, 3rd ed., McGraw-Hill, New York, 1979.

Words and Terms

Bidding
Building Program
Design Phases: programming, conceptual, development, construction documents, bidding
Engineering
Life-cycle costs
Planning
Tare

Significant Relations, Functions, and Issues

Two steps of building design: planning and engineering.
Sequential phases of design—from programming to construction.
Functions of the architect as agent of the owner.
Building construction contract awarding process.
Cost control for construction related to the owner/contractor contractural agreement.
Separate effects of cost control measures on design, construction, maintenance, and operation costs.

1.3 SYSTEMS DESIGN APPROACH TO BUILDING

The General Critique of Sec. 1.2 indicates that the basic traditional building procedure could be improved by

1. More questioning of the cost effectiveness of proposed building components and greater efforts to obtain better alternatives.
2. Coordinating the work of various design and construction specialists to achieve more cost-effective designs; for example, use of multipurpose building components in which the products of two or more specialties are integrated.
3. Placing relevant emphasis on both construction and life-cycle costs.
4. Having construction experts contribute their knowledge of construction and costs to the design process.
5. Use of techniques that will reduce the number of mistakes and omissions in design that are not discovered until after construction starts.

The systems design approach described briefly in this section and in more detail in following chapters offers opportunities for such improvements.

Operations Research

Development of the technique known as operations research or systems analysis began early in the 20th Century but became more intense after 1940. Many attempts have been made to define it, but none of the definitions appears to be completely satisfactory. They either are so broad as to encompass other procedures or they consist merely of a listing of the tools used in operations research. Consider, for example, the definition proposed by the Committee on Operations Research of the National Research Council:

Operations research is the application of the scientific method to the study of the operations of large complex organizations or activities.

The *scientific method* comprises the following steps:

1. Collection of data, observations of natural phenomena
2. Formulation of an hypothesis capable of predicting future observations
3. Testing the hypothesis to verify the accuracy of its predictions and abandonment or improvement of the hypothesis if it is inaccurate

Operations research does satisfy the definition; but architects and engineers also can justifiably claim that the design procedures they have been using are also covered by the definition. A major difference, however, between traditional design procedures and operations research, or systems analysis, is that the traditional design steps are vague. As a result, there usually is only a fortuitous connection between the statement of requirements, or program, and the final design. Systems analysis instead marks a precise path that guides creativity toward the best decisions.

Definition of a System

Before the systems design method can be explained in full, a knowledge of terms used is necessary. Similarly, before the method can be applied to building design, a knowledge of basic building components is essential. Following are some basic definitions:

A system is an assemblage of components formed to serve specific functions or to meet specific objectives and subject to constraints, or restrictions.

Thus, a system comprises two or more essential, compatible and interrelated components. Each component contributes to the performance of the system in serving the specified functions or meeting the specified objectives. Usually, operation, or even the mere existence, of one component affects in some way the performance of other components. In addition, the required performance of the system as a whole, as well as constraints on the system, impose restrictions on each component.

A building satisfies the preceding definition and description of a system. Even a simple building, with only floor, roof, walls, doors and windows, is a system. The components can be assembled to provide the essential functions:

1. Surface on which activities can take place and furnishings or materials can be stored.
2. Shelter from the weather
3. Access to and from the shelter
4. Light within the shelter
5. Ventilation within the shelter

The components are essential, compatible and interrelated. The combination of floor, roof and walls, for example, meets the requirement of shelter from the weather, because these components fully enclose the spaces within the building. Floor, roof and walls also must be compatible, because they must fit tightly together to exclude precipitation. In addition, they are interrelated, because they interconnect, and sometimes the walls are required to support the roof. Similar comments can be made about walls, doors and windows.

Systems Analysis

In systems analysis, a system is first resolved into its basic components. Then, it is investigated to determine the nature, interaction and performance of the components and of the system as a whole.

Components also may be grouped into smaller assemblages that meet the definition of a system. Such assemblages are called *subsys-*

tems. Systems analysis also may be applied to subsystems.

A complex system can be resolved into many sizes and types of subsystems. For example, it may not only be possible to separate subsystems into subsubsystems but also to recombine parts taken from each subsubsystem into a new subsystem. Hence, a system can be analyzed in many different ways.

Consider, for instance, a building wall. In some buildings, a wall can be composed of a basic component. In other buildings, a wall may consist of several components: exterior surfacing, water-resistant sheathing, wood studs, insulation and interior paneling. In the latter case, the wall may be considered a subsystem. In such buildings, power may be supplied by electricity, in which case the electrical equipment and wiring may be considered a subsystem. Suppose now that the wiring is incorporated in the wall, between the wood studs. Then, the wiring and other wall components may be considered a subsystem.

Systems Design

Systems design is the application of the scientific method to selection and assembly of components or subsystems to form the optimum system to attain specified goals and objectives while subject to given constraints, or restrictions.

Applied to buildings, systems design must provide answers to the following questions:

1. What precisely should the building accomplish?
2. What conditions exist, or will exist after construction, that are beyond the designers' control?
3. What requirements for the building or conditions affecting system performance does design control?
4. What performance requirements and time and cost criteria can be used to evaluate the building design?

Value Analysis

An additional step, aimed at reduction of life-cycle costs of buildings, was introduced into the traditional building procedure about the middle of the 20th Century. This new step required a study, generally by separate cost estimators and engineers, of ways to reduce cost, in addition to normal design considerations of cost and function. The technique used in such studies became known as value analysis, or value engineering.

When used, value analysis often was permitted or required by a clause in the construction contract. This clause gave the general contractor an opportunity to suggest changes in the working drawings and specifications. As an incentive for so doing, the contractor was given a share of the resulting savings, if the owner accepted the suggestions. Thus, the design changes were made during the construction phase, after the designers had completed their work, and the construction contract had to be amended accordingly. Despite the late application of value analysis in the building procedure, the technique generally yielded appreciable savings to owners.

Nevertheless, the technique is applicable in the other phases of the building process. Furthermore, experience has shown that cost-reduction efforts are more effective in the earlier phases of design. Consequently, it is logical to start value analysis in the conceptual phase and continue it right through the contract documents phase.

As practiced, value analysis is an orderly sequence of steps with the goal of lowering life-cycle costs of a proposed system. In the search for cost reductions, the analysts question the choice of systems and components and propose alternatives that are more cost effective. Based on observations of current value-analysis practice, the following definitions are proposed:

Value is a measure of benefits anticipated from a system or from the contribution of a component to system performance. This measure must be capable of serving as a guide in a choice between alternatives in system evaluation.

Value analysis is an investigation of the relationship between life-cycle costs and values of a system, its components and alternatives to these. The objective is attainment of the lowest life-cycle cost for acceptable system performance.

Note that the first definition permits value to be negative; for example, an increase in cost to an owner because of a component characteristic, such as strength or thickness.

Note also that the second definition sets the goal of value analysis as the least cost for acceptable system performance. This differs from the goal of systems design, as previously defined, which is to produce the optimum solution. Thus, systems design may seek either the least cost, as does value analysis, or the best performance for a given cost. Since there is a common goal, it appears well worthwhile to integrate value analysis into the overall design process.

Sometimes, value can be expressed in terms of money; for example, profit resulting from a system change. If this measure is used, value analysis is facilitated, because life-cycle cost and value, or benefit, can be directly compared in monetary terms. Often, however, in value analysis of buildings, value is based on a subjective decision of the owner. He must, for example, decide how much more he is willing to pay for an increase in attractiveness of the building exterior, or conversely, whether the savings in construction cost from a decrease in attractiveness is worthwhile. Thus, value analysis must be capable of evaluating satisfaction, prestige, acceptability, morale, gloom, glare, draftiness, noise, etc.

Value analysis is further discussed in Chap. 3, Sec. 3.12.

Systems Design Procedure

In accordance with the definition of systems design and the description of the scientific method, the systems design procedure has three essential parts: analysis, synthesis and appraisal. These may be carried out in sequence or simultaneously.

Analysis is the process of giving the designers and value engineers an understanding of what the system should accomplish. Analysis includes collection of data, identification of the objectives and constraints, and establishment of performance criteria and relationships between variables.

Synthesis is the process of selecting compo-

nents to form a system that meets the design objectives while subject to the constraints.

Appraisal is the process of evaluating system performance. Value analysis is part of the appraisal phase, to insure cost effectiveness of components. Data obtained in the appraisal are used to effect improvements in the system, through feedback of information to analysis and synthesis.

Thus, the procedure is repetitive, or cyclical. Information feedback should insure that the system produced in each cycle is better than its predecessor. Consequently, the design should converge on the optimum system. Whether that system can actually be attained, however, depends on the skills of the designers and the value engineers. Each cycle consumes time; costs mount with time, and in addition, the design must be completed by a deadline. Hence, design may have to be terminated before the optimum system has been achieved.

Optimization

For complex systems, such as buildings, it usually is impractical to optimize a complete system by simultaneous synthesis of optimum components. It may be necessary to design some components or subsystems in sequence, while, to save time, other parts are designed simultaneously. In practice, therefore, synthesis of a system can develop by combination of components or subsystems that can be realistically optimized rather than by direct optimization of the complete system. Nevertheless, to obtain the best system by this procedure, the effects of each component or subsystem on other components must be taken into account.

Thus, in building design, floor plans and exterior views of the building are produced first. Then, structural framing, heating and cooling subsystems and electrical subsystems usually are synthesized simultaneously. The effects of the various subsystems on each other may result in changes in each subsystem. Also, value analysis may suggest improvements to reduce life-cycle costs. At some stage in the design, however, the system can be studied as a whole with optimization of the total system as the goal.

That optimization of a building's subsystems does not necessarily lead to optimization of the building can be demonstrated by a simple example. Assume that a tennis court is to be enclosed in a building. Suppose that initially a design is proposed with four vertical walls and a flat horizontal roof (see Fig. 1.6a) and with other appurtenances essential for a tennis court. Suppose also that the building is resolved into three subsystems: the walls, the roof and the other appurtenances. Then, each subsystem is designed for the lowest possible cost. For that condition, the cost determined for the building is the sum of the subsystem costs. There may now be possible, however, a lower-cost building achieving the same results. For example, a curved, cylindrical enclosure may be constructed between the sides of the court, with vertical walls at the two ends (see Fig. 1.6b). The curved enclosure may be less costly than the two walls and the roof it replaces. Hence, the alternative building (see Fig. 1.6b) would cost less than the one with four walls and roof.

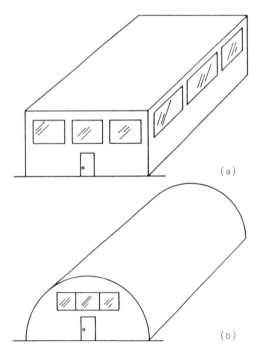

(a)

(b)

Fig. 1.6. Two possible building shapes for enclosing a tennis court. (a) Building with four walls and a roof; (b) building with cylindrical enclosure and two end walls.

Adaptation of the Traditional Procedure to Systems Design

In the application of the systems-design procedure to buildings, it is desirable to retain as much of the basic traditional building procedure as is feasible, because of its advantages. In fact, with some revision of the procedures in the various phases, the steps of the traditional procedure indicated in Fig. 1.5 can be retained in systems design of buildings. In brief, the changes are as follows:

To have the greatest impact on design efficiency, systems design application should commence at the very earliest stages of the building process. Programming should be done in anticipation of the use of value analysis as a design tool; value analysis challenges should be made of proposals in the conceptual stage, where design changes can be made with little or no cost involved in effecting the changes.

Systems design procedures should continue through the design development phase, in which major features of the design are synthesized. In that phase also there is little cost for making design changes.

During the contract documents phase, however, major system design changes become costly in terms of time lost and redesign work required. Consequently, complete application of the systems design procedure may not be desirable. But for work that originates in this phase, such as that involving detailed design and writing of specifications, value analysis may be profitably applied.

During the construction phase, changes are very expensive and should preferably be limited to corrections of mistakes or adjustments for unanticipated situations. If changes are proposed by the contractor, they should be subjected to value analysis to assure their fit with the original design objectives.

General Critique

The systems design procedure, as outlined in this section, has nearly all the advantages of the basic traditional building procedure. Systems design, in addition, offers the five desired improvements that were described at the beginning of this section.

Two of the improvements are readily discerned. Incorporation of value analysis in the procedure provides the desired questioning of design proposals. Though aimed primarily at cost reduction, value analysis, as a part of systems design, will also encourage innovation, because use of new materials, equipment or methods is one means to reduce costs. System design also provides for the desired inclusion of life-cycle costs in evaluations of design proposals. To the extent that data are available, maintenance and operating costs can be incorporated in cost estimates and value analyses.

Systems design does offer the other three improvements (coordination of design specialists, contributions of construction experts, and reduction of errors), although how it does may not be evident at this time. Those improvements come about as a consequence of the organization of personnel required for effective execution of systems design. This is discussed in Sec. 1.4.

Systems design, however, does have disadvantages. It takes more time and effort than the traditional procedure. Consequently, design costs are higher. To offset the higher costs, the owner should pay a higher design fee. It may be difficult, though, to persuade an owner to do this, since some design firms providing less desirable design services may offer to do the design for less money.

Several difficulties are likely to be encountered in application of systems design:

Needed data may not be available.

Information supplied by the owner for compilation of the building program may be incomplete initially or misunderstood by the designers.

Requirements, as stated by the owner, may not be the actual requirements. Knowledge of existing conditions; for example, subsurface conditions at the building site, may be erroneous.

Many of these difficulties, however, are also experienced with the traditional procedure. They are mentioned here to preclude the impression that systems design is a cure-all for the problems of building design.

In addition, the appraisal process is more difficult and may be inaccurate in systems design. As with the traditional procedure, the means used in cost estimating may be faulty, but also in some cases, in value analysis, the means used in determining values may be erroneous.

In summary, the systems design approach to buildings is superior to the traditional procedure. But higher design fees are required to offset higher design costs.

References

F. Merritt, *Building Design and Construction Handbook*, 4th ed., McGraw-Hill, New York, 1982.

A. Mudge, *Value Engineering*, Society of American Value Engineers, 1981.

D. Meredith et al., *Design and Planning of Engineering Systems*, 2nd ed., Prentice-Hall, Englewood Cliffs, NJ, 1985.

Words and Terms

Analysis	Synthesis
Appraisal	System
Component	Systems Design
Operations research	Subsystem
Optimization	Value
Scientific method	Value analysis

Significant Relations, Functions and Issues

Steps in systems design procedure: analysis, synthesis, and appraisal.

Composition and resolution of systems into subunits: subsystems and components.

Suboptimization: the perfecting of parts that does not necessarily improve the whole system.

1.4. DESIGN BY BUILDING TEAM

Systems design of buildings requires, as does traditional design, the skills of diverse specialists. These may be the same specialists as those required for traditional design. In addition, for the purposes of systems design, additional specialists, such as value engineers, cost estimators, construction experts and custodians or plant engineers, are needed. But for systems design to be effective, the specialists must operate differently from the independent manner in which they did for traditional design.

In systems design, account must be taken of the interaction of building components and the effects of each component on the performance of the system. For better performance of the system and for cost effectiveness, unnecessary components should be eliminated and, where possible, two or more components should be combined. For these tasks to be accomplished with facility when the components are the responsibility of different specialists, those concerned should be in direct and immediate communication. They should work together as a team.

Thus, it is highly desirable that those responsible for design and construction of a building form a building team, to contribute their skills jointly. Working together, the various specialists provide a diversity of approach to synthesis, a multitude of paths to creative design. The diversity of skills available for analyses insures that all ramifications of a decision will be considered. With several experienced designers with broad backgrounds reviewing the design output, mistakes and omissions become less likely.

The Team Leader

Just as for athletic teams, a building team needs a leader to direct the team effort and to insure that the owner's objectives are met at minimum life-cycle cost and in the least time.

The team leader should be a generalist, familiar with all aspects of building design and construction, cost estimating and value analysis. This person should not only have leadership abilities but also architectural or engineering training, artistic talents, business expertise, public relations capabilities, management skills and a professional attitude. As a professional, the leader must abide by the highest standards of conduct and provide faithful services to clients and the public, just as doctors and lawyers are required to do. Though acting legally as an agent of the owner, the professional must be fair and objective in dealing with contractors, especially when called on for interpretations of provisions of construction contract documents

or approval of payments. In addition, the leader should be skillful in maintaining good relations with public officials, including representatives of local building departments and zoning commissions.

By education and experience, the team leader may be an architect, a structural, mechanical, electrical, value or industrial engineer, or a construction manager or other professional with the required capabilities. Under state laws established for protection of the public, however, the leader should be registered as either an architect or a professional engineer. In either case, state registration is achieved after completing architectural or engineering courses, years of architecture or engineering experience and passing a written examination given by the state.

The leader and the other members of the team may be employees of a single firm or representatives of different firms participating in a joint venture for design of a project. The leader may also be the prime contractor with the owner and engage consultants to serve on the team.

The leader provides liaison between the owner, members of the team and contractors. With responsibility for all design activities, the leader coordinates and expedites the work, motivates the team to the highest level of performance and communicates clearly and accurately all necessary information to all concerned with the project.

Other Team Members

Preferably, all team members should have the same characteristics required of a team leader. Such characteristics are needed to execute subsystem design efficiently and to discharge responsibilities to clients and the public. In particular, architects and engineers should be licensed to practice their professions. Following is a brief description of each of the specialists usually included on the building team:

An *architect* is a professional with a broad background in building design. This background should be sufficient to permit design of a simple building, such as a one-family house, without the help of specialists. The architect should be trained to analyze the needs and desires of clients and to transform those requirements into buildings. The training should also include study of the human factors involved in building use and operation. In addition, the architect should be familiar with the influence on buildings and their occupants of natural factors, such as geography, climate, material resources, site and orientation; the influence of economic, technological and sociological factors; and the influence of allied arts. Furthermore, this professional should have artistic talents, appropriate to making buildings attractive in appearance, inside and outside.

Thus, as part of a building team, the architect may be delegated responsibility for any or all of the following plus any other design tasks for which he may have capabilities:

1. Preparation of the program
2. Arrangement and location of the building on the site
3. Control of traffic and access to the site and the building
4. Use of natural features of the site
5. Climate considerations in building design
6. Proper relationship between the building, its neighbors and the community
7. Aesthetics
8. Compliance of the building and the site with health, safety and zoning ordinances and building codes
9. Determination of the size and shape of interior spaces for human needs and the relationship of such spaces to each other
10. Interior and exterior surface finishes, doors, windows, stairs, ramps, building hardware and, if required, interior decoration
11. Inspection of construction

A *structural engineer* is a specialist trained in the application of scientific principles to design of load-bearing walls, floors, roofs, foundations and skeleton framing needed for the support of buildings and building components. As part of the building team, this engineer may be delegated responsibility for structural design required for the building project and inspection of structural members and connections during construction.

A *mechanical engineer* is a specialist trained in the application of scientific principles to design of plumbing and plumbing fixtures; heating, ventilation and air conditioning; elevators; escalators; horizontal walkways; dumbwaiters and conveyors. This engineer also may have capabilities for designing machines and planning their location for such buildings as factories and hospitals. As part of the building team, the mechanical engineer may be delegated the responsibility for design and inspection of the installation of the aforementioned elements.

An *electrical engineer* is a specialist trained in the application of scientific principles to design of electric circuits, electric controls and safety devices, electric motors and generators, electric lighting and other electric equipment. As part of the building team, the electrical engineer may be delegated responsibility for design and inspection of the installation of the aforementioned elements.

A *construction manager* is a specialist with considerable experience in building construction. This expert may be a general contractor, or a former project manager for a general contractor, or an architect or engineer with practical knowledge of construction management. The construction manager should have the knowledge, experience and skill to direct construction of a complex building, though he may not be engaged for erection of the building the team is to design. He must be familiar with all commonly used construction methods. He must be a good judge of contractor and subcontractor capabilities. He must be a good negotiator and expediter. He must be capable of preparing or supervising the preparation of accurate cost estimates during the various design phases. He must know how to schedule the construction work so that the project will be completed at the required date. During construction, he must insure that costs are controlled and that the project is kept on schedule. As part of the building team, the construction manager may be assigned any or all of the following tasks:

1. Advising on the costs of building components
2. Providing cost estimates, when needed, for the whole building

3. Indicating the effects of selected components on construction methods and costs
4. Recommending cost-reducing measures
5. Assisting in selection of contractors and subcontractors
6. Negotiating construction contracts with contractors and subcontractors
7. Scheduling construction
8. Cost control during construction
9. Expediting deliveries of materials and equipment and keeping the project on schedule
10. Inspection of the work as it proceeds

A *value engineer* is a specialist trained in value analysis. As part of the building team, the value engineer may head a group of value analysts, each of whom may be a specialist, for example, in structural systems, plumbing systems, electrical systems, or cost estimating. The value engineer provides liaison between the building team and the value analysis group.

The building team may also include architectural consultants, such as architects who specialize in hospital or school design; landscape architects; acoustics consultants and other specialists, depending on the type of building to be designed.

Design, in the sense used in the preceding descriptions, means analysis, synthesis and appraisal; preparation of schematic, preliminary and working drawings; and development of outline and final specifications.

References

Architect's Handbook of Professional Practice, American Institute of Architects, Washington, DC.
W. Caudill, *Architecture by Team*, Van Nostrand Reinhold, New York (out of print).

Words and Terms

Architect
Construction manager
Electrical engineer
Mechanical engineer
Structural engineer
Value engineer

Significant Relations, Functions and Issues

Need for communication and coordination in the design team.
Responsibilities and skills of the design team leader.
Functions of the design team members: construction manager and structural, electrical, mechanical, and value engineers.

General References and Sources for Additional Study

These are books that deal comprehensively with several topics covered in this chapter. Topic-specific references relating to individual chapter sections are listed at the ends of the sections.

F. Merritt, *Building Design and Construction Handbook*, 4th ed., McGraw-Hill, New York, 1982.
Architect's Handbook of Professional Practice, (Publ. No. A511), American Institute of Architects. (Volumes 1, 2, and 3).
D. Merdith et al., *Design and Planning of Engineering Systems*, 2nd ed., Prentice-Hall, Englewood Cliffs, NJ, 1985.
S. Andriole, *Interactive Computer Based Systems Design and Development*, Van Nostrand Reinhold, New York, 1983.
A. Gheorghe, *Applied Systems Engineering*, Wiley, New York, 1982.
P. O'Connor, *Practical Reliability Engineering*, Wiley, New York, 1985.
A. Dell'Isola, *Value Engineering in the Construction Industry*, Van Nostrand Reinhold, New York, 1983.
L. Zimmerman and G. Hart, *Value Engineering: A Practical Approach for Owners, Designers, and Contractors*, Van Nostrand Reinhold, New York, 1981.

EXERCISES

The following questions and problems are provided for review of the individual sections and chapter as a whole.

Section 1.1

1. What events made skyscrapers desirable and practical?
2. What do general contractors do in the building process?

3. What are the purposes of:
   (a) building codes?
   (b) zoning codes?
4. What do construction managers do in the building process?

Section 1.2

5. Describe two major ways of selecting a general contractor.
6. Name the major steps in the traditional building procedure.
7. What documents are produced by the building designers to form a part of the construction contract.
8. Describe some of the disadvantages of the basic traditional building procedure.

Section 1.3

9. What is accomplished by:
   (a) systems analysis?
   (b) systems design?

10. What is the purpose of value analysis?
11. Describe the three essential parts of systems design.

Section 1.4

12. Who is best qualified to be the leader of the design team for implementation of systems design? Why?
13. What responsibilities and tasks may be assigned to the construction manager?

General

14. What are the disadvantages of awarding a contract to lowest bidder for:
    (a) design of a building?
    (b) construction of a building?
15. What provision is made in systems design to insure that each design cycle is an improvement over the preceding one?
16. Compare the objectives of analysis, synthesis, and appraisal.

Chapter 2

Basic Building Elements and Their Representation

Overall optimization of the building system is the goal of systems design. Buildings, however, usually are too complex for immediate, direct optimization of the total system. Instead, it is first necessary to synthesize subsystems that, when combined, form the building system. After normal design studies and value analysis of these subsystems, they may be replaced partly or entirely by better subsystems. This cycle may be repeated several times. Then, the final subsystems may be optimized to yield the optimum building system.

The subsystems usually are composed of basic elements common to most buildings. For the preceding process to be carried out, a knowledge of these basic elements and of some of the simpler, commonly used subsystems in which they are incorporated is essential. This information is provided in this chapter.

This chapter also describes the means by which designers' concepts of buildings, building elements to be used and the manner in which they are to be assembled are communicated to others, in particular to owners, contractors and building department officials. Subsystem design is discussed in later chapters.

To simplify terminology, a building as a whole is called a building system in this book, or simply a building. Major subsystems of buildings are called systems; for example, floor systems, roof systems, plumbing systems, etc. Two or more components of such systems may form a subsystem.

2.1. MAIN PARTS OF BUILDINGS

Nearly all buildings are constructed of certain basic elements. For illustrative purposes, several of these are indicated on the cross section of a simple, one-story building, with basement, shown in Fig. 2.1.

Structure

To provide a flat, horizontal surface on which desired human activities can take place, all buildings contain at least one *floor*. In primitive buildings, the ground may be used as the floor. In better buildings, the floor may be a *deck* laid on the ground or supported above ground on structural members, such as the *joist* indicated in Fig. 2.1.

To shelter the uppermost floor, buildings are topped with a *roof*, usually waterproofed to exclude precipitation. Often it is necessary to support the roof over the top floor on structural members, such as the *rafter* shown in Fig. 2.1. For further protection against wind, rain, snow and extreme temperatures, the outer perimeter of the floors are enclosed with an *exterior wall* extending from ground to roof (see Fig. 2.1). If the building extends below the ground surface, for example, to provide a *basement* as

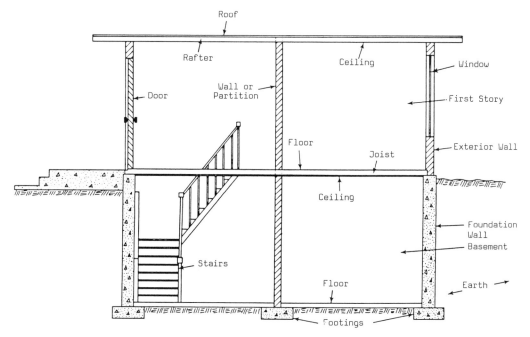

Fig. 2.1. Vertical cross section of a one-story building with basement.

does the structure in Fig. 2.1, *foundation walls* must be furnished to carry the exterior walls and to keep the earth outside from collapsing into the basement. Unless the foundation walls can be seated on strong rock, some sort of support must be furnished to keep them from sinking into the soil. For this purpose, *spread footings*, such as those shown in Fig. 2.1, are often used. These distribute the load of the walls over a large enough area that settlement of the soil under the walls is inconsequential.

In most buildings, spaces for various activities are enclosed, to separate them from each other, to form *rooms*. The enclosures are called *interior walls* or *partitions*.

Circulation

At least one partition or wall around a room has an opening to permit entry or to exit from the room. Such openings usually are equipped with a *door*, a panel that can be moved to fill the opening, to bar passage, or to clear the opening. Exterior walls also have openings equipped with doors, to permit entry to and exit from the building interior.

In multistory buildings, because there is one floor above another, *stairs* are provided, for normal or emergency use, to permit movement from one floor to the next. Sometimes, stairways with moving steps, driven by electric power, called *escalators*, are installed to move people from floor to floor. In buildings with many floors, *elevators*, powered lifting devices, are provided for vertical transportation. In some buildings, such as parking garages and stadiums, sloping floors, or *ramps*, are used for movement between floors.

Environmental Control

To admit daylight to the building interior and to give occupants a view of the outdoors, the exterior walls usually contain openings in which *windows* glazed with a transparent material are inserted. The windows, like the exterior walls in which they are placed, must exclude wind, rain, snow and extreme temperatures. Also, the windows often are openable so that they can be used to ventilate the building interior.

For maintenance of desirable indoor temperatures, equipment usually must be installed for *heating* or *cooling*, or both. Often, this equipment is supplemented by *ducts* or *pipes* that conduct warmed or chilled air or liquid to various rooms in the building. In addition, *chimneys* are provided to vent to the outdoors smoke and gases produced in burning fuel for heating.

Plumbing

In most buildings, certain pipes referred to as plumbing, must be installed. Some of these pipes are necessary for bringing *water* into the building and distributing it to points where needed. Other pipes are essential for collecting *wastewater*, roof rainwater drainage and sometimes other wastes and conducting those substances out of the building, to an external sewage disposal system. Still other plumbing may be used to bring *heating gas* into a building and distribute it to points where needed; and other plumbing is needed for *venting* air or gases from some of the pipes, when necessary, to the outdoors.

Also considered as part of the plumbing system are associated *valves*, *traps* and other controls and fixtures. The plumbing fixtures include *sinks*, *lavatories*, *bathtubs*, *water closets*, *urinals* and *bidets*.

Electrical Systems

In most buildings, *electric equipment and wiring* are provided to bring electric power into the interior and distribute the power where needed, for lighting, heating, operating motors, control systems and electronic equipment. *Lighting fixtures* also are considered part of the electrical system. Other wiring also is installed for communication purposes, such as *telephone*, *paging and signal and alarm systems*.

2.2. FLOORS AND CEILINGS

As mentioned in Sec. 2.1, floors provide the flat, horizontal surfaces on which desired human activities take place in a building. Primarily then, a floor is a deck on which people walk, vehicles ride, furniture is supported, equipment rests and materials are stored.

Floor-Ceiling Systems

Often, for aesthetic reasons, for foot comfort, for noise control or to protect the deck from wear, a *floor covering* is placed atop the deck. In such cases, the deck is called a *subfloor*.

When a floor is not placed directly on the ground; for example, when a floor extends above a room below, some means must be provided for supporting the deck in place. For this purpose, the deck may be propped up on such supports as walls, partitions or columns (posts). If the deck is made strong and stiff enough, it can span unassisted between those supports. Usually, however, supports are placed far apart so as not to interfere with the room layout below. As a result of such spacing, the deck would have to be too thick and too heavy to be self supporting. In such cases, horizontal structural members, called *beams*, have to be provided to carry the weight of the deck and the loads on it to the vertical supports.

Figure 2.2 illustrates two of many types of floor construction in use. Figure 2.2*a* shows a floor system often used in houses. The plywood subfloor is covered with carpeting. On the underside, the subfloor is supported on

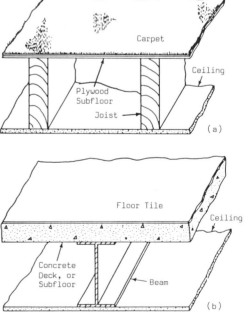

Fig. 2.2. Floor construction. (*a*) Plywood subfloor on wood joists. (*b*) Concrete deck on steel beams.

wood structural members, called joists. (*Joist* is a term generally applied to very light, closely spaced, floor-beams.) Because the plywood is thin, the joists are closely spaced, usually 16 or 24 in. center to center, to provide adequate support. Figure 2.2*b* shows a floor sometimes used in office buildings. The subfloor is strong and thick, often made of concrete. It may be covered with linoleum, asphalt or vinyl tile, or carpeting. Beams for supporting this floor may be placed relatively far apart and have to be strong and stiff. They may be steel beams, as shown in Fig. 2.2*b*, or concrete or timber beams.

The underside of the floor, including the floorbeams, and decorative treatment that may be applied to that side is called a *ceiling*. Alternatively, a ceiling may be a separate element, or membrane, placed below the subfloor and beams and usually supported by them. Figure 2.2 shows flat, horizontal ceilings.

The *plenum*, or space, between deck and ceiling below in Fig. 2.2 need not be wasted. It can be put to use for housing recessed lighting fixtures and as a passageway for ducts, pipes and wiring. Otherwise, space for these elements, except the lighting fixtures, might have to be provided above the floor, where space is much more valuable.

Fire Protection

Beams, whether heavy members or joists, are critical members. If they should be damaged, they might bend excessively or break, causing collapse of the floor and serious injury to building occupants. Damage to beams might be caused by overloading the floor, cutting holes in improper places in a beam for passage of pipes or ducts, or by fire or high heat. Overloading, however, usually is very unlikely. Structural engineers design beams for much heavier loads than those likely to be imposed. Holes, though, sometimes are cut in the wrong places by ignorant, improperly trained or careless construction personnel. Proper supervision and inspection can prevent this or at least institute corrective measures before an accident results. Fire damage, like overloading, can become a rare occurrence by good building design.

Beams usually are made of concrete, wood, or steel. These materials have different fire resistances. Concrete, if thick enough, can withstand fire for hours. But wood structural members are slow burning at best and combustible at worst. Steel structural members, though incombustible, can be damaged by fire, if the fire is hot enough and lasts long enough.

Both wood and steel members, however, can be protected from fire. A common method is to enclose such members with a suitable thickness of an insulating, incombustible material, such as concrete or plaster. As an alternative, wood members can be impregnated with fire-retardant chemicals.

Tests have been made to determine for how long a time specific thicknesses of various materials can protect structural components from a rapid buildup of heat, called a *standard fire*. Based on the tests, these thicknesses and components have been assigned *fire ratings*. The ratings give the time, in hours, that the various types of construction so protected can withstand a standard fire. Building codes, in turn, indicate the minimum fire ratings that building components should possess, depending on type of building and how the building is used.

Concrete is an incombustible material with good resistance to heat flow. When concrete floors are constructed with the thickness required for structural purposes, the floors usually are assigned a high-enough fire rating to protect wood or steel beams below from a fire above the floors. In such cases, however, fire protection still may be required for the bottom and sides of the beams. For steel beams, this protection may be furnished by complete embedding of the beams in concrete, with a minimum cover of 1 or 2 in.; but this type of construction is heavy and therefore often undesirable. When this is the case, the sides and, if necessary, the bottom of the beams can be sprayed with a lightweight, protective material to a thickness of about 1 or 2 in. (see Fig. 2.3*a*), or the fire protection can be boxed out with 1- or 2-in.-thick plaster or concrete (see Fig. 2.3*b*).

In many buildings, however, for aesthetic reasons as well as for fire protection, the beam bottoms and sides are protected with a continu-

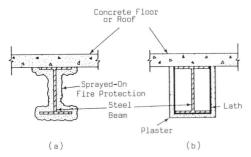

Fig. 2.3. Fire protection for beams supporting a concrete floor or roof. (*a*) With a sprayed-on insulating material. (*b*) With boxing-in by insulating construction.

ous ceiling, as shown in Fig. 2.2. Gypsum plaster, gypsumboard, or insulating, acoustic tiles often are used for such fire protection.

Thus, a floor-ceiling system often may consist of a floor covering, subfloor, beams, fire protection, plenum and ceiling. Floor-ceiling systems are further discussed in Chap. 16.

2.3. ROOFS

The purpose of a roof, as indicated in Sec. 2.1, is mainly to shelter the uppermost floor of a building. Thus, the roof must exclude wind, rain and snow. Generally, it is desirable also that the roof resist passage of heat, to keep out solar heat in warm weather and to prevent heat from escaping from the building in cold weather. In addition, the roof must be strong and stiff enough to support anticipated loads, including wind, ponded rainwater, collections of snow and weight of repairmen.

Roof construction resembles floor construction. Usually, a roof, like a floor, has a top covering. For a roof, however, the covering generally is wind resistant and waterproof, and unless intended to serve also as a promenade or a patio, the roof covering is not so wear resistant as a floor covering. Called *roofing*, this waterproof layer usually is thin. Therefore, it is laid on a *roof deck*, which is similar to the subfloor in a floor system. Also, as in a floor, *beams* often have to be furnished to support the roof deck, which has to span over the interior spaces of the building. In addition, a *ceiling* may be placed under and supported from the roof and beams. Unlike a floor, however, a roof

often incorporates a layer of *thermal insulation* to resist passage of heat.

Figure 2.4 illustrates two of many types of roof construction in use. Figure 2.4*a* shows a cross section through a sloping roof often used for houses. A deck is needed to support the roofing, which may be roofing paper or felt covered by protective shingles, tile or similar, relatively small, overlapping elements. These are usually also part of the aesthetic treatment of the building, because a sloping roof is visible from the ground. In Fig. 2.4*a*, the deck is shown supported on wood rafters, which are laid along the slope of the roof and rest on the exterior walls of the building. (*Rafter* is a term generally applied to a light roof beam.) The rafters are closely spaced, usually 16 or 24 in. on centers. Because of the close spacing, a thin deck can be used; for example, plywood. A deck this thin often is called *sheathing*, as indicated in Fig. 2.4*a*, to denote its primary role as an enclosure.

Fig. 2.4*b* shows a flat roof, often used for industrial or office buildings. The roof deck is strong and thick, frequently made of concrete. It usually is covered with a continuous, bituminous, waterproofing membrane. Structural members, called *purlins*, for supporting this deck may be placed relatively far apart and may be steel beams, as shown in Fig. 2.4*b*, or concrete or timber beams.

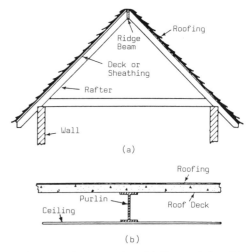

Fig. 2.4. Roof construction. (*a*) Plywood sheathing on sloping rafters. (*b*) Flat concrete deck on steel beams.

Just as is done for floorbeams, protection against injury must be provided for roof beams. Fire protection, in particular, can be furnished for roof beams in the same way as for floorbeams, as indicated in Sec. 2.2. Building codes indicate the minimum fire ratings that roof construction should possess, depending on type of building and how the building is to be used.

Thus, a roof-ceiling system often consists of roofing, roof deck, beams, thermal insulation, fire protection, plenum and ceiling. Roof systems are further discussed in Chap. 15.

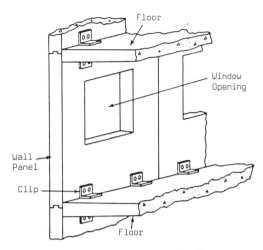

Fig. 2.6. Concrete panel wall.

2.4. EXTERIOR WALLS AND OPENINGS

For many reasons, buildings are enclosed by walls along their perimeters. The most important reason is to shelter the building interior from wind, rain, snow and extreme temperatures.

An exterior wall may be a single element or it may consist of several elements. In the latter case, a typical wall may be built with an exterior facing, a backing, insulation and an interior facing.

In general, a wall, interior or exterior, may be built in one of the following ways:

Unit Masonry. One basic way is to assemble a wall with small units, such as clay brick, concrete block, glass block, or clay tile, held together by a cement, such as mortar. Figure 2.5 shows a wall consisting of two vertical layers, or *wythes*, of clay brick.

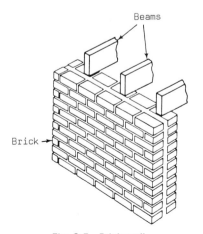

Fig. 2.5. Brick wall.

Panel Wall. A second basic way is to form a wall with large units. A panel, for example, may be large enough to extend from floor to ceiling and to incorporate at least one window. Figure 2.6 illustrates such a panel. Sometimes, however, a panel need be only deep enough to extend from a floor to a window above or below.

Framed Wall. A third basic way is to construct a wall with thin, closely spaced structural members to which interior and exterior facings are attached and between which insulation may be placed. Figure 2.7 is an example of a wood-framed exterior wall, viewed from the inside, often used for small houses. The vertical structural members, called *studs*, are tied together at top and bottom with horizontal members, called *plates*. A continuous bracing member, called *sheathing*, which may be plywood or a gypsum panel, is attached to the outer face of the studs. If the sheathing is not waterproof, a waterproofing sheet is fastened to its outer surface. Then, the exterior is covered by a facing, which may be brick, wood siding, asbestos-cement shingles or other finish desired by the architect and the owner. Thermal insulation may be installed between the studs. An interior finish, such as gypsum plaster or gypsumboard, usually is attached to the interior face of the studs, to complete the wall.

Combination Walls. Because metals, brick, concrete and clay tile are strong, durable and water and fire resistant, one of these materials

Fig. 2.7. Load-bearing wood-framed wall.

often is used as an outer facing. To reduce wall costs, a less-expensive material may be used as a backup. Often, for example, unit masonry may be used as the exterior facing with wood framing, or unit masonry may be the backup with a panel facing.

Curtain Walls

An exterior wall may serve primarily as an enclosure. Such a wall is known as a nonload-bearing, or curtain, wall. The wall in Fig. 2.6 is a curtain wall. Supported by the floors above and below, the wall need be strong enough to carry only its own weight and wind pressure on its exterior face.

Load-bearing Walls

An exterior wall also may be used to transmit to the foundations loads from other building components, such as other walls, beams, floors and roof. Such a wall is known as a load-bearing wall, or, for short, a bearing wall. Figure 2.5 shows a brick-bearing wall, while Fig. 2.7 illustrates a wood-framed bearing wall.

Openings

In Sec. 2.1, the necessity of doors for entrance to and exit from a building and the desirability of windows are indicated. Openings must be provided for these in the exterior walls.

Where such openings occur, structural support must be provided over each opening to carry the weight of the wall above as well as any other loads on that portion of the wall. In the past, such loads were often supported on masonry arches. Currently, the practice is to carry the load on straight, horizontal beams. For masonry walls, the beams, often steel an-

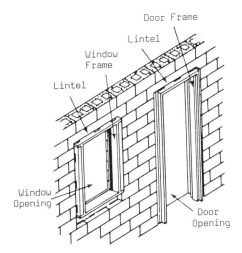

Fig. 2.8. Lintels support the wall above openings.

gles or rectangular concrete beams, are called *lintels* (see Fig. 2.8). In wood-framed walls, the beams are called *top headers*.

Windows

In exterior walls, openings equipped with windows substitute a transparent material for the opaque walls. Such openings offer occupants a view of the outside (see Fig. 2.9) or, for retail stores, provide passersby a view of merchandise on display inside. The transparent material usually is glass, but plastics also may be used. In either case, the material, called *glazing*, generally is held in place by light framing, known as *sash*. The combination of sash and glass is usually referred to as a window.

An important function of windows is transmission of daylight for illuminating the adjacent building interior. When windows are openable, the opening may also be used to provide interior ventilation. Many types of windows are available.

Supports, called a *window frame*, usually are provided around the perimeter of the opening and secured to the wall (see Fig. 2.8). For sliding windows, the frame carries guides in which the sash slides. For swinging windows, the frame contains stops against which the window closes.

In addition, *hardware* must be provided to enable the window to function as required. The hardware includes locks, grips for moving the window, hinges for swinging windows, and sash balances and pulleys for vertically sliding windows.

Fig. 2.9. Windows in an exterior wall.

Exterior-Wall System

As indicated in the preceding, an exterior wall may have many components. It is not unusual for a wall system to include interior and exterior facings, backup, thermal insulation, windows, doors, and lintels and other framing around openings. Exterior-wall systems are further discussed in Chap. 15.

2.5. PARTITIONS, DOORS, AND INTERIOR-WALL FINISHES

As indicated in Sec. 2.1, interior walls or partitions are used to separate spaces in the interior of buildings. The term interior walls often is reserved for load-bearing walls, whereas the term partitions generally is applied to nonload-bearing walls.

Neither interior walls nor partitions are subjected to such strenuous conditions as exterior walls. For example, they usually do not have to withstand outside weather or solar heat, but they do have two surfaces that must meet the same requirements as the interior faces of exterior walls, as described later.

Because load conditions generally are not severe, partitions may be constructed of such brittle materials as glass (see Fig. 2.10a), weak materials as gypsum (see Fig. 2.10b and c), or thin materials as sheet metal (see Fig. 2.10a). Some light framing, however, may be necessary to hold these materials in place.

Some partitions may be permanently fixed in place. Others may be movable, easily shifted. Still others may be foldable, like a horizontally sliding door.

Load-bearing walls must be strong enough to transmit vertical loads imposed on them to supports below. Such interior walls may extend vertically from roof to foundations (see Figs. 2.10d and 2.12).

Often, interior walls and partitions are required to be fire resistant as well as capable of limiting passage of sound between adjoining spaces or both.

Doors

Exterior walls are provided with openings for permitting entrance to and exit from buildings (see Fig. 2.8). These openings are equipped with doors that open to allow entry or exit and close to bar passage. Similarly, openings are provided in interior walls and partitions to permit movement of people and equipment between interior spaces. These openings also are usually equipped with doors to control passage and also for privacy.

Many types of doors are available for these purposes. They may be hinged on top or sides, to swing open or shut. They may slide horizontally or vertically. Or they may revolve about a vertical axis in the center of the opening.

A *lintel* is required to support the portion of the wall above the door. Additional framing, called a *door frame*, also is needed for supporting the door and the *stops* against which it closes (see Fig. 2.11).

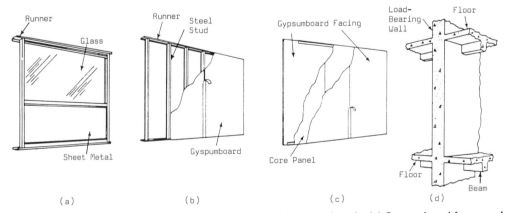

Fig. 2.10. Partitions. (a) Nonload-bearing. (b) Gypsumboard on metal studs. (c) Gypsumboard face panels laminated to gypsum core panel. (d) Load-bearing concrete interior wall.

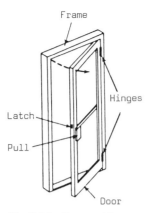

Fig. 2.11. Door and frame.

In addition, hardware must be furnished to enable the door to function as required. For example, a swinging door must be provided hinges on which to swing. Also, a lock or latch usually is needed to hold the door in the closed position. A knob or pull is desirable for opening and closing the door and controlling its movement.

Fig. 2.12. Load-bearing concrete wall supports concrete floors in a multistory building. Floors and walls were prefabricated away from the building site. (Courtesy Formigli Corp.)

Hardware

Builders' hardware is a general term covering a wide variety of fastenings and devices, such as locks, hinges and pulleys. It includes finishing and rough hardware.

Finishing hardware consists of items that are made in attractive shapes and finishes and are usually visible as an integral part of the completed building. Door and exposed window hardware are examples of finishing hardware.

Rough hardware applies to utility items that are not usually finished for attractive appearance. Rough hardware includes nails, screws, bolts, and window sash balances and pulleys.

Interior-Wall Finishes

The inside faces of exterior walls and faces of interior walls and partitions that are exposed to view in rooms, work areas or corridors should usually satisfy such requirements as attractive appearance, easy to clean, durable under indoor conditions and inexpensive maintenance. Preferably, the facings should be fire and water resistant and also should have acoustic properties appropriate to the space enclosed.

A wide variety of finishes are used for interior walls. In residential and commercial construction, plaster and gypsumboard, with paint or wallpaper decorative treatment, are often used because of good fire resistance. Sometimes, however, plywood, fiberboards or plastics are chosen for aesthetic reasons. For factories or schools, where harder or perhaps chemical-resistant finishes are desired, unit-masonry or tile surfaces often are left exposed or given a tough, decorative coating. In restaurants and theaters, in contrast, acoustic requirements are given high priority, though fire resistance and aesthetics also are important.

Interior-Wall System

As indicated in the preceding, an interior wall or partition may have several components. Often, an interior-wall system may include a facing on one or two sides, a backup, means of attachment to floors and ceilings, doors and lintels or other framing around openings. Interior-wall systems and doors are further discussed in Chap. 16.

FOR SECTIONS 2.1 THROUGH 2.5

References

C. Ramsey and H. Sleeper, *Architectural Graphic Standards*, 8th ed., Wiley, New York, 1988.

F. Merritt, *Building Design and Construction Handbook*, 4th ed., McGraw-Hill, New York, 1982.

Words and Terms

Beam	Plenum
Ceiling	Purlin
Curtain wall	Rafter
Deck	Roofing
Ducts	Sash
Fire rating	Sheathing
Glazing	Spread footing
Joist	Stud
Lintel	Subfloor
Masonry	Thermal insulation
Partition	Vents
Piping	Wiring
Plate	Wythe

Significant Relations, Functions and Issues

Components of building elements: structure, surfacing (structural decks, sheathing), finishes (roofing, flooring, etc.), enhancements (thermal insulation, weather seals, fire protection, etc.).

Floor functions and features: horizontal surface, facilitation of activities, support of suspended items (ceiling, equipment, etc.), creation of plenum, fire separation.

Roof functions: drainable surface, exclusion of precipitation, insulation of building exterior, support of suspended items, facilitation of openings (chimneys, vents, ducts, skylights, etc.).

Exterior wall functions: insulation of building exterior, major exterior building appearance, facilitation of openings for windows and doors.

Interior wall functions: interior space and circulation control, separation for fire, acoustics and security, ease of rearrangement if nonstructural.

Circulation elements: doors, stairs, elevators, escalators and ramps.

2.6. STRUCTURAL FRAMING AND FOUNDATIONS

Sections 2.2 and 2.3 indicate that floors and roofs must be strong and stiff enough to span alone over spaces below or else beams must be provided to support them. In either case, decks or beams must be propped in place. For this purpose, additional structural members must be provided.

Sometimes, load-bearing walls can be used, as pointed out in Secs. 2.4 and 2.5 (see also Fig. 2.12). In other cases, especially when beams support the floors or roof, strong, slender, vertical members, called *columns*, are used. If, however, columns were used under every beam, the building interior might become objectionably cluttered with them. So, instead, the beams often are supported on strong cross beams, called *girders*, which then are seated on the columns (see Fig. 2.13). This type of construction is called *skeleton framing*.

Foundations

The vertical supports for floors and roof must carry all loads to foundations situated at or below ground level. The ground is the ultimate support for the building.

Foundations are the structural members that transmit building loads directly to the ground. Usually, foundations are built of concrete, because this material is strong and durable.

When a building has a basement, it is enclosed in continuous foundation walls, to exclude the surrounding earth. In that case, perimeter, or exterior, walls and columns of the upper part of the building (*superstructure*) may be seated on the foundation walls.

When there is no basement, foundations should extend into the ground at least to the *frost line*, the depth below which the ground is not likely to freeze in cold weather. Freezing and thawing of the soil can cause undesirable movements of foundations seated on that soil.

Ordinarily, soil will settle excessively if called on to support a column or wall directly, so walls are spread out at the base to distribute the loads they carry over large enough areas that settlement is inconsequential. The spread-out base under a wall is known as a *continuous spread footing* (see Fig. 2.14a). Similarly, if column loads are to be distributed directly to the ground, each column is seated on a broad, thick pad, called an *individual spread footing* (see Fig. 2.14b). Sometimes, however, soil is

Fig. 2.13. Skeleton framing of structural steel for a multistory building. Inclined columns are used to increase spacing of exterior columns in the lower part of the building. (Courtesy United States Steel Corp.)

so weak that the spread footings for columns become so large that it becomes more economical to provide one huge spread footing for the whole building. Such a footing is called a *raft*, or *mat*, *footing* and occasionally a *floating foundation*.

When the soil is very weak, spread footings may be impractical. In such cases, it may be necessary to support the columns and walls on *piles*. These are structural members very much like columns, except that piles are driven into the ground. Usually, several piles are required to support a column or a wall. Consequently, a thick *cap*, or *pile footing*, of concrete is placed across the top of the group of piles to distribute the load from the column or wall to the piles (see Fig. 2.15).

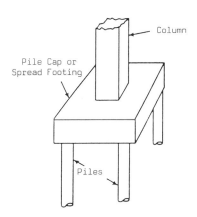

Fig. 2.14. Spread footings. (*a*) Continuous footing for wall. (*b*) Individual footing for column.

Fig. 2.15. Pile footing for column.

Lateral Stability

Walls and columns by themselves have good capability for supporting the weight of a building and its contents (*gravity loads*). Not all building loads are vertical, though. Wind or earthquakes, for example, may impose horizontal forces on the building. Walls or columns alone may not be adequate to withstand these lateral loads, which, if not resisted, could destroy the building in either of two ways: Horizontal loads may overturn the building or they might collapse it like a house of cards. If adequate precautions are not taken, the horizontal loads might rack rectangular beam-column or beam-wall framing into a flattened parallelepiped.

To prevent overturning, height-width and height-thickness ratios of buildings must be kept within reasonable limits. Also, column bases must be anchored to prevent uplift.

To prevent a racking failure, the structural framing must be designed to transmit the horizontal forces to the ground. Several means are available for doing this.

One way is to provide diagonal members, called *bracing*. These work with beams and columns or other structural members in transmitting horizontal forces to the foundations and from them to the ground (see Fig. 2.16a).

Another way is to make *rigid connections* between beams and columns, to restrict rotation of these joints (see Fig. 2.16b). Then, the lateral loads cannot distort the rectangular beam-column framing into a parallelepiped.

Still another way is to provide long walls, called *shear walls*, in two perpendicular directions (see Fig. 2.16c). Because a wall by itself has low resistance to horizontal forces acting perpendicular to its faces although it has high resistance to such forces acting parallel to its faces, one wall alone cannot resist wind or earthquake forces that may come from any direction. But no matter in what direction the forces may act, two perpendicular walls can resist them.

Fire Protection

Sections 2.2 and 2.3 point out that fire protection may be required for floors and roofs, and especially for beams, and describe how such protection generally is provided. Similarly, fire protection may be required for columns and bracing.

Bracing in buildings where fire protection is required often is encased in floors, roof or walls. In such cases, the encasement usually provides adequate fire protection.

Columns also may be encased in walls that provide adequate fire protection. Otherwise, columns may be encased in concrete or enclosed in boxed-out fireproofing, much like the beam in Fig. 2.3b.

Structural System

From an overall view, the structural system may be considered to consist of load-bearing walls, skeleton framing (beams and columns),

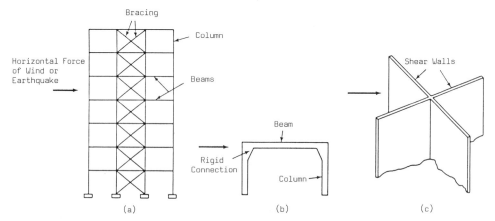

Fig. 2.16. Lateral bracing of buildings to resist horizontal loads. (*a*) X bracing. (*b*) Rigid-frame construction. (*c*) Shear walls.

bracing, shear walls and foundations. Because foundation design has become a specialty, the structural system is sometimes partitioned into two systems: *foundations*, or *substructure*, and *superstructure*, the walls and framing above the foundations. Structural systems are further discussed in Chap. 8.

FOR SECTION 2.6

References

M. Salvadori, *Structure in Architecture: The Building of Buildings*, 3rd ed., Prentice-Hall, Englewood Cliffs, NJ, 1986.
J. Ambrose, *Building Structures*, Wiley, New York, 1988.
F. Merritt, *Building Design and Construction Handbook*, 4th ed., McGraw-Hill, New York, 1982.

Words and Terms

Beam	Rigid connections
Bracing	Shear walls
Column	Skeleton framing
Frost line	Spread footings: continuous,
Foundations	individual
Girders	Substructure
Lateral stability	Superstructure
Piles	

Significant Relations, Functions, and Issues

Nature of structural system: skeleton framing versus bearing wall.
Foundation issues: depth below grade (to bottom of construction, to good soil, below frost line), type (spread footing or deep—pile or caissons), size for load magnitude.
Lateral stability issues: type of bracing, critical load—wind or seismic, three-dimensional stability.
Relation of superstructure to substructure.

2.7 PLUMBING

The main functions of plumbing are twofold:

1. To bring water and also heating gas, if desired, from sources outside a building to places inside where they are needed
2. To collect wastewater and stormwater at points inside the building, or on the roof, or elsewhere on the site and to deliver these wastes to sewers outside the building

Execution of these functions of plumbing primarily requires water, air, gas and pipes. Also needed, however, are the following:

1. Fixtures for utilizing water, such as lavatories, drinking fountains, bathtubs and showers
2. Fixtures for receiving wastewater and stormwater, such as water closets, urinals and drains
3. Control and safety devices, such as valves, faucets and traps
4. Storage tanks and pumps
5. Vents for removal of gases generated in the wastewater system or by combustion
6. Fire fighting devices, such as detection devices, alarms, sprinklers, hoses and hose valves

Water

Availability of good water in adequate quantity is a prime consideration in locating, designing and constructing any building. Usually, there must be a potable supply ample to meet the needs of all who will reside, work or visit in the building. In addition to this basic domestic need, there must be water for heating, air conditioning, fire protection and wastewater disposal. Also, for industrial buildings, there are numerous process uses plus a vast equipment-cooling job for water. It is the function of the plumbing system to transport the needed water from points of entry into the building to points of use.

Heating Gas

This is an optional fuel often selected for buildings. Because the gas can form explosive mixtures when air is present, gas piping must be absolutely airtight, not only to prevent gas from escaping but also to prevent air from entering.

Wastewater Disposal

The ability to get rid of wastes is as important a consideration in building design as water supply. Even for a small house with the normal small flow of domestic sewage, early determination is necessary as to whether sewers are available and can be connected to easily; or if not, whether local regulations or physical conditions permit other economic means of disposal, such as cesspools or septic tanks. For big industrial buildings requiring large quantities of water for cooling and processes, site selection may well hinge on available and allowable means of wastewater disposal. The cost of bringing water into a building may prove small compared to the cost of discharging the water after use in a condition acceptable to those responsible for preventing pollution of the environment. When wastewater cannot be discharged untreated into a public sewer, the alternative usually is provision by the building owner of sewage treatment facilities.

The responsibility of a building's plumbing system for wastewater removal extends from points of reception inside the building to a public sewer or other main sewer outside.

Plumbing Code

Because improper functioning of any plumbing in a building can impair the health or safety of occupants and possibly others in the community, state and municipal regulations have been established to govern plumbing design and installation. These regulations often incorporate or are based on the "National Plumbing Code," which has been promulgated by the American National Standards Institute as a standard, designated A40.8. This code gives basic goals in environmental sanitation and is a useful aid in designing and installing plumbing systems in all classes of buildings.

Plumbing systems are further discussed in Chap. 9.

2.8. HEATING, VENTILATING, AND AIR CONDITIONING (HVAC) SYSTEMS

Two issues regarding the building interior are usually combined for design purposes. The first is the need for fresh, clean air, described as the need for *ventilation*, and the second is the need for *thermal control*. Systems design to achieve control of these two conditions are often combined, but it is also possible to do the two tasks separately.

Ventilation is needed in the interior of a building to supply clean air for breathing and to remove odors, tobacco smoke, carbon dioxide and other undesirable gases. Ventilation, however, also is useful for drawing warm or cool air into a building from outdoors to make the interior more comfortable. For this purpose, large quantities of outdoor air often are needed, whereas much smaller quantities of fresh air usually are essential for the prime objective of ventilation.

If an interior space has windows that are openable, the simplest way to ventilate it is to open the windows. This method, however, often is impractical or cannot be used (see Sec. 10.6).

When natural ventilation cannot be used, *mechanical ventilation* is necessary. In such cases, fans are used to draw fresh air into the building and to distribute the air to interior spaces. Often, the air is filtered to remove particles in it before it is distributed within the building. When it is necessary to ventilate remote or windowless spaces, fresh air can be distributed through conduits or ducts to those spaces.

In many cases, fresh air can be introduced directly into interior spaces only in mild weather. In cold weather, the air must first be heated, and in hot weather, the air must first be cooled or the occupants will be made uncomfortable.

General conditioning of ventilating air may involve many concerns, depending on the nature of building activities and that of the climate and general environment outside the building. In warm climates, where cooling is generally the more critical problem—instead of heating—it is usual to combine ventilation and general thermal conditioning in a single operation. In very cold climates, however, adequate heating of large masses of cold air is usually not practical. Hence only the minimum volume of air required for ventilation is heated. General building heating in cold climates is mostly achieved by other means. In many cases, in fact, unless the building exterior is very tightly sealed, ventilation in very cold weather is assumed to be adequately achieved by the leaking of air into the interior through cracks around doors and windows and other construction joints. In the latter case, however, interior spaces at some distance from the exterior walls may still need some form of mechanical ventilation.

HVAC Systems

A wide variety of systems are available for heating, ventilation and air conditioning (HVAC). Basically, they may be divided into two classes: central plant and unit.

A central-plant system concentrates heating or cooling sources in one area to serve a substantial portion of a building or one or more buildings.

A unit system has two or more heating or cooling sources throughout a building.

For example, a house with a furnace in the basement has a central-plant system, whereas one heated by a fireplace or stove in each of several rooms has a unit system. An industrial building heated with steam from a boiler in a boiler room has a central-plant system, whereas a building heated by direct-fired heaters in strategic locations throughout production and storage areas has a unit system.

The two classes differ not only in sizes and capacities of equipment required but also in methods of delivering heating and cooling to points where needed. Central-plant systems generally require conduits, pipes or ducts for distribution of heating or cooling media. Unit systems, in contrast, usually can supply heating and cooling directly to the spaces requiring them. Often, however, central-plant systems give better distribution and are more economical to operate, though initial costs may be higher.

Humidity

An important factor affecting human comfort or, in some cases, a desirable industrial environment is humidity. Building air almost always contains humidity, some water in vapor form. The relative amount of this vapor influences the comfort of building occupants, depending on the temperature. In some cases, humidity is necessary for manufacturing processes and in other cases, it is undesirable, for example, for some storage spaces.

When a building is heated, the relative humidity decreases unless moisture is added to the air. If the air becomes too dry, occupants will become uncomfortable. Hence, it is often necessary to add moisture to building air during the heating season.

In hot weather, high humidity will make occupants of a building uncomfortable. In such cases, removal of moisture from the air is desirable.

Consequently, a HVAC system should not only provide appropriate temperatures within a building but also control the humidity.

HVAC is discussed in more detail in Chap. 10.

2.9. LIGHTING

Illumination is a necessity in a building. Without light, humans cannot see and are unable to perform many essential activities. Furthermore, moving about would be hazardous within a building, because of potential collisions with unseen objects, the peril of tripping and the danger of falling down stairs.

Good lighting, for a specific building function, requires an adequate quantity of light, good quality of illumination and proper colors. These characteristics are interrelated; each affects the others. In addition, effects of lighting are significantly influenced by the colors, textures and reflectivities of objects illuminated.

Illumination of a building interior may be accomplished by natural or artificial means.

Natural illumination is provided by daylight. It is brought into a building through fenestration, such as windows in the exterior walls or monitors or skylights on the roof.

Artificial illumination usually is accomplished by consumption of electric power in incandescent, fluorescent, electroluminescent or other electric lamps and occasionally by burning of candles, or oil or gas lamps. For artificial lighting, a light source usually is enclosed in a housing, called a *luminaire* or *lighting fixture*, which may also contain devices for directing and controlling the light output.

Electric power is conducted to the light

sources by wires. Manual switches for permitting or interrupting the flow of electric current or dimmers for varying the electric voltage to light sources are incorporated in the wiring and installed at convenient locations for operation by building occupants.

Luminaires are mounted on or in ceilings, walls or on furniture. The fixtures may be constructed to aim light directly on tasks to be performed or objects to be illuminated or to distribute light by reflection off walls, ceilings, floors or objects in a room. Electric wiring to the fixtures may be concealed in spaces in walls or floors or between ceilings and floors or extended exposed from electric outlets in walls or floors.

Thus, a lighting system consists of fenestration (windows, monitors, skylights, etc.), artificial light sources, luminaires, mounting equipment for the lighting fixtures, electric wiring, ceilings, walls, floors and control devices, such as switches, dimmers, reflectors, diffusers and refractors. Lighting systems are further discussed in Chap. 11.

2.10. ACOUSTICS

Acoustics is the science of sound, its production, transmission and effects. (Sound and vibrations are closely related.) Acoustic properties of an enclosed space are qualities that affect distinct hearing.

One objective of the application of acoustics to buildings is reduction or elimination of noise from building interiors. Noise is unwanted sound. Acoustical comfort requires primarily the absence of noise. In some cases, noise can be a health hazard; for example, when sound intensity is so high that it impairs hearing. If production of noise in a building is unavoidable, transmission of the noise from point of origin to other parts of the building should be prevented. Accomplishment of this is one of the purposes of acoustics design.

Another objective of acoustics applications is provision of an environment that enhances communication, whether in the form of speech or music. Such an environment generally requires a degree of quiet that depends on the purpose of the space. For example, the degree of quiet required in a theater may be much different from that acceptable in a factory.

In many building interior situations a major concern is for the establishment of some degree of acoustic privacy. This may relate to keeping conversations from being overhead by persons outside some private space, or to a need for freedom from the intrusion of sounds—the latter being a form of noise control. Separations between adjacent apartments, hotel rooms, classrooms, and private offices commonly present concern for these matters.

Installations in a building for sound control may be considered parts of an acoustical system. But it generally will be more efficient if acoustical installations and measures are integrated in other major building systems or subsystems. For example, design of partitions, walls, ceilings and floors should, from the start, have the objective of meeting acoustical requirements. Tacking on acoustical corrections after design or construction has been completed can be costly and not nearly as effective. Acoustical design is discussed in more detail in Chap. 12.

2.11. ELECTRIC SUPPLY

Electric power for buildings usually is purchased from a utility company, publicly or privately owned. Sometimes, however, batteries or a generating plant are provided for a building to supply power for emergency use. Occasionally, a generating plant is installed for normal operation. This may be necessary for buildings in remote locations or for industrial buildings with large or special needs for power that make generation in their own plants economical or essential.

In a building, electric power finds a wide range of uses, including space heating; cooking; operation of motors, pumps, compressors and other electric equipment and controls; operation of electronic devices, including computers; transmission of communication signals; and provision of artificial illumination and other radiation, such as ultraviolet, infrared and X ray.

Electric power is brought from a generating source through cables to an entrance control point and often to a meter in a building. From there, electricity is distributed throughout the building by means of additional conductors. Where needed, means are furnished for withdrawing electric power from the system for operation of electric equipment. Also, controls for permitting flow of electricity or shutting it off, and devices for adjusting voltages, are provided at points in the distribution system. In addition, provision must be made to prevent undesirable flow of electricity from the system.

Accident Prevention

Even a relatively low-power system such as that for a small house can deliver devastating amounts of power. Hence, extraordinary measures must be taken to insure personal safety in use of electricity. Power systems must be designed and installed with protection of human life as a prime consideration.

Also, should electricity be unleashed in unwanted places, perhaps as the result of electrical breakdowns, not only may electric components be destroyed but, in addition, other severe property damage may result, including fire damage. In industrial plants, production equipment may be put out of commission. As a result, replacement and related delays may be costly. In some cases, merely shutting down and restarting operations because of an electrical breakdown may be expensive. Consequently, safety features must be incorporated in the power system for protection of property.

For safety reasons, therefore, state and municipal regulations have been established to govern system design and installation. These regulations often incorporate or are based on the "National Electrical Code," sponsored by the National Fire Protection Association. This code, as well as the legal regulations, however, contain only minimum requirements for safety. Strict application will not insure satisfactory performance of an electrical system. More than minimum specifications often are needed.

Electrical Systems

Generally, buildings may be considered to incorporate two interrelated electrical systems. One system handles communications, including telephone, video monitoring, background music, paging, signal and alarm subsystems. The other system meets the remaining electrical power needs of the building and its occupants.

Both systems have as major elements conductors for distribution of electricity, outlets for tapping the conductors for electricity and controls for turning on or shutting off the flow of electricity to any point in the systems. The conductors and outlets may be considered parts of an electrical subsystem; but it generally will be more efficient if these subsystems are integrated in other major building subsystems. For example, design of partitions, walls, ceilings and floors should, from the start, consider these elements as potential conduits for the electric conductors and possible housing for outlets.

Electrical systems are further discussed in Chap. 13.

2.12. VERTICAL-CIRCULATION ELEMENTS

Very important components of multistory buildings are those that provide a means for movement of people, supplies and equipment between levels.

Ramps

A sloping floor, or ramp, is used for movement of people and vehicles in some buildings, such as garages and stadiums. A ramp also is useful to accommodate persons in wheelchairs in other types of buildings.

Usually, however, a ramp occupies more space than stairs, which can be set on a steeper slope. People can move vertically along a much steeper slope on stairs than on ramps. Stairs, then, are generally provided, for both normal and emergency use.

Stairways

A stair comprises a set of *treads*, or horizontal platforms, and their supports. Each tread is placed a convenient distance horizontally from and vertically above a preceding tread to permit people to walk on a slope from one floor of a building to a floor above (see Fig. 2.17). Often, a vertical enclosure, called a *riser*, is placed between adjacent treads. A riser and the tread

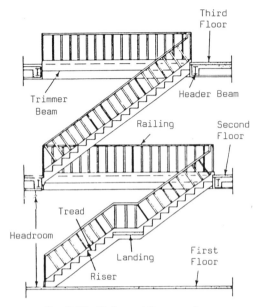

Fig. 2.17. Stairs and floor openings.

above often are referred to as a *step*. The steps of a stairway can be made self supporting but generally are supported on structural members. Stairs are usually provided with *railings* along the sides, for safety reasons.

Where two floors are connected by stairs, an opening at least as wide as the stairs must be provided in the upper floor over the stairs to permit passage to that level. The opening must extend far enough from the top of the stairs out over the steps to prevent persons using the stairs from injuring their heads through collision with the ceiling, floor or structural members at or near the edges of the opening. For this purpose, adequate clearance, or *headroom*, must be provided between every tread and construction above.

Structural framing usually is required around the perimeter of the opening to support the edges of the floor. Also, railings or an enclosure must be provided to prevent people or things from falling through the opening (see Fig. 2.17). The enclosure also may be required for fire protection.

Escalators

In buildings in which there is very heavy pedestrian traffic between floors, for example, department stores, moving stairs, powered by electric motors, may be provided for convenience and rapid movement. Called escalators, these stairs consist basically of a conveyor belt with steps attached, motor, controls and structural supports.

Elevators

For speedier vertical transportation, especially in tall buildings, or for movement of supplies and equipment between levels, elevators usually are installed. They operate in a fire-resistant, vertical shaft. The shaft has openings, protected by doors, at each floor served. Transportation is furnished by an enclosed car suspended on and moved by cables (see Fig. 2.18*a*) or supported atop a piston moved by hydraulic pressure (see Fig. 2.18*b*). The cable-type elevator, driven by electric motors, is suitable for much taller buildings than is the hydraulic type.

Movement of Goods

When elevators are available, they may be used to move freight to the various levels of multistory buildings. For movement of small items, small cable-suspended elevators, called *dumbwaiters*, may be installed. For handling a large flow of light supplies, such as paper work, *vertical conveyors* may be provided. *Belt convey-*

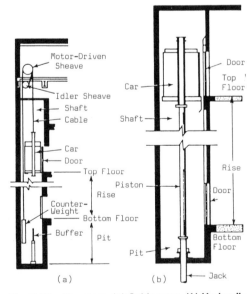

Fig. 2.18. Elevators. (*a*) Cable type. (*b*) Hydraulic.

ors often are used in factories and storage buildings for moving goods both horizontally and vertically.

Vertical circulation elements are discussed in more detail in Chap. 14.

FOR SECTIONS 2.7 THROUGH 2.12

References

B. Stein et al., *Mechanical and Electrical Equipment for Buildings*, 7th ed., Wiley, New York, 1986.
J. Flynn and W. Segil, *Architectural Interior Systems: Lighting, Air Conditioning*, Acoustics, 2nd ed., Van Nostrand Reinhold, New York, 1987.

Words and Terms

Acoustics	Luminaire
Elevator	Plumbing
Escalator	Ramp
HVAC	Stair
Illumination: natural, artificial	Ventilation: natural, mechanical

Significant Relations, Functions and Issues

Functions of plumbing: supply and waste removal.
Functions of HVAC systems: air change and quality, thermal control.
Aspects of lighting: visual tasks, natural illumination (daylight), artificial illumination (electrical), components of lighting systems (power, wiring, fixtures, controls).
Concerns for acoustics: hearing, privacy, noise control, acoustic isolation, and separation of interior spaces.
Aspects of electrical systems: power source, distribution, power level control, general flow control (switches, circuit-breakers, etc.), delivery devices, usage, communication systems, signaling.
Vertical circulation components: ramps, stairs, elevators, escalators, devices for movement of goods.

2.13. WHY DRAWINGS ARE NECESSARY

An architect or engineer designing a building may have a fairly complete picture of the required structure in his mind, but a mental picture at best cannot be entirely accurate nor absolutely complete. Too many items are involved, and there are too many details that are impossible to design and correlate with mental pictures alone. Consequently, the designers' mental pictures must be converted to drawings on paper, film or cloth, where concepts can be developed and completed.

Even if architects and engineers were able to visualize accurately and completely in their minds a picture of the required building, they would find it impossible to transmit exactly the same mental picture to the building owner, consultants, contractors, financiers and others interested in the building. The concepts must be conveyed from the designers to others concerned through construction drawings, which make clear exactly what the designers have in mind for the building.

Construction drawings (also called contract drawings or working drawings) are picture-like representations that show how a building that is to be constructed will appear. They are also called *plans* or *prints*. The latter term refers to reproductions that are used for study, review, fabrication and construction, to preserve the original drawings.

The drawings must show the builders what to do in every phase of construction. In effect, they constitute graphic instructions to the builders. Every detail of construction from foundations to roof must be indicated, to show what has to be placed where and how attached. This must be done in such a manner as to avoid any confusion or misunderstanding.

2.14. DRAWING CONVENTIONS

Construction drawings have to be made in a size for convenient handling by those who have to use them. Hence, elements depicted are usually shown much smaller than actual size. Also, to give an accurate depiction of elements and their positioning in the building, the drawings nearly always are prepared *to scale*.

Each dimension of an element on a drawing bears the same ratio to the actual dimension of the element as does every other dimension shown to the corresponding actual dimension.

Drawings, therefore, are miniature as well as picture-like representations of the building, an exact reproduction of the building on a small scale. (Scales are discussed further in Sec. 2.17.)

Because of the relatively small size of drawings, however, many building components cannot be shown on some drawings exactly as

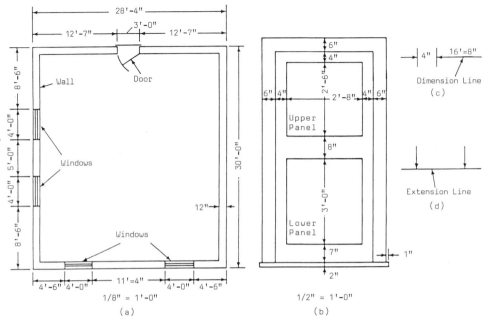

Fig. 2.19. Arrows are used to give dimensions of drawings. (*a*) Floor plan for a one-room building. (*b*) Elevation view of a door. (*c*) Alternate ways of showing dimensions. (*d*) The way to show the extent of a dimension line.

they will look when installed in the building. Consequently, designers have to use a special kind of graphic language to indicate the many items that they cannot actually picture. This language employs *symbols* to represent materials and components that cannot be reproduced exactly. Note, for example, how windows and doors are indicated in Fig. 2.19.

2.15. TYPES OF DRAWINGS

Several different types of drawings are required to show all the information needed for construction of a building. They form a set of construction plans. Following are some of the types that might be included:

Perspective drawings look like pictures and often are drawn to show a building owner a picture of the building before construction begins.

Elevation views show what the exterior of a building will look like. Usually, four such views are required for an ordinary building. Elevation views are discussed further in Sec. 2.18.

Plan views show what a building or its horizontal components, such as floors and roofs, look like when viewed from above. A typical architectural plan view shows the building interior and indicates sizes, shapes, and arrangement of rooms and other spaces, doors, windows, toilet fixtures, kitchen equipment, and other needed information. A structural framing plan indicates the location, orientation, and sizes and gives other pertinent information for floor or roof structural members, such as beams, girders, and columns. An air-conditioning plan gives similar data for equipment, pipes, and ducts. An electrical plan provides information on wiring, power-using equipment, controls, and outlets that supply electric power. Plan views are further discussed in Sec. 2.19.

Section views are used to show the interior construction of various building parts. Section views are discussed in Sec. 2.21.

Detail views are used to provide required information about structural assembly, trim, and various special equipment. Such views often are given to supply information that cannot be shown in the elevation, plan, or section views. Detail views are discussed in Sec. 2.22.

Survey plans supply information concerning the site on which a building is to be constructed. Survey plans are discussed in Sec. 2.23.

Plot plans show where a building is to be placed on a site, how it is to be oriented in that location, and how the ground around the building is to be graded. Plot plans are discussed in Sec. 2.23.

Originals and Prints

The single set of working drawings prepared by designers is called an original. This one set could not normally serve the purposes of all persons concerned with construction of the building. It is not practical, however, for the designers to draw several sets of identical plans. Hence, to provide the many sets of plans needed, the original set is reproduced by a duplicating process. The reproductions are referred to as prints. (In years past, reproductions were called blueprints, because they were made with white lines on a blue background. Now, prints usually are made with black or brown lines on a white background.)

2.16. SPECIFICATIONS

It usually is impossible to provide on drawings all the information necessary for construction of a building. Some types of information, such as the type of brick to be used to face a wall or the type of windows to be incorporated in the wall, are best provided in written form; but if such data were to be written in notes on the construction drawings, they would become so cluttered and confusing that building construction would be hindered.

As a result, construction drawings are almost always accompanied by separate written instructions, called specifications. These provide all information concerning materials, methods of construction, standards of construction, and the manner of conducting the work that is not furnished on the drawings. Thus, specifications

supplement the drawings. Both are equally important to construction of the building.

Specifications are discussed in Chap. 5, Sec. 5.4.

2.17. SCALES AND DIMENSIONS ON DRAWINGS

The process of drawing the parts of a building to a proportionate size that can be contained on convenient-size sheets of paper is called drawing to scale. The drawings must be in exact proportion to the actual dimensions of the components they represent. For example, for most buildings, the scale used makes the drawings $\frac{1}{48}$ the actual size. Thus, instead of drawing a window opening 3 ft wide, the designer draws it $\frac{1}{48}$ of 3 ft (36 in.), or $\frac{3}{4}$-in. long. As a result, the drawing looks like the full-size component but is only $\frac{1}{48}$ the size.

Selection of a scale for a drawing depends both on the size of the sheet of paper to be used and the size of the building or components to be drawn. For an ordinary building, elevations and plans often are prepared to a scale of $\frac{1}{4}'' = 1'\text{-}0''$. (One-quarter inch equals one foot. On construction drawings, prime marks are used to indicate feet, and double-prime marks to indicate inches.) For detail drawings on which types of construction and materials are shown, $\frac{1}{2}'' = 1'\text{-}0''$ or $\frac{3}{4}'' = 1'\text{-}0''$ may be used; but if detail parts are very small, and an easy-to-read drawing is desired, $3'' = 1'\text{-}0''$ may be chosen. Very small scales, such as $\frac{1}{8}'' = 1'\text{-}0''$ and $\frac{1}{16}'' = 1'\text{-}0''$, are generally used for exceptionally large elevation views or for plot and survey plans, to keep the overall size of drawings within reasonable limits.

Title Block

Each of the several drawings comprising a set of construction plans is provided with a title block. It usually is placed at the lower right-hand corner of the sheet. The title block shows the name of the building, names of designers, type of drawing and name of component shown. The title block also gives other information, such as scale used, revisions and date revisions were made. For example, a title block might indicate in large letters that the drawing shows the First-Floor Plan. If a single scale were used for that drawing, that scale might be indicated under the type of drawing; for example, under First-Floor Plan. If a drawing contains parts drawn to different scales, each part should have a title given in large letters directly under it, and the scale should be indicated under each title. Title blocks are further discussed in Chap. 5.

Dimensions

Construction drawings would not permit construction of the building intended by the building owner and the designers if the drawings were merely a drawn-to-scale picture of the structure. They must also show the dimensions of the building and its parts. Everyone concerned wants to know the length, width and ceiling height of each room. Builders want to know the wall thicknesses, foundation depth and thickness, sizes and locations of window and door openings, and numerous other size stipulations. Cost estimators also need to know sizes because most of the costs they calculate involve sizes of various materials.

Size or space stipulations on drawings usually are indicated by a system of lines, arrows, and numbers, called dimensions.

Despite the scale used for the drawing, dimensions give actual or full sizes or distances.

Figure 2.19*a* shows a plan view of a space enclosed by four walls. The walls contain four windows and a door. The drawing was made to a scale of $\frac{1}{8}'' = 1'-0''$. (Reproduction of a drawing in this book is done for illustrative purposes by a photographic process and is unlikely to be to the scale indicated.) The drawing shows that the overall dimensions of the enclosure are 30 ft by 28 ft 4 in. (Feet are denoted by prime marks, and inches by double prime marks.) Arrows indicate that the wall is 12 in. thick. Windows are 4 ft wide and the door, 3 ft wide.

The limits of each dimension are indicated by a pair of arrows. An arrow is called a *dimension line* (see Fig. 2.19*c*). Each arrow terminates at an *extension line* (see Fig. 2.19*d*). Thus, a pair of extension lines shows where each dimension, indicated by a set of arrows, ends.

Numbers between each pair of arrows give the actual size or distance between the extension lines.

Figure 2.19*b* shows an elevation view that might be used as a picture-like representation of a door. This method of showing dimensions of the door and its parts is typical for doors and other items for which special millwork is required. Thus, if a door or other item of other than stock (standard) size is required, the designer prepares a detail like the one in Fig. 2.19*b*, to show exactly what he has in mind. The size of each part of the door can be determined from the horizontal and vertical rows of dimensions.

Sometimes, different variations are used for indicating dimensions. For example, when a single dimension line with an arrowhead at each end is used to give a dimension, the number giving the size may be shown above or alongside the line. In some cases, arrowheads may be replaced by dots.

2.18. ELEVATION VIEWS

An elevation view of an object is the projection of the object on a vertical plane. Thus, an elevation view shows what a vertical side of the object looks like when viewed by someone facing that side.

To visualize an elevation of a building, imagine that you can stand outside it so as to face squarely one side of it. The face will appear to lie in a vertical plane. The image, to scale, of the building in the vertical plane is an elevation. Because a rectangular building has four faces, it also has four elevations.

Figure 2.20*a* is a perspective drawing of a one-story house, on which the four directions in which the walls face are indicated as north, south, east and west. Thus, the house will have four elevations correspondingly named, as shown in Fig. 2.20*b* to *e*.

West Elevation

Imagine that you are standing at a distance from the building and facing its west side squarely. You will then see the west elevation (see Fig. 2.20*b*). It was drawn by projecting

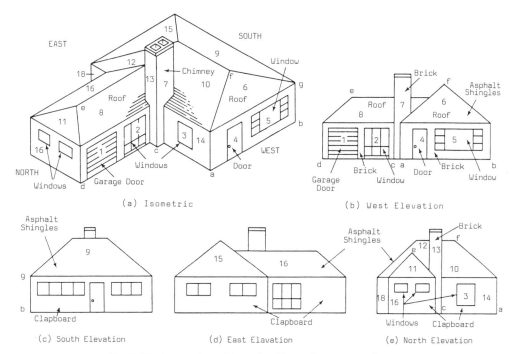

Fig. 2.20. Isometric and four elevations of a one-story house.

the west sides of the building on a vertical north-south plane.

The roof areas marked 6 and 8 on the elevation are the same areas shown in the perspective marked 6 and 8. The chimney area, labeled 7 on the elevation, is the same as the chimney side marked 7 in the perspective. Similarly, the roof points marked e and f are identical in both perspective and west elevation. The doors and windows labeled 1, 2, 4 and 5 also are the same in perspective and elevation. Points a and c, at ground level, appear close together in the elevation, though they are shown relatively far apart in the perspective.

The wall marked 14 in the perspective and shown in the north elevation (see Fig. 2.20e) does not appear as an area in the west elevation, because the wall is perpendicular to the plane of projection. Instead, wall 14, when viewed from the west, is seen as a vertical line and thus is indicated by a vertical line above a on the west elevation.

East, North and South Elevations

The east elevation is obtained much like the west elevation, by projecting the east sides of the building on a vertical north-south plane. But the view is drawn as seen from the east (Fig. 2.20d).

Similarly, the north and south elevations are drawn by projecting the north and south sides, respectively, on a vertical east-west plane. For the north elevation, the north sides are viewed from the north (see Fig. 2.20e). For the south elevation, the south sides are viewed from the south (see Fig. 2.20c).

Note that no dimensions are shown in any of the elevations in Fig. 2.20. In an actual construction drawing, dimensions of the walls and openings in them would have been given.

2.19. PLAN VIEWS

A plan view shows what the interior of a building looks like when viewed from above. While an elevation view is actually a projection on a vertical plane, a plan view is a projection on a horizontal plane. For a floor plan, the view is usually obtained by making an imaginary horizontal cut through the building and then projecting the exposed parts on a horizontal plane. For a roof plan, however, visualizing a horizontal cut is generally unnecessary, because roofs are exposed to view from above.

Several types of plan are used for construction of a building. When a plan is drawn to show the size, shape and location of rooms, it is called a floor plan. When a plan shows the structural framing supporting the floor, the view is called a floor framing plan. Similarly, other plan views may show ductwork for HVAC, electric equipment and wiring, and other information.

To visualize a plan view, imagine that the building is cut horizontally and the top is removed so that you can look straight down at the cut surfaces. These surfaces and the floor below will appear to lie in a horizontal plane. The image, to scale, of the lower part of the building is a plan view. (In some cases, the view may be drawn as seen from below; for example, for a ceiling plan, to show lighting fixtures and air-conditioning outlets.)

Figure 2.21a is a perspective of a one-story office building, containing a reception room, office and toilet (water closet, w.c.). Imagine the structure cut through horizontally, as indicated by the dashed line from x to y, as if by a large saw. Imagine also that the top part can be lifted so that you can look down squarely at the surfaces cut along the lines abcdef (see Fig. 2.21b). The cut surfaces, shown in heavy black, and the floor between them, shown in white, constitute the plan view of the ground (first) floor (see Fig. 2.21c). Doors, windows and rooms, as well as partitions and exterior walls, are all shown.

Note that no dimensions are given for the floor plan in Fig. 2.21. In an actual construction drawing, dimensions of walls, openings in the walls and rooms would have been given.

For a multistory building, a plan view would be drawn for each floor, including basement, if present, and roof.

Cut surfaces are not always represented by heavy black lines. Often, it is desirable to show the boundaries, in detail, of a cut object. In those cases, the cut surfaces may be indicated by a symbol. One commonly used symbol is cross hatching, closely spaced light lines, generally drawn at a 45° angle with a main boundary. Sometimes, the symbol selected represents a specific material, such as brick or concrete.

2.20. LINES

Designers use different types of lines on drawings. Each type of line is applicable to a specific purpose.

Solid lines usually represent *edges* of objects.

Short dashes are used to indicate *invisible edges*, boundaries covered by a part shown in the view being drawn. For example, in the perspective in Fig. 2.22a, which shows the exterior of a building, a dashed line indicates the location of the ceiling, because, in that view, it is hidden by the exterior wall.

Dash-and-dot lines, made up of alternating long and short dashes, generally are used as a *centerline*. This is a line drawn to mark the middle of a building or a component. A centerline, in addition, is labeled with an inter-

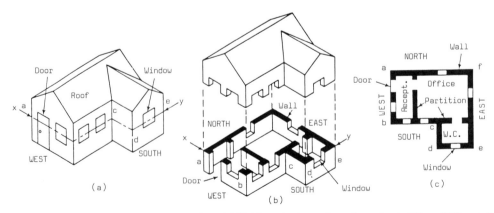

Fig. 2.21.   Guide to visualization of a plan view. (*a*) Making a horizontal cut through a building. (*b*) Removing the part above the cut. (*c*) Floor plan of the building.

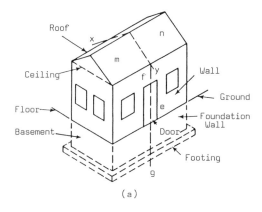

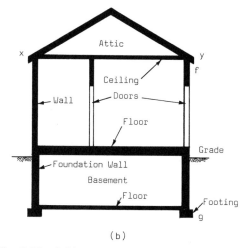

Fig. 2.22. Guide to visualization of a vertical section. (a) Making a vertical cut through a building. (b) Vertical section of the building.

secting *c* and *L*. This representation is used for the centerline of the construction, called a truss, in Fig. 2.23.

A heavy line, made up of long dashes, often is used to indicate where a building is to be imagined cut to obtain a view, called a section. (The plan views discussed in Sec. 2.19 are horizontal sections.) Arrows are drawn at the ends of a *section line* to indicate the direction in which to look to obtain the view. Note that section lines need not be continuous, straight lines but may have abrupt changes to show, in one view, different cuts or levels.

Broken solid lines, with a wavy break at intervals, are used to indicate that parts of a drawing have been omitted or that the full length of some part has not been shown. Such lines are

used along the right side of the truss in Fig. 2.23 to show that almost half the truss has been omitted from the drawing.

2.21. SECTIONS

Seldom do elevation and plan views alone show sufficient information to enable a builder to determine exactly how the various parts of a building are to be assembled and connected. For example, an elevation view in Fig. 2.20 shows at what height above ground a window is to be set, while the corresponding floor plan would show how far the window is to be placed from one end of the wall. Also, the specifications for the building would define the materials and their quality to be used for the window. There are, however, many good ways of constructing such a wall and window, and some undesirable ways. Unless the builder is shown exactly what the designer has in mind, a type of wall may be built that will not please the designer. To preclude such an event, designers provide additional drawings, called sections.

A section shows the interior construction of a part of the building. This type of drawing indicates how various structural components are to be assembled and connected. Usually, sections are drawn to a larger scale than plan and elevation views, because sections are intended to show more detail.

In the explanation of how to visualize a plan view, you were asked to imagine a building to be cut through horizontally. A plan view actually is one type of section. In general, however, to obtain a section needed for specific illustrative purposes, a designer may imagine the building cut through at any angle. Usually, however, sections are taken horizontally, vertically or sometimes perpendicular to an inclined surface. This is convenient for drawing purposes, because building parts, such as floors and walls are horizontal and vertical, respectively, and structural parts used in inclined surfaces, such as the framing in sloping roofs, generally are perpendicular to each other.

To visualize a vertical section, for example, imagine a building or one of its parts cut through vertically and one part removed so that you can look squarely at the cut surfaces of the remaining portion. The image, to scale, of the remaining portion is a section view; that is, a vertical section is a projection on a vertical plane of the parts exposed by a vertical cut.

Figure 2.22a shows a perspective of a small, one-story building. Imagine that the structure can be cut through vertically along the lines between x and y. The dashed line fg shows the path of the cut, which goes through an opening for a door. Imagine now that the part of the building labeled m is moved away so that you can look squarely at the cut surfaces of the n part. Figure 2.22b shows the resulting section view.

The door is not shown in this view, because it is imagined not installed at the time of cutting. The cut surfaces in this drawing are shown in black, but the doorway is left white, because no surfaces were cut.

In section views, including plans, cut surfaces usually are indicated by some symbol. In Figs. 2.21c and 2.22b, the cut surfaces are shown in solid black. Sometimes, however, use is made of a symbol that represents the material that has been cut. For example, concrete generally is represented by small triangles in a matrix of dots; plaster by dots; brick by closely spaced lines at a 45° angle with the horizontal lines of a drawing; and other materials by similar conventional symbols.

If the designer does not intend to indicate a specific material, he may use *cross hatching*, closely spaced lines at a 45° angle. For different but adjacent parts, cross hatching is sloped in opposite directions, to distinguish the parts.

Symbol lines are drawn much lighter than lines representing edges.

The vertical section in Fig. 2.22b does not give any dimensions. In an actual construction drawing, dimensions would be shown for all the parts in the section.

2.22. DETAILS

A detail view supplies information about structural assembly, trim, and various special equipment that cannot be given in elevations, plans, or full-building sections. This has largely to do with the scale of drawings. For example, ele-

vations, plans, and full-building sections are usually drawn at $\frac{1}{4}$ in. equals one ft ($1:48$) or smaller, so that they will fit on a reasonably sized sheet when printed for use. Details, on the other hand, especially when drawn of very small portions of the whole building, can be drawn as large as full size—although scales of $1:4$ (3 in. equals one ft), $1:8$ (1.5 in. equals one ft), or $1:16$ (0.75 in. equals one ft) are more often used.

Many details are drawn as vertical sections, although any form of drawing can be made large in size for explanation of particular details of the assemblage or the form of individual parts. In some cases, where ordinary orthographic projection (x-y-z, right angle views, such as plans, elevations and vertical sections) does not fully suffice to explain the assemblage, isometric or even perspective views may be required for clarity.

Details views may generally be classified as placement or assembly types. *Placement details* are used to show the desired arrangement of objects in the finished construction. Plans and elevations are of this class, and detail plans or elevations may be used to explain objects in greater detail, such as the arrangement of fixtures in a bathroom or the details of a single window. *Assembly details* are used to explain individual components (such as parts of the structural system or the piping system). Figure 2.23a is an example of an assembly detail, illustrating how a wood truss for the roof framing of a house is to be assembled. The drawing shows a partial elevation of the truss. As the truss is symmetrical, it is not necessary to show the whole truss for the purposes of the detail assembly drawing.

For detail drawings it is common to show only just as much of the whole part as is necessary to clarify the desired detail information. Specifications would establish the type of wood and other detailed information on materials, construction tolerances, and so on. Notation on the detail view should be limited to the identifications and dimensions that are specifically required to clarify the work. General building dimensions should be indicated on plans, elevations or full-building sections, and detailed material information should be given in the

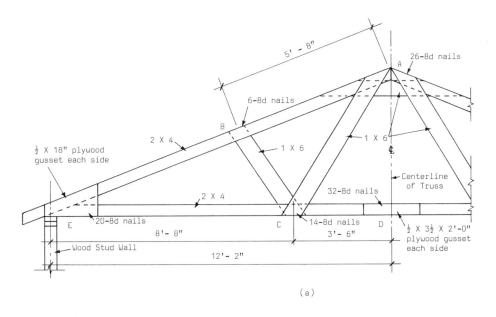

(a)

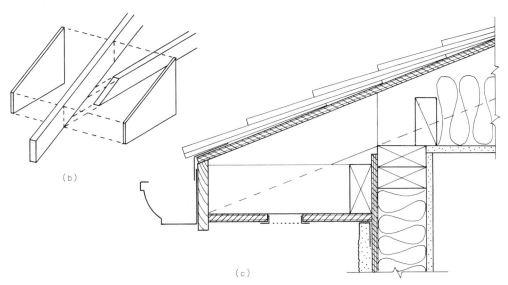

(b)

(c)

Fig. 2.23. Construction detail drawings. (a) Elevation of a truss. (b) Exploded view of a truss joint—joint E in the truss in (a). (c) Comprehensive architectural detail section, incorporating joint E of the truss.

specifications. Repetition of such information allows for the possibility of confusion and conflict when the information does not agree in different locations due to error.

The exploded isometric view of Joint E, shown in Fig. 2.23b, more clearly indicates the form of the parts and the assembly of the joint; for example, the fact that there are gussets on both sides of the joint. The note on the drawing

actually calls for this, but the isometric view drives the idea home.

Some of the most important drawings for building construction are the detail sections of the form shown in Fig. 2.23c. These show the complete construction assemblage, with arrangement of parts and the identities of the individual materials made as clear as possible. For persons trained to read such drawings, they are very informative. Standardized graphic symbols are used to indicate materials and the notation uses terms carefully chosen to agree with those used elsewhere on the construction drawings or in the specifications.

While the comprehensive detail section in Fig. 2.23c is most useful for understanding the total nature of the construction, the forms of individual parts are often obscured by the complexity of such views. Thus, to explain the truss assemblage—although the truss appears in partial view in Fig. 2.23c—it is really necessary to remove the truss and show it alone in the views in Figures 2.23a and b. For the workers who perform the single task of making the trusses, this is useful and sufficient. For this reason, the full set of building construction contract drawings normally contains both fully detailed architectural drawings and separate drawings showing only the structure, or the electrical components, or the plumbing, and so on.

2.23. SURVEY AND PLOT PLANS

Before land is purchased, the purchaser should have a land surveyor survey the lot, for several reasons. One reason is that the survey will indicate the exact boundaries of the lot. Another reason is that most municipalities require a survey plan of a lot for establishment of ownership. Such a plan also is required by a bank before it will make a loan for land purchase or building construction. In addition, a survey plan is needed by architects and engineers for analysis of the property, to determine location of buildings, access roads, walks, parking lots and equipment.

A survey plan shows how a building site looks when viewed from above. The plan should show boundaries and exact dimensions and give elevations (heights) of the land. Also, the drawing should show boundary streets and highways; utilities available, such as water, gas and electricity; directions of the compass; and topographic features, such as trees, brooks and lakes. Survey plans are usually drawn to a scale of $0.1'' = 1.0'$.

Figure 2.24 shows a survey plan for a city lot. The drawing indicates that the northern

border of the lot is an avenue in which there are a water main and two sewers. The drawing also shows that the lot is rectangular, 98×158 ft in size, and contains several trees.

The proposed building is not shown on a survey plan but instead is drawn on a separate drawing, called a plot plan. Developed from the survey plan, the plot plan also is a view of the building site from above; but the plot plan is used to show the location and orientation of the proposed buildings on the lot. This plan should also indicate how the grounds around the buildings are to be graded and where walks, parking lots and storage areas are to be located. In addition, it should provide other information that builders need before they can stake out the buildings and excavate for foundations. Plot plans are often drawn to a scale of $\frac{1}{16}'' = 1'\text{-}0''$.

Figure 2.25 shows a plot plan developed from Fig. 2.24 to show the location of a house to be built on the lot. The area covered by cross hatching (closely spaced 45° lines) represents the house. The drawing indicates that the finished floor level of the building must be at Elevation 277. The front of the house is to be 20 ft from the northern boundary. The driveway is to be 18 ft wide, and the sidewalk between porch and street, 4 ft 9 in. wide. Steps are shown between porch and sidewalk.

Lines

On survey and plot plans, boundaries are represented by a line consisting of a repeated set of a long dash and two short dashes. Land elevations are indicated by curved lines, called contours, drawn somewhat lighter than boundary lines. Utilities, such as sewers and water mains, are represented by dashed lines. Solid lines are used to indicate internal boundaries, such as those of buildings and walls. Various symbols are used to represent topographic features, such as trees, swamps and waterways.

Elevations

Heights of points on a building site are determined relative to a *datum*, or reference level, established by the municipality or other legal authority. The datum is assigned a specific elevation, such as 0 or 100 ft.

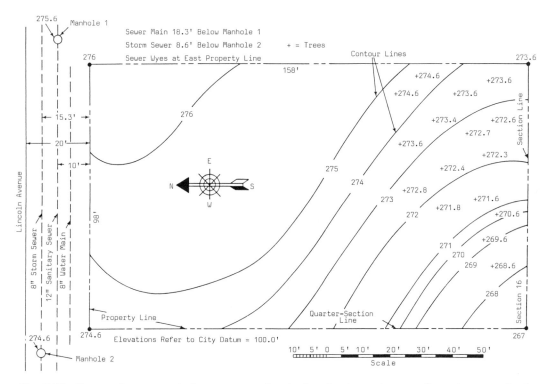

Fig. 2.24. Survey plan prepared after a survey of a city lot. Section and quarter lines shown are legal reference lines for locating property boundaries.

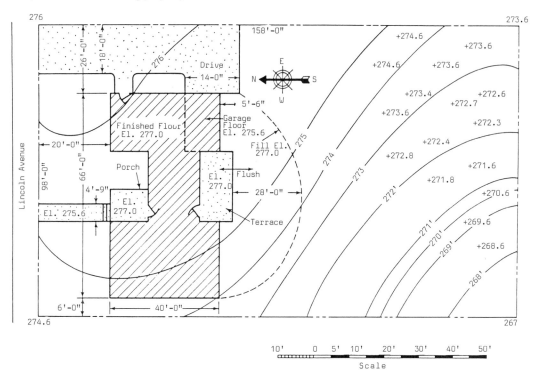

Fig. 2.25. Plot plan prepared from the survey plan in Fig. 2.24, showing the location of the building and regrading with fill to the dashed-line contour.

For convenience in surveying, other points of known elevation, relative to the datum, may be established in the region. These are known as *bench marks*. A bench mark 20 ft above a datum of 100 would be assigned an elevation of 100 + 20, or 120 ft. A bench mark 10 ft below this datum would be given an elevation of 100 − 10, or 90 ft. From bench marks, surveyors determine with their instruments the elevations of various points on a building site. These points are used to plot contours.

A *contour* is a curved line that connects all points of the same level. For convenience, a contour may be imagined as the waterline that would be formed on the shore if a lake were to be created at the site with water up to the level assigned to the contour. For example, with the datum assigned an elevation of 100 ft, a contour marked 100 connects all points on the site that are at the same elevation as the datum. A contour marked 150 connects all points that are 50 ft above the datum.

The closer contours are, the steeper is the slope of the land. Note that contours can meet only at a vertical cliff.

In Figures 2.24 and 2.25, elevations are shown relative to a datum of 100. Thus, the contour marked 276 is 176 ft above the city datum. Contours are drawn for 1-ft intervals of elevation. (If the property were to be flooded by a lake to the 276-ft level, the 276 contour would represent the waterline along the shore throughout the site. Each time the lake were to be lowered 1 ft, the waterline would lie along another contour.) Between contours, elevations are given in tenths of a foot to indicate the heights of topographic features, such as trees.

The plot plan in Fig. 2.25 indicates that a fill is required on the south side of the house. The boundary of the fill is represented by the dashed line 28 ft from the terrace.

FOR SECTIONS 2.13 THROUGH 2.23

References

C. Ramsey and H. Sleeper, *Architectural Graphic Standards*, 8th ed., Wiley, New York, 1988.

T. French and C. Vierck, *Engineering Drawing and Graphic Technology*, 13th ed., McGraw-Hill, New York, 1986.

R. Liebing and M. Paul, *Architectural Working Drawings*, 2nd ed., Wiley, New York, 1983.

F. Ching, *Architectural Graphics*, 2nd ed., Van Nostrand Reinhold, New York, 1985.

Words and Terms

Bench mark	Plot plan
Construction drawings	Scale
Contour line	Section drawing
Datum	Specifications
Elevation drawing	Survey plan
Perspective drawing	Symbols
Plan drawing	

Significant Relations, Functions, and Issues

Purposes of drawings: communication of design information, relation to specifications.
Relation of scale to level of detail possible in drawings.
Need to use conventional drawing techniques and symbols and notation with terms that are compatible with specifications, for clarity of communications.

GENERAL REFERENCES AND SOURCES FOR ADDITIONAL STUDY

These are books that deal comprehensively with several topics covered in this chapter. Topic-specific references relating to individual chapter sections are listed at the ends of the sections.

C. Ramsey and H. Sleeper, *Architectural Graphic Standards*, 8th ed., Wiley, New York, 1988.

F. Merritt, *Building Design and Construction Handbook*, 4th ed., McGraw-Hill, New York, 1982.

J. Ambrose, *Building Structures*, Wiley, New York, 1988.

B. Stein et al., *Mechanical and Electrical Equipment for Buildings*, 7th ed., Wiley, New York, 1986.

F. Ching, *Architectural Graphics*, 2nd ed., Van Nostrand Reinhold, New York, 1985.

EXERCISES, CHAP. 2

The following questions and problems are provided for review of the individual sections of the chapter:

Section 2.1

1. Name the following basic building elements:

(a) A horizontal structural member that supports a deck above ground.

(b) A vertical wall that prevents earth from coming into a basement.

(c) A horizontal element that keeps a foundation wall from sinking into the ground.

2. What are the purposes of the following building elements?
   (a) Roof
   (b) Windows
   (c) Partitions
   (d) Doors
   (e) Chimney
   (f) Plumbing

3. Describe two different ways of providing light inside a building.

Section 2.2

4. What are the purposes of:
   (a) floor covering?
   (b) ceilings?

5. Where are joists used and for what purpose?

6. The weight of a subfloor, floor covering, and ceiling as well as loads from people and furnishings are supported on a set of beams. What may be used to support the beams?

7. From what source of information should you obtain the fire rating required for an interior wall of a building?

8. A floor system has been tested and assigned a fire rating of 2 hr. What does this signify?

9. How is the fire rating of a building component determined?

10. How can a wood beam be made fire resistant?

11. Describe two methods for protecting a steel beam from fire.

Sections 2.3 and 2.4

12. Describe the purposes of:
    (a) A roof
    (b) Roofing
    (c) Thermal insulation incorporated in a roof system
    (d) Rafters

13. What is the main purpose of an exterior wall?

14. Why are exterior walls usually a combination of different materials?

15. What is the primary difference between a curtain wall and a bearing wall?

16. How does unit masonry construction differ from:
    (a) Panel construction?
    (b) Framed construction?

17. Where is a lintel used?

18. Describe three important functions of a window.

19. What provision is made in a wall opening to receive a window?

Section 2.5

20. What are the major purposes of interior walls?

21. What prevents a door that is being closed from swinging past the wall opening?

22. Builder's hardware is classified either as finishing or as rough. Which type are the following?
    (a) Doorknob
    (b) Locks
    (c) Nails
    (d) Hinges
    (e) Windows sash balances

23. What are the usual means of directly supporting a door?

Section 2.6

24. What types of members are used in skeleton framing?

25. What is the purpose of the following?
    (a) Column
    (b) Girder
    (c) Bracing
    (d) Rigid connection
    (e) Foundation

26. Why is it desirable that footings be placed below the frost line?

27. What must be done for safety reasons besides just transmitting building loads to the ground?

28. For what soil conditions are the following suitable?
    (a) Spread footings

(b) Mat

(c) Piles

29. Describe some methods of protecting beams from fire.

Section 2.7

30. What is the purpose of water supply plumbing?

31. Why must gas plumbing be a sealed system?

32. What is the purpose of wastewater plumbing?

33. Why is the presence of both water-supply plumbing and wastewater plumbing in a building a potential health hazard?

34. What consideration should be given to water-supply and wastewater disposal in selection of a site for a building?

Section 2.8

35. Why is ventilation necessary?

36. Why is mechanical ventilation used?

37. What effect on relative humidity does heating of air have?

38. Why is it desirable to add humidity to a building when it is heated in cold weather?

39. What effect does humidity have on building occupants in hot weather?

40. Describe briefly the two types of HVAC systems.

Section 2.9

41. What are the three factors that determine good lighting?

42. Why should color of emitted light be considered in selection of a light source?

43. What are the two methods used for illumination of the interior of a building?

44. Explain why walls, floors and ceilings should be treated as parts of the lighting system.

Section 2.10

45. When does sound become noise?

46. Which requires greater sound control: an office building or a factory? Why?

47. Why should acoustics be considered in design of ceilings, walls and floors?

Section 2.11

48. How is electricity distributed in a building?

49. Why is placement of electrical conductors within walls, floors or ceilings desirable?

50. What should controls do in an electrical distribution system?

51. Why is safety a prime consideration in design of an electrical system?

Section 2.12

52. Define a ramp.

53. What advantages do:
    (a) Stairs have over ramps?
    (b) Elevators have over escalators?

54. What effect on the slope of stairs does the following have?
    (a) Decreasing the width of the treads with no change in risers.
    (b) Decreasing the height of risers with no change in the treads.

55. What is the purpose of headroom?

56. Describe two commonly used methods for propelling elevator cars.

57. What is the purpose of a dumbwaiter?

Sections 2.13 to 2.15

58. What is the purpose of construction drawings? Of specifications?

59. Give two reasons why drawings are drawn to scale.

60. Why are symbols necessary for construction drawings?

61. How does a perspective drawing of a building differ from an elevation view of the same building?

62. What does a floor plan show?

63. What does a section view show?

64. In what views in construction drawings would you find information on each of the following?
    (a) Height of windows above a floor
    (b) Arrangement of rooms
    (c) Location of doors in partitions

(d) Arrangement of floor beams for the second floor of a building

(e) Construction of window framing

65. How does a survey plan differ primarily from a plot plan?

Sections 2.16 and 2.17

66. What information do specifications provide?

67. Why are specifications necessary?

68. How can you tell what scale was used for a drawing?

69. What distance is represented by a line 2-in. long when the scale is:
    (a) $\frac{1}{4}$ in. = 1 ft?
    (b) $\frac{1}{8}$ in. = 1 ft?
    (c) 3 in. = 1 ft?

Section 2.18

70. How is the elevation view of the side of a building facing northeast obtained?

71. In Fig. 2.20, in what elevation would you look to find:
    (a) The number of windows in wall 14?
    (b) The location of garage door 1?
    (c) Material used for the outer facing of wall 16?

Section 2.19

72. How is a floor plan of a building obtained?

73. How is a ceiling lighting plan obtained?

74. In Fig. 2.21, what information is given by the floor plan that is not given by the isometric or any of the elevations?

Section 2.20

75. What type of line should be used in an elevation to represent an opening for a window in a wall?

76. What type of line should be used to represent in a floor plan the opening in a partition for a door?

77. What type of line should be used in an elevation to show where a cut is to be made for a ceiling plan?

78. What are the edges of the foundations in Fig. 2.22a represented by dashed lines?

Sections 2.21 and 2.22

79. How is a vertical section obtained?

80. How is a horizontal section obtained?

81. In Fig. 2.22, what information is given by the vertical section that is not provided by the isometric or the elevations?

82. Why is a detail usually drawn to a large scale?

Section 2.23

83. Can contour lines cross? Explain your answer.

84. In Fig. 2.24:
    (a) What utilities are available close to the lot?
    (b) What is the frontage along the avenue of the lot?
    (c) What part of the lot is steepest?
    (d) Is the northern part of the lot flat or steeply sloped?
    (e) How far is the storm sewer from the north property line?

85. In Fig. 2.25:
    (a) How far is the west side of the house from the west boundary line?
    (b) What is the elevation of the garage floor?
    (c) How far above the ground is the finished floor at the northeast and northwest corners of the building?

Chapter 3

Systems Design Method

Building design, being as much an art as a science, requires creativity, imagination and judgment. These talents can be inspired and assisted by systems design to produce better and less costly buildings. The big advantage of systems design over traditional building design is that systems design marks clearly the precise path for production of optimum results.

In Sec. 1.3, systems design is defined as the application of the scientific method to selection and assembly of components or subsystems to form the optimum system to attain specified goals and objectives while subject to constraints or restrictions. The scientific method requires observation and collection of data, formulation of an hypothesis and testing of the hypothesis.

Section 1.3 also points out that the systems design procedure requires three essential steps: analysis, synthesis and appraisal. The purpose of analysis is to indicate what the system is to accomplish. Synthesis is the formulation of a system that meets objectives and constraints. Appraisal evaluates systems performance and costs. To insure cost effectiveness of components, value analysis is included in the appraisal phase. Value analysis investigates the relationship between life-cycle costs and the values of a system, its components and alternatives to these, to obtain the lowest life-cycle cost for acceptable system performance. In practice, these steps may overlap.

In this chapter, the systems design method is explored in detail. Chapter 4 discusses the practical application of the method to building design.

As proposed in this chapter, systems design is applicable to the whole building as a system, to each of its systems and subsystems, and to component systems. The method requires that, at the start, the characteristics required of the system be described. Then, a system with these characteristics is developed. Various methods may be used to refine the system, to attain acceptable performance at least life-cycle cost, or the best performance for a given cost, or some intermediate performance and cost. Next, value analysis is applied to see if costs can be reduced. Alternative systems are investigated in a similar manner. Finally, all systems are compared and the optimum system is chosen.

Execution of the method is expedited by the use of models. These are discussed in Sec. 3.1.

3.1. MODELS

As used in systems analysis, a model is a representation of an actual system for the purposes of optimization and appraisal. A prime requisite for a model is that it be able to predict the behavior of the system within the range of concern.

For each condition imposed on the system and each reaction of the system to that condition, there must be a known corresponding condition that, when imposed on the model, evokes a determinable response that corresponds to the system reaction.

The correlation need not be perfect but should be close enough to serve the purposes for which the model is to be used.

For practical reasons, the model should be a simple one, consistent with the role for which it is chosen. In addition, the cost of formulating and using a model should be negligible compared with the cost of assembling and testing the actual system.

A model may be formulated from among a wide range of possibilities. A system or a typical portion of it, for instance, may serve as its own model. For example, a few piles for supporting a foundation are sometimes driven full depth into the ground and then tested in place with gradually increasing loads to determine their safe load-carrying capacity. The data obtained in the test are used to establish the safe loads for other piles to be driven nearby under the same soil conditions.

Sometimes, a model may be essentially a replica of a system to a small scale. For example, a model sometimes is constructed about $\frac{1}{100}$ or less the size of an actual building for tests in a wind tunnel to determine the effects of wind on the building or of the building on air movements.

But many other models are possible and can be used in systems analysis. A model, for example, may be a set of mathematical relationships, graphs, tables or words. Regardless of form, however, a model, to be useful, must behave like the real system. Consequently, a model should be tested continuously during its formulation for correlation with the real system.

Types of Models

Despite the variety of models that may be used in systems analysis, models may be classified as one of only three types: iconic, symbolic or analog.

Iconic models bear a physical resemblance to the real system. They differ from the real system in scale and often are simpler. The previously mentioned models of buildings used in wind tunnel tests are iconic.

Symbolic models represent by symbols the conditions imposed on the real system and the reactions of the system to those conditions. With such models, the relationships between imposed conditions and reactions, or performance, can be generally, and yet compactly, represented. For example, the maximum safe load on a steel hanger can be represented by $P = AF_y/k$, where P is the load, kips (thousands of pounds); A is the cross-sectional area of the hanger, sq in.; F_y is the yield strength of the steel, ksi (kips per sq in.); and k is a load factor that provides a safety margin. Symbolic models generally are preferred for systems analysis when they can be used, because they take less time to formulate, are less costly to develop and use, and are easy to manipulate.

Analog models are real systems but with physical properties different from those of the actual system. If mathematical formulas could be written to represent the behavior of a system and its analog, the formulas would be identical in form, although the symbols used might be different. For example, a slide rule is an analog model for representing numbers by distances. When lengths on a slide rule are made proportional to logarithms of numbers, addition of lengths on a slide rule is equivalent to multiplication of the corresponding numbers. Similarly, electric current can be used to determine heat flow through a metal plate; a soap membrane can be used to determine torsional stress in a shaft; and light can be used to determine bending stresses in a beam.

It sometimes may be necessary to represent a system by more than one model. The models in such cases may be used in combination, much like a set of simultaneous linear algebraic equations; or they may be used in sequence, the output of one model serving as the input of another.

Regardless of the type of model, systems and models must meet the following conditions:

1. It must be possible to construct the model from a knowledge of the known characteristics of the system. Only those known characteristics that are essential, however, need be considered. Many properties of a model or system may be irrelevant.

2. It must be possible to predict the response of the system from a knowledge of the response of the model.
3. Accuracy of the response of the system obtained by use of the model must be assured, through tests of the model, to be within acceptable tolerances. The tests may be made physically or by mathematical computations, or both.

Elements of a Model

A model relates imposed conditions, which usually can be expressed numerically and hence can be represented by variables, and corresponding responses, which also usually can be expressed numerically and can be represented by variables. Sometimes, the variables are known within reasonable accuracy with certainty; that is, they are deterministic. Often, however, only a probable value of each variable can be assumed; that is, the variables are random. The variables of concern in building design and their relationships, however, are usually so intricate that it is impractical to treat them as random, with probabilistic or statistical methods. Hence, variables usually are treated as if they were deterministic, sometimes assigned a mean value, sometimes an extreme value and often what is considered, by consensus, an acceptable value.

Variables representing imposed conditions and properties of the system usually may be considered *independent variables*. These are of two types:

1. Variables over which the designer has complete control: $x_1, x_2, x_3, \ldots$
2. Variables over which the designer has no control: $y_1, y_2, y_3, \ldots$

Variables representing the response or performance of the system may be considered *dependent variables*: $z_1, z_2, z_3. \ldots$. They are functions of the independent variables.

These functions also contain parameters, such as coefficients, constants and exponents. When a form of model is selected, these parameters have to be set to match the response of the real system; that is, the model must be calibrated.

As an example, consider a symbolic model representing the cost of operating a heating sys-

tem for a building. Assume that the shape of the building and the material in the exterior walls have been determined. Then, the cost can be shown to be a function of wall thickness and difference between interior and outdoor temperatures:

$$C = f(t, T_i, T_o) \qquad (3.1)$$

where

C = cost of heating
t = wall thickness
T_i = indoor temperature
T_o = outdoor temperature

C is the dependent variable, the response of the system that the designers are interested in determining. T_i in this case is a parameter, a design value established for the comfort of building occupants. T_o, although actually a random variable, may be taken as the expected value for the location of the building, date and time of day. In any event, T_o is an uncontrollable variable. In contrast, t is a value that can be chosen by the designer and therefore is a controllable variable.

In Sec. 3.8, where the various steps of systems design are discussed, one step is given succinctly as "Model the system and apply the model." This step, however, requires several actions, illustrated in Fig. 3.1:

1. Formulation of a model and its calibration, setting of values for the parameters.
2. Values have to be estimated for the uncontrollable variables.
3. Values have to be determined for the controllable variables from constraints and conditions for optimization.
4. Finally, the sought-after response of the system can be found, from the relationship of the variables, through use of the model.

Cost Models

Costs are such an overriding consideration in design and construction of many buildings that cost models deserve special attention. They are discussed briefly in the following paragraphs and other parts of the book.

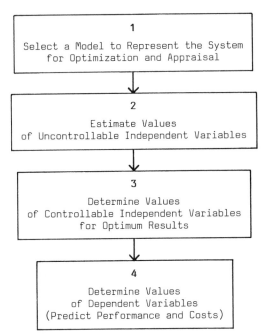

1

Select a Model to Represent the System
for Optimization and Appraisal

2

Estimate Values
of Uncontrollable Independent Variables

3

Determine Values
of Controllable Independent Variables
for Optimum Results

4

Determine Values
of Dependent Variables
(Predict Performance and Costs)

Fig. 3.1. **Steps in "Model the system and apply the model." See Fig. 3.4.**

As indicated in Chap. 1, during the various design phases, concepts are generated, changed, developed and then worked out in detail. In the early design stages, systems and some components may be selected tentatively, and systems and subsystems may be specified only in general form. With the considerable uncertainty that exists in those stages, reliability of cost estimates is not likely to be good. Cost models used in those stages, therefore, may be very simple. For example, at a very early stage, cost C of the whole building may be represented by

$$C = Ap \qquad (3.2)$$

where

A = floor area, sq ft, provided in the building
p = unit cost, dollars per sq ft

The unit cost may be based on past experience with similar buildings.

Equation (3.2) may be interpreted differently for specific buildings. For example, for a school, A may be taken as the number of pupils and p as cost per pupil. For a hospital, A may be chosen as the number of beds and p as cost per bed. Similarly, for an apartment building, A may be selected as the number of apartments and p as cost per apartment.

As design advances, more information becomes available for cost estimating. The reliability of cost estimates can then be improved. At some design stage, for example, building cost C can be expressed as the sum of the cost of its systems:

$$C = A_1 p_1 + A_2 p_2 + \cdots + A_n p_n = \Sigma A_i p_i \quad (3.3)$$

where

A_i = convenient unit for the i-th system
p_i = cost per unit for the i-th system

As design develops further, still more information becomes available. Even greater reliability is feasible for the cost estimate. Building costs may still be expressed as the sum of the costs of the systems, but those costs should then be given with greater accuracy than obtained with the terms in Eq. (3.3). For example, system costs may be expressed as the sum of subsystem costs, with those costs in the general form of Eq. (3.2). In that case,

$$C = \Sigma A_j p_j \qquad (3.4)$$

where

A_j = convenient unit for the j-th subsystem
p_j = cost per unit for the j-th subsystem

Eventually, enough information becomes available that costs may be estimated in detail, as a contractor would do in preparing a bid. The cost of systems may then be obtained as the sum of the purchase price of components delivered to the site, wages for construction workers, handling and construction equipment costs, and contractor's overhead and profit.

Similar cost models may be formulated for maintenance and operating costs for a building.

Optimization

Optimum means best. Optimization, therefore, is the act of producing the best.

In systems design, the objective is to find the single best system. If there is only one criterion, such as least cost, for judging which system is best, it may be feasible to generate a system that is clearly better than all others, but if there are more than one criterion, for example, least cost and best performance, then an optimum solution may or may not exist. Hence, in establishment of objectives for a system, if optimization is desired, preferably only one criterion for selection of the best system should be chosen.

This criterion may be expressed in the form:

$$\text{Optimize } z_r = f_r(x_1, x_2, \cdots y_1, y_2 \cdots) \quad (3.5)$$

where

z_r = dependent variable to be optimized (maximized or minimized)
x = controllable variable, identified by a subscript
y = uncontrollable variable, identified by a subscript
f_r = the *objective function*.

The system, however, generally is subject to one or more constraints; for example, a building code may specify a minimum thickness or a minimum fire rating, or building geometry may require a minimum clearance, thus imposing limits on system dimensions. These constraints may be expressed in the form

$$f_1(x_1, x_2, \cdots y_1, y_2, \cdots) \geqslant 0$$
$$f_2(x_1, x_2, \cdots y_1, y_2, \cdots) \geqslant 0$$
$$\cdots \cdots \cdots \cdots$$
$$f_n(x_1, x_2, \cdots y_1, y_2, \cdots) \geqslant 0 \quad (3.6)$$

Thus, Eqs. (3.5) and (3.6) must be solved simultaneously to produce the optimum solution. The solution yields values of the controllable variables x as functions of the uncontrollable variables y, which, when substituted in Eq. (3.5), optimizes z_r.

Many techniques are available for finding a solution. Sometimes, calculus can be used. When Eqs. (3.5) and (3.6) are linear, linear programming can be used. When time is a variable, dynamic programming may be applicable. In such cases, use of a high-speed electronic computer usually is necessary for practical computation.

For a building as a whole, its systems and larger subsystems, direct application of Eqs. (3.5) and (3.6) is impractical, because of the large number of variables and constraints. The difficulties that may be encountered are perhaps best illustrated by an example.

Consider a building with skeleton framing. In such a building, columns are usually spaced along rows in two perpendicular directions. The quadrangle, such as *ABCD* in Fig. 3.2a, formed by four columns, is called a *bay*. The area of the bay is a controllable variable that may affect construction and operating costs of several systems. Bay area, in particular, has a considerable effect on structural costs and on production, or activity, costs. The latter may be measured by the loss of revenue to the owner because of the effect on production, or activity,

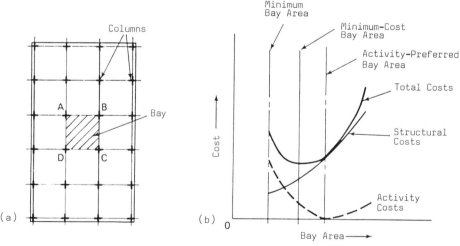

Fig. 3.2. (*a*) Plan view of building showing location of columns. (*b*) Curves showing variation with bay area of structural and activity costs and of the sum of those costs.

of making the bay area smaller or larger than desired for efficient operation, flexibility and future expansion.

Because of the large number of variables involved in optimization of bay area, an exact solution generally is impractical. Any of several strategies for choosing bay size consequently may be adopted. These include selection of: the efficient, or activity-preferred, area, making the loss of revenue zero; the minimum area considered essential; the area making the sum of structural and activity costs a minimum; or some arbitrary area chosen by the owner.

Logic appears to favor the minimum-cost strategy. The influence of bay area on structural and activity costs is very large compared with its effect on other system costs. The strategy, for example, may be carried out as illustrated in Fig. 3.2*b*. Structural costs would be computed and plotted for gradual increases in bay area from the minimum considered essential. Similarly, the increase in activity cost would be computed and plotted for gradual changes in bay area from the activity-preferred size. Then, the sum of the costs would be plotted. The bay area corresponding to the low point of the resulting curve would be the bay area to specify.

Optimization of the sum of structural and activity costs, however, also involves many variables and may be impractical for several other reasons. It may, for example, be impracticable for the owner to predict the loss in future revenue because of variations in bay area. In some cases, activity cost may not be calculable; for example, for a residence.

Structural costs are affected by bay area through its influence on column spacing. This influence is not easy to predict. For example, column spacing can determine choice of structural materials and type of framing. For a specific decision on these, increasing column spacing may increase floor-framing costs but, in some cases, decrease foundation costs. Consequently, many alternative structural systems may have to be investigated to determine the bay area for minimum cost.

Even if an optimum bay area is not determinable, such studies are well worthwhile because they indicate the general range of costs with variations in bay size. Sometimes, owners demand much larger bays than are essential to meet objectives but then either accept smaller bays or drop the project when confronted with the difference in estimated structural costs.

Suboptimization. In some cases, it may be practicable to optimize a system by a process called suboptimization, in which smaller, simpler components are optimized in sequence. The process is discussed in Sec. 3.13.

Simulation. Systems subject to change may sometimes be optimized by a process called simulation, which may also involve trial and error. For the purpose, the actual system or a model may be used. In the latter case, high-speed electronic computers may be very useful. The actual system may be used when it is readily accessible and the changes to be made do not affect other systems and have little or no effect on cost of the system after installation. For example, a HVAC duct system, after completion, may be operated for a variety of conditions to determine the optimum damper position for each condition.

References

F. Jelen and J. Black, *Cost and Optimization Engineering*, 2nd ed., McGraw-Hill, New York, 1982.

R. Stark and R. Nicholls, *Mathematical Foundation for Design*, McGraw-Hill, New York, 1972.

Words and Terms

Model: iconic, symbolic, analog
Objective function
Optimization
Simulation
Suboptimization
Variables: dependent, independent

Significant Relations, Functions, and Issues

Elements of a model: variables, responses, measurement, evaluation.
Methods of modeling or types of models: iconic, symbolic, analog.
Relation of optimization to objectives.
Optimization through observation of successive simulations.

3.2 VALUE MEASURES
FOR COMPARISONS

For the purposes of comparing systems and selecting the best one, some criteria or values must be established as a guide in making the decision. Each must correspond to a measurable system characteristic or to a response (output) to an imposed condition (input). This requirement implies that characteristics or responses must be distinguishable. Thus, it must be feasible to assign different identification marks or numbers to those criteria or values that are different.

It is desirable, but not essential, that criteria and values be quantifiable; that is, that it be possible to arrange assigned numbers in an order that is significant in a comparison. Selection of the best system is easier when comparisons are made on the basis of quantifiable criteria or values, such as costs and revenue.

Any one of four types of measurement scales may be used for criteria and values in systems design: ratio, interval, ordinal or nominal.

Ratio Scales

Engineers generally prefer to use measures that are well defined, such as costs, distances and weights. If a value of $6 is assigned to system *A* and of $3 to system *B*, it can be accepted that *A* costs twice as much as *B*. If *A* weighs 12 lb and *B* 3 lb, it can be accepted that *A* weighs four times as much as *B*. Such scales are called ratio scales.

A ratio scale has the property that, if any characteristic of a system is assigned a value number k, any characteristic that is n times as large must be assigned a value number nk. The absence of the characteristics is denoted by zero.

Interval Scales

For some characteristics, a well-defined measure may not be available; however, it may be possible to use a scale that would at least give a numerical measure of differences in characteristics. The Celsius scale for measuring temperatures is an example. Temperature is a measure of the heat in an object. On the Celsius scale, zero is arbitrarily set at the temperature at which water freezes but does not indicate the absence of heat. Consequently, 80°C does not indicate the presence of twice as much heat in an object as would be present at 40°C, although 80 is twice 40. Relative measurements, however, are still possible. The scale may indicate, for example, that four times as much heat is required to raise the temperature of the object from 40°C to 80°C, an increase of 40°C, as would be needed to raise the temperature from 40°C to 50°C, an increase of 10°C.

An interval scale has the property that equal intervals between assigned value numbers represent equal differences in the characteristic being measured. Zero on the scale is established arbitrarily.

Interval scales are often used in making comparisons of systems that are the same except for a few characteristics or responses. Calculations for selection of the best system in such cases need only take into account the characteristics or responses that are different. Also, only the differences need be compared.

Ordinal Scales

For some characteristics, the measure may be based on a purely subjective decision or the characteristic to be measured may not be precisely defined. Beauty is an example. While aesthetics may be a prime consideration in the design of some buildings, how much more beautiful is one building than another? Generally, if a decision can be reached in a comparison of two objects, it can at best be that one is more, or less, beautiful than the other. If, then, one of these objects is compared with a third object, a decision might be reached that either is more, or less, beautiful than the other. By such a process, it may be possible to assign numbers ranking the objects in order of beauty. But the numbers would not measure how much the objects differ in beauty. The numbers would form an ordinal scale.

An ordinal scale has the property that the magnitude of value numbers assigned to a characteristic indicate whether an object has more, or less, of the characteristic than another object or is the same with respect to that characteristic.

Ordinal scales are useful in comparisons of systems where criteria cannot be expressed in strictly economic measures, such as costs or revenues in dollars. For example, an owner may seek a low-cost building but also want it to be aesthetically appealing. He may also require low maintenance. With several objectives and criteria to be met, it may be necessary to trade-off higher construction costs for more attractive and more durable materials than a least-cost building would permit. Selection of the best system to meet these objectives may have to be based on an ordinal scale.

To illustrate how an ordinal scale may be set up for comparison of systems with several objectives when only some values are quantifiable, the following example presents a scale that has been used for value analysis. In the example, the scale is applied to a comparison of two partitions, one all metal, the other, glass and metal. Calculations are shown in Table 3.1.

The characteristics of concern in the comparison are listed in the first column. These characteristics are assigned a weight, in accordance with the relative importance of the design objectives, as judged by the analysts. The weight assigned to each characteristic may range from

1 for low priority to 10 for highest priority and is shown in the second column.

Next, a relative value is assigned each alternative partition for each characteristic. For example, for construction cost, the all-metal partition is given a value of 10 in Table 3.1 and the glass-and-metal partition a value of 8, because the all-metal partition costs somewhat less. Also, the glass-and-metal partition is assigned a value of 9 for appearance, because the analysts considered it to be slightly more attractive than the other, which is given a value of 7.

For each characteristic, then, the weights and values are multiplied and the products are entered in the table as weighted values. Finally, the ratio of the sum of the weighted values to the partition cost is computed for each alternative. The all-metal partition would be recommended because its ratio is larger.

Nominal Scales

For some characteristics, the measure may be capable of doing no more than indicating that two characteristics are different. No value, however, is assigned to the difference. Such a measurement scale is called a nominal scale.

The measures of such a scale, for example, may indicate the presence or absence of a char-

Table 3.1. Comparison of Alternative Partitions

Characteristics	Relative importance	Alternatives				
		1 All metal		2 Glass and metal		
		Relative value	Weighted value	Relative value	Weighted value	
Construction cost	8	10	80	8	64	
Appearance	9	7	63	9	81	
Sound transmission	5	5	25	4	20	
Privacy	3	10	30	2	6	
Visibility	10	0	0	8	80	
Movability	2	8	16	8	16	
Power outlets	4	0	0	0	0	
Durability	10	9	90	9	90	
Low maintenance	8	7	56	5	40	
Total weighted values			360		397	
Cost			$12,000		$15,000	
Ratio of values to cost			0.0300		0.0265	

acteristic. The measure, for instance, might be that a fan is or is not needed; or space is or is not available for electric wiring; or all components are or are not factory assembled.

3.3 COMPARISONS OF SYSTEMS

The discussion of ordinal scales makes an important point: In evaluation of systems where many factors in addition to cost have to be considered, analysts have to determine the relative importance of design objectives to the owner, building users and the public and the weight to be assigned to values. How important is initial cost? Aesthetics? Maintenance? Flexibility?

While the owner would like to optimize all of the values, optimization of more than one value may not be possible. Consequently, systems analysts may have to determine the psychological value to the owner of system characteristics or responses, the intensity of his feeling for them. Such values then can be used in system comparisons. Psychological measurement, however, is crude compared with economic measurement. Nevertheless, in some cases, it may form the only basis for making a decision.

Economic comparisons are preferred for several reasons: When properly and accurately made, they are much more reliable than comparisons based on subjective values. Also, comparisons expressed in monetary units are likely to be more easily understood by owners. In addition, comparisons may be facilitated by use of money units, because the many different system characteristics to be evaluated in a choice between alternatives may be made commensurable by transformation into money. Consequently, where possible, selection of a system should be based on an economic comparison.

Basis for Decisions

In the choice between alternative systems, only the differences between system values are significant and need be compared.

For example, suppose an exterior wall considered for a building is initially planned to contain 4 in. of insulation. Suppose also that studies indicate that the insulation would save $200 annually in HVAC costs. Why not then for economy use 5 in. of insulation? The question should be answered by subtracting the additional equivalent annual cost of 1 in. of insulation from the corresponding decrease in annual HVAC costs. (The effects of a thicker wall on reduction of interior space and on interfacing with other building components should also be taken into account if relevant.) If the difference is positive, 5 in. of insulation would be better than 4 in. If the difference is negative, the added insulation is not an improvement. It is the difference in savings that should control the decision.

Maximization of Profit

Costs are very important in building design because they usually are incorporated in criteria, or measures, that determine whether a building meets the owner's objectives. Generally, for example, an owner would like to recover his investment and maximize the profit, or return, on his investment in the building. *Return* is the difference between revenue from use of the building and total costs. The last is the sum of initial investment, maintenance and operating costs. With the objective of maximum return, therefore, it is the difference between revenue and costs that should be maximized.

Sometimes, instead, costs are minimized. This could lead to an erroneous decision in choosing between alternatives. Minimum cost yields a maximum return only if revenue is unaffected by the choice of systems or does not decrease as rapidly as cost. Similarly, maximizing revenue could lead to a poor decision unless costs are unaffected by the decision or do not increase as rapidly as revenue.

Also, sometimes, only initial investment, or construction, cost is minimized. This, too, could lead to a bad decision, even though revenue would not be adversely affected. Life-cycle costs should be used in computing profit, not just initial investment cost. Life-cycle costs include maintenance and operating costs. Sometimes also, depreciation and taxes may be important.

Maximum profit, however, may not be sufficient to meet an owner's objectives in some cases. The owner may want profit to be commensurate with the risk involved in making the investment and with the return available from other investment opportunities. Thus, he might require that the rate of return, the ratio of return to investment, be larger than all of the following:

Interest rate for borrowed money
Rate for government bonds or notes
Rate for highly rated corporate bonds
Rate of return expected from a business

Consequently, the decision whether to proceed with construction of a building may well hinge on whether a maximum return can be realized that is large enough to make the rate of return appealing to the client.

Time Value of Money

The preceding discussion should make evident the importance of the time value of money in economic comparisons. All costs represent money that must be borrowed or that could otherwise be invested at a current interest rate, depending on the risk considered acceptable. Consequently, in economic comparisons, interest rates should be used to convert costs of different types, such as initial investment and annual costs, to a common base. For example, initial investment may be changed to an equivalent annual cost, or annual cost may be converted to present worth. Use of interest rates for these purposes is discussed in Sec. 3.4.

3.4. RETURN ON INVESTMENT

A typical economic comparison of alternative systems involves evaluations of initial capital investments, salvage values after several years, annual disbursements and annual revenues. For the comparison, it is necessary to make these different types of costs and revenues commensurable. This is usually done in either of two ways:

1. Conversion of all costs and income to equivalent uniform annual costs and income.
2. Conversion of all costs and income to present worth as of time zero for the annual series of disbursements and income.

Present worth is the amount of money that, invested at time zero at a specified rate of return, would yield annually the required series of disbursements and income.

The conversions should assume a rate of return that is attractive to the owner. It should be at least equal to the interest rate that would have to be paid if the amount of the investment had to be borrowed. Consequently, the desired rate of return usually is referred to in conversion calculations as the interest rate.

The conversions should also be based on the actual time periods involved, or reasonable estimates of them. For example, salvage values should be specified as the expected return on sale of an item after a specific number of years that the item has been in service. To simplify calculations, interest is computed for the end of each year.

Compound interest formulas should be used for the conversions. Thus, a sum invested increases over a specific number of years to

$$S = P(1 + i)^n \qquad (3.7)$$

where

S = future amount of money, equivalent to P, at the end of n periods with interest i
i = interest rate per interest period
n = number of interest periods
P = present sum of money = present worth of investment at time zero

Present worth of a future sum of money can be obtained by solving Eq. (3.7) for P:

$$P = S(1 + i)^{-n} \qquad (3.8)$$

A capital investment P can be recovered in n years with interest i through a series of annual payments R. The amount of the annual payment for capital recovery is given by

$$R = P\left[\frac{i}{1 - (1 + i)^{-n}}\right] = P\left[\frac{i}{(1 + i)^n - 1} + i\right] \qquad (3.9)$$

The present worth of an annual series of payments R can be obtained by solving Eq. (3.9) for P:

$$P = R\left[\frac{1 - (1 + i)^{-n}}{i}\right] \qquad (3.10)$$

The present worth of an annual series of payments continued indefinitely then is

$$P = \frac{R}{i} \qquad (3.11)$$

When equipment has salvage value V after n years, capital recovery can be computed by subtraction of the present worth of the salvage value from the capital investment:

$$R = [P - V(1 + i)^{-n}] \left[\frac{i}{(1 + i)^n - 1} + i\right] \qquad (3.12)$$

Example 3.1. Annual Cost Comparison

Alternatives: Two heating units are being considered for an office building. Estimates for the units are as follows:

	UNIT A	UNIT B
Initial cost	$30,000	$50,000
Life, years	10	20
Salvage value	$5,000	$10,000
Annual costs	$3,000	$2,000

The annual costs include operation, maintenance, property taxes and insurance. Which unit would be more economical if the rate of return is chosen at 8%?

Comparison: Annual costs are computed as follows:

UNIT A

By Eq. (3.12),

$R = [\$30,000$

$\quad - \$5,000(1.08)^{-10}] \left[\dfrac{0.08}{(1.08)^{10} - 1} + 0.08\right]$

	= $4,125
Annual costs	= $3,000
Total annual cost	= $7,125

UNIT B

By Eq. (3.12),

$R = [\$50,000$

$\quad - \$10,000(1.08)^{-20}] \left[\dfrac{0.08}{(1.08)^{20} - 1} + 0.08\right]$

	= $4,874
Annual costs	= $2,000
Total annual cost	= $6,874

Conclusion: Unit B is more economical because its annual cost is lower.

Example 3.2. Present Worth Comparison

Compare Units A and B of Example 3.1 by use of present worths.

Comparison: Whereas the alternatives have different service lives, conversion of all costs and income to present worth must be based on a common service life. A convenient simple assumption for doing this is that replacement assets will repeat the investment and annual costs predicted for the initial asset. In accord with this assumption, a common service life to be used in the comparison of present worths must be selected. Sometimes, it is convenient to choose for the common service life the least common multiple of the lives of the alternatives. In other cases, annual costs may be assumed to be perpetual. The present worths of such annual costs are known as *capitalized costs*.

For this example, assume a common service life of 20 years. Hence, Unit A will presumably be replaced at the end of 10 years by a similar unit at a cost of $30,000, less the salvage value. The new unit will be assumed to have a salvage value of $5,000 at the end of 20 years.

UNIT A

Initial investment	= $30,000
Present worth of replacement cost in 10 years [Eq. (3.8)]	
= ($30,000 - $5,000)(1.08)^{-10}	= 11,580
Present worth of annual costs for 20 years [Eq. (3.10)]	
= $3,000 \left[\dfrac{1 - (1.08)^{-20}}{0.08}\right]$	= 29,454
Present worth of all costs	= $71,034

Income:

Present worth of salvage value after 20 years [Eq. (3.8)]	
= $5,000(1.08)^{-20}	= 1,073
Present worth of net costs for 20 years	= $69,961

UNIT B	
Initial investment	= $50,000
Present worth of annual costs for 20 years [Eq. 3.10)]	
$= \$2,000 \left[\dfrac{1 - (1.08)^{-20}}{0.08} \right]$	= 19,636
Present worth of all costs	= $69,636
Income:	
Present worth of salvage value after 20 years [Eq. (3.8)]	
$= \$10,000(1.08)^{-20}$	= 2,145
Present worth of net cost for 20 years	= $67,491

Conclusion: Unit B is more economical because it will cost less.

Benefit-Cost Comparisons

As indicated previously, the objective of an economic comparison may be selection of a system yielding the maximum return; i.e., the largest difference between revenues and costs. In the preceding examples, however, least cost is the criterion for selection of a system rather than maximum return, because revenue is assumed to be unaffected by the decision, except for salvage values of the equipment. In other cases, revenue may be affected by the decision and should be taken into account.

Revenue, though, may be thought of as more than monetary income. Revenue may also include intangible gains or prevention of losses. For example, the decision whether to waterproof a basement should take into account the damage that would result were the basement to be flooded. Nonoccurrence of such losses would be a financial benefit accruing from waterproofing. Another example is the decision to enclose acoustically a noisy machine. The benefits would be worker comfort, improved worker efficiency and possibly also the ability to obtain workers at lower wages. Benefits may be a better term to use than revenues in such cases.

Consequently, in economic comparisons, the objective may be to maximize the difference between benefits and costs.

Example 3.3. Benefit-Cost Comparison

A step in a manufacturing process requires impact forming of a product. The noise produced by the impact, while not likely to impair the hearing of workers, is unpleasant and will cause a loss of worker efficiency. Three alternatives are under consideration.

Alternatives:

Plan 1. Select Machine A and normal operation.

Plan 2. Select Machine A and isolate it with an acoustical enclosure, thus improving worker efficiency.

Plan 3. Select quieter Machine B, which costs considerably more, but thus free workers from the restrictions of the enclosure.

Estimates for the plans are as follows:

	PLAN 1	PLAN 2	PLAN 3
Initial machine cost	$10,000	$10,000	$20,000
Life, years	5	5	7
Salvage value	0	0	0
Annual costs for machine operation	$ 2,000	$ 2,000	$ 1,500
Acoustical protection cost	0	$ 2,000	0
Annual value of improved efficiency	0	$ 800	$ 1,000

Which plan would be the most economical for the assumption of a 10% rate of return?

Comparison: The following annual costs are computed:

	PLAN 1	PLAN 2	PLAN 3
Capital recovery of machine cost [see Eq. (3.12)]	$ 2,638	$ 2,638	$ 4,108
Capital recovery of acoustical protection	0	528	0
Annual costs for machine operation	2,000	2,000	1,500
Total annual cost	$ 4,638	$ 5,166	$ 5,608

Benefits:

Annual value of improved efficiency	0	800	1,000
Net annual cost	$ 4,638	$ 4,366	$ 4,608
Benefit-cost ratio	0	0.155	0.178

Conclusion: Plan 2 is the most economical because its annual cost is lowest.

Benefit-Cost Ratios

Note that in Example 3.3, Plan 3 has a higher ratio of annual benefit to annual cost than Plan 2. Yet, Plan 2 is more economical. Benefit-cost ratios, the example indicates, are not a reliable measure of the relative economy of alternative systems—at least not when the ratios are based on total costs.

Reliable results can be obtained, however, if the ratios are taken as that of increment in benefit to the increment in cost that goes to produce the benefit. Alternatives may then be compared in pairs in the order of increasing costs. The incremental benefit-cost ratio of the system selected should exceed unity.

For example, in the comparison of Plan 2 with Plan 1, Plan 2 costs $528 more than Plan 1, has a benefit increment of $800, and therefore has a benefit-cost ratio of $800/528 = 1.51$. In the comparison of Plan 3 with Plan 2, Plan 3 costs $442 more, has a benefit increment of $1,000 - $800 = 200, and therefore has a benefit-cost ratio of 0.452. Consequently, the extra cost of Plan 3 is not warranted because the incremental benefit-cost ratio is less than unity.

For Sections 3.2–3.4

References

C. Churchman, *Prediction and Optimal Decision: Philosophical Issues of a Science of Values*, Greenwood, 1982.

E. Grant et al., *Principles of Engineering Economy*, 7th ed., Wiley, New York, 1985.

W. Fabrycky and G. Thuesen, *Engineering Economy*, 6th ed., Prentice-Hall, Englewood Cliffs, NJ, 1984.

Words and Terms

Alternatives
Benefit-cost comparison
Investment
Measurement scales: ratio, interval, ordinal, nominal
Present worth
Profit
Return

Significant Relations, Functions and Issues

Process of use and establishing of values: defining objectives, measurement systems, comparisons, conclusions. Time value of money.

3.5. CONSTRAINTS IMPOSED BY BUILDING CODES

States and communities establish regulations governing building construction under the police powers of the state, to protect the health, welfare and safety of the community. These regulations comprise a building code, which applies a multitude of constraints on building design.

Building codes are administered by a building department. In many communities, the building department not only enforces the building code but also the zoning code, subdivision regulations and other laws affecting buildings.

If a state has a building code, its provisions usually take precedence over municipal codes if the state regulations are more stringent.

The requirements of building codes generally are the minimum needed to protect the public. Architects and engineers therefore must use judgment in applying codes, to protect fully the interests of both clients and the public. Often, more than minimum criteria must be satisfied if a building is to serve efficiently and if personal injuries are to be prevented in use of the building.

Sometimes, code requirements are not adequate to protect the interests of either the client or the public. For example, a building code may specify a minimum thickness of concrete floor, which may be adequate for the client's immediate needs. But the client's needs may change or he may sell the building to a new owner with different needs, in either case making the floor thickness unsafe without expensive alterations. As another example, code requirements for fire resistance may not be enough for public safety. An oven is completely fire resistant but unsafe for humans when the heat is on. Past fires have demonstrated that fire-resistant buildings may actually be huge ovens! When the owner's interests conflict with the public interest, the public interest must prevail.

Code Enforcement

The building department enforces regulations under its purview by checking building plans before construction starts and then inspecting the work during construction. If the department approves the plans, it issues a building permit authorizing construction to start. If, while work is under way, a building inspector finds a violation of a regulation, he issues an order for removal of the violation. Failure to correct a violation subjects a contractor or owner to fines and even to imprisonment. Decisions of the building department, however, may be appealed to a Board of Appeals or to the courts, whether design or construction is concerned. When the building has been completed and approved by an inspector, a certificate permitting occupancy is issued.

Building codes, in general, apply only to work within lot boundaries. (Exceptions include relatively short overhangs, bridges between adjacent buildings, or under-sidewalk vaults.) Construction affecting sidewalks or streets, curb elimination for driveways, water and sewer connections, and other types of work on public property usually are controlled by regulations under the jurisdiction of other departments, such as a department of highways and sewers or a water department. Contractors often have to obtain permits from such departments.

Types of Codes

Attempts have been made in the past to classify building codes as specification type or performance type.

A *specification-type code* specifies specific materials for specific uses. It gives minimum or maximum thickness, height, or length, or com-binations of these. For example, this type of code may specify that an exterior wall must be made of brick or concrete. It may also require that one-story walls must be at least 8 in. thick.

In contrast, a *performance-type code* specifies the performance requirements of buildings and their components. It leaves materials, methods, and dimensions to the option of the designer so long as the performance requirements are satisfied. For example, this type of code specifies that an exterior wall must be:

1. Strong enough to resist all loads that may be imposed on it
2. Stiff enough that loads will not cause permanent deformations or cracking
3. Durable
4. Capable of achieving a stipulated fire rating
5. Resistant to passage of heat, sound, and water.

The code may apply quantitative values, like the fire rating, to many of these desired characteristics.

Performance-type codes have many supporters, because this type gives designers more freedom in selecting materials and methods, readily permits use of new materials and methods, and does not become obsolete as quickly. With specification-type codes, new legislation often is required before new materials or methods may be used. Even when such action is not necessary, building officials may be slow in approving new things, to be certain that their use is safe. In practice, however, performance-type codes have not shown the advantages over specification-type codes that have been expected. The principal reason is that as materials are demonstrated to meet performance requirements they are placed on a list of approved materials. If materials planned for a project are not on the list, extensive investigations may be necessary to obtain approval of those materials. By the time the investigations are completed, it may be too late to use those materials on the project for which they were proposed.

Actually, performance-type codes are an idealization. A purely performance-type code has never been written. Sufficient information for the purpose is not available. Consequently, all codes are partly performance type and partly specifications type. Whether a building code is considered to be of either type depends on the degree to which it relies on performance requirements.

Forms of Codes

Building codes often vary in form with locality. In general, however, they consist of two parts, of which,

One part deals with administration and enforcement, including:

1. Licenses, permits, fees, certificates of occupancy
2. Safety
3. Projections beyond street lines
4. Alterations
5. Maintenance
6. Applications, approval of plans, stop-work orders
7. Posting of buildings to indicate permissible live loads and occupant loads

The second part contains the regulations directly affecting building design and construction, and is, in turn, subdivided to deal separately with:

1. Occupancy and construction-type classifications, limitations on these classes, fire protection, and means of egress
2. Structural requirements
3. Lighting and heating, ventilating, air conditioning, and refrigeration (HVAC) regulations
4. Plumbing and gas piping
5. Elevators and conveyors
6. Electrical code
7. Safety of public and property during construction operations

The form of subdivision depends on the municipality.

Adoption of Standards

Building codes generally consist of a mixture of good practices and minimum standards of adequacy. To obtain building regulations suitable to local conditions, a community may develop a completely new building code for its own use and adopt it by legislative action. By similar action, the community may adopt the latest version of a model code, such as those promulgated by associations of building officials or the American Insurance Association, or a state code, or any of these codes with modifications. The legislation need simply indicate that a specific *code of given date* is adopted, except for certain listed modifications. This action is called *adoption by reference.*

It is common practice also for building codes to adopt by reference existing standards of various types. For example, a building code may, in this manner, incorporate the latest version of ANSI A40.8, ''The National Plumbing Code,'' American National Standards Institute; or a code may adopt by reference any of the many standard specifications for materials or methods of ASTM; or a code may adopt by reference the standard building code requirements for structural design and construction promulgated by the American Institute of Steel Construction, American Institute of Timber Construction, and the American Concrete Institute.

Code Constraints on Design

Many of the architectural and structural constraints imposed by building codes depend on various classifications of buildings defined in the codes. In general, a building may be classified according to:

Fire zone in which it is located

Occupancy group, depending on building use

Type of construction, as a measure of fire protection offered

Fire zones usually are shown on a community's fire-district zoning map. The building code indicates what types of construction and occupancy groups are permitted or prohibited in each zone.

Occupancy group is determined by the building official in accordance with the use or character of occupancy of the building. Typical classifications include:

Places of assembly, such as theaters, concert halls, auditoriums, and stadiums

Schools

Hospitals and nursing homes

Industrial buildings with hazardous contents

Buildings in which combustible materials may be stored

Industrial buildings with noncombustible contents

Hotels, apartment buildings, dormitories, convents, monasteries

One- and two-story dwellings

Type of construction is determined by the building official in accordance with the degree of public safety and resistance to fire offered by the building and its components. These characteristics are measured by the fire ratings assigned to building walls and partitions, structural frame, shaft enclosures, floors, roofs, doors and windows. Fire ratings of various constructions used in buildings are determined by a standard test (usually ASTM E119, ''Standard Methods of Fire Tests of Building Construction and Material,'' promulgated by ASTM, formerly the American Society for Testing and Materials), and measured in hours.

Some building codes give fire-resistance requirements, in addition, for exterior walls and protection of wall openings in accordance with location of a building on a site and distances to property lines and other buildings. The objective is to prevent or delay spread of fire from one building to another.

To prevent or delay spread of fire over very large areas on any level of a building, codes usually specify the *maximum allowable floor area* enclosed within walls of appropriate fire resistance on any level. The areas permitted depend on occupancy group and type of construction.

Maximum building height and number of stories also are specified in building codes for fire safety. These limits, too, depend on occupancy group and type of construction.

Similarly, *occupant load*, or number of persons permitted in a building or room, is specified. The objective is to enable rapid and orderly egress in emergencies, such as fire, smoke, gases, earthquake or any event that might cause panic. Occupant load for any use is determined by dividing the floor area as-

signed to that use by a specified number of square feet per occupant. Building codes list permitted occupant loads in accordance with the type of use of the area or the building. Associated with these loads is a specified number of exits of adequate capacity and fire protection that must be provided.

The structural subdivision of a building code lists the minimum loads for which a building or its components must be designed. The subdivision may also indicate the minimum structural capacities required, allowable unit stresses or maximum permitted deflections. Sometimes, in addition, minimum thicknesses of materials are specified, as well as maximum spacing of bracing.

In a similar manner, building codes apply constraints to mechanical, electrical and other components of buildings. Also, the codes contain rules governing construction of buildings including use of equipment, such as cranes.

References

Architect's Handbook of Professional Practice, American Institute of Architects.

Uniform Building Code, International Conference of Building Officials.

The BOCA Basic National Building Code, Building Officials and Code Administrators International.

Standard Building Code, Southern Building Code Congress International.

Words and Terms

Adoption of standards

Building code

Fire zone

Occupancy group

Occupant load

Performance-type code

Specification-type code

Type of construction (code classification)

Significant Relations, Functions and Issues

Administration of building code: granting of permits for construction, inspections.

Types of codes: performance, specification.

Adoption of documented standards by reference.

Code constraints on design work.

3.6. ZONING CODES

Buildings in a community may be regulated under the police powers of the state, to protect the health, welfare and safety of the community. The regulations promulgated for this purpose generally comprise a zoning code, which applies numerous constraints on building design.

Zoning codes are usually administered by a planning commission or by a building department. The commission also may establish related subdivision regulations, to control subdivision of large parcels of land by developers. Subdivision regulations also act as constraints on designers, who are not completely free, as a result, to maximize economic or aesthetic effects of a building, because design must comply with the regulations.

Zoning is an important planning tool in guiding growth and other changes in a community or region. Planning goals include better living conditions, safety, sanitation, quiet and provision for growth and population increases. In endeavoring to achieve these goals, zoning may restrict the right of a property owner to use his property as he sees fit. But at the same time, zoning protects the property owner from being injured by improper use of nearby property.

Zoning attempts to achieve the planning objectives through control of land use, building height, lot or building area and population density. For the purpose, the planning commission divides the community into a number of districts, in which limits are placed on the features to be controlled.

Land-Use Regulations

These determine the type of occupancy permitted in each district, such as industrial, commercial or residential (single- or two-family dwellings or apartment buildings, for example).

Building-Height Regulations

Height may be controlled practically in any of several ways. One way is to place a limit on the number of stories or the height, in feet, from street to roof (see Fig. 3.3a). Another way is to

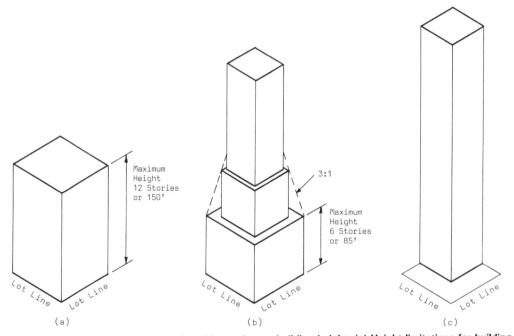

Fig. 3.3. Illustrations of limitations placed by zoning on building height. (a) Height limitations for building constructed up to lot lines. (b) Setback required by a 3:1 sky exposure plane. A tower with floor area at any level not exceeding 40% of the lot area may project above that plane, but the floor-area ratio of the building may not exceed 15. (c) A sheer tower may be allowed a floor-area ratio of 15 if the floor area at any level is 55% or less of the lot area or a floor-area ratio of 18 if floor area is 40% of the lot area.

require that a building lie within specified sloping planes defined with respect to lot lines (see Fig. 3.3*b*). These envelopes are known as *sky-exposure planes*. Such a regulation not only places a practical limit on overall building height, for economic reasons, because a building gets smaller as it gets higher, but the planes also set a limit on height of portions of the building, because they must fit within the specified enveloping planes. Shape of building (appearance), consequently, is considerably influenced by this type of zoning regulation.

Still another way to control building height in a zoning regulation is to specify a maximum floor-area ratio, the ratio of the maximum floor area permitted within a building to the area of the lot. This type of regulation controls bulk and trades off additional floor area in a building for additional unused space on the lot. For instance, for a floor-area ratio of 10, each square foot by which a lot is expanded permits an addition of 10 sq ft of floor area within the building to be constructed on the lot. The effect of floor-area ratio, for practical reasons, is also to limit the overall height of a building and portions of it, because, in congested city districts, lots are very expensive. Land cost increases rapidly with lot area.

Figure 3.3*c* illustrates a case where a considerable portion of a lot is devoted to a plaza for public use. A building without setbacks, a sheer tower, may be erected to a considerable height in accordance with local floor-area-ratio zoning regulations. For example, one city assigns a floor-area ratio of 15 if the cross-sectional area of the tower does not exceed 55% of the lot area or a ratio of 18 if the tower area does not exceed 40% of the lot area. Consequently, on a 10,000-sq ft lot, with a tower area of 40% of 10,000, or 4,000 sq ft, a building would be permitted a total of 18 × 10,000 = 180,000 sq ft. Thus, the tower could be built 180,000/4,000 = 45 stories high.

Area Regulations

One type of regulation on lot area specifies the minimum distances that must be provided between a building exterior and the nearest lot line on all sides. Also, the regulation specifies

the minimum frontage the lot must have along a street.

Another type establishes minimum area of lots for single-family houses and minimum lot areas per family for apartment buildings. The objective of this type of regulation is to control population density. An alternative is to specify the maximum number of families permitted per acre and allow the developer the option of selecting the types of buildings he prefers for satisfying that criterion.

Zoning Map

Current land-use controls are usually indicated on a drawing called a zoning map. It is primarily based on existing land use when it was prepared, modified by granting of variances by the planning commission and changed by rezoning legislation.

Master Plan

In addition to the zoning map, the planning commission usually prepares a master plan as a guide to the growth of the community. An important part of the master plan is a future land-use plan. The objective is to steer changes in the zoning map in the direction of the future land-use plan.

Other Types of Zoning

The following legal regulations also may constrain building design. Aimed at accomplishing specific purposes, they are superimposed on the standard zoning patterns.

Airport zoning is one example. Its objectives are to maintain obstruction-free approach zones and to provide noise-attenuating distances around an airport. The approach zones are maintained by establishment of limits on building heights. These limits vary with distance from and orientation with respect to the airport.

Fire zones are another example. They prohibit certain types of construction that otherwise might be permissible. The restrictions depend on congestion in each zone, population density and proximity and height of buildings.

Land subdivision regulations are still another example. The local zoning ordinance specifies minimum lot area and minimum frontage a lot may have along a street. Subdivision regulations, in contrast, specify the level of improvements to be installed in new land-development projects. These regulations contain criteria for location, grade, width and pavement of streets, length of blocks, open spaces to be provided and right of way for utilities.

References

Architect's Handbook of Professional Practice, American Institute of Architects.

J. Sweet, *Legal Aspects of Architecture, Engineering, and the Construction Process*, West Publishing Co., 1970.

Words and Terms

Area regulations
Building height regulations
Fire zones
Land subdivision
Land-use regulations
Master plan
Zoning
Zoning code
Zoning map

Significant Relations, Functions, and Issues

Zoning as a community planning tool.

3.7. OTHER CONSTRAINING REGULATIONS

In addition to building and zoning codes, there are other legal requirements affecting building design and construction. Local departments of highways, streets, sewers and water have regulations with which building construction must comply. Also, local utility companies have standards that must be met if a building is to be serviced. Designers and construction contractors must be alert to the possibility of these and other constraints and to apply them if they are applicable. In particular, buildings are likely to be subject to requirements of the following agencies.

Health

State or local health departments may have jurisdiction over conditions in buildings that could affect the health of occupants or visitors. Food-handling establishments, hospitals and nursing homes are especially likely to be subject to health department regulation. But health departments may also have the responsibility for enforcing such regulations as those requiring maintenance of suitable indoor temperatures in cold weather.

Labor

For industrial and office buildings and retail stores, there may be laws for employee safety and health established by the state department of labor. Designers must insure that buildings they design provide conditions that are in accordance with the law. The law may require that building plans be submitted to the department of labor for review before construction starts. Failure to comply with these laws will subject an owner to fines. During construction of a building, contractors, as employers, also must comply with the labor laws.

Occupational Safety and Health Administration

For occupational safety and health, the U.S. Congress passed in 1970 the Occupational Safety and Health Act (OSHA). This act contains regulations governing conditions under which employees work. In particular, OSHA contains detailed standards for construction. Contractors and subcontractors must comply with these regulations during construction.

Designers must insure that buildings they design provide conditions that are acceptable under OSHA. There is no provision in the law, however, for reviewing plans before construction starts. Inspections usually are made by the administrating agency only after complaints have been received. Consequently, owners and their design and construction agents should be thoroughly familiar with the law and interpretations of it and should insure compliance.

Housing

For residential buildings for which government-insured mortgages are to be secured, the standards of the Federal Housing Administration or the Veterans Administration apply. In particular, housing must comply with FHA "Minimum Property Standards."

Military

Materials used in military construction must conform with Federal Specifications. Each military department may have regulations affecting construction performed for it.

3.8. SYSTEMS DESIGN STEPS

The preceding sections provide much of the background information needed for systems design of buildings. In this section, the steps required for execution of systems design are outlined.

The procedure proposed has nine basic steps (see Fig. 3.4). These are generally taken in sequence; but Steps 3 through 8, synthesizing, analyzing and appraising alternative systems, may be repeated as many times as costs and deadlines permit or until the designers and analysts are unable to generate new alternatives worth considering.

In showing the steps in sequence, Fig. 3.4 has been simplified for the purposes of illustrating and explaining the design procedure.

Though not shown in Fig. 3.4, several alternatives may be acted on concurrently in actual practice, rather than in sequence. This would make possible comparisons of a group of alternatives simultaneously.

Also, though not shown in Fig. 3.4, some steps may start before earlier ones have been completed. In addition, though not indicated in Fig. 3.4, there may be feedback of information from some steps to earlier ones. These possible feedbacks would create loops—return to an earlier step, revisions, and repeat of a sequence of steps.

Implied but not shown in Fig. 3.4 is a very important action—data collection. To show this would complicate the flow diagram. Data collection is likely to be almost continuous from the start in systems design. Information is needed to prepare objectives and constraints, develop criteria, select and calibrate models,

and evaluate alternatives. Early in design, much of the information that will be needed may not be available or the need may not be recognized. Hence, data may have to be collected throughout most of the design process.

In brief, system design comprises these stages: data collection and problem formulation, synthesis, analysis, value analysis, appraisal and decision. The steps of these stages are as follows:

Step 1. Define briefly what is needed. Indicate what the system is to accomplish. Describe the effects the environment or other systems will have on the performance of the required system. Also, indicate the effects the system will have on its surroundings.

Step 2. In view of what is needed and the expected interaction of proposed systems with the environment and other systems, develop a set of objectives that must be met. Also, compile a set of constraints indicating the range within which values of controllable variables must lie.

Step 3. Conceive a system that potentially could meet all the objectives and constraints of Step 2.

Step 4. Model the system and apply the model. This requires the actions indicated in Fig. 3.1 and explained in Sec. 3.1.

Step 5. Evaluate the system. Determine if it meets all objectives and constraints satisfactorily. In particular, see if construction costs lie within the owner's budget.

Step 6. Apply value analysis. If potential improvements are possible, eliminate components that are not essential, make simplifying or cost-saving changes, or integrate components so that one component can do the work of several.

Step 7. Because the changes made in Step 6 result in a new or modified system, model the system and apply the model, as in Step 4. This may require a new model, recalibration of the former model or just substitution in the former model of new values of the controllable variables.

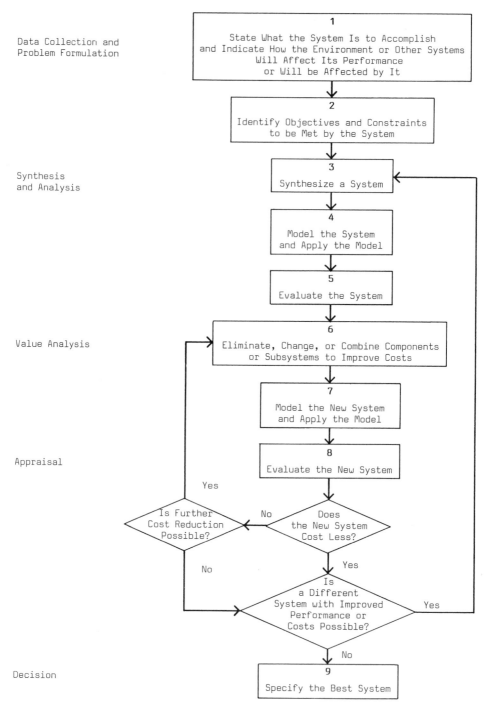

Fig. 3.4. Steps in Systems Design.

Step 8. Evaluate the new or modified system. Compare it with any other alternative systems that have been evaluated. If the new system is more expensive than any of the others, see if additional changes can reduce costs. If they can, make the changes (in effect, return to Step 6) and repeat Steps 7 and 8. Next, try to generate an alternative system that will cost less or will perform better. If this can be done, model the improved system and apply the model (in effect, return to Step 4); then, repeat Steps 5 through 8 with the improved system. If a better or less costly system does not appear feasible, proceed to Step 9.

Step 9. Select and specify the best system from among the alternatives investigated.

The following example is presented to illustrate the systems design procedure. The conditions described and the proposed solution have been simplified, perhaps over-simplified, for the purposes of the example.

Example 3.4. Selection of Shape and Size of an Office Building

At the start of the conceptual phase of design of an office building, the following information is provided:

The owner is a federal government agency, exempt from zoning-code requirements. The owner wants to build an office building for the sole use of the agency. A total of 350,000 sq ft of office floor area is needed. But preliminary studies of office layout indicate that no floor should provide less than 13,000 sq ft of office area. Also, studies show that the service core, containing stairs, elevators, toilets and service rooms, is likely to require about 2,500 sq ft of floor area per story. Budget: $41,500,000, exclusive of land cost.

The owner wants the building erected on a 23,000-sq-ft lot owned by the agency. Located in a congested, central business district of a big city, the lot has a frontage on the south of 200 ft along an avenue (see Fig. 3.5). On the east, the lot has a frontage of 130 ft along a street. And on the west, the lot has a frontage of 100 ft along another street. The lot may be considered, for convenience, to consist of two rectangular areas: Area 1 with 20,000 sq ft, and Area 2 with 3,000 sq ft.

Adjacent to the lot, on the northwest, is a museum, famous as a landmark. This building is about 40 ft high. Other buildings nearby, however, are skyscrapers, mostly about 400 ft tall, or higher.

Along the avenue, buildings usually are set back from the property line, to permit wider-than-usual sidewalks or to provide plazas for public use. In contrast, along the streets, buildings usually are constructed along the property line.

The owner, being a government agency, wants

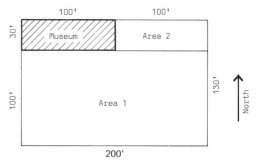

Fig. 3.5. Building site for Example 3.4.

to set a good example for other builders. Consequently, the owner would like the building to be constructed to enhance the community, to provide a public service, if possible. Though exempt from local building ordinances, the owner, for the preceding purpose, requests that the building be designed and constructed in accordance with the city building code and zoning ordinance.

Step 1. Goal Provide, on a lot owned by a federal government agency, a building for the agency with at least 350,000 sq ft of office area at a cost not exceeding $41,500,000. The building should be an asset to the community.

Step 2. Objectives and Constraints
Objectives:
1. Design a building for the 23,000-sq ft lot shown in Fig. 3.5. Area 1 is 100 × 200 ft, and Area 2, 30 × 100 ft.
2. Provide a total office floor area of 350,000 sq ft minimum.
3. Provide, on each level of the building, office floor area of at least 13,000 sq ft, plus service-core area of about 2,500 sq ft.
4. As a public service, provide as much open space at street level for public use as possible, consistent with the preceding objectives.
5. Relate the building to and harmonize it in appearance and position on the lot with neighboring office buildings, many of which rise 400 ft or more above street level. Hence, if possible, align the building with those on the avenue that are set back from the property line.
6. As a public service, provide open space around the landmark museum.
7. For the public good, abide by the city building and zoning codes.

Constraints:

1. Construction cost of building must not exceeding $41,500,000. Cost estimates at this stage of design may be based on a construction cost of $100 per sq ft of gross floor area, the average reported for recently constructed office buildings in the central business district.

2. Maximum possible floor area per story = lot area = 23,000 sq ft; minimum required floor area per story = 13,000 + 2,500 = 15,500 sq ft.

3. Zoning regulations (Note: Floor-area ratio is the ratio of total gross floor area to the lot area.)

   (1) Buildings constructed up to property lines along streets or avenues may rise 85 ft above street level without a setback. Parts of buildings more than 85 ft high must lie within a sky exposure plane starting at the 85-ft level at the property line and with an upward slope of 3:1 away from the street (see Fig. 3.3b), until the floor area in any story does not exceed 40% of the lot area. Maximum permissible floor-area ratio = 15.

   (2) Buildings set back from the property line to provide a wider sidewalk or a plaza are permitted a sky exposure plane as in regulation (1) but with a slope of 4:1. Maximum permissible floor-area ratio = 18.

   (3) Sheer towers set back from the property line to provide a wider sidewalk or a plaza (see Fig. 3.3c) and with a gross floor area per story not exceeding 50% of the lot area are permitted a floor-area ratio = 17.

Step 3. Alternative 1 A building satisfying zoning regulation (1) would provide no open space at street level. Such a building, therefore, would not meet objectives 4 to 6. It cannot be considered an acceptable system.

Instead, for Alternative 1, consider a building satisfying zoning regulation (2). The building then would consist of a tower rising from a broader base. The base, in turn, being set back from the property lines along the streets, would be smaller than the lot (see Fig. 3.6). The service core would be placed in Area 2 of the lot.

Step 4. Alternative 1 Model For a floor-area ratio of 18, the total gross floor area permitted is 18 × 23,000 = 414,000 sq ft.

For a sky exposure plane with slope 4:1, the setback from the property line and height of

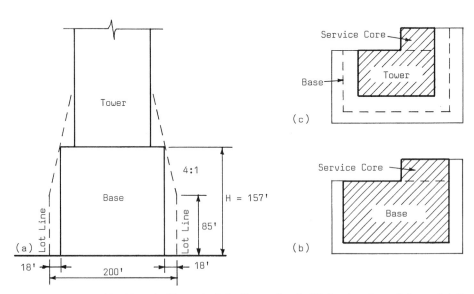

Fig. 3.6. (*a*) Elevation of building for Example 3.4, Alternative 1. (*b*) Section through base. (*c*) Section through tower.

base H to provide a base with maximum floor area can be determined mathematically. On the assumption of a 25-ft-high first story and 12-ft-high stories above, use of differential calculus indicates that the base should be 12 stories (157 ft) high (see Fig. 3.6a).

Set back from the property line 18 ft, the service core in the base would be 30 × (100−18) ft, thus providing a floor area of 2,460 sq ft. Office floor area in the base would be 13,450 sq ft per story, or a total in the 12 stories of 161,400 sq ft. Total service-core floor area in the base would be 29,500 sq ft. Consequently, the base would provide a total floor area of 190,900 sq ft.

The tower then would have to provide office areas totaling at least 350,000−161,400 = 188,600 sq ft. Floor area per story in the tower, however, may not exceed 40% of the lot area, or 0.40 × 23,000 = 9,200 sq ft. Because some elevators and perhaps also some stairs need not be extended above the base, a smaller service-core floor area than 2,500 sq ft, say 2,000 sq ft, may be assumed for the tower. In that case, the maximum office area per story that can be provided in the tower is 9,200 − 2,000 = 7,200 sq ft. To furnish the total office area required, the tower therefore would have to extend above the base 188,600/7,200 = 26 stories.

The building would have a total height of 12 + 26 = 38 stories. The base would provide a gross floor area of 190,900 sq ft and the tower, 26 × 9,200, or 239,200 sq ft. The building would then furnish a total floor area of 430,400 sq ft.

Construction cost of the building would be 430,400 × $100 = $43,040,000.

Step 5. Evaluation of Alternative 1 The estimated cost exceeds the $41,500,000 budget. The gross floor area exceeds the 414,000 sq ft permitted for a floor-area ratio of 18. Furthermore, the floor area per story in the tower is much less than the 13,000 sq ft desired by the owner and therefore is too small to be useful. Alternative 1 consequently is unsatisfactory and cannot be improved by value analysis. Loop back to Step 3.

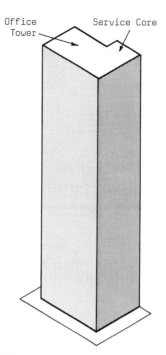

Office Tower Service Core

Fig. 3.7. Sheer tower for Example 3.4, Alternative 2.

Step 3. Alternative 2 Consider a sheer tower satisfying zoning regulation (3) (see Fig. 3.7). As for Alternative 1, the service core would be placed in Area 2 of the lot.

Step 4. Alternative 2 Model For a floor-area ratio of 17, the total gross floor area permitted is 17 × 23,000 = 391,000 sq ft.

Under zoning regulation (3), the tower is permitted an area of only 50% of the lot area, or 0.50 × 23,000 = 11,500 sq ft. This would require a building 391,000/11,500 = 34 stories high. With a service-core floor area of 2,500 sq ft, the floor area available in each story for offices is 11,500 − 2,500 = 9,000 sq ft.

Step 5. Evaluation of Alternative 2 The floor area per story in the sheer tower, being much less than 13,000 sq ft, is too small to be useful. The system cannot be improved by value analysis. Therefore, loop back to Step 3.

Step 3. Alternative 3 It appears to be impossible to meet the owner's objectives of a total office floor area of at least 350,000 sq ft and office floor area per story of at least 13,000 sq ft and also satisfy the zoning code. Therefore, the owner's objectives must be changed or the project will have to be abandoned.

The requirements that the presently owned lot be used and for minimum floor areas appear to be essential. The owner, however, need not abide completely by the zoning code. Since the owner is exempt from the city code requirements, objective 7 (see Step 2) could be relaxed. A more general objective would be:

7. Abide by the local building code and respect the intent of the zoning code.

This change in objective would make possible a trade-off of additional space in the building and greater building bulk for more open space for public use at street level. As a result also, objectives 5 and 6 could be more readily met.

To provide a large open area to serve as a plaza at street level for public use, consider a sheer tower with a major portion of it raised up above street level on stilts, substantial columns (see Fig. 3.8). The service core would extend from the ground to the roof. The office floors would

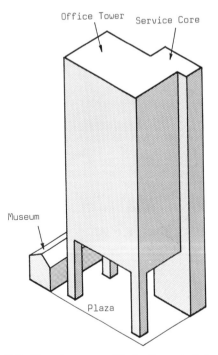

Fig. 3.8. Shear tower on stilts for Example 3.4, Alternative 3.

start at a level, say about 100 ft above the street, that would be sufficiently high above the landmark museum not to cut off light and air movements and to give the plaza a feeling of openness.

Step 4. Alternative 3 Model The floor area per story in the tower can be selected in several ways, since the constraints of the zoning code on tower area have been relaxed. For example, the area can be chosen mathematically by equating the product of the unit cost, $100 per sq ft, number of stories and floor area per story to the owner's budget, $41,500,000. But the resulting structure might be bulkier than necessary. So instead, select the minimum floor area consistent with objective 3.

Thus, each tower floor would be assigned an office area of 13,000 sq ft. The service core would have an area of 2,500 sq ft, as for the other alternatives considered. In addition, a floor area of about 700 sq ft would be provided to connect the service core to the office area. Total floor area per story would then be 16,200 sq ft.

To furnish a total office area of 350,000 sq ft, the number of stories required is 350,000/13,000 = 27.

Construction cost of the building is estimated at $100 × 27 × 16,200 = $43,740,000.

Step 5. Evaluation of Alternative 3 The sheer tower provides the minimum total office area of 350,000 sq ft, with at least 13,000 sq ft of office per story. But the $41,500,000 budget would be exceeded.

Step 6. Value Analysis To bring the construction cost within the budget, the floor area must be reduced.

Alternative 3A. Floor-area requirements for offices can be decreased by treating the 700-sq ft area connecting the service core to the office portion as office space. The former rectangular office area can then be reduced to 12,300 sq ft and the total floor area per story to 15,500 sq ft. The building then would have a total area of 27 × 15,500 = 418,500 sq ft.

Alternative 3B. Floor-area requirements for the service core can be decreased by reducing the number of stories in the building. If, for example, the 700 sq ft were added to, rather than subtracted from the 13,000 sq ft, the number of stories required would be 350,000/13,700 = 25.5, say 25. More accurately then, a 25-story tower would provide 350,000/25 = 14,000 sq ft of office area per story. Total floor area per story would be 16,500 sq ft, and the total floor area in the building would be 16,500 × 25 = 412,500 sq ft.

Step 7. Recalibration of Model
Alternative 3A. Construction cost would be $100 × 418,500 = $41,850,000.
Alternative 3B. Construction cost would be $100 × 412,500 = $41,250,000.

Step 8. Evaluation of Alternatives
Alternatives 3A and 3B provide both the required total office area and the required office area per story. Both alternatives provide a plaza at street level with an area of nearly 20,000 sq ft, less the space required for the stilts. With the office floors starting 100 ft above the plaza, Alternative 3A would be about 420-ft high and Alternative 3B about 400-ft high. These heights would be about the same as those of neighboring office buildings. Also, both alternatives would be set back from the joint property lines with the museum. Thus, the alternatives meet the first six objectives listed in Step 1.

Estimated construction costs of the alternatives are close to the $41,500,000 limit. While the estimated cost of Alternative 3A slightly exceeds the budget, the excess is small and acceptable at this early stage of design.

The alternatives differ principally in bulk and height. The relationship between these factors and the zoning code must be taken into account in evaluation of the alternatives.

Alternative 3A has a ratio of floor area per story to lot area of about 67%. Floor-area ratio is 418,500/23,000 = 18.

Alternative 3B has a ratio of floor area per story to lot area of about 72%. Floor-area ratio is 412,500/23,000 = 18. The principal difference in the alternatives then is that Alternative 3B is two stories lower but occupies 5% more of the lot.

Step 9. Decision Alternatives 3A and 3B meet objectives and constraints about equally well but construction cost of either is very close to the budgetary limit. Unit cost will have to be kept below $100 per sq ft. A lower building will facilitate this, because it will have less wall area, shorter pipe and wiring runs, shorter stairs, less costly elevators and lower structural framing costs.

Recommend Alternative 3B, the 25-story sheer tower on stilts, to the owner.

3.9. SYSTEM GOALS

Before design of a system can proceed, it is necessary to have a definite design program, a list of requirements to be satisfied by the system and of conditions that exist before the system is built. From information in the program, as required by Step 1 of systems design (see Sec. 3.8), goals to be met by the system must be defined.

Applied to system design, goals are desired results expressed broadly. They should encompass all the design objectives, guide generation of alternative designs and control selection of the best alternative.

Goals may be classified generally as service or interactive.

Service Goals

Service goals indicate what the system is to accomplish. They apply to such factors as function, or use, of the system, strength, aesthetics, safety, and initial, maintenance and operating costs.

Since Step 2 of systems design provides the details of required system performance, the statement of goals should be brief and to the point. For example:

Given: Lot C in City D and construction budget of $6,000,000

Design: A factory for the Widget Company for production of 1,000,000 widgets annually and an attached office area for three high-level executives and 20 office workers.

Also, the statement of goals should be broad. Goals that are too narrow may lead designers to overlook favorable alternatives. For example, suppose that the system to be designed is an exterior wall. Stating that the goal is a brick curtain wall would be too restrictive and might rule out suitable alternatives, such as limestone or precast-concrete walls. A goal calling for a curtain wall without restrictions, or either a curtain wall or a load-bearing wall, generally would be better.

Interactive Goals

Interactive goals indicate how the system will affect the environment and other systems.

Environmental interactive goals are those concerned with the response of the system to human needs and feelings. Buildings are for people; hence, a building and its components should be constructed to appropriate human scale. It should be built with concern for the view of the building from outside and the view of outside from within the building. Also, while design should recognize the importance of the client's needs and desires, the primary concern of design should be the health, welfare and safety of building users, whether they be occupants, visitors or the client's employees. Another prime concern should be the good of the community. The building should not contribute unduly to pedestrian and vehicular congestion or cause shortages of resources. Discharges from the building should neither pollute the air nor bodies of water. Nor should the building excessively restrict movement of air, block passage of light, or interfere with communication signals, such as radio and television, to neighboring buildings. For these purposes, environmental interactive goals, as applicable, should be specified to supplement the service goals.

Other interactive goals deal with the desired effects of the system on other systems. For example, a goal for an exterior wall might be light weight, to lighten the load on the structural frame. Or a goal for an electric lighting system might be low power consumption, not only for energy conservation directly but also to decrease the heat gain from the lights and thus to lighten the load on the building cooling system.

Other Interaction

In addition to the statement of goals, Step 1 also should describe how the environment or other systems will affect the system to be designed. The descriptions might provide such information as lot location and size, land surface and subsurface conditions, construction budget, type of community, type of neighboring buildings, streets and utilities.

When known at the start of the design, the descriptions should be included in given information. They need not be only verbal; maps and photographs could be used.

When the information is not available, the effects may have to be assumed or estimated in synthesizing alternative systems and in developing models. For example, during the conceptual design phase, when a building site might not yet have been purchased, design might proceed on the assumption that the lot will be flat and of ample size. When a site is purchased later, the design might have to be revised to accord with actual conditions. Similarly, during this design phase, when subsoil explorations have not yet been completed, design of the structural frame might proceed on the assumption that ordinary spread footings can be used. If the foundation investigations indicate otherwise, the design would have to be revised. In both cases, for the revised designs, the new information becomes part of given information.

Example Goal

Given: Construction budget of $50,000 and lot A, 100 × 100 ft, in a middle-class, residential section of Suburb B. Well-drained land slopes slightly toward the street side of the lot. Municipal water and sewers are available on the street side. Gas and electricity also are available on the street side from underground lines of A Gas & Electric Company. There are existing houses on both sides of the lot and at the rear. All are one-story high and have red brick walls and hipped roofs. Entrances face the street. (This information is given verbally here but in practice would be provided on a survey map, supplemented by photographs.)

Design: A house for Mr. and Mrs. Will B. Homeowner and their two sons and daughter. Children's ages are 17, 15 and 12.

The house should harmonize with the adjoining houses on the street.

Design must be completed within 90 days. Construction must be completed within 180 days thereafter.

3.10. SYSTEM OBJECTIVES

An essential phase of Step 2 of systems design is identification of system objectives (see Sec. 3.8). These are similar to goals. But whereas goals are broad, objectives are specific.

An objective is a desired result achieving or assisting in the achievement of one or more specified goals.

Associated with an objective must be at least one *criterion* or a range of values that indicates that the objective has been met and that can serve as a guide in evaluations of alternative systems.

Expressed another way, an objective is a statement of the response (output) required of a system to specific conditions imposed on the system (input). A criterion then is the range in which the measure of the response must lie. Thus, the response must be measurable but not necessarily quantifiable. Any convenient measurement scale may be used to measure it (see Sec. 3.2).

Objectives for systems design usually may be listed starting with broad generalizations and then developed at more detailed levels to guide design of the system.

Basic Objectives

There are several basic objectives that are generally imposed on building design. They specify that requirements of building codes, zoning ordinances, subdivision regulations, utility companies, fire marshalls, health departments, labor departments, Occupational Safety and Health Administration, etc., must be met. Since these objectives occur so frequently, they may be considered imposed by implication and not listed with other objectives in Step 2. If they were to be waived, however, an objective should be listed to indicate the intent of a replacement. Also, where there may be some ambiguity in applicability of a code or there are some other reasons for specifying a code, an objective should be given to indicate which code building design must satisfy.

Another set of basic objectives that should be stated explicitly deals with costs and time. These objectives should comply with the client's requirements and reflect the seriousness with which he views his proposed budget. The objectives, for example, should indicate whether initial, maintenance or operating costs, or any combination of them, are to be minimized. Also, the objectives should note whether construction time is to be minimized. Energy conservation may be an implied objective if covered by a legal regulation, or implied by an objective concerned with minimization of operating cost, or required by a specific objective.

In accord with the classification of goals as service or interactive, objectives may be similarly classified.

Service Objectives

A primary objective of building design is to serve the needs of the client and building users. Accordingly, a set of objectives must be provided to insure that those needs will be met. If, for example, the building is a factory, objectives must indicate the size, nature and relationship of facilities needed for production; power, water and other resources required; wastes, smoke and heat that must be disposed of; and environments that must be provided. Similarly, if the building is a school, objectives must indicate the size, nature and relationship of classrooms, lecture halls, study rooms, auditoriums, gymnasiums, offices, library and other educational facilities needed; power, water and other resources required; waste disposal; and environments that must be provided. For building components, similar objectives insuring that functional requirements will be met must be compiled.

Other service objectives should deal with specific characteristics of a building and its components: appearance, strength, durability, stiffness, operation, maintenance and fire resistance. Still other objectives should be concerned with human aspects: safety; convenience in moving about and in locating and using facilities; and comfort, including thermal and acoustical. Additional objectives usually are needed to specify controls needed for operation of systems provided to meet the preceding objectives.

Interactive Objectives

Those objectives that specify how the system to be designed will affect the environment and other systems are secondary to the preceding objectives but nevertheless are important.

Environmental interactive objectives should be specified to attain environmental interactive goals, as explained in Sec. 3.9. In the interests of public health, welfare and safety, these objectives seek to avoid pollution, to respect the rights of neighbors, and to enhance community life.

Additional interactive objectives are necessary to attain goals concerning the effects of the system on other systems, as explained in Sec. 3.9.

Sources of Criteria

As mentioned previously, at least one criterion should be associated with each objective, to be used as an indication that the objective has been met. The criterion should apply to a measure of an appropriate system response. Criteria may be chosen from any of numerous sources, depending on the particular objectives. Note that one criterion may be applicable to more than one objective, while one objective may be associated with more than one criterion.

Criteria for objectives related to legal regulations, such as building codes, zoning ordinances, health laws and labor department rules, usually may be obtained from those regulations. Criteria dealing with quality of materials and methods of testing materials often may be secured from specifications of ASTM (formerly American Society for Testing and Materials) or Federal Specifications. Criteria for such characteristics of systems as strength and resistance to deformation and for fabrication and construction methods generally may be found in industry codes of practice, such as those of the American Institute of Steel Construction, American Concrete Institute and American Institute of Timber Construction. Also, criteria may be obtained from recommendations of professional societies, such as the American Institute of Architects, American Society of Heating, Refrigerating and Air-Conditioning Engineers and Institute of Electrical and Electronic Engineers.

In some cases, it may be necessary to develop criteria based on the owner's feelings or estimates of values to him of various system responses. Criteria applicable to aesthetics are of this type. Alternatively, such criteria may be derived from a consensus of the members of the building team or of building users or others who will be affected by the objectives. In other cases, the only source of criteria may be the experience and judgment of the designers.

Relative Importance of Objectives

In addition to identifying the objectives of a system, the designers and analysts also must determine the relative importance of the objectives. If money could be used as a measure of importance, ranking of objectives would be easy. Many system values, or benefits, however, are not quantifiable. Appearance is one example. Comfort of building users is another. Consequently, some means must be adopted for weighting system values in accordance with importance to the client, building users and the public (see Sec. 3.3).

One method that has been used for doing this is described in Sec. 3.2 (see Table 3.1). Other methods have also been tried. (See, for example, C. E. Osgood, G. J. Suci and P. H. Tannenbaum, "The Measurement of Meaning," University of Illinois Press, Urbana, Ill., and L. L. Thurstone, "The Measurement of Values," University of Chicago Press, Chicago, Ill.)

3.11. SYSTEM CONSTRAINTS

As indicated in Sec. 3.10, objectives and criteria are related to system responses. When a system is modeled, responses are represented by dependent variables. The independent variables, which represent the input to the system and system properties, may be controllable by the designer or uncontrollable (see Sec. 3.1). The designer, however, may not be completely free to select any values he desires for the controllable variables. There may be restrictions—legal, economic, physical, chemical, temporal, psychological, sociological, aesthetic, etc.—that either fix the values of these variables or establish a range in which they must lie.

Constraints are restrictions on the values of controllable variables that represent properties of the system.

Associated with a constraint must be at least one standard. A *standard* is a specific desired value of a controllable variable. A minimum standard is a value below which the variable should not fall. A maximum standard is a value that the variable should not exceed.

An example of a constraint is a building-code requirement that the thickness of a one-story, load-bearing, brick wall shall not be less than 6 in. In this case, 6 in. is a minimum standard.

Another example of a constraint is a health-department regulation that when the outdoor temperatures between October 1 and April 1 fall below 65°F buildings must be heated to maintain a temperature of at least 68°F. In this case, 68°F is a minimum standard. If the client were to require that at no time should temperatures in occupied areas, other than entranceways, of a building exceed 77°F, that temperature would be a maximum standard.

Sometimes, it may be difficult to distinguish between objectives, which are related to responses, and constraints, which are related to system properties. For example, costs may be considered a response of a system or a property of the system, depending on circumstances. A restriction on cost may then be imposed accordingly either as a criterion for an objective or a standard for a constraint. More specifically, suppose an owner wished to minimize costs. That would be an objective. If, instead, the owner established a budget that must not be exceeded but did not care how much less than the budget would be spent, there would be no necessity to minimize cost. Thus, the budget amount would be a standard. Similarly, the maximum permissible completion date for construction would be a standard if the owner did not care how much earlier the project were to be completed. Another example is beauty, which sometimes may be considered to be a response of a system and sometimes, a system property.

For Sections 3.8–3.11

References

D. Meredith et al., *Design and Planning of Engineering Systems*, 2nd ed., Prentice-Hall, Englewood Cliffs, NJ, 1985.

S. Andriole, *Interactive Computer Based Systems Design and Development*, Van Nostrand Reinhold, New York, 1983.

A. Gheorghe, *Applied Systems Engineering*, Wiley, New York, 1982.

Words and Terms

Constraints
Criteria
Feedback
Goals: service, interactive
Loop
Objectives: basic, service, interactive
Standards

Significant Relations, Functions and Issues

Steps in systems design, from data collection to specification.
Analysis, evaluation, and comparison of alternatives.
Anticipation of feedback and looping in the design process.
Purposes and relations of goals and objectives.
Importance of criteria in defining of objectives.
Sources of criteria.

3.12. VALUE ANALYSIS

As defined in Sec. 1.3, value analysis is an investigation of the relationship between life-cycle costs and values of a system, its components and alternatives to these, to obtain the lowest life-cycle cost for an acceptable performance.

Value is a measure of benefits anticipated from a system response or from the contribution of a component to a system response. This measure is used as a guide in a choice between alternatives. Scales that may be used for value measures are discussed in Sec. 3.2.

Life-cycle costs may be taken as the sum of the present worth of initial, maintenance and operating costs; or they may equally well be taken as the sum of equivalent annual initial, maintenance and operating costs. In either form, life-cycle costs encompass money measures of such system characteristics as quality, energy consumption and efficiency. Furthermore, initial, or construction, costs include tax, wage, handling, storing, shipping, fabrication, erection, finishing and clean-up costs, in addition to the purchase price of materials and equipment. Requiring life-cycle costs to be minimized, therefore, is equivalent to requiring that the sum of all the previously mentioned component costs be minimized.

Value analysis requires that system values (benefits) be balanced against the costs of providing them. Thus, value analysis is not merely a device for paring down costs. Its aim is to go as far as possible toward relevant goals and objectives while minimizing total cost. Consequently, other values than economic ones often must be considered. This requires determination of the relative importance of objectives (see Sec. 3.10) and weighting of values accordingly (see Sec. 3.2).

Weighting of values should reflect the seriousness with which the client views the construction budget. Often, the budget is established as the maximum permissible construction cost. If costs cannot be kept within the budget, the project may be canceled. When the constraint of the budget is a governing factor, minimization of life-cycle costs, as desirable as it may be as an objective, may not be realizable. In such cases, a prime concern of value analysis is keeping construction costs within the budget. As a result, space, quality, reliability and low energy consumption may be traded off for initial cost savings, with consequent higher maintenance and operating costs and poorer aesthetic results. The client may plan on correcting these at a later date, if possible, when funds become available; or the client may be willing to accept the adverse effects indefinitely as the price he has to pay for current lack of suitable funds.

Steps in Value Analysis

Regardless of the design phase in which value analysis is applied, the value analysts must be thoroughly acquainted with the building program, system goals, system objectives and criteria, and system constraints and standards. Through study of design drawings and specifications and proposed construction contracts, if available, the analysts should also familiarize themselves with the system or systems to be analyzed. From information supplied by the client and the system designers, the analysts then should determine the relative importance of the system objectives and weight values ac-

cordingly. The weighted values are to be compared with estimated costs.

For the purposes of cost analyses and comparisons, cost estimators should develop and calibrate cost models from records of costs of previous similar systems. When a cost estimate is obtained for construction of a system, the result should be compared with the budget and an indication obtained as to how much cost cutting is needed. When the estimate is smaller than the budget, further cost studies may not be necessary but they still are desirable, to insure that the client will be getting his money's worth.

With the aid of the cost models, the search for ways to cut costs is facilitated. With complicated systems, however, it may not be practical to investigate every component or even every subsystem. Instead, the analysts should identify target items for study. These may be discovered through comparisons with previously recorded costs of similar items. For example, a subsystem whose cost represented a high percentage of the total system cost in a previous system would be a good target. As another example, a subsystem whose cost differed substantially from the cost of a similar subsystem in a previous system would be a suitable target.

To cut costs, the analysts may seek to eliminate components, substitute more efficient or less costly components, or combine components so that one component can serve the purposes of two or more. The effects of the changes on costs can be estimated with the aid of the cost models. Because the changes are also likely to affect system values, the analysts should determine new weighted values for the revised system. Costs and weighted values should then be compared to provide a basis for the decision as to whether the changes are warranted.

The cost-cutting investigation should not be restricted just to the system itself. The analysts should review the building program, specifications and construction contracts as well as criteria and standards to determine whether they are essential, too restrictive, or in other ways add unnecessarily to costs. If the analysts should find that a change is necessary, they should report this to the designers.

System changes recommended by the value

analysts will result in a new system. It should be treated as indicated in Step 7 and subsequent steps of systems design (see Fig. 3.4).

References

L. Miles, *Techniques of Value Analysis and Engineering*, 2nd ed., McGraw-Hill, New York, 1972.

Words and Terms

Benefit
Benefit-cost analysis
Life-cycle costs
Tradeoffs

Significant Relations, Functions and Issues

Relative importance of separate objectives.
Relative flexibility of criteria and constraints.
Reconsideration of objectives based on value analysis.

3.13. OPTIMUM DESIGN OF COMPLEX SYSTEMS

Section 3.1 points out that if optimization of a system is an objective, preferably only one criterion for the selection of the best system should be chosen. The criterion may be expressed in the form of an objective function [see Eq. (3.5)]. The system, however, also has to satisfy constraints, which may be expressed in the form of Eqs. (3.6).

For a complicated system, such as a building, a direct solution generally is impractical. There are too many variables and constraints. Also, it may be necessary that other objectives, which, while not the prime concern, also must be met.

Selection of Previously Used Systems

To meet the many objectives of a complicated system, some designers recommend, without thorough study, a system that has been used previously and worked satisfactorily. They usually offer any or all of the following reasons for this:

1. Design costs mount with time spent on design, and the design fee that the client is willing to pay is not sufficient to cover design costs for a thorough study.
2. The deadline for completion of design is too close to permit a thorough study.

3. Contractors quote lower prices for constructing systems that have been built before and build them faster and better than systems with which they are not familiar.
4. The probability of getting successful results with a system that has been successful in the past is very high.

These are good reasons and generally true. As a result, the system chosen may sometimes be the best for the client. It will actually be the best if the system had been developed through study and experience to meet certain objectives and constraints and these all happen to be the same as those of the client. Often, however, the client has objectives that differ from those of other owners or the constraints, such as budget, building codes, zoning ordinances, foundation conditions, or climate, differ from those imposed on previous systems. In such cases, if a system is selected without adequate systems analysis, the client either will not attain his goals or will pay too much for what he gets.

Trial and Error

An alternative that often is used to try to attain optimization of a system is trial and error. In many cases, this is the only feasible method, because of system complexity. (Simulation may be considered a form of trial and error. See p. 63.) Trial and error involves selection of a tentative system and a sequence of attempts to improve it by changing controllable variables while observing the effects on the dependent variables. (See Example 3.4, p. 79.) The procedure has at least the two following disadvantages:

1. It may have to be terminated before optimization has been achieved, because of design time and cost limitations.
2. The nature of the initial system selected may be such that, even if the system were to be optimized, it still would not be the optimum. For example, if a long-span structural system was to be designed for lowest cost and the initial system selected was a concrete frame, optimization by trial and error would lead to the lowest-cost concrete frame. The true optimum, however, might be the lowest-cost steel

frame or the lowest-cost thin concrete shell, and not a frame at all.

Recognizing these disadvantages, the designers must rely on experience, skill, imagination and judgment in using trial and error to attain optimization. The aim should be to approach the optimum if design costs and time have to halt the design effort at any stage.

Suboptimization

The procedure most often used for a complicated system, such as a building, is to try to attain optimization of the system by suboptimization; that is, by first optimizing subsystems. This procedure, however, for several reasons, may not yield a true optimum. Usually, because of the interaction of system components, design of a subsystem affects the design of other subsystems. Hence, a subsystem cannot be optimized until the others have been designed and their effects evaluated. This usually makes necessary a trial-and-error procedure for design, which has the disadvantages previously mentioned.

For example, in optimizing structural costs, minimum costs will not always be obtained if, first, costs of roof and floor framing are minimized and then column and foundation costs are minimized. For, though column and foundation sizes are determined by the load from the roof and floor framing, the minimum-cost roof and floor framing may be heavier than other alternatives and thus require more costly columns and foundations than would the alternatives. The total cost of the framing, therefore, may not be the optimum.

Sometimes, it may be possible to optimize a system by suboptimization directly when components influence each other in series. For example, consider a system with three subsystems (see Fig. 3.9a). Subsystem 1 is assumed to have a known input. This subsystem affects only subsystem 2; that is, the output of subsystem 1 equals the input to subsystem 2. Similarly, subsystem 2 provides input only to subsystem 3, whereas subsystem 3 does not affect any other subsystem. Hence, the subsystems are in series.

Suboptimization may be started with the end component, subsystem 3, because optimization of that component has no effect on input to preceding components.

Subsystem 3, however, cannot be selected immediately, because the input to it depends on the design of the other subsystems and therefore is not known at this stage. To provide the needed input information, preliminary designs of possible optimum subsystems may be made in sequence, beginning with the first subsystem, subsystem 1, to obtain estimates of their outputs. With a potential input or a range of inputs assumed, one or more optimum designs may be prepared for subsystem 3, the end subsystem. Next, subsystems 2 and 3 can be optimized together for an assumed input or range of inputs, with no effect on subsystem 1. Then, the process can be repeated with subsystem 1, the three subsystems being optimized in combination. Since the input to subsystem 1 is known, the optimum system can be selected from the alternatives considered.

The procedure may be illustrated by a hypothetical example. Assume that the system is a one-story structural frame (see Fig. 3.9b) and that inputs and outputs are loads (see Fig. 3.9c). The roof would correspond to subsystem 1 in Fig. 3.9a, columns to subsystem 2 and footings to subsystem 3. As shown in Table 3.2, p. 92, and Fig. 3.9c, the load (input) on the roof is 400 lb. Load is transmitted in sequence from the roof through the columns to the footings. Cost of the whole structural system is to be minimized.

Suboptimization therefore can be started with the footings. The load on the footings, however, is not known initially, because the weight of roof and columns to be added to the 400-lb roof load cannot be determined until they have been designed. So preliminary designs of roof and columns are made to obtain estimates of the probable weights. As indicated in Table 3.2, three alternative designs are prepared for the roof, 1A, 1B and 1C (Step 1). With the weights of those alternatives added to the roof load, three alternative designs are prepared for the columns, 2A, 2B and 2C (Step 2). A set of loads that might be expected to be imposed on the footings is now determined by the output of Step 2.

Suboptimization, starting with the footings,

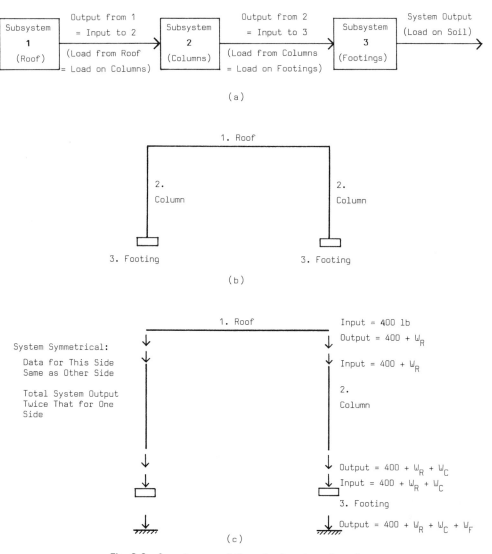

Fig. 3.9. A system consisting of subsystems in series.

can now begin. As shown in Table 3.2, optimum footing designs, $3A_o$, $3B_o$ and $3C_o$ are prepared for the range of loads that might be expected (Step 1 of suboptimization). Costs are estimated for each design. Next, for the range of loads anticipated, columns are selected to make the cost of columns and footings a minimum (Step 2). Finally, for the load imposed on the roof, which is given as 400 lb, the roof and framing are selected to make the cost of roof, columns and footings a minimum (Step 3). In this example, subsystem $1C_o$ which has the highest cost ($2,700 – $1,000 = $1,700), is selected for the roof, because, when combined with optimum columns and footings, that roof yields the lowest-cost structural frame.

Reference

F. Jelen and J. Black, *Cost and Optimization Engineering*, 2nd ed., McGraw-Hill, New York, 1982.

Words and Terms

Suboptimization
Subsystem
Trial and error

Table 3.2. Suboptimization of a Simple Frame

	Preliminary Design					
	Step 1			Step 2		
Load (input)	Subsystem type	Load + weight (output)	Load (input)	Subsystem type	Load + weight (output)
400	1A	2,700	2,700	2A	3,000
400	1B	1,800	1,800	2B	2,000
400	1C	800	800	2C	1,000

	Suboptimization									
	Step 3			Step 2			Step 1			
Input	Optimum subsystems	Lowest cost	Input	Optimum subsystems	Lowest cost	Input	Optimum subsystem	Lowest cost	Output
400	$1A_O + 2A_O + 3A_O$	$2,800	2,700	$2A_O + 3A_O$	$1,800	3,000	$3A_O$	$1,000	4,000
400	$1B_O + 2B_O + 3B_O$	$2,900	1,800	$2B_O + 3B_O$	$1,400	2,000	$3B_O$	$800	2,700
400	$1C_O + 2C_O + 3C_O$	$2,700	800	$2C_O + 3C_O$	$1,000	1,000	$3C_O$	$600	1,500

Note. The lowest-cost system consists of the optimum subsystems of types 1C, 2C, and 3C.

Significant Relations, Functions and Issues

Problems of selecting previously used systems.
Difficulties of optimization of complex systems.
Give and take of suboptimization; need for interactive analysis.

GENERAL REFERENCES AND SOURCES FOR ADDITIONAL STUDY, CHAP. 3

These are books for general reference, grouped under the five categories shown. References relating to the individual chapter sections are listed at the ends of the sections.

Models and Optimization

F. Jelen and J. Black, *Cost Optimization Engineering*, 2nd ed., McGraw-Hill, New York, 1982.
R. Stark and R. Nicholls, *Mathematical Foundation for Design*, McGraw-Hill, New York, 1972.

Comparisons of Alternatives

C. Churchman, *Prediction and Optimal Decision: Philosophical Issues of a Science of Values*, Greenwood, 1982.
E. Grant et al., *Principles of Engineering Economy*, 7th ed., Wiley, New York, 1985.
W. Fabrycky and G. Thuesen, *Engineering Economy*, 6th ed., Prentice-Hall, Englewood Cliffs, NJ, 1984.

Value Analysis

M. Macedo et al., *Value Management for Construction*, Wiley, New York, 1978.
L. Miles, *Techniques of Value Analysis and Engineering*, 2nd ed., McGraw-Hill, New York, 1972.
A. Mudge, *Value Engineering*, Society of American Value Engineers, 1981.
P. O'Connor, *Practical Reliability Engineering*, Wiley, New York, 1985.
A. Dell'Isola, *Value Engineering in the Construction Industry*, Van Nostrand Reinhold, New York, 1983.
L. Zimmerman and G. Hart, *Value Engineering: A Practical Approach for Owners, Designers, and Contractors*, Van Nostrand Reinhold, New York, 1981.

Systems Design

M. Sanders and E. McCormick, *Human Factors in Engineering and Design*, 6th ed., McGraw-Hill, New York, 1987.
D. Meridith, et al., *Design and Planning of Engineering Systems*, 2nd ed., 1985.
S. Andriole, *Interactive Computer Based Systems Design and Development*, Van Nostrand Reinhold, New York, 1983.

A. Gheorghe, *Applied Systems Engineering*, Wiley, New York, 1982.

Building Codes

Uniform Building Code, 1988 ed., International Conference of Building Officials. (New edition every three years.)

The Standard Building Code, 1988 ed., Southern Building Code Congress International. (New edition every three years.)

EXERCISES

The following questions and problems are provided for review of the individual sections and Chapter 3 as a whole.

Section 3.1

1. What is the most important requirement for a model? What should be done to insure that a model meets this requirement?

2. Compare the role of models in systems analysis with that of hypothesis in the scientific method.

3. What do iconic models and analog models have in common?

4. What are the principal advantages of symbolic models?

5. A beam with length L is attached with a bolt to the top of each of two columns. One column is placed at one end of the beam, and the second column at a distance a ($a < L$) from that end. A load P is set on the unsupported end of the beam. Construct a symbolic model that gives, for every load P, the loads imposed on the columns. Define the symbols used in the model. Test the model to verify its validity, by (1) setting $a = L/2$ and (2) letting a approach L in magnitude.

6. (1) At the start of design of an industrial building to produce 1,000,000 widgets annually, the owner establishes a budget for the project of $3,000,000. Studies of existing widget factories indicate that construction costs, adjusted for time and regional differences, ranged from $2,500 to $3,500 per thousand widgets produced annually. Is the project likely to be feasible?

(2) The owner establishes an objective of 150,000 sq ft of floor area for the proposed $3,000,000 widget factory. Studies show that adjusted construction costs for similar buildings range from $25 to $30 per sq ft of floor area. Is the project likely to be economically feasible?

7. During design of an office building with a proposed floor area of 100,000 sq ft and budget of $3,000,000, cost is estimated at $16 per sq ft of floor area for architectural components, $5 per sq ft for the structural system and $12 per sq ft for mechanical and electrical systems. To how many square feet should the floor area be changed to meet the budget?

8. Excavation for a 9 × 6-ft by 1-ft thick concrete footing is predicted to take 2 hr for a workman with payroll cost of $15 per hour. Formwork for the footing concrete is estimated to cost $40. Concrete and reinforcing steel in place is expected to cost $50 per cu yd of concrete. Overhead and profit is assumed at 20% of material and labor costs. How much will the footing cost?

9. Cost records for an existing factory indicate that building maintenance costs have averaged $30 per year per 1,000 sq ft of floor area. A similar proposed factory will have 1,000,000 sq ft of floor area. Estimate the average annual maintenance cost for the proposed building on the assumption that costs will increase 100% during the service life of the factory.

Sections 3.2 to 3.4

10. If a businessman can get a rate of return of 15% annually by investing $10,000 in his business, how much money will he have at the end of 10 years if he reinvests the return every year? Assume annual income tax at 5% of the total invested each year.

11. A $200 blower purchased today is estimated to have a salvage value of $20 after 5 years of service.

(1) For a rate of return of 10%, what is the present worth of the blower?

(2) What is the present worth of the salvage value for a 10% return?

12. How much should capital recovery be each year for 20 years if a building costs $1,000,000 and the desired rate of return is 10%?

13. An industrialist is contemplating installing in his factory a labor-saving device at a cost of $10,000. Annual savings of $5,000 will be sufficient to enable the industrialist to recover his investment in 3 years, though the device has a life of 10 years. Show that the rate of return in the first 3 years will be about 23%.

14. (1) What is the present worth of annual revenues of $117,460 for 20 years if the rate of return is 10%?

(2) What is the present worth of annual revenues of $117,460 continued indefinitely if the rate of return is 10%?

15. How much should capital recovery be each year for a $5,000 industrial crane if the salvage value after 5 years is $1,000? Assume a 10% rate of return.

16. A factory is being designed with plans for doubling its size after 10 years. The owner wants an emergency electric power-generating plant installed. Two plans are being considered:

Plan 1. Purchase equipment initially for the planned future size of factory. The equipment will cost $100,000 initially and will have a probable life of 25 years, but no salvage value. Annual maintenance and operating costs are estimated at $12,000.

Plan 2. Purchase equipment for the initial size of factory and add more generating equipment when the factory is expanded 10 years later. Initial equipment will cost $60,000. It is estimated that the added equipment will cost $80,000. In both cases, equipment is estimated to have a life of 25 years with no salvage value; but at the end of 15 years, salvage value is estimated at $20,000. Annual operating and maintenance costs are estimated to be $7,200 for the initial equipment and $14,400 for the final installation.

Which plan will be more economical with an 8% rate of return?

17. A $50,000 house is initially designed with no roof insulation. The HVAC installation will cost $4,000. Maintenance and operating costs will be $700 annually for HVAC. Addition of insulation in the roof will result in the following costs:

INSULATION THICKNESS, IN.	COST INSTALLED	HVAC COST INSTALLED	ANNUAL HVAC COSTS
2	$100	$3,700	$600
3	$125	$3,700	$590
4	$150	$3,500	$585
5	$175	$3,500	$580
6	$200	$3,300	$575

What is the most economical insulation thickness? Assume a 10% rate of return.

18. Incremental benefit-cost ratio for Alternative 1 over Alternative 2 is 0.86, and of Alternative 2 over Alternative 3, 2.10. Which alternative is best?

Section 3.5

19. Who enforces building codes?

20. A building code states that the minimum size of copper electrical conductor permitted is No. 14. An electrical engineer specifies a minimum size of No. 12, which is larger and more costly, for a residence, although calculations indicate that No. 14 is more than adequate in some cases. What justification does the engineer have for his specification?

21. What document does a contractor need from a building department before construction of a building may start?

22. What document does an owner need to show that he has building-department permission to occupy a new building?

23. What are the principal differences between specification-type and performance-type codes?

24. Where would you find information as to whether a wood school building may be built along a specific city street?

25. What provisions do building codes contain for prevention of spread of fire in any story of a large building?

26. What provisions do building codes have to insure egress for occupants in emergencies?

Section 3.6

27. What are the purposes of a zoning code?
28. What is the relationship between zoning and subdivision regulations?
29. What two types of zoning regulations should be checked to determine if a wood-frame factory may be built on a lot fronting on a specific street?
30. A builder plans a 60-story building in a city. The city zoning ordinance will ordinarily permit this height on the size of lot owned by the building and the lot location. What other zoning regulations should the builder check?
31. A developer plans to erect 100 houses on land zoned by a county for residential construction. (1) What ordinance should the developer consult for limits on minimum lot size? (2) What regulations govern street layout? (3) What regulations specify how far each house must be from its lot lines.
32. A developer owns a 500-ft-long strip of land, wide enough for only one row of houses, along a street. The zoning code requires a minimum frontage of 20 ft for lots along that street. What is the maximum number of lots into which the land may be subdivided?
33. A builder owns a 10,000-sq ft lot. If he provides a plaza at street level, the zoning code permits a floor-area ratio of 15 if the average area of each floor does not exceed 55% of the lot area. How many stories high may a sheet tower be constructed on the lot?

Section 3.8

34. What are the purposes of Steps 1 and 2 of the systems design procedure?
35. What does Step 4 of the systems design procedure accomplish?
36. What is the purpose of value analysis?
37. What action specifically does Step 9 call for?

38. A one-story building with 2,500 sq ft of floor area is to be enclosed with a 10-ft-high exterior wall. If the wall is built in straight sections, it will cost $8 per sq ft of wall area; if built in curved sections, $10 per sq ft. Corners cost $250 each to build. What shape should the building have in plan and what should its dimensions be to minimize construction cost?

Sections 3.9 to 3.11

39. Explain the relationship between system goals and objectives.
40. How does a system objective differ from a system constraint?
41. What purposes do criteria serve with respect to system objectives?
42. What purposes do standards serve with respect to system constraints?
43. When a model of a system is formulated, to what do criteria and standards, respectively, apply?
44. A manufacturer of heavily advertised, consumer products is the client for design of an office building to be built along a heavily traveled highway. He requires that the building be a showpiece, because of the advertising value to his products. In this case, would beauty be an objective or a constraint?
45. A factory is being designed for construction in an industrial park. The owner states that it must be built for the least possible cost and sets a tight budget. Management of the industrial park, however, will not permit buildings in the park that are not sufficiently handsome to obtain approval of its architectural committee. In this case, would beauty be an objective or a constraint?

Section 3.12

46. Basically, what is a system value?
47. Name at least three components of construction cost of an installed window.
48. What kind of costs are included in life-cycle costs besides initial cost?

49. Describe two alternative ways of converting the components of life-cycle costs to the same basis so that they can be added.
50. A change is being considered in a system under design. If value is expressed in monetary terms, what should the minimum ratio of value added by the change to the resulting cost increase be to justify an improvement in the system? What should the maximum ratio of value lost by the change to the resulting cost saving be to justify the change?

General

51. Define systems design.
52. What steps in systems design are called for by "Model the system and apply the model"?
53. What is the purpose of the objective function?
54. Why should interest rates be used in making economic comparisons of alternatives?
55. A client owns a building that he expects to sell in 5 years for $100,000. If the rate of return is 10%, what is the present worth of the building?
56. A client anticipates that he will have to replace his $500,000 building in 5 years. Salvage value is estimated at $100,000. How much money should the client put aside annually at 6% interest to have $500,000 in 5 years for purchase of a new building?
57. Maintenance costs of a building are averaging $30,000 per year. If the interest rate is 6%, what is the present worth of these costs for a 10-year period? What would the

present worth be if the costs continued indefinitely?
58. An owner is considering two types of buildings for a proposed factory. Revenues from use of the buildings will not be affected by his choice. Estimates for cost of the alternatives are as follows:

	BUILDING 1	BUILDING 2
First cost	$100,00	$240,000
Life, years	20	40
Salvage value	$20,000	$40,000
Annual disbursements	$18,000	$12,000
Rate of return, %	8	8

Which building will be more economical?
59. In what legal documents should you look for requirements for:
   (a) Number of street-level exits from a building?
   (b) Minimum distance of a building from a rear lot line?
   (c) Height of building and number of stories?
   (d) Minimum width of streets in a new development?
   (e) Electrical conduit to be used in a building?
60. Why should building designers be familiar with the requirements of OSHA for factory conditions? Why should contractors be familiar with OSHA requirements?
61. Describe the advantages and disadvantages of using standard plans and specifications for several buildings of the same type for the same owner but to be constructed on different sites.

Chapter 4

Application of Systems Design to Buildings

In Chap. 3, systems design is proposed as a precise procedure for development of an optimum system. The method consists of six stages: data collection and problem formulation, synthesis, analysis, value analysis, appraisal and decision. In addition to data collection, which could be a continuous activity feeding information to every design step, there are nine basic design steps (see Fig. 3.4). These generally should be executed in sequence; however, the procedure also calls for loops from advanced steps back to earlier steps and then ahead again, as new information that can be used to improve the system is generated.

In brief, systems design requires designers to start with a list of goals, objectives and constraints. Criteria must then be established, as a measure of system response, to indicate whether or not objectives have been met. Also, standards must be set as a measure of the constraints on properties of the system. Next, designers must propose one or more designs that will satisfy the objectives and constraints. With the aid of models, the designers should analyze the proposed systems and attempt to obtain an optimum design for each. The alternative systems should be evaluated and compared. After evaluation by the designers, one or more of the best systems should be subjected to value analysis. In this process, the systems may be changed to improve their cost effectiveness or suggestions

for alternative designs may be proposed. In either case, the designers should analyze the alternatives, evaluate them and seek new improvements. With this procedure, the design should improve as new information develops and therefore should converge on the optimum for the given objectives and constraints.

Application of systems design to buildings is made difficult by the following factors:

1. A building is a very complicated system. Design of any of its component systems may affect the design of many, perhaps all, others.
2. Design costs generally mount rapidly with additional investigations of alternative systems. Design fees may not be sufficient to cover the costs of numerous studies.
3. Time available for design often is limited and thus restricts the number of investigations that can be made.

Systems design, therefore, must be adapted to building design with these factors in mind. This chapter describes one way of applying the method to buildings.

The design process requires the designers to make tentative decisions as various requirements are considered and various parts of the system are tackled. The results of these decisions must then be tested for validity against the results of previous decisions. The resulting facility must be a well integrated unit, not just an assembly of solutions to individual objectives. Thus, every component or subsystem should exist only to serve the purposes of the whole system.

4.1. CONSIDERATIONS IN ADAPTATION OF SYSTEMS DESIGN

For complex buildings, the probability is very small that direct application of systems design to whole systems will be completely successful. There is so much interaction to such a high degree among building components that usually designers can attain an optimum building only by suboptimization. This optimization procedure, however, because of the effects component systems have on each other, usually has to be executed by further suboptimization or, more likely, by trial and error. The latter process requires that one or more alternatives be synthesized, analyzed and evaluated, then discarded, improved or replaced by other alternatives, and the process repeated continuously in a search for the best solution.

The trial-and-error approach to design has long been accepted practice in traditional design. In fact, the traditional building procedure described in Sec. 1.2 evolved over a long period of time to handle the trial-and-error process effectively. For the purpose, the traditional building procedure provides a phased approach to selection of a final system. The phases, varying from the general to the specific in sequence, comprise:

1. A *conceptual phase*, in which alternative building systems are synthesized and investigated.
2. A *design development phase*, in which alternative component systems are synthesized and investigated.
3. A *contract documents phase*, in which details are worked out.
4. A *construction phase*, in which the building is erected.

The procedure gives designers an opportunity in each phase to submit results to owners for approval and to start subsequent phases with previous results approved.

There appear to be at least two good reasons for adapting systems design to the traditional building procedure. One reason is that, despite the defects discussed in Sec. 1.2, it has worked well in practice. The second is that building designers are familiar with it and are more likely to adopt modifications of it than to discard it for something completely new with probable higher design costs. Hence, the rest of this chapter will be concerned with considerations in adapting systems design to the traditional procedure.

Design by Building Team

The greatest change in traditional practice required by systems design, in addition to the orderly step-by-step convergence to an optimum design, is design by a building team, as discussed in Sec. 1.4.

In traditional practice, a prime professional, generally an architect, assumes responsibility for building design. He is assisted by consultants, each of whom works individually in applying his specialty to meet design objectives. The prime professional correlates the work of the specialists. Usually, however, there is little or no effort to integrate their work to produce economies by making one component serve several functions.

In design by a building team, there still is a prime professional but his prime task is to serve as the team leader. Throughout design, all members of the team contribute their knowledge, experience, skill and imagination. Their work is not only correlated but also guided and integrated. Also, since construction experts and building operators may be members of the team, the results of team design should be better designs and buildings with lower life-cycle costs.

Suboptimization in Building Design

Because of the complexity of buildings, optimization usually is feasible only by suboptimization. This process, discussed in Sec. 3.13, is fraught with pitfalls. The most treacherous pitfall is that use of the process may give the impression that an optimum has been attained when the result actually may not be a true optimum. Experience and judgment are the only means of avoiding the pitfalls.

The technique of suboptimization of a system with subsystems in series is likely to have limited application in building design. Such suboptimization may be useful only for small subsystems or subsubsystems, because most building components affect or are affected by many other components.

The general principle, however, may be adapted to suboptimization of larger subsystems where the effects of interaction are small enough that the components may be treated as if they were in series. Errors introduced by this assumption then may be corrected, if substantial, after the interaction effects have been evaluated.

When construction duration and costs are important considerations, greater economies may result from integration of subsystems than from suboptimization of individual subsystems. Consider, for example, a floor system composed of a deck and beams, with lowest cost as the design objective. Optimization of the floor cost conceivably could result in a more costly building. The design might be such that costs of installing HVAC ducts, placing electric wiring, lengthening vertical pipe runs and building higher exterior walls would be larger than with other types of floors. Results closer to the true optimum are more likely to be attained by incorporating ducts, wiring conduit and piping in the floor system and optimizing that system. Since knowledge of several specialties is required for design of the system, this example points to the desirability of design by a building team.

Construction Considerations

Whether traditional or systems design is used, designers should take into account construction conditions that not only normally exist but also special conditions that are likely to exist when the building is constructed.

For example, designers should insure that it is feasible to fabricate and erect building components as drawn and specified. For this, designers need a knowledge of fabrication and erection methods. In addition, a construction expert should check the designs as soon as possible, certainly before bids are requested. This is only one of many useful services that can be performed by the construction consultant on the building team.

Another useful service that the construction consultant can perform is to advise the other team members of construction market conditions that are likely to exist when the building is constructed. He should forecast the availability of materials and equipment that the designers are considering specifying. There is no sense, for example, in specifying a window type that the manufacturer will be unable to deliver when it is needed for the building. Also, the construction consultant should predict the availability of contractors, subcontractors and labor that might be required. Shortages of one type or another not only might require substitutions for specified materials and equipment but also rescheduling of construction contract awards.

When the owner needs a building in a hurry and design and construction time consequently must be minimized, the construction consultant should, in addition, assist in scheduling all phases of the work to insure that the deadline will be met at minimum cost.

Phased Construction (Fast Track)

Construction cost and project duration are interrelated.

Cost increases when construction time surpasses the optimum. That happens because the decrease in wages through use of fewer workers is more than offset by constant overhead costs, which continue as long as the building is under construction, and because of the cost of delays due to inefficiency, bad weather or other causes.

Also, cost tends to increase when construction time is shorter than optimum. That happens generally because wages rise due to use of more workers or overtime payments and bonuses.

Usually, therefore, contractors strive to optimize construction time.

Design, though, also influences construction time, for designers can speed construction by calling for systems that can be erected quickly.

Also, designers can shorten project duration by cooperating with the construction consultant and construction contractors in a speed-up technique known as phased construction, or fast track.

In this process, construction starts before design has been completed. Early phases of construction are begun while later phases are still being designed. Contracts are awarded for subsequent phases as rapidly as design is completed. For example, as soon as foundation drawings have been completed, site work on the foundations commences. As the floor framing plans are finished, the contractor orders the structural steel and concrete reinforcing bars needed. Structural members for the lower floors of a tall building are erected while the upper floors are still being designed. Similarly, an early start can be made on placement of exterior walls and windows, construction of partitions and even finishing operations.

Whether construction be normal or phased, design and construction must be integrated. The building team must work together as a unit from project inception to completion to insure that the owner attains his goals. Balancing design quality, construction cost and project duration, the team should aim at production of an optimum product.

Prefabrication and Industrialized Building

As mentioned previously, designers can speed construction by specifying components that can be erected quickly. Designers have many options for doing this.

One option is to specify systems that have been coordinated so that they can be assembled swiftly in the field without the necessity of cutting them to fit.

Another option is to specify systems that have been preassembled in a factory. Such systems are likely to have also the advantage of better quality, because of assembly under controlled conditions and close supervision. They are likely to be lower cost too, because of mass production, use of fast, powerful machines and lower wages than those paid construction workers. Preassembled systems are also known as prefabricated components or, in the case of concrete, as precast concrete.

Buildings formed with large preassembled systems are often referred to as industrialized buildings. The goal generally is to employ to the greatest extent possible, in shop and field, the mass-production techniques that have proved successful in factories. Before specifying such buildings, however, designers should take precautions to insure that use of preassembly will not be counterproductive. For one thing, designers should check that the owner's objectives will be met and that constraints, especially building-code requirements, will be satisfied. For another, they should verify that shipping, handling, storing and erection costs will not exceed savings in purchase price and preassembly and that the systems will be delivered when needed.

Designers should bear in mind that traditional building is difficult to compete with because it often employs a form of mass, or assembly-line, production. In factories, a product being assembled usually moves past stationary workers, who perform a task on it. In building construction, in contrast, the product is stationary and the workers move from product to product to perform the same task. The major disadvantages of this type of production are limited use of machines and uncontrolled environment (climate), which can halt work or affect quality.

Design Priorities

With many options open to them to speed construction and cut costs, designers should logically consider the options in order of potential for achieving objectives. A possible sequence would be the following:

1. Selection of an available industrialized building
2. Design of an industrialized building (if the client needs many buildings of the same type)
3. Forming a building with prefabricated components or systems
4. Specification of as many prefabricated and standard components as possible.

5. Repetition of elements of the design as many times as possible. This may permit mass production of some components. Also, as workers become familiar with those elements, erection will be speeded.

6. Design of elements for erection so that trades will be employed continuously. For example, suppose designers of a multistory building were to call for a brick interior wall to support steel beams supported at the other end by steel columns. Bricks are laid by masons, whereas steel beams and columns are placed by ironworkers. Since the steel columns usually can be erected faster than a brick wall can be constructed, the ironworkers will be idle while waiting for the masons to finish. Thus, this type of construction would be slower than all-steel framing.

When to Apply Systems Design

In general, systems design may be used in all phases of the building procedure. Systems design should start in the conceptual phase and should be used continuously thereafter. The procedure is especially advantageous in the early design phases because design changes then involve little or no cost.

In the contract documents phase, systems design preferably should be applied only to the details being worked out in that phase and not to revisions of major systems or subsystems. Such changes are likely to be costly in that phase. Value analysis, however, could be cost effective when applied to the specifications and owner-contractor agreements.

In the construction phase, systems design should be applied only when design is required because of changes that have to be made in the plans and specifications. Time may be too short, however, for thorough studies, but at least value analysis should be used.

References

Architects Handbook of Professional Practice, American Institute of Architects.

W. Caudill, *Architecture by Team*, Van Nostrand Reinhold (out of print).

Words and Terms

Building design team
Design priorities
Fast-track (phased construction)
Industrialized buildings
Prefabrication
Suboptimization

Significant Relations, Functions and Issues

Phases of the traditional design procedure: conceptual, development, contract documents, construction.
Use of suboptimization for complex systems; problems of effective integration.
Overlapping of design and construction phases: fast-track.
Utilization of predesigned components and industrialized buildings for faster design and construction.

4.2. ROLE OF OWNER

When systems design is used, the duties and responsibilities of the owner during the building process are substantially the same as for the traditional building procedure. There are some differences, however, in the initial steps.

Generally, the basic steps taken by an owner in the process of having a building designed and constructed are as follows:

1. Recognizes the need for a new building.
2. Establishes goals and determines project feasibility.
3. Establishes building program, budget and time schedule.
4. Makes preliminary financial arrangements.
5. Selects construction program manager or construction representative to act as authorizing agent and project overseer, unless the owner will act in that capacity.
6. Selects prime professional, construction manager and other members of the building team.
7. Approves schematic drawings and rough cost estimate.
8. Purchases a building site and arranges for surveys and subsurface explorations to provide information for building placement, foundation design and construction, and landscaping.
9. Develops harmonious relations with the

community in which the building will be constructed.

10. Assists with critical design decisions and approves preliminary drawings, outline specifications and preliminary cost estimate.
11. Makes payments of fees to designers as work progresses.
12. Approves contract documents and final cost estimates.
13. Makes final financial arrangements, obtains construction loan.
14. Awards construction contracts and orders construction to start.
15. Obtains liability, property and other desirable insurance.
16. Inspects construction as it proceeds.
17. Makes payments to contractors as work progresses.
18. Approves completed project.

In many cases, there may be additional steps. For example, the owner may have to make such decisions as to whether or not phased construction must be used, whether separate contracts or multiple contracts should be used or whether the general contractor should also be the construction manager, serving on the building team. Note also that the steps may not all be exactly in the order listed and that some steps may overlap.

Some of the steps require additional comment.

Selection of Construction Representative

An early decision the owner has to make is whether he will personally manage the construction program—act as authorizing agent and project overseer, provide information needed for design and construction, make decisions, approve plans, specifications and contracts, engage surveyors and consultants, approve payments for all work and expedite design and construction, if necessary—or will designate one or more representatives to assume those duties. If he decides to assign a construction representative, the owner nevertheless retains the power to set goals, establish a construction budget and completion date, and make final decisions, approvals and changes.

For large and complex buildings, involving expenditures of large sums of money, appointment of a construction program manager, who manages both design and construction, or at least a construction representative, usually is desirable. Large corporations and public agencies with a big construction program, for example, often have a construction department, from which a staff engineer is assigned to serve as owner's representative. For less complicated buildings, the owner may not require a representative but may rely instead for assistance on the members of the building team. There usually are additional fees, however, when the consultants provide services in addition to their basic services.

Selection of Building Team

The skills, knowledge, experience and imagination of the members of the building team are critical to the results. Poor design can produce a defective and inefficient building, high maintenance and operating costs or unnecessarily high construction costs, or a combination of these. The fees paid the consultants usually are a relatively small percentage of the total construction cost, and in any event, competent designers can easily save the owner more than the amount of the fees. Consequently, the owner will find it advisable to engage the best talents to serve on the building team.

In selecting the prime professional, who assumes responsibility for complete design, the owner should evaluate the firm's technical qualifications, experience, reputation, financial standing, past accomplishments in related fields and ability to absorb an additional work load. In addition, the owner should learn whom the firm will place in charge of the project to serve as leader of the building team. The owner should verify that this manager has the experience and the capabilities required for a team leader, as outlined in Sec. 1.4.

Similar considerations should apply in selection of a construction consultant, construction manager, or general contractor, or any other member of the building team; but, in addition, it is important that the owner learn whether these professionals have demonstrated on past projects a capability and personality suitable for teamwork. Those who have not should be avoided.

Community Relations

Efforts to establish good relations with the community in which a building is to be constructed should start before a site for the building is purchased. This is especially important if changes in the zoning ordinance will have to be requested to permit the type of building contemplated. Before a variance will be granted, the planning commission will hold public hearings and solicit opinions from the community. Hence, it is desirable to inform the public of the nature and purpose of the proposed building and to indicate the benefits to and potential harmful effects on the community. The report should discuss objectively the environmental impact anticipated, including effects on local and regional economics, recreation, ecology, aesthetics, housing and resources. Such information, however, should be provided the public even when zoning changes are not needed, because it will promote good public relations.

A public relations program is the responsibility of the owner; however, the building team should assist the owner, and if necessary, suggest and guide the program, because poor public relations can halt or delay the project or produce other adverse effects.

Goals and Program

Design cannot start until the owner establishes the goals and objectives for the building. In addition, the owner must give some indication as to the relative importance of each objective for use in evaluations of alternative systems.

Preparation of the building program, or list of requirements for space, services and environment, is also the responsibility of the owner. The building team will assist with the program, but there usually is an additional fee for such service.

4.3. CONCEPTUAL PHASE OF SYSTEMS DESIGN

The conceptual phase of design is the start of the search for the best system for a specific set of objectives and constraints.

The purpose of the phase is to convert the building program, goals, objectives, constraints, data on site conditions and other relevant information into a building system that has high potential for client approval and that, if approved by the client, will be the basis for design development.

The results of the phase should be schematics—floor plans, simple elevations and sections, sketches, renderings, perspective drawings or models—and project descriptions that will give the client a broad picture of how the building and its site will look when completed. Accompanying these illustrations should be a rough cost estimate to indicate the approximate cost of the facility planned.

During the conceptual phase, many alternative designs may have to be investigated by the building team. To begin, each member of the team will generate initial concepts for his specialty. Most likely, these concepts will have to be adjusted or discarded as new information is developed. Changes especially will occur on interaction with concepts developed by other members of the team. Eventually, some concepts will stand out as being worth developing in detail in the next design phase.

Since it is the purpose of the conceptual phase to present a broad picture of the proposed building, the design effort usually concentrates only on important features, such as those effecting the goals, or functions, of the building, those representing a high proportion of the total construction or life-cycle costs and those having significant effects on aesthetics, the environment and the community.

The effort involves all members of the team, but one member usually plays a major role. For ordinary buildings, such as houses, hospitals, schools, office buildings and churches, an architect has this role. He has responsibility for aesthetics, environmental impact and planning the functional spaces, and the areas used for activities and services. For more special types of buildings, an engineer most likely will be assigned the role. For example, responsibility for environmental impact and planning the functional spaces may be assigned to a mechanical engineer for an industrial plant, to an electrical engineer for a power plant and to a civil engineer for a sewage or water treatment plant.

This team member initiates the design effort by synthesizing one or more functional-space systems to meet relevant objectives, constraints and site information. His initial concepts are provided to the other team members in the form of floor plans and simple elevations and sections, with other pertinent information. After studying them, the consultants first offer suggestions for improvement and later develop schematic designs for their own specialties, such as HVAC, structural and electrical systems, to correlate with the proposed functional-space system.

In the following discussions of the conceptual phase, an ordinary building will be assumed as the goal of design, in which case an architect will play the major role in design. The procedure when another team member plans the functional spaces would not differ significantly.

Preliminary Information Needed

Before conceptual design can proceed, essential basic design data must be obtained from the owner. The building team may have to call the owner's attention to the need for this information and assist him in providing it. Predesign information should include the following:

Feasibility Study. A feasibility study may be made with the owner's own staff or a building committee, or with outside specialist consultants or the prime professional for building design. The last alternative, often requiring payment of an additional fee for the study, has the advantage of familiarizing the prime professional with the owner's operations and general requirements for the building. The study should:

1. Anticipate facilities needed.
2. Estimate construction and operating costs.
3. Make economic comparisons of proposed facilities and of alternatives such as renovation of existing buildings or leasing instead of constructing a new building.
4. Anticipate capital financing requirements and the feasibility of providing the funds as needed.
5. Estimate future personnel requirements.
6. Indicate resources or services, such as electricity, gas and transportation, required.

7. Indicate potential locations for the facilities and pertinent requirements for markets, environment and legislation affecting construction and operation.
8. Recommend a specific course of action.

If the feasibility study recommends a new building, design may proceed.

Building Program. The list of requirements for the building should include the following:

1. Scope and type of project.
2. Relationship of the building to other buildings on the site or adjacent to the site.
3. Characteristics of the occupants.
4. Special requirements for the building.
5. Functional requirements, including circulation of people, material handling and work flow.
6. Priority of requirements.
7. Relationship of activities to be carried out in the building pertinent to the location of spaces for those activities with respect to each other and the amount of flexibility permitted in assigning the spaces.
8. Site development requirements.
9. Equipment to be supplied by the owner and equipment to be supplied by the construction contractor.

Budget. The owner should generally indicate the maximum amount of money he is willing to pay for design and construction of the building. He may also require that construction cost be minimized or that life-cycle cost be minimized.

The budget usually is based on the sum of the following estimated costs based on cost records for previously built similar buildings:

1. Cost of spaces for activities and services, as required by the program;
2. Foundation costs;
3. Cost of site preparation and improvement;
4. Equipment costs;
5. Contingency costs, including design fees, inspection fees, costs of site surveys and subsurface explorations, legal fees, financing costs, administration costs and costs of changes during construction.

Contingency costs generally are estimated at about 15% of the sum of the other costs.

In addition, since design and construction may take one or two years or more, the budget should make an additional allowance for rising wages and prices and other inflationary effects during that period.

Completion Date. The owner may set a deadline for completion of the project and occupancy of the building. This deadline, in turn, may impose restrictions on the time available for building design. The deadlines may be necessary because of commitments made by the owner for use of the building, because of high interest costs for financing over the period of design and construction or because of revenues desired from use of the building on its early completion. In any event, a practical schedule should be prepared as soon as possible for both design and construction. The design schedule preferably should allot reasonable amounts of time to each of the design phases. The construction schedule, if possible, should permit construction at normal speed to keep construction costs at the optimum. If necessary, however, phased construction may have to be used if time allotted for construction is too short.

Management Decisions. The owner should make as soon as possible some basic decisions affecting the execution of design and construction:

1. Whether to appoint a construction representative or a construction program manager.
2. Whether to engage a construction manager or a construction consultant to serve on the building team.
3. Whether phased or normal construction is to be used.
4. Whether the construction contract will be awarded to a general contractor, who will engage all subcontractors, or whether separate contracts will be given to several prime contractors.
5. Type of contract to be used—lump sum or cost plus fixed fee.
6. Form of general conditions of the contract to be used.
7. Whether design is to be executed by a building team or by an architect assisted by consulting engineers.

Predesign Activities. The prime professional may organize a building team before preparing a proposal for design of a building, after being asked to do so by the client, or after signing the design agreement with the client, depending on particular circumstances. After being assembled, the building team assigns personnel to the project.

The first task tackled by the team is review of the building program, construction budget and construction schedule. Besides familiarizing themselves with the requirements, the designers also insure that it is feasible to comply with them. Next, from the program and other information elicited from the owner, the prime professional compiles the goals and objectives for the project and provides the list to the team members for study. Also, the team assembles the constraints on design. From the objectives and constraints, the designers develop criteria and standards that the system must meet. In the process, the team assembles and reviews building codes, zoning ordinances, health department regulations, OSHA rules, etc. In addition, the prime professional informs the team of the design schedule to be met. Finally, a preliminary report of the environmental impact of the project may be prepared by a team member.

If a site for the building has not been purchased, the prime professional may assist the owner in selecting a site. If one has been purchased, the owner should provide site information, if necessary engaging for the purpose surveyors and soil consultants. Members of the building team also should visit the site to become personally acquainted with conditions there.

Design of Functional-Space System

As pointed out previously, an architect usually has responsibility for planning the spaces required for activities and services. He should find it worthwhile to apply the systems-design approach to this task.

From the building program and other information supplied by the owner, the architect should compile a list of spaces that will be needed and the approximate floor area that will be required for each space. The other members of the building team should supply additional information on spaces needed for their specialties. The architect also should allow space for horizontal and vertical circulation, reception of visitors, lounges, etc. Then, the architect should compile other objectives and list constraints, for which he should establish criteria and standards. Next, following the systems-design procedure, he should generate alternative systems, model and evaluate them. Systems may be judged by how well they meet objectives and constraints, construction cost, operating efficiency and space efficiency, as indicated by the ratio of useful floor area to gross floor area. The result should be schematic floor plans that the architect should submit to the other members of the building team and value analysts for study and recommendations.

Example. As an example of the development of a floor plan in the conceptual phase, consider a one-story house with basement for a family with two small children. The program indicates that the main floor will have to contain spaces for the 11 elements represented in Fig. 4.1(a).

The family requires three bedrooms, BR 1 for the parents and BR 2 and BR 3 for the two children. Two bathrooms are needed, with access to B 1 only from BR 1. The family also requires a foyer at the front entrance, with access to stairs to the basement. A kitchen is wanted, next to the dining room and close to an enclosed porch. In addition, the family would like a large living room accessible directly from the foyer.

From anticipated furniture, closet and activity requirements given in the program, areas of the elements are estimated and shapes are assumed as shown in Fig. 4.1(b). Then, the desired relationships between spaces are noted in table form (see Fig. 4.2).

In Fig. 4.2, the relative closeness desired and the relative importance of proximity are indicated, in the order of importance, by the letters A, E, I, O, U and X. The reason for the decision is indicated by a number. For example, the requirement that B 1 be next to BR 1 is indicated by an A at the intersection of the row labeled Bathroom 1 and the column marked BR 1. The reason for the requirement is indicated by the number 2 at the same intersection. In the summary of reasons at the right in Fig. 4.2, 2

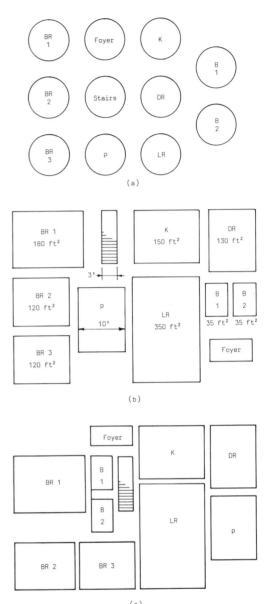

Fig. 4.1. (a) Schematic drawing indicating owner's basic needs for a one-story house with a basement. BR = bedroom, B = bath, P = porch, LR = living room, DR = dining room, K = kitchen. (b) Schematic drawing made to indicate probable floor-area requirements and room shapes to meet needs shown in (a). (c) Schematic drawing showing desired relative locations of spaces to meet the needs shown in (a).

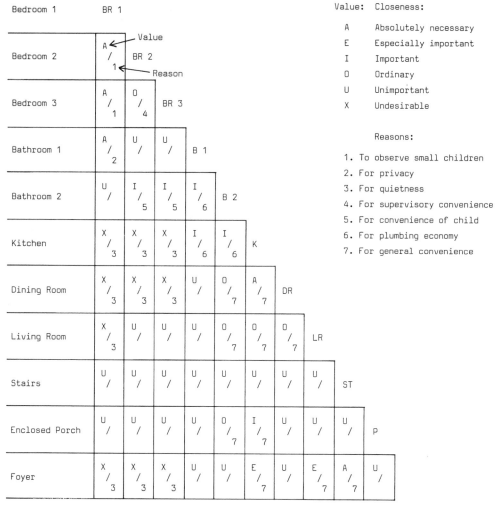

Fig. 4.2. Activity relationships for a small house.

corresponds to "For privacy." Similarly, the requirement that the dining room be next to the kitchen is noted by an A at the intersection of the row for dining room and the column marked K. The reason is "7. For general convenience."

Figure 4.1(c) represents an early attempt to place the elements shown in Fig. 4.1(b) in positions that satisfy the proximity requirements of Fig. 4.2. With the elements in these places, however, the room shapes do not lend themselves to formation of a regular shape for the building. With modifications of the room shapes, the floor plan shown in Fig. 4.3 results. If the rough cost estimate is within the owner's budget and the owner approves the

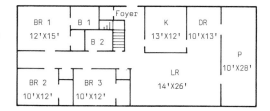

Fig. 4.3. Schematic floor plan with rooms located, stairs positioned, corridors shown. Window locations have not yet been determined.

floor plan, it may be developed in greater detail.

For multistory buildings, a floor plan may be developed in a similar manner for each floor. If several floors will be identical, however, a typical plan may be prepared for them, and the title of the drawing should indicate to what floors the plan applies.

In the development of the floor plans, consideration must be given to entrances to and exits from the building, access to each floor and internal circulation, or traffic flow. Also, the floor plans must be developed in conjunction with considerations of site conditions. For this to be done, the placement of the building on the site must be taken into account. Positions of walks, driveways and parking areas must be included in these considerations.

Information Flow

The flow of information in the conceptual phase when the architect plays the major role is shown in Figs. 4.4 to 4.7. In all cases, information passes from the owner to the prime professional and other members of the building team. Also, recommended concepts flow from the members of the building team to the prime professional and the owner.

Figure 4.4 shows the flow of information to the architect for execution of his main tasks. The diagram indicates that information given the architect also is given to the other members of the building team and that he confers with them for comment and suggestions. He then develops schematic architectural drawings, which are submitted to value analysts for comment and suggestions. Next, the drawings, modified as required by the analysis, are reviewed by the building team. Finally, the drawings, again modified as required by the re-

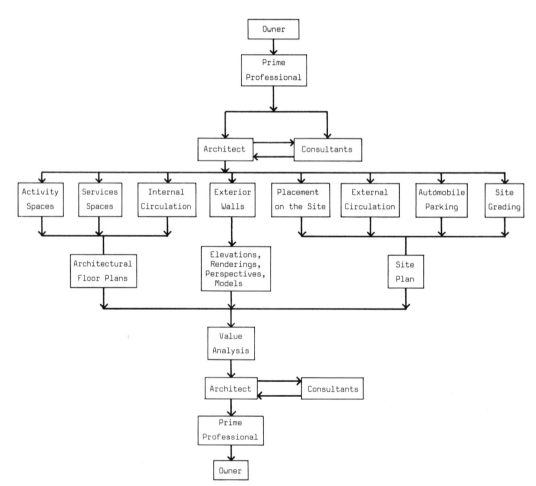

Fig. 4.4. Flow of information to and from the architect during conceptual design for an ordinary building.

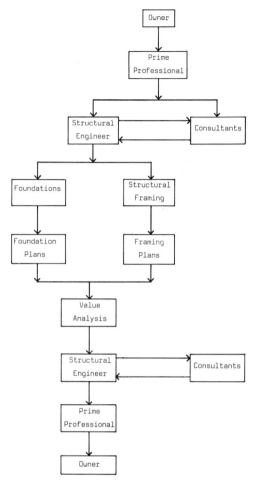

Fig. 4.5. Flow of information to and from the structural engineer during the conceptual design phase for an ordinary building.

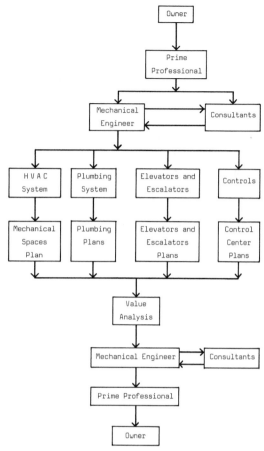

Fig. 4.6. Flow of information to and from the mechanical engineer during the conceptual design phase of an ordinary building.

view, are forwarded to the prime professional and the owner for approval.

Figure 4.5 shows the flow of information to the structural consultant for execution of his main tasks. This diagram also indicates the information given the structural engineer is given to the other members of the building team, too. He then develops schematic structural drawings, which are subjected to value analysis and to review by the other consultants. Next, the drawings, modified as required by the studies, are forwarded to the prime professional and the owner for approval.

Similarly, Figures 4.6 and 4.7 show the flow of information to the mechanical and electrical consultants, respectively.

Design of Systems

Each of the members of the building team should apply systems design to the systems for which he is responsible.

For illustrative purposes, Fig. 4.8 shows possible steps in systems design of a structural system for a multistory building. The diagram follows closely the steps in Fig. 3.4 for general systems design.

FOR SECTIONS 4.2 AND 4.3

References

AIA, *Architect's Handbook of Professional Practice*, American Institute of Architects.

W. Caudill, *Architecture by Team*, Van Nostrand Reinhold (out of print).

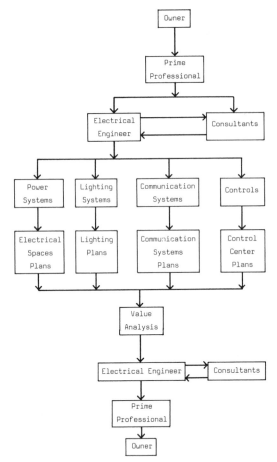

Fig. 4.7. Flow of information to and from the electrical engineer during the conceptual design phase for an ordinary building.

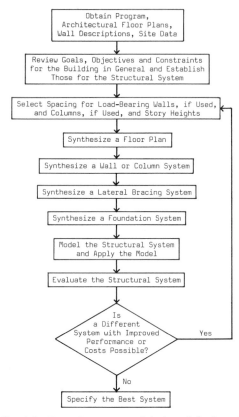

Fig. 4.8. Steps in conceptual design of the foundations and structural framing.

Space analysis techniques for development of building plans.

Flow of information in conceptual design phase.

Words and Terms

Budget
Building program
Feasibility study
Predesign activities
Prime design professional

Significant Relations, Functions and Issues

Owner's decision to maintain personal control or assign management of design and construction to others.

Need to manage community relations for large projects.

Owner's responsibility to establish goals and objectives for the project.

Owner's selection of prime design professional; related to nature of project.

Critical management decisions of owner.

4.4 DESIGN DEVELOPMENT PHASE OF SYSTEMS DESIGN

After the client approves the schematic drawings, project descriptions and cost projections, the desirability of the building concepts proposed is established. The technological feasibility, however, is still open to question. Can the system be made to function as presently conceived? Can it be constructed with currently available methods and equipment? Can it be constructed speedily, efficiently and reliably at expected costs and with low maintenance? If the answers to these questions are negative, the development of alternative concepts may be required, depending on what the designers learn as they develop the design in greater detail.

The purposes of the design development phase therefore are:

1. To bring the proposed system into clearer focus by determining materials to be used for, and sizes of components of, the important features synthesized in the conceptual phase.
2. To develop further the concepts of the conceptual phase by proposing concepts of other essential features and the materials and sizes to be used.
3. To determine the technological feasibility of the developed design.
4. To prepare design drawings and project descriptions that will be the basis for preparation of contract documents.

The results of this phase should be preliminary design drawings—floor plans, elevations, sections, some details, renderings and perhaps also a model—giving building dimensions and showing locations of equipment, pipes, ducts, wiring and controls; outline specifications and a more refined, although still preliminary cost estimate and construction schedule.

At the start of design development, the owner and the building team should review the program, objectives and constraints to insure that they are all still valid. The owner should at that time impose additional or special requirements previously overlooked, if any are necessary. The designers should verify compliance of the schematics with the program and building and zoning codes and with all other legal regulations. In addition, the owner should supply more detailed information concerning the site, if required, especially information on subsurface conditions and soil types and properties.

In the conceptual phase, effort is concentrated only on important features. In design development, all systems must be determined and analyzed. Additional elements to be specified include fire protection and other life-safety systems; security systems; lighting; telephone; paging systems; intercommunication systems; sound control; conveyors, cranes and other material-handling equipment; closed-circuit and cable television; clocks; and such supplies as vacuum, steam, heating gas, compressed air, oxygen and distilled water.

Optimization

Each member of the building team develops the design for his specialty (see Sec. 1.2), but because of the consequences of decisions by each specialist, the team members must confer frequently with each other and advise the others of the current status of their designs. The designers also should compare drawings of proposed systems, to determine space requirements and tolerances and to eliminate incompatible or undesirable situations. Flow of information during the design development phase is similar to that shown in Figs. 4.4 to 4.7.

The prime professional, with an overall view of all team accomplishments, should insure integration of components and subsystems to form the optimum system for the given objectives and constraints. Initially, each team member should apply systems design and endeavor to obtain the optimum systems for his specialty. The team should then examine the results to see if the component systems when combined form a true optimum.

At the start, the building system may be treated as though it were composed of subsystems in series. The functional-space system and systems having small effects on the others (for example, partition, electrical and plumbing systems) may be tentatively considered true optimums. Then, the exterior-wall system may be combined with those systems and optimized. Next, the HVAC system may be combined with the preceding combination and optimized. After that, the structural system may be combined with all the others and optimized. Finally, the optimized result should be restudied as a system to determine if integration of components would produce a better system.

If an alternative system is not evident, the cycle should be repeated. This cycle, however, to make all the systems compatible, should start with changes necessary in the functional-space system, including internal circulation, and the component systems that had been tentatively treated as true optimums. These adjusted systems should then be optimized. Next, the effects of the new optimized systems on the others should be determined and, if necessary, new optimum systems should be designed. If changes are substantial, the cycle may have to be repeated again.

At the termination of the second cycle and subsequent cycles, if any, the results should be subjected to value analysis. The building team should then evaluate and act on suggested modifications and alternatives.

Costs and Time

Also, at the end of each cycle, an estimate should be made of construction and life-cycle costs and construction time. If any of these do not meet objectives and constraints, the system should be modified accordingly. If, for example, construction costs are too high, the building may have to be made smaller, less expensive materials or equipment may have to be specified, or some requirements may have to be changed. If shortages of materials, equipment or certain types of construction workers will delay construction, substitutes should be specified.

With more detailed information on the composition of the building and its equipment available than at the end of the conceptual phase, a more accurate estimate of construction cost is now feasible. The cost, for instance, may be based on historical unit costs for each of the systems comprising the building. Typical elements of the estimate would include items covered by the general conditions of the contract and the contractor's fee, sitework, foundations, masonry, concrete, structural steel, ornamental metal, carpentry, roofing, windows, doors, hardware, glass, curtain walls, plaster and gypsumboard, metal partitions, tile work, ceilings, HVAC, elevators, plumbing, electrical system and painting.

Specifications

Each member of the building team during the design development phase should keep notes of decisions made, the date they were made and the reasons for them. The notes dealing with materials and equipment to be installed should be compiled as outline specifications. These will form the basis for the final specifications.

Approvals

Important questions should be submitted to the owner for decision as they arise. Also, the owner's approval of resources, especially fuels, to be used should be obtained. Preliminary drawings, outline specifications and preliminary estimates of costs and construction time should be submitted to the owner for approval at the end of the second cycle and later cycles when convergence or near convergence to the optimum system is evident.

As drawings take final shape, designers should obtain tentative approval of regulatory agencies concerned, especially the building department and zoning commission, state and local fire departments and health department. In some cases, opinions may be desirable from a state labor department and the Occupational Safety and Health Administration.

Mechanical and electrical engineers should obtain the approval of all utility companies concerned for service connections.

Significant Relations, Functions and Issues

Basic purposes of the design development phase.
Cyclic integration by suboptimization in series and sequential combinations of subsystems.
Transitions to contract documents phase: outline specifications, preliminary drawings, preliminary approvals.

GENERAL REFERENCES AND SOURCES FOR ADDITIONAL STUDY

These are general references for the chapter; see also the references listed at the ends of chapter subsections.

Architect's Handbook of Professional Practice, American Institute of Architects.
W. Caudill, *Architecture by Team*, Van Nostrand Reinhold (out of print).
A. Gheorge, *Applied Systems Engineering*, Wiley, 1982.
F. Merritt, *Building Design and Construction Handbook*, 4th ed., McGraw-Hill, 1982.

EXERCISES

The following questions and problems are provided for review of the individual sections and the chapter as a whole.

Section 4.1

1. When systems design is used with the traditional building procedure, what are the major changes required in traditional practice?
2. What are the advantages of phased design?
3. Why is teamwork desirable in systems design?
4. What are the duties of the prime professional designer on the building team?
5. Why is it desirable that structural, mechanical and electrical engineers cooperate in design of a floor system?
6. What are the duties of a construction consultant during design?
7. Explain the variation of construction costs with construction duration.
8. What are the advantages of phased construction?
9. Why must design and construction of buildings be integrated?
10. In what way does traditional field assembly of buildings resemble factory assembly-line production?
11. Discuss the advantages and disadvantages of prefabrication.
12. What are the advantages of repetition of components in building construction?
13. Why is it desirable that different building components be dimensionally coordinated?
14. A grocery store chain wants to construct 10 large identical market buildings in a region. If business then goes well, the chain plans to erect more such markets. Use of an industrialized building appears to have good potential for this application. But investigation indicates that no design currently available is suitable for the client's needs. What should the building team recommend as the best alternative? Justify your answer.

15. A 20-story building has 18 floors with identical structural framing. Framing for the roof, however, could be lighter and less costly than that for the floors. Discuss the advantages of using the floor framing also for the roof.

Section 4.2

16. What are the advantages to an owner of engaging a construction representative to assist him in administering design and construction of a building?
17. What is the difference between duties of a construction program manager and those of a construction manager?
18. What information should a building program provide?
19. Who is responsible for providing information concerning the building site?
20. An owner employs a construction manager for a building to be constructed. The manager negotiates a contract for construction with a general contractor. Who should sign the contract and assume legal responsibility for payments for construction?
21. What are the advantages to an owner of assigning to one member of the building team the duties of prime professional?
22. What are some of the most important characteristics that members of the building team should have?
23. Why is a public relations program aimed at the community where a building is to be constructed important to the owner? To the building team? When should the program start? What should be its purposes?

Section 4.3

24. What is accomplished in the conceptual phase of building design?
25. What are the primary design concerns in the conceptual phase?
26. Describe briefly the basic predesign data needed.
27. What design information should the building team provide for the start of the conceptual phase?

28. A one-story school building requires six classrooms, each with a floor area of 700 sq ft, a 2,000-sq ft auditorium, a 2,500-sq ft gymnasium, a 500-sq ft library, a 600-sq ft cafeteria, two 60-sq ft toilets and a 1,000-sq ft administration area. Compile a closeness table and draw a schematic plan for the main floor.

29. At completion of the conceptual phase of design, a building team has produced an office building with a floor area of 100,000 sq ft.
   (a) Estimate the construction cost if similar buildings in the same city constructed recently have averaged $40 per sq ft.
   (b) What would the estimate be if a pile foundation is required and will cost about $100,000 more than the spread-footing foundations used for the other office buildings?
   (c) What would the estimate be if costs are expected to increase at a rate of 10% per year during the 2 years the building (with pile foundations) will be under design and construction?

30. Which member of the building team is responsible for drawings for:
   (a) Exterior walls?
   (b) Foundations?
   (c) Plumbing?
   (d) Telephone wiring?
   (e) Site grading?

Section 4.4

31. What are the purposes of the design development phase?

32. A construction consultant estimates that a multistory office building with a structural steel frame can be erected in 15 months at a cost of $4,000,000. He also estimates that the building with a concrete frame can be constructed in 18 months at a cost of $3,800,000. The owner anticipates a net revenue of $100,000 per month when the building is occupied. On the basis of this information alone, which type of frame should the building team recommend?

Chapter 5

Contract Documents and Construction Methods

After owner approval of the preliminary drawings, outline specifications, and preliminary cost estimate and construction schedule, the contract documents phase begins. In this phase, the building team develops working drawings, specifications, and final cost estimate and construction schedule. Design in this phase differs from the traditional principally in closer coordination of the work of the various specialists and tighter integration of the building systems. Changes of major systems are undesirable in this phase because they will be time consuming and costly. Systems design and especially value analysis, however, may still be profitably applied to details being worked out, final specifications, and general and special conditions of the contract.

After the owner approves the contract documents, construction contracts may be awarded and construction may proceed. (In phased construction, construction may begin before all the working drawings and specifications have been completed.) Design changes may be made after construction starts, but they will be more costly than if made before award of the construction contract.

The contract documents are graphic and written means of conveying concepts of the structure to the builders and assigning duties and responsibilities during the construction phase. The documents enable the owner to obtain the building portrayed in them. They allow the designers to indicate what is to be constructed. They specify to the selected contractors the materials to be used, the equipment to be installed and the assemblage of the materials to produce the desired building. Also, the documents detail the payments to be made for this work.

5.1. RESPONSIBILITIES ASSIGNED BY THE CONSTRUCTION CONTRACT

The contract for construction is solely between the owner and the contractor. The prime professional (responsible for execution of design) unless also the owner of the building, is not a party to the contract. Nevertheless, he prepares some of the contract documents—the working drawings and specifications—and assists the owner in preparing the owner-contractor agreement and conditions of the contract. For preparation of the agreement and conditions of the contract, legal counsel is at least desirable and generally necessary. Contract law differs from state to state, making it necessary to obtain information from the owner's legal counsel regarding the law of the state in which the building is to be erected.

While it is feasible for an owner to make an oral contract for construction of a building, in general, this is very risky for both the owner and the contractor. No construction should be undertaken without a written contract. Similarly, all changes in the contract before bidding and all modifications after the contract has been awarded, including change orders, should be in writing. Otherwise, costly disputes may arise.

Contractor Responsibilities

A construction contract generally assigns the following responsibilities to the general contractor for a project to be built:

1. Performance of all work in accordance with working drawings and specifications and change orders issued by the owner. Thus, the contractor also is responsible for the performance of all subcontractors and workmen.
2. Starting and completing the project on the dates specified in the contract.   ·
3. Quality of workmanship. The general contractor is required to correct any work that does not conform with plans and specifications.
4. Payments of all taxes, fees, licenses and royalties and for all labor, materials, equipment, tools, utilities and other services necessary. The general contractor also is responsible for reimbursement of all subcontractors he engages for work they perform.
5. Securing all permits and fees necessary and compliance with all legal regulations.
6. Checking and submitting for approval to the owner or his agent all samples and shop drawings as required by the plans and specifications.
7. Use of safety measures and good housekeeping on the building site, plus provision of insurance coverage, to protect the owner, building designers and other owner agents against financial losses from property damage or personal injuries to employees, visitors or the general public during construction.
8. Cooperation with other contractors, if any, engaged by the owner.
9. Providing access to the work to the owner and his agents.

Owner Responsibilities

A construction contract usually assigns the following responsibilities to the owner of a project to be constructed. (The owner may delegate authority for carrying out these responsibilities to one or more agents.)

1. Preparation of working drawings and specifications that clearly define what is to be built and either:
   (a) Stipulate materials to be used and their quality and the equipment to be installed or:
   (b) Present performance requirements for the building, its structure and installed equipment, but not both.
2. Approving work as completed, making decisions that become necessary as work proceeds, approving subcontractors, approving samples and shop drawings, and issuing instructions to the contractor. The contract should indicate who has authority to act as the owner's agents during construction and the limits on the authority given to each. Specifically, the contract should make clear who has authority to issue on the owner's behalf instructions to the contractor.
3. Payments to the general contractor for all work, including changes and extra work ordered by the owner. The owner also may be required to pay for extra work arising from unexpected conditions, such as subsurface conditions on the site that were not disclosed because of inadequate or inaccurate information the owner provided.
4. Furnishing surveys and subsurface information concerning the site.
5. Securing and paying for easements in permanent structures or permanent changes in existing facilities.
6. Inspection of the work to insure compliance with the contract documents and rejection of nonconforming work. When necessary, the owner should especially advise the contractor of the likelihood of cost overruns or late completion, when

such conditions are discovered, and of the need for better control of costs and time.

7. Supplying materials and equipment and installation labor not covered by the working drawings and specifications.
8. Provision of insurance against financial loss from property damage and personal injuries before, during and after construction.

Responsibilities of Owner's Agents

The owner, while retaining the right to exert his authority under the contract at any time, may assign complete authority to act on his behalf during construction to one agent or may divide this authority among several agents.

Public agencies and corporations that have their own construction departments, for example, may assign complete authority to a staff architect or engineer. Some owners may give the authority to a construction program manager or to the prime design professional.

In some cases, authority may be divided between the prime professional and a construction representative or a construction manager.

The prime design professional, in any case, has responsibility for preparation of working drawings and specifications. The contract usually also assigns him the responsibility for interpreting these documents when the contractor has questions or a dispute arises between the contractor and the owner. The contract may, in addition, oblige the prime professional to inspect construction to insure compliance with plans and specifications, approve samples and shop drawings submitted by the contractor, and to design work required by change orders during construction. (These duties, which often require payment by the owner of additional fees, must also be covered in a separate owner-designer agreement.) Responsibility for assessing construction progress and authorizing periodic payments to the contractor for work in place, in some cases, may be assigned to the prime professional or, in other cases, to a construction representative or a construction manager.

General administration of the construction contract may, at the owner's option, be assigned to the prime professional, a construction representative, or a construction manager. The contract administrator is given authority to issue instructions, including change orders, to the contractor on the owner's behalf. He also is assigned responsibility for approving subcontractors. In addition, he is charged with responsibility for insuring that costs are controlled, that proper insurance coverage is maintained, that the contractor complies with all legal regulations, and that the project is kept on schedule. If inspection duties are not assigned to the prime professional, the contract administrator will have to engage inspectors. Furthermore, he or the prime professional may be given authority to settle all claims or disputes between the owner and the contractor; but under the contract, such decisions may be subject to arbitration by outside parties named in the contract.

While some of these responsibilities may involve a conflict of interest, many years of experience have indicated that professionals can execute these duties responsibly and with fairness to both parties to the contract.

What but Not How to Build

One aspect of the contract documents is worthy of special note. They always endeavor to specify precisely what the designers intend to have built. They avoid, whenever possible, instructions to the contractors concerning methods to be used for construction. The reason for this is that if the contractor uses the specified methods and the results are unsatisfactory, the responsibility for the unacceptable work falls on the designer. On the other hand, if the contract documents indicate only the results to be obtained and leave the methods to be used at the option of the contractor, the responsibility for the outcome rests on him.

References

AIA, *Architect's Handbook of Professional Practice*, Vol. 1, American Institute of Architects.
J. Sweet, *Legal Aspects of Architecture, Engineering, and the Construction Process*, West Publishing Co., 1970.

Words and Terms

Prime design professional
Specifications
Working drawings

Significant Relations, Functions and Issues

Responsibilities of the contractors.
Responsibilities of the owner.
Owner's assigned agents: prime design professional, construction representative, construction manager.
Basic function of contract documents: control of *what* is built, not *how* it is built.

5.2 COMPONENTS OF THE CONTRACT DOCUMENTS

Basically, the construction contract documents consist of:

1. Owner-contractor agreement
2. General conditions
3. Supplementary conditions
4. Drawings
5. Specifications
6. Addenda
7. Modifications

The owner-contractor agreement indicates what the contractor is to do, for how much money and in what period of time.

The general conditions contain requirements generally applicable to all types of building construction.

The supplementary conditions extend or modify the general conditions to meet the requirements of the specific project.

The drawings show graphically the building to be constructed.

The specifications list the materials to be used and equipment to be installed in the structure and provide necessary information about them that cannot easily be given in the drawings.

If it is necessary to make changes in the preceding documents before execution of the owner-contractor agreement, the prime professional, who is responsible for design, issues on behalf of the owner *addenda* incorporating the revisions. These addenda should be given simultaneously to all bidders so that all bids can be prepared on an equal basis. Addenda are part of the contract documents.

If changes become necessary after execution of the agreement, the owner and the contractor must agree on the *modifications*, which include *change orders* and *interpretations* of drawings and specifications. (See also Secs. 5.3 and 5.4.)

Project Manual

For the convenience of those concerned with the contract documents, all the documents *except the drawings* may be bound in a volume, called a project manual, to provide an orderly, systematic arrangement of project requirements. In addition, bidding requirements, though not contract documents, are desirably incorporated in the manual for the convenience of bidders. Bidding requirements, which govern preparation and submission of proposals by contractors, are described in Sec. 5.5.

Project manuals are generally organized as follows:

1. Table of contents
2. Addenda
3. Bidding requirements
4. Owner-contractor agreement
5. General conditions
6. Supplementary conditions
7. Schedule of drawings
8. Specifications

For large projects, however, a single volume may be inconvenient. In such cases, some of the divisions of the specifications, such as the mechanical, or the electrical, or specialty items, may be bound as separate volumes. Each volume should also contain the addenda, bidding requirements, conditions of the contract, and the division of the specifications that presents general requirements.

Construction Contract Forms

A typical owner-contractor agreement is presented for illustrative purposes only. An agreement is a legal document and therefore advice of an attorney in its preparation is advisable.

AGREEMENT

made this ___ day of _____ in the year _____

BETWEEN

ABC Company, the owner, and
IMA Building Corp., the contractor.
The owner and the contractor agree as follows:

Article 1. The Contract Documents

The contract documents consist of this agreement, conditions of the contract (general, supplementary and other conditions), drawings, specifications, all addenda issued before execution of this agreement and all modifications issued afterward. All the documents form the contract, and all are as fully a part of the contract as if attached to this agreement or repeated in it.

Article 2. The Work

The contractor shall perform all the work required by the contract documents for the ABC Office Building to be located at _____ Street and _____ Avenue, _____ City, State.

Article 3. Prime Professional

The prime professional for this project is _____.

Article 4. Times of Commencement and Completion

The work to be performed under the terms of this contract shall begin not later than _____ and be completed not later than _____.

Article 5. Contract Payments

The owner shall pay the contractor for the performance of the work, subject to additions and deductions by change order as provided in the conditions of the contract, in current funds, the contract sum of _____.

Article 6. Progress Payments

Based upon applications for payment submitted to the prime professional by the contractor and certificates for payment issued by the prime professional, the owner shall make progress payments on account of the contract sum to the contractor as provided in the conditions of the contract as follows: _____

Article 7. Final Payment

The entire unpaid balance of the contract sum shall be paid by the owner to the contractor _____ days after substantial completion of the work unless otherwise stipulated in the Certificate of Substantial Completion, if the work has then been completed, the contract fully performed and a final Certificate for Payment has been issued by the prime professional.

Article 8. Miscellaneous Provisions

Terms used in this agreement and defined in the conditions of the contract shall have the meanings designated in those conditions.

The contract documents that constitute the entire agreement between the owner and the contractor are listed in Article 1 and, except for modifications issued after execution of this agreement, are as follows: [Documents should be listed with page or sheet number and dates where applicable.]

This agreement executed the day and year first written above.

_____	_____
Owner	Contractor

The owner-contractor agreement specifies the method of payment to the general contractor for constructing the building. Consequently, the payment method selected strongly influences the terms of the agreement and the conditions of the contract. Contracts, therefore, may be classified in accordance with payment method as lump-sum, guaranteed-upset-price, cost-plus-fixed-fee, or management contracts.

Standard forms are available for some of these types. The American Institute of Architects, for example, publishes standard forms for lump-sum and cost-plus contracts. Some government agencies and large corporations with extensive construction programs have developed their own standard forms. The advantages of such forms are that contractors become familiar with them and readily accept them, and the chances of omitting important requirements are reduced. If modifications of a standard form are required, the owner's legal counsel should draft the agreement.

General Conditions of the Contract

Applicable to building construction in general, the general conditions are made a part of each construction contract by reference in the owner-contractor agreement (see preceding subsection **Construction Contract Forms**). Because of variations in local and project requirements, extension and modification of the general conditions usually are necessary. These are accomplished by also making special, or supplementary, conditions part of the same contract (see next subsection).

If separate prime contracts are awarded, the general conditions should be made a part of each prime contract. For example, if the owner should engage directly an electrical contractor and a plumbing and heating contractor, as well as a general contractor, the general conditions should be incorporated into each contract.

A major portion of the general conditions is devoted to descriptions of the rights, responsibilities, duties and relationships of the parties to the construction contract and their authorized agents, generally as listed in Sec. 5.1.

The first article of the general conditions, however, usually is broad in scope. It defines the contract documents, the work and the project. The article also points out that the contract documents form the contract and indicates how the contract may be amended or modified. In addition, the article notes that the contract documents are complementary, and what is required by any one shall be as binding as if required by all. (This clause is the reason why requirements in the drawings should not be inconsistent or conflict with those in the specifications.)

Other requirements usually included in the first article are that:

The owner and the contractor should sign at least three copies of the contract documents.

The owner will furnish the contractor without charge all copies of drawings and specifications reasonably necessary for execution of the work. The drawings and specifications, being the property of the prime design professional, may not be used on any other project and should be returned to him on request on completion of the work.

By executing the contract, the contractor represents that he has visited the site. Consequently, it is presumed that he has familiarized himself with conditions under which work is to be performed and has correlated his observations with the requirements of the contract documents.

On request, the prime design professional will deliver, in writing or in the form of drawings, interpretations necessary for proper execution or progress of the work.

The final article of the general conditions usually deals with circumstances under which the contract may be terminated by either party, other than by completion of the work, and describes the means for so doing.

Supplementary Conditions of the Contract

Because requirements differ from project to project, supplementary conditions are generally needed to extend or modify the general conditions of the construction contract. The supplementary conditions are made a part of each construction contract by reference in the owner-contractor agreement (see preceding subsection, **Construction Contract Forms**). They usually are prepared by the prime design professional with the aid of the owner's legal and insurance counsels.

Nothing should be incorporated into the supplementary conditions that can be covered in the specifications. Being a well organized listing of requirements, the specifications make it easy for the various trades to determine what work is to be done and what their responsibilities and duties are. A requirement placed in the supplementary conditions when it should be in the specifications runs the risk of being overlooked during construction.

The supplementary conditions may consist of modifications of the standard form of general conditions as required for a project and of additional conditions. Any of many additional conditions may be included. Among the more common are provisions for substitution of materials and equipment for those specified, accident prevention, allowances for unpredictable items, and payments of bonuses to the contractor for early completion of the project or cutting costs or of liquidated damages by the contractor for late completion of the project. Also often included are provisions for bracing and shoring, project offices and other temporary facilities, posting notices and signs, and provision of water, electricity, temporary heat, scaffolding, hoists and ladders during construction. In addition, the supplementary conditions may deal with the influence of weather on construction.

5.3. CONTRACT DRAWINGS

The construction, or working, drawings the contractor uses to determine what is to be built are given legal status by being made part of the contract by reference in the owner-contractor agreement agreement (see Sec. 5.2, **Construction Contract Forms**). The purpose of the drawings, which are often also called the plans, is to depict graphically the extent and characteristics of the work covered by the contract. The drawings are complemented by the specifications, also part of the contract, in which information is compiled concerning the building and its components that cannot be shown graphically.

Changes made in the drawings before the owner-contractor agreement is signed are incorporated in the contract as addenda. Later changes are included as modifications of the contract.

The drawings show the site to scale and the location of the building on the site. Sufficient dimensions are given to enable the contractor to position the structure precisely where the designers intend it to be and to orient it properly. The drawings also show how the building will look on the outside when viewed from various angles. Plan views are included for each level, from the lowest basement to the highest roof, to show the arrangement of the interior. Other drawings show the foundations, structural framing, electrical installation, plumbing, stairs, elevators, HVAC, and other components.

On every drawing, sufficient dimensions are given to enable the contractor to locate every item, observe its size, and determine how it is to be assembled in the building. Overall dimensions also are included. Where necessary, details are shown to a large scale.

Numbering of Drawings

All sheets should be numbered for identification. The numbers also are useful in referring the plan reader from one sheet to related information contained on another sheet.

The first sheet of the set of working drawings is the title sheet. It contains the name of the project, location, name of owner, project identification number and names of designers. Usually, it also provides a table of contents for the drawings. It may also provide a list and explanation of the symbols and abbreviations used in the drawings.

The following sheets are grouped in accordance with the type of practice of the designers who prepared them. The architectural sheets are assembled in sequence and often are assigned a number prefixed with the letter A. They generally are followed by the structural drawings, each given a number with the prefix S. Next come the mechanical drawings, often with each plumbing sheet numbered in sequence with the prefix P and with each HVAC sheet numbered in sequence with the prefix HVAC. After that come the electrical drawings, each assigned a number with the prefix E. If other drawings are necessary, they follow and are similarly identified by letters and numbers. A typical sequence is indicated in Table 5.1.

Title Block

Each sheet carries at the bottom, usually at the lower right-hand corner, a title block (see Fig. 5.1) that contains the sheet identification and general information about the project and the sheet. The title block prominently displays the name and location of the project, the name and address of the design firm responsible for the drawing, and the sheet number. The block also contains the project identification number, the

Table 5.1. Suggested Sequence of Contract Drawings

Title sheet
Table of contents, symbols, abbreviations

Architectural drawings
Topographical survey, site plan, landscaping plan
Elevations
Floor plans, starting with lowest basement
Roof plan
Sections
Details
Schedules

Structural drawings
Soil test borings
Foundations
Floor plans, starting with lowest floor
Roof plan
Sections
Details
Schedules

Mechanical drawings
Site plan
Plumbing plans
Plumbing details, schedules, and stack diagrams
Heating, ventilation, and air-conditioning plans
HVAC details and schedules

Electrical drawings
Site plan
Electrical power and lighting plans
Details and schedules

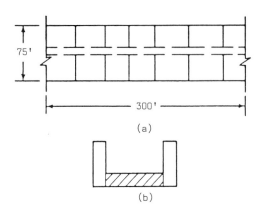

Fig. 5.2. (a) Plan view of part of a building (shown here to a much smaller scale than that used on the construction drawings). (b) Block plan of the building, with cross-hatching indicating the location in the building of the plan view in (a). (The block plan is shown to about the same scale as might be used on the drawing.)

ious times on the sheet. Consultants may be listed in the title block or nearby.

Additional necessary information is provided near the title block. If the drawing contains a plan view, a north arrow is shown. If a single scale is used for the whole drawing, the scale, if not given below the title of the sheet, should be indicated near the title block. If the sheet shows a plan that is only part of the overall view, a block plan may be drawn to indicate the location of the part shown on the drawing relative to the rest of the plan (see Fig. 5.2). When necessary, a key may be provided to explain notations used for identifying and locating sections and details. In the same area of the drawing as the title block, other information required by the local building department or the state boards of architecture and engineering may be included, such as signatures and registration seals of the architectural or engineering firm responsible for the drawing. The signature of the owner or his representative accepting the drawing should be adjacent to those signatures.

initials of the draftsman, the date of completion of the drawing, the initials of the person who checked the drawing, the date the checking was completed, and the date of issuance of the sheet. In addition, the title of the sheet, such as North Elevation, First Floor Plan, or Details, is prominently displayed. Space is also provided for listing revisions made at var-

Notes and Schedules

The working drawings also contain notes and listings of materials and equipment, called schedules.

The notes support and explain some items shown in the drawings. Because writing clutters the drawings and because the information the

Fig. 5.1. Example of title block for a drawing.

notes contain may be overlooked during preparation of a cost estimate or, even worse, during construction, notes should be kept to a minimum. Preferably, the information they would provide should be incorporated in the specifications. (It is not advisable to put written information both on the drawings and in the specifications, for emphasis, to prevent oversight or for any other reason. Repetition can be the source of inconsistencies and the cause of conflicts between the contractor and the owner.)

The schedules provide information that is conveniently tabulated, such as a listing of doors and their types, sizes and hand; windows and their types and sizes; room finishes; builders' hardware; and structural columns and their components and dimensions.

Relationship of Drawings and Specifications

The working drawings and specifications, equal components of the contract documents, complement each other. They serve different purposes.

The drawings are a diagrammatic representation of the building to be constructed. The specifications are a written description (see Sec. 5.4). They present requirements that cannot be readily shown graphically but can be conveniently expressed in words. Thus, specifications prescribe the type and quality of materials required, the performance characteristics of equipment to be installed, and workmanship desired in installation of materials and equipment. The specifications may also name acceptable sources from whom the contractor may purchase the required materials and equipment.

Information provided by words on drawings, such as notes, should be brief and general. Notes should describe a type of construction and its location and quantity required. The specifications should expand on the characteristics of the materials involved and the quality of workmanship required for their installation. For example, a note on a drawing may read "Insulation." The specifications will completely describe the insulation, either by naming several acceptable proprietary products and their manufacturers or by giving a desired thermal coefficient, indicating the desired physical state, such as board, granular, reflective, or blanket, and specifying quality requirements. The specifications also will dictate the method to be used in fastening the insulation in the positions indicated in the drawings and describe the results to be achieved.

Reference

AIA, *Architect's Handbook of Professional Practice*, American Institute of Architects.

Words and Terms

Addenda
Change orders
Contract documents
Contract drawings
Contract specifications
General conditions
Interpretations
Modifications
Owner-contractor agreement
Project manual
Supplementary conditions
Title block

Significant Relations, Functions, and Issues

Components of the contract documents.
Articles of the owner-contractor agreement.
Relationship of drawings and specifications.

5.4 SPECIFICATIONS

The specifications for a building are made part of the contract by reference in the owner-contractor agreement (see Sec. 5.2, **Construction Contract Forms**). A written description of construction requirements, the specifications complement the working drawings. Neither takes precedence over the other.

Because specifications are a legal document, specification writers tend to use language as precise as lawyers use. Specifications, however, are primarily intended for the use of prime contractors and subcontractors and should, however, be written so that they can easily understand the requirements. Hence, specifications should be brief, clear, and precise. They should be organized in an orderly and logical manner, and generally accepted practices should be followed.

Master Specifications

Design organizations that have been in existence for several years build up a file of specifications. From these, they can develop a generally applicable type, called a master specification. They adapt this to a specific project by deleting inapplicable portions and adding appropriate requirements.

Use of a master specification may be expedited with computers. The master specification can be stored in the computer memory. Parts applicable to a specific project may be recalled for viewing on a monitor, revised if desired or expanded, then printed or stored for later printing.

The American Institute of Architects established in 1969 a nonprofit corporation, Production Systems for Architects and Engineers, Inc., which developed a computerized master specification, called MASTERSPEC. Available to all professionals on an annual subscription basis, the program enables a central facility to receive, maintain, evaluate and transmit specification information in concise form. After editing the master specification, the subscriber can secure a computer printout from which he can obtain copies needed, without any intermediate typing steps.

Basic Principles

A fundamental concept is that *specifications should be in accordance with the general prevailing practices in the construction industry.*

Specification organization therefore should correlate with the common practice in which prime contractors prepare their proposals from bids submitted by subcontractors. These bids may represent as much as 85% or more of the work for a project. Consequently, the specifications should be written and organized for the convenience of subcontractors as well as for prime contractors.

All items of work covered by the contract should be specified in the specifications.

It follows therefore that every item shown on the drawings should be prescribed in the specifications. This precaution will reduce the chances of a subcontractor overlooking an item that is required only by graphical depiction on

the drawings. As indicated in Sec. 5.3, **Relationship of Drawings and Specifications**, however, the specifications should supply complementary information, such as quality and workmanship desired, not repeat the information provided in the drawings, such as size, shape, location and quantity required.

Specifications should be divided into sections, each applicable to part or all of the work of only one subcontractor.

Each item of work covered by the contract should be treated once, and only once, in the specifications, and only in the appropriate section.

Each section should give the scope of and fully describe the work to be performed by the subcontractor. Separate sections also should be provided for work to be performed by the prime contractors.

When a subcontractor may perform different construction operations, a separate section should be devoted to each operation. For example, a masonry subcontractor may lay brick, concrete block, glass block, structural clay tile and similar materials. A single section dealing with all of these would be too complex and omissions and duplications might go undiscovered. Preferably, a separate section should be provided for each type of work.

Divisions

For convenience, related sections are grouped into divisions. The divisions, in turn, usually are organized in accordance with the Uniform Construction Index favored by the Construction Specifications Institute and others. Recommended divisions and their sequence are given in Table 5.2.

The first division, general requirements, deals with items of the contractor's work that are general in nature. The division should provide a summary of the work to be done by the contractor, work to be done by other contractors, and work to be postponed. Materials and equipment to be provided by the owner should be listed. Whether construction will be performed under a single general contract or under separate contracts should be indicated. Also, a description of the site should be included. In addition, the division should indicate how tests,

Table 5.2. Recommended Divisions of Specifications

1. General requirements
2. Site work
3. Concrete
4. Masonry
5. Metals
6. Wood and plastics
7. Thermal and moisture protection
8. Doors and windows
9. Finishes
10. Specialties
11. Equipment
12. Furnishings
13. Special construction
14. Conveying systems
15. Mechanical
16. Electrical

reports and construction progress schedules should be handled, and how allowances and alternates should be treated. Furthermore, reference should be made to applicable building standards, such as those of the American National Standards Institute, ASTM, American Institute of Steel Construction, and American Concrete Institute.

There may be some difficulty in deciding whether an item should be dealt with in supplementary conditions to the contract or in the specifications. In general, if the requirements can logically be included in the specifications, preferably they should be placed there. If, however, the item is of a legal nature, closely related to the general conditions, especially an extension of them, it should be incorporated into the supplementary conditions. If the item concerns the work of the contractor at the site, the item should be treated in Division 1 of Table 5.2.

Sections

Each division is composed of sections. Those in the divisions after Division 1 deal with specific construction operations and are often referred to as technical, or trade, sections. Every section should be assigned a number indicating its sequence in its division and a title indicating the work the contractor or subcontractor is to perform.

It is advantageous to standardize the format

Table 5.3. Recommended Format for Technical Sections

Preface: reference to conditions of the contract and Division 1

General provisions
Scope of the work: materials or equipment to be furnished and installed under this section; materials or equipment furnished but not installed; materials or equipment installed under this section but not furnished under this section

Notes
Quality control: necessity for prior approval of materials or equipment; industry standards to be met
Delivery and storage of materials or equipment
Protection and cleaning

Materials or Equipment
Acceptable manufacturers
Substitutions: prohibition or procedure for obtaining approval
Specifications for materials or equipment

Fabrication, Installation, and Testing
Closeout and Continuing Requirements
Schedules
Inspections
Guarantees, warranties, and bonds
Closeout

Submissions: samples, shop drawings, test reports, maintenance and operating instructions

Alternates

of technical sections, because then contractors become familiar with the arrangement and are likelier to make fewer mistakes in preparing proposals and in performing the work. Because of the wide variety of work involved in building construction, however, variations from a standard format often are desirable for simplification, clarity, and convenience. Table 5.3 gives a recommended format.

For legal reasons, it is desirable that each technical section be prefaced with the statement: *The general provisions of the contract, including the conditions of the contract and Division 1, as appropriate, apply to the work specified in this section.*

Since a separate subcontract is concerned with each section, this statement insures that the subcontractor has been informed that the conditions of the prime contract and Division 1 are part of his subcontract. Thus, those requirements are given only once in the project manual and need not be repeated for each section.

Types of Specifications

The material or equipment specifications in the technical sections may be written in any of several different ways. Types of specifications that are used include performance, descriptive, reference, proprietary, and base-bid specifications.

Performance Specifications

This type defines the work by specifying the results desired. It does not give dimensions, specific materials, finishes, nor methods of manufacture. It does not tell the contractor how to do the work. The specification delegates to the contractor the responsibility for the design or selection of the product and determination of the method of installation. He has complete freedom to employ his knowledge and experience to achieve the results itemized in the specifications. The prime design professional has the task of evaluating in detail each bidder's proposal to determine whether the items proposed by the bidders will meet the performance requirements and of recommending the best proposal to the owner. Because of the difficulty of evaluating results and the undesirability of withholding payment from the contractor until results have been determined, the contract should obligate each prime contractor to supply the owner with a written guarantee that labor and materials furnished and work performed are in accordance with the requirements of the contract. The guarantee should apply for at least 1 year. In a similar fashion, the prime contractor should get guarantees from fabricators, manufacturers, and subcontractors who supply products or perform the work.

To illustrate the type of requirements that might be incorporated in a *performance specification*, the following is an example of what might be written for insulation for a hot pipe:

Pipe insulation shall be completely in contact with the pipe, fully enclosing it, and firmly fastened in place. Thickness of the insulation

shall not exceed 1 in. The insulation material shall have the following properties: Passage of heat through the material shall not exceed 0.30 Btu per hr per sq ft per °F. It shall be suitable for use at temperatures up to 1,200°F without mechanical failure. The material shall be incombustible, insoluble in water, odorless, verminproof, rotproof, mildewproof, strong enough to resist light wear and light accidental blows without permanent deformations or damage, and capable of retaining its shape in normal usage.

The architect or engineer who writes such a specification knows of at least one material that meets the specification requirements and also knows the generally accepted practice of installing it. If the contractor proposes that material and installation method, he will obtain ready approval. If he proposes a different material or method, the architect or engineer will have to determine whether specification requirements will be satisfied and will make a comparison of properties and costs with those for the material that was contemplated when the specification was written.

Because of the difficulty of writing and evaluating performance specifications, many specifications are a combination of this type and another type, usually the descriptive type.

Descriptive Specifications

This type describes the components of a product and how they are to be assembled. Every material is identified; the structure is fully described; the method of fastening is specified; and the sequence of assembly is prescribed. The contractor is required to furnish and install the product in accordance with the description. If the installation passes inspection, responsibility for functioning and performance of the product rests on the specifier. Consequently, unless the specifier is certain that the product, when properly installed, will function properly, he should not use a descriptive specification.

The following is an example of a *descriptive specification* for insulation for a hot pipe:

Pipe insulation shall be of the block (sectional and segmental) type, molded of a chemically reacted hydrous calcium silicate consisting of at least 75% hydrous calcium silicate and between 15 and 20% graded asbestos fiber. The fiber shall be well distributed so that neither material shall be in excess in samples taken at random. The insulation material shall be suitable for use at temperatures up to 1,200°F without mechanical failure. Average density shall be about 11 lb per cu ft, oven dried. Modulus of rupture shall average 50 psi for three samples 6 in. wide and $1\frac{1}{2}$ in. thick supported on a 10-in. span and carrying a midspan concentrated load distributed over the width of the block. Maximum linear shrinkage should not exceed $1\frac{1}{2}\%$ when the material is heated to 1,000°F for 24 hr. The material shall be insoluble in water. Conductivity at 200°F should average 0.44.

Reference Specifications

This type may be basically a performance or descriptive type but may refer to a standard specification to indicate properties and quality desired and methods of test required. Standard specifications usually adopted by reference include those of the American National Standards Institute, ASTM, and the federal government. Following is an example of a reference specification:

Cement shall be portland cement conforming to ASTM C150-86, "Standard Specification for Portland Cement," Type I.

Many companies manufacture products to conform to such standard specifications and will furnish, on request, independent laboratory reports substantiating the performance of their products. Such products can be specified with confidence that minimum requirements will be met.

Proprietary Specifications

This type specifies products by trade name, model number, and manufacturer. It eliminates the task of determining whether a product meets specification requirements. Use of this type of specification is risky, however, because a lengthy period may elapse between writing of a specification and ordering of the product, during which time the manufacturer may make undesirable changes in the product. Another disadvantage is that proprietary specifications may permit use of alternative products that are not equal in every respect. Hence, when such specifications are used, the specifier should be familiar with the products and their past performance and with the reputations of the manufacturers and the subcontractors in servicing those products.

Naming only one proprietary product as acceptable in a specification is very risky. If the product should not be available when needed for construction, the work may be delayed until a suitable substitute is obtained. Also, with only one product considered acceptable, there is no price competition for furnishing it, and costs therefore may be unduly high. Consequently, two or more names should be provided for each product to insure competition and availability. Permitting the use of a product "or equal" is not satisfactory because the proposed equal may not actually be so.

Base-Bid Specifications

This type indicates acceptable products by listing at least three trade names and corresponding model numbers and manufacturers, but the bidder is permitted to offer substitutes. The bidder is required to prepare his proposal with prices from the named suppliers as base bids and to indicate for each proposed alternate the price and properties. The owner selects the product to be used. Base-bid specifications offer the greatest control of product quality.

References

Architect's Handbook of Professional Practice, American Institute of Architects.

H. Meier, *Construction Specifications Handbook*, Prentice-Hall, 1983.

C. Dunham et al., *Contracts, Specifications, and Law for Engineers*, 3rd ed., McGraw-Hill, 1979.

Words and Terms

Base-bid specifications
Descriptive specifications
Master specifications

Performance specifications
Proprietary specifications
Reference specifications
Uniform Construction Index

Significant Relations, Functions, and Issues

Standard divisions of the specifications, based on the Uniform Construction Index.
Types of specifications: master, performance, descriptive, reference, proprietary, base-bid.

5.5 BIDDING REQUIREMENTS

Bidding documents and requirements are incorporated in the project manual for a building to be constructed to inform prospective bidders of all provisions for submission of proposals. These documents are not part of the contract documents. The intent is to provide fair competition, that is, to have all bidders invited to submit proposals compete on an equal basis.

The documents and requirements include:

1. An advertisement for bids if any competent contractor will be considered, or an invitation to bid if only prequalified contractors will be considered
2. Instructions to bidders
3. Proposal form
4. Sample form for the owner-contractor agreement
5. Contractor's qualification statement, if required
6. Requirements for various types of bonds
7. Consent of surety
8. Noncollusion affidavit, if required

Advertisement for Bids

This is a printed notice in newspapers or other periodicals. It advises that proposals will be received by the owner for construction of the building. If the advertisement is required by law, the statute usually indicates how many consecutive times the notice must be published. The notice should contain the following:

1. Name of owner, name of contract and location of project
2. Time and place for receiving bids

3. Brief description of the project
4. Places and times for examination of drawings and specifications or indication from whom they may be borrowed and deposit required
5. Information on required guarantees, such as a bid bond
6. Information on a performance and payment bond, if one is required

Invitation to Bid

An invitation to bid contains practically the same information as an advertisement for bids and is used to invite proposals from prequalified contractors, selected for experience, qualifications and financial ability. Usually, the invitation is in the form of a letter, signed by the owner, construction representative or prime design professional.

Bonds

A performance and payment bond is a guarantee to the owner, equal to the total amount of the bid, that everything required by the construction contract will be faithfully done. Also, the contractor is required to pay all lawful claims of subcontractors, material suppliers and labor for all work done and all materials supplied in performance of work under the terms of the contract. In addition, the owner must be protected against suits by those persons, infringement by the contractor of patents and copyrights, and claims for property damage or personal injury incurred by anyone during performance of the work.

This bond usually is provided by a surety company on behalf of a contractor, based on knowledge of his competency and financial condition. Liability protection against suits and other claims may be provided by an insurance company.

A bid bond is a guarantee to the owner that the bidder, if offered the construction contract at the prices he bid, will sign it. This bond also may be supplied by a surety company; however, the company usually will issue a bid bond only after the performance bond for the project has been underwritten and approved. Sometimes, a contractor may be permitted to submit a certified check instead of a bid bond. He should not do so unless he has assurances that his surety will approve a performance bond for his execution of the contract.

Instructions to Bidders

The bidding instructions should describe procedures for preparation, submission, receipt, opening, withdrawal and rejection of bids. The instructions also should indicate who will answer questions concerning the drawings and specifications. Usually, to be certain that all bidders are treated fairly, the owner's representative will answer all questions in writing by issuing addenda. These documents, part of the contract, change, modify or clarify the drawings, specifications or contract conditions and are sent to all known bidders.

The bidding instructions, in addition, should require bidders to visit the site to ascertain pertinent local conditions.

For projects for which bid advertisements are published, the bidding instructions should cover qualifications of bidders, submission and return of bidders' guarantees, payment of taxes and wage rates to be paid. Sometimes, also, the general contractor may be informed of the need to submit a list of subcontractors and material suppliers to be engaged for the project. In addition, for public works, a noncollusion affidavit may be required from each bidder, to discourage agreements between the bidders.

Proposal Form

For uniformity in presentation of bidders' proposals, the bidding requirements provide a form to be used by all bidders in submitting bids. In signing the form, the bidder acknowledges that he is familiar with the contract documents, has examined the building site and has received the addenda issued. The bidder also states the price for which he agrees to furnish all materials and perform all work required by the construction contract. He further agrees to complete the project in the stipulated time and to execute, within a specified period, a contract for the project if his proposal is accepted.

The proposal form usually also contains blank spaces for prices that may be added to or deducted from the total sum quoted for construction of the project if the owner elects to make changes or select an alternate given in the drawings or specifications. If unit prices apply to changes, the form should have spaces for the unit prices for each type of work to which they apply.

The proposal should be submitted sealed to the owner or designated representative at the specified location and within the required time.

Opening of Bids

Sealed proposals should be accepted by the owner up to the time specified in the advertisement for bids or invitation to bid. At the specified time and place, the owner or his representative should open the sealed bids in the presence of the bidders and disclose the complete contents of each to those present. For public works, the owner's representative should announce publicly the lump sums or unit prices bid. Final tabulation of the results, however, need not be made at the bid opening nor at award of the contract. Often, the tabulation requires considerable study, especially when the owner has the option of selecting various alternates given in the contract documents.

Unless required to do so by law, the owner need not award the contract to the lowest bidder. For reasons peculiar to a specific project or because of special conditions, the owner may choose another bidder, despite his higher price.

Shortly after the bid opening, the owner should return the bid guarantees submitted by bidders not likely to be selected. He may retain some, say three, of the bid guarantees while bid studies continue. These three should be returned later, after the owner and the selected contractor sign the construction contract.

The agreement between the owner and contractor must be signed by the owner or a duly authorized representative and by an authorized officer of the contracting company.

References

Architect's Handbook of Professional Practice, American Institute of Architects.

J. Sweet, *Legal Aspects of Architecture, Engineering, and the Construction Process*, West Publishing Co., 1970.

Words and Terms

Advertisement for bids
Bid proposal form
Bonds: performance and payment, bid
Instructions to bidders
Invitation to bid

Significant Relations, Functions, and Issues

Type of bidding procedure: by advertisement, by invitation.
Bonds as protection for owner.
Need for uniformity of information to all bidders.
Process of receiving, opening, and evaluating of the bids.
Use of bids as a basis for the construction contract (agreement between owner and contractor).

5.6 CONTRACTORS DRAWINGS

It is typical on building projects for some of the subcontractors to be required to submit drawings for the completion of their work. These drawings may in some cases consist of a final level of completion of the design work, as is the case for subcontractors whose work involves custom designed installations—special cabinetry and works of art, for example. The basic contract drawings will allow for this work, but leave the actual details to be developed by the installers, with a formal approval process spelled out in the contract. The drawings submitted by the subcontractor will constitute both a submittal of the detailed work for approval and the description of the actual work for completion of the contract.

More frequently, contractors drawings consist of those required for clear indication of *how* the work is to be done. In these cases the final detail of the work is still required to conform with that shown on the earlier contract drawings prepared by the building designers. These contract drawings are also submitted for approval—usually meaning approval by the professionals who prepared the contract drawings. Approval, however, does not relieve the contractors from responsibility for completion of the work as spelled out on the contract drawings and the specifications. If the contractors intend to make any changes or substitutions, these must be clearly indicated and specifically negotiated with the owner.

Many subcontractors work essentially as installers; merely obtaining the items indicated on the contract drawings and described in the specifications and proceeding to place them properly in the building. In other cases, however, subcontractors must also fabricate or produce the components of the systems they install, which may involve some amount of final detail development of the components. Arrangements for this vary in terms of the process for approval and the roles of the owner, design professionals, general contractor, and the particular subcontractors involved.

5.7. CONSTRUCTION AND OCCUPANCY PERMITS

Before construction of a building may start, the local building department usually requires that the drawings and specifications be approved and a building permit be obtained from the department. Obtaining the permit generally is the responsibility of the owner; his signature often is required on the application for the permit. The drawings, however, may be submitted on his behalf by the prime design professional, or the construction contract may require the general contractor to obtain the building permit and pay required fees.

Note that a building department will approve only drawings prepared by an architect or engineer licensed by the state and so certified by seal and signature.

Because the building department may require revisions of the drawings so that the building will comply with all building-code and zoning-code regulations, it generally is advisable for the owner to have the drawings submitted to the department before bids are requested. Then, the revisions can be made before receipt of bids or signing of the construction contract. If the changes are made later, the prime professional will have to issue addenda before bids are received or, after the contract has been signed, issue change orders, an even more costly practice.

The owner also is generally responsible for obtaining any other permits required for construction of the project under the contract, such as those for temporary closing of a street, trucking of very long loads, or temporary shutoff of utilities that have to be relocated. The contractor, however, has the responsibility of securing the necessary permits for work done at his option on the project but not required by the contract. For example, if he elects to use a crane operating from a street for erection of steel or placement of concrete and a permit is required for that purpose, the contractor should obtain the permit and pay the required fee.

After construction starts, the building department periodically sends inspectors to inspect the building to insure conformance with building-code and zoning-code regulations. The department also may require the contractor to notify it when certain critical stages have been reached or critical items are ready so that an inspector may be dispatched to check compliance.

In addition other municipal and state agencies may send inspectors to insure that items under their jurisdiction comply with legal requirements. The state fire marshall or local fire department, for example, may inspect construction; so may the state labor department, state housing division, or a federal housing agency, if it is concerned. If violations are observed, the contractor is informed of them and required to correct them. He is subject to penalties if he does not remove the violations.

When construction has been completed, the contractor notifies the building department, as well as the owner, that the project is ready for final inspection. The department then sends inspectors to the building for that purpose. Any violations discovered must be removed. When no violations are found, the department issues a certificate of occupancy. This gives to the owner the department's permission to occupy the premises.

5.8 CONSTRUCTION PROCEDURES

Construction usually starts shortly after the owner and the general contractor have signed the owner-contractor agreement. The owner should send the contractor a written *notice to proceed*. The notice should also advise the contractor when he can enter the property and begin work. The contractor has the number of days stipulated in the contract in which to start construction.

When the project has a tight construction schedule, the owner may want an early construction start. He may ask the contractor to begin work before the formal signing of the contract. In this case, the owner may issue to the contractor a *letter of intent*. This letter indicates that the contract will be awarded to the contractor and gives him notice to proceed with construction. The letter should provide for compensation to the contractor if the owner should not award the contract to the contractor. Both parties must sign the letter of intent to make it effective.

Construction Supervision

The head of the construction company may personally take charge of construction operations or if the company is doing several jobs simultaneously, he may assign a company officer or project manager for that purpose. To assist the manager, a field superintendent usually is assigned to the building site with responsibility for all activities there.

The owner may assign a representative to the site for surveillance of the work. The representative would have the responsibility of keeping progress records, supervising inspectors, and in some cases, keeping cost records. The representative may have the title *clerk of the works*, *architect's superintendent*, *engineer's superintendent* or *resident engineer*. When a construction manager is engaged, he generally assumes responsibility for construction surveillance. Members of the building team may be required to visit the site only occasionally. Alternatively, the prime design professional may be assigned the responsibility for surveillance and engagement of the project representative.

The basic steps in construction are summarized in Table 5.4.

Subcontract Awards

On receipt of the notice to proceed or letter of intent, the contractor mobilizes his forces and equipment to start work. In accordance with a construction schedule he has planned, he notifies subcontractors that he is ready to sign contracts with them and, if necessary, issues letters of intent to those subcontractors required for the early stages of construction. After he signs

Table 5.4. Basic Steps in Construction

1. Obtaining of permits and issuance by owner of letter of intent or written notice to proceed
2. Planning and scheduling of construction operations in detail
3. Mobilization of equipment and personnel for project
4. Notification to subcontractors of contract award, issuance of letters of intent, awarding of subcontracts, advance ordering of materials and equipment, issuance to subcontractors of notice to proceed
5. Survey of adjacent structures and terrain
6. Survey for construction layout
7. Establishment of field offices
8. Erection of fences and bridges
9. Demolition, site preparation, and excavation, including bracing of earth sides, drainage, and utility relocation
10. Construction of foundations
11. Erection of structural framing and stairs
12. Placement of temporary flooring, if needed
13. Installation of pipes, ducts, and electric conduit
14. Erection of material hoists, if needed
15. Construction of permanent floors
16. Installation of elevators in tall buildings
17. Placement of exterior walls and windows
18. Fireproofing of steel framing, if required
19. Construction of fixed partitions
20. Construction of roof and placement of roofing
21. Finishing operations
22. Removal of temporary structures and clean up
23. Landscaping
24. Final inspections and project acceptance
25. Issuance by building department of certificate of occupancy
26. Final payment to contractor and occupation of premises

contracts with the others, he advises them of the dates on which they are scheduled to begin work on the project.

Scheduling

A subcontractor who arrives on the job ready to start work before the project has reached the appropriate stage generally is unable to begin. One who starts late may delay the project. Consequently, work must be so scheduled that subcontractors report for work exactly when needed. Similarly, materials must be delivered close to the time when they are needed. If they arrive late, work is delayed. If they arrive too soon, sufficient storage space may not be avail-

able. Consequently, the contractor must plan his operations well in advance, carefully and accurately.

Surveying

One of the first steps is to survey the building site and surrounding terrain and property. The survey of existing conditions is needed to determine conditions of adjacent structures that may possibly be damaged by the contractor's operations. This survey must be supplemented by readings of elevations at the foundations of the nearby buildings for use in later determinations of the occurrence of settlement or lateral movement.

Also, a survey is needed to provide a general layout for construction, including base lines, offset lines, and reference points, such as bench marks (points of known elevation). These lines and points are used for geometric control during construction. Measuring from them, the surveyors locate and orient structural members, such as columns and walls, and maintain verticality of vertical components. Also, the surveyors determine the elevations of foundations, floors and roofs. In addition, they establish control points inside the building, from which they make other measurements of distance and elevation.

Construction Offices

Another early step the contractor takes is establishment of construction offices on or adjacent to the building site. The space is needed for housing records, permits, and construction documents, administrative and supervisory personnel when not in the field, and clerical and secretarial help. The offices also are useful for communication and meeting purposes. Job meetings are held periodically with the owner's representatives and subcontractors for dissemination of information, to settle controversies, and to avoid problems. Similarly, subcontractors set up offices, usually close to those of the general contractor, but only for their own staff. The owner too may set up offices for his representatives. Temporary buildings may be used for all of these offices.

Fences and Bridges

Another early step the contractor takes is to fence in the property, especially if excavation will extend near the lot boundaries. Fencing, however, often is omitted if the contractor believes that there is practically no chance that the public will be injured by construction activities.

As construction advances above ground, the contractor usually erects a shed, or bridge, along the fence above streets or walkways that adjoin the fence. The purpose of the bridge is to protect passersby from objects accidentally dropped from the structure during its erection.

Site Preparation

If the lot already contains buildings or other structures not to be incorporated in the new building, they must be removed. For this purpose, a demolition subcontractor is employed. If the site is heavily wooded, the area to be occupied by the building and other facilities must be cleared. Often, the land has to be graded.

Also, substantial excavation may be required; for example, when the building has a basement or deep foundations are needed. This work may be performed by the foundation subcontractor or a separate earthwork subcontractor. In addition, the earthwork subcontract may call for removal or relocation of existing underground utilities. If special provisions must be made for draining the excavation to prevent water that seeps into it from hindering construction, a special subcontract may be let for pumping out the water, for instance, to a wellpoint subcontractor. If extensive trenching is needed for water supply and sewage lines, the earthwork contract also may include that work; otherwise, it may be covered by the plumbing subcontract.

Foundations

As excavation proceeds, it usually becomes necessary to support the earth around the boundaries to prevent it from caving in or from moving laterally. Earth movements might cause nearby structures to settle or move laterally and damage them. A common method of providing support is to drive sheetpiles around the excavation, and then, in some manner, brace the sheeting.

If piles are needed to support the building, they are driven to required depths into the bottom of the excavation and capped with concrete footings. If spread footings are specified, soil is excavated to required depths and concrete is placed to form the footings. Also, foundation walls are concreted along the perimeter of the excavation. Earth backfill then is placed behind the walls, if necessary, to prevent movement of adjoining soil when the sheeting supporting it is removed.

Structural Framing

If the framing of the building is to be structural steel, the general contractor may award a contract for fabrication and erection of the steel to a single subcontractor, or one contract to a steel fabricator and a second to an erection subcontractor. If the framing is to be concrete, a concrete subcontractor will be employed. A concrete subcontractor will be engaged in any event if the floors, walls, or other parts of the structure are made of concrete.

Erection of walls and columns can start as soon as the concrete in the foundations has gained sufficient strength, usually at least a week after the concrete was cast. Beams and girders, if required, or concrete floors are placed between the vertical structural members as soon as possible. The horizontal members are needed to brace the verticals, to prevent them from toppling over. Additional diagonal bracing also may be installed, to keep columns in vertical alignment. The beams and girders, in addition, provide support for temporary flooring for workmen. (To avoid interference with erection of structural steel framing, permanent floors usually are not installed in multistory buildings until the framing is in place several stories above. Building codes, however, for safety reasons, often place a limit on the number of stories the framing may be advanced above the floors.) Load-bearing walls must be constructed before the skeleton framing (beams or floor slabs and columns), because the walls have to support the beams or floor slabs.

Other Components

Installation of stairs generally follows closely behind erection of the framing to enable workmen to reach the levels at which they have to work. Piping, ductwork, and electric conduit to be embedded in permanent construction also are installed early.

The general contractor usually provides hoists for use by the subcontractors to raise materials and equipment to the floors where they will be needed. The subcontractors also may use cranes or derricks for hoisting materials and equipment. Also, for tall buildings, when the framing becomes high enough, elevators are installed to lift workmen to working levels.

As the framing rises, permanent floors are installed at successive levels. Placement of exterior walls follows closely behind. Windows are set in the walls, often without glazing to prevent the glass from being broken accidentally while construction proceeds. Meanwhile, electrical, plumbing, and HVAC subcontractors continue with installation of wiring, piping, and ductwork. If the framing is structural steel and the underside requires fireproofing, fire-resistant material should be placed to protect the framing, unless the ceiling will serve that purpose. Fixed partitions may be constructed next. By this time, usually, the roof can be installed.

Finishing Operations

Before the building can be considered completed, however, there are numerous operations still to be performed. Ceilings have to be placed, roofing and flashing laid down, wallboard or panelling attached to interior wall surfaces and tile set. Electric lighting fixtures have to be mounted and switches, electrical outlets and electrical controls installed. Plumbing fixtures have to be seated in place. Furnaces, air-conditioning equipment, permanent elevators and escalators, heating and cooling devices for rooms, electric motors and other items called for in the drawings and specifications must be installed. Glass must be placed in the windows, floor coverings laid down, movable partitions set in place, doors hung and finishing hardware installed. All temporary construction, such as field offices, fences and bridges, must be removed. The site must be landscaped and paved. Finally, the building interior must be painted and cleaned.

When the general contractor believes the building has been completed, he notifies, in writing, the owner and his site representative. The owner then should apply to the building department for a certificate of occupancy and determine by inspection whether the work has, in fact, been completed in accordance with the drawings and specifications. If it has not, the owner should report, in writing and in detail, to the contractor the additional work required. The owner should make a careful final inspection and not rely on the building department inspection to protect his interests. Building department inspectors are primarily concerned with discovering violations of the building code.

When the work has been satisfactorily completed, the owner must determine accurately the amount due the contractor, including the value of work done under change orders and extra work authorized during construction. Within a period stipulated in the contract, usually 30 days, the owner must pay the contractor the amount due, less any money the owner, under the contract, may be permitted to withhold tentatively. He also should promptly furnish the contractor and his surety a statement of acceptance of the project or of exceptions. On receipt of the certificate of occupancy, the owner may occupy the building.

GENERAL REFERENCES AND SOURCES FOR ADDITIONAL STUDY

These are books that deal generally with topics covered in the chapter. Topic-specific references relating to the individual chapter sections are listed at the ends of the sections.

The Architect's Handbook of Professional Practice, Vol. 1, American Institute of Architects.

J. Sweet, *Legal Aspects of Architecture, Engineering, and the Construction Process*, West Publishing Co., 1970.

R. Hershberger, *Programming for Architecture*, Van Nostrand Reinhold, 1987.

C. Dunham et al., *Contracts, Specifications, and Law for Engineers*, 3rd ed., McGraw-Hill, 1979.

J. Clark, *Understanding and Using Engineering Service and Construction Contracts*, Van Nostrand Reinhold, 1986.

R. McHugh, *Working Drawing Handbook*, 2nd ed., Van Nostrand Reinhold, 1982.

H. Meier, *Construction Specifications Handbook*, Prentice-Hall, 1983.

EXERCISES

The following questions and problems are provided for review of the individual sections and the chapter as a whole.

Section 5.1

1. Who besides the owner and members of the building team should help prepare the contract documents?

2. Name and describe briefly the contract documents.

3. Who are the parties (signatories) to the construction contracts?

4. If, during construction of a building, the owner asks the contractor to perform work not covered by the drawings and specifications, what should the contractor do if he is willing to do the extra work?

5. Why should contracts and change orders be in writing?

6. Why are modifications of the contract during construction undesirable?

7. Besides maintenance of a good reputation, what incentive does a contractor have to provide good workmanship?

8. What risks are incurred when the specifications list in detail the procedures the contractor must follow and the required results?

9. Who is responsible for securing construction permits?

10. Under what circumstances should a contractor use substitute materials not called for in the contract documents?

SECTION 5.2

11. How do addenda and modifications of the contract differ?

12. What is the purpose of the project manual?

13. (a) Which of the documents in the project manual are not contract documents?

    (b) Which of the contract documents are not normally incorporated in the manual?

14. What requirements usually are contained in the owner-contractor agreement?

15. What are the advantages of using a standard construction contract?

16. Is it necessary to repeat in the specifications a requirement given in the conditions of the contract? Explain your answer.

17. What are the purposes of:

    (a) General conditions of the contract?

    (b) Supplementary conditions of the contract?

Sections 5.3 and 5.4

18. If working drawings show to scale and label all items comprising a building, why are specifications still necessary?

19. A contractor preparing a construction proposal discovers a conflict between the working drawings and the specifications. The prime professional corrects the conflict and notifies the observent contractor. What else should the prime professional do before bids are received?

20. During construction of a building, the general contractor notifies the prime professional that a window type specified is no longer being manufactured and is not available for purchase. What should the prime professional do?

21. In what contract document for a 10-story building should you look to determine:

    (a) Beam spacing for the fifth floor?

    (b) The tolerances permitted in fabrication of steel beams for the fifth floor?

    (c) Layout of air ducts for the fifth story?

    (d) Quality of material used for the air ducts for the fifth story?

22. What information is provided by a door schedule?

23. Why should every item shown in the drawing also be specified in the specification?

Why is repetition of a requirement in both documents undesirable?

24. What are the advantages of a master specification?

25. What is the relationship of the organization of specifications and the subcontract method of construction?

26. What are the advantages to the building owner of a product specification that gives the contractor a choice of several products?

27. What type of information should be given by notes on working drawings and what type by specifications?

28. A drawing of a foundation wall does not show a drain at the base of the wall, but a note on the drawing states: "Install 4 in. cast-iron drain pipe completely around foundation and connection to sewer." Is the contractor required to furnish and install the pipe without additional compensation? Justify your answer.

Section 5.5

29. Describe briefly the usual bidding requirement documents.

30. Compare the advantages and disadvantages of an advertisement for bids and an invitation to bid.

31. What should be the value of the performance and payment bond a contractor is required to post for a building he seeks to construct?

32. What is the purpose of a bid bond?

33. The prime professional for a project receives a telephone call from a prospective bidder who questions some items in the drawings and specifications. What should the prime professional do?

34. In what document should a builder look to find information on the opening of bids?

35. Shortly after bids have been requested for building construction, the prime professional receives a telephone call from a material supplier requesting additional information. How should the prime professional handle the request?

Sections 5.6 to 5.8

36. Why must an owner or a competent representative inspect construction despite frequent building department inspections?

37. After a building permit has been issued, a company president, on recommendation of the building architect, decides to request bids from six general contractors:
    (a) What means should the president use to request bids?
    (b) How does the president insure that the selected bidder will sign the construction contract?
    (c) Are bids usually examined as they are received?
    (d) Must the company president sign the contract with the low bidder?

38. A city public works department has obtained a building permit for a proposed building. The city engineer, in accordance with city law, must accept proposals from all interested contractors and engage the contractor who submits the lowest bid.
    (a) What means should the engineer use to request bids?
    (b) How does the engineer insure that the bidder selected will complete the work after signing the contract?

39. A company president engages for design and construction of a factory a general contractor who is neither a registered architect nor a professional engineer.
    (a) Will the building department issue a building permit for plans drawn by the contractor that satisfy the building code? Explain your answer.
    (b) How will the building department know from inspection of the drawings whether or not the person who prepared them is an architect or an engineer?

40. After selecting a general contractor, how can an owner get construction started immediately without signing a construction contract?

41. Compare the responsibilities and duties of the contractor's field superintendent with those of the clerk of the works.

Chapter 6

Life Safety Concerns

Buildings must be designed for both normal and emergency conditions. Building designers should, in initial design of buildings, take precautions to protect property from major damage, and especially from collapse, due to accidents or disasters; but designers must also provide for life safety of occupants, neighbors and passersby in emergency situations. Such situations may be caused by high winds, earthquake, intruders or fire. Cost-effective protection against their adverse effects can be achieved with the systems-design approach, applied from the start of conceptual design.

Building codes contain many requirements for prevention of major property damage and for life safety. But codes do not always cover extreme conditions or special cases. Safety requirements in codes generally are minimum standards applicable to ordinary buildings, those not unusually large or tall and those not used for purposes with which there has been little experience, such as production of nuclear power. Building designers therefore must use judgment in adopting code provisions and apply more stringent requirements when specific conditions warrant them.

Economic and sometimes sociological factors often rule out provisions for full protection against extreme conditions that are possible but highly unlikely to occur. For example, it is possible but very costly to build a one-family house that can withstand a violent tornado. Furthermore, because such a house would have

small or no windows, a family preferring large windows would choose such windows and risk the possibility of injuries from tornadoes. Consequently, building designers should weigh the possibility of extreme conditions occurring and balance risks and costs.

From statistical studies of such natural phenomena as snowfalls, high winds and earthquakes, probabilities of extreme conditions being exceeded in a year in various parts of the United States have been determined. For example, the probability is 0.04, or four chances in one hundred, of a wind faster than 80 mph blowing through New York City, or 0.02, two chances in one hundred, of a wind faster than 60 mph blowing through Los Angeles.

The reciprocal of the probability is called the *mean recurrence interval*. This gives the average time in years between the occurrence of any condition that exceeds the specific extreme condition. Thus, the mean recurrence interval of a wind faster than 80 mph in New York City is $\frac{1}{0.04}$, or 25 years, and of a wind faster than 60 mph in Los Angeles, 50 years.

Designers can use mean recurrence intervals to establish reasonable design values. For example, if the expected life of a building to be erected in Los Angeles is 50 years, it would be logical to design the building for a 60-mph wind, which would be unlikely to be exceeded in the next 50 years. (Mean recurrence intervals for snowfalls, winds and earthquakes are given in ANSI "Building Code Requirements for Minimum Design Loads in Buildings and Other Structures," A58.1-1982, American National Standards Institute.)

6.1. WINDSTORMS

Every year, high winds in the United States cause property damage costing many millions of dollars and kill and injure many persons. Deaths from tornadoes alone average about 100 persons annually. Many of these deaths and injuries result from collapse of buildings. Better building design and construction therefore not only could prevent much of the wind damage but also could save many lives.

Furthermore, with the systems-design approach, designers could incorporate adequate wind resistance in buildings and protect lives with little or no increase in costs over former inadequate measures.

Wind Characteristics

Experience has shown that probably no area in the United States is immune from incidence of winds of 100 mph. Furthermore, tornadoes have been reported in all states except for four or five western states. Tornado winds have been estimated as high as 600 mph. Hurricanes have struck areas of the country with winds as high as 200 mph.

Straight Winds. Generally, wind damage appears to be caused by severe straight winds, although air in the storms may be rotating about a nearly vertical axis. In the case of hurricanes with rotating winds, the radius of curvature is so large that the path of the wind may be considered straight. In the case of tornadoes with winds rotating in a narrow funnel, damage appears to be caused by severe winds in the same general direction as the funnel movement.

Gusts. Wind velocity, however, usually does not remain constant for long. The wind often strikes a building as gusts. Wind velocity in such cases rises rapidly and may drop off just as fast. Hence, wind actually imposes dynamic loads on buildings.

Effects of Friction. Because of natural and man-made obstructions along the ground, wind velocity is lower along the ground than higher up. Ground characteristics within a range of at least 1 mile of a building are likely to affect velocity of the wind striking the building. The rougher the terrain the more the air will be slowed. Building codes often take this effect into account by permitting lower design wind loads for buildings in the center of a large city than for buildings in a suburban area and woods, and allowing lower loads for buildings in suburban areas than for buildings in flat, open country.

The effects of ground roughness on wind velocity diminish with height above ground and eventually become negligible. In the center of a large city, wind velocity above an elevation of about 1,500 ft relative to the ground may be unaffected by the ground surface and buildings. For suburban areas and woods, wind velocity may be considered nearly constant above an elevation of about 1,200 ft. For flat, open country, the limiting elevation for roughness effects may be about 900 ft, and for flat, coastal areas, 700 ft.

Velocity Measurement. For standardization purposes, wind velocities are reported for an elevation of 10 m (32.8 ft) above ground. If winds are not measured at the level, the wind velocities recorded at another elevation are converted to velocities at the 10-m level. Building codes often require buildings up to 30 to 50 ft high to be designed for wind velocities at the 10-m level.

Variation with Height. With the velocity known at the 10-m height, the velocity at any height above ground up to the limiting elevations previously mentioned may be estimated from Eq. (6.1).

$$v_z = v_{10} \left(\frac{z}{32.8} \right)^n \qquad (6.1)$$

where

v_z = velocity at height z above ground

v_{10} = velocity 32.8 ft above ground

z = elevation above ground, ft
  ($32.8 \leq z \leq 900$, 1,200 or 1,400)

n = exponent with value depending on roughness of terrain

For centers of large cities and very rough, hilly terrain, n may be taken equal to $\frac{1}{3}$. For suburban areas, towns, city outskirts, wooded areas and rolling terrain, n may be taken equal to $\frac{1}{4.5}$. For flat open country and grassland, n may be taken equal to $\frac{1}{7}$ and for flat, coastal areas, $\frac{1}{10}$.

Shielding. A building may be shielded from the wind from certain directions by adjacent buildings or hills. But it would not be conservative to design a building for lesser wind loads because of such shielding. In the future, the shielding buildings may be removed or the hill modified, with resulting increases in wind buffeting.

Channeling Effects. Man-made or natural obstructions might channel wind toward a building or increase the intensities of gusts. In such cases, if the increased wind loads can be estimated, they should be provided for in design of the building. Often, wind tunnel tests of a model of a building, neighboring buildings and the nearby terrain are useful in predicting wind behavior.

Orientation. In many parts of the United States, high winds generally may come from a specific direction. Nevertheless, it is possible for high winds to come from other directions too. Consequently, it is advisable in building design to assume that wind may come from any direction.

Inclination. In many cases, it is reasonable to assume wind velocities as horizontal vectors. This assumption may be adequate for design of vertical walls and structural framing other than in roofs; but because of possible turbulence in the vicinity of a building or because of the slope of the terrain, wind velocity may be inclined 10° or 15° or more from the horizontal, up or down. The vertical component of the wind in such cases may impose severe loading on roofs, balconies, eaves and other overhangs.

Design Loads. The preceding description of wind characteristics should make it evident that wind loads are uncontrollable, random variables. They also are dynamic rather than static. Nevertheless, for ordinary buildings, it is usual practice to assume probable maximum wind velocities and to treat the associated pressures on the buildings as constant loads. For unusually tall or slender buildings, detailed structural analyses aided by wind tunnel tests of models generally is advisable, with the wind treated as a dynamic load.

Wind Pressures and Suctions

For a wind velocity v, mph, the basic velocity pressure p, psf, on a flat surface normal to the velocity is defined by

$$p = Kv^2 \qquad (6.2)$$

where $K = 0.00052$ when $n = \frac{1}{3}$

$\qquad = 0.0013$ when $n = \frac{1}{4.5}$

$\qquad = 0.0026$ when $n = \frac{1}{7}$

$\qquad = 0.0036$ when $n = \frac{1}{10}$

Table 6.1 lists some basic pressures for winds with 50-year recurrence interval for various regions of the United States and for various heights above ground. These pressures can serve as a guide in the absence of building-code requirements. Note, however, that Table 6.1 does not allow for tornado winds or extreme hurricanes.

The basic total force P, lb, due to a basic wind pressure p is given by

$$P = Ap \qquad (6.3)$$

where A = area of perpendicular surface, sq ft.

The effects of gusts may be taken into account by applying an appropriate gust factor G. With gusts, the wind force equals

$$P = GAp = Aq \qquad (6.4)$$

where $q = Gp$.

Dependent on type of exposure and dynamic response characteristics of the obstruction, G is probably best determined from wind tunnel tests of models or from observations of similar existing structures.

The effective pressure acting on a building or a building component depends on the building geometry. The effect of geometry is generally taken into account by multiplying q by a pressure coefficient C. For example, external pressure on the windward wall of a building may be

Table 6.1. Basic Wind Pressures for Design of Framing, Vertical Walls and Windows of Ordinary Rectangular Buildings, psf[a]

Height zone, ft above curb	Types of exposures								
	A^b	B^c	C^d	A^b	B^c	C^d	A^b	B^c	C^d
	Coastal areas, N.W. and S.E. United States[e]			Northern and central United States[f]			Other parts of United States[g]		
0–50	20	40	65	15	25	40	15	20	35
51–100	30	50	75	20	35	50	15	25	40
101–300	40	65	85	25	45	60	20	35	45
301–600	65	85	105	40	55	70	35	45	55
Over 600	85	100	120	60	70	80	45	55	65

[a]For winds with 50-year recurrence interval. For computation of more exact wind pressures, see ANSI Standard A58.1-1982.
[b]Centers of large cities and very rough, hilly terrain.
[c]Suburban areas, towns, city outskirts, wooded areas, and rolling terrain.
[d]Flat open country, flat open coastal belts, and grassland.
[e]100-mph basic wind speed.
[f]90-mph basic wind speed.
[g]80-mph basic wind speed.

computed from

$$P_e = C_p A q_e \qquad (6.5)$$

where

A = projected area of the structure on a vertical plane normal to the wind direction

C_p = external-pressure coefficient

Building codes give recommended minimum values for pressure coefficients for ordinary buildings. For unusual buildings, they may be determined from wind tunnel tests of models.

Pressure coefficients are given a positive sign when the pressure tends to push a building component toward the building interior. They are given a negative sign when the pressure tends to pull a building component outward. Nega-

tive pressures also are called suctions or, for a roof, uplift.

Figure 6.1a illustrates wind flow over a sloping roof of a low building. As indicated in Fig. 6.1b, the wind creates a positive pressure on the windward wall normal to the wind direction and a negative pressure on the leeward wall. For the roof slope shown, it is likely that the wind will create an uplift over the whole roof. Pressure coefficients for the windward wall therefore will be positive and those for the leeward wall and the roof, negative.

If there are openings in the building walls, internal pressures will be imposed on the walls, floors and roofs.

The net pressure acting on a building component then is the difference between the pres-

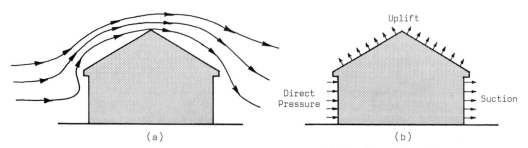

Fig. 6.1. Effect of wind on a low building with sloping roof. (a) Wind flow. (b) Wind pressures.

sures acting on opposite faces (vector sum of forces acting).

Design for Wind

Every building and its components should be designed to withstand, without collapsing, tearing away, breaking or cracking, maximum winds that are likely to occur within the anticipated service life of the building. Generally, a 50-year mean recurrence interval should be the basis for selection of a maximum basic wind velocity for a permanent building. But for unusual buildings or for those presenting an unusually high hazard to life and property in case of failure, a 100-year mean recurrence interval should be used.

For economic reasons, it is impractical to design buildings to resist violent tornadoes without considerable building damage. Conservative designs, however, usually incorporate a safety factor to provide reserve strength against unexpected loads, poor quality materials and low-grade workmanship that escapes attention. Hence, a building properly designed to withstand probable wind loads without damage should have sufficient reserve strength to resist much stronger winds with little or no damage.

For light construction, such as one- and two-story houses, which tend to collapse when in the path of a tornado, it is advisable to incorporate a well-protected shelter. It may be located in the basement or on the ground floor of basementless buildings. The shelter should be enclosed on all sides with a strong material, such as thick, reinforced concrete. The entrance preferably should be inside the building and should provide a 90° turn to prevent flying debris from entering the shelter. In a tornado, debris, such as wood beams, can become flying missiles and penetrate several ordinary building walls.

The systems-design approach can be helpful in reducing damage from high winds, especially when systems design is used from the start of the conceptual phase of design. For example, though a building may be designed for a probable maximum wind from any direction, the building may be oriented to resist extremely high winds in the direction from which they are

likely to come. In the midwestern region of the United States, for instance, tornadoes generally move from southwest to northeast. Hence, in that region, a building could be placed on its site and shaped to have low exposure to the southwest and strengthened to have high resistance to such tornadoes.

As another example, all connections, from foundation to roof, between building components could be given adequate strength to withstand extreme winds without failing at only a slight increase in construction cost.

As still another example, measures can be taken to reduce wind pressures. This may be accomplished with appendages on the walls or roofline irregularities; or vents may be placed in roofs to relieve uplift. Wind tunnel tests may give clues.

Wind resistance should be an integral part of every building system. Designing a system initially only for gravity loads and then adding strengthening elements for wind resistance is likely to be more costly and not so structurally effective as providing strength and stiffness for both gravity and wind loads from the start of design.

Designers should consider the possibilities of different failure modes under wind loads and provide against them.

Overturning. Wind loads are often referred to as lateral forces, because they act against the sides of buildings as substantially horizontal forces, compared with gravity loads, which act vertically.

Note that a wind striking the sides of a rectangular building obliquely may be resolved into two components, each component perpendicular to a windward side. In analysis of the effects of wind loads then, the response of the building to each component can be studied separately and the effects of both components determined.

Considered as a rigid body, a building subjected to horizontal forces W may be overturned. It would tend to rotate about the edge of its base on the leeward side (see Fig. 6.2a). The tendency to overturn is resisted by the weight M of the building. Building codes usually require that the resistance to overturning be at least 50% greater than the overturning force.

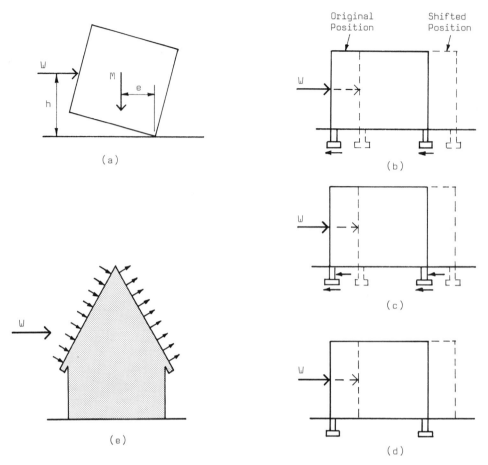

Fig. 6.2. Some potential modes of failure for buildings subjected to high winds. (*a*) Overturning. (*b*) Sliding, unresisted. (*c*) Sliding, resisted by weak soil.(*d*) Sliding off foundations. (*e*) Roof uplift.

If *Wh* is the overturning moment about the base and *Me* is the resisting moment about the leeward edge of the base,

$$Me \geqslant 1.5Wh \qquad (6.6)$$

The resistance to overturning can be augmented by anchoring the building firmly to its foundations. The weight of earth atop footings then may be included with the weight of the building in computation of *Me*.

Sliding. In addition to tending to overturn a building, wind forces also tend to push a building horizontally. This movement is resisted by friction, earth pressure and connections between superstructure and substructure.

Like overturning resistance, sliding resistance due to friction depends on the weight of

the building. If a building subjected to high winds is firmly connected to foundations that are located near the ground surface, the foundations may slide in the direction of the wind unless there is sufficient friction between them and the soil or unless the foundations are anchored to the ground (see Fig. 6.2*b*). For a building with deep foundations, earth between footings and ground surface will assist the friction forces in resisting sliding (see Fig. 6.2*c*). (With some soils, however, resistance to movement may decrease when the ground gets wet. The possibility of this occurring should be considered by the designers.)

To insure development of required sliding resistance, whether foundations are shallow or deep, it is essential that designers call for strong connections between superstructure and foundations. In the absence of such connections, strong winds have pushed many small buildings off their foundations, with disastrous consequences to occupants and property (see Fig. 6.2d). Buildings should be securely anchored to prevent both sliding and overturning.

Roof Uplift or Sliding. Flat roofs and roofs with slopes up to about 45° may be subjected to suction over the whole area. The uplift may be severe enough to draw the roof, or parts of it, away from the rest of the building, unless the roof is firmly anchored to the building frame and its components are securely attached to each other and to the frame. Often, when high winds peel a roof from a building with load-bearing walls, one or more of these walls also topples.

The weight of a roof cannot be relied on to hold it in place in strong winds. Positive anchorage should be provided between the roof and its supports.

Steeply sloped roofs may be subjected to positive pressures or suctions, depending on the direction of inclination (see Fig. 6.2e). The resulting forces may slide a roof from its supports or suck components loose. Such damage can also be prevented by positive anchorage.

Sway and Collapse. Strong winds may collapse a building, without overturning it or causing it to slide, unless adequate means are provided to transmit the wind loads through foundations into ground strong enough to resist the loads.

Load-bearing walls have to be braced against caving in or being sucked outward by winds. Floors and roof, if securely attached to the walls, can serve as bracing, but some means must be present to transmit the wind loads to the ground. If not, the floors and roof will shift under the horizontal forces and permit the walls to topple.

Curtain walls should be anchored to the structural frame of the building or to floors and roof attached to the frame. Connections should be strong enough to transmit wind loads from the walls to the frame. Then, some means must be provided to transmit the loads from the frame through the foundations to the ground. If this is not done, the building may topple like a house of cards (see Fig. 6.3a).

Any of several structural devices may be used to carry wind loads to the ground. Figures 6.3b to d illustrate some of the most commonly used ones.

Figure 6.3b shows a shear wall, which may be used to brace load-bearing walls directly or to withstand wind loads on floors and roof. The wall has high resistance to horizontal forces parallel to its length. If two such walls are placed perpendicular to each other, they can resist wind from any direction, since any wind force can be resolved into components parallel to each of the walls.

Figure 6.3c shows a structural frame with diagonal structural members to carry the wind loads W to the ground. The diagonals are called X bracing. The arrows in Fig. 6.3c show the paths taken by wind forces until they reach the ground. Note that the diagonals and the girders (major beams) transmit the horizontal forces W to the leeward columns, which carry the forces vertically to the ground. The windward columns, in contrast, carry vertically upward forces from the ground that keep the building from overturning.

Figure 6.3d shows a rigid frame subjected to horizontal wind forces. The wind tries to topple the building in the manner indicated in Fig. 6.3a; however, in the rigid frame in Fig. 6.3d, the girders and columns are rigidly connected to each other. Any tendency for the ends of the girders to rotate is resisted by the columns, because the connections maintain the right angle at each joint. The frame may shift a little in the direction of the wind, but the frame cannot collapse until the strength of the members and connections is exhausted. As in the X-braced frame, the leeward columns transmit the horizontal forces vertically to the ground. The windward columns carry upward the forces from the ground that prevent overturning.

Designers must bear in mind that the objective of shear walls, bracing and rigid frames is to convey loads to ground that can withstand those loads. Any gap in the load path to such ground, i.e. any failure to transmit load, can lead to disaster. Consequently, not only must designers provide a continuous load path but also they must make every element along the path strong enough to carry imposed loads.

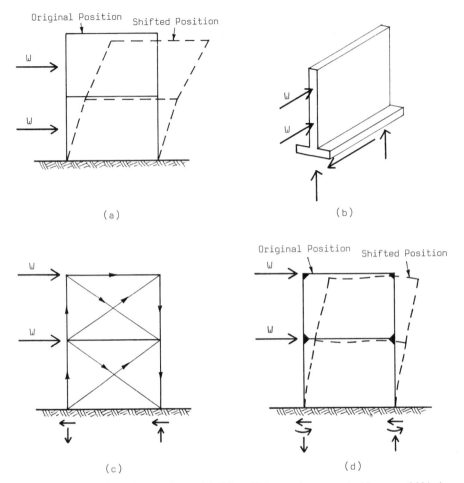

Fig. 6.3. (*a*) Wind loads *W* topple an unbraced building. This may be prevented by use of (*b*) shear wall, (*c*) diagonal bracing, or (*d*) rigid-frame action.

This means that connections as well as girders, columns, bracing, foundations and soil must have adequate capacity.

Sway or Drift. Strength, however, cannot be the sole consideration. Potential movements of the building also must be considered. Wind forces, even when static, cause sidesway, or drift, a shifting of the upper part of a building in the direction of the wind. Winds, though, are dynamic loads and they can cause a building to sway violently back and forth until it falls apart, unless the building is made stiff enough to resist such movements. So design against wind must have two other objectives besides provision of sufficient strength. Total drift of a building must be limited to prevent damage to building components, especially cracking of

brittle materials, such as plaster or concrete walls. In addition, vibration of buildings must be controlled so as not to damage building components or annoy occupants.

These objectives can be attained by proper shaping of a building, arrangements of structural components to resist drift, and selection of members with adequate dimensions and geometry to withstand changes in dimensions. For example, low, squat buildings have less sidesway than tall, slender buildings. Hence, decreasing the ratio of building height to least base dimension, width or length, will reduce drift. As another example, thin rectangular buildings have more sidesway than square or circular buildings with the same floor area per story. Thus, making buildings more compact will reduce drift. But thin rectangles can be used with reduced drift if they are arranged in perpendicular wings, to brace each other. Buildings T or H shaped in plan can consequently be efficient in resisting sidesway because, regardless of the direction from which the wind blows, they have long walls or long lines of columns with high resistance to wind in the direction of wind components.

Design of a building as a system requires that gravity and lateral loads be considered simultaneously, to achieve optimum results. Often, it becomes possible to provide wind resistance through this approach with no increase in cost over that for supporting gravity loads alone.

Structural framing is discussed further in Sec. 6.2 and Chap. 8.

6.2 EARTHQUAKES

Earthquakes may occur anywhere in the United States. Therefore, all buildings should be designed to withstand them. Proper aseismic design should produce buildings capable of surviving minor temblors with no damage. With good systems design, this should be done with no increase in construction cost over that for gravity and wind loads. Also, proper aseismic design should produce buildings capable of surviving major earthquakes without collapsing. Good systems design should be helpful in minimizing the cost of achieving this objective.

The probability of a violent temblor occurring at the same time as a high wind appears to be very small. Hence, building codes generally do not require buildings to be designed for simultaneous occurrence of wind and seismic loads. As a result, the full strength and stiffness provided a building for resistance to seismic loads are also considered available to resist wind loads.

If a strong earthquake should occur, sidesway of buildings is likely to be severer than for winds. As a result, even if structural components are made strong enough to prevent collapse, buildings may suffer considerable damage. Nonstructural components especially may be vulnerable. For example, walls may be stiffer than the structural components but not so strong. Being stiffer, the walls will be subjected to greater forces, which can cause severe cracking of the walls or their collapse. Also, the walls may interfere with planned actions of the structural components and thus cause additional damage. Consequently, aseismic design requires thorough knowledge of structural engineering and building material properties and also calls for exercise of good judgment to save lives and minimize property damage.

Characteristics of Earthquakes

Earthquakes occur because of sudden movements inside the earth, with simultaneous release of tremendous amounts of energy. The location at which the temblor originates is called the *hypocenter*. The point on the surface of the earth directly above the hypocenter is called the *epicenter*.

The shock produces both longitudinal and transverse vibrations in the earth's crust. The shock waves travel at different velocities away from the hypocenter, some traveling through the earth's crust and some along the ground surface. The waves consequently arrive at distant locations at different times. Hence, at points away from the hypocenter, seismic vibrations are a combination of longitudinal, transverse and surface waves. The effects are made even more complicated by reflection of waves from dense portions of the crust and consequent magnification or reduction of vibration amplitudes where waves meet.

Normally, an earthquake starts with faint vibrations of the ground surface, which last only a short time. These usually are followed by severe shock waves, which continue for a longer period. Then, the vibrations gradually vanish. The initial faint vibration registers arrival of the first longitudinal waves. The shocks occur because longitudinal, transverse and surface waves arrive simultaneously.

Movements of the earth at any point during an earthquake may be recorded with seismographs and plotted as seismograms. These diagrams show the variation with time of components of the displacements. Seismograms of earthquakes that have occurred indicate that seismic wave forms are very complex.

Measurements of ground accelerations that occur during a temblor also are important. Newton's law states:

$$F = Ma = \frac{W}{g} a \qquad (6.6)$$

where

F = force, lb

M = mass accelerated

a = acceleration, ft per sec^2

W = weight accelerated, lb

g = acceleration due to gravity = 32.2 ft per sec^2

Hence, inertial forces resisting earthquake accelerations are proportional to those accelerations. Accelerations may be plotted as accelerograms, which show the variation with time of components of the ground accelerations.

Seismic Severity. Several scales are in use for measuring the severity of earthquakes.

A scale commonly used in the United States for indicating seismic intensity is the Modified Mercalli Scale, which is based on subjective criteria. The scale has twelve divisions. The more severe an earthquake is, the higher the number assigned to it. Mercalli intensity I indicates vibrations detected only by sensitive instruments. Intensity V denotes waves felt by nearly everyone. Intensity IX marks occurrence of considerable damage to well designed structures. Intensity XII registers total damage. Thus, the Mercalli scale indicates the severity of an earthquake at a specific location.

Another scale used in the United States is the Richter scale, which measures the magnitude of an earthquake. The scale is based on the maximum amplitude of ground motion and distance of the point of measurement of the amplitude from the epicenter. Richter magnitudes range from zero to 8.9. The smallest values correspond to the smallest Mercalli intensities and the value of 8 approximately to Mercalli intensity XI.

Influence of Ground Conditions. Investigations of earthquake damage indicate that there is a marked difference in the degree of damage in similar structures at different points at the same distance from an epicenter. The difference in damage appears to be due to types of soil at those points. (Sometimes, though, variations in damage may be due to magnification or reduction of vibration amplitudes as a result of wave reflections.)

Soil type affects intensity and wave form of motion. Furthermore, some soils may suffer a loss of strength in a temblor and allow large, uneven settlements of foundations, with large consequent property damage. Not only soils near the surface but also earth deep down may have these effects.

Observations indicate that movements are very much larger in alluvial soils (sands or clays deposited by flowing water) than in rocky areas or diluvial soils (material deposited by glaciers). Behavior of reclaimed land (fills) appears to be even poorer than alluvial soils when subjected to earthquakes. Seismic intensity seems to increase in the following order: hard ground, sand and gravel, sand, clay.

It seems, therefore, that disasters could be averted by not placing buildings on sites with soils that will have large displacements in earthquakes.

Ground Motions. Seismic waves may reach a building site from any direction. The ground motions are vibratory in three dimensions—up and down, back and forth horizontally. A building supported by the ground subjected to an earthquake has to move with the ground and therefore also moves up and down and back and forth horizontally. In accordance with Newton's law [see Eq. (6.6)], the accelerations are accompanied by inertial forces equal to the product of mass being accelerated and the acceleration. The inertial forces act in the same directions as the accelerations of the building.

Consequently, buildings should be designed to resist seismic forces from any direction. These forces are uncontrollable, random variables. Varying in intensity and direction with time, they also are dynamic loads.

Design Loads

Seismic loads can be resolved into vertical and horizontal components. Vertical components, however, usually are of little concern in building design. Buildings are designed for gravity loads with a conservative safety factor and therefore have considerable reserve for resisting additional vertical loads. Also, the added strength and stiffness provided for withstanding high winds is available for resisting earthquakes.

Major damage usually is caused by the horizontal component of the seismic loads. Consequently, buildings should be designed to resist the maximum likely horizontal component. (Note that the horizontal component can be resolved into two perpendicular components for convenience in design and analysis.)

Seismic loads can be determined from the accelerations of the various parts of the building. These motions depend on the ground motions and the dynamic properties of the building.

With the aid of computers, probable seismic design loads can be computed from historical earthquake records and dynamic structural analysis of the building. The calculations, however, are complex and their accuracy may be questionable, because the historical records may not be applicable to the site conditions and future earthquakes may be completely different from previous ones.

Building codes may permit use of an alternative static loading for which structural analysis is much simpler. This loading applies forces to the parts of a building in proportion to their weight.

To begin with, a total lateral force is specified. This load is determined by multiplying the total weight of the building by various coefficients. The coefficients account for the seismic history of the zone in which the building will be erected, the type of structural framing and the dynamic properties of the building.

The static seismic loads are assumed to act horizontally at each floor level. For buildings more than two-stories high, a part of the total lateral load is distributed to each floor in proportion to the weight of building parts attributable to that level. The roof, however, in recognition of the dynamic behavior of buildings under seismic loads, is assigned a force that depends on the building height-width ratio. For one-story and two-story buildings, a uniformly distributed seismic loading may be specified because of their relatively large stiffness.

Response of Structures

Seismic resistance should be an integral part of every building system. As for wind loads,

seismic loads must be transmitted along continuous paths from the various parts of a building to ground strong enough to withstand those loads. In addition, the building should be made stiff enough to keep the amplitude of sidesway within acceptable limits. Furthermore, since the response of a building to seismic loads is a vibratory motion, provision must be made to damp the vibrations through absorption of the energy of motion. For economy, systems design should utilize the lateral-force-resisting system for both wind and earthquake resistance.

Designers should consider the possibilities of different modes of failure in earthquakes and provide against them. The failure modes possible generally are overturning or sliding, as for wind loads; collapse like a house of cards; severe twisting and excessive sidesway. Destructive sway may occur not only because of the magnitude of the seismic forces but also, since they are transient dynamic loads, because of build up of vibrations.

Design Measures

A primary concern in aseismic design should be to transmit seismic loads to ground strong enough to resist them. Structural members provided for this purpose should be strong enough to transmit the imposed forces and should be capable of controlling sidesway. Also, the members should be ductile, so they can absorb large amounts of energy without breaking. Connections between members also should be strong and ductile.

As for wind resistance, many devices, including rigid frames, X bracing and shear walls, may be used to transmit seismic loads to the ground and to resist twisting of the building. Ductile rigid frames, however, generally are advantageous because of large energy-absorption capacity.

Floors and roofs are usually relied on to transmit the lateral forces to the resisting elements. In this role, a floor or roof may act as a diaphragm, or deep horizontal beam. (Horizontal bracing, however, may be used instead.) Diaphragms with openings, for stairs or elevators, should be reinforced around the openings to bypass the horizontal forces.

Overturning and sliding can be resisted, as for

wind, by utilizing the weight of the building and anchoring the building firmly to its foundations. In addition, it is desirable that individual footings, especially pile and caisson footings, be tied to each other to prevent relative movement.

As for wind loads, sidesway can be controlled by proper shaping of a building, arrangements of structural components to resist drift, and selection of members with adequate dimensions and geometry to withstand changes in dimensions. No precise criteria placing limitations on sidesway are available. Some engineers have suggested that, for buildings over 13 stories high and with ratios of height to least base dimension exceeding 2.5, drift in any story should not be more than 0.25% of the story height for wind or 0.5% of the story height for earthquakes (computed for the equivalent static load previously described).

Curtain walls and partitions should be capable of accommodating building movements caused by lateral forces or temperature changes. Connections and intersections should allow for a relative movement between stories of at least twice the drift per story. Also, sufficient separation should be provided between adjacent buildings or between two elements of an irregular building to prevent them from striking each other during vibratory motion.

Structural framing is further discussed in Chap. 8.

SECTIONS 6.1 and 6.2

References

American National Standard Minimum Design Loads for Buildings and Other Structures, American National Standards Institute, 1982.

Uniform Building Code, International Conference of Building Officials, 1988.

Standard Building Code, Southern Building Code Congress International, 1988.

J. Ambrose and D. Vergun, *Design for Lateral Forces*, Wiley, 1987.

C. Arnold and R. Reitherman, *Building Configuration and Seismic Design*, Wiley, 1982.

Words and Terms

Aseismic	Bracing
Braced frame	Damping
Diaphragm	Richter scale
Drift	Overturn
Dyanamic loads	Rigid frame
Earthquake	Shear wall
Gust	Torsion
Modified Mercalli Scale	Uplift

Significant Relations, Functions, and Issues

Influence of building size, form, weight, and location on wind and earthquake effects.

General nature of critical wind and earthquake effects on building components (roof, walls, bracing) and the building as a whole.

Computation of wind and earthquake effects for design.

Basic types and details of lateral bracing systems.

6.3. FIRE

Loss of life, injuries and property damage in building fires in the past have been tragically large. In an effort to curtail these losses, building officials devote far more than half of the usual building codes to fire protection. As a result, owners must spend considerable sums of money to provide fire protection in buildings to meet code requirements. Designers therefore are professionally obligated not only to abide by the word of the law but also by its spirit. Also, obligated to the economic welfare of the owners, designers, in addition, should seek ways to provide life safety in buildings and to avoid or minimize property damage due to fires at least cost to the owners.

An owner pays for fire protection in several ways. Initially, he pays for installation of fire protection when a building is constructed. Then, he pays for maintenance and operation of the fire-protection system. Also, usually as long as he maintains ownership, he pays fire-insurance premiums every year to cover possible fire losses. The last payments may amount to a considerable sum over a long period.

Building designers can help lower those costs by providing fire protection that will secure for the owner lower fire-insurance premiums. This, however, may result in higher construction and operation costs. The design effort nevertheless should aim at optimizing life-cycle costs, the sum of construction, operation, maintenance and insurance costs.

Because of insurance companies' concern with fire protection, they have promulgated many standards for the purpose that are widely used. Many have been adopted by reference in building codes and are specified by government agencies. Generally, insurance-oriented standards, such as those of the National Fire Protection Association and Factory Mutual System, are primarily concerned with avoiding property losses by fire, whereas municipal building codes mainly aim at life safety. Building designers therefore should consider both standards for life safety and those protecting the owners' economic interests by preventing property damage. Standards, however, usually present minimum requirements. Often, public safety and the owners' special needs require more stringent fire protection and emergency measures than those specified in building codes and standards.

The multivolume "National Fire Codes" of the National Fire Protection Association, Quincy, MA 02269, contains more than 200 standards, which are updated annually.

The Factory Mutual Engineering Corporation, Norwood, MA 02062, publishes standards applicable to properties insured by the Factory Mutual System. FM also has available a list of devices it has tested and approved.

Underwriters Laboratories, Inc., 333 Pfingsten Road, Northbrook, IL 60062, makes fire tests in its laboratories and reports the fire resistance found for various types of constructions. UL reports the devices and systems it approves in "Fire Protection Equipment List," which is updated bimonthly and annually. Also, UL lists approved building components in "Building Materials List."

For federal government buildings, requirements of the General Services Administration must be observed.

Many other government agencies also promulgate standards that must be adhered to, even for nongovernmental buildings. Many standards of the federal Occupational Safety and Health Administration, for example, are concerned with life safety in fires. Also, many states have safety codes applicable to commercial and industrial buildings. These codes may be administered by a state Department of Labor, Fire Marshal's office, Education Department or Health Department.

The American National Standards Institute, Inc., also promulgates standards affecting life safety in buildings. In particular, ANSI A117.1, "Specifications for Making Buildings and Facilities Accessible to and Usable by the Physically Handicapped," is applicable to building design for both normal and emergency conditions.

Fire Loads and Resistance Ratings

Fires occur in buildings because they contain combustibles, materials that burn when ignited. The potential severity of a fire depends on the amount and arrangement of these materials.

Combustibles may be present within a building or in the building structure. Contents of a building are related to the type of occupancy, whereas combustibility of structure is related to type of construction. Accordingly, building codes classify buildings by occupancy and construction, as described in Sec. 3.5.

Fire load, measured in pounds per square foot (psf) of floor area, is defined as the amount of combustibles present in a building. Heat content liberated in a fire may range from 7,000 to 8,000 Btu per lb for materials such as paper or wood to more than twice as much for materials such as petroleum products, fats, waxes and alcohol.

Fire load appears to be closely related to fire severity. Burnout tests made by the National Institute of Science and Technology indicate the relationship shown in Table 6.2.

Fire resistance of building materials and assemblies of materials is determined in standardized fire tests. In these tests, temperature is made to vary with time in a controlled manner. Figure 6.4 shows a standard time-temperature curve usually followed. The ability of constructions to withstand fire in these tests is expressed as a fire rating in hours. Fire ratings determined by Underwriters Laboratories, Inc., are tabulated in the UL "Building Materials List."

Building codes classify types of construction in accordance with fire ratings of structural members, exterior walls, fire divisions, fire

Table 6.2. Relation between Weight of Combustibles and Fire Severity[a]

Average Weight of Combustibles, Psf	Equivalent Fire Severity, Hr
5	$\frac{1}{2}$
10	1
20	2
30	3
40	$4\frac{1}{2}$
50	6
60	$7\frac{1}{2}$

[a]Based on National Bureau of Standards Report BMS92, "Classifications of Building Constructions," U.S. Government Printing Office, Washington, D.C.

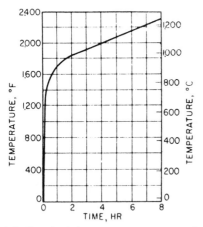

Fig. 6.4. Standard time-temperature curve for fire tests of building components.

separations and ceiling-floor assemblies. Codes usually also specify the ratings required for interior finishes of walls, ceilings and floors. Methods for determining such ratings are described in standards of ASTM, formerly American Society for Testing and Materials, such as E84 and E119. The UL "Building Materials List" also reports such ratings.

Building codes, however, do not relate life-safety hazards directly to fire load. Instead, codes deal with hazards through requirements for interior finishes, ventilation and means of egress in event of fire.

Height and Area Restrictions

To limit the spread of fire and the length of travel of occupants to places of refuge, build-ings may be compartmented horizontally and vertically. Fire-resistant floors and ceilings are used to prevent fire from spreading from story to story. Fire-resistant walls, called *fire walls*, are used to prevent fire from spreading horizontally. Openings in these fire barriers for passage of occupants in normal or emergency circumstances also must be fire protected.

Building codes may restrict building height and floor areas included between fire walls in accordance with potential fire hazards associated with type of occupancy and type of construction. Usually, the greater the fire resistance of the structure the greater the permissible height and floor area. Because of the excellent past record of sprinklers in early extinguishment or control of fires, greater heights and larger floor areas are often permitted when automatic sprinklers are installed.

Classes of Fires

Methods used for extinguishing some burning materials may not be suitable for others. Hence, for convenience in indicating the effectiveness of extinguishing media, such as water, powders, gases or foam, fires may be classified in accordance with the type of combustible material burning. A classification system developed by Underwriters Laboratories, Inc., defines the following four types of fires:

Class A fires. Ordinary combustibles. Extinguishable with water or by cooling or by coating with a suitable chemical powder.

Class B fires. Flammable liquids. Extinguishable by smothering or careful application of a cooling agent.

Class C fires. Live electrical equipment. Extinguishable with a nonconducting medium. A conducting agent can be used if the circuit is interrupted.

Class D fires. Metals, such as magnesium, powdered aluminum and sodium, that burn. Extinguishable by specially trained personnel applying special powders.

6.4 FIRE EXTINGUISHMENT

Writers of building codes and concerned building designers generally take the position that a fire *will* occur in any building and then proceed to consider what can be done about it. For preservation of the building, as well as the safety of the occupants, a major concern is for the rapid extinguishing of the fire. The means for achieving this vary, depending on the building form and construction, the occupancy and the nature of the combustible materials that fuel the fire. This section discusses some of the ordinary means for extinguishing building fires.

Sprinklers

Automatic sprinklers have proven very effective in early extinguishment of fires. In fact, that is their main purpose; but they are also useful in curtailing the spread of fire and hot gases by cooling the environment around a fire. Sprinklers are suitable for extinguishing Class A fires. Sprinklers also may be used for some Class B and Class C fires.

A sprinkler system basically consists of fire detectors, water for extinguishing fires, heads for discharging the water when actuated by the detectors, and piping for delivering the water to the heads. Heads should be located at ceiling and roof levels to completely cover the interior of the building. Intervals between heads on the piping should be small enough to provide desired concentration of water on every square foot of floor.

Requirements governing design and installation of sprinkler systems are given in building codes and in standards of the National Fire Protection Association and Factory Mutual System. Generally, the requirements of the local code will govern, but designers should check with the owner's insurance carrier to determine if other standards may also apply. If such standards are ignored, the owner may have to pay higher than necessary fire-insurance premiums.

Standpipes

A standpipe is a water pipe within a building to which hoses may be attached for fire fighting. Standpipes are required in buildings in which fires may occur too high to be reached by ground-based fire-department equipment. These pipes also may be necessary in low buildings with large floor areas, the interiors of which may be difficult to reach with hose streams from the outside.

Sprinklers and standpipes are further discussed as part of the plumbing system in Sec. 9.4.

Chemical Extinguishing Systems

Small fires in buildings in ordinary materials, such as paper, wood and fabrics, when first starting, often may be rapidly extinguished with water, propelled by compressed gases, from hand-held extinguishers. Building codes may require such extinguishers to be located at convenient places in buildings. Occupants should be taught to operate the extinguishers. There is a risk in their use, however, in that the attempt to fight a fire with an extinguisher may delay notification of the fire department or other better-equipped fire fighters of the presence of the fire.

Instead of plain water for extinguishing fires, chemicals or water plus chemicals may be used. Applied by automatic sprinklers, hoses, hand-held extinguishers, portable wheeled equipment or larger devices, chemicals may be desirable or necessary for fires in certain materials.

Foams. For flammable liquids, such as gasoline, a foamed chemical, mostly a conglomeration of air- or gas-filled bubbles, may be useful. Three types are suitable for fire extinguishment: chemical foam; air, or mechanical, foam; and high-expansion foam. Chemical foam is formed by the reaction of water with powders. Usually, sodium bicarbonate and aluminum sulfate are used, forming carbon-dioxide bubbles. Air, or mechanical, foam is produced by mixing water with a protein-based chemical concentrate. High-expansion foam is generated by passage of air through a screen constantly wetted by a chemical solution, usually with a detergent base. The volume of foam produced by this method relative to the volume of water used is a great many times the volume produced by the other methods. The foams extinguish a fire by smothering it and cooling the surface.

Carbon Dioxide. For flammable liquids or live electrical fires, carbon dioxide may be useful. It is also suitable for equipment fires, such as those in gasoline or diesel engines, because the gas requires no cleanup. Stored in containers under pressure, it is immediately ready for discharge when a valve is opened. Heavier than air, the gas tends to drop into the base of a fire and extinguish it by reducing the oxygen concentration.

Halon 1301. For use in the same circumstances as carbon dioxide, bromotrifluoromethane ($CBrF_3$), or Halon 1301, acts much faster. This gas also requires no cleanup. It extinguishes fires by interfering with the chain reaction necessary to maintain combustion.

Dry Chemicals. For Class B and C fires, dry chemicals, such as sodium bicarbonate, may be suitable. They tend to extinguish fires by breaking the chain reaction for combustion. When dry chemicals are used, cleanup after a fire may be difficult.

Dry Powders. For combustible metals, dry powders, different from the dry chemicals previously mentioned, usually are the most suitable extinguishing agent. Specific metals require specific dry powders. Fires in metals should be fought only by properly trained personnel.

6.5. EMERGENCY EGRESS

For life safety in buildings in event of fire or other emergencies, provisions must be made for safe, rapid egress of occupants, at least from the dangerous areas and preferably also from the buildings. The escape routes must be fire protected and smoke free to allow safe passage of occupants.

An *exit* is a means of egress from the interior of a building to an open exterior space beyond the reach of a building fire. The means of egress may be provided by exterior door openings and enclosed horizontal and vertical passageways.

Section 6.3 points out the desirability of using fire walls to compartment buildings, to limit the spread of fire and the length of travel to places of refuge. It is also necessary within compartments to use on floors, ceilings and walls interior finishes that will not spread flames.

In addition, structural members should have sufficiently high fire ratings to prevent collapse, for a few hours at least. The objectives of this are to allow all occupants to be evacuated and to give fire fighters time to extinguish the fire. If structural members are inadequate for the purpose, they may be fire protected with other materials. For example, beams and columns may be encased in concrete, enclosed with plaster, gypsum blocks or gypsumboard, or sprayed with insulating material.

Section 6.4 discusses the use of automatic sprinklers to extinguish fires as soon as they start and to cool surrounding areas. Also, it is important, as soon as fire is detected, to sound an alarm and notify the fire department. In addition, a communications system should instruct occupants on the evacuation procedure to be followed or other precautionary measures.

There is great danger of panic in emergency situations. Panic, however, seldom develops if occupants can move freely toward exits that they can see clearly, that are within a short distance and that can be reached by safe, unobstructed, uncongested paths. Thus, the objective of life-safety design should be to provide such rapid, safe egress from all areas of buildings that will preclude development of panic. Moreover, more than one path to safety should be provided in case one safe means of escape becomes unavailable. All paths must be accessible to and usable by handicapped persons, including those in wheelchairs, if they may be occupants.

To permit prompt escape of occupants from danger, building codes specify the number, size, arrangement and marking of exit facilities, in addition to other life-safety measures. The requirements depend on the types of occupancy and construction.

Generally, building codes require a building to have at least two means of egress from every floor of a building. These exits should be remote from each other, to minimize any possibility that both may become blocked in an emergency.

Codes usually also specify that exits and other vertical openings between floors of a building be fire protected, to prevent spread of fire, smoke or fumes between stories.

In addition, codes limit the size of openings

Table 6.3. Maximum Sizes of
Openings in Fire Walls

Protection of adjoining spaces	Max area, sq ft	Max dimension, ft
Unsprinklered	120[a]	12[b]
Sprinklers on both sides	150[a]	15[a]
Building fully sprinklered	Unlimited[a]	Unlimited[a]

[a]But not more than 25% of the wall length or 56 sq ft per door if the fire barrier serves as a horizontal exit.
[b]But not more than 25% of the wall length.
Based on New York City Building Code.

Table 6.4. Typical Occupant Load
Requirements for Buildings

Occupancy	Net floor area per occupant, sq ft
Bowling alleys	50
Classrooms	20
Dance floors	10
Dining spaces (nonresidential)	12
Garages and open parking structures	250
Gymnasiums	15
Habitable rooms	140
Industrial shops	200
Institutional sleeping rooms	
Adults	75
Children	50
Infants	25
Kindergartens	35
Libraries	25
Offices	100
Passenger terminals or platforms	1.5C[a]
Sales areas (retail)	
First floor or basement	25
Other floors	50
Seating areas (audience) in places of assembly	
Fixed seats	D[b]
Movable seats	10

[a]C = capacity of all passenger vehicles that can be unloaded simultaneously.
[b]D = number of seats or occupants for which space is to be used.
Based on New York City Building Code.

in fire walls (see Table 6.3). Furthermore, openings must be fire protected. For example, a door used for an opening in a fire wall should be a fire door, one that has a fire rating commensurate with that of the wall, as required by the building code.

Required Exit Capacity

Means of egress in event of fire or other emergencies should have sufficient capacity to permit rapid passage of the anticipated number of escapees. This number depends on a factor called the occupant load.

Occupant load of a building space is the maximum number of persons that may be in the space at any time. Building codes may specify the minimum permitted capacity of exits in terms of occupant load, given as net floor area, sq ft, per person, for various types of occupancy (see Table 6.4). In such cases, the number of occupants per space can be computed by dividing the floor area, sq ft, by the specified occupant load.

The occupant load of any space should include the occupant load of other spaces if the occupants have to pass through that space to reach an exit.

With the occupant load known, the required opening width for exits can be determined by dividing the number of occupants per space by the capacity of the exit.

Capacity of exits is measured in units of 22 in. of width. (Fractions of a unit of width less than 12 in. should be ignored, but 12 in. or more added to a full unit may be counted as one-half unit.) Building codes may specify the

maximum design capacity of an opening as the number of persons per 22-in. unit, for various types of occupancy (see Table 6.5).

When occupant load is divided by unit capacity to determine the minimum required exit width, a mixed fraction may result. In such cases, the next larger integer or integer plus one-half should be used to determine the exit dimensions.

Building codes, however, also specify a minimum width for exits (see Table 6.5) and may require at least two separated exits. These requirements govern. Generally, building codes set the minimum width of corridors at 44 in. and exit door openings at 36 in. (See also Sec. 16.2.)

Example. Determination of Door Width

An office has 20,000 sq ft of open floor area. The building code requires at least two exits,

Table 6.5. Capacity of Exits, Persons per 22-in. Unit of Exit Width

Occupancy type	To outdoors at grade	Other doors	Min corridor width, in.
High hazard	50	40	36
Storage	75	60	36
Mercantile	100	80	44
Industrial	100	80	44
Business	100	80	44
Educational	100	80	66
Institutional			
For detention	50	40	44
For handicapped	30	30	96
Hotels, motels,			
apartments	50	40	44

	From assembly place	From safe area	
Assembly			44
Theaters	50	100	
Concert halls	80	125	
Churches	80	125	
Outdoor structures	400	500	
Museums	80	125	
Restaurants	50	100	

each protected by 2-hr fire doors. The exits lead to stairways. How wide should each door opening be?

Table 6.4 gives the occupant load for offices as 100 sq ft per occupant. Therefore, the space may be occupied by 20,000/100, or 200 persons. (If the designer knows that the owner plans to employ more than 200 persons in that office area, calculations should be based on the actual number to be employed.) Table 6.5 gives the allowable exit capacity per unit for business occupancies as 80 persons. The number of units of width required then is 200/80, or 2.5 units. If these are divided equally into two openings, each exit would be 2.5/2, or 1.25 units wide. Width of each opening required, therefore, is 1 X 22 + 12 = 34 in. Use the minimum permitted opening of 36 in.

Travel Distance and Dead-End Limits

To insure that occupants will have sufficient time to escape from a dangerous area, building codes limit the travel distance from the most remote point in any room or space to a door that opens to an outdoor space, stairway or exit passageway. The maximum distance permitted depends on the type of occupancy and whether the space is sprinklered. For unsprinklered spaces, for example, maximum permitted travel may range from 100 ft for storage and institutional buildings to 150 ft for residential, mercantile and industrial occupancies. For sprinklered spaces, maximum permitted travel may range from 150 ft for high-hazard and storage buildings to 300 ft for businesses, with 200 ft usually permitted for other types of occupancy.

Lengths of passageways or courts that lead to a dead end also are restricted or prohibited (for high-hazard occupancies). For example, a code may set the maximum length to a dead end as 30 ft for assembly, educational and institutional buildings, 40 ft for residential buildings and 50 ft for all other occupancies, except high hazard.

Location of Exits

All exits and access facilities should be placed so as to be clearly visible to occupants who may have to use them, or their locations should be clearly marked. If an exit is not immediately accessible from an open floor area, a safe continuous passageway should be provided directly to the exit. The path should be kept unobstructed at all times. Furthermore, it should be so located that occupants will not have to travel toward any high-hazard areas not fully shielded.

Types of Exits

Building codes generally indicate what types of facilities may qualify as exits. These usually include:

Corridors—enclosed public passageways, which lead from rooms or spaces to exits. Minimum floor-to-ceiling height is 7 ft 6 in., although 7 ft may be permitted for short stretches. Minimum width depends on type of occupancy (see Table 6.5). Building codes may require subdivision of corridors into lengths not exceeding 300 ft for educational buildings and 150 ft for institutional buildings. The subdivision should be accomplished with noncombustible partitions incorporating smoke-stop doors. Codes also may require the corridor enclosures to have a fire rating of 1 or 2 hrs.

Exit Passageways—horizontal extensions of vertical exits, or a passage leading from a yard or court to an outdoor space. Minimum floor-to-ceiling height is the same as for corridors. Width should be at least that of the vertical exit. Building codes may require the passageway enclosures to have a 2-hr fire rating. A street-floor lobby may serve as an exit passageway if it is sufficiently wide to accommodate the occupant load of all contributing spaces on the lobby floor.

Exit Doors—doors providing access to streets (these doors need not have a fire rating) and doors to stairs and exit passageways ($\frac{3}{4}$-hr fire rating). (See also Sec. 16.2.)

Horizontal Exit—access to a refuge area. The exit may consist of doors through walls with 2-hr fire rating, balcony offering passage around a fire barrier to another compartment or building, or a bridge or tunnel between two buildings. Doors should have a fire rating of $1\frac{1}{2}$ hr, except that doors in fire barriers with 3- or 4-hr fire rating should have a $1\frac{1}{2}$-hr rated door on each face of the fire division. Balconies, bridges and tunnels should be at least as wide as the doors opening on them and their enclosures or sides should have a fire rating of 2 hr. Exterior-wall openings below any open bridge or balcony, or within 30 ft horizontally of such construction should have $\frac{3}{4}$-hr fire protection.

Interior Stairs—stairs within a building that serve as an exit. Building codes generally require such stairs to be constructed of noncombustible materials but may except one-story or two-story, low-hazard buildings. Stair enclosures should have a 2-hr fire rating, except in low dwellings, where no enclosure may be required. (See also Sec. 14.2.)

Exterior Stairs—stairs that are open to the outdoors and that serve as an exit to ground

level. Building codes limit the height of such stairs, often to not more than 75 ft or six stories. The stairs usually should be constructed of noncombustible materials and topped with a fire-resistant roof. Openings in walls within 10 ft of the stairs should have $\frac{3}{4}$-hr fire protection.

Smokeproof Tower—a continuous fire-resistant enclosure protecting a stairway from fire or smoke in a building. Passage between building and tower should be provided on every floor by vestibules or balconies directly open to the outdoors. Enclosures should have a 2-hr fire rating. Access to the vestibules or balconies and entrances to the tower should be through doorways at least 40 in. wide, protected by self-closing fire doors. The vestibules or balconies should be at least as wide and long as the required doorway width.

Escalators—moving stairs. These may be used as exits instead of interior stairs if they meet applicable requirements of such stairs and if they move in the direction of exit travel or stop gradually when an automatic fire detection system signals a fire.

Moving Walks—horizontal or inclined conveyor belts for passengers. These may be used as exits if they meet the requirements for exit passageways and move in the direction of exit travel or stop gradually when an automatic fire detection system signals a fire.

Fire Escapes—exterior stairs, with railings, that are open to the outdoors, except possibly along a building exterior wall. These formerly were permitted but generally no longer are.

Elevators are not recognized as a reliable means of egress in a fire.

Refuge Areas

A refuge area is a space safe from fire. The refuge should be at about the same level as the areas served and separated from them by construction with at least a 2-hr fire rating. Fire doors to the refuge area should have at least a $1\frac{1}{2}$-hr fire rating.

Size of the refuge area should be adequate for the occupant load of the areas served, in addition to its own occupant load, allowing 3 sq ft of open space per person (30 sq ft per person for hospital or nursing-home patients). There should be at least one vertical exit and, in locations over 11 stories above ground, one elevator for evacuation of occupants from the refuge area.

6.6. FIRE PROTECTION

Preceding sections have considered two of the primary concerns with regard to building fires: the rapid control and extinguishing of the fire and the egress of the building's occupants in a safe manner. There are many other factors relating to potential damage or injury from fires that may have some bearing on the building design. This section discusses some of the other major concerns for general protection from the hazards of fires.

Fire-Detection Devices

The next best thing to preventing a fire from occurring is to detect it as soon as it starts or in an incipient stage. Many devices are available for early detection of fires. When a fire occurs near one, the device can perform automatically serveral important operations, such as sound an alarm locally; notify a central station and the fire department; open automatic sprinklers; start and stop fans, industrial processes, escalators and elevators; shut fire doors.

Underwriters Laboratories, Inc. (UL) has tested and reported on many fire-detection devices. On approving a device, UL specifies the maximum distance between detectors giving area coverage. Often, however, building conditions may make closer spacing advisable.

Detectors may be classified into five types, depending on method of operation: fixed-temperature, rate-of-rise, photoelectric, combustion-products and flame.

Fixed-Temperature Detectors. These devices are set to signal a fire when one element is subjected to a specific temperature. There may, however, be a delay between the time when ambient (room) temperatures rise beyond this temperature and the element attains it. For example, ambient temperature may reach about 200°F by the time the detector reaches its rated temperature of 135°F. Several different types of fixed-temperature detectors are available. They usually are designed to close an electric circuit when the rated temperature is reached.

Rate-of-Rise Detectors. Operating independently of heat level, these detectors signal a fire when temperature rises rapidly. For example, a detector may operate when it registers a temperature rise at the rate of 10°F or more per min. Rate-of-rise detectors do not have the disadvantage of thermal lag as do fixed-temperature devices. Several different types of rate-of-rise detectors are available.

Photoelectric Detectors. These are actuated when visibility is decreased by smoke. In a photoelectric detector, a light ray is directed across a chamber so as not to strike a photoelectric cell. If smoke particles collect in the chamber, they deflect the ray so that it impinges on the cell, thus causing an electric current to flow in a warning circuit. Photoelectric detectors are useful where a potential fire may generate considerable smoke before much heat develops or flames can be observed.

Combustion-Products Detectors. As the name implies, these devices signal a fire when they detect products of combustion. They may be ionization or resistance-bridge types. The ionization type employs gases ionized by alpha particles from radioactive material to detect a change in the composition of ambient air. The resistance-bridge type operates when combustion products change the electrical impedance of an electric bridge grid circuit. Both types are useful for giving early warning of a fire, when combustion products are still invisible.

Flame Detectors. These devices signal a fire when they detect light from combustion. One type detects light in the ultraviolet range, whereas another type detects light in the infrared range.

Smoke and Heat Stops and Vents

A fire gives off heat and often a considerable amount of smoke. Both products can build up rapidly to lethal concentrations and spread the fire, if confined within the building. Consequently, in addition to immediate application of large quantities of water or chemicals to smother the fire or cool the fire source and surrounding space, speedy removal of the heat and smoke from the building is necessary. Methods of doing this depend on the size of buildings and whether they are one story high or multistory.

Small buildings can release heat and smoke through open or broken windows or through roof vents.

Large, one-story buildings, such as those used for manufacturing and storage, may have interior areas cut off by fire walls or too far from exterior walls for effective venting through windows. Often, such buildings are impracticable to vent around the perimeter because they are windowless. Hence, for large, one-story buildings, the only practical method for removing heat and smoke from a fire usually is through openings in the roof. (Venting is desirable as an auxiliary safety measure even when buildings are equipped with automatic sprinklers.)

Generally, smoke and heat should be vented from large, one-story buildings by natural draft. The discharge apertures of the vents should always be open or otherwise should open automatically when a fire is detected. Vents that may be closed should be openable by fire fighters from the outside. Venting may be done with monitors (openable windows that project above the main roof), continuous vents (narrow slots with a weather hood above), unit-type vents or sawtooth skylights.

As a guide, Table 6.6 gives an approximate ratio for determination of vent area. In deciding on the area to be used, designers should consider the quantity, size, shape and combustibility of building contents and structure. They should provide sufficient vent area to prevent dangerous accumulations of smoke during the time necessary for evacuation of the floor area to be served, with a margin of safety to allow for unforeseen situations.

Table 6.6. Approximate Areas and Spacings for Roof Vents

Type of contents	Ratio of vent area to floor area	Maximum spacing, ft
Low heat release	1:150	150
Moderate heat release	1:100	120
High heat release	1:30	75

Unit-type vents come in sizes from 4 × 4 ft to 10 × 10 ft. The maximum distance between vents usually should not exceed the spacing given in Table 6.6. Generally, a large number of closely spaced, small vents is better than a few large vents. The reason for this is that with close spacing the probability is greater that a vent will be close to any location where a fire may occur.

In multistory buildings, only the top story can be vented through the roof. Often, the windows are normally closed, and even when openable, they are not operable automatically. Consequently, heat and smoke in lower-story fires must be collected at the source, ducted through the stories above and discharged above the roof. Shafts should be provided for this purpose.

Each smoke shaft should be equipped with an exhaust fan. In buildings with air-conditioning ducts, return-air ducts, which will pick up smoke, should be controlled with dampers to discharge into a smoke shaft when smoke is detected. A smoke detector installed at the inlet to each return-air duct should actuate the smoke exhaust fan and the dampers. When smoke is detected, the smoke exhaust fan should start and supply-air blowers should stop automatically. Manual override controls, however, should be installed in a location that will be accessible under all conditions. Smoke-detector operation should be supervised from a central station.

To prevent spread of fire from one part of a building to another and to confine the smoke and heat of a fire to one area from which they can be exhausted safely, building codes require compartmentation of a building by fire divisions. The floor area permitted to be included between fire divisions depends on types of occupancy and construction and whether the building is sprinklered.

A *fire division* is any construction with the fire-resistance rating and structural stability under fire conditions required for the types of occupancy and construction of the building to bar spread of fire between adjoining buildings or between parts of the same building on opposite sides of the division.

A fire division may be an exterior wall, fire window, fire wall, fire door, floor, ceiling or firestop.

A *firestop* is a solid or compact, tight closure incorporated in a concealed space in a building to retard spread of flames or hot gases. Every partition and wall should be firestopped at each floor level, at the top-story ceiling level and at the level of support for roofs. Also, every large unoccupied attic space should be subdivided by firestops into areas of 3,000 sq ft or less. In addition, any large plenum or space between a ceiling and floor or roof should be subdivided. Firestops extending the full depth of the space should be placed along the line of supports of the structural members and elsewhere to enclose spaces between ceiling and floor with areas not exceeding 1,000 sq ft nor 3,000 sq ft when between ceiling and roof.

For life safety of occupants during evacuation from multistory buildings through smokeproof towers, it is desirable to pump fresh air into the towers to pressurize them. Maintenance of a higher-than-normal air pressure is intended to prevent smoke from entering the towers through openings in the enclosure that may not be completely closed. The procedure, however, has some disadvantages. One is that the pressure may make opening doors to leave the tower difficult. Another is that in many buildings standpipe connections are located in the towers and fire fighters have to open the door to the fire floor to move a hose toward the fire. This disadvantage can be overcome by placing the hose valves within the building at the tower doors, if permitted by the building code, while leaving the standpipe, as customary, in the smokeproof tower.

Systems Design for Fire Protection

Sections 6.3–6.6 described the elements necessary for life safety and protection of property in event of fire or other emergencies in buildings. In summary, these elements are:

1. Limitation of potential fire loads, with respect to both combustibility and ability to generate smoke and toxic gases.
2. Compartmentation of buildings by fire divisions to confine a fire to a limited space.
3. Provision of refuge areas and safe evacuation routes to outdoors.
4. Prompt detection of fires, with warning to occupants who may be affected and notification of presence of fire to fire fighters.
5. Communication of instructions to occupants as to procedures to adopt for safety, such as to stay in place, proceed to a designated refuge area or evacuate the building.
6. Early extinguishment of any fire that may occur, primarily by automatic sprinklers but also by trained fire fighters.
7. Provision, for fire fighting, of adequate water supply, appropriate chemicals, adequate-sized piping, conveniently located valves, hoses, pumps and other equipment necessary.
8. Removal of heat and smoke from the building as rapidly as possible without exposing occupants to them, with the HVAC system, if one is present, assisting in venting the building and by pressurizing smokeproof towers, elevator shafts and other exits.

Emergency Power

In addition, not discussed before, a standby electric power and light system should be installed in large buildings. The system should be equipped with a generator that will start automatically on failure of normal electric service. The emergency electric supply should be capable of operating all emergency electric equipment at full power within 60 seconds of failure of normal service. Emergency equipment to be operated includes lights for exits, elevators for fire fighters, escalators and moving walks designated as exits, exhaust fans and pressurizing blowers, communications systems, detectors, and controls needed for fire fighting and life safety during evacuation of occupants.

Emergency Elevators

The vertical transportation system should make available at least one elevator for control by fire fighters, to give them access to any floor from the street-floor lobby. Elevator controls should be designed to preclude elevators from stopping automatically at floors affected by fire. In the past, lives have been lost when fires damaged elevator signaling devices, stopping elevators with passengers at the fire floor and opening the elevator doors.

Systems Design for Life Safety

For maximum protection of life and property in event of fire or other emergency at least cost, all the preceding elements should be integrated into a single life-safety system so that they work in unison to meet all objectives.

Some of the elements may be considered permanent. They require no supervision other than that necessary for ordinary maintenance. These elements include the various fire divisions, structural members and exits. With the systems design approach, cost of the fire-resistance functions of these building components can be offset by assigning them additional functions, where feasible.

Other elements, such as detectors, automatic sprinklers and the emergency HVAC system, require at least frequent observation of their condition, if not constant supervision.

Supervision can be efficiently provided by personnel at a properly equipped control center, which may include an electronic computer, supplemented by personnel performing scheduled maintenance. The control center can continuously monitor alarms, gate valves, temperatures, air and water pressures and perform other pertinent functions. In addition, in emergencies, the control center can hold two-way conversations with occupants throughout the building and notify the fire and police departments. Furthermore, the control center can dispatch investigators to sources of potential trouble or send maintenance personnel to make emergency repairs, when necessary.

For more efficient operation of the total building system and greater economy, the control center can also be assigned many other functions. The center can become the key element of a system that, for example:

1. Meets life-safety objectives
2. Warns of intruders
3. Controls HVAC to conserve energy
4. Switches on emergency power
5. Turns lights on and off
6. Communicates with building occupants, when necessary
7. Schedules building maintenance and repair
8. Puts elevators under manual control for emergencies

SECTIONS 6.3 TO 6.6

References

Uniform Building Code, International Conference of Building Officials, 1988.
Standard Building Code, Southern Building Code Congress International, 1988.
Life Safety Code, National Fire Protection Association, 1988.
Fire Protection Handbook, NFPA

Words and Terms

Class of fire, A to D
Combustible
Egress in emergency
Extinguishment
Exit and occupant load
Fire division
Fire load
Fire severity, in hours
Occupant load
Refuge area
Smoke and heat stops
Smokeproof tower
Sprinklers
Standpipe
Venting
Width of exits, individual and total

Significant Relations, Functions and Issues

Fire/cost relations: long term benefits of design to lower risk and reduce insurance premiums.
Rating of building construction components and systems for fire resistance.
Building height and floor area restrictions related to occupancy and fire resistance of construction.
Means for fire extinguishing related to type (class) of fire.

Exit requirements related to occupant load.
Fire control by use of stops, divisions and vents.

6.7. SECURITY

Means for prevention of theft and vandalism in buildings after the owners occupy them should be an integral part of the building system.

Provision should be made from the start of design for control of access to buildings and to specific areas, if desired by the owner. For some buildings, tight security may be essential for certain sections, such as rooms housing valuable materials or expensive equipment, like a large computer.

For detection of intruders, television monitors and intrusion alarms may be installed. For control of access, doors may be equipped with locks operated by keys or by devices that read identification cards, hand prints or voice vibrations. For protection of valuables, thick steel safes may be provided.

For a small building, alarm systems may be rigged to sound an alarm and to notify the police the instant an intruder attempts to enter the locked building or security area.

For a large building, guards are needed to monitor the various devices or to patrol the building. Therefore, a control center should be provided for observation purposes. In addition, communications should be established between the center, various parts of the building and police and fire departments. Also, a guard room should be provided for guards not on duty and for files and lockers.

With the use of electronic devices, security systems can be installed to do the following:

1. Sound an alarm when an intruder attempts to enter.
2. Identify the point of intrusion.
3. Turn on lights.
4. Display the intruder on television and record observations on video tape.
5. Call police automatically.
6. Restrict entry to specific areas only to properly identified personnel and at permitted times.
7. Change locks automatically.

Costs for security can be cut if the systems-design approach is used to combine security measures with other controls. For example, the control center and its equipment, including a computer, if desired, can be used not only for security but also for HVAC controls and fire-safety equipment. In addition, personnel, television monitors and sensors as well as electric wiring can share tasks related to security, HVAC and fire detection, extinguishment and communications.

6.8. BARRIER-FREE ENVIRONMENTS

Ordinary building safety concerns are based primarily on an assumption that the building occupants are able-bodied and in full possession of the faculties of a normal adult. It is assumed that occupants can walk (use stairs), see (read exit signs), hear (be alerted by fire alarms), use their hands (open doors), and generally function adequately in panic situations (not retarded, not very young, not marginally senile, etc.). However, in almost all types of building occupancies there will be some persons who do not have all of these faculties intact. In recent times, the building codes have been made to recognize this situation, and most buildings must now be designed with some recognition of the need for barrier-free environments. Barriers are anything that interfere with use of the building—in particular, devices and components involving entrances, exits, warning systems, rest rooms, and general vertical and horizontal movement through the building.

Special efforts to create barrier-free environments must be made for buildings that have a large number of occupants with special needs: day care centers, convalescent hospitals, and health care facilities, for example. However, the same accommodations are also generally required for any building that involves use by the public or houses employees. These concerns may be principally addressed to usage and access but at some level may involve safety when hazardous conditions are at issue.

It is virtually impossible to produce a physical environment that is optimally accommodating to a range of people that includes normal, healthy adults, small children, enfeebled and easily disoriented elderly persons, and persons who are blind or wheelchair confined. In some instances, what is best for one group is bad for another. The elaborate facilities required to provide access and egress for persons in wheelchairs may in effect represent confusing barriers for blind, very young, or elderly persons. Where occupancy is more specific, some level of optimization may be feasible, but where the public as a whole must be facilitated, considerable compromise must be anticipated.

Building codes, design practices, and the development of construction components—notably hardware, signage, elevator controls, paving and floor finishes—are steadily being designed with a concern for a wider range of occupant capabilities. Hard design data are being developed from research and the experience deriving from experimentation and design implementations.

SECTIONS 6.7 AND 6.8

References

P. Hopf, *Handbook of Building Security Planning and Design*, McGraw-Hill, 1979.

P. Hopf and J. Raeber, *Access for the Handicapped*, Van Nostrand Reinhold, 1984.

M. Valins, *Housing for Elderly People*, Van Nostrand Reinhold, 1987.

Specifications for Making Buildings and Facilities Accessible to and Usable by the Physically Handicapped, ANSI A117.1, American National Standards Institute, New York.

Words and Terms

Detection and alarm systems
Entrance control
Handicapped access
Selective reduction of barriers

Significant Relations, Functions, and Issues

Access control related to degree of security required.
Need for entrance by controlled means.
Need for reduction of barriers and hazards for selected groups of persons with diminished faculties.

6.9. TOXIC MATERIALS

There is a great range of materials used for building construction. Some materials are used essentially in raw, natural form as in the case of wood used for structural purposes. In most cases, however, building products are produced from synthesized, processed, materials. For example, wood is often used as an ingredient in a synthesized material for paper, cardboard, and particleboard products. In some cases wood is also processed by being impregnated or coated with materials, thus involving a composite material in its finished form.

General experience together with extensive medical research has produced a long list of potentially dangerous materials, posing the possibility of sickness, injury or death upon exposure to them. Some cases are long-standing as in the case of lead, which has virtually been eliminated as an ingredient in paints. More recent cases involve construction products containing asbestos, formaldehyde, and chlorine. Publicity from legal actions and the work of advocacy groups has brought pressure on manufacturers, builders, designers, and the administrators of building codes to respond by restricting, eliminating, or otherwise controlling the use of such materials.

Danger may be present merely in exposure to some toxic materials on a continuing basis. Thus, when many house paints were lead-based, the chipping and flaking of the painted surfaces over time led to an accumulation of particles containing lead that sometimes were picked up and ingested by the building occupants. Another danger is that occurring during a fire when products of combustion may include highly toxic materials—particularly, lethal gases. These have been the main cause of death in most fires in recent times, the major culprit being various plastic materials used for furnishings, decorations, and building construction products such as piping and insulation.

Although danger to building occupants is a major design concern, potential danger to construction and maintenance workers also should be investigated to determine the need for modification or elimination of hazardous building products. One such product is asbestos, the hazard of which was dramatized by massive law suits brought by workers. Only as a secondary effect did the public become alarmed about the hazards represented by the dormant presence of the material in many existing buildings.

As is usually the case, the original reason for

use in construction of a particular material is positive. For example, one factor that led to widespread use of asbestos was its high resistance to fire. Thus, an optimized design process, with fire resistance as a major value, could easily serve as justification for use of the material. Add the other plus factors for use of the inert mineral material (water-resistive, non-rotting, etc.) and the result (as it actually developed) quickly produced quite popular, widespread use of the material. Only much later did the danger of lung infection from ingestion of asbestos fibers become evident.

Both public awareness and industry caution concerning liability are steadily growing in this area. This will hopefully both rectify some of the errors of the past and allow for some confidence in the use of new products. However, development of new materials and products and the slow feedback of medical research results call for considerable restraint in acceptance of unproven items for building construction. This is unavoidably inhibiting to creative, pioneering designers, but is an ethical issue of major proportions.

6.10. CONSTRUCTION SAFETY

Pressures brought by trade unions and various advocacy groups have greatly increased concerns for the reduction of hazards during the construction process. This form of pressure—resulting in legal actions and the creation of legislation and agencies for enforcement—falls most heavily on manufacturers and contractors, adding to the overhead expense for various types of work. As this causes some shifts in the relative cost of certain types of construction, such influence bears on designers who make basic choices of materials, products, and entire systems.

Direct-cost factors involving required safety measures are routinely reflected in unit prices used for cost estimating. Less easy to deal with are more subtle effects such as the general reluctance of workers or contractors to deal with some forms of construction because of the complexity or general annoyance of complying with the actions or documentations required because of safety requirements. The latter can in

effect sometimes result in a form of boycott, which may be quite regional or only short-lived, but can have major influence in bidding on particular forms of construction. This can be a major factor in establishing what is defined as "local practice".

GENERAL REFERENCES AND SOURCES FOR ADDITIONAL STUDY

These are books that deal comprehensively with several topics covered in this chapter. Topic-specific references relating to individual chapter sections are listed at the end of each individual section.

American National Standard Minimum Design Loads for Buildings and Other Structures, ANSI A58.1-1982, American National Standards Institute, New York, 1982.

Uniform Building Code, International Conference of Building Officials, Whittier, CA, 1988 (new editions every three years).

Standard Building Code, Southern Building Code Congress International, Birmingham, AL.

Life Safety Code, NFPA 101, National Fire Protection Association, Quincy, MA, 1988.

J. Lathrop, *Life Safety Code Handbook*, National Fire Protection Association, Quincy, MA, 1988.

J. Ambrose and D. Vergun, *Design for Lateral Forces*, Wiley, New York, 1987.

P. Hopf, *Handbook of Building Security Planning and Design*, McGraw-Hill, New York, 1979.

P. Hopf and J. Raeber, *Access for the Handicapped*, Van Nostrand Reinhold, New York, 1984.

EXERCISES

The following questions and problems are provided for review of the individual sections and the chapter as a whole.

Sections 6.1 and 6.2

1. Why should lateral loads on buildings be treated as dynamic loads? Why might dynamic loads have severer effects on buildings than static lateral loads of the same magnitude?

2. The walls of a building face north-south and east-west. The maximum wind may blow from any direction. Explain why the building should be designed to withstand full design wind pressures against the north

and south walls and also separately against the east and west walls.

3. Wind pressure on a building 32.8 ft (10 m) above ground is 20 psf. What is the pressure 240 ft above grade if the building is located in the center of a large city?

4. A factory building, 30 ft high, is 20 × 100 ft in plan. For wind pressure of 20 psf, what is the lateral wind force on each wall?

5. At what value should the mean recurrence interval be taken for design of a permanent ordinary building?

6. A symmetrical building 40 ft wide weighs 200 tons. Lateral forces may total 100 tons and their resultant is 40 ft above grade. If the building relies only on its weight for stability, can it be considered safe against overturning by the lateral forces. Justify your answer.

7. A 30 × 60-ft roof weighs 10 tons and relies for stability on its own weight. If basic wind pressure on the roof may average 40 psf and the pressure coefficient is −0.5, is the roof stable? Justify your answer.

8. Wind velocity is measured at 50 mph at a station 32.8 ft (10 m) above ground. The station is located in rough, hilly terrain.
   (1) What would the velocity have been 240 ft above ground?
   (2) What would the basic velocity pressure have been 240 ft above ground?

9. A building code requires that a 30-ft-high building be designed for a minimum effective velocity pressure, including gust effects, of 10 psf. The building is 100 sq ft in plan and has a flat roof. The building code specifies an external pressure coefficient C_p of 0.8 for windward walls and −0.5 for leeward walls. For what minimum lateral wind forces should the building be designed to prevent overturning and sliding at the base?

10. Explain the importance of anchoring a roof to its supports.

11. Why are seismic forces assumed to be proportional to weight?

12. The probability of an earthquake of Modified Mercalli Scale intensity V at City A is 0.01. What is the mean recurrence interval of earthquakes of that intensity?

13. A developer is considering two sites for a high-rise apartment building in southern California, a state where severe earthquakes have occurred in the past. One site is the remains of an ancient river bed and has deep layers of clay. The other site is on high ground and has deep layers of sand with some clay. Which site should be selected? Why?

14. Why is ductility important in aseismic design?

15. Describe a shear wall and explain its purpose.

16. Describe a rigid frame and explain its purpose.

Sections 6.3 to 6.6

17. Why should standards for fire protection specified by insurance companies be applied in design of a building for a private or governmental owner?

18. Why is type of occupancy important in determining fire-protection requirements for a building?

19. Why is type of construction important in determining fire-protection requirements for a building?

20. Define fire load.

21. What is meant when a building component is reported to have a 4-hr fire rating?

22. What is a fire wall?

23. How should a burning liquid be extinguished?

24. What are the basic components of a sprinkler system?

25. An office has a fire load of 10 psf. What is the equivalent fire severity?

26. What means should be used to prevent fire from spreading:
   (a) Vertically from story to story?
   (b) Horizontally throughout a complete story?

27. What is the main advantage of automatic sprinklers?

28. In multistory buildings, where are standpipe risers usually placed?

29. What may be used to extinguish a fire around an electric motor?

30. What elements may be incorporated in a building to reduce chances of panic if a fire occurs?
31. A restaurant has a 2,400-sq ft dining area.
    (a) From Table 6.4, determine the maximum number of persons permitted in the dining room.
    (b) From Table 6.5, determine the minimum number and size of exits, if maximum door size is limited to 44 in.
32. How much floor area, as a minimum, should be allotted to a refuge area for 100 persons?
33. Compare advantages and disadvantages of fixed-temperature and rate-of-rise detectors.
34. What types of fire detectors are useful for detecting a smoldering fire?
35. A one-story factory contains 30,000 sq ft of floor area. Materials and equipment handled or installed may be classified as low heat release. What is the minimum vent area that should be provided in the roof?
36. Where must firestops be used?
37. What are the advantages of a multipurpose control center?
38. What is the objective of life-safety design for emergencies demanding evacuation of occupants from dangerous areas?
39. How do foams extinguish a fire?
40. What is the purpose of exits?
41. Name and describe briefly at least three facilities that building codes generally recognize as a reliable exit.

42. Under what conditions can an escalator be considered an exit?
43. A two-story industrial laboratory building has 5,000 sq ft of floor area on each level. It will have exit stairs at each end of the building with 2-hr fire doors.
    (1) On the second floor, how wide should each door opening be?
    (2) What is the minimum width permitted for the corridors to the exits?
44. Why must structural members be fire protected?
45. How can a communication system be used to prevent occupant panic?
46. What is the purpose of a refuge area?

Sections 6.7 to 6.10

47. Describe at least four tasks a security system should perform to prevent theft and vandalism.
48. With regard to establishment of barrier-free environments, what is meant by the term ''barrier''?
49. Why is it difficult to optimize barrier-free facilities for all types of building users?
50. How does undue concentration on a single favorable property of a material or product sometimes result in major use of a toxic material?
51. How may concerns for worker safety affect designer's choices for materials or products?

Chapter 7

Building Sites and Foundations

The building site is subject to design manipulation in a limited way. Site boundaries constitute the most fixed set of conditions. In addition, the nature of adjacent properties and other boundary conditions provide major constraints. Surface contours and existing site landscaping may be altered to some degree, but must conform to site drainage, erosion, zoning, and environmental impact considerations on a general neighborhood or regional basis.

Site materials can be a resource for construction and must be dealt with as such. Surface materials, however, may be unsuitable for the site and may need to be replaced. The general excavation and recontouring of the site should be accomplished as much as possible without requiring excess removal or importing of ground materials. Excavation work and support of the building will involve subsurface materials that present simple and uncomplicated or challenging and expensive problems.

This chapter discusses the general problems and design factors relating to specific building sites and to the development of site and foundation constructions.

7.1. SITE CONSIDERATIONS

Building designers frequently are confronted with situations where the owner has already purchased a site and they are required to design a building for that site. In such cases, they will adapt a building to the site, if it is practical to do so, and will endeavor to keep construction costs as low as possible under existing conditions. Sometimes, however, it is necessary or desirable to choose another site. In one case, for example, soil investigations at a site indicated that foundation construction would cost several hundred thousand dollars more than if good ground were available. The owner decided to buy another site.

The decisions as to which site to purchase, when to buy it, how much land to include, and how much money to pay for it are strictly the owner's. It is to his advantage, though, to have the advice of consultants, especially his design consultants, in site selection. They, however, usually will charge for this service, because site evaluations or comparisons are not ordinarily included in the basic services provided by architects and engineers.

The best time for selecting a site occurs after the design program has been established and a good estimate has been made of the owner's space needs. Schematic studies then can indicate how many buildings will be required, how much land will be needed for each building, how much space will be needed around each building and characteristics desired of the site and surrounding property. Estimates also can be made of utilities needed. The data derived from these studies are basic considerations in selecting a site.

Land Costs

Final selection of a site, however, is likely to be determined by other considerations than just the suitability of the lot for the building. Initial cost of the lot, for example, may be a significant factor.

It is important, though, that the consultants call the owner's attention to the fact that the purchase price of a lot is not the only initial cost. While there are broker fees, legal fees, registration fees, title insurance premium and other costs that the owner may be aware of, there are likely to be other considerably larger costs that he may not expect, unless advised of them by his consultants. These costs arise from zoning or subdivision regulations, provision of access to the building, obtaining of utilities, site preparation, foundation construction and other conditions, depending on the type of building. Such costs vary from site to site. In comparing prices of proposed building sites, therefore, the owner should be advised to add or subtract cost differentials to account for these costs.

Site Selection

Table 7.1 provides a partial check list as a guide in site selection. The check list is partial in the sense that the factors included should always be investigated, but, in addition, other factors affecting a specific project also should be investigated.

For example, for an industrial or commercial project, availability of a labor supply and housing for executives and labor should be determined and taken into account. For a residential development, the distances to schools, shopping, medical facilities and religious institutions may be critical factors. For a shopping center, the number of potential customers and the range of incomes are crucial.

Inclusion of an item in the check list is not an implication that its presence or absence at a site is favorable or unfavorable to a decision to purchase the site. Each factor must be evaluated with respect to the specific project being considered.

Physical Features

Various physical characteristics of the site may exert influence on the design of buildings for the site. If site space is restricted, it may be

Table 7.1. Check List for Site Selection Considerations

Internal Site Characteristics
  Area and shape of lot—need for parking, storage areas, future expansion
  Topology—slopes, surface water, trees, drainage, rock outcroppings
  Geological conditions—surface soils, subsoils, water table, risk of landslide, flood, earthquake

Location
  Owner preference for region, urban, suburban, or rural area
  Distance from population centers and facilities for education, recreation, medical service

Transportation
  Accessibility of site—easements and rights of way needed
  Highways, airports, railroads, waterways, surface transportation

Costs and Legal Concerns
  Initial price, fees, taxes, insurance, permits
  Clear title, easements, rights of way granted to others
  Building codes, zoning, subdivision ordinances
  Site work, access roads, services, utilities

Utilities and Services Required
  Water, sewer, electricity, gas, telephone
  Mail service and fire and police protection

Environmental Impact on Proposed Project
  Business and political climate, local labor-management relations, local employment conditions, available labor
  Character of neighborhood, attitude of nearby residents
  Proposed development or highway construction in area
  Congestion, noise, trends of neighborhood, proximity to airports
  Views from site, appearance of approaches to site
  Climate, prevailing winds, fog, smog, dust storms, odors

Environmental Impact of Project
  Congestion, pollution, noise, parking, housing, schools
  Services required—utilities, police, fire, transportation
  Taxes and assessments
  Economic, educational, sociological, cultural

necessary to have a high-rise building, instead of a less costly low-rise building. If the site boundaries are not rectangular, it may be necessary to mold the building plan to the site shape, which is likely to add complexity and some compromising to the general planning of the building. For tight sites it is often necessary to provide some parking in lower levels of the building, the planning for which adds constraints to the layout of vertical structural elements in the upper levels of the building.

Slope of the terrain also is an important consideration. Moderately sloping or rolling terrain is preferable to flat or steep land. Flat land is difficult to drain of rainwater. For steep slopes (surfaces rising or falling more than 10 ft vertically in 100 ft horizontally), improvement costs rise rapidly. Heavy grading to flatten slopes not only adds to land costs but also creates the risk of later uneven building settlement or land erosion. In addition, fast runoff of rainwater from slopes, as well as collection of water in marshes, swamps or wet pockets, must be prevented, and this type of work is costly.

Rock at a convenient distance below the ground surface often is advantageous for foundations of buildings, but rock outcroppings that interfere with building or road construction may have to be removed with explosives, at considerable expense.

The difference in foundation construction costs for good land and bad land may be sufficient for rejection of an otherwise suitable site. One of the earliest steps in site evaluation, therefore, should be an investigation of subsurface conditions at the site.

Transportation

For most types of buildings, easy access to a main thoroughfare is a prime requisite. For commuting of employees from home to work, receipt of supplies, and dispatch of output, building operations usually depend heavily on transportation by automobile. Note, however, that it is easy access to a major highway, not its nearness, that is important. A building fronting on a limited-access highway, for example, may not be able to discharge traffic to it without inconvenient, lengthy detours. Similarly, a building near a major interchange may be undesirable, because of congested traffic, difficulty of access and egress, confusion caused by the interchange layout, and noise and vibration from the traffic.

For some industrial plants, the type of product to be shipped may be such that shipment by railroad is economically necessary. In such cases, not only must the site selected be located along a railroad, but also an agreement should be reached with the railroad for provision of freight service at acceptable rates and intervals of time. For companies oriented to air transportation, a location near an airport may be a prime requirement.

Zoning

Zoning or subdivision regulations often may determine how a site under consideration may be used, the types of buildings that may be constructed, the types of occupancies that may be permitted and the nature of the construction (see Sec. 3.6). In some cases, the type of building contemplated for the site may be prohibited, or land costs may be too high relative to the expected return on investment when the parcel of land is subdivided as required by law. Similarly, limits on building height or floor area may make a contemplated building uneconomical. Sometimes, however, with the help of legal counsel, the owner may be able to obtain a variance from zoning requirements that will permit the proposed building.

Building codes are not likely to have such drastic effects on a proposed project. They may, however, be a factor in site comparisons when the sites lie in different jurisdictions, inasmuch as building-code requirements are likely to be different in different communities.

References

H. Rubinstein, *Guide to Site and Environmental Planning*, 3rd ed., Wiley, New York, 1987.
J. Simonds, *Landscape Architecture: A Manual of Site Planning and Design*, 2nd ed., McGraw-Hill, New York, 1987.

Significant Relations, Functions and Issues

Site selection criteria.
Building-to-site relations: elevation, site plan, grading, access, neighborhood, environment.
Site recontouring: drainage, topology, construction, site boundaries.
Zoning, rights-of-way, easements, subdivision ordinances.
Feasibility of site and foundations, regarding site topology, access for equipment, subsurface conditions, water.

7.2. SITE SURVEYS

When a site is being considered for purchase, a site survey is conducted to provide information needed for making a decision regarding that purchase. The information provided should be that necessary for evaluation and comparison of alternative sites and in determination of the suitability of a specific lot for the building and its uses. After a site has been selected and purchased, the purposes of site surveys are to provide information needed for planning the use of the land in detail, locating the building and other facilities on the lot, installing utilities and constructing foundations. In either case, the information is given diagrammatically and, to a limited extent, by notes on two or more maps. Additional information is provided in written reports on surface and subsurface conditions and their significance in design and construction of the proposed building.

One type of map is used primarily to indicate the location of the site with respect to population centers or other points of interest to the building owner. Another type of map, the survey plan described in Sec. 2.23, shows property lines, topography and utility locations. The plot plan described in Sec. 2.23 shows the proposed location and orientation of the building to be constructed, site grading to be done, parking areas to be provided, driveways and other installations planned. The plot plan is developed from the survey plan.

Site Location Map

This map is useful, in the early stages of site-selection studies as well as after purchase of a lot, in showing where the new building would be located relative to existing facilities. The map may be drawn to a very small scale, compatible with provision of the following information:

1. location of site relative to nearby population centers
2. jurisdictional boundaries
3. major highways and streets
4. principal approaches to the site
5. transportation lines
6. employment centers
7. shopping centers, schools, religious institutions, recreational facilities
8. appropriate zoning regulations, such as those governing land use, for example, nearby parcels restricted only to residential, or only to industrial construction

Survey Plan

This map delineates the boundaries of the lot. The map need not be drawn to a large scale unless considerable detail must be shown. For large parcels, a scale of 1 in. = 100 ft may be adequate. The map should show the following:

1. Lengths, bearings (directions), curve data and angles at intersections of all boundary lines
2. Locations and dimensions of streets along the boundaries and of streets, easements and rights of way within the parcel, with deed or dedication references
3. Location of intersection lines of adjoining tracts and any encroachment on boundaries of the lot
4. Names of owners of adjoining property or reference to recorded subdivision of that property
5. Position and description of physical boundary markers and of official bench marks, triangulation stations and surveying monuments within or near the property
6. Area of the site and each parcel comprising it
7. Topography of the site—contours, lakes, marshes, rock outcroppings, etc.
8. True and magnetic meridian (north arrow) on the date of survey
9. Utility installations adjoining or passing through the site

Topographic Map

A separate topographic map should be prepared when considerable detail must be shown for a site or when a lot has steep slopes. Topographic maps show the nature of the terrain and locate natural features, such as lakes, streams, rock outcroppings, boulders and important trees, as well as structures and other man-made items existing on the site. The scale should not be smaller than 1 in. = 100 ft.

Slopes are indicated on such maps by contour lines. Each line represents a specific level above a base elevation, or datum, as explained in Sec. 2.23. The steeper the slope of the terrain, the closer will be the contour lines. For relatively flat land (slopes up to about 3%; that is, 3 ft vertically in 100 ft horizontally), contours may be drawn for height intervals of 1 ft. For slopes up to about 15%, the contour interval may be 2 ft, and for steeper slopes, 5 ft.

The location of test pits or borings for soil investigations of the site may be added to the topographic map.

Utilities Map

A separate map showing type, location and sizes, if appropriate, of utilities adjacent to or within the site should be prepared if the amount of data to be given would make a single map with combined information confusing or difficult to read.

The utilities map should be drawn to a scale not smaller than 1 in. = 100 ft. In addition to location, it should provide the following information:

1. Sizes and invert elevations of existing sewers, open drainage channels, catchbasins and manholes
2. Sizes of water, gas and steam pipes and underground electrical conduit
3. Widths of railroad tracks and rights of way
4. Police and fire-alarm call boxes and similar devices
5. Dimensions of utility easements or rights of way

Surveying Methods

Site surveys are not included in the basic services provided by architects and engineers. The owner, or the architect or engineer on his behalf, should engage a licensed surveyor to make land surveys and draw the maps. Geotechnical consultants should be engaged for subsurface investigations and reports.

Property-line lengths and directions and setting of boundary markers require land surveys of relatively high accuracy. These surveys are usually made with a transit or a theodolite and tape. Topographic and utility surveys may be made with a transit or a plane table and a stadia rod. Also, such surveys may be made with electronic instruments, such as a tellurometer, which uses microwaves to determine distances; an electrotape, which uses radio-frequency signals; or a geodimeter, which employs light. For large parcels, aerial surveys offer economy and speed in obtaining topographic information. By photogrammetric methods, contours and natural and artificial features can be plotted on an aerial photograph of the site.

References

C. Ramsey and H. Sleeper, *Architectural Graphic Standards*, 9th ed., Wiley, 1988.
J. DeChiara and L. E. Koppelman, *Time-Saver Standards for Site Planning*, McGraw-Hill, 1984.
F. Merritt, *Building Design and Construction Handbook*, 4th ed., McGraw-Hill, 1982, Sec. 23.

Words and Terms

Site Location Map
Survey Plan
Topographic Map
Utilities Map

Significant Relations, Functions and Issues

Site development, regarding: existing topology, location of streets and utilities, control of water runoff.
Building positioning on site, regarding: grading, utility connections, excavation for construction.

7.3. SOIL CONSIDERATIONS FOR SITE AND FOUNDATION DESIGN

Surface and near-surface ground materials are generally composed of combinations of the following:

1. Rock, solid or fractured
2. Soil, in naturally formed deposits

3. Fill materials of recent origin
4. Organic materials in partially decomposed form
5. Liquids, mostly water

A number of considerations must be taken into account in design of the building foundations and the general site development. With respect to site ground conditions, some typical concerns are the following:

1. The relative ease of excavation
2. Site water conditions: ease of and possible effects of any required site dewatering for construction
3. Feasibility of using excavated site materials for fill and site finish grading
4. Ability of the soil to stand on a relatively vertical cut in an excavation
5. Effects of construction activities—notably the movement of workers and equipment—on surface soils
6. Reliability and structural capacity of near-surface materials for foundation support
7. Long-time effects of site changes: paving, irrigating, recontouring
8. Necessary provisions for frost protection, soil shrinkage, subsidence, consolidation, expansion, erosion
9. Need for dampproofing and/or waterproofing of subgrade construction of occupied spaces
10. Ease of installation of buried services: water, gas, sewer, utilities, phones
11. Special provisions for existing features: large trees, buildings (abandoned or remaining), existing underground services or easements for same

Ground conditions at the site constitute a given condition which must be dealt with in some feasible manner. If building and site design requirements do not mesh well with given conditions, a lot of adjustment and compensation must be made. Investigations of the site conditions must be made to inventory the existing ground materials, with special attention to properties critical to building and site design concerns. The following material in this section deals with a discussion of various ground materials, their significant properties, how they behave with respect to effects of building and site construction, and the problems of establishing criteria for design.

Soil Properties and Identification

Of the various ground materials previously described, we are concerned primarily with soil and rock. Fill materials of recent geological-formation origin and those with a high percentage of organic materials are generally not useful for site construction or foundation support, although they may have potential for backfill or finish grading to support plantings. A precise distinction between soil and rock is somewhat difficult, as some soils are quite hard when dry and cemented, while some rock formations are highly weathered and decomposed or have extensive fractures. At the extreme, the distinction is simple and clear: for example, loose sand versus solid granite. A precise definition for engineering purposes must be made on the basis of various responses of the materials to handling and to investigative tests. Some of these are described in other portions of this chapter, relating to specific materials, property definitions, and excavation and construction problems.

Soil is generally defined as material consisting of discrete particles that are relatively easy to separate by moderate pulverizing action or by saturation with water. A specific soil mass is visualized as consisting of three parts: solid, liquid and air. By either volume or weight, these are represented as shown in Fig. 7-1. The nonsolid portion of the volume is called the *void* and is typically filled partly with liquid (usually

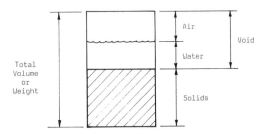

Fig. 7.1. Three-part composition of typical soil mass.

water) and partly with air, unless the soil is saturated or is baked totally dry.

Soil weight (density) is determined by the weights of the solid and liquid portions. The weight of the solids may be determined by weighing an oven-dried sample. The weight of the water present before drying is given by the difference in weight between that sample and the weight of the sample before drying. The specific gravity of primary soil materials (sand, silt, clay, rock) vary over a short range—from about 2.60 to 2.75—so that the volume of the solids is easily predicted from the weight; or, if the volume is known, the weight is easily predicted.

Various significant engineering properties of soils are defined in terms of the proportions of materials as represented in Fig. 7.1. Major ones are the following:

1. *Void ratio* (*e*). This generally expresses the relative porosity or density of the soil mass, and is defined as follows:
$$e = \frac{\text{volume of the voids}}{\text{volume of the solids}}$$

2. *Porosity* (*n*). This is the actual percentage of the void, defined as
$$n = \frac{\text{volume of the voids}}{\text{total soil volume}} (100)$$
This generally defines the rate at which water will flow into or out of soils with coarse grains (sand and gravel), although water flow is measured by tests and expressed as relative *permeability*, which is also affected by actual particle size and gradation of particle sizes.

3. *Water content* (*w*). This is one means of expressing the amount of water, defined as
w (in percent)
$$= \frac{\text{weight of water in the sample}}{\text{weight of solids in the sample}} (100)$$

4. *Saturation* (*S*). This expresses the amount of water as a ratio, similar to the void ratio, thus:
$$S = \frac{\text{volume of water}}{\text{volume of voids}}$$
Oversaturation, with *S* greater than one, is possible in some soils, when some of the soil particles are made to float.

Particle size is a major factor in soil classification, as the major types of soils are essentially defined on this basis. Figure 7.2 indicates the general form of the graph that is used for classifications on the basis of particle size. Size is displayed horizontally on the graph, using a log scale, and indicating the usual boundaries for common soil types. Distinctions become less clear for very fine materials, so that other factors must be used to clearly distinguish between sand and silt and between silt and clay. Soil deposits typically consist of mixtures of a range of particles, and the vertical scale in the graph in Fig. 7.2 indicates the percentage of the soil volume that is represented by various particle sizes. The curved lines on the graph indicate typical displays of size analyses for soil samples; the forms of the curves represent different soil types, as follows:

1. *Well-graded soil*. This is indicated by a smooth curve, spanning a considerable range of size.
2. *Uniform soil*. This is indicated by a curve that is mostly vertical in a short range of size.
3. *Gap-graded soil*. This curve is significantly flexed, indicating a significant lack of middle-sized soil particles.

Particle size alone, together with grading evaluations, will provide indications of major engineering properties.

Particle sizes of coarser materials are determined by passing the loose soil materials through increasingly finer sieves, as indicated at the top of the graph in Fig. 7.2. For fine-grained materials, size is measured by the rate of settlement of particles in an agitated soil-water mixture.

Particle shape is also significant, being mostly bulky in rounded or angular form, although flaky and needlelike shapes are also possible. Soil mobility, compaction, and settlement may be affected by shape—this often being critical with specific types of soils.

A major distinction is made between *cohesive* and *cohesionless* soils. Sand and gravel generally represent cohesionless soil materials, while clay is cohesive. Soils, though, usually

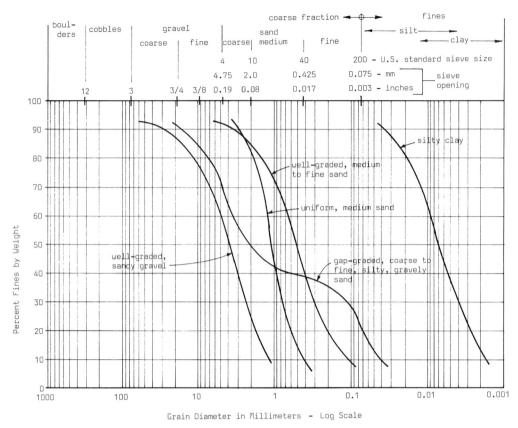

7.2. **Grain size measurement and plot for soil particles. Standard-size sieves are used to determine grada-tion of size by percent. Both size and gradation are critical to soil properties.**

are mixtures of various materials, taking on a general character on the basis of the types of soil materials as well as their relative amounts in the overall soil volume. A quite small percentage of clay, for example, can give considerable cohesive character to predominantly sandy soils.

At the extreme, cohesive and cohesionless soils are quite different in many regards, the properties of critical concern being quite different. For consideration of structural capacity, we may compare sand and clay as follows:

Sand. Has little compression resistance without some confinement; principle stress mechanism is shear resistance; important data are penetration resistance (measured as number of blows (N) for advancing a soil sampler), density (measured as weight), grain shape, predominant grain size and nature of size gradation; some loss of strength when saturated.

Clay. Principal stress resistance in tension; confinement is of concern only when clay is wet (oozes); important are the unconfined compressive strength (q_u), liquid limit (w_L), plastic index (I_p) and relative consistency (soft to hard).

On the basis of various observed and tested properties, soils are typically classified by various systems. The systems used are based on user concerns—major ones being those used by the highway construction interests and by agriculturists. For engineering purposes, the principal system used is the Unified System (ASTM Designation D-2487), abbreviated in Fig. 7.3. This system defines fifteen soil types, each represented by a two-letter symbol, and establishes the specific properties that identify each type. Most building codes use this as the basic reference for establishing foundation soil requirements. Codes often provide tables of lim-

Major Divisions				Group Symbols	Descriptive Names
Coarse-Grained Soils More than 50% retained on No. 200 sieve	Gravels 50% or more of coarse fraction retained on No. 4 sieve	Clean Gravels		GW	Well-graded gravels and gravel-sand mixtures, little or no fines
				GP	Poorly graded gravels and gravel-sand mixtures, little or no fines
		Gravels with Fines		GM	Silty gravels, gravel-sand-silt mixtures
				GC	Clayey gravels, gravel-sand-clay mixtures
	Sands More than 50% of coarse fraction passes No. 4 sieve	Clean Sands		SW	Well-graded sands and gravelly sands, little or no fines
				SP	Poorly graded sands and gravelly sands, little or no fines
		Sands with Fines		SM	Silty sands, sand-silt mixtures
				SC	Clayey sands, sand-clay mixtures
Fine-Grained Soils 50% or more passes No. 200 sieve	Silts and Clays Liquid limit 50% or less			ML	Inorganic silts, very fine sands, rock flour, silty or clayey fine sands
				CL	Inorganic clays of low to medium plasticity, gravelly clays, sandy clays, silty clays, lean clays
				OL	Organic silts and organic silty clays of low plasticity
	Silts and Clays Liquid limit greater than 50%			MH	Inorganic silts, micaceous or diatomaceous fine sands or silts, elastic silts
				CH	Inorganic clays of high plasticity, fat clays
				OH	Organic clays of medium to high plasticity
Highly Organic Soils				Pt	Peat, muck and other highly organic soils

Fig. 7.3. Unified Soil Classification System. Used generally for basic identification of soil type. Various criteria (not shown here) based on tests are used to establish classification. (After ASTM D2487-85.)

iting values for foundation design (called presumptive values) based on some amount of prescribed soil identification. Table 7.2 is reprinted from the Uniform Building Code and indicates the Unified System as the basic classification reference.

The following are some of the most common ground materials. These are commonly named for our purpose here, although some have quite specific engineering definitions in the various classification systems.

Loam, or topsoil, is a mixture of humus, or organic material, and sand, silt or clay. It generally is not suitable for supporting foundations.

Bedrock is sound, hard rock lying in the position where it was formed and underlain by no other material but rock. Usually, bedrock is capable of withstanding very high pressures from foundations and therefore is very desirable for supporting buildings. When bedrock is found near the ground surface, excavation of overlying soil to expose the rock and set footings on it often is the most economical alternative. When bedrock is deep down and overlain by

Table 7.2. Allowable Foundation and Lateral Pressure[a]

CLASS OF MATERIALS[2]	ALLOWABLE FOUNDATION PRESSURE, LB/SQ. FT.[3]	LATERAL BEARING LB/SQ./FT./ FT. OF DEPTH BELOW NATURAL GRADE[4]	LATERAL SLIDING[1]	
			COEF-FICIENT[5]	RESISTANCE LB/SQ. FT.[6]
1. Massive Crystalline Bedrock	4000	1200	.70	
2. Sedimentary and Foliated Rock	2000	400	.35	
3. Sandy Gravel and/or Gravel (GW and GP)	2000	200	.35	
4. Sand, Silty Sand, Clayey Sand, Silty Gravel and Clayey Gravel (SW, SP, SM, SC, GM and GC)	1500	150	.25	
5. Clay, Sandy Clay, Silty Clay and Clayey Silt (CL, ML, MH and CH)	1000[7]	100		130

[1]Lateral bearing and lateral sliding resistance may be combined.

[2]For soil classifications OL, OH and PT (i.e., organic clays and peat), a foundation investigation shall be required.

[3]All values of allowable foundation pressure are for footings having a minimum width of 12 inches and a minimum depth of 12 inches into natural grade. Except as in Footnote 7 below, increase of 20 percent allowed for each additional foot of width and/or depth to a maximum value of three times the designated value.

[4]May be increased the amount of the designated value for each additional foot of depth to a maximum of 15 times the designated value. Isolated poles for uses such as flagpoles or signs and poles used to support buildings which are not adversely affected by a 1/2-inch motion at ground surface due to short-term lateral loads may be designed using lateral bearing values equal to two times the tabulated values.

[5]Coefficient to be multiplied by the dead load.

[6]Lateral sliding resistance value to be multiplied by the contact area. In no case shall the lateral sliding resistance exceed one half the dead load.

[7]No increase for width is allowed.

[a]Source: Reproduced from the Uniform Building Code, 1988 edition, with permission of the publishers, International Conference of Building Officials.

weak soils, however, it may be more economical to drive supports, such as piles, from the ground surface through the weak soil to the rock to carry a building. Care should be taken in soil investigations not to mistake weathered rock or boulders for bedrock.

Weathered rock is the name applied to materials at some stage in the deterioration of bedrock into soil. This type of rock cannot be trusted to carry heavy loads.

Boulders are rock fragments over about 10 in. in maximum dimension. They too cannot be trusted to carry heavy loads, because, when embedded in weak soils, they may tip over when loaded.

Gravel consists of rock fragments between 2 mm and 6 in. in size. When composed of hard, sound rock, it makes a good foundation material.

Sand consists of rock particles between 0.05 and 2 mm in size. The smallest particles may be classified as fine sand, the largest as coarse sand, and the intermediate sizes as medium sand. Dense sands usually make a good foundation material. Fine sands may be converted by water into quicksand, which may flow out from under even a very lightly loaded foundation.

Silt and clay consist of particles so small that individual particles cannot readily be distinguished with the unaided eye. In one classification system, silt comprises particles larger than those in clay. In another classification system, silt is defined as a fine-grained, inorganic soil that cannot be made plastic by adjustment of water content and that exhibits little or no strength when air-dried. Clay is defined as a fine-grained, inorganic soil that can be made plastic by adjustment of water content and exhibits considerable strength when air-dried. Thus, clay loses its plasticity when dried and its strength when wetted. It may make a satisfactory foundation material under certain conditions. Silt is not a desirable foundation material, because when it gets wet its strength cannot be relied on.

Sand and gravel are considered cohesionless materials, because their particles do not adhere to each other. They derive their strength from internal friction. In contrast, silt and clay are considered cohesive materials, because their particles tend to adhere when the water content is low.

Hardpan consists of cemented material containing rock fragments. Some hardpans consist of mixtures of sand, gravel and clay or silt. Glacial hardpans may be composed of particles, ranging in size from clay to boulders, that were at least partly cemented together by high pressures from glaciers. Some hardpans, depending on the degree of consolidation, make very good foundation material.

Till is a glacial deposit of mixtures of clay, silt, sand, gravel and boulders. If highly compressed in the natural state, till may serve as a good foundation material. Loose tills vary in character and may cause uneven settlement of buildings supported on them.

Muck, or mud, is a sticky mixture of soil and water. Because of its lack of stability, muck seldom can be used as a foundation material or as a fill to build up ground to a desired level.

Foundation Design Criteria

Investigation of site conditions is aimed partly at establishing data for the building foundation design and planning of the site work and foundation construction. Information and recommendations must be obtained that address the following concerns.

Allowable Bearing Pressure. This is the limiting value for the vertical pressure under shallow bearing foundation elements. It will be affected by the type of soil materials encountered, by seasonal fluctuations of the ground water level, by any deep frost conditions, by the depth of the footing below the ground surface (called *surcharge*), by the sensitivity of the type of building construction to settlements, and—in special situations—by numerous other possible data and circumstances. In simple situations, for modest sized buildings where considerable previous construction has been in place for some time, design values may be primarily based on recommended presumptive values; often stipulated by local building codes. For large projects, or where unusual conditions exist, it is common to seek recommendations from experienced geotechnical engineers, supplied with considerable investigative data.

Settlement. Downward movement of foundations, as the building is progressively stacked upon them, is an unavoidable eventuality, except for foundations bearing directly on massive bedrock. The precise magnitude of movements of complex constructions on multilayered soil masses is quite difficult to predict. In most cases, movements will be small, and the primary concern may be for a *uniformity* of the settlement, rather than a precise prediction of the magnitude. Again, for modest structures, bearing on firm soils with relatively low imposed vertical pressure, settlement is seldom a major concern. However, if any of the following situations occur, settlements should be very carefully studied:

1. When soils of a highly unstable or compressible nature are encountered
2. When vertical pressures are considerable and bearing footings are used—especially

when any of the following situations 3 and 4 exist

3. When the construction is sensitive to movements (notably to differential settlements of separate foundation elements), as are concrete rigid frames, tall towers and masonry or plastered walls

4. When nonuniform settlements may cause serious misalignment of sensitive equipment, or even of tall elevators, large doors, or other building elements requiring careful fit or joining

In some instances, the design vertical pressure may be reduced to limit the magnitude of settlements.

Water Effects. Water—is typically present in all soils, except for those in very dry, desert conditions. At building sites, the effects of precipitation plus irrigation for plantings will often keep a notable magnitude of moisture in soils near the ground surface. A special concern is that for the free-water level (sometimes called the water table) in the ground, below which relatively porous soils will be essentially saturated. This level normally fluctuates over time as precipitation amounts vary—especially in areas where long periods occur with no precipitation. Repeated changes in the moisture content of soils, from saturated to near dry, can be the source of various problems—most notably in fine-grained soils subject to erosion, flotation or high magnitude of volume change due to shrinkage and expansion. A high free-water level can also be a problem during construction where considerable deep excavation is required. Regrading of the site, covering of major portions of the site surface with buildings and paving and provision of extensive irrigation are all effects constituting major adjustments of the previous natural site environment and may result in major changes in some soil materials near the ground surface.

Horizontal Force Effects. Horizontally directed force effects are usually of one of the following origins.

1. *Horizontal stresses from vertical forces.* When a large vertical force is imposed on soil the resulting stresses in the soil are three-dimensional in nature. The soil mass tends to bulge out horizontally. This effect can be a major one in some situations—most notably in soils with a high clay content. Adjacent foundations may experience horizontal movements or nearby excavations may be pushed outward.

2. *Active lateral pressure.* This is the horizontal force effect exerted by a soil mass against some vertical retaining structure, such as a basement wall. This is generally visualized by considering the soil to behave like a fluid, exerting pressure in proportion to the distance below the top of the fluid mass (ground level). If the ground slopes upward behind the retaining structure (as with a hillside retaining wall), or some additional load (such as a heavy vehicle) imposes additional vertical load on the ground surface, this pressure will be further increased. Water conditions and the type of soil will also cause variations in both the magnitude and nature of distribution of this type of pressure.

3. *Passive lateral pressure.* This represents the resistance developed by a soil mass against the horizontal movement of some object through the soil. This is the basic means by which the actual horizontal forces caused by wind and seismic action are transferred to the ground—by soil-mass pressures against the basement walls or sides of foundation elements. As with active pressures, this effect varies in its potential magnitude with depth below the ground surface, although some limiting total magnitude exists.

Frictional Resistance. When combined with vertical forces, horizontal stresses will also be resisted by friction on the soil in the case of bearing type foundations (friction on the bottom of footings resisting lateral sliding). For coarse-grained soils (sand and gravel) the potential friction resistance varies with the vertical force and is generally independent of the contact area. For clays, the resistance will vary with the cohesion per unit of contact area, with the vertical force being considered only in terms of a certain minimum amount to develop the friction effect.

Both passive lateral bearing and frictional sliding resistances as well as presumptive vertical bearing pressures are often stipulated in building codes (see Table 7.2).

Stability. The likelihood for a soil mass to remain in its present structural state depends on its relative stability. Significant lack of stability may result in erosion, subsidence, lateral movement, viscous flow or liquefaction. All of these actions can have disastrous effects on supported structures and their potential occurrence is a major concern in soil investigation. During excavation work as well as in final site grading, a major concern is that for the stability of slopes, as discussed in Sec. 7.8. Principal destabilizing effects are those due to fluctuations in water content, unbalancing of the equilibrium of pressures caused by deep excavations or heavy surface loads and dynamic shocks such as those caused by earthquakes. Loose sands, highly plastic soft clays and cemented soils with high voids are examples of potentially unstable soils. Modification of some soils may be necessary and various techniques are employed, as described in Sec. 7.7.

Excavation and Construction Concerns. Performance of necessary excavation and site grading and the general advancement of site and building construction work must be planned with consideration of various factors relating to soil conditions. Need for bracing and possibly dewatering of large excavations is a major task; this is discussed in Sec. 7.9. Surface materials that can be used for backfill, pavement subgrades or for finish grading as topsoil should be stockpiled for future use before they are lost during the excavation and construction processes. In some cases, existence of large boulders, tree roots, old wells, cesspools, underground tanks or buried construction for utility tunnels and vaults may present major tasks during excavation as well as possibly requiring some reconsideration in siting of buildings and design of foundation elements. For urban sites, these matters may be of heightened concern because they may require protection of adjacent buildings, streets, buried utilities, and other structures.

Pile Foundations. These are discussed in general in Sec. 7.5. A critical factor is the determination at an early stage of design of the *need* for deep foundations and the type to be used. For piles, critical early decisions must be made as to the likely required length, the use of friction versus end-bearing piles and any special problems that might be encountered in advancing the piles. Heavy equipment must be used for pile driving and the movement of the equipment to and on the site may be a problem in some cases. Pile driving is also disturbing to the neighbors and may present problems in this regard.

Pier (Caisson) Foundations. These are also discussed in general in Sec. 7.5. They represent a need for a very deep foundation, two primary concerns being for the effects of water and the potential collapse of the sides of the excavation. The excavation must be successfully advanced and then filled up with concrete; in some cases requiring the lining of the walls of the dug shaft and a dewatering process. The term ''caisson'' derives from a technique used primarily for bridge piers, in which an airtight chamber is sunk by digging out from under it; then is filled with concrete once in place. The deep foundation that is not driven as a pile is more generally described as a pier. In some cases, it may be advanced a great distance below grade, and its design (and development of the construction planning) may require quite deep soil explorations. Piers—like end-bearing piles—often have their safe load capacities verified by actual load tests; however, as the size of the pier increases, this becomes less feasible. If load tests are not performed, the necessity for reliable and complete soil information becomes essential.

References

J. Ambrose, *Simplified Design of Building Foundations*, 2nd ed., Wiley, 1988.

J. Bowles, *Foundation Analysis and Design*, 3rd ed., McGraw-Hill, 1982.

G. Sowers, *Introductory Soil Mechanics and Foundations: Geotechnic Engineering*, 4th ed., Macmillan, 1979.

Words and Terms

Clay
Cohesionless
Cohesive
Density
Fill
Grain size
Gravel
Penetration Resistance
Permeability
Porosity
Presumptive bearing pressure
Rock
Sand
Settlement: allowable, differential
Silt
Surcharge
Unified System of soil classification
Void

Significant Relations, Functions, and Issues

Excavation: extent, ease, dewatering for, bracing for.
Effects of construction activity on the site and excavated soils for bearing.
Site development in general related to building foundation design.
Soil identification and evaluation for use.
Establishment of design criteria for site and foundation systems.
Settlement: computation of, control, effects on building.

7.4. SHALLOW BEARING FOUNDATIONS

In situations where reasonably stable, bearing-resistive soils occur near the ground surface, the commonly employed foundation system is that using shallow bearing footings. The most common forms of such footings are the simple strip footing used beneath bearing walls and the rectangular pad under individual columns. There are, however, various other forms of footings for different elements of building construction or special situations. Some of the most common types of footings are shown in Fig. 7.4.

The principal function of bearing foundations is mainly transfer of vertical force through contact pressure on the bottom of the footings. A primary design decision is selection of the maximum permitted bearing pressure, which is important in determining the area of contact (plan size of the footing). This area may be calculated from

$$A = \frac{P}{q} \qquad (7.1)$$

in which A is the required footing area, P the total load including the weight of the footing and q is the unit of allowable soil pressure. Design of the footing may proceed as for a reinforced concrete flexural member: a single-direction cantilever for a wall footing and a two-way cantilever for a column footing.

For small to medium-size projects, construction of footings is often quite crude, involving a minimum of forming—especially in soils where a vertical cut for the footing sides can be made for a shallow excavation. In such cases, construction consists essentially of casting concrete in a hole in the ground. Economy is generally obtained by using a relatively low grade of concrete, a bare minimum of reinforcing, and a minimum of forming—often only that required to obtain a reasonably true top surface for the beginning of the construction of the supported object. From a construction detailing point of view, this latter function is of primary concern: the providing of a platform for the building construction. Concerns in this regard are for the accurate location of the top of the footing, the centering of the footing beneath the supported object and the accurate installation of any anchorage devices, such as anchor bolts or dowels for reinforcement.

Wall and column footings are used so repetitively in common situations that their designs are mostly achieved by using tabulated data, such as those in the *CRSI Handbook* (see References at end of this Section). Complete structural design is usually limited to special foot-

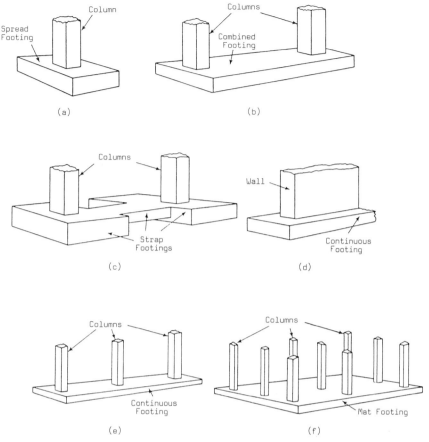

Fig. 7.4. Shallow bearing foundations (also called spread footings). (a) Single footings for column. (b) Combined footings for closely-spaced columns. (c) Cantilever, or strapped, footing, used at building edge on tight sites. (d) Continuuous strip footing for wall. (e) Continuous footing for a row of columns. (f) Large single footing for number of columns or a whole building, called a mat or raft.

ings, such as combined column footings or rectangular footings for individual columns.

A major concern for shallow bearing footings is the anticipated vertical movement caused by the loads on the footing, called *settlement*. Some amount of settlement must be expected if the footing bears on anything other than solid rock, and solid rock is not often available. The magnitude of settlement is frequently the principal factor in determining the limiting soil pressure for a footing. Settlements can quite often be predicted with acceptable accuracy on the basis of the soil materials, the thickness of individual strata (layers) of different soil materials, the magnitude of the vertical loads and bearing area of the footings. The prediction, however, should preferably be made by a qual-

ified, experienced, geotechnical engineer, as it requires expert interpretation of investigative data and collation of many factors.

Settlement mechanisms develop differently in various types of soils and in response to various actions. The initial settlement caused by the weight of building and contents, however, may be most significant, especially in loose sands and sand-gravel mixtures. In soft wet clays, on the other hand, settlement may occur over time as the clay mass readjusts to the changes in pressure—oozing in three dimensions in the directions of less restraint. The potential critical nature of settlements is largely predictable from investigative data, if the data are properly obtained and carefully analyzed.

How serious settlement effects are depends

not only on the magnitude of the settlements but also on characteristics of the construction. Tilting of tower structures, cracking of plaster or masonry, and misalignment of elevators or other sensitive equipment are examples of the effects of movements of the building supports. Especially critical is the effect of differential movements of supports of stiff, rigid-frame structures. Differential settlements are, in fact, of more frequent concern than the magnitude

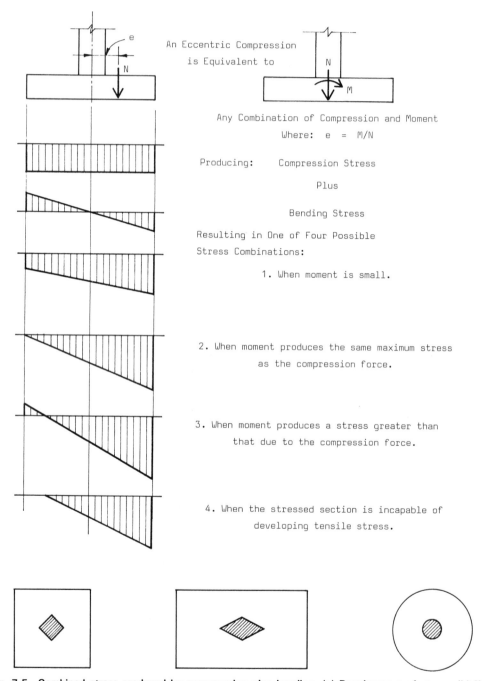

Fig. 7.5. Combined stress produced by compression plus bending. (a) Development of stress. (b) Kern limits for common sections; indicates limit for eccentric load without tension stress or uplift.

of the vertical movements. If all of the footings for a building settle the same amount, there will be essentially no damage to the construction. If the uniform settlement is small, or if adequate provision can be made to compensate for it (such as simply building the footings a bit high), the magnitude of the settlement may be inconsequential.

In some situations differential settlements are partly controlled by designing specifically to control them. This involves an analysis of the nature of settlements in terms of both the soil mechanisms and the loads that cause them. For settlements that occur mostly at the time of loading, footing sizes may be proportioned on the basis of the loads, with some emphasis on the dead loads which are more predictable. For long-time settlements of considerable magnitude, it is sometimes necessary to place adjustable elements between the footings and the supported construction, with adjustments made periodically as settlements are monitored.

In addition to their resistance to vertical loads, bearing footings are often required to develop resistance to the effects of lateral, uplift, or overturning actions. The problems of horizontal soil pressures and general resistance to uplift are discussed in Sec. 7.6. Footings subjected to combinations of vertical compression and overturning moment, such as the supports for freestanding walls, towers, and isolated shear walls, must resolve the combined effects of compression and bending as shown in Fig. 7.5. As it is not feasible to develop tension resistance at the contact face between the footing and the soil, the total resistive effort must be achieved with compression stress. Eccentric loadings may result in development of a partly loaded contact face, or *cracked section*, when the magnitude of bending stress exceeds that of the direct compression. This stress distribution is not very desirable for a footing because of the implications of rotational settlement. Thus, the usual conservative design limit is the condition for which maximum tensile bending stresses and direct compressive stresses are equal. Visualizing the combined actions as equivalent to those produced by a mislocation (eccentricity) of the compression force, it is possible to derive the maximum eccentricity for

the limiting stress condition. This is the basis for establishing the *kern* limit for an area; the form of such a limit for simple areas is shown in Fig. 7.5.

References

J. Ambrose, *Simplified Design of Building Foundations*, 2nd ed., Wiley, 1988.

J. Bowles, *Foundation Analysis and Design*, 3rd ed., McGraw-Hill, 1982.

F. Merritt, *Building Design and Construction Handbook*, 4th ed., McGraw-Hill, 1982, Sections 6 and 10.

Design Aids

C. Ramsey and H. Sleeper, *Architectural Graphic Standards*, 9th ed., Wiley, 1988.

CRSI Handbook and *CRSI Manual of Standard Practice*, Concrete Reinforcing Steel Institute, 1984.

Words and Terms

Shallow bearing foundations

Footings: wall, column, cantilever, rectangular, combined, mat

Lateral pressure

Soil friction

Kern limit

Geotechnical engineer

Significant Relations, Functions and Issues

Vertical bearing: magnitude of, dispersion in soil mass, settlement from.

Lateral, uplift and overturning moments on footings.

7.5. DEEP FOUNDATIONS

In many situations the ground mass immediately below the bottom of the building is not suitable for use of direct bearing of footings. For tall buildings this may simply be due to the magnitude of the loads. In most cases, however, there is some problem with the soil itself or with some potential destabilizing effect, such as washout erosion in waterfront locations. For such situations it becomes necessary to go deeper into the ground for the transfer of the bearing loads. (See Fig. 7.6.)

If the distance to good bearing material is rel-

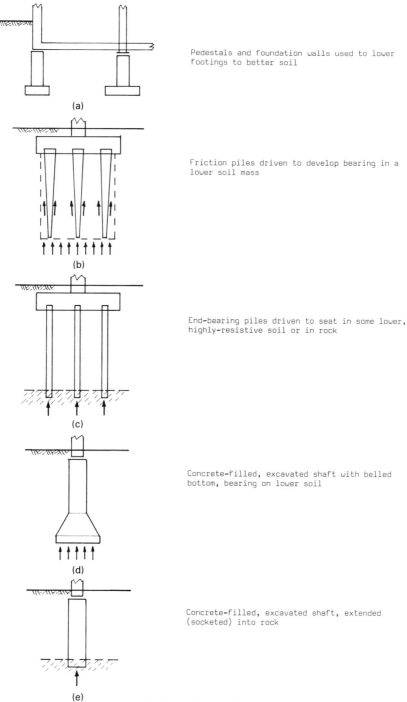

Pedestals and foundation walls used to lower footings to better soil

(a)

Friction piles driven to develop bearing in a lower soil mass

(b)

End-bearing piles driven to seat in some lower, highly-resistive soil or in rock

(c)

Concrete-filled, excavated shaft with belled bottom, bearing on lower soil

(d)

Concrete-filled, excavated shaft, extended (socketed) into rock

(e)

Fig. 7.6. Types of deep foundations.

atively short, it may be possible to simply excavate into that soil, construct the usual bearing footings, then build short columns (called piers or pedestals) up to the bottom of the superstruc-

ture (Fig. 7.6a). In some situations, there are other motivations for the use of such transitional elements, such as cases where supported elements (wood or steel columns, for example) must be kept out of contact with soil. If short pedestals are used, the additional cost may be minor, consisting mostly only of the additional excavation and the construction of the pedestals.

When it becomes necessary to lower the bearing transfer a considerable distance (say 15 ft or more), the usual solution consists of the use of either piles or piers. This decision is not lightly made, as the cost of such foundation systems is usually much more than that of simple footings. Use of piles or piers consists essentially of erecting the building on stilts, the stilts being used to transfer the vertical bearing to some point significantly distant from the bottom of the building. The distinction between piles and piers has to do with the means for placing them in the ground: piles are dynamically inserted (much like pounding a nail into a board) while piers are essentially concrete columns, the concrete being cast in excavated shafts.

Because of the means of their installation, the precision of the location of piles is difficult to control. It is thus typical to use groups of piles for support of loads that require precise location, such as single building columns. Piers, on the other hand, are mostly used singly, except where a very large single platform must be supported. Building loads may be placed directly on top of piers, while a transitional concrete cap (not unlike a thick footing) is required between a group of piles and a column base.

Piles may consist of timber poles (stripped tree trunks), rolled steel sections (H-shaped), thick-walled or fluted-walled steel pipe, or precast concrete. These elements are driven to one of two forms of resistance development: simple skin friction (Fig. 7.6b) or end point bearing (Fig. 7.6c). For friction piles, the load capacity is ordinarily established by the difficulty of driving it the last few feet. With the aid of a calibrated driving device, the number of blows required to advance a pile the last foot or so can be converted to an extrapolated static force resistance. Building codes usually have empirical formulas that can be used for this, although more complex analyses may be possible using additional factors and a computer-aided investigation.

When piles are closely clustered in a group, the group capacity determines the load that may be supported. It may be calculated by treating the pile cluster as a large single block, equivalent to a bearing footing with a plan size of that of the pile group.

End bearing piles driven into rock present a much different situation. This type of foundation is generally feasible only with steel piles and capacities must usually be determined by load tests. The load tests are more for the purpose of determining the proper seating of the piles in the rock; the actual load capacity is usually that for the steel pile acting as a column.

Piles are installed by specialty contractors, often using patented equipment or special pile systems. As the necessary heavy equipment is difficult to move over great distances, the type of pile foundation used is often restricted by the local availability of individual contracting organizations. Since a particular type of pile or pile-driving technique is usually best suited to particular soil conditions, local marketing of services will typically favor particular systems. The following discussion deals with some typical types of piles and driving methods and some general considerations for their use.

Pile Types

Timber Piles. Timber piles consist of straight tree trunks, similar to those used for utility poles, that are driven with the small end down, primarily as friction piles. Their length is limited to that obtainable from the species of tree available. In most areas where timber is plentiful, lengths up to 50 or 60 ft are obtainable, whereas piles up to 80 or 90 ft may be obtained in some areas. The maximum driving force, and consequently the usable load, is limited by the problems of shattering either the leading point or the driven end. It is generally not possible to drive timber piles through very hard soil strata or through soil containing large rocks. Usable design working loads are typically limited to 50 to 60 k (1 k = 1 kip = 1,000 lb).

Decay of the wood is a major problem, especially where the tops of piles are above the groundwater line. Treatment with creosote will prolong the pile life but is only a delaying measure, not one of permanent protection. One technique is to drive the wood piles below the waterline and then build concrete piers on top of them up to the desired support level for the building.

For driving through difficult soils, or to end bearing, wood piles are sometimes fitted with steel points. This reduces the problem of damage at the leading point, but does not increase resistance to shattering at the driven end.

Because of their relative flexibility, long timber piles may be relatively easily diverted during driving, with the pile ending up in something other than a straight, vertical position. The smaller the pile group, the more this effect can produce an unstable structural condition. Where this is considered to be a strong possibility, piles are sometimes deliberately driven at an angle, with the outer piles in a group splayed out for increased lateral stability of the group. While not often utilized in buildings, this splaying out, called battering, of the outer piles is done routinely for foundations for isolated towers and bridge piers in order to develop resistance to lateral forces.

Timber piles are somewhat limited in their ability to accommodate to variations in driven length. In some situations the finished length of piles can only be approximated, as the actual driving resistance encountered establishes the required length for an individual pile. Thus the specific length of the pile to be driven may be either too long or too short. If too long, the timber pile can easily be cut off. However, if it is too short, it is not so easy to splice on additional length. Typically, the lengths chosen for the piles are quite conservatively long, with considerable cutting off tolerated in order to avoid the need for splicing.

Cast-in-Place Concrete Piles. Various methods are used for installing concrete piles for which the shaft of the pile is cast in place in the ground. Most of these systems utilize materials or equipment produced by a particular manufacturer, who in some cases is also the installation contractor. The systems are as follows:

1. *Armco system.* In this system a thin-walled steel cylinder is driven by inserting a heavy steel driving core, called a mandrel, inside the cylinder. The cylinder is then dragged into the ground as the mandrel is driven. Once in place, the mandrel is removed for reuse and the hollow cylinder is filled with concrete.

2. *Raymond Step-Taper pile.* This is similar to the Armco system in that a heavy core is used to insert a thin-walled cylinder into the ground. In this case the cylinder is made of spirally corrugated sheet steel and has a tapered vertical profile, both of which tend to increase the skin friction.

3. *Union Metal Monotube pile.* With this system the hollow cylinder is fluted longitudinally to increase its stiffness, permitting it to be driven without the mandrel. The fluting also increases the surface area, which tends to add to the friction resistance for supporting loads.

4. *Franki pile with permanent steel shell.* The Franki pile is created by depositing a mass of concrete into a shallow hole and then driving this concrete "plug" into the ground. Where a permanent liner is desired for the pile shaft, a spirally corrugated steel shell is engaged to the concrete plug and is dragged down with the driven plug. When the plug has arrived at the desired depth, the steel shell is then filled with concrete.

5. *Franki pile without permanent shell.* In this case the plug is driven without the permanent shell. If conditions require it, a smooth shell is used and is withdrawn as the concrete is deposited. The concrete fill is additionally rammed into the hole as it is deposited, which assures a tight fit for better friction between the concrete and the soil.

Both length and load range is limited for these systems, based on the size of elements, the strength of materials, and the driving techniques. The load range generally extends from timber piles at the lower end up to as much as 400 kips for some systems.

Precast Concrete Piles. Some of the largest and highest-load-capacity piles have been built of precast concrete. In larger sizes these are usually made hollow cylinders, to reduce both the amount of material used and the weight for handling. These are more generally used for bridges and waterfront construction. A problem with these piles is establishing their precise in-place length. They are usually difficult to cut off as well as to splice. One solution is to produce them in modular lengths with a typical splice joint, which permits some degree of adjustment. The final finished top is then produced as a cast-in-place concrete cap.

In smaller sizes these piles are competitive in load capacity with those of cast-in-place concrete and steel. For deep water installations huge piles several hundred feet in length have been produced. These are floated into place and then dropped into position with their own dead weight ramming them home. Precast concrete piles often are prestressed with high-strength steel bars or wires to limit tensile stresses during driving.

Steel Piles. Steel pipes and **H**-sections are widely used for piles, especially where great length or load capacity is required or where driving is difficult and requires excessive driving force. Although the piles themselves are quite expensive, their ability to achieve great length, their higher load capacity, and the relative ease of cutting or splicing them may be sufficient advantages to offset their price. As with timber piles of great length, their relative flexibility presents the problems of assuring exact straightness during driving.

Pile Caps. When a group of piles support a column or pier, load transfer is accomplished through a pile cap. The piles are driven close together to keep cap size to a minimum. The exact spacing allowable is related to the pile size and the driving technique. Ordinary spacings are 2 ft 6 in. for small timber piles and 3 ft for most other piles of the size range ordinarily used in building foundations.

Pile caps function much like column foot-

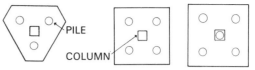

Fig. 7.7. Caps used for groups of three, four and five piles.

ings, and will generally be of a size close to that of a column footing for the same total load with a relatively high soil pressure. Pile layouts typically follow classical patterns, based on the number of piles in the group. Typical layouts are shown in Fig. 7.7. Special layouts, of course, may be used for groups carrying bearing walls, shear walls, elevator towers, combined foundations for closely spaced columns, and other special situations.

Although the three-pile group is ordinarily preferred as the minimum for a column, the use of lateral bracing between groups may offer a degree of additional stability permitting the possibility of using a two-pile group, or even a single pile, for lightly loaded columns. This may extend the feasibility of using piles for a given situation, especially where column loads are less than that developed by even a single pile, which is not uncommon for single-story buildings of light construction and a low roof live load. Lateral bracing may be provided by foundation walls or grade beams or by the addition of ties between pile caps.

Drilled-in Piers

When loads are relatively light, the most common form of pier is the drilled-in pier consisting of a vertical round shaft and a bell-shaped bottom, as shown in Fig. 7.6d. When soil conditions permit, the pier shaft is excavated with a large auger-type drill similar to that used for large post holes and water wells. When the shaft has reached the desired bearing soil strata, the auger is withdrawn and an expansion element is inserted to form the bell. The decision to use such a foundation, the determination of the necessary sizes and details for the piers, and the development of any necessary inspection or testing during the construction should all be done by persons with experience in this type of construction.

This type of foundation is usually feasible only when a reasonably strong soil can be reached with a minimum-length pier. The pier shaft is usually designed as an unreinforced concrete column, although the upper part of the shaft is often provided with some reinforcement. This is done to give the upper part of the pier some additional resistance to bending caused by lateral forces or column loads that are slightly eccentric from the pier centroid.

The usual limit for the bell diameter is three times the shaft diameter. With this as an upper limit, actual bell diameters are sometimes determined at the time of drilling on the basis of field tests performed on the soil actually encountered at the bottom of the shaft.

Where subgrade rock is within a practical depth, the bell may be eliminated. Reinforced with a structural shape, such as an H-beam, socketed in the rock (Fig. 7.6e), a drilled in pier can support very large loads.

One of the advantages of drilled piers is that they may usually be installed with a higher degree of control on the final position of the pier tops than is possible with driven piles. It thus becomes more feasible to consider the use of a single pier for the support of a column load. For the support of walls, shear walls, elevator pits, or groups of closely spaced columns, however, it may be necessary to use clusters or rows of piers.

References

J. Bowles, *Foundation Analysis and Design*, 3rd ed., McGraw-Hill, 1982.

H. Winterkorn and H. Fang, *Foundation Engineering Handbook*, Van Nostrand Reinhold, 1975.

Sweets Architectural File, for various priority systems for piles and drilled piers.

Words and Terms

Pile: friction, end-bearing, (see next Section 7.6), cap for
Pier: caisson, belled, drilled.
Kips

Significant Relations, Functions and Issues

Need for deep foundation.
Selection of foundation type and construction method.
Provisions for lateral and uplift forces.
Determination of vertical load capacity.
Required testing, before, during, and after installation.

7.6. LATERAL AND UPLIFT FORCES ON STRUCTURES

While resistance to vertical force is the primary function of foundations, there are many situations in which horizontal and uplift loads develop. The following are some types of structures and situations involving such actions.

Basement Walls

Basement walls are vertical load-bearing walls, but they must also resist inward soil pressures on their outside surfaces. The horizontal soil pressure is usually assumed to vary in magnitude with the distance below grade, as shown in Fig. 7.8, with the soil acting in the manner of a fluid. For investigation the equivalent fluid soil is assumed to have a unit density of approximately one third of its actual weight. It is also common to assume some surcharge effect, due to either an overburden of soil above the surface, a sloping ground surface, or a wheel load from some vehicle near the building. The typical horizontal pressure loading for a basement wall is therefore that represented by the trapezoidal distribution shown in Fig. 7.8b. In addition to these functions, basement walls may also serve as beams when they must support columns directly or must span between isolated footings or pile caps. Finally, they must serve as exterior walls for any subgrade occupied spaces, and must prohibit water penetration and limit thermal transmission.

Freestanding Walls

These are walls supported only by their foundation bases. They may occur inside buildings as partition walls, but occur more often as ex-

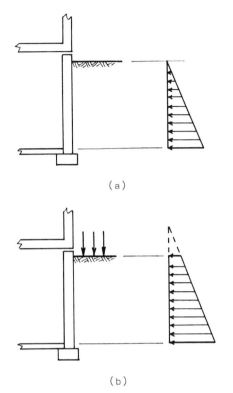

(a)

(b)

Fig. 7.8. Horizontal soil pressure on basement wall (a) without surcharge. (b) with surcharge.

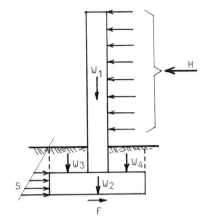

Fig. 7.9. Total effect of gravity and lateral forces on a freestanding wall. W₁ = weight of wall, W₂ = weight of footing, W₃ and W₄ = weight of soil, S = passive soil pressure, F = friction.

7.10*a*. If the shear wall is an interior wall, it may be built as a freestanding wall, with the combination of active and resistive forces shown in Fig. 7.10*b*. The single interior shear wall is seldom actually free, however, and may resolve horizontal forces through elements of the floor or basement construction, as shown in Fig. 7.10*c*.

Retaining Walls

Changes in ground elevations that occur gradually can be achieved by simply sloping the soil. When abrupt changes must be made, however, some type of soil-retaining structure is required; the type used depending largely on the height difference to be achieved. Small changes of a foot or so can be accomplished with a simple curb, but for greater heights, a cantilevered retaining wall often is used. For heights from a few feet up to 10 ft or so, a common form is that shown in Fig. 7.11*a*. The wall may be built of masonry or solid concrete or core-grouted, concrete block. For very tall walls it is common to use some form of bracing, as shown in Fig. 7.11*b*.

Low walls may be designed for the equivalent fluid pressure described for basement walls, although a more rigorous investigation relating to specific properties of the retained

terior walls or fences. The foundations must support the weight of the wall, but must also develop resistance to the horizontal forces of wind or earthquakes. The combination of vertical and lateral forces results in pressures on the bottom of the foundation that vary as illustrated in Fig. 7.5. The lateral loads will be resisted by a combination of passive soil pressure on the side of the footing and pressure on the buried portion of the wall, plus sliding friction on the bottom of the footing. The action of the active and resistive forces is shown in Fig. 7.9.

Shear Walls

Shear walls (Sec. 8.16) often occur as portions of exterior walls, with their support provided by either a continuous basement wall or a grade beam (a beam at ground level). In such cases, horizontal forces on the wall will produce shear and bending in the supports as shown in Fig.

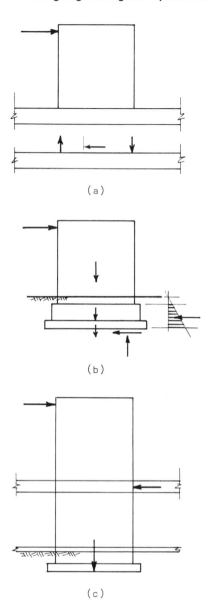

Fig. 7.10. Actions of shear wall foundations. (a) Wall on continous foundation. (b) Freestanding wall. (c) Wall restrained by the building construction.

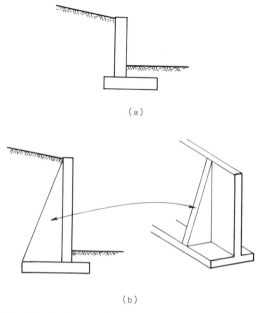

Fig. 7.11. Forms of retaining walls. (a) Low wall, cantilevered from footing. (b) Tall wall with counterfort braces.

Abutments

It is occasionally necessary to provide a form of foundation for a permanent combination of vertical and horizontal forces, such as at the base of an arch. This structure is called an abutment, and the simplest form is that shown in Fig. 7.12a. Whereas footings that are subject to lateral loads from wind or earthquakes may be designed for the uneven soil pressures shown in Fig. 7.5, such pressures, when the lateral load is permanent, will result in some tilting of the foundation. It is therefore desirable to have the line of action of the resultant load pass through the centroid of the footing bearing area, as shown in Fig. 7.12b. When the horizontal force is large, or the resultant load on the abutment is at a very low angle, or the load application occurs a considerable distance above the footing, it may be necessary to use an off-center footing or one with a non-rectangular bearing area, to get the footing centroid in the proper location. The structure in Fig. 7.12c indicates the use of a T-shaped footing for this purpose, and also shows the use of an intermediate grade beam to reduce the bending in the footing. For structures such as arches or ga-

soil is usually made for high walls. When the ground surface slopes significantly (at more than 1:5 or so) there is some added pressure which is usually accounted for by using a surcharge effect, as shown in Fig. 7.8. A major objective for retaining walls is to prevent collection of water in the retained soil behind the wall, usually achieved by installing through-the-wall drains and a coarse-grained, porous fill behind the wall.

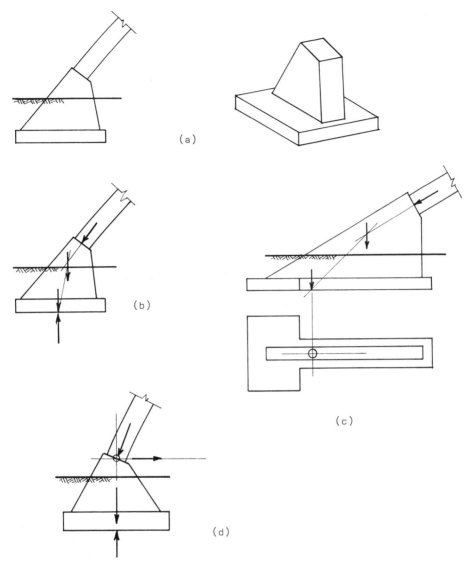

Fig. 7.12. Forms of abutments for arches. (a) Simple abutment with rectangular footing. (b) Force resolutions to obtain uniform soil pressure for (a). (c) Abutment with T-shaped footing. (d) Abutment for a tied arch; develops only vertical resistance with a horizontal tie at the hinge at the base of the arch.

bled frames, it may be possible to provide a cross-tie that resists the horizontal force without involving the footing, as is shown in Fig. 7.12d; in which case the footing is simply designed for the vertical load.

Lateral Loads on Deep Foundations

Piles and piers offer very limited resistance to lateral forces at their tops. For buildings, the usual solution is to use ties and struts in the foundation construction to transfer the horizontal forces to basement walls or grade beams. Where this is not possible, it may be necessary to use battered piles (driven at an angle), drilled-in tiebacks, or other means to establish significant lateral resistance. A special structure sometimes used in waterfront or hillside conditions is that shown in Fig. 7-13c, in which piles or piers are developed to provide a rigid frame action with a horizontal concrete frame system; called a *downhill frame*. Additional

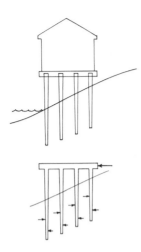

Fig. 7.13. Action of a downhill frame, functioning as a fixed-base rigid frame.

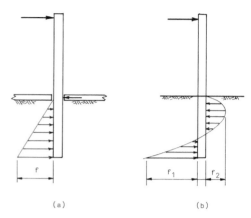

Fig. 7.14. Lateral resistance of buried poles. (a) with ground-level restraint. (b) with soil resistance only.

stiffness of such a structure is obtained if short drilled-in piers can be inserted a sufficient distance into solid rock to provide fixity at their lower ends. If piles or piers must develop lateral resistance without any of these measures, they must usually do so in the manner of pole structures, described as follows.

Pole Structures

Where good timber poles are plentiful, they may be used for a building foundation. The poles may be driven-in piles—mostly in waterfront locations—but may instead be partly buried in excavated holes. In one form of construction—called pole platform construction—the poles extend up to provide a platform on which the building is erected. The other basic system—called pole frame construction—uses the poles as building columns, with floors and roofs framed by attachment to the extended poles. For lateral forces, the buried poles function in one of the two ways shown in Fig. 7.14. In Fig. 7.14a the pole is restrained at ground level—for example, by a concrete floor slab. This results in a rotation about ground level and development of the lateral soil pressure indicated. If no restraint exists, the poles behave as shown in Fig. 7.14b, and their resistance is determined by empirical formulas. (See the Uniform Building Code or specific references on pole construction.)

Uplift Forces

Resistance to uplift forces may be achieved in various ways. The magnitude of the force is a major consideration, resulting in modest response for a simple tent stake and monumental construction for the end anchorage of suspension bridge cables. The resisting force may be developed by engaging sufficient soil mass or by creating a constructed dead weight. Concrete foundations themselves tend to be sufficient to anchor most light structures. Large piers may be used for the value of their own dead weight, with some more resistance offered if they have belled bottoms, as shown in Fig. 7.6d. Major problems occur mostly with very tall structures, or single-footed structures, such as signposts or light towers, or tall shear walls for which the overturning moment exceeds the gravity restoring moment.

Relatively long piles will develop significant resistance to pullout, which may be used for tension resistance. For these, the reliable design load capacity must usually be determined by load tests. For concrete piles or piers, tension development will require considerable vertical reinforcement, representing a major increase in cost.

References

J. Ambrose, *Simplified Design of Building Foundations*, 2nd ed., Wiley, 1988.

J. Bowles, *Foundation Analysis and Design*, 3rd ed., McGraw-Hill, 1982.

Words and Terms

Abutment
Active soil pressure
Battered pile
Downhill frame
Overturning moment
Passive soil pressure
Pole frame construction
Soil friction
Surcharge
Drilled-in tieback

Significant Relations, Functions and Issues

Active pressure on retaining structures: affected by height, soil type, water, surcharge.
Development of resistance to horizontal movement in soils: affects of soil type, depth below grade.
Selection of type of retaining structure related to height of retained soil.

7.7. SITE DEVELOPMENT CONSIDERATIONS

General development of the building site may be a simple matter or a major part of the project. Extensive site construction or need for major corrective efforts on difficult sites may make the site work a considerable design and construction planning undertaking on its own. As site design and construction is not the major topic of this book, we will consider only a few of the typical situations that relate to the building planning and the design of foundations.

Finish Grading—Cut and Fill

Most sites require some degree of finish ''trimming'' to accommodate the building and develop necessary walkways, drives, landscaping, and so on. Economy and ease of the construction is generally served best if there is a minimum requirement to either take away or bring to the site significant amounts of soil materials. This translates to a desire to balance the cuts (below existing grade) with the fills (above grade) where possible. In this regard the excavation for the building foundations usually represents a major cut, unless the finish level of the grade is to be raised a significant amount.

The ability to do any cross-trading of soil materials on the site will of course depend on the nature of the existing materials. Existing surface soils may not make good backfill and excavated materials may not be good for landscaping work.

Removal of Objectionable Materials

Site materials not useable as backfill, paving subgrades, or planting fill may have to be removed—and, if necessary, replaced with imported materials. This can only be determined after considerable information about the site is obtained and the site and building designs are carried to a relatively complete stage of design. Removal and replacement of soil are an expense to be avoided if possible and may affect the siting of the building and general plans for site development. It is also possible that site materials objectionable for one project may be useful for another, making trading of materials possible. Some form of soil modification may also be possible, as described in the following.

Soil Modification

The work of excavating and grading is a form of modification of a sort; materials may be basically unchanged, but soil structures are altered. In the process of moving the soils, some new materials may be introduced, resulting in some new character for the soil. Modifications may also be made of unexcavated soils. Some possibilities for this are as follows:

1. *Consolidation* by vibration or overburden (stacking soil on the site) or by flooding to dissolve the bonds in highly voided, cemented soils.
2. *Surface compaction* for better pavement bases.
3. *Infiltration of fine materials*, such as cement or bentonite clay, to reduce voids and lower the permeability of fine, loose sands.

Modifications may have the basic purpose to improve the soil conditions, or may be done to cause certain unstable effects to occur in advance of construction. Flooding to cause collapse in cemented void soils may be done to prevent large settlement after construction, when extensive irrigation for plantings may cause the ground to sink. This is a common occurrence in arid climates where irrigation is used around buildings.

Surface Drainage

Control of surface water runoff is a major concern in developing the finished site contours and site construction. This also needs to be coordinated with the building roof drainage and measures undertaken to prevent water buildup behind retaining walls and on the outsides of basement walls. Care must be taken not to create channels of water flowing onto neighboring properties, unless they existed as established streams before site development in which case damming or otherwise obstructing them may be objectionable.

Site Construction

Construction of pavements, retaining walls, ditches, planting structures, pools and other site structures may be essentially separate from or simply an extension of the building construction. Although overall integrity of the construction is to be desired, less conservativism may be exercised in the ancillary design work unless safety may be adversely affected. Exterior exposure conditions for this work make detailing of the construction quite responsive to local climate conditions. Concrete admixtures or types of special cements, depth of footings below grade, and control joints in pavements are some of the variables in this regard. Local codes may provide some guidance, but this is more a matter of evolved local practices.

Utilities

Electric power, telephone lines, gas, water, and sewer services may all be delivered underground. The presence of existing main distribution or collection lines—especially gravity-flow sewers—may offer important constraints to site planning or to the general siting of the

building. Consideration in planning should also be given to the following: Underground tunnels or vaults may require vents or manholes for access which must be allowed for in the site planning. Penetration of service lines into the building must be accommodated by the building foundations and basement walls. General access for modification, maintenance or repair should also be considered.

Landscaping

Except for very tight urban locations, most building sites will have some form of landscaping with some amount of plantings. The following should be investigated with regard to possible effects on the building:

1. *Adequate provision for plant growth.* Plantings may occur over underground structures. Hence, adequate depth for plant growth and space for the necessary waterproofing are needed. Siting and vertical positioning of building spaces must make adequate provisions for planting requirements.
2. *Provision for effects of irrigation.* Frequent watering may saturate the soil and cause uneven settlement of retaining walls, basement walls or buried utilities, or may cause sinking of cemented void soils.
3. *Roots of trees and shrubs.* These may intrude into basements or utility tunnels, or may get beneath some light foundations and push them up or sideways.

Provision for these, or for other potential adverse conditions, may well offer opportunities for systems integration during the design stages.

Reference

J. DeChiara and L. E. Koppelman, *Time-Saver Standards for Site Planning*, McGraw-Hill, 1984.

Words and Terms

Site grading
Soil Modification

Significant Relations, Functions, and Issues

Balancing of cut and fill in grading.
Effects of construction activity on surface and subsurface soils.
Modifications to improve soil properties.
Intrusion of buried utilities in site development.
Provisions for and effects of landscaping and irrigation.

7.8 COFFERDAMS AND FOUNDATION WALLS

Depending on the type of soil, shallow basements or cellars may be excavated and foundation walls and footings constructed in stiff soils with no lining of the earth perimeter or, in weak soils, with braced wood sheeting to prevent collapse of the earth sidewalls. Deep cellars and high foundation walls require that excavation be carried out within an enclosure to keep out water and to prevent earth sidewalls from collapsing. A cofferdam generally is used for the purpose.

A cofferdam is a temporary wall used for protecting an excavation. One of its most important functions usually is to permit work to be carried out on a dry, or nearly dry, site.

There are many different types of cofferdams, including simple earth dikes, cells filled with earth, braced single walls and double walls with earth between the walls. For excavations for buildings, braced single-wall cofferdams are generally used. Such walls though must be carefully constructed, especially if there are streets or other buildings nearby. Small inward movement of such cofferdams may cause cave-ins and damage to nearby construction.

Such movements and consequent damage can be prevented only by adequate bracing of the cofferdams. Not only must the bracing be strong enough to sustain imposed loads, such as earth and hydrostatic pressures and the weight of traffic outside the excavation, but it must also be seated on practically immovable footings, anchors or walls.

Single-Wall Cofferdams

Cofferdams for building excavations may be constructed in any of many different ways. Some of these are illustrated in Fig. 7.15.

Figure 7.15a shows a type of single-wall cofferdam that is often used when dry conditions are expected during excavation. Structural steel piles, called *soldier piles*, are driven vertically into the ground, usually at intervals of 5 to 10 ft, around the perimeter of the planned excavation. Meanwhile, excavation proceeds in the central portion of the enclosure toward the depth required for the cellar and interior wall and column footings. As the soldier piles are placed with their bottoms embedded below the required depth of excavation, excavation starts between the piles. To support the earth sidewalls as work proceeds, wood boards, called *lagging*, are set horizontally between the soldier piles. Small gaps are left between the lagging to permit water to drain through and prevent buildup of hydrostatic pressures against the cofferdams. At intervals, as depth of the excavation increases, horizontal braces, called *wales*, are attached to the soldier piles. Also, the wales are braced with rigid struts to an opposite cofferdam wall or with diagonal struts, called *rakers*, extending to rigid supports in the ground.

When wet conditions may occur during excavation, the single-wall cofferdam may preferably be constructed with sheetpiles (see Fig. 7.15b to f). *Sheetpiles* are thin structural steel shapes fabricated to interlock with each other along their edges when they are driven into the ground. They are driven in the same way as other piles. In cross section, sheetpiles may be straight, channel (C) shaped, or zees. Special sections are fabricated for special purposes, for example, for forming cofferdam corners. The sheetpiles, which form a continuous wall, may be braced, as excavation proceeds, in the same way as cofferdams with soldier piles and lagging; for example, with wales and rakers, as in Fig. 7.15b, or with cross-lot bracing, as in Fig. 7.15e and f.

These types of bracing, however, have the disadvantage that they tend to interfere with construction operations within the cofferdam. To avoid this disadvantage, means have been developed for placing bracing, other than wales, outside the cofferdam. For shallow excavations, for example, the top of the cofferdam may be tied back to a concrete anchor, or *dead man*, buried in the soil outside the enclo-

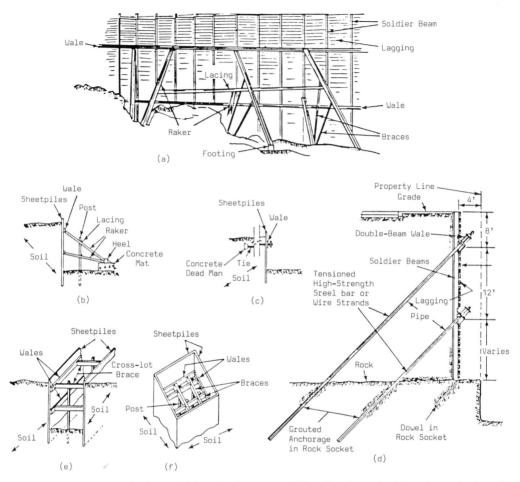

Fig. 7.15. Single-wall cofferdams. (a) Lagging between soldier piles, braced with wales and rakes. (b) Sheet pile with braces. (c) Sheet piles with tieback to deadman. (d) Wall with drilled-in tiebacks. (e) Cross-braced walls in narrow excavation (trench). (f) Walls with two-way cross-lot bracing.

sure (see Fig. 7.15c). For deep excavations, for which bracing is required at several vertical intervals, wales at those levels may be restrained by tensioned, high-strength steel bars or wire strands anchored in rock (see Fig. 7.15d). For the purpose, holes are drilled on a diagonal through the soil outside the cofferdam until rock is penetrated. Next, a pipe is placed in each hole to maintain the opening. A steel bar or several wire strands are inserted in the pipe, and one end is anchored with grout in the rock socket while the opposite end is fastened to hydraulic jacks set on a wale. The jacks apply a high tension to the bar or strands, which then are anchored to the wale. The resulting forces restrain the cofferdam against inward movement under earth and hydrostatic pressures.

A cofferdam also may be constructed with precast-concrete panels or by forming continuous walls by casting in place concrete piles in bored holes.

Another method that may be used is the slurry-trench method, which permits construction of a concrete wall in a trench. The trench is excavated in short lengths. As excavation proceeds, the trench is filled with a slurry of bentonite, a mixture of water and fine inorganic particles. The fluid pressure of the slurry prevents the sidewalls of the trench from collapsing. Concrete is then placed in the trench, replacing the slurry.

After excavation has been completed within a soldier-pile or sheetpile cofferdam, formwork can be erected around the perimeter for the

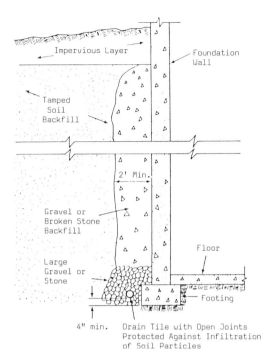

Fig. 7.16. Provision for drainage at a foundation wall.

foundation walls. Finally, concrete is placed within the framework to form the walls. After the concrete has hardened, the formwork and cofferdam can be removed, for reuse elsewhere. Drains should be placed behind the walls along wall footings, to conduct away water, and a porous backfill, such as gravel, should be placed against the wall to allow water to seep down to the drain (see Fig. 7.16).

References

J. Bowles, *Foundation Analysis and Design*, 3rd ed. McGraw-Hill, 1982.
G. Sowers, *Introductory Soil Mechanics and Foundations: Geotechnic Engineering*, 4th ed., Macmillan, 1979.
L. Zeevaert, *Foundation Engineering for Difficult Subsoil Conditions*, 2nd ed., Van Nostrand Reinhold, 1982.

Words and Terms

Cofferdam
Dead man
Drilled-in rock anchor
Lagging
Raker
Sheetpiling
Soldier beam or pile
Tieback
Wale
Slurry-trench method

Significant Relations, Functions and Issues

Bracing for excavation related to height of cut, construction in excavation, and protection of property or buildings adjacent to cut.
Drainage of backfill along cofferdams

7.9. DEWATERING OF EXCAVATIONS

Several construction operations must be carried out within an excavation for a building. These include erection of formwork and placing of concrete for footings, walls, piers, columns and floors and perhaps also erection of steel columns and beams. These operations can be executed more efficiently if the excavation is kept dry.

Provision for dewatering therefore usually has to be made for excavations for buildings. Dewatering, however, also has other advantages than just permitting construction to be carried out in the dry. Removal of water makes excavated material lighter and easier to handle. Dewatering also prevents loss of soil below slopes or from the bottom of the excavation, a loss that can cause cave-ins. In addition, removal of water can avoid a *quick* or *boiling* bottom in the excavation; for example, prevent conversion of a fine sand to quicksand.

Often, an excavation becomes wet because the water table, or level of groundwater, is above the bottom of the excavation. To keep the excavation dry, the water table should be at least 2 ft, and preferably 5 ft, below the bottom of the excavation in most soils.

Any of several methods may be used for lowering the water table, when necessary, and for draining the bottom of the excavation. Information obtained from site exploration should be useful for deciding on the most suitable and economical dewatering method. This information should cover types of soil in and below the excavation, probable groundwater levels during construction, permeability of the soils and quantities of water to be removed. Pumping tests are useful in obtaining data for estimating capacity of pumps needed as well as indicating the drainage characteristics of the soils.

Dewatering Methods

When the groundwater table lies below the excavation bottom, water may enter the excavation only during rainstorms, or by seepage through side slopes or through or under cofferdams. In many small excavations, or where there are dense or cemented soils, water may be collected in ditches or sumps at the excavation bottom and pumped out. This is the most economical dewatering method.

Where seepage from the excavation sides may be considerable, it may be cut off with a sheet-pile cofferdam, grout curtains or concrete-pile or slurry-trench walls. For sheetpile cofferdams in pervious soils, water should be intercepted before it reaches the enclosure, to avoid high pressures on the sheetpiles. Deep wells or wellpoints may be placed outside the cofferdams for the purpose.

Deep wells, from 6 to 20 in. in diameter, are placed around the perimeter of the excavation to intercept seepage or to lower the water table. Water collecting in the wells is removed with centrifugal or turbine pumps at the well bottoms. The pumps are enclosed in protective well screens and a sand-gravel filter.

Wellpoints often are used for lowering the water table in pervious soils or for intercepting seepage (see Fig. 7.17). Wellpoints are metal well screens, about 2 to 3 in. in diameter and up to about 4 ft long, that are placed below the bottom of the excavation and around the perimeter. A riser connects each wellpoint to a collection pipe, or header, above ground. A combined vacuum and centrifugal pump removes the water from the header.

References

G. Sowers, *Introductory Soil Mechanics and Foundations: Geotechnic Engineering*, 4th ed., Macmillan, 1979.
L. Zeevaert, *Foundation Engineering for Difficult Subsoil Conditions*, 2nd ed., Van Nostrand Reinhold, 1982.

Words and Terms

Deep wells
Dewatering
Quick or boiling effect
Seepage
Well points

Significant Relations, Functions, and Issues

Conditions requiring need for dewatering.
Dewatering methods related to soil type, water level in soil and nature of excavation and bracing of cuts.

7.10. INVESTIGATION AND TESTING

Some investigation of site and subsurface conditions must be made for any building project. The extent of investigation and its timing varies

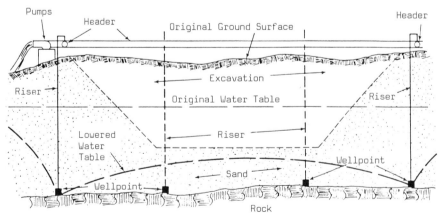

Fig. 7.17. Wellpoint installation for an excavation.

considerably, depending on the size and nature of the construction and the site conditions. Cost of investigative work must be considered, and its control may be quite important for small building projects. However, if serious problems exist, they must be sufficiently investigated, regardless of the project size.

Site surveys will provide considerable information for the general site development. For the building design, however, some investigation must be made of the subsurface conditions that affect the foundations and any subgrade construction. For small buildings with shallow foundations, a simple soil exploration may be sufficient, possibly conducted with very minor equipment, such as hand augers or post-hole diggers. Such minor investigations performed by persons experienced in geotechnical work can reveal considerable data, which may be significant for site development and for preliminary design of foundations.

For most building projects, however, permit-granting agencies will require some investigation consisting of deep soil sampling and the performance of minimal testing. Such soil explorations should be performed by experienced persons who will usually also make some recommendations for foundation design criteria and any special conditions that can be anticipated during the site work and excavation.

The form and extent of subsurface investigations, the equipment used to achieve them and the type of tests to be performed, all vary considerably. When the site is fully unexplored, the first investigations may be done primarily to establish what type of investigation is really required. In many situations it is possible to predict from past experience or previous construction on the site the likelihood of encountering specific conditions in the region of the site. Information is also forthcoming from various sources such as:

1. Government engineering, building or highway departments.
2. Government agricultural agencies or agricultural industry organizations.
3. Various agencies that conduct studies for water resources, such as erosion control, and seismic activity.

4. U.S. Armed Forces studies for nonclassified projects.

Most geotechnical engineering organizations in the soil investigation and testing business will make use of any such information that is available for regions they regularly serve, and they can thus work from a considerable database to predict conditions on any site.

In the end what is required is whatever investigation is necessary to assure the adequacy of the foundation design and to provide a base for planning of the site work and excavation for the building construction. This may present some problems in timing, as some soil exploration must be made before design can be done, while the investigation required must be based partly on knowing the building location, the type of construction, and the anticipated magnitude of foundation loadings. The result of this can sometimes produce some interaction between design and investigation—such as when the subsurface investigation shows that the desired location for the building is the worst possible place on the site for foundations.

Investigation and testing must generally address the establishing of three categories of information:

First is the identification of types of soils in the various typical layers that occur at different distances below the ground surface. Using the Unified System (Fig. 7.3) for classification, pertinent data are obtained. Simply identifying soils as one of the 15 categories in the Unified System does not yield all of the desired information for the engineering design work, however.

Second, depending on the soil type, additional data may be required. For sand—in addition to information for the classification only—structural behavior will be determined by relative density (from loose to dense), penetration resistance (measured as the number of blows required to advance a standard soil sampler), grain shape and water content. For clay the principal tested structural property is unconfined compressive strength, q_u. This property may be approximated by simple field tests, but must be more accurately determined by laboratory tests.

Third, information may be needed to evaluate the potential for using excavated soil materials for site-development work—for pavement subgrade bases, for plantings, for retaining wall backfill, for example. This often requires more information on surface and near-surface materials, which are not of major concern in the deep-soil investigation for foundation design criteria.

For large projects, for deep excavations or whenever major problems are anticipated during construction, it is often necessary to perform some tests during the performance of the excavation work. These may be done primarily only to verify data from previous investigations or to confirm assumptions made for the engineering design. For some types of foundations, however, there are investigations that are a normal part of the work. Foundations on rock, typically using very high levels of bearing pressure, require tests on the rock at the time of excavation, involving drilling to some length into the rock material encountered. End-bearing piles seated in rock must usually be load tested, as the usual empirical formulas using pile driving data do not work for this situation.

Soil properties are interrelated, so that a single tested property—such as unit weight—should relate to other tested properties, such as penetration resistance, permeability, consolidation and water content. This allows the inference of some properties from the identification of others, but more importantly permits cross-checking to verify reliability of investigative data.

The following is a description of various explorations and tests that are commonly made to support site and foundation design work.

Explorations

Visual inspection is an essential preliminary step for building designers, foundation consultants, construction managers, and prospective construction bidders. Visual inspection should provide information on surface soils, rock outcroppings, surface water, slopes, accessibility for equipment for subsurface exploration, grading and excavating, availability of water for drilling equipment, existing structures on the site, former structures on the site and adjacent structures. If possible, inspection should determine whether underground utilities may be passing through the site. Small truck-mounted auger drills, may be used for quick visual soil analysis and location of groundwater below grade.

Test pits are holes dug on a site to investigate soil conditions. They permit visual examination of soil in place and provide information on the difficulty of digging. They also make it possible to obtain an undisturbed sample of the soil manually. Cost of digging a test pit, however, increases with depth. Hence, this method is limited for economic reasons to relatively shallow depths below grade.

Borings usually are resorted to for sampling soils at greater depths than those desirable with test pits. Boring is a drilling process in which a hole is formed in the ground for soil sampling or rock drilling. The hole may be protected by insertion of a steel casing or with drilling mud, a slurry of water and clay. Borings may be carried out in several different ways, but the most satisfactory results are usually obtained by ''dry'' sampling.

''Dry sampling has the objective of obtaining a complete sample of the natural soil. For the purpose, a hole is drilled; for example, with a hollow-stem auger. Samples are obtained by lowering a drill rod with a sampler on the bottom end through the hollow-stem auger to the bottom of the hole. Then, the sampler is driven beyond the lead point of the auger to secure a sample of the soil.

Any of several different types of samplers, also called spoons, may be used to obtain a dry sample from a borehole. Thin-walled types cause less soil disturbance, but thick-walled types are preferable for sampling stiff, nonplastic soils. Some samplers have sectional linings for collecting samples, to permit delivery of samples to a laboratory without manual handling of the soil.

A sampler is driven into the bottom of a hole with a free-falling weight. Standard practice is to use a 140-lb weight falling 30 in. on a spoon with 2-in. outside diameter. The number of blows required to drive the spoon are recorded for each foot of penetration into the soil. This record is useful as a measure of the soil resistance encountered and may be used for soil classification. If undisturbed soil samples can be obtained, however, unconfined compression tests on them in a laboratory can be of greater value to engineers than the number of spoon blows. Shearing strength, for example, equals one-half the unconfined compressive strength of the soil.

Rock samples generally are obtained in the form of rock cores. For the purpose, rotary drilling with shot or diamond bits is used. A complete core, however, may not always be obtainable. Hence, investigators should report the percentage of recovery of rock, the ratio of length of core obtained to distance drilled. Generally, the higher the percentage of recovery, the better the condition of the rock. Note, however, that recovery also depends on care taken in sampling and on type of bit used in drilling.

Water-table depth and its variation over a period of time should be reported when a site investigation is made, because the presence of groundwater can affect foundation design and construction. One method is to dig a permanent observation well and take weekly or monthly readings of the water level. Another method is to take readings in boreholes.

Local building codes may specify the minimum number of borings or test pits in terms of building area, for example, one for every 2,500 sq ft of ground area. These codes also may specify the depth for the holes and pits. At least one boring, though, should extend into bedrock.

Soil Tests in the Field

Properties of soils from which predictions of their behavior under building loads may be made sometimes can be determined directly in the field. Such tests may measure soil density, permeability, compressibility or shearing strength. The tests may be made at the bottom of test pits or of casings driven into the soil or inserted in boreholes. After the casings have been cleaned out, any of several different kinds of tests may be carried out on the exposed bottom. Common tests include loading of cone-shaped plungers or tight-fitting bearing plates to measure resistance of the soil to penetration or compression. Sometimes, a vane shear test may be made. In this test, a rod with two to four vertical plates, or vanes, at the tip is inserted into the soil and rotated. The torques required to start and maintain rotation can be correlated with shear resistance and internal friction. The data can be used to estimate soil bearing capacity and pile friction resistance and hence pile length required.

Bearing capacity of a soil to support footings often is estimated from load tests made in the field. The tests are made on a small area of soil in a pit at the level of the footings to be built. In the tests, loads are applied in increments to a bearing plate resting on the soil. Hydraulic jacks or weights may be used to load the plate. Settlement of the plate after each increment has been in place 24 hr is recorded, but sometimes sizable settlements continue for a long time, in which case it is necessary to wait until settlement stops. The data recorded usually are plotted to form a load-settlement curve.

In load tests of a typical footing for a building, loads are continually increased to 150 or 200% of the expected footing design load. The design load may be considered acceptable if, when applied in the test, the load does not cause a settlement exceeding a specified amount, for example, $\frac{3}{4}$ in. Also, under the maximum applied load, settlement must be nearly proportional to that under the design load. Building codes generally prescribe the procedure for making a load test.

Soil Tests in Laboratories

Any of many different laboratory tests may be made on soils to identify those present, determine their properties and predict their behavior under building loads. Usually, however, only a few different tests are necessary. More tests may be needed though for foundations with heavy or dynamic loads and those on weak or unreliable soils. Among the more commonly made tests are the following:

Mechanical analyses are performed to determine the percentages of different size particles in a soil sample. Different sizes of sieves are used to separate coarse particles. Fine particles are separated by sedimentation, usually by hydrometer test. The gradations measured in such analyses can be used to indicate the type of soil and a wide variety of properties, such as permeability, frost resistance, compactability and shearing strength.

Density determinations are made to measure the compressibility of soils. Loosely packed soils are more compressible than compact ones.

Compaction tests are made to determine the maximum density that can be achieved for a soil. These tests provide data for later use in the field to insure that the desired degree of compaction of a fill is achieved with compaction equipment.

Moisture-content determinations are made for use in estimating soil compactability and compressibility and to predict the shearing strength of clays at varying water contents.

Consistency tests may be made on fine-grained soils to predict their shearing strength at varying water contents. Atterberg-limit tests, for example, are made to determine the water contents that change fine-grained soils from solids to semisolids, then to plastic materials, and finally to liquids.

Permeability tests are conducted to estimate subsurface water flow, such as artesian flow and flow under sheetpiling.

Compression tests of various types are made to determine the compressibility and shearing strength of soils.

Consolidation tests are made to obtain information for predicting anticipated settlement of soils under building loads.

Direct shear tests are conducted to determine the bearing capacity of soils under building loads and the stability of ground slopes.

The value of information obtained from laboratory tests depends greatly on the care with which samples are extracted, locations from which samples are taken, and the care in storing, handling and delivering samples to the laboratory, as well as the execution of the tests. Intelligent interpretation of test results also is important.

References

J. Bowles, *Foundation Analysis and Design*, 3rd ed., McGraw-Hill, 1982.
G. Sowers, *Introductory Soil Mechanics and Foundation: Geotechnic Engineering*, 4th ed., Macmillan, 1979.

Words and Terms

Borings
Rock cores
Soil sampling (dry)
Test pits
Density
Penetration resistance
Compaction
Consistency
Consolidation

Significant Relations, Functions and Issues

Types of investigations and tests related to soil conditions and nature of construction.
Timing of soil investigations related to schedule of design work and construction.
Importance of reliability of investigation and interpretations of data.
Field tests for soil density, permeability, compressibility, shearing strength, unconfined compressive strength
Laboratory tests for particle size and gradation, density, water content, consolidation

7.11. SYSTEMS-DESIGN APPROACH TO SITE ADAPTATION

Two cases must be considered in systems design of a building. One is the situation where a site has already been purchased and the building must be adapted to it. The second is the situation where a site has not yet been selected before design starts. In the latter case, the building designers may be able to influence purchase of a site that will permit use of the most economical foundations and that will have mostly beneficial effects on building design and construction.

The lot and foundations may be treated as a major subsystem of the building, but for brevity will be referred to as the site-foundation system. The systems-design steps illustrated in Fig. 3.4 should be applied to this system.

There are complications, however, because the site-foundation system plays conflicting roles in building design. The lot is both the initial and final subsystem in a sequence of subsystems comprising a building. The lot affects design of the basic subsystems, such as the building envelope (shape, plan dimensions and number of stories, or height), orientation of the building on the site and access to the building from streets and highways. In contrast, as the end subsystem, the site-foundation system does not affect the design of the building structural system, after the type and geometry of the building system has been selected. In view of the associated complications of the conflicting roles, a practical procedure is to prepare schematics of the building superstructure, with due consideration to site characteristics, and then to apply systems design to site adaptation and preliminary design of foundations.

Data Collection and Problem Formulation

In selection of a site, reconnaissance, surveys and preliminary soil sampling, as well as purchase-cost comparisons, should provide information to guide decision making. After a site has been bought, data should be collected to guide site adaptation and foundation design, as described in Sec. 7.2 and 7.3. Site surveys and soil investigations should provide information on size and shape of lot, surface conditions, slopes, rock outcroppings, underlying soils, water table, access to site, utilities available, possible interferences with construction operations, and neighbors and adjoining construction. In addition, the possible effects of build-

ing construction on neighbors and adjoining construction should be determined.

The goal may be stated succinctly as: to adapt the building to the site or the site to the building to meet the owner's goal for the building.

Objectives

Start of a listing of objectives depends on whether or not a site has been selected. When a site has not been selected, one objective is to select a site that gives designers freedom to design a building that meets the owner's goal efficiently and economically. Given a site, one objective is to choose sizes, shape and orientation of the building to make the most efficient use of the site. Other objectives then may be listed to give details on what is required of the building and site. Included should be the important objective of selecting the most economical foundation system for the building size and the site surface and subsurface conditions and for meeting requirements for supporting superstructure walls and columns.

Constraints

Numerous constraints may be imposed on site adaptation and foundation design. Among the most important are the lot size and shape, surface and subsurface conditions that make building design and construction difficult or costly, locations of building walls and columns, site grading and drainage, depth of excavation required and cofferdams needed, and provision of access to the lot and the building. Also important are considerations that have to be given to community relations, neighboring construction, building-code requirements and zoning regulations.

Synthesis and Analysis

Schematics of the superstructure should be prepared to meet site-foundation system objectives as well as superstructure objectives. At the same time, the schematics should satisfy the constraints imposed on both the site-foundation system and the superstructure. Analysis of the proposed design should verify that the objectives and constraints are indeed met.

If the owner approves the schematics, preliminary designs of alternative foundation systems should be developed. Costs of these systems should then be estimated.

Value Analysis and Appraisal

The benefits and costs of the alternative foundations should be compared. The evaluation should lead to selection of the optimum type of foundations for the building site. Then, the optimum foundations of the chosen type should be designed. Inasmuch as the foundations are the final subsystem in the sequence of subsystems composing the building system, the type of foundations selected for the preliminary design will be unaffected by the superstructure preliminary design, unless it is changed from that shown in the schematics.

GENERAL REFERENCES AND SOURCES FOR ADDITIONAL STUDY

These are books that deal comprehensively with several topics covered in this chapter. Topic-specific references relating to individual chapter sections are listed at the ends of the sections.

F. Merritt, *Building Design and Construction Handbook*, 4th ed., McGraw-Hill, 1982.

G. Sowers, *Introductory Soil Mechanics and Foundations: Geotechnic Engineering*, 4th ed., Macmillan, 1979.

J. Bowles, *Foundation Analysis and Design*, 3rd ed., McGraw-Hill, 1982.

J. Ambrose,*Simplified Design of Building Foundations*, 2nd ed., Wiley, 1988.

J. De Chiara andd L. E. Koppelman, *Time-Saver Standards for Site Planning*, McGraw-Hill, 1984.

C. Ramsey and H. Sleeper, *Architectural Graphic Standards*, 9th ed., Wiley, 1988.

EXERCISES

The following questions and problems are provided for review of the individual sections of the chapter.

Section 7.1

1. Why do construction plans for a building on one site have to be revised for use for an identical building on a different site?
2. An owner has purchased a 20,000-sq ft site for an office building. Studies show the need for at least 25,000 sq ft of floor area for the building. What effect will this have on design of the building?
3. What site of each of the following pairs should a designer recommend for a large factory?
   (a) Triangular lot or square lot, both with the same area.
   (b) Level lot or one with a 2% slope.
   (c) Lot with rock at the surface or one with rock about 20 ft below the surface.
   (d) Lot on a two-way service road one half mile from a freeway entrance/exit or a lot on a one-way service road midway between entrance/exits one mile apart.
4. Describe at least three ways in which zoning affects selection of a site for:
   (a) A one-family dwelling.
   (b) A factory.
   (c) A high-rise office building.

Section 7.2

5. An architect is recommending to a client purchase of a ten-acre parcel for development as a shopping center. What map should be examined to determine:
   (a) The potential market to be served by the shopping center?
   (b) Availability of water, sewers, electricity and gas?
   (c) Access roads?
   (d) Who owns adjoining property?
   (e) Current occupancy usage of nearby properties?
   (f) Whether zoning would permit a shopping center to be built?
6. Who should prepare the survey plan for a site?

Section 7.3

7. What two materials ordinarily take up the void in a soil?

8. What are the principal differences between cohesionless and cohesive soils?
9. If the specific gravity of the solid particles in a soil is 2.65 and the tested void ratio is 0.3, what is the unit weight of the dry soil in lb/ft^3?
10. What is the difference between a well-graded soil and one that is:
    (a) uniformly graded?
    (b) gap-graded?
11. What are the various factors to be considered in establishing the design unit bearing pressure for bearing footings?
12. What particular conditions make settlements of increased concern in foundation design?
13. What is the difference between active and passive lateral soil pressure?
14. How is frictional resistance determined differently for sand and clay soils?

Section 7.4

15. A wall footing is to support a thick, reinforced masonry wall. Besides consideration of allowable bearing pressure, what should be noted in establishing the footing dimensions?
16. A one story building with shallow footings is to have footings bear on silty sand at approximately 2 ft below natural grade. Based on data from Table 7.2, find:
    (a) Required width for a wall footing; load is 2.4 kips/ft; footings 12 in. thick.
    (b) Side dimension for a square column footing; column load is 80 kips; assume footing 16 in. thick.

Section 7.5

17. What are the principal considerations that influence a decision to use deep foundations instead of a shallow bearing foundation?
18. Why are piles usually placed in groups?
19. What soil conditions make the installation of piles or piers difficult?
20. What is the purpose of a belled bottom on a drilled pier?

Section 7.6

21. What is meant by the term "equivalent fluid pressure"?
22. Why is it desirable to have the load resultant coincide with the plan centroid of an abutment?
23. What are some means for bracing of pile and pier foundations for lateral load effects?
24. What is the difference between pole frame and pole platform construction?

Section 7.7

25. In developing plans for site grading, why is it desirable to balance the cuts and fills?
26. What are some positive changes that may be effected by soil modification?
27. What considerations must be made in developing the site with regard to water runoff?
28. What site design considerations may be affected by accommodation of service lines for utilities?
29. Describe some of the relations between development of landscaping and design of the building and its foundations.

Section 7.8

30. What are the purposes of foundation walls?
31. What are the purposes of grade beams?
32. What are the purposes of cofferdams?
33. Describe the steps in forming a deep foundation with a sheetpile cofferdam.
34. Describe the slurry-trench method of constructing a foundation wall.
35. How can a buildup of hydrostatic pressure in permeable soil against a foundation wall be prevented?
36. Under what conditions can horizontal bracing between parallel foundation walls be considered safe?

Section 7.9

37. What is the most economical method, when it works, for dewatering an excavation?

38. Where should deep wells be placed to dewater an excavation?
39. Describe the installation of a wellpoint system for dewatering an excavation.

Section 7.10

40. What soil properties can usually be adequately determined by field tests?
41. What soil properties require laboratory tests for reliable determination?

42. What data are most significant for establishing the engineering design criteria for:
    (a) sand?
    (b) clay?
    (c) bedrock?
43. Describe the problems involved in deciding what kinds of investigations to make and how and when to make them for the foundation design of a large building project on an extensive site in an area with no history of building construction.

Chapter 8

Structural Systems

Structural systems are major subsystems incorporated to resist the loads in and on a building. The prime function of the systems is to transmit safely the loads from the upper portion, or superstructure, of the building to the foundations and the ground.

Floors, ceilings and roofs may serve simply as working surfaces or enclosures, which transmit loads to a structural system. More economically, these building components may also serve as part of the structural system, participating in the load-carrying function. Similarly, partitions and walls may serve simply as space dividers and fire stops. Again, these components may serve also economically as part of the structural system. The systems design approach encourages such multipurpose use of building components.

Comprehension of the role of structural systems requires a knowledge of:

1. Types and magnitudes of loads that may be imposed on a building
2. Structural materials and their characteristics
3. Structural analysis and design theory and practice
4. Types of structural systems, their behavior under load and probable life-cycle cost
5. Methods of erecting structural systems

Structural analysis and design lie in the province of specialists, called structural engineers. They may serve as independent consultants to the prime design professional for a building or,

preferably, as part of the building team. The scope of structural engineering is broad and complex. This chapter therefore can only describe briefly the most significant aspects of structural systems.

8.1. BUILDING LOADS

Loads are the external forces acting on a building or a component of a building. They tend to deform the structure and its components, although in a properly designed structure, the deformations are not noticeable. There are many ways in which loads are classified. The more important classifications are described below.

Types of Stress

One method classifies loads in accordance with the deformation effect on the components resisting them. Thus, the type of load depends on the way in which it is applied to a component: Tensile forces tend to stretch a component. Compressive forces squeeze a component. Shearing forces tend to slide parts of a component past each other (a cubical element becomes a parallelepiped). (See Fig. 8.1.)

These forces occur because motion produced by loads is required to be negligible. Hence, when a load is applied to a structural component, an equal and opposite reaction must also be developed to maintain static equilibrium. The function of a structural component is to re-

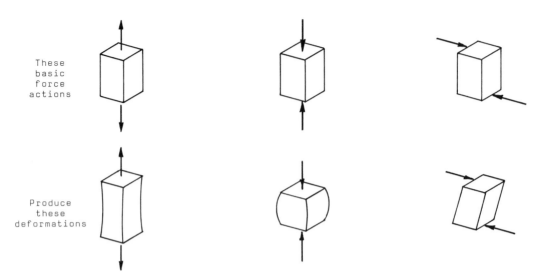

These
basic
force
actions

Produce
these
deformations

Fig. 8.1. Force actions and deformations.

sist the load and its reaction, and in so doing, the component is subjected to tension, compression or shear, or a combination of these. The reaction may be supplied by other structural members, foundations or supports external to the building.

Laws of Equilibrium. Because the reaction prevents translatory motion, the sum of all the external forces acting on a structural member in equilibrium must be zero. If the external forces are resolved into horizontal components H_1 $H_2, \ldots$ and vertical components $V_1, V_2, \ldots$, then

$$\Sigma H_i = 0 \quad \Sigma V_i = 0 \quad i = 1, 2, \ldots \quad (8.1)$$

Because the reaction to a load also prevents rotation, the sum of the moments of the external forces about any point must be zero:

$$\Sigma M = 0 \quad (8.2)$$

Now imagine a structural member cut into two parts but with each part restrained from motion by reactions with the other part. These reactions are called **stresses**. They are the internal forces in a structural member that resist the loads and external reactions. Because each of the two parts are in static equilibrium, the laws of equilibrium expressed by Eqs. (8.1) and (8.2) hold for each part.

Because loads are known, these laws often

may be used to determine internal stresses and external reactions. For example, Eqs. (8.1) and (8.2) may be used to determine three unknowns in any nonconcurrent coplanar force system. The equations may yield the magnitude of three forces for which the direction and point of application are known, or the magnitude, direction and point of application of a single force.

As an illustration of the use of the laws of equilibrium, the reactions of a simple beam will be computed. Line AB in Fig. 8.2a represents the beam, which is shown to have a 20-ft span. A support at A cannot resist rotation nor horizontal movement. It can supply only a vertical reaction R_1. A support at B cannot resist rotation. It can supply a reaction with a horizontal component H and a vertical component R_2. A 10-kip load (10,000 lb) is applied at a 45° angle to the beam at C, 5 ft from A. The load has horizontal and vertical components equal to 7.07 kips (see Fig. 8.2b). There are therefore three unknown reactions to these components, R_1, R_2 and H, to be determined from Eqs. (8.1) and (8.2).

Because the sum of the horizontal components of the external forces must be zero, H must be equal to the 7.07-kip horizontal component of the load but oppositely directed (see Fig. 8.2b). To make the sum of the vertical components of the external forces vanish: R_1 +

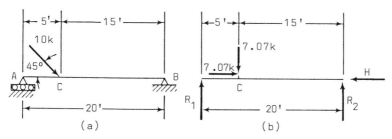

Fig. 8.2. Simple beam with inclined load.

$R_2 - 7.07 = 0$, where the negative sign is assigned to the downward acting force. To apply Eq. (8.2), moments are taken about B, which makes the moment arms of R_2 and the horizontal forces equal to zero. The result is $20R_1 - 15 \times 7.07 = 0$, from which $R_1 = 5.30$ kips. Substitution in the equation for the sum of the vertical forces yields $R_2 = 7.07 - 5.30 = 1.77$ kips.

If moments are taken about another point, say A, the result will not be another independent equation. The calculations, however, will be a check on the preceding results: $20R_2 - 5 \times 7.07 = 0$, from which again $R_2 = 1.77$ kips.

Static and Dynamic Loads

Another classification method for loads takes into account the rate of variation of load with time.

Static loads are forces that are applied slowly to a building and then remain nearly constant. Weight of building components, such as floors and roof, is one example.

Dynamic loads are forces that vary with time. They include *moving loads*, such as automobiles in a garage; repeated, and impact loads. *Repeated loads* are forces that are applied many times and cause the magnitude and sometimes also the direction of the stresses in a component to change. Forces from a machine with off-balance rotating parts are one example. *Impact loads* are forces that require a structure or a component to absorb energy quickly. Dropping of a heavy weight on a floor is one example.

Distributed and Concentrated Loads

Another classification method for loads takes into account the degree to which a load is spread out over a supporting member and the location of the load relative to an axis passing through the centroid of sections of the member.

Distributed loads are forces spread out over a relatively large area of a supporting member. Uniformly distributed loads are those that have constant magnitude and direction. Weight of a concrete floor slab of constant thickness and density is one example.

Concentrated loads are forces that have a small contact area on a supporting member relative to the entire surface area. One example is the load from a beam on a girder supporting it. Concentrated loads and loads that for practical purposes may be considered concentrated may be further classified as follows:

An *axial load* on a section of a supporting member is a force that passes through the centroid of the section and is perpendicular to the section.

An *eccentric load* on a section of a supporting member is a force perpendicular to the section but not passing through its centroid. Such loads tend to bend the member.

A *torsional load* on a section of a supporting member is a force offset from a point, called the *shear center*, of the section. Such loads tend to twist the member.

Design Loads

Still another classification method is one generally employed in building codes. It takes into account the nature of the source of the load.

Dead loads include the weight of a building and its components and anything that may be installed and left in place for a long time.

Table 8.1. Minimum Design Dead Loads for Buildings

Type of Construction	Psf
Ceilings	
Plaster (on tile or concrete)	5
Suspended gypsum lath and plaster	10
Concrete slabs	
Stone aggregate, reinforced, per inch of thickness	12.5
Lightweight aggregate, reinforced, per inch of thickness	9
Floor finishes	
Cement, per inch of thickness	12
Ceramic or quarry tile, 1-in.	12
Hardwood flooring, $\frac{7}{8}$-in.	4
Plywood subflooring, $\frac{1}{2}$-in.	1.5
Resilient flooring (asphalt tile, linoleum, etc.)	2
Glass	
Single-strength	1.2
Double-strength or $\frac{1}{8}$-in. plate	1.6
Insulation	
Glass-fiber bats, per inch of thickness	0.5
Urethane, 2-in.	1.2
Partitions	
Gypsum plaster, with sand, per inch of thickness	8.5
Gypsum plaster, with lightweight aggregate, per inch	4
Steel studs, with plaster on two sides	18
Wood studs, 2 × 4-in., with plaster on one side	11
Wood studs, 2 × 4-in., with plaster on two sides	19
Roof coverings	
Composition, 4-ply felt and gravel	5.5
Composition, 5-ply felt and gravel	6
Shingles	
Asbestos-cement	4
Asphalt	2
Wood	3
Walls	
Clay or concrete brick, per 4-in. wythe	33–46
Concrete block, 8-in. hollow, with stone aggregate	55
Concrete block, 8-in. hollow, lightweight aggregate	35
Gypsum block, 4-in. hollow	12.5
Waterproofing	
5-ply membrane	5

Materials	Lb per cu ft
Ashlar masonry	
Granite	165
Limestone	135–165
Marble	173
Sandstone	144
Cement, portland, loose	90
Concrete, stone aggregate, reinforced	150
Steel	490
Wood	
Douglas fir	40
Pine	33–50

Live loads include occupants and installations that may be relocated, removed or apply dynamic forces.

Impact loads are dynamic forces applied by live loads. Because they are considered related to live loads, impact loads generally are taken as a fraction of the live loads causing them.

Wind, snow and seismic loads are forces caused, respectively, by wind pressure, weight of snow and inertia in earthquakes. Snow loads usually are treated as additional dead load, whereas the other two types may be considered dynamic loads or may be taken into account by use of approximately equivalent static loads.

In design, structural engineers apply the maximum probable load that may occur or the load required by the applicable building code, whichever is larger. Tables 8.1 through 8.4 illustrate the type of data that might be given in building codes. More general comprehensive design data on loads are given in the American National Standards Institute *Building Code Requirements for Minimum Design Loads in Buildings and Other Structures*.

8.2. DEFORMATIONS OF STRUCTURAL MEMBERS

Section 8.1 points out that loads cause a structural member to deform. In a properly designed member, the deformations produced by design loads are very small. In tests, though, large deformations can be produced.

Tension and Elongation

Figure 8.3*a* shows a straight structural member in static equilibrium under the action of a pair of equal but oppositely directed axial tensile

Table 8.2. Minimum Design Uniformly Distributed Live Loads, Impact Included[a]

Occupancy or use	Psf
Auditoriums with fixed seats	60
Auditorium with movable seats	100
Garages, for passenger cars	50
Hospitals	
Operating rooms, laboratories and service areas	60
Patients' rooms, wards and personnel areas	40
Libraries	
Reading rooms	60
Stack areas (books and shelving 65 lb per cu ft)	150
Lobbies, first floor	100
Manufacturing	125–250
Office buildings	
Corridors above first floor	80
Files	125
Offices	50
Residential	
Apartments and hotel guest rooms	40
Attics, uninhabitable	20
Corridors (multifamily and hotels)	80
One- and two-family	40
Retail stores	
Basement and first floor	100
Upper floors	75
Schools	
Classrooms	40
Corridors	80
Toilet areas	40

[a]See local building code for permitted reductions for large loaded areas.

Table 8.3. Pressures, psf, for Winds with 50-Year Recurrence Interval

Height zone, ft above curb	Exposures								
	110-mph basic wind speed[a] (Coastal areas, N.W. and S.E. United States)			90-mph basic wind speed[a] (Northern and central United States)			80-mph basic wind speed[a] (Other parts of the United States)		
	A^b	B^c	C^d	A^b	B^c	C^d	A^b	B^c	C^d
0–50	20	40	65	15	25	40	15	20	35
51–100	30	50	75	20	35	50	15	25	40
101–300	40	65	85	25	45	60	20	35	45
301–600	65	85	105	40	55	70	35	45	55
Over 600	85	100	120	60	70	80	45	55	65

[a]At 30-ft height above ground surface.
[b]Centers of large cities and rough, hilly terrain.
[c]Suburban regions, wooded areas and rolling ground.
[d]Flat, open country or coast, and grassland.

Table 8.4. Roof Design Loads, psf, for Snow Depth with 50-Year Recurrence Interval

Regions (other then mountainous)[a]	Roof angle with horizontal, degrees			
	0–30	40	50	60
Southern states	5	5	5	0
Central and northwestern states	10	10	5	5
Middle Atlantic states	30	25	20	10
Northern states	50	40	30	15

[a]For mountainous regions, snow load should be based on analysis of local climate and topography.

forces T. Note that the arrows representing T are directed away from the ends of the member. As indicated in Fig. 8.3a, the forces stretch the member. The elongation occurs in the direction of the forces and its magnitude is shown as e. Tests and experience indicate that for a specific member, the larger the magnitude of T the larger e will be.

Compression and Buckling

Figure 8.3b shows a straight structural member in static equilibrium under the action of a pair

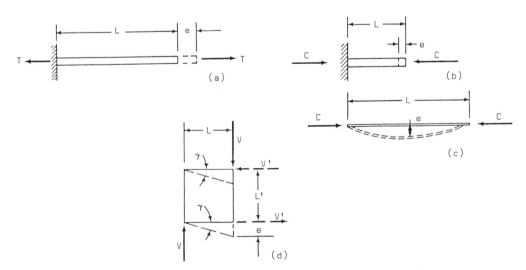

Fig. 8.3. Deformation effects of force actions. (a) Tension. (b) Compression of short element. (c) Compression of slender element. (d) Shear.

of equal but oppositely directed axial compressive forces C. Note that the arrows representing C are directed toward the ends of the member. Two cases must be recognized: short compression members and columns.

A short compression member is one with length L in the direction of C not much larger than the dimensions of the member perpendicular to the length. As shown for the short member in Fig. 8.3b, C causes the member to shorten an amount e in the direction of C. Tests and experience indicate that for a specific member, the larger the magnitude of C, the larger e will be.

A column is a compression member with length L in the direction of C much larger than the dimensions of the member perpendicular to C. For small values of C, the column may behave in the same way as a short compression member; but if C is made larger, the central portion of the member will move perpendicular to the length (see Fig. 8.3c). This sort of movement is called *buckling*. A specific value of C, called the *Euler load*, will hold the column in equilibrium in the buckled position. Larger loads will cause an increase in buckling until the column fails. Tests and experience indicate, however, that a relatively small force applied normal to the length of the column at an appropriate point can prevent buckling. This observation indicates that proper bracing can stop columns from buckling.

Shear

The rectangle in Fig. 8.3d may represent a structural member, such as a short bracket, or an element isolated from the interior of a structural member. The vertical arrow directed downward represents a shearing force V. For equilibrium, an equal but oppositely directed vertical force must be provided, by an external reaction in the case of the bracket or by a shear stress in the case of the internal element. These forces tend to make vertical sections of the member slide past each other. As a result, one edge of the rectangle moves a distance e relative to the other edge. Tests and experience indicate that the larger the magnitude of V, the larger e will become.

In Fig. 8.3d, the load V and its reaction are shown a distance L apart. Thus, they form a couple with a moment VL in the clockwise direction. For equilibrium, a counterclockwise moment also equal to VL must be provided by external forces or by stresses. These are indicated in Fig. 8.3d by the dashed-line horizontal arrows V'. From this it may be concluded that if shears act on an element in one direction, shears must also act on the element in a perpendicular direction.

Load-Deformation Curves

To study the behavior of a member, tests may be performed on it to measure deformations as loads are increased in increments. For graphic representation of the results, each measured deformation may be plotted for the corresponding load producing it. If the points are connected with a line, the result is a load-deformation curve, also known as a load-deflection curve. While the curve provides information on the structural behavior of the member, the results are applicable only to that specific member. For more generally applicable results of such tests, the concepts of stress and strain are more useful. These are discussed in Sec. 8.3.

8.3. UNIT STRESSES AND STRAINS

Deformation as described in Sec. 8.2 is the total change produced by loads in the dimension of a member in the direction of the loads. Deformation is also referred to as strain.

Unit strain, or unit deformation, in any direction at any point in a structural member is the deformation per unit of length in that direction.

Types of Unit Strain

Consider the structural member in Fig. 8.3a subjected to axial tensile forces, which cause an elongation e of the member. If the unit strains can be considered constant along the member, then the unit strain at every point can be obtained by dividing e by the length L; that is, the tensile unit strain ϵ_t in the direction of T

equals

$$\epsilon_t = \frac{e}{L} \qquad (8.3)$$

Consider now the structural member in Fig. 8.3b subjected to axial compressive forces, which cause a shortening e of the member. Again, if the unit strains ϵ_c can be considered constant along the length L of the member, the compressive unit strain in the direction of C equals

$$\epsilon_c = \frac{e}{L} \qquad (8.4)$$

Finally, consider the rectangular element in Fig. 8.3d subjected to shearing forces. If the deformation e is divided by the distance L over which it occurs, the result is the angular rotation γ, radians, of the sides of the rectangle when it is distorted into a parallelogram. Thus, the shearing unit strain is given by the angle

$$\gamma = \frac{e}{L} \qquad (8.5)$$

In general, unit strains are not constant in a loaded structural member. They actually represent the limiting value of a ratio giving deformation per unit length.

Unit Stresses

Section 8.1 defines stress as the internal force in a structural member that resists loads and external reactions.

Unit stress is the load per unit of area at a point in a structural member and in a specific direction.

Consider the structural member in Fig. 8.3a subjected to horizontal axial tensile forces T. Imagine the member cut into two parts by a vertical section and still maintained in equilibrium. Thus, each cut end must be subjected to a stress equal but opposite to T that is supplied by the reaction of the other part. Assume now that the unit stresses are constant over each cut end (see Fig. 8.4a). Then, by the definition of unit stress, the product of A, the area of the cut end, and the unit tensile stress f_t must equal T, for equilibrium. So, for constant unit tensile stress.

$$f_t = \frac{T}{A} \qquad (8.6)$$

Imagine now the structural member in Fig. 8.3b, subjected to horizontal axial compressive forces C, cut into two parts by a vertical section and still maintained in equilibrium. Assume that the unit stresses are constant over each cut end (see Fig. 8.4b). Then, from the requirement of equilibrium, the unit compressive stress equals

$$f_c = \frac{C}{A} \qquad (8.7)$$

where A is the area of the cut end.

Tensile and compressive stresses are sometimes referred to as normal stresses, because they act on an area normal to the loads. Under this concept, tensile stresses are considered as positive

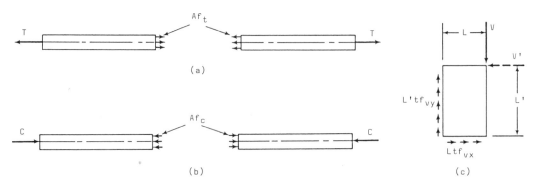

Fig. 8.4. Development of unit stresses. (a) Tension stress. (b) Compressive stress. (c) Shearing stress.

normal stresses and compressive stresses as negative normal stresses.

Shearing unit stress acts differently. The area over which this type of stress acts is the sliding area and therefore must be taken in the direction of the shear force. Consider, for example, the element of a structural member represented by a rectangle with sides of length L and L' in Fig. 8.4c. A downward shearing force V must be counteracted, for equilibrium, by upward shearing unit stresses f_{vy}. If the sliding area $A = L't$, where t is the thickness of the member and the unit shearing stresses are considered constant over A, then for equilibrium $L'tf_{vy} = V$. So for constant unit shear,

$$f_{vy} = \frac{V}{L't} \qquad (8.8)$$

Also, for equilibrium, the element must be subjected to a horizontal shear V', which is counteracted by horizontal unit shearing stresses f_{vx}. These stresses act over an area Lt. Hence,

$$f_{vx} = \frac{V'}{Lt} \qquad (8.9)$$

In addition, for equilibrium, the couple VL must equal the couple $V'L'$, or $(L'tf_{vy})L = (Ltf_{vx})L'$. Division by LtL' yields

$$f_{vy} = f_{vx} \qquad (8.10)$$

Consequently, the unit shearing stresses in perpendicular directions are equal. They therefore can be represented simply by f_v.

One other type of unit stress should be considered at this point. This type of stress, called bearing stress, is the same type discussed in Chap. 3 as the pressure under a spread footing. Figure 8.5a shows a load P applied to a structural member 1, which, in turn, transmits the

load to a second structural member 2 over a bearing area A. As indicated in Fig. 8.5b, the reaction of 2 on 1 is a bearing stress f_b, assumed constant over A. For equilibrium, $Af_b = P$. Then, the bearing stress on 1 is

$$f_b = \frac{P}{A} \qquad (8.11)$$

Also, for equilibrium, the reaction of 1 on 2 (see Fig. 8.5c) is a bearing stress f_b, oppositely directed, given by Eq. (8.11).

In general, unit stresses are not constant in a loaded structural member but vary from point to point. The unit stress at any point in a specific direction is the limiting value of the ratio of the internal force on any small area to that area, as the area is taken smaller and smaller.

Stress-Strain Curves

To study the behavior of a structural material, load tests are performed on a specimen of standard size and shape. For materials that have been in use a long time, the specimen size and shape generally are taken to accord with the requirements of an applicable method of test given in an ASTM specification. In these tests, loads are increased in increments and the deformation is measured for each load. Then, unit stresses are computed from the loads, and unit strains from the deformations. For graphic representation of the results, each strain may be plotted for the corresponding unit stress. If the points are connected by a line, the result is a stress-strain curve.

While a load-deflection curve provides information on the behavior under load of the specific member tested, a stress-strain curve supplies information on the mechanical properties of the material tested. This information is applicable to practically any size and shape of structural member made of the material. Stress-strain curves will be discussed in more detail later.

SECTIONS 8.1–8.3

References

ANSI, *American National Standards Minimum Design Loads for Buildings and Other structures*, American National Standards Institute, New York.

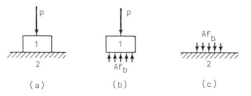

Fig. 8.5. Bearing stress. (a) Load presses member 1 against member 2. (b) Unit bearing stress on member 1. (c) Unit bearing stress on member 2.

H. Parker and J. Ambrose, *Simplified Mechanics and Strength of Materials*, 4th ed., Wiley, New York, 1986.

R. Gytkowski, *Structures: Fundamental Theory and Behavior*, 2nd ed., Van Nostrand Reinhold, New York, 1987.

Words and Terms

Buckling
Deformation
Elongation
Force, types of, actions
Load: dead, live, static, dynamic, wind, snow, seismic, impact, distributed, concentrated
Strain
Stress
Unit stress

Significant Relations, Functions and Issues

Function of structure; structural role of building construction components.

Aspects of understanding of structures.

Types of stresses; nature of unit stress; forms of deformation.

Loads: sources, effects, measurement.

8.4. IDEALIZATION OF STRUCTURAL MATERIALS

Stress-strain curves obtained from a standard load test of a structural material are indicative of the behavior of structural members made of that material. Several mechanical properties of importance can be deduced from such curves.

Material Properties and Stress-Strain Curves

Some examples of stress-strain curves for different materials are shown in Fig. 8.6. These were developed from tension tests in which a specimen was loaded in increments until it fractured.

Curve *OA* in Fig. 8.6*a* is indicative of the behavior of a brittle material. For the material tested, stress is proportional to strain throughout the loading. Fracture occurs suddenly at point *A*. The ultimate tensile strength, or unit stress at failure, is represented by F_u.

Ductility. The curve in Fig. 8.6*b* is indicative of the behavior of a ductile material. Ductility is the ability of a material to undergo large deformations before it fractures. Initially, for the material tested, as indicated by line *OB*, stress is proportional to strain. Between points *B* and *C*, the stress-strain curve may be irregular or nearly horizontal. Beyond *C*, strains increase rapidly with little increase in stress, or a nominal decrease, until fracture occurs. The large deformations before fracture give ample warning of the imminence of failure. Consequently, ductility is a very desirable characteristic of structural materials.

Elastic Behavior. If a material, after being subjected to a load, returns to its original size after it has been unloaded, it is said to be *elastic*. If the size is different, the material is called

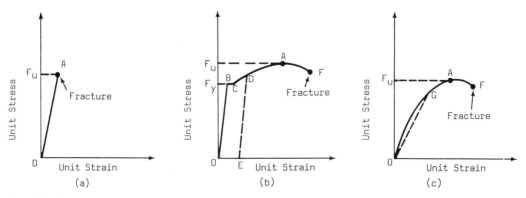

Fig. 8.6. Stress–strain curves for various materials. (*a*) Curve for an elastic but brittle material. (*b*) Curve for an elastic, ductile material. (*c*) Curve for a material with no proportional limit.

inelastic. The material for which the stress-strain curve in Fig. 8.6*b* was developed is elastic up to a stress called the *elastic limit.* If the material is loaded to a larger stress, it will not return to its original size when unloaded. It has become inelastic. The curve for slow unloading is nearly parallel to line *OB*, the initial portion of the stress-strain curve. Thus, as indicated in Fig. 8.6*b*, if the material is loaded beyond the elastic limit until the point *D* on the stress-strain curve is reached, the unloading curve will be *DE.* The material will then have a permanent set, or residual unit strain, *OE.* If the material is now reloaded, the stress-strain curve will lie along *ED*, back to *D.* It has again become elastic, but with the permanent set *OE.*

The stress at which strain and stress cease to be proportional is known as the *proportional limit.*

The stress F_y beyond which there appears to be an increase in strain with no increase or a small decrease in stress is called the *yield point.*

The elastic limit, proportional limit and yield point, if they exist, are located close together on the stress-strain curve. For some materials, determination of the values of these stresses is very difficult. Furthermore, some materials do not have a proportional limit or a recognizable yield point or elastic limit. Figure 8.6*c* shows the stress-strain curve for a material with no proportional limit and with a yield point that is poorly defined.

For such materials, an arbitrary stress, called the *yield strength*, also denoted by F_y, may be used as a measure of the beginning of plastic strain, or inelastic behavior. The yield strength is defined as the stress corresponding to a specific permanent set, usually 0.20% (0.002 in. per in.).

The yield point and yield strength are important in structural design because they are used as the limit of usefulness of a structural material. At higher stresses, the material suffers permanent damage, undergoes large deformations, which may damage supported construction, and is close to failure. It has become customary, consequently, to apply safety factors to F_y in the determination of allowable unit stresses or safe loads for ductile materials.

Some additional structural properties of note are the following:

Poisson's ratio (μ) is the ratio of lateral unit strain to longitudinal unit strain in a material. Under tension, for example, a member lengthens in the longitudinal direction and shortens laterally. For steel this ratio is about 0.3 and for concrete it is about 0.25.

Modulus of elasticity (E) is the ratio of normal stress (tension or compression) to strain within the proportional limit. On the stress-strain graph, this is measured as the tangent of the angle of the curve (such as line *OB* in Fig. 8.6*b*).

Toughness is the ability of a material to absorb large amounts of energy (dynamic loading) without failure. This is often affected by temperature, toughness being reduced at low temperatures.

Modulus of rigidity or shearing modulus (G) is the constant of proportionality when shearing unit strain is proportional to unit shear stress:

$$f_v = G\gamma \qquad (8.12)$$

where

f_v = shearing unit stress
γ = shearing unit strain
G = modulus of rigidity, or shearing modulus of elasticity

It is possible to determine G from a linear stress-strain curve when Poisson's ratio is known, because G is related to the modulus of elasticity in tension and compression:

$$G = \frac{E}{2(1 + \mu)} \qquad (8.13)$$

where

μ = Poisson's ratio for the material

Idealized Structural Materials

To simplify structural analysis and design, structural materials usually are represented by a simple mathematical model. The model often assumes that a material is homogeneous; that is, there is no change in the material from point to point in a structural member. Also, the model generally assumes that the material is isotropic, that it has the same properties in all directions.

For some materials, additional assumptions are made. Some common ones for structural steel, for example, is that the material has the same modulus of elasticity in compression and tension and is ductile and tough.

Hooke's Law. Figure 8.7 shows the stress-strain curve in tension for an idealized ductile, linearly elastic structural material. For this material, Hooke's law applies up to the yield stress F_y.

Hooke's law states that unit strain is proportional to unit stress. The law can be represented by the equation

$$f = E\epsilon \qquad (8.14)$$

where

f = unit stress
ϵ = unit strain
E = modulus of elasticity
  (also called Young's
  modulus)

Accordingly, line OB, the initial portion of the stress-strain curve, in Fig. 8.7 is a straight line with slope E. At point B, the stress is F_y.

Plastic Behavior. The portion of the stress-strain curve beyond B often is taken as a hori-

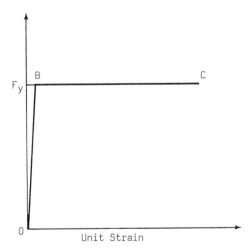

Fig. 8.7. Stress–strain curve for an idealized ductile, linearly elastic material.

zontal line, such as BC in Fig. 8.7. This is a conservative assumption for ductile materials, since they require at least a small increase in stress to produce a large increase in strain. In the range BC, the idealized material is said to be *plastic*.

This property of a material is important, because it affects the ultimate strength of a structural member. When a portion of the member is stressed under load to F_y, the member does not necessarily fail when the load is increased. That portion of the member yields; that is, undergoes large deformations, with the increase in load. Nonetheless, if the rest of the member is not stressed to F_y and is so constructed that the large deformations do not cause failure, the member can sustain larger loads than that causing F_y in only a few places. Thus, plastic behavior can increase the load-carrying capacity of the structural member beyond that for local departure from the elastic range.

Departure of Actual Materials from Ideal

Few actual materials can be accurately represented by the idealized structural material. Consequently, for these materials, the mathematical model must be modified.

Sometimes, the modification may be minor. For example, for a ductile material with no proportional limit, such as a material with a stress-strain curve like that shown in Fig. 8.6c, a secant modulus may be substituted for the modulus of elasticity in the assumption of a stress-strain curve observing Hooke's law. The *secant modulus* is the slope of a line, such as OG in Fig. 8.6c, from the origin to a specified point on the actual stress-strain curve. The point is chosen to make the shape of the linear stress-strain curve for the idealized material approximate that in Fig. 8.6c.

Design Bases for Structural Materials

Either of two methods is generally used in design of structures. In one method, *allowable* or *working stresses* are established. Under design loads, these stresses may not be exceeded. In the second method, called *ultimate-strength design, limit design* or *load-factor design*, design loads are multiplied by appropriate factors and the structure is permitted to be stressed or strained to the limit of usefulness under the factored loads.

Allowable stresses for a specific material usually are determined by dividing the yield strength or the ultimate strength by a constant safety factor greater than unity. A smaller safety factor is permitted for such combinations of loading as dead and live loads with wind or seismic loads.

When ultimate-strength design is used, load factors greater than unity are so chosen as to reflect the probability of occurrence of excessive loadings. For example, a larger factor is applied to live loads than to dead loads, because of the greater probability of maximum design live loadings 'being exceeded than of design dead loads being exceeded.

8.5. STRUCTURAL MATERIALS

The idealized structural material described in Sec. 8.4 is homogeneous and isotropic. It exhibits linearly elastic behavior up to a yield strength F_y. If the material also is ductile, it exhibits plastic behavior under loads greater than that causing F_y. For structural analysis and design, mathematical models predict the behavior of structures made of the idealized structural material.

Actual materials, however, may not be represented accurately by the idealization. Consequently, the idealized material and the mathematical models must be revised to improve the accuracy of predictions of structural behavior. For this to be done, a knowledge of the properties of structural materials in use is required.

In this section, the properties of some commonly used structural materials are described briefly. In following sections, mathematical models for predictions of the behavior of these materials are discussed.

Material and Design Specifications

After a material has been tested thoroughly and has been used for several years, a standard specification is written for it. Material and test specifications of ASTM (formerly American Society for Testing and Materials) are often used for structural materials. Building codes often adopt such specifications by reference. Standard *design specifications* also are available for commonly used structural materials. Design specifications usually are developed and promulgated by trade associations or engineering societies concerned with safe, economical and proper uses of materials. Examples of such specifications are: for structural steel—the American Institute of Steel Construction; for reinforced concrete—the American Concrete Institute; and for wood construction—the American Institute of Timber Construction.

Structural Steel

Steel is a highly modifiable material due to the process of production of the raw material plus the various processes of production of finished products. Steel is basically an alloy of iron and carbon. For structural steel the carbon content is kept quite small, generally less than 0.35%, since the carbon tends to decrease ductility and weldability, although it increases the strength of the iron. Other metals are often alloyed to the carbon and iron to produce particular properties of the finished steel, such as high strength, corrosion resistance, weldability, and toughness.

Structural steels used in building construction are often specified by reference to ASTM specifications. For example, steels meeting the requirements of ASTM A36, Standard Specification for Structural Steel, are called A36 steel, which is considered as sufficient identification. Some ASTM specifications apply to steel with common characteristics but with different yield points. In such cases, steels with the same minimum yield point are grouped in grade, with the grade designation being the value of the yield point in ksi.

The plates and shapes used in structural applications are produced by hot rolling material produced in steelmaking. This hot rolling tends to improve grain structure in the direction of rolling. Consequently, ductility and bendability are better in the direction of rolling than in transverse directions. Also, because more rolling is needed to produce thin steel products than thick ones, thin material has larger ultimate tensile strength and yield stress than thick material. Consequently, ASTM specifications often specify higher minimum yield stresses as thickness is decreased (see Table 8.5). This, in turn, permits use of larger allowable unit stresses or load-carrying capacity in design as thickness is decreased.

In accordance with the chemical content, structural steels usually used in building construction are classified as carbon steels; high-strength, low-alloy steels; and high-strength, heat-treated, low-alloy steels. As indicated in Table 8.5, high-strength, low-alloy steels have greater strength than carbon steels. Also, the heat-treated steel A514 has greater strength than the other two classes of steel. The greater the strength, however, the greater the cost per unit weight of the steel.

The structural steels have stress-strain curves similar to that shown in Fig. 8.6c. Figure 8.8 shows to an enlarged scale the portions of the curves for some structural steels in the elastic range and somewhat beyond. The shapes of these curves indicate that the stress-strain curve in Fig. 8.7 for the idealized structural material

Table 8.5. Properties of Structural Steels

ASTM specification	Thickness, in.	Minimum tensile strength, ksi	Minimum yield point or strength, ksi	Relative corrosion resistance[a]
		Carbon Steels		
A36	To 8 in. incl.	58–80[c]	36	1[b]
A529	To $\frac{1}{2}$ in. incl.	60–85[c]	42	1
		High Strength, Low-Alloy Steels		
A441	To $\frac{3}{4}$ incl.	70	50	2
	Over $\frac{3}{4}$ to $1\frac{1}{2}$	67	46	2
	Over $1\frac{1}{2}$ to 4 incl.	63	42	2
	Over 4 to 8 incl.	60	40	2
A572	Gr 42: to 4 incl.	60	42	1
	Gr 45: to $1\frac{1}{2}$ incl.	60	45	1
	Gr 50: to $1\frac{1}{2}$ incl.	65	50	1
	Gr 55: to $1\frac{1}{2}$ incl.	70	55	1
	Gr 60: to 1 incl.	75	60	1
	Gr 65: to $\frac{1}{2}$ incl.	80	65	1
A242	To $\frac{3}{4}$ incl.	70	50	4–8
	Over $\frac{3}{4}$ to $1\frac{1}{2}$	67	46	4–8
	Over $1\frac{1}{2}$ to 4 incl.	63	42	4–8
A588	To 4 incl.	70	50	4
	Over 4 to 5	67	46	4
	Over 5 to 8 incl.	63	42	4
		Heat-Treated, Low-Alloy Steels		
A514	To $\frac{3}{4}$ incl.	115–135	100	1–4
	Over $\frac{3}{4}$ to $2\frac{1}{2}$	115–135	100	1–4
	Over $2\frac{1}{2}$ to 4 incl.	105–135	90	1–4

[a]Relative to carbon steels low in copper.
[b]A36 steel with 0.20% copper has a relative corrosion resistance of 2.
[c]Minimum tensile strength may not exceed the higher value.

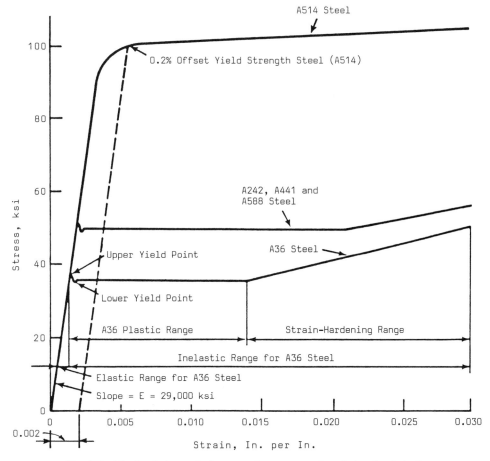

Fig. 8.8.   Idealized stress-strain curves for some grades of structural steel.

can represent, with acceptable accuracy for design purposes, the structural steels generally used in building construction.

Steel mills use a different classification method for steel products. Included are *structural shapes (heavy)* and *(light)*. The former applies to shapes with at least one cross-sectional dimension of 3 in. or more, whereas the latter applies to shapes of smaller size, such as bars.

Shapes are identified by their cross-sectional geometry; for example, wide-flange or H-shapes, I-beams, bearing piles, miscellaneous shapes, structural tees, channels, angles, pipe and structural tubing (see Fig. 8.9). For convenience, these shapes usually are designated by letter symbols, as listed in Table 8.6.

A specific shape is specified by a listing in the following order: symbol, depth and weight.

For example, W12 × 36 specifies a wide-flange shape with nominal depth of 12 in. and weight of 36 lb per lin ft. Actual dimensions are listed in the AISC Steel Construction Manual and steel producers' catalogs. (The X symbol in the designation is a separator and is read "by.")

Plates are designated differently. The designation lists in the following order: symbol, thickness, width and length. For example, PL $\frac{1}{2}$ × 15 × 2'-6" specifies a plate $\frac{1}{2}$ in. thick, 15 in. wide and 2 ft 6 in. long.

Properties of Structural Steels. As indicated by the stress-strain curves of Fig. 8.8, structural steels are linearly elastic until stressed nearly to the yield point. Beyond the yield point, they may be considered plastic.

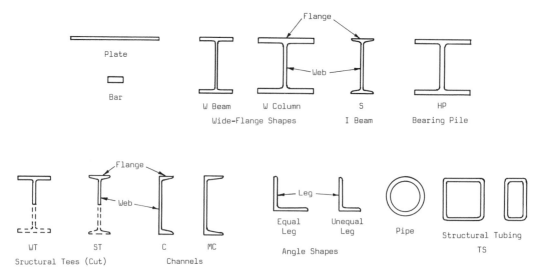

Fig. 8.9. Rolled structural steel shapes and their symbols.

Table 8.6. Structural-Steel Shape Designations

Section	Symbol
Plates	PL
Wide-flange (H) shapes	W
Standard I-shapes	S
Bearing-pile shapes	HP
Similar shapes that cannot be included in W, S or HP	M
Structural tees made by cutting a W, S or M shape	WT, ST, MT
American standard channels	C
All other channel shapes	MC
Angles	L
Tubing	TS

Stiffness. Structural steel is the stiffest of the commonly used structural materials. Stiffness may be measured by the modulus of elasticity, which may be considered the unit stress required to produce a unit strain of 1 in. per in. if the material were to stay in the elastic range. Structural steels have a modulus of elasticity of about 29,000 ksi (30,000 ksi often is used in design). The modulus of elasticity is taken as the same value in tension and compression. Poisson's ratio for structural steels is about 0.3.

Stiffness is important because it is an indication of the resistance offered by a loaded structural member to deformations and deflections.

Weight. ASTM specification A6 specifies that rolled steel shall be assumed to weigh 0.2833 lb per cu in.

Volume Changes. Steel is a good conductor of heat and electricity. It has a thermal coefficient of expansion of 0.000 0065 in. per in. per °F. This is about the same as that for concrete but much larger than that for wood.

Corrosion. Steel tends to form an iron oxide in the presence of oxygen and water. Unless special alloying elements are present, the oxide is very weak and is called rust. The process of forming the weak oxide is called corrosion.

The rate of corrosion of a steel depends on the type and amount of alloying elements incorporated in the material. A copper-bearing steel with at least 0.20% copper has about twice the corrosion resistance of ordinary carbon steels. Some steels, such as A441, A242 and A588, are known as *weathering steels* because they can offer even greater resistance to corrosion (see Table 8.5). These steels form an oxide, but it adheres strongly to the base metal and prevents further corrosion. Other steels should be protected against corrosion by coats of paint or concrete or by cathodic protection.

Fire Protection. Structural steels also need protection against high temperatures, because they tend to lose strength under such conditions. Consequently, steel members should be protected against the effects of fire if exposure to fire is possible. For this purpose, the steel may be coated with insulating, fireproof materials, such as concrete, plaster, mineral fibers and special paints. (See also Secs. 2.2 and 6.3.)

Steel Cables. When considered a permanent part of structural steel framework, steel cables are considered to be in the classification of structural steel. Cables may be used as vertical hangers or they may be strung between two points in a curve to support other building components. The types of cables used for these purposes are known as bridge strand or bridge rope.

A *strand* consists of steel wires coiling helically in a symmetrical arrangement about a center wire. A *rope* is formed similarly but with strands instead of wire. The wires used in forming these products are cold drawn and do not have a definite yield point. Safe loads on strands and rope, therefore, are determined by dividing the specified minimum breaking strength for a specific nominal diameter and type of cable by a safety factor greater than unity.

Design Rules for Structural Steels. Structural steel design often is based on the AISC Specification for the Design, Fabrication and Erection of Structural Steel for Buildings. This specification may be adopted as a whole by local building codes or with some modifications.

The design rules in this specification apply to elastic and plastic behavior of structural steels. The rules generally assume that the steels have a stress-strain curve similar to that of the idealized material in Fig. 8.7 and that the modulus of elasticity is the same in tension and compression. The specification is included in the AISC Manual of Steel Construction but separate copies also are available.

Fabrication. The AISC Specification also presents requirements for fabrication and erection of structural steel framing. Fabrication is the operation, usually conducted in a shop, of cutting steel plates and shapes to specified sizes and assembling the components into finished members, ready for shipment to the building site and for erection.

The intent of steel designers is conveyed to the fabricating shop in *detail drawings*. They usually are prepared by shop detailers, employed by the fabricator, from the steel designers' drawings. The detail drawings are generally of two types: shop working drawings and erection drawings. Called details, shop drawings are prepared for every member of the framework. They provide all information necessary for fabricating each member. Erection drawings guide the steel erector in constructing the framework. They show the location and orientation of every member, or assembly of components, called shipping pieces, that will be shipped to the building site.

Fasteners for Steel Connections. Components or members may be connected to each other with rivets or bolts or by welding. For connections with rivets and bolts, holes must be provided for the fasteners in the fabricating shop. The holes must be accurately located and of the proper size for the fasteners to be used. The holes may be formed by punching or drilling, the former being faster and less expensive. Punching, however, is suitable only for thin material, usually up to about 1-in. thickness for carbon steels and $\frac{1}{2}$-in. for heat-treated steel.

A *rivet* consists of a cylindrical shank with an enlarged end, or head. For making a connection between steel members, the rivet is heated until it is cherry red, placed in aligned holes in the members and hammered with a bull riveter or a riveting gun to form a second head. When the rivet cools, it shrinks, and the two heads force the connected members into tight contact.

A *bolt* consists of a cylindrical shank with a head on one end and threads on the opposite end. For making a connection between steel members, the bolt is inserted in aligned holes in the members and a nut is turned on the threaded end of the bolt shank to hold the members in tight contact. In some cases, washers may be required under the nut and sometimes also under the bolt head. Bolts generally are used in field connections as well as in shop connections, but rivets usually are used only in shop connections, for economic reasons.

The AISC Specification requires that special high-strength bolts be used for major connections in tall buildings and for connections subject to moving loads, impact or stress reversal. These bolts should conform to ASTM specification A325, Specification for High-Strength Bolts for Structural Steel Joints, or A490, Specification for Quenched and Tempered Alloy Steel Bolts for Structural Steel Joints (for use with high-strength steels). For such connections, the high-strength bolts are highly tensioned by tightening of the nuts, and the parts of the connection are so tightly clamped together that slippage is prevented by friction.

A *weld* joins two steel components by fusion. It is economical, because it reduces the number of holes and the amount of connection material needed from that required with fasteners. Also, welding is less noisy than bolting or riveting. Use of welding in steel construction is governed by the American Welding Society Structural Welding Code, AWS D1.1. The most commonly used welding methods employ a metal electrode to strike an electric arc that supplies sufficient heat to melt the metal to be joined, or base metal, and the tip of the electrode. The electrode supplies filler metal for building up the weld.

Welds used for structural steels are either fillet or groove. Fillet welds are used to join parts at an angle with each other, often 90°. In the process, molten weld metal is built up in the angle (see Fig. 8.10a). Groove welds may be used to connect parts lying in the same plane

or at an angle with each other. For this purpose, one edge to be connected is cut on a slope, so that when the edge is placed against another edge or surface in the connection, a gap, or groove, is formed. In the welding process, molten metal is built up in the gap (see Fig. 8.10b and c). Groove welds may be either complete (see Fig. 8.10b) or partial penetration (see Fig. 8.10c), depending on the depth of gap and solid weld metal.

Clearances in Steel Erection. Steel designers should select member sizes and shapes and arrange the components so that fabrication and erection operations can be easily and properly performed. There should be ample clearances, for example, for application of riveting machines, tightening of nuts on bolts with wrenches, and welding with electrodes and welding machines. Another important example is provision of ample clearances for erection of beams between columns. Consequently, designers should be familiar with fabrication and erection methods, and their designs should anticipate the methods likely to be used. In addition, designers should anticipate conditions under which pieces are to be shipped from the fabricating shop to the building site. Lengths and widths of trucks or railroad cars, or height limitations on shipments by highway or railroad, may restrict the size of pieces that can be moved. The size restrictions may determine whether members must be shipped in parts and later spliced, and if so, the location of the splices.

Erection Equipment for Steel Framing. When a piece arrives at the building site, the steel member may be moved to storage or immediately erected in its final position. Often, the handling is done with a crane (see Fig. 8.11). Depending on the terrain, the crane may be mounted on wheels or tractor treads. It carries a long boom, sometimes with an extension on the end, called a jib, over which steel lines are passed for picking up building components. Operating at ground level, the crane can rotate and raise or lower the boom and jib, while a drum winds or unwinds the lines.

Stiffleg derricks also are frequently used for steel erection. Such derricks consist of a rotat-

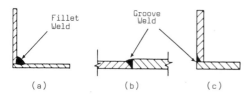

Fig. 8.10. Welds used for structural steel. (*a*) Fillet weld. (*b*) Complete penetration groove weld. (*c*) Partial penetration groove weld.

Fig. 8.11. Crane erecting structural steel. (Courtesy of American Hoist and Derrick Company)

able vertical mast, held in position by two stiff-legs, a boom pinned at the bottom of the mast, and steel lines passing over the top of the boom for picking up building components. In erection of a tall building, a stiffleg derrick usually is set on the top level of steel to erect the next tier and is jumped upward as erection progresses. Thus, while use of a crane is limited to framework with height less than that of boom and jib—usually about 200 ft—a stiffleg derrick can be used for framework of any height. Other equipment, such as guy derricks, also may be used in a similar fashion.

Cold-Formed Steels

As mentioned previously, certain steel items, though used as structural components, are not considered structural steel if not defined as such in the AISC Code of Standard Practice for Steel Buildings and Bridges. An important classification of structural items made of steel but not considered structural steel is cold-formed steel. Made generally of sheet or strip steel or of bars with small cross section, or of combinations of these materials, cold-formed steel members offer substantially the same advantages as structural steel, although price per pound may be greater, but are intended for use for light loads and short beam or panel spans. These members can be used as floor or roof deck or curtain walls as well as beams and columns.

Like hot-rolled steel, cold-formed members must be protected against corrosion. This usually is done by painting or galvanizing. They also must be protected against fire, and this is usually done with concrete floor or roof decks and plaster or fire-resistant acoustic ceilings.

Cold-formed steel members may be classified as structural framing formed from sheet or strip steel; deck and panels; or open-web steel joists.

Production of Cold-formed Shapes. Cold-formed framing, deck and panels are made of relatively thin steel formed by bending sheet or strip steel in roll-forming machines, press brakes or bending brakes. Shapes so produced may also be used in door and window frames, partitions, wall studs, floor joists, sheathing and molding.

Thickness of cold-formed sheet steel often is designated by a gage number, but decimal parts of an inch are preferable. Table 8.7 lists the equivalent thickness in inches for gage numbers and approximate unit weight. Thickness of strip steel is always given as decimal parts of an inch.

Stress-strain curves for cold-formed steels are similar to that of the idealized structural material (see Fig. 8.7). Modulus of elasticity is about 30,000 ksi in tension and compression. Hence, cold-formed steel shapes may be designed by the same procedures used for hot-rolled structural steel shapes. Local buckling, however, is a greater possibility with cold-formed shapes, because of the thinner material used. To account for this effect, the width or depth of a shape used in design is reduced from the actual dimensions, the reduction being

Table 8.7. Manufacturers' Standard Gage for Steel Sheets

Gage No.	Equivalent thickness, inc.	Weight, psi
3	0.239	10.00
4	0.224	9.38
5	0.209	8.75
6	0.194	8.13
7	0.179	7.50
8	0.164	6.88
9	0.150	6.25
10	0.135	5.63
12	0.105	4.38
14	0.075	3.13
16	0.060	2.50
18	0.048	2.00
20	0.036	1.50
22	0.030	1.25
24	0.024	1.00

greater the larger the ratio of width or depth to thickness.

Design of cold-formed steel members generally is based on the American Iron and Steel Institute (AISI) Specification for the Design of Cold-Formed Steel Members. This specification may be adopted by building codes as a whole or with modifications.

Cold-Formed Steel Shapes. Because cold-formed steel is thin, it can be bent readily into desired configurations or built up from bent shapes for specific architectural and structural applications. Some shapes used structurally are similar to hot-rolled structural shapes. Channels, zees and angles can be roll-formed in a single operation from one piece of material (see Fig. 8.12a to c). They can also be provided with lips along flange edges to stiffen the flanges against local buckling (see Fig. 8.12d to f). I sections may be produced by welding two angles to a channel (see Fig. 8.12g and h) or by welding two channels back to back (see Fig. 8.12i and j). As indicated in Fig. 8.12h and j, I sections also may be provided with stiffening lips.

Other structurally useful sections made with cold-formed steel include the hat, open box and U shapes shown in Fig. 8.12k to m. Made with

two webs, these shapes are stiffer laterally than the single-web shapes in Fig. 8.12a to j.

Deck and panels may be produced by forming out of one piece of material a wide shape with stiffening configurations. The configurations may be rounded corrugations or sharply bent ribs, such as the series of hat shapes shown in Fig. 8.12n, sometimes used for roof deck. Often, deck shapes are built up by addition of a flat sheet on the underside of the hats (Fig. 8.12o), for greater stiffening, or by attaching a series of inverted hat shapes, to form parallel lines of hollow boxes, or cells, which offer both greater strength and stiffening.

Steel roof deck consists of ribbed sheets with nesting or upstanding-seam side joints, for interlocking adjoining panels, as shown in Fig. 8.12n and o. The deck should be designed to support roof loads applied between purlins, roof beams or trusses. Floor deck may also be used for roofs.

Cellular panels generally are used for *floor deck* (see Fig. 8.12p). The advantage of the cellular type is that the cells provide space in which electric wiring and piping may be placed. This avoids increasing the floor depth to incorporate the wiring and piping and also conceals the unsightly network from view. The cells also may be used for air distribution in air-conditioning systems. Consequently, cellular deck is always a competitive alternative to other types of floor systems in systems design of buildings.

Connections for Cold-Formed Steels. Components of cold-formed steels may be interconnected with bolts, rivets or welds in the same way as hot-rolled structural components; but because cold-formed steels are thin, other methods often may be conveniently used. Arc welding, for example, may be used to join parts with spot welds. As another example, in fabricating shops, resistance welding may be used because of speed and low cost. Spot welds are formed by this process by clamping the parts between two electrodes through which an electric current passes, to fuse the parts. Projection welding is another form of spot welding that may be used. A projection or protuberance is formed on one of the mating parts, and when

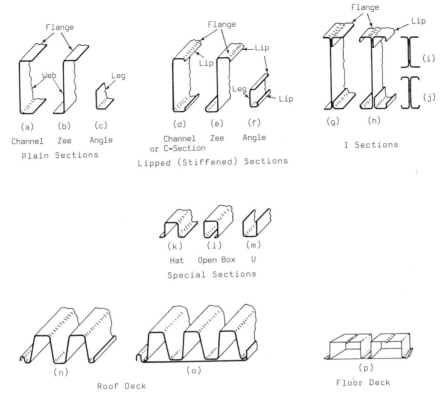

Fig. 8.12. Cold-formed, light-gage steel shapes.

the parts are brought together and current is passed through, a weld is formed.

Bolts are convenient for making connections in the field when loads have to be transmitted between connected parts. Tapping screws may be used for field joints that do not have to carry calculated gravity loads. Tapping screws used for connecting cold-formed siding and roofing are generally preassembled with neoprene washers to control leaking, squeaking, cracking and crazing.

Open-Web Steel Joists. These are truss-like, load-carrying members suitable for direct support of floors and roofs in buildings. The joists are usually fabricated from relatively small bars, which form continuous, zigzag web members between chords, and the top and bottom chords may be made of bar-size angles or shapes formed from flat-rolled steel. (Cold working in rolling chords of sheet or strip steel increases the strength of the metal. Yield strengths exceeding 150% of the minimum yield point of the plain material may be obtained.) Components of open-web steel joists may be joined by resistance or electric-arc welding.

Joists used for short spans are simply called open-web steel joists. Such joists, however, may span up to about 60 ft. Those used for long spans, up to about 144 ft, are commonly known as long-span joists. Both types of joists may be specified by reference to specifications adopted jointly by AISC and the Steel Joist Institute (SJI).

While SJI has standardized many aspects of design, fabrication and erection of open-web steel joists, exact details of specific joists vary with different manufacturers. Some typical arrangements in elevation and cross section are shown in Fig. 8.13. Some joists are produced with top chords that have provisions for nailing of wood deck to them.

Open-web joists are very flexible laterally. Consequently, they should be braced as soon as possible after erection and before construction loads may be applied to them. Often, bracing

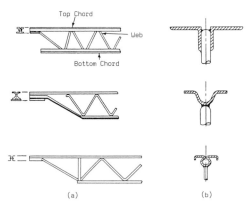

Fig. 8.13. Typical forms for open-web steel joists (prefabricated light trusses).

of the top chord may be provided by a floor or roof deck immediately attached and by struts between the bottom chord and adjacent joists, or by bridging, continuous rigid struts between the chords of one joist and the corresponding chords of an adjacent joist. Also, wall-bearing joists should be firmly secured at the supports with masonry anchors. Where joists rest on steel beams, the joists should be welded, bolted or clipped to the beams.

Wood

Wood is an organic product of nature. Often, except for drying, wood is used in its natural state.

This product of nature offers numerous advantages in structural applications as well as in such architectural applications as interior and exterior wall facings and floor coverings. Wood has high strength but low cost per unit of weight. It is ductile and resilient (high shock-absorption capacity). It can be easily sawn to desired dimensions and bent to sharp curvature. It can readily be shop fabricated into structural members, ready for shipment to the building site for erection. Often, the light weight of wood makes possible erection without the aid and cost of mechanical hoisting equipment.

Wood, however, has some disadvantages in structural applications. For example, for the same load-carrying capacity, bulkier wood members are needed as opposed to structural steel or concrete. Where space is critical and

heavy loads must be supported, wood is at a disadvantage, despite lower cost. Wood also has the disadvantage of being combustible. This disadvantage, however, can be overcome to some extent by use of bulky members, which are slow burning and therefore permitted in low buildings with nonhazardous contents, by use of fire-retardant treatments and by enclosure with fire-resistant materials. But the two disadvantages generally make use of wood structural members impractical for high-rise buildings. In addition, consideration must be given to the possibility that wood can decay or may be attacked and destroyed by insects.

Just as there are different types of structural steels, there are also different types of wood. For one thing, wood from different species of trees have different characteristics. For another, wood cut from two trees of the same species that grew side by side probably would not have the same strength; and even if all characteristics were initially the same, two pieces of wood may develop in a relatively short time different defects that would influence strength differently. Research and experience, though, have shown how these differences can be taken into account. Consequently, wood is a useful, reliable and economical structural material.

Structure of Wood. Wood has a cellular, fibrous structure that is responsible for many of its characteristics. The cell walls, made essentially of cellulose, are cemented together by lignin, another organic substance. Positioned vertically in trees before they are cut down, the cells range in length from 0.25 to 0.33 in. and are about 1% as wide. A cross section also has horizontally positioned bands of cells called rays. In addition, because of the manner in which trees grow, differences in cell thickness occur between seasons and are displayed in the cross section as annular growth rings. Because of this composition, wood is neither homogeneous nor isotropic; that is, it has different properties in different directions.

Moisture Content. Unlike other structural materials, wood undergoes little dimensional change directly due to temperature changes; however, it may develop significant dimensional changes due to increase or decrease of moisture content.

Wood from a newly felled tree is called *green* because the interior is wet. The wood may be allowed to dry naturally (seasoned) or may be dried in a kiln. The first step in the drying process is exodus of free water from the cavities in the wood. Eventually, the cavities will contain only air, but the cell walls, or fibers, will still be saturated. At this fiber-saturation stage, the moisture content of the wood may be between 25 to 30% of the weight of the oven-dry wood.

Except for change in weight, very few of the properties of wood, including dimensions, change during removal of free water. If drying continues past the fiber-saturation point, however, the cell walls lose water, and the wood begins to shrink across the grain (normal to the direction of the fibers). Shrinkage continues nearly linearly as moisture content decreases to zero. In ordinary usage, however, moisture content will stabilize in accordance with the humidity of the environment. If humidity increases, the wood fibers will absorb moisture and the wood will swell (see Table 8.8). Many properties of wood are affected by its moisture content.

Defects. Wood also contains or develops defects that influence its properties. Knots, for example, are formed when a branch, embedded in the tree, is cut through in the process of lumber manufacture. Another example is slope of grain, cross grain or deviation of fiber from a line parallel to the sides of a piece of wood.

Table 8.8. Properties of Douglas Fir and Southern Pine

Property	Douglas fir, coastal	Douglas fir, inland	Southern pine, longleaf	Southern pine, shortleaf
Moisture content when green, %	38	48	63	81
Weight, lb per cu ft				
When green	38.2	36.3	50.2	45.9
With 12% moisture	33.8	31.4	41.1	35.2
Add for each 1% moisture increase	0.17	0.14	0.18	0.15
Shrinkage from green dimensions when dried to 20% moisture, %[a]				
Radial direction	1.7	1.4	1.6	1.6
Tangential direction	2.6	2.5	2.6	2.6
Volumetric	3.9	3.6	4.1	4.1
Shrinkage from green dimensions when dried to 12% moisture, %[a]				
Radial direction	2.7	2.2	2.7	2.7
Tangential direction	4.1	4.1	4.1	4.1
Volumetric	6.2	5.8	6.6	6.6
Modulus of elasticity, ksi				
When green	1,550	1,340	1,600	1,390
With 12% moisture	1,920	1,610	1,990	1,760
Proportional limit, compression parallel to grain, ksi				
When green	3.4	2.5	3.4	2.5
With 12% moisture	6.5	5.5	6.2	5.1
Proportional limit, compression perpendicular to grain, ksi				
When green	0.5	0.5	0.6	0.4
With 12% moisture	0.9	1.0	1.2	1.0
Proportional limit, bending, ksi				
When green	4.8	3.6	5.2	3.9
With 12% moisture	8.1	7.4	9.3	7.7

[a]Total longitudinal shrinkage of normal species from fiber saturation to oven dry is minor, usually ranges from 0.17 to 0.3% of the length when green.

Other examples are shakes and checks, lengthwise grain separations between or through growth rings.

To some extent, the deleterious effects on strength of all these defects may be overcome by grading of wood in accordance with the type, number and size of defects present, so that appropriate material can be selected for specific tasks, with appropriate allowable stresses assigned.

Also, different wood pieces can be combined in such a way as to average out the effects of solid material and material with defects. For example, several pieces of lumber may be laminated, with glue or nails, to form the equivalent of a large timber member. Because of the low probability of defects occurring in several components at the same cross section, the laminated member will have much greater strength than if it were made of a single ordinary wood piece of the same size. Similarly, thin sheets, or veneers, of wood can be bonded together to form plywood, with a reduction in the probability of defects being concentrated in any section.

Hardwoods and Softwoods. Trees whose wood is used in construction may be divided into two classes, hardwoods and softwoods. Hardwoods have broad leaves, which are shed at the end of each growing season. Softwoods, or conifers, have needlelike or scalelike leaves, and many of the species in this category are evergreen.

The fact that a tree is classified as a softwood does not mean that its wood is softer than that from a hardwood tree. Some softwoods are harder than medium density hardwoods. Two of the most commonly used woods for structural purposes, Douglas fir and southern pine, are softwoods. Hardwoods, such as oak, maple, birch, beech and pecan, usually are used for flooring or interior trim.

Softwood Lumber Standards. Softwood lumber is generally produced to meet the requirements of Product Standard PS 20-70, a voluntary standard developed by the National Institute of Standards and Technology and wood producers, distributors and users. This standard sets dimensional requirements for standard sizes, technical requirements, inspection and testing procedures, and methods of grading and marking lumber. The standard includes a provision for grading of lumber by mechanical means. Lumber so graded, called machine-stress-rated lumber, is distinguished from lumber that is stress graded visually in that machine-graded lumber is nondestructively tested and marked to indicate: Machine Rated, rated stress in bending and the modulus of elasticity.

Glued-laminated timber is generally produced to conform with Product Standard PS 56-73. This is a voluntary standard that gives minimum requirements for production, testing, inspection, identification and certification of structural glued-laminated timber. In addition, structural members should be fabricated to conform with AITC 117, Standard Specification for Structural Glued-Laminated Timber of Douglas Fir, Western Larch, Southern Pine and California Redwood, developed by the American Institute of Timber Construction.

Design Specification for Structural Lumber. Practice followed in design of visually stress-graded lumber, machine-stress-rated lumber, structural glued-laminated timber and lumber used in repetitive-member systems generally conforms with the National Design Specification for Stress-Grade Lumber and Its Fastenings, recommended by the National Forest Products Association. (*Repetitive-member systems* consist of three or more framing members, spaced not more than 24 in. center-to-center, that are joined by floor, roof or other load-distributing members so that the framing members share the load.)

Classification of Structural Lumber. The design specification requires that lumber grades be specified by commercial grade names. Structural lumber consists of lumber classifications as follows: dimension, beams and stringers, and posts and timbers. The specification assigns allowable unit stresses for each grade in these classifications.

Each lumber grade comprises pieces of lumber that may be slightly different from each other but all suitable for the use for which the grade is intended. Grading rules applied by generally accepted grading agencies describe the pieces that may be accepted in each grade. For those use and size classifications for which stress values are assigned, the grade rules limit strength-reducing characteristics.

Dimension lumber denotes pieces at least 2 in. but less than 5 in. thick, and at least 2 in. in nominal width. The following classes are included: framing, special dimension, and joists and planks, each with several grades. The framing classification covers studs 10 ft or less in length. The special-dimension classification covers framing for which appearance is important, machine-stress-rated lumber and decking.

Beams and stringers denote pieces of rectangular cross section, at least 5 in. thick and 8 in. wide, and graded with respect to strength in bending when loaded on the narrow face.

Posts and timbers comprise lumber of square or nearly square cross section, graded primarily for use as posts or columns carrying longitudinal loads but adapted also for miscellaneous uses in which strength in bending is not especially important.

Boards are lumber pieces not more than $1\frac{1}{2}$ in. thick and at least 2 in. wide. The classification includes appearance grades, sheathing and form lumber.

Methods for Establishing Allowable Stresses. Because of variability in characteristics of wood, as previously described, it is impractical to establish allowable design stresses by applying safety factors to ultimate strengths, yield strengths or proportional limits, as is done for the more uniform structural steels. Instead, tests are made on small specimens substantially free of defects to determine strength data and then factors are applied to determine basic allowable stresses and modifications to account for the influences of various characteristics.

The National Design Specification for Stress-Grade Lumber and Its Fastenings, mentioned previously, presents allowable unit stresses based on rules of the various agencies that write grading rules. If these stresses are used, each piece

of lumber must be identified by the grade mark of a competent lumber grading or inspection agency.

Standard Sizes of Lumber. As mentioned previously, Product Standard PS 20-70 establishes dimensional requirements for standard sizes of lumber. These standard sizes apply to rectangular cross sections and are specified by their nominal dimensions. Actual dimensions differ from the nominal to allow for dressing the lumber to size and for moisture content. PS 20-70 lists minimum dressed sizes for lumber in both the dry and green conditions. In this case, *dry lumber* is defined as lumber seasoned to a moisture content of 19% or less, and *green lumber* as containing more than 19% moisture.

Generally, dry lumber with a nominal dimension up to 7 in. is $\frac{1}{2}$ in. smaller than the nominal size. For example, a 2 × 4 actually is $1\frac{1}{2}$ × $3\frac{1}{2}$. For nominal dimensions of 8 in. or more, actual size is $\frac{3}{4}$ in. less than nominal. For boards, actual thickness is $\frac{1}{4}$ in. less than nominal. Actual sizes for green lumber are slightly larger than for dry lumber to allow for shrinkage when moisture content drops below 19%.

Glued-laminated timbers are generally fabricated with nominal 2-in.-thick lumber, unless they are to be sharply curved (see Fig. 8.14). For sharply curved members, nominal 1-in.-thick lumber is usually used. Standard sizes for glued-laminated timbers are based on the minimum dressed sizes for the laminations and are given in AITC 113, Standard for Dimensions of Glued-Laminated Structural Members.

Feet Board Measure. Payment for wood is generally based on volume, measured in feet

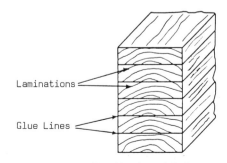

Fig. 8.14. Glued-laminated timber.

board measure (fbm). A board foot of a piece of lumber is determined by multiplying the nominal thickness, in., by the nominal width, in., and by the length, ft.

Structural Behavior of Wood. Wood is nonhomogeneous and anisotropic (has different properties in different directions) because of its cellular structure, the orientation of its cells and the way in which diameter increases when trees grow. Properties of wood usually are determined in three perpendicular directions: longitudinally, radially and tangentially (see Fig. 8.15). The difference between properties in the radial and tangential directions is seldom significant for structural purposes. In general, it is desirable to differentiate only between properties in the longitudinal direction, or parallel to the grain, and the transverse directions, perpendicular to, or across, the grain.

Wood has much greater strength and stiffness parallel to the grain than across the grain, the difference in tension being much larger than in compression. Table 8.8 (p. 227) lists proportional limits for Douglas fir and southern pine in compression in the two directions for comparison. The compressive strength of wood at an angle other than parallel or perpendicular to grain may be computed from

$$C_\theta = \frac{C_1 C_2}{C_1 \sin^2 \theta + C_2 \cos^2 \theta} \qquad (8.15)$$

where

C_θ = strength at desired angle θ with the grain

C_1 = compressive strength parallel to the grain

C_2 = compressive strength perpendicular to the grain

The stress-strain curve for wood in each principal direction resembles that in Fig. 8.6b (p. 214), except that it has no recognizable yield point. The modulus of elasticity, as well as the proportional limit, is different in each direction. The modulus given in Table 8.8 (p. 227) is for static bending loads.

Wood is designed so that nowhere does the stress produced by design loads exceed allowable unit stresses that will keep the material in the elastic range. Consequently, wood can be treated in structural analysis and design in the same way as the idealized structural material with the stress-strain curve of Fig. 8.7 (p. 216), for the elastic range and for each principal direction.

Fabrication may involve the following operations in shop or field: boring, cutting, sawing, trimming, dapping, routing and otherwise shaping, framing and finishing lumber, sawn or laminated, to assemble the pieces in final position in a structure. Shop details should be prepared in advance of fabrication and submitted to the structural engineer for approval. The details should give complete information necessary for fabrication and erection of components, including location, type, size and extent of all connections, fastenings and amount of camber (upward bending of beams to compensate for expected deflections).

Glued-laminated timbers are fabricated by bonding lumber laminations with adhesive so that the grain directions of all laminations are essentially parallel (see Fig. 8.14). To form a wide, deep member, narrow boards may be edge-glued, and short boards may be endglued, in each layer. The resulting laminations can then be face-glued to form the large timber.

The strength, stiffness and service life of glued-laminated timbers depend on the grade of lumber used for the laminations and the glue joint produced. Selection of the adhesive to use

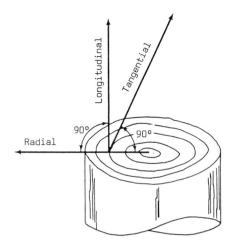

Fig. 8.15. Wood log with directions indicated for measuring properties.

should take into account the wood species, type of treatment, if any, given the wood, and whether condition of use of the timber will be wet or dry. Casein is generally used for dry-use timber. Resorcinol or phenol-resorcinol resins often are used for wet-use or preservative-treated timber. Glued joints may be cured by heat by any of several methods.

Fasteners for Structural Lumber. Fabrication of wood members and their erection is easily carried out, not only because of the light weight of wood and the ease with which it can be cut to size and shape in shop and field, but also because there are numerous easy ways for joining wood parts.

Small, lightly loaded wood parts can readily be joined with *nails* or *spikes*. The latter are available in longer lengths and in larger diameters for the same length than nails. Most nails and spikes used in construction are made of common wire steel, with a head on one end and a point on the other end of the shank. They also are available in hardened steel, alloy steels, aluminum and copper. Shanks may be smooth bright, cement coated, blued, galvanized, etched or barbed. Some shanks come with annular grooves or helical threads, for improved holding power.

The unit of measure for specification of nails and short spikes is the penny, represented by the letter d. Penny measure indicates length, usually measured from under the head to the tip of the point. A two-penny (2d) nail is 1-in. long. From this length, lengths increase in $\frac{1}{4}$-in. increments per penny, to 3 in. for a 10d nail. For longer nails, generally lengths increase in $\frac{1}{8}$-in. increments per penny to 4 in. for a 20d nail, then in $\frac{1}{2}$-in. increments for every 10d, to 6 in. for a 60d nail. Diameters are standardized for each penny size.

Nails and spikes may be driven directly with a hammer through wood to connect two or more pieces. If, however, nails have to be placed closer together than half their length, holes smaller than the nail diameter should first be drilled at the nail locations and then the nails may be driven in the prebored holes.

Nails and spikes may be driven at any angle with the grain if they are to be loaded in com-pression; however, if they are to be loaded in tension or subjected to withdrawal forces, nails and spikes should not be placed in end grain or parallel to the grain. On the other hand, some reliance can be placed on toenailing (driving nails diagonally) where two members meet at a sharp angle and the loads are primarily com-pressive. Best results with toenailing are obtained when each nail is started at one-third the nail length from the end of the piece being joined and the nail is driven at an angle of about 30° to the face of the piece.

Among the many factors determining the withdrawal resistance of a nail or spike, one of the most important is the length of penetration into the piece receiving the point. Load tables for withdrawal resistance generally are based on a penetration of 11 nail or spike diameters. Penetration into the piece receiving the point should be a minimum of $3\frac{2}{3}$ diameters.

Wood screws are an alternative to nails and spikes but with greater holding power, because the threads project into the wood. Common types of screws come with flat heads, for use in countersunk holes when a flush surface is desired, or with oval or round heads, for appearance or to avoid countersinking. Also, screws are available in steel, brass and other metals. They should be placed in wood pieces only perpendicular to the grain, preferably in pre-drilled holes, to prevent splitting. Embedment into the piece holding the point should be at least seven times the shank diameter.

Lag screws are large screws capable of resisting large loads. They are sometimes used instead of bolts, especially where tightening a nut or bolt would be difficult or where a nut would be objectionable for appearance reasons.

Common, or machine, bolts, usually with square heads and nuts, are often used to connect load-carrying wood pieces. Holes with a diameter that permits easy placement of bolts have to be drilled in the wood. The holes must be accurately located. Metal washers should be placed under nuts and bolt heads to protect the wood when the nuts are tightened and to distribute the pressure over the wood surface.

Tightening of the nuts holds the pieces together, while loads are resisted by pressure of the wood against the bolt shank.

Exposed metal fasteners may be subjected to corrosion or chemical attack. If so, they should be protected by painting, galvanizing or plating with a corrosion-resistant metal. In some cases, hot tar or pitch may provide a suitable protective coating.

Timber connectors often are used with bolts to provide a more efficient joint with fewer bolts. Connectors are metal devices that transmit load between parts to be joined, while bolts function mainly to hold the parts in contact. Generally, a connector is placed in a groove in the wood.

Split rings (see Fig. 8.16a) are efficient for joining wood pieces. Made of steel, the rings are $2\frac{1}{2}$ or 4 in. in diameter. To make a connection, a groove for each ring is cut with a special tool to a depth of half the ring depth in the contact surface of each piece to be joined. Also, a hole for a bolt is drilled at the center of the

circle and through each piece. The lower part of the split ring is inserted in the groove of one piece and then the groove in the second piece is placed on the upper part of the ring. The ring is provided with a tongue-and-groove split, to obtain a tight fit. Placement of a bolt in the center hole and tightening of a nut on the bolt holds the pieces of the connection together (see Fig. 8.16b).

Shear plates (see Fig. 8.16c) are efficient for joining wood to steel components. Circular in shape, they come with a smooth back face, for contact with the steel part, and a circumferential lip, for insertion in a round groove in the wood contact surface. A bolt hole must be provided in both the steel and wood parts of the joint at the center of the circle. As with split rings, the bolt holds the parts in contact (see Fig. 8.16d). Shear plates, used in pairs, back to back, may also be used to connect wood components (see Fig. 8.16e).

Glued Lumber Joints. Still another alternative for making connections between wood pieces is

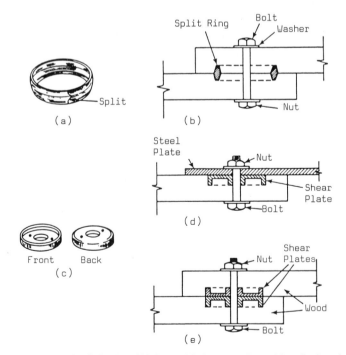

Fig. 8.16. Timber connectors. (a) Split ring. (b) Assembled connection with split ring. (c) Shear plate. (d) Steel member connected to wood member with shear plate and bolt. (e) Wood members connected with a pair of shear plates and a bolt.

glueing. Adhesives described previously for fabricating glued-laminated timbers are also suitable for other glued joints. In all cases, consideration must be given in selection of an adhesive as to whether service conditions will be wet or dry.

When a strong glued joint is required, pieces should be placed with the grain direction parallel. Only in special cases, such as fabrication of plywood, may lumber be glued with the grain direction at an angle in pieces to be joined.

Joints between end-grain surfaces of wood pieces are not likely to be reliable. Consequently, joints between ends of pieces with fibers parallel are usually made by cutting sloping, mating surfaces at the ends and glueing those surfaces [*scarf joints* (see Fig. 8.17a)]. Or fingers may be cut into the ends and joined with glue by interlocking [*finger joints* (see Fig. 8.17b and c)].

Preservative Treatments for Wood. If wood cannot be kept dry or permanently submerged in water, it will decay. This, however, can be avoided by application of a preservative treatment. There are several types of preservative that may be used. Selection of an appropriate preservative depends on the service expected of the wood member for the specific conditions of exposure.

Among the commonly used preservatives are creosote and creosote solutions; oil-borne chemicals, such as pentachlorphenol; and waterborne inorganic salts, such as certain chromated zinc compounds, copper compounds and fluorchrome compounds. Preservatives may be applied in any of many ways. In some cases, brushing or dipping may provide adequate protection. For maximum effect, however, pressure application is necessary. Effectiveness depends on depth of penetration and amount of retention of preservative.

Wood also may be impregnated to make it fire retardant. For such purposes, salts containing ammonium and phosphates are often used. They are recommended, however, only for interior or dry-use service conditions, because the salts may leach out. With proper preparation, the impregnated surfaces may be painted. Fire-retardant surface treatments also are available. Their effectiveness generally depends on formation, under heat, of a blanket of inert gas bubbles, which retards combustion and insulates the wood below.

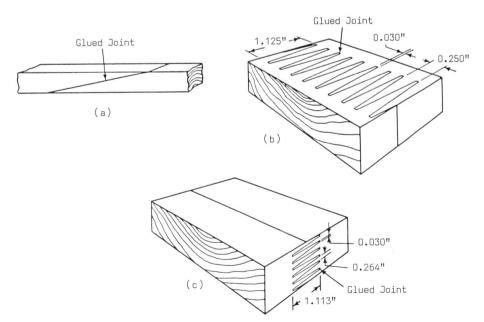

Fig. 8.17. Wood joints. (*a*) Scarf joint for connecting ends of wood members. (*b*) Finger joint with fingers along the width of the members. (*c*) Finger joint with fingers over the depth of the members.

Plywood

Plywood essentially is a wood product, just as is glued-laminated timber. Plywood, however, is fabricated as sheets, or panels, often 4 ft wide by 8 ft long and $\frac{1}{4}$ to $1\frac{1}{8}$ in. thick for construction applications, although other-size panels also are available. Principal advantages over lumber are its more nearly equal strength properties in the length and width directions, greater resistance to checking, less shrinkage and swelling from changes in moisture content, and greater resistance to splitting.

Production of Plywood. Plywood is built up of thin wood sheets, or veneers. It contains an odd number of layers of these veneers. Each layer, or ply, consists of one or more veneers. The plies are glued together with the grain of adjacent layers at right angles to each other (see Fig. 8.18).

The veneers are peeled from logs by rotating them against a long knife. After being dried to a moisture content of about 2 to 5%, each veneer is graded and then coated on one face with glue. Next, another veneer may be placed on the glued surface to form a ply, which also is then coated on one face with glue. Finally, plies are combined about a central ply so that plies at the same distance from the central ply have their grain direction parallel. This symmetry is desirable to reduce warping, twisting and shrinking. The glued plies are usually cured by hot pressing in a large hydraulic press. The manufacturing process generally conforms with the latest edition of the voluntary product standard PS 1, for Softwood Plywood—Construction and Industrial.

Design Standard for Plywood. Design practice usually complies with the Plywood Design Specification of the American Plywood Association.

This specification classifies wood species into five groups in accordance with modulus of elasticity in bending and important strength properties. It also distinguishes between two types of plywood, interior and exterior, and its various grades. In addition, the specification presents design methods and allowable design stresses. Various supplements to the specification cover design of such structural components as beams, curved panels, stressed-skin panels and sandwich panels.

Classification Systems for Plywood. All woods within a species group are assigned the same allowable stress. The group classification is usually determined by the species in the face and back veneer of the plywood panel. Unless the grade classification requires all plies to be of the same species, inner plies may be made of a different species than the outer plies.

Southern pine and Douglas fir from northern areas fall in Group 1. Douglas fir from southern states are included in Group 2, which is assigned lower stresses than Group 1.

Some types of allowable stress also are lower for interior-type plywood than for exterior-type used in dry conditions. Shear strength, however, depends on the type of glue used in the plywood.

The classification into interior and exterior types is based on resistance of the plywood to moisture. Exterior plywood is made with waterproof glue and high-quality veneers, incorporating only small, minor defects, such as small

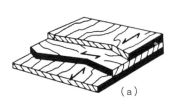

(a)

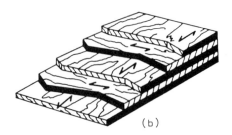

(b)

Fig. 8.18. Assemblage of plywood panels. (*a*) Three layer, three ply. (*b*) Five layer, five ply. Arrows show direction of grain in the plies.

knots, knotholes and patches. Interior plywood also may be made with waterproof glue but the veneers may be of lower quality than that permitted for exterior plywood. Interior plywood may be used where its moisture content in service will not continuously or repeatedly exceed 18% or where it will not be exposed to the weather. For wet conditions, exterior plywood should be used.

Veneer is classified into the following grades:

N and *A*—no knots, restricted patches, *N* being used for natural finishes and *A* for paintable surfaces

    B—small round knots, small patches, round plugs (often used as outer facing of plywood forms for concrete)

    C—small knots, knotholes and patches (lowest grade permitted for exterior plywood; often used for a facing on sheathing). *C plugged* is a better quality *C* grade, often used in floor underlayment.

    D—large knots and knotholes, used for inner plies and back ply of interior plywood

Plywood panels, depending on the wood species and veneer grades in the plies, may be considered to be in an *engineered grade* or an *appearance grade*. Engineered grades consist mostly of unsanded sheathing panels designated *C-D Interior* or *C-C Exterior*, the letters referring to the veneer grades in front and back panels. These grades are specified by giving thickness and an identification index, which is discussed later. *C-D Interior* and *C-C Exterior* may additionally be graded *Structural I* or *II*, intended for use where strength properties are important. All plies of *Structural I* must be one of the Group 1 species. Groups 2 and 3, which are assigned lower design stresses, are permitted in *Structural II*. Both structural grades are made only with exterior glue and have additional restrictions on knot size and patches. Appearance grades are specified by thickness, veneer grades in face and back plies, and species group.

The *identification index* used on sheathing is a measure of plywood stiffness and strength parallel to the grain of the face plies. The index consists of two numbers, written left to right, with a slash line between them. The number on the left indicates the maximum spacing, in., for roof supports for uniform live loads of at least 35 psf. The number on the right indicates the maximum spacing, in., for floor supports for uniform live loads of at least 160 psf and concentrated loads, such as pianos and refrigerators. For example, $\frac{1}{2}''$ 32/16 C-C INT-APA specifies a $\frac{1}{2}$-in. thick C-C interior-type plywood panel, which could be used as roof sheathing supported by rafters spaced not more than 32 in. center-to-center or as subflooring with joists spaced not more than 16 in. center-to-center.

Plywood Construction. Plywood can be used mainly as paneling or built up into structural members, much as structural steel plates are used to build up structural steel members.

Paneling may serve as facings for interior and exterior surfaces of building walls, as subflooring, or as roof or wall sheathing.

Subflooring is a structural deck, which is supported on joists and on which is placed carpeting, floor tile, linoleum or other decorative, wearing surface.

Sheathing is an enclosure, supported by rafters in roofs or by studs in walls, and on the outer surface of which is usually placed waterproofing and a decorative, weather-resistant facing.

In built-up members, plywood may be used as the web of beams, such as the I beam in Fig. 8.19*a* and the box beam in Fig. 8.19*b*, with lumber top and bottom flanges. Plywood also may be used as the faces, or skins, of sandwich panels. In such panels, the skins are separated by, but glued to, a structural core capable of resisting shear (see Fig. 8.19*c*). The core may be closely spaced lumber ribs, resin-impregnated-paper honeycomb or foamed plastic. The honeycomb or plastic, however, should not be assumed capable of resisting flexural or direct stresses.

As another alternative, plywood may be used as one or both faces of a stressed-skin panel (see Fig. 8.19*d*). In such a panel, plywood fac-

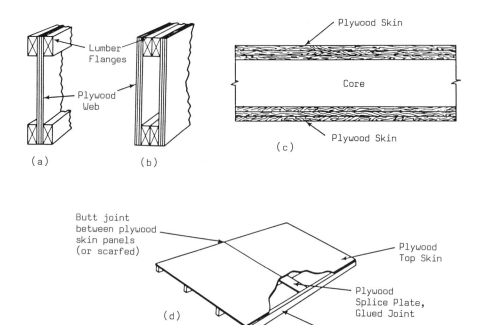

Fig. 8.19. Plywood structural members. (*a*) I beam. (*b*) Box beam. (*c*) Sandwich panel. (*d*) Stressed-skin panel.

ings are glued to lumber stringers, for use as floor or roof members or as wall members subjected to bending. The skins and stringers act together as a series of **T** or **I** beams. The skins resist nearly all the bending forces and also serve as sheathing or exposed facings. The stringers resist shear on the panel. Cross bracing, such as headers at the panel ends and interior bridging, may be placed between the stringers to keep them aligned, support edges of the skin and help distribute concentrated loads.

Concretes

Just as there are many structural steels, there are many concretes. The concretes that are commonly used for structural purposes in buildings consist basically of a mixture of portland cement, fine aggregate, coarse aggregate and water. Temporarily plastic and moldable, the mixture soon forms a hard mass, usually in a few hours. Just as chemicals are added to steels to form alloys to achieve specific results, chemicals, called *admixtures*, may be added to

the concrete mixture to secure desired properties. A specific proportioning of contents of concrete is called a *mix*.

A short time after the solid components of concrete are mixed with water, the mass stiffens, or sets. The time required for setting is controlled by ingredients in the portland cement and sometimes also by admixtures to allow time for placement of concrete in molds, usually called *forms*, and for producing desired surfaces. Then, additional chemical reactions cause the mass to strengthen gradually. The concrete may continue to gain strength for more than one year, although most of the strength gain takes place in the first few days after hardening starts.

The chemical reactions are accompanied by release of heat, which must be dissipated to the atmosphere. If this heat cannot be removed during the early stages of hardening, as may happen in hot weather or when the concrete is cast in large masses, the concrete will get hot and crack. Precautions must be taken to avoid such situations, as will be discussed later.

Designed by knowledgeable structural engineers and produced and handled by reliable, competent contractors, concrete is an excellent, economical structural material. It offers high strength relative to its cost. While plastic, it can be cast in forms on the building site to produce almost any desired shape. Also, it can be precast with strict quality control in factories, then shipped to the building site and erected in a manner similar to that for structural and cold-formed steel members. In addition, concrete is durable and can serve as a wearing surface for floors and pavements or as an exposed surface for walls. It does not need painting. Furthermore, it has high resistance to fire and penetration of water.

Concrete, however, has disadvantages, especially when it is cast in place on the building site. For one thing, quality control may be difficult. Often, many subcontractors are involved in supplying ingredients, designing the mixes, producing the mixes, placing the concrete and curing it, inspecting the process and testing the results. If any step should be faulty and cause production of poor concrete, responsibility for the undesirable results may be impossible to determine.

Another disadvantage is that concrete is brittle. While its compressive strength is substantial, tensile strength is small and failure in tension is sudden. As a result, plain concrete is used only where it will be subjected principally to compression and in members, such as pedestals, that are bulky. Concrete, however, can be used in members subjected to tension if reinforcing materials, such as steel, capable of resisting the tension, are incorporated in it. Concrete can also sustain tension if it is prestressed, held in compression permanently by external forces applied before the tensile loads. Reinforced, or prestressed, concrete serves economically in a wide variety of structural applications in buildings, from footings to roofs.

Types of Concrete. The main types of concrete used in building construction may be classified as normal, air-entrained or lightweight concrete. Heavyweight concrete may be used where shielding from high-penetration radiation, such as that from nuclear reactors, is needed.

Normal concrete is generally used for structural members, including foundations. It is made with portland cement, sand as the fine aggregate, gravel or crushed stone as the coarse aggregate, and water.

Air-entrained concrete is used where the material will be subjected to cycles of freezing and thawing in service. While normal concrete contains some entrapped air, this air does not provide adequate protection from damage from freezing and thawing. So air in small, disconnected bubbles is purposely entrained in the concrete. This may be done by incorporation of chemicals in the portland cement or by addition of admixtures to the concrete mix. The tiny air bubbles in the concrete provide reservoirs into which ice formed in concrete in cold weather can expand and thus prevent spalling of the concrete. A slight loss in strength results from air entrainment, but the penalty is small compared with the benefits from improved durability.

Lightweight concrete is used where less weight is needed than that imposed by normal or air-entrained concrete or where weight reduction will cut construction costs. Lightweight concrete may be made by substitution of light aggregates for sand and gravel, by expansion of the mix before it sets by addition of a chemical, such as aluminum powder, or by incorporation of a stabilized foam. Concrete into which nails can be driven can be produced by replacement of the coarse aggregate with pine sawdust.

Heavyweight concrete is made with heavy aggregates, such as iron ore, barite, or iron shot and iron punchings. This type of concrete may weigh more than twice as much as an equal volume of normal concrete.

Methods of selecting mixes for the various types of concrete are described in ACI 211.1, Recommended Practice for Selecting Proportions for Normal and Heavyweight Concrete, and ACI 211.2, Recommended Practice for Selecting Proportions for Structural Lightweight Concrete, American Concrete Institute.

Production, placement, finishing and testing of concrete are described in ACI 301, Specifications for Structural Concrete for Buildings. This standard also incorporates by reference applicable ASTM specifications. Admixtures are discussed in the ACI Guide for Use of Admixtures in Concrete. The preceding recommended practices and many others are incorporated in the ACI Manual of Concrete Practice but also are usually available as separate publications.

Portland Cements. There are several types of portland cement, but only a few are frequently used in building construction. All are made by blending a carefully proportioned mixture of calcareous, or lime-containing, materials and argillaceous, or clayey, materials. Burned at high temperatures in a rotary kiln, the mixture forms hard pellets, called clinker. The clinker is then ground with gypsum to the fine powder called portland cement. The gypsum is used to control the rate of setting of concrete mixes. For some types of cement, an air-entraining agent is added.

The most effective ingredients of portland cement in the formation of normal concrete are tricalcium silicate (C_3S), dicalcium silicate (C_2S), tricalcium aluminate (C_3A) and tetracalcium aluminum ferrite (C_4AF). The proportions of these chemicals and requirements placed on physical properties distinguish the various types of portland cements from each other.

Portland cements are usually produced to meet the requirements of ASTM C150, Standard Specification for Portland Cement. This specification details requirements for general-purpose cements, usually used for structural applications; cements modified for use where exposure to sulfate attack will be moderate or where there will be somewhat more than normal heat of hydration during hardening of the concrete; high-early-strength cements; and air-entraining cements.

Aggregates. Fine and coarse aggregates for concretes should be treated as separate ingredients. Aggregates for normal concrete should conform with ASTM C33, Specifications for Concrete Aggregates. Those for lightweight concrete should comply with ASTM C330, Specifications for Lightweight Aggregates for Structural Concrete.

Water. Concretes should be made only with clean, freshwater. If the water is drinkable, it is generally acceptable. If not, specimens of concrete made with the water should be tested to verify attainment of desired concrete properties.

Desired Characteristics of Normal Concrete. Within specific inherent limitations, properties of concrete can be changed to achieve desired objectives by changing the ingredients of the mix and the proportions of the ingredients used. The main variations that are usually made are:

1. Type of cement
2. Ratio of weight of water to weight of cement (water-cement ratio)
3. Ratio of weight of cement to weight of aggregates
4. Size of coarse aggregate
5. Ratio of weight of fine aggregate to weight of coarse aggregate
6. Gradation, or proportioning of sizes, of aggregates
7. Use of admixtures

Workability. For complete hydration of cement, about $2\frac{1}{2}$ gal of water combines chemically with each sack (94 lb) of cement. But a mix with this water-cement ratio would be too stiff for uniform mixing, proper handling and easy placement. Consequently, additional water is used to make the concrete flow more readily. Workability thus is a measure of the ease with which the ingredients of concrete can be mixed and the resulting mix handled and placed. The concept of workability embraces other characteristics of the plastic mix, such as consistency, adhesiveness and plasticity.

To evaluate the workability of a mix, a slump test is often made, as described in ASTM C143, Standard Method of Test for Slump of Portland Cement Concrete. In this test, which actually measures consistency, a mold with the shape of the frustum of a cone is filled with a specimen of the mix. The mold is removed and the change in height, or slump, of the mass is measured. The slump is used as an indication of the workability of the mix. Slump for structural concrete may range from zero for a very stiff mix to 4 in., one-third the cone height, for a plastic mix.

Durability of concrete is generally achieved by production of a dense, high-strength concrete, made with hard, round, chemically stable aggregates. Such a concrete has a high resistance to abrasion. For resistance to freezing and thawing, air-entrained concrete is used.

Watertightness. Water in excess of that needed for hydration of the cement evaporates eventually, leaving voids and cavities in the concrete. If these become connected, water can penetrate the surface and pass through the concrete. Consequently, water added to the concrete mix should be kept to the minimum needed for acceptable workability.

Volume Changes. From the plastic state on, concrete is likely to undergo changes in volume. While in the plastic state, concrete may settle before it sets. To reduce the amount of settlement, concrete, after placement in formwork, often is vibrated with mechanical vibrators or pushed with a spade to consolidate the mix and insure complete filling of the forms. After the concrete starts to harden, it may shrink as it dries, and cracking may occur as a consequence. Drying shrinkage may be limited by keeping the amount of cement and mixing water to the minimum necessary for attaining other desired properties of the concrete and by moist curing of the concrete as it hardens. The cracks may be kept small or closed by use of reinforcing steel or prestress.

After the concrete has hardened, it will expand and contract with temperature changes in its environment. The coefficient of expansion of concrete depends on many factors but averages about 0.000 0055 in. per in. per °F, nearly the same as that of steel.

Another cause of volume change is chemical reaction between ingredients of the concrete. Such changes can be avoided by selection of nonreactive aggregates or by addition of pozzolanic material, such as fly ash, to the mix.

Strength. After concrete sets, it gains strength rapidly at first and then much more slowly as it ages. Because concrete has a low tensile strength, engineers seldom rely on this property. Instead, concrete strength is measured by the ultimate compressive strength of a sample. In the United States, it is common practice to measure this property by testing a cylindrical sample in compression when it is 28 days old. The sample may be taken from the mix being placed in the forms or by taking drilled cores from hardened concrete. Handling and testing of the specimens are prescribed in various ASTM specifications and often done by independent testing laboratories.

Concrete strength is related linearly to the water-cement ratio. The lower the ratio, the higher the strength of a workable mix. Hence, strength can be increased by decreasing the amount of water or increasing the amount of cement in the mix. Strength also may be raised by use of higher-strength aggregates, grading the aggregates to reduce the percentage of voids in the concrete, moist curing the concrete after it has set, vibrating the concrete in the forms to densify it, or sucking out excess water from the concrete in the forms by means of vacuum. High strength in the first few days can be achieved by use of high-early-strength cement, addition of appropriate admixtures or high curing temperatures, but long-term strength may not be significantly influenced by these measures.

Ultimate compressive strength of normal structural concrete at 28 days usually specified ranges from about 3,000 psi to about 10,000 psi. Strengths exceeding 15,000 psi have been achieved in building construction, however, by use of special admixtures and cement substitutes. Superplasticizers, which are cement-dispersion agents, are used to reduce water requirements and increase strength without impairing workability. Similarly, silica fume, or microsilica, a waste product of electric-arc furnaces, is substituted for a portion of the required portland cement to achieve substantial strength increases.

Weight. Normal concrete weighs about 145 lb per cu ft. For reinforced concrete, 150 lb per cu ft is generally assumed for design purposes, to include the weight of steel reinforcing.

The weight of lightweight concretes depends on aggregates used or amount of expansion or foaming, in accordance with the technique used for reducing weight. With vermiculite or perlite, weight may be about 60 lb per cu ft; with scoria, pumice or expanded clay or shale, about 100 lb per cu ft; and with cinders, 115 lb per cu ft with sand and 85 lb per cu ft without sand.

Heavyweight concretes made with steel shot for fine aggregate and with steel punchings and sheared bars as coarse aggregates may weigh from 250 to 290 lb per cu ft.

Creep. As do other materials, concrete when subjected to static load deforms, and the amount of initial deformation depends on the magnitude of the load. If the load remains on the concrete, the deformation will increase. This change with time of deformation under constant load is called creep. The rate of creep gradually diminishes with time and the total creep approaches a limiting value, which may be as large as three times the initial deformation.

Stress-Strain Curve for Concrete. When tested in compression, concrete has a stress-strain curve similar to that shown in Fig. 8.6c, p. 214. Consequently, the idealized structural material with the stress-strain curve shown in Fig. 8.7, p. 216, is at best only a rough approximation of concrete in compression.

The portion of the stress-strain curve up to about 40% of the ultimate load is sufficiently close to a straight line that a secant modulus of elasticity may represent that portion. The value of the modulus may be taken, in psi, for normal concrete as

$$E_c = 57,000 \sqrt{f'_c} \tag{8.16}$$

where f'_c = 28-day compressive strength of the concrete, psi.

Poisson's ratio for concrete is about 0.25.

Manufacture of Concretes. The mix for normal concrete is specified by indicating the weight, lb, of water, sand and stone to be used per 94-lb bag of cement. Preferably also, type of cement, gradation of the aggregates and maximum size of coarse aggregate should be specified. Mixes are often briefly but not completely described by giving the ratio of cement to sand to coarse aggregate by weight; for instance, $1 : 1\frac{1}{2} : 3$.

The proportion of components should be selected to obtain a concrete with the desired properties for anticipated service conditions and at lowest cost. Although strength and other desirable properties improve with increase in cement content of a mix, cement is relatively expensive. Hence, for economy, cement content should be kept to a minimum. Also, because strength increases with decrease in water-cement ratio, the amount of water should be kept to the minimum necessary to produce acceptable workability. Attainment of these objectives is facilitated by selection of the largest-size coarse aggregates consistent with job requirements and by good gradation of the aggregates to keep the volume of voids in the concrete small. Optimum mixes for specific job requirements usually are determined by making and testing trial batches with different proportions of ingredients.

Batching and Mixing. While concrete may be prepared on or near the building site, it is usually more convenient to purchase ready-mixed concrete batched at a central plant and delivered to the site in a mechanical mixer mounted on a truck. The mix ingredients are stored at the plant in batching equipment before being fed to the mixer. The equipment includes hoppers, water tank, scales, and controls for adjusting weights of ingredients to be supplied to the mixer. Since mixing takes only a few minutes, best control of mixing time can be achieved by adding water to the mixer after the truck arrives at the site, where the operation can be inspected.

Placement of Concrete. There are many ways of conveying concrete from the mixer to its final position in the building. In all cases, precautions should be taken to prevent segregation of the concrete ingredients after discharge from the mixer. Usually, the mixer discharges into a chute. To avoid segregation, the concrete should be fed into a length of pipe inserted at the end of the chute to carry the mix directly down to the forms or into buckets, hoppers, carts or conveyor belts that will transport it to the forms.

In multistory buildings, concrete may be lifted to upper levels in buckets raised by cranes, hoists, or elevators. There, it is usually discharged into hoppers and then fed to barrows or motorized carts for delivery to forms (see Fig. 8.20a). Alternatively, concrete may be pumped through pipes from ground level directly to the forms (see Fig. 8.20b). Near ground level, concrete may be speedily placed with conveyor belts (see Fig. 8.20c).

Concrete also may be sprayed onto a backup surface and built up to desired thicknesses. Sprayed concrete, also called *shotcrete* or *gunite*, is applied by an air jet from a gun, or mechanical feeder, mixer and compressor. Shotcrete can be made with a low water-cement ratio to obtain high strength.

After the concrete has been placed in final position in the forms, it should be immediately spaded or vibrated with electric or pneumatic

(a)

(b)

Fig. 8.20. Placement of concrete. (a) Motorized cart transports ready-mixed concrete to forms for floor of a multistory building. (b) Pipes carry concrete from a pump to forms. (c) Conveyor carries concrete to column forms. (a) Courtesy of H. H. Robertson Co.; (b) and (c) courtesy of Morgan Manufacturing Co.

(c)

Fig. 8.20. (*Continued*)

vibrators. The objectives are to eliminate voids and insure close contact of the concrete with forms, reinforcing and other embedded objects.

Construction Joints. A construction joint is formed when new concrete must be bonded to concrete that has already hardened. Steps should be taken to insure a secure bond.

First, the hardened surface to be bonded should be cleaned, washed with a jet of air and water at 100 psi, and sandblasted or brushed vigorously with fine-wire brooms. Next, the surface should be washed and allowed to dry. Then, before the new concrete is placed against the surface, it should be coated with $\frac{1}{2}$ in. of mortar, a mixture of sand, cement and water. The new concrete should be placed before the mortar dries.

Finishing of Concrete Surfaces. After concrete has been consolidated in the forms, the top surface should be brought to the desired level and given the specified shape, smoothness or texture. The surface usually is leveled or shaped by screeding. In this process, a straightedge or a board with the specified shape is moved along *screeds*, rigid guides set at appropriate elevations. Then, if desired, as would be the case for floors and roofs, the top surface may be smoothed with power floats or by hand with wood floats. If a finer surface is desired, it may be steel-troweled. Floating may begin as soon as the surface has hardened sufficiently to bear a worker's weight without indentation. Troweling may be done when the surface is hard enough that excess fine material in the mix will not be drawn to the top and, for an extra-hard finish, again when the surface has nearly hardened. Excessive manipulation, however, can cause the surface to check, craze and dust later.

Curing of Concrete. Although more water than needed for hydration of the cement is incorporated in a concrete mix, concrete nevertheless must be kept moist after it has set. If the water evaporates, hydration of the cement may be delayed or prevented, and a weak concrete may result. Curing is any operation performed after concrete has set that improves hydration. Generally, curing is achieved by maintaining a moist environment by addition of water. For this purpose, water may be continuously sprinklered or ponded on the surfaces exposed or the surfaces may be covered with wet burlap. Normal concrete should be cured this way for at least 14 days. An alternative method is coating of the surfaces with a sealer to prevent evaporation.

Precast concrete and concrete cast in cold weather may be steam-cured in enclosures to speed hydration. Temperatures maintained usually range between 100 and 165°F. The result is a product with high early strength. Higher 24-hr strengths can be attained, however, if start of steam curing is delayed 1 to 6 hr, to allow early cement reactions to take place and sufficient hardening to occur so that the concrete can withstand the rapid temperature changes when curing begins. Another process that may be used for factory-produced concrete products is *autoclaving*, or high-pressure steam curing, with temperatures above the boiling point of water. The process may provide high early strength, low volume change on drying and better chemical resistance.

Formwork for Concrete. Forms are used to support and shape concrete until it gains sufficient strength to be self supporting. They also may be provided with coatings or liners to produce desired surface textures.

Forms for horizontal or inclined concrete members usually must be supported on falsework, temporary construction, of adequate strength and sufficient rigidity to keep deflections within acceptable limits (see Fig. 8.21a). Forms for vertical members must be braced to keep them plumb.

The forms themselves must also be strong and rigid to satisfy dimensional tolerances. In addition, they must be tight, to prevent water or mortar from leaking out. Because forms usually are only temporary construction devices, they must be low cost, and when possible, they should be designed and scheduled for repeated reuse, for economical reasons. Figure 8.21b illustrates the assembly of wall forms with identical, prefabricated panels. Figure 8.21c shows a large, prefabricated form, called a *flying form*, used repeatedly for the floors and roof of a multistory building. Because forms are usually temporary construction and reuse, or at least salvage of the materials, is desirable, forms should be lightweight and also easily and speedily erected and removed. Materials generally used for forms include lumber, plywood, cold-formed steel, reinforced plastic and precast concrete.

Parallel vertical forms, such as those for walls, often are kept at the proper distance apart by struts or ties, the former usually being placed at the top of the forms. Form ties that will pass through the concrete should have as small a cross section as possible, because the holes they form may permit water to leak through.

Forms should be treated before concrete is placed in them to prevent the concrete from sticking to them. Paraffin-base mineral oils generally are used to coat wood forms, and marine-engine oils for steel forms.

While forms usually are kept immovable until they are to be removed, continuously movable forms, called *slip forms*, also may be used in appropriate situations. In building construction, slip forms sometimes are used for concrete walls of multistory buildings. The forms are slowly jacked up on steel bars embedded in the concrete. Climbing rates range from about 2 to 12 in. per hr.

In multistory buildings, floor and roof forms usually are supported on vertical falsework, called *shores*, that extend to and are supported on the floor below (see Fig. 8.21a). Because construction must proceed speedily, that floor seldom is old enough and therefore sufficiently strong to withstand the load of falsework and newly placed concrete at the upper level. Consequently, as soon as the forms for a floor have been removed, the floor should be *reshored*, or again supported on falsework extending to the floor beneath. It may be necessary to reshore

(a)

(b)

Fig. 8.21. Forming systems for sitecast concrete. (*a*) Falsework temporarily supports forms for floors and roofs. (*b*) Prefabricated panels used for walls. (*c*) Flying form for repetitive use. (Courtesy Symons Manufacturing Co.)

(c)

Fig. 8.21. (*Continued*)

several floors in series to support the new concrete adequately.

Cold-Weather Concreting. New concrete should be protected against freezing during cold weather for at least 4 days after placement in the forms. For this purpose, the mix may be made with water heated to at least 140°F, and if necessary, the aggregates may also be heated. An enclosure should be placed around the forms or the entire construction space, and the space within the enclosure should be heated to maintain temperatures above freezing. Danger of freezing and thawing may be reduced by production of concrete with high early strength.

Hot-Weather Concreting. In hot weather, concrete may set undesirably fast. Addition of water to counteract this is poor practice for many reasons.

At high outdoor temperatures mixing water evaporates swiftly, the concrete shrinks fast and the rate of cooling of the concrete from the high initial temperatures may be high. As a result, the concrete will crack. Such conditions may be avoided by keeping the temperature of concrete during casting below 90°F. For this purpose, water and aggregates may be chilled. If necessary, work may be carried out at night. Also, forms may be sprinkled with cool water. In addition, set-retarding admixtures may be used in the mix to counterattack the set-accelerating effects of hot weather and to lessen the need for additional mixing water to maintain workability. Curing should be started as soon as the concrete has hardened sufficiently to resist surface damage. Continuous water curing gives best results.

Contraction Joints. As pointed out previously, concrete shrinks when it dries after setting or whenever the temperature drops. If the shrinkage is restrained, whether by friction or obstructions, the concrete will crack. Shrinkage and restraint are difficult to prevent. Consequently, structural engineers design concrete components to prevent or control cracking. Sufficient prestress, keeping the concrete continuously under compression after it hardens, can prevent cracks from opening. Use of reinforcing steel can keep the size of cracks small.

Contraction joints are another crack control measure. They are used in concrete floors, roof slabs and long walls mainly to control location of cracks, by creating a plane of weakness along which cracks are likely to occur. A contraction joint is an indentation in the concrete, in effect, a man-made crack. Width may be $\frac{1}{4}$ to $\frac{3}{8}$ in. and depth one-sixth to one-quarter the slab thickness. The joint may be formed by grooving the surface of the slab during finishing or with a saw cut shortly after the concrete hardens.

Expansion Joints. Temperature changes can cause large changes in length of long slabs and walls. Provision must be made for these movements or the components may crack or buckle. For this purpose, expansion joints are used. They provide complete separation of the parts of a long component. The opening usually is sealed with a compressible material to prevent the gap from becoming jammed with dirt and, consequently, inoperative. Where watertightness is required, a flexible water stop of rubber, plastic or noncorrosive metal should be placed between the separated parts. If transfer of load between the parts is desirable, steel bars, called *dowels*, should connect the parts, with provision for one end of each dowel to slide in a close-fitting cap or thimble.

Reinforced Concrete

The disadvantages of the low tensile strength of concretes can be largely overcome by incorporating in them, in proper locations and with appropriate anchorage lengths, steel bars or wires designed to withstand tensile forces. The steel may serve entirely as reinforcing against tension, in which case, the combination of steel and concrete is called reinforced concrete; or the steel may be used as tendons, to prestress the concrete.

Reinforcing steel may also be used in beams and columns to withstand compressive forces and thus permit use of smaller concrete members. In addition, steel may be used to control crack openings due to shrinkage and temperature change, to distribute load to the concrete and other reinforcing steel, or both.

Design of reinforced concrete should conform with requirements of the local building code or with ACI 318, Building Code Requirements for Reinforced Concrete. A *Commentary on ACI 318* also is available from the American Concrete Institute. It contains explanations and interpretations of the requirements in ACI 318.

Reinforcing Steel. Reinforcing bars and wires come with smooth surfaces or with protuberances, or deformations, for gripping the concrete. Deformed bars are generally used in preference to smooth bars because of superior bond with the concrete. Wires are usually provided as welded mesh, or fabric. The wires may be smooth, because the cross wiring of fabric gives excellent bond with concrete. Bars and wires are usually produced to conform with applicable ASTM specifications.

Reinforcing bars are generally used in beams, columns, walls, footings and long-span or heavily loaded floor or roof slabs. Welded-wire fabric is frequently used in short-span or lightly loaded floor or roof slabs. The bars or wires are placed in forms, as required by design drawings, before concrete is placed. They are held in their specified positions by *bar supports*. These are commercially available in four general types of material: wire, precast concrete, molded plastic and asbestos cement.

Under ASTM specifications, reinforcing bars may be made of billet steel (Grades 40, 60 and 75), axle steel (Grades 40 and 60) and rerolled rail steel (Grades 50 and 60). The grade numbers indicate the yield strengths of the steel in ksi. Grade 60 is generally used in building construction. It is provided in eleven diameters, ranging from a nominal $\frac{3}{8}$ in. to $2\frac{1}{4}$ in. Sizes are denoted by integers that are about eight times the nominal bar diameters. Thus, a No. 3 bar has a nominal diameter of $\frac{3}{8}$ in., and a No. 8 bar, 1 in.

Stress-strain diagrams for reinforcing steels permit the assumption that design may be based on the idealized structural material with stress-strain curve shown in Fig. 8.7, p. 216.

Reinforcing bars, often referred to as *rebars* for short, may be fabricated in a shop or in the field, the choice often depending on local union restrictions. Fabrication consists of cutting the bars to required lengths and performing specified bending. The rebar supplier often *details* the work (prepares shop drawings and diagrams for placing the bars in the forms), fabricates the bars and delivers them to the building site. Sometimes, the supplier also places the rebars in the form.

Rebars may be placed in concrete members singly at specified spacings or in bundles at specified locations. Bundling is advantageous where space available for reinforcing in a concrete member is very tight. A bundle is assembled by wiring in tight contact up to four bars, none larger than No. 11.

Because of limits on shipping lengths of rebars, they have to be spliced when lengths longer than shipping lengths are required. Splices may be made by welding. Bars up to No. 11 in size, however, may simply be overlapped and tied together. Welded-wire fabric also may be spliced by overlapping.

For protection against fire and corrosion, reinforcing steel must be embedded deeply enough that the concrete will serve as a protective cover. Except for slabs and joists, depth of cover between steel extreme surface and concrete exposed surface should at least equal the bar diameter. For extreme exposures, such as exposure to seawater or contact with the ground, cover should be 3 to 4 in. In unexposed concrete, minimum cover should be $\frac{3}{4}$ in. for joists spaced not more than 30 in. center-to-center, slabs and walls and $1\frac{1}{2}$ in. for beams and girders.

Precast Concrete

When concrete products are manufactured in other than their final position and then assembled on the structure being erected, they are considered precast. They may be unreinforced, reinforced or prestressed. Precast products include blocks, pipes, slabs, beams, columns and piles.

Precasting is advantageous when it permits economical mass production of concrete members with strict quality control. Because the final products must be handled several times, shipped to the site and erected, low weight is important. Consequently, precast products generally are made of high-strength concrete to permit production of thin members. Also, they are often prestressed to withstand handling stresses as well as loads in service.

While precasting of concrete members may be more economical than production of the members by casting them in place, care must be taken that precasting savings are not offset by the cost of handling, transporting, erecting and making satisfactory connections in the field.

Lift-slab construction is a special type of precasting performed on the building site. In this method, floor and roof slabs are cast one atop the other at or near ground level and then lifted into place with jacks set atop the building columns. After each slab, beginning with the topmost one, reaches its final position, steel collars embedded in the slab are welded to the columns to secure the slab in place. Lift-slab construction thus offers many of the advantages of precasting while eliminating many storing, handling and transportation disadvantages.

Tilt-up wall construction is another special type of precasting performed on the building site. For low buildings, it provides advantages similar to those for lift-slab construction. In tilt-up construction, panels are cast horizontally at ground level at their final position. After the panels have gained sufficient strength, they are tilted into position, braced, then connected to each other and joined to floors and roof.

Prestressed Concrete

In the design of reinforced concrete members, structural engineers assume that the reinforcing steel and only the concrete in compression are effective in supporting loads. Use of concrete, however, can be made much more efficient by applying a compressive prestress to the concrete. If the prestress is large enough, the concrete need never be in tension. In that case, the whole cross section of the member becomes effective in supporting loads. Furthermore, the prestress will either prevent formation of cracks or will hold them closed if they should form. With the whole section thus effective, it becomes more economical to use higher-strength concrete for prestressed concrete than for reinforced concrete, where much of the cross section is not considered capable of supporting loads.

The usual procedure in prestressing concrete is to tension high-strength steel cables or bars, called *tendons*, and then anchor them to the concrete. Because the concrete resists the tendency of the stretched steel to shorten, the concrete becomes compressed. Prestressing often is classified in accordance with the sequence in which concrete is cast and hardened and the tendons tensioned and anchored to the concrete.

Pretensioning is a method in which tendons are tensioned between external anchorages and then concrete is cast to form a member and embed the tendons throughout its length. After the concrete attains sufficient strength to withstand prestress, the tendons are released from the anchorages. Being bonded to the concrete, the tendons impose compressive forces on it.

Posttensioning is a method in which concrete is cast around but not bonded to unstressed tendons, which may be sheathed in protective ducts. After the concrete attains sufficient strength, the tendons are tensioned with jacks acting against the hardened concrete. Then, the stressed tendons are anchored to the concrete, imposing compressive forces on it.

Prestress need not completely eliminate tension in the concrete. Sometimes, it is economical to permit small tensile stresses or to add reinforcing steel to resist substantial tension. The latter method is called *partial prestressing*.

The initial prestress applied by the tendons to the concrete decreases for several reasons. For one thing, when prestress is transferred from the steel to the concrete, the concrete shortens

elastically, allowing some shortening of the tendons. Also, there are losses due to friction if the tendons are curved and due to some slip at anchorages. In addition, there are losses that increase with time, such as those due to creep or drying shrinkage of the concrete or relaxation of the steel. The total of these losses can be substantial. Hence, it is desirable to use high-strength tendons and apply a large tension to them in prestressing, to make the prestress losses a small percentage of the applied prestress force.

For pretensioning, spaced wires are usually used as tendons. As with deformed rebars, deformations on the wires improve bond with the concrete for transfer of prestress.

For posttensioning, where the tendons are placed in ducts to prevent bond with the concrete initially, bars, strands or groups of parallel wires are generally used for prestressing. *Grout*, a fluid mix of cement, fine aggregate and water, is usually pumped into the ducts after the tendons have been anchored, to establish bond between the steel and the concrete.

To protect the tendons from fire and corrosion, sufficient concrete cover should be provided the steel, as is done for reinforced concrete.

Anchorage devices of many types are available for transfer of prestress from tendons to concrete in posttensioning. The methods used depend on the type of tendons. For example, bars may come with threaded ends on which a nut may be tightened until it bears against steel bearing plates embedded in the concrete. Strands may be supplied with threaded swaged fittings on which a nut may be tightened. Single wires may be provided with button heads to bear against an anchorage. Also, wires may be anchored by wedging them with a conical wedge against the sides of a conical opening.

Unit Masonry and Mortars

Structural members and some self-supporting enclosures often are constructed of inorganic, nonmetallic, durable, fire-resistant components, called masonry units, that are small and light enough for a worker to handle manually. While these units may be fabricated into a structural member or a panel in an off-site shop, they usually are assembled in their final position with a cementitious material.

Unit masonry. This is a built-up construction or combination of masonry units bonded together with mortar or other cementitious material. *Mortar* is a plastic mixture of a cementitious material, usually portland and masonry cements or portland cement and lime, fine aggregates and water. Sometimes, however, masonry units are bonded together with sulfur or an organic cement (plastics).

Masonry Units. These may be solid or hollow. Solid units have a net cross-sectional area, in every plane parallel to the bearing surface, equal to 75% or more of the gross cross-sectional area measured in the same plane. Hollow units are those with a smaller net cross-sectional area, because of open spaces in them.

A line of unit masonry one unit wide is called a *wythe*. A line of masonry one unit deep is referred to as a *course*. In masonry walls, a masonry unit laid with length horizontal and parallel with the wall face is known as a *stretcher*. A *header* or *bonder* is a masonry unit laid flat across a wall with the end surface exposed, to bond two wythes. Separate wythes also may be interconnected with metal ties. Joints may be made by mortaring together overlapping and interlocking units.

Because unit masonry and mortar each have a low tensile strength, masonry generally is used for members subjected principally to compression. Units with substantial compressive strength, however, also can be reinforced with steel or prestressed to resist tension, as is done with reinforced and prestressed concrete. Principal uses for masonry, however, are for walls and partitions.

Unit masonry may be made of various materials. Some are suitable for load-bearing construction; some can be exposed to the weather; some are used principally for appearance reasons; and some serve mainly as fire protection for structural members and shafts. Commonly used materials include brick, concrete block, ceramic veneer, stone, gypsum block and glass block. Materials generally used for structural purposes include brick, concrete block and stone.

Brick is a rectangular masonry unit, at least 75% solid, made from burned clay, shale or a mixture of these materials.

Concrete block is a machine-formed masonry unit composed of portland cement, aggregates and water.

Stone is a masonry unit composed of or cut from natural rock. Ashlar stone is a rectangular unit usually larger than brick. Rubble stone is a roughly shaped stone.

Mortar. The strength of a unit-masonry structural member depends on the strengths of units and mortar and on the bond between units and mortar. The strength and bond of the mortar, however, usually govern. Allowable unit stresses for unit masonry generally are based on the properties of the mortar.

Structural Masonry

There is a considerable variety of constructions that may be called by the general term masonry, but a critical distinction is that made between structural and nonstructural masonry. Both may be made with the same units, the distinction being the intended purpose and general character of the construction. Masonry used to produce bearing walls, supporting piers, and shear walls must be structural—the principal design concern being for structural properties of the units, mortar, and other construction details. Whereas appearance of the finished construction is of major importance with nonstructural masonry, and while not to be ignored with structural masonry, it is not the overriding concern of the structural designer.

A second structural-masonry distinction is between reinforced and unreinforced masonry. All forms of structural masonry use some form of joint reinforcement to improve the resistance of the construction to cracking and joint separations. Reinforced masonry, however, is developed much the same as reinforced concrete, with steel rods used to develop major tensile forces. When carefully designed and constructed, reinforced masonry structures can attain strengths competitive with reinforced concrete and levels of raw strength and toughness considerably beyond those capable with the equivalent forms of unreinforced construction. For structural applications, reinforced masonry construction is the only form permitted by codes in regions of high seismic risk.

Composite Materials and Composite Construction

In previous sections, some examples of composite materials, such as reinforced and prestressed concretes, and some examples of composite construction, such as drilled-in caissons, were discussed. In the following, some additional examples are described.

The distinction between composite materials and composite construction is vague. In general, production of composite materials is a manufacturing process, carried out in an off-site plant, whereas composite construction is an assembly process done with the components in their final position in the building.

Composite materials may be basically classified as matrix, laminate or sandwich systems.

Matrix systems (see Fig. 8.22a) consist of a discontinuous phase, such as particles, flakes or fibers, embedded in a continuous phase or matrix. Steel-fiber-reinforced concrete and glass-fiber-reinforced plastics are examples of such systems.

Laminates (see Fig. 8.22b) are formed by bonding together two or more layers of mate-rials, all of which share the load-carrying function. The layers may all be made of the same material, as is the case with plywood and glued-laminated timbers, or they may be made of different materials, such as steel and plastic. For example, bearing plates for beams are sometimes made of layers of Teflon and steel, a composite that permits movement of a supported member and also provides strength.

Sandwich systems (see Fig. 8.22c) comprise at least two load-carrying layers, or skins, between which is a core not relied on for carrying substantial proportions of the loads but which serves to separate and brace the skins. Sandwich panels with metal or plywood skins and insulation cores, for example, are often used in wall and roof construction.

Composite construction is primarily used to take advantage, for economic reasons, of specific properties of different materials. Concrete is one material often used because it can serve simultaneously as an enclosure, for fire protection and as a wear- or weather-resistant surface, while advantage can be taken of its compressive strength. Thus, concrete often is used as the exposed surface of a floor, ceiling or roof. Because concrete is weak in tension or because additional compressive strength is needed, it is advantageous to form a composite of concrete and a material strong in tension, such as structural steel.

One example of composite construction is the drilled-in caisson, in which a pile is formed by embedding a structural steel shape in concrete. Another example is a column formed by filling a steel pipe with concrete. Still another example is a beam in which the top flange is a rein-

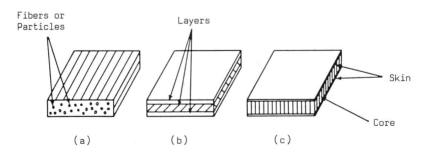

Fig. 8.22. Composite materials. (a) Matrix system. (b) Laminate. (c) Sandwich.

forced concrete floor or roof slab and the web and bottom flange are structural steel. In this case, because the concrete is intended to resist compressive stresses while the steel at the same cross section is resisting tensile stresses, some means must be provided to bond the two materials together. Connectors, such as headed steel studs or steel channels, with web vertical, often are welded to the top of the steel for embedment in the concrete to tie the materials together.

A characteristic of composites of materials that share the load-carrying function is that at any point where two materials or layers are integrated, the unit strain must be the same in both.

In a homogeneous material in the elastic range, the centroid of a cross section lies at the intersection of two perpendicular axes so located that the moments of the areas on opposite sides of an axis, taken about that axis, are zero. For a composite material in the elastic range, in contrast, the centroid is located at the intersection of two perpendicular axes so located that

the moments of the products of each area and its modulus of elasticity on opposite sides of an axis, taken about that axis, are zero.

To illustrate, consider a prism of uniform cross section composed of two materials and with a load of 200 kips in compression (see Fig. 8.23a). Suppose one material is concrete, 12-in. thick and 10 in. wide, with a cross-sectional area of 120 sq. in. and a modulus of elasticity of 3,000 ksi. Assume that the other material is structural steel, $\frac{1}{2}$-in. thick and 10-in. wide, with an area of 5 sq in. and a modulus of elasticity of 30,000 ksi. The materials are bonded together along the 10-in. wide faces for the full length of the prism.

To find the centroid of the section, moments may be taken about the outer face of the concrete. Let x be the distance from that face to the centroidal axis parallel to the face (see Fig. 8.23b). The distance from the concrete face to the centroid of the concrete is $12/2 = 6$ in. and to the centroid of the steel $12 + 0.5/2 = 12.25$ in. The product of the area A_c of the concrete and its modulus of elasticity E_c is

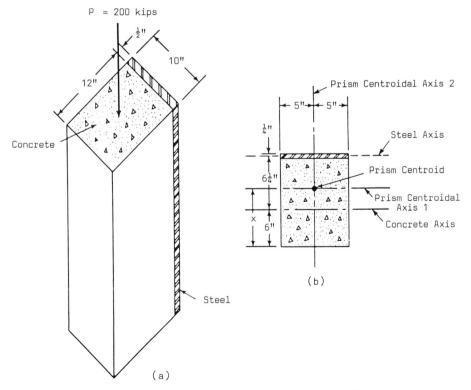

Fig. 8.23. (a) Short composite column. (b) Cross section of column.

$$A_c E_c = 120 \times 3,000 = 360,000$$

The product of the area A_s of the steel and its modulus of elasticity E_s is

$$A_s E_s = 5 \times 30,000 = 150,000$$

The sum of these products is $A_c E_c + A_s E_s = 510,000$. Then, to balance the products of areas and elastic moduli about the centroidal axis,

$$510,000\, x = 360,000 \times 6 + 150,000 \times 12.25$$
$$= 3,997,500$$

Solving for x yields $x = 7.84$ in.; that is, the centroidal axis is 7.84 in. from the outer face of the concrete and $12.5 - 7.84 = 4.66$ in. from the outer face of the steel. The 200-kip load on the prism must pass through the intersection of this axis with the perpendicular axis at the midpoint of the prism or it will cause the prism to bend.

Since at the interface of the concrete and the steel, the unit strain in each material is the same, then from Hooke's law [see Eq. (8.14)] and Eq. (8.7), the load taken by each material is proportional to the product of its area by its modulus of elasticity. Also, the sum of the loads on each material must be equal to the total load $P = 200$ kips on the prism. From these relationships, the load on each material can be calculated and then the stress determined by dividing that load by the area of the material. Thus, for the concrete and steel prism, the stress in the concrete is

$$f_c = \frac{PE_c}{A_c E_c + A_s E_s}$$

$$= \frac{200 \times 3,000}{360,000 + 150,000} = 1.18 \text{ ksi}$$

and the stress in the steel is

$$f_s = \frac{PE_s}{A_c E_c + A_s E_s}$$

$$= \frac{200 \times 30,000}{360,000 + 150,000} = 11.76 \text{ ksi}$$

Division of the expression for f_c by that for f_s indicates that the unit stress in each material is proportional to the modulus of elasticity. This also follows from Hooke's law, because the unit strain is the same in both materials. Consequently,

$$f_s = n f_c \qquad \text{(Eq. 8.17)}$$

where

n = modular ratio = E_s / E_c

SECTIONS 8.4 AND 8.5

References

H. Parker and J. Ambrose, *Simplified Mechanics and Strength of Materials*, 4th ed., Wiley, New York, 1986.

F. Merritt, *Building Design and Construction Handbook*, 4th ed., McGraw-Hill, New York, 1982.

H. Rosen, *Construction Materials for Architects*, Wiley, New York, 1985.

R. Smith, *Materials of Construction*, 3rd ed., McGraw-Hill, New York, 1979.

D. Watson, *Construction Materials and Practices*, 3rd ed., McGraw-Hill, New York, 1986.

E. Allen, *Fundamentals of Building Construction: Materials and Methods*, Wiley, New York, 1985.

Words and Terms

Allowable (working) stress
Brittle: fracture, material
Composites:
  Composite construction
  Composite material
  Laminates
  Matrix systems
  Modular ratio (n)
  Sandwich systems
Concrete:
  Admixture
  Aggregate
  Air-entrained concrete
  Curing
  Forms
  Portland cement, types
  Precast concrete
  Prestressed concrete: pretensioned, posttensioned
  Rebar
  Specified compressive strength (f_c')
  Water-cement ratio
  Workability
Design specifications
Ductility
Elastic: limit, stress/strain behavior
Hooke's law
Masonry:
  Course
  Header
  Masonry unit

Mortar
Reinforced masonry
Stretcher
Structural masonry
Wythe
Modulus of elasticity
Plastic stress/strain behavior
Poisson's ratio
Proportional limit
Steel:
    Cable, strand, rope
    Cold-formed members
    Forming: drawing, rolling
    Gage, of sheet steel, of wire
    Open-web joist
    Rivet
    Structural shapes
    Weathering steels
    Welding
Strain hardening
Toughness:
Ultimate strength
Ultimate-strength design (limit design, load-factor design)
Yield: point, strength
Wood:
    Defects
    Fasteners: bolts, connectors (shear), nails, screws, sheet
      metal
    Glued-laminated timber
    Grade (of structural lumber)
    Hardwood/softwood
    Moisture content
    Plywood: grade, identification index
    Species (tree)

Significant Relations, Functions and Issues

Stress/strain behavior: elastic, inelastic, plastic, ductile, brittle.
Measurements of stress/strain behavior: elastic limit, yield strength, ultimate strength, modulus of elasticity, modulus of rigidity (shear), Poisson's ratio.
Design (stress-based): allowable stress, safety factors
Design (strength-based): design loads, load factors
Standard design specifications
Steel: basic metallurgy, production methods, standard products, fasteners, fabrication and erection.
Wood: material identity and grade classification, standard structural products, design specifications, fasteners.
Plywood: production, classification, identification index.
Concrete: components of mix, control of properties, placing and forming, finishing, curing, reinforcing.
Precast and prestressed concrete: construction procedures and applications.
Masonry: units, mortar, construction controls, structural, reinforced and unreinforced.
Composite materials and constructions: interactive structural behaviors, construction applications.

8.6. TYPICAL MAJOR CONSTRAINTS ON STRUCTURAL SYSTEMS

Building codes and nationally recognized specifications and standards govern certain aspects of the design of a structural system, such as minimum design loads, acceptable materials, load factors, allowable unit stresses and often methods for computing load-carrying capacity of components. As a result, corresponding design variables are controlled or constrained, depending on whether the designer is restricted to standard values or to a range of values for the variables.

Loads

Structural systems must be designed to support the maximum loads that are likely to occur during construction and their service life or the minimum design loads specified in building codes, whichever is larger. Design loads on buildings include dead, snow, live, impact, wind and seismic loads, and sometimes also earth pressures (see Sec. 8.1).

For a specific component, only part of the dead load, the weight of building contents, such as furniture or equipment, and the weight of components previously designed are known at the start of design. The weight of the component being designed is unknown until a material is selected for the component and its size determined. This part of the dead load is a controllable variable. It is, however, subject to several constraints, typical of which are least cost, space limitations, and requirements imposed by other building systems.

Snow load depends primarily on the climate. A local building code, however, may specify a minimum design load. The load used, in any event, is a partly controllable variable, because it can be reduced by use of sloping roofs.

Live and impact loads are uncontrollable variables. They are determined by loads related to the function of the building, the type of occupancy and by minimum values specified by building codes. Some building components, furnishings and equipment may be considered dead rather than live loads if their location is permanent. Partitions, however, if they may be shifted in the future, should be treated as live loads. Building codes though, often prescribe a uniformly distributed floor load, to be added to the dead load, to take into account the uncertainty of partition locations. Also, codes may specify a concentrated load that should be located to produce maximum effects on the system and its components, because of the possibility of unanticipated heavy loads.

Wind, seismic and earth loads are partly controllable variables. Basically, these loads depend on local environmental conditions; but they also vary with such characteristics of the structural system as exposed surface areas, mass and stiffness.

Building designers thus have an opportunity to reduce building costs, by adjusting values of those variables that are partly or completely controllable, to reduce the loads on the structural system as much as possible.

Stability

A prime requisite of a structural system is that it be stable, unable to move freely and permit damage to property or injury to occupants or the public, despite the maximum loads that are likely to occur during construction and the service life of the building. Stability is provided by proper arrangement and interconnection of building components that have adequate strength and rigidity. These components must provide a continuous path along which loads are transmitted to the ground.

Trusses, for example, are made stable by assembling a sequence of structural members arranged in a triangular configuration. A horizontal beam supported at its ends and subjected to vertical loads is made stable by restricting points of uncontrolled rotation, or *hinges*, only to the two end supports. If, however, a third hinge is inserted in the beam, for instance, at midspan, the beam would become a mechanism and collapse under the vertical loads. In contrast, an arch with hinges at two end supports and midspan is stable.

While only a single path of load transmission is necessary for each load, it is prudent to provide redundancy; that is, at least two paths. Then, if one path is destroyed by an unforeseen accident, as has happened in the past in storms or explosions, the structural system may basically remain stable. Often, redundancy can be provided with little or no increase in initial construction cost.

Strength

This is the ability of a structural system and its components to withstand without excessive deformation or collapse the maximum loads that may be imposed during construction and the service life of the building. Both strength under static loads and strength under dynamic loads, which includes fatigue strength under cyclic loads, are important.

Static strength of a component may be measured in either of two ways. One measure is the load that causes excessive deformation or collapse. This load, for adequate strength, must be equal to or greater than the service load multiplied by a factor greater than unity, as prescribed in a building code or nationally recognized standard.

A second measure is the maximum load under which nowhere in the component will allowable unit stresses assigned by a building code or nationally recognized standard be exceeded. This load, for adequate strength, must be equal to or greater than the service load.

Fatigue strength also may be measured in either of two ways. One measure is the load determined by the maximum allowable unit stress assigned for fatigue, to avoid sudden failure under repetitive loads. A second measure is the allowable range of stresses at any point as the unit stress fluctuates with change in load. For stress reversal, in which stresses alternate between tension and compression, the stress range should be taken as the numerical sum of the maximum repeated tensile and compressive stresses. Fatigue strength is likely to be of concern only for components subjected to frequently moving loads, such as crane runways, or for supports for machinery. While wind and seismic loads cause fluctuating stresses, occurrence of full design loads is usually too infrequent to govern strength design.

Strength under dynamic loads in general is measured by the ability of the structural system and its components to absorb energy. Except for tall, slender buildings and special structures, however, structural systems are often permitted by building codes to be designed for an equivalent static loading. Determination of energy-absorbing capacity, however, may be required for systems to be constructed in regions known or suspected to be subject to heavy seismic shocks.

Rigidity

This is the ability of a structural system and its components to withstand without excessive deformation the maximum loads that may be imposed during construction and the service life of the building. Of special concern are control of beam deflections; drift, or sway, of a building under horizontal loads; and prevention of buckling, either locally or overall, of components subjected to compression.

Rigidity is necessary for components subjected to dynamic loads to control vibrations and their transmission.

Cost

The objective of structural engineering is a stable structural system with required strength and rigidity that will have the lowest life-cycle cost. It is not sufficient that the materials selected for the system have the lowest cost or that the system contain the smallest amount of materials of all possible alternatives.

Life-cycle cost is the sum for the whole structural system of the costs of raw materials, fabrication, handling, storage, shipping, erection and maintenance after erection. Because this sum is difficult or impracticable to estimate during the design stage, other measures are often used in cost comparisons.

For example, cost of a structural steel frame may be measured by multiplying the weight, lb, of steel by a price per lb. Cost of a wood frame may be measured by multiplying feet board measure by a price per ft. Cost of a concrete system may be measured as the sum of the product of volume of concrete, cu yd, by price per cu yd and the product of weight of reinforcing steel, lb, by a price per lb. In each case, the unit price is taken as greater than that for the raw material to cover fabrication, erection and other costs. In selection of the unit price, care should be taken to incorporate the effects of fabrication and erection characteristics of each structural system.

SECTION 8.6

References

American National Standard Minimum Design Loads for Buildings and Other Structures, American National Standards Institute, New York, 1982.

F. Merritt, *Building Design and Construction Handbook*, 4th ed., McGraw-Hill, New York, 1982.

J. Ambrose, *Building Structures*, Wiley, New York, 1988.

Words and Terms

Cost (structural): components of, life-cycle
Rigidity
Stability
Strength: static, dynamic, fatugue, adequate service
Loads: dead, design (service)

Significant Relations, Functions, and Issues

Control of loads as a design variable.
Components of control of rigidity and stability.
Redundancy as a safety factor.
Types of strength.
Structural cost as a design variable.

8.7. TENSION MEMBERS

This section deals only with members subjected to axial tension (no bending or torsion). Design of such members requires selection of a material and determination of dimensions of a cross section normal to the load. Also, the type of connections to be made to other members and the type, number and size of fasteners to be used in the connections must be decided. This decision affects the net cross-sectional area at the connection and may govern the design of the member.

The required cross-sectional area may be calculated by recasting Eq. (8.6) in the form:

$$A = \frac{T}{f_t} \qquad (8.18)$$

If ultimate-strength design is used, T represents the factored load and f_t, the yield stress of the material. If working-stress design is used, T is the service load and f_t, the allowable unit-tensile stress. In either case, A is the critical net cross-sectional area, over which failure is likely to occur.

The *net section* equals the gross cross-sectional area, or the area included between the outer surfaces of the member, less the area of any openings or holes. Generally, tension members have a constant cross section, except at connections. Consequently, the critical design section occurs at connections, where openings may be provided for bolt holes (see Fig. 8.24a, b and d) or for connectors (see Fig. 8.24e, f and h) or where the area may be reduced by threads to receive nuts.

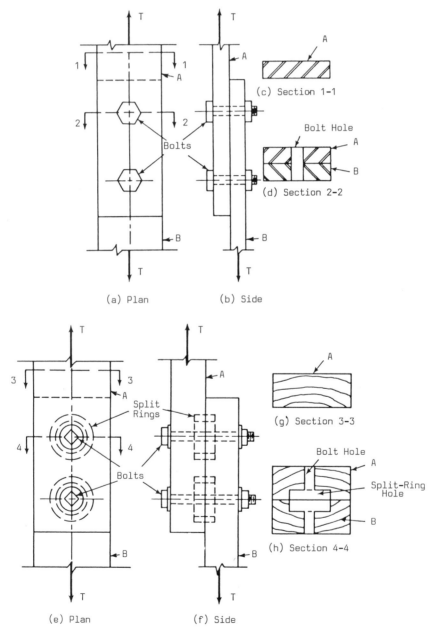

Fig. 8.24. Bolted joints in tension. (a) to (d) Steel bars connected with two bolts. (e) to (h) Wood members connected with bolts and split rings.

Figure 8.24*a* shows a plan view and Fig. 8.24*b* a side view of a bolted connection between two rectangular steel bars under tension. Between connections, bar *A* has the solid cross section (normal to the tensile force *T*) shown in Fig. 8.24*c*. At the upper bolt, however, the cross-sectional area is reduced by the area of the bolt hole, the product of the material thickness and sum of bolt diameter and $\frac{1}{8}$ in. clearance (see Fig. 8.24*d*).

Figure 8.24*e* shows a plan view and Fig. 8.24*f* a side view of a split-ring connection between two pieces of lumber in tension. Between connections, piece *A* has the solid cross section (normal to the tensile force *T*) shown in Fig. 8.24*g*. At the upper bolt, however, the cross-sectional area is reduced by the sum of the projected area of the split ring on the section and the portion of the bolt-hole area not included in that projected area (Fig. 8.24*h*).

The critical section need not be normal to the tensile stress. The critical area may occur in a diagonal plane or along a zigzag surface where there are two or more holes near each other. The AISC Specification for the Design, Fabrication and Erection of Structural Steel for Buildings and the NFPA National Design Specification for Stress-Grade Lumber and Its Fastenings present methods for computing the net section of steel and wood tension members, respectively. Similar methods may be used for other materials.

8.8. COLUMNS

This section deals with members subjected to axial compression. For short compression members, dimensions of cross sections normal to the load may be computed with Eq. (8.7), by solving for the area, as is done for tension members. No reduction in area need be made for bolt holes, however, because fasteners are assumed to fill the holes and to be capable of withstanding the compression.

For some materials, such as concrete, that are weak in tension, provision should be made to resist tensile stresses, computed with Poisson's ratio for the material, in directions normal to the compressive stresses. For this purpose, concrete compression members usually are reinforced around the perimeter, with steel-bar ties or spirals encircling longitudinal reinforcing bars.

Columns are long compression members. Dimensions of their cross sections normal to the axial load are relatively small compared with their length in the direction of the load. Although a long compression member may be straight when the load is applied (see Fig. 8.25*a*), it may bend, or buckle, suddenly and collapse when a certain load is exceeded (see Fig. 8.25*b*). This load is called the *critical*, or *Euler*, *load*.

Buckling may occur long before the yield stress is reached at any point in the column. The strength of a long column, therefore, is not determined by the unit stress in Eq. (8.7), as is the strength of short compression members, but by the maximum load the column can sustain without becoming unstable and buckling. In members intermediate in length between short and long columns, however, the yield stress may be exceeded at some points before buckling occurs.

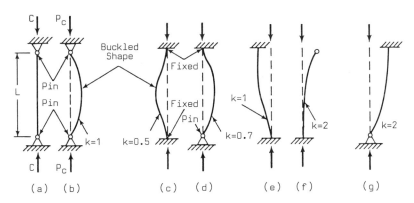

Fig. 8.25. Effective lengths of columns for various end conditions.

Stable Equilibrium

The column with ends free to rotate shown in Fig. 8.25a is initially straight. It will remain straight as long as the load C is less than the critical load P_c. If a small transverse force is applied, the column will deflect, but it will become straight again when the force is removed. This behavior indicates that when $C < P_c$, internal and external forces are in stable equilibrium.

Unstable Equilibrium

If C acting on the column in Fig. 8.25a is increased to P_c and a small transverse force is applied, the column will deflect, as with smaller loads, but when the force is removed, the column will remain in the bent position (see Fig. 8.25b). Repeated application and removal of small transverse forces or application of axial loads greater than P_c will cause the column to fail by buckling. This behavior indicates that when $C = P_c$, internal and external forces are in unstable equilibrium.

Euler Loads

Application of bending theory to analysis of column behavior indicates that, if stresses throughout the member do not exceed the yield stress, the smallest value of the Euler load for a pin-ended column is given by

$$P_c = \frac{\pi^2 EI}{L^2} \qquad (8.19)$$

where

E = modulus of elasticity of the material in the column

L = length of column

I = moment of inertia about an axis through the centroid of the column cross section

The axis for which moment of inertia is smallest should be chosen, because buckling will occur in the direction normal to this axis. (*Moment of inertia* is the sum of the products of each area comprising the cross section by the square of the distance of the area from the axis.)

Note that the Euler load, which determines the strength of a column, depends not on the strength of the material but, as indicated by Eq. (8.19), on the stiffness of the material, as measured by E.

As mentioned previously, column behavior depends on the ratio of length to cross-sectional dimensions. This relationship may be expressed more precisely by representing cross-sectional dimensions by the radius of gyration r of the cross section.

Radius of gyration is defined by

$$r = \sqrt{\frac{I}{A}} \qquad (8.20)$$

where

A = cross-sectional area of the column

For a rectangular section, $r = t/\sqrt{12}$ about an axis through the centroid and in the direction of the width, where t is the thickness. For a circular section, $r = d/4$ for every axis through the centroid, where d is the diameter of the circle. The AISC Steel Construction Manual lists the moment of inertia and radius of gyration for structural steel shapes.

Equation (8.19) for the Euler load can be expressed in terms of the radius of gyration if both sides of the equation are divided by A and Eq. (8.20) is used to eliminate I. The result is

$$\frac{P_c}{A} = \frac{\pi^2 E}{(L/r)^2} \qquad (8.21)$$

The left side of the equation gives as a measure of column strength a unit-compressive stress. The right side of the equation indicates that the stress is proportional to modulus of elasticity and is inversely proportional to the square of the ratio of length to least radius of gyration.

This important ratio is known as the *slenderness ratio* of the column.

Equation (8.21) applies only to pin-ended columns with stresses within the elastic limit. For other end conditions, the column formula may be written as

$$\frac{P_c}{A} = \frac{\pi^2 E}{(kL/r)^2} \qquad (8.22)$$

where

k = factor determined by end-support conditions for the column; for example:

Both ends fixed against translation and rotation (see Fig. 8.25c), $k = 0.5$

One end pinned, one end completely fixed (see Fig. 8.25d), $k = 0.7$

Both ends fixed against rotation, but one end can drift (Fig. 8.25e), $k = 1.0$

One end completely fixed, but the other end can drift and rotate (see Fig. 8.25f), $k = 2.0$

One end pinned, but the other end can drift while fixed against rotation (see Fig. 8.25g), $k = 2.0$

The product of k and the actual column length is called the *effective length* of the column. For the conditions shown in Fig. 8.25, effective length ranges from half the actual column length to double the column length. *Slenderness ratio*, in general then, is the ratio of effective length to least radius of gyration of the column cross section.

In columns with a slenderness ratio below a certain limiting value, the elastic limit may be exceeded before the column buckles. In such cases, E can no longer be considered constant. It may be more accurate to substitute for E in Eq. (8.22) a *tangent modulus*, the slope of the stress-strain curve for the material at a point corresponding to actual unit strains in the column.

Column Curves

These are the lines obtained by plotting critical unit stress with respect to the corresponding slenderness ratio. A typical column curve (see Fig. 8.26) consists of two parts: a curve showing the relationship between Euler loads and slenderness ratios, which applies for large slenderness ratios, and a curve showing the relationship between tangent-modulus critical loads and slenderness ratios, which applies for smaller slenderness ratios.

The curve for the smaller slenderness ratios is greatly influenced by the shape of the stress-strain curve for the column material. Figure 8.26a shows the column curve for a material that does not have a sharply defined yield point; for instance, a material with a stress-strain curve such as that in Fig. 8.6c, p. 214. For such material, the tangent-modulus critical loads increase with decrease in slenderness ratio. Figure 8.26b shows the column curve for a material with a stress-strain curve that approximates that shown in Fig. 8.6b, with a sharply defined yield point. For such material, the tangent-modulus critical loads become nearly constant at small slenderness ratios, because the tangent modulus of the material is very small.

For large slenderness ratios, column curves have a steep slope. Consequently, critical loads are very sensitive to end conditions, as measured by the factor k. Thus, the effect of end conditions on the stability of a column is much larger for long columns than for short columns.

Local Buckling

The preceding discussion of column instability considers only buckling of a column as a whole. Instead, a column may fail because of buckling of one of its components; for example, a thin flange or web. Hence, in determination of the

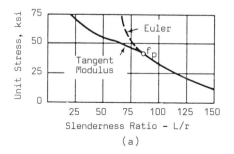

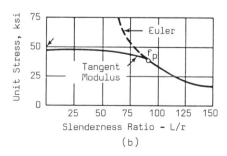

Fig. 8.26. Column curves for materials. (*a*) Without a sharply defined yield point. (*b*) With a sharply defined yield point.

strength of a column, the stability of components should be investigated as well as that of the whole column. To decrease the possibility of local buckling, design standards generally limit the ratio of unsupported length or width to thickness of components in compression.

Behavior of Actual Columns

Equation (8.22) for column strength is derived from theoretical considerations. Columns in structures, however, behave differently from the idealized column on which the equation is based. Actual column behavior is affected by many factors, including the effects of nonhomogeneity of materials, initial stresses, initial crookedness and eccentricity of load. Also, effects of end conditions may be difficult to evaluate accurately. Consequently, columns generally are designed with the aid of empirical formulas. Different equations are used for different materials and for short, intermediate and long columns.

Structural-Steel Column Formulas

For axially loaded structural steel columns, the allowable compressive stress on the gross cross section is given by formulas selected in accordance with the range in which the slenderness ratio kL/r of the columns lies. One formula is used for short columns; another formula is used for long columns; and still another formula is used for slender bracing and secondary members. The division between short and long columns is determined by the slenderness ratio C_c corresponding to the maximum stress for elastic buckling failure:

$$C_c = \sqrt{2\pi^2 E/F_y} \qquad (8.23)$$

where

E = the modulus of elasticity of the steel
= 29,000 ksi
F_y = the specified minimum yield stress, ksi, of the steel

For $kL/r < C_c$, the allowable compressive stress, ksi, is

$$F_a = \frac{[1 - (kL/r)^2/2C_c{}^2]F_y}{F.S.} \qquad (8.24)$$

$$F.S. = \frac{5}{3} + \frac{3kL/r}{8C_c} - \frac{(kL/r)^2}{8C_c{}^2} \qquad (8.25)$$

$F.S.$ is a safety factor. It varies from 1.67 when $kL/r = 0$ to 1.92 when $kL/r = C_c$.

When $kL/r > C_c$, the allowable compressive stress, ksi, is given by Eq. (8.22), with a safety factor of 1.92 and E taken as 29,000 ksi:

$$F_a = \frac{149,000}{(kL/r)^2} \qquad (8.26)$$

Since F_y does not appear in this formula, the allowable stress is the same for all structural steels.

For bracing and secondary members with $L/r > 120$, the allowable compressive stress, ksi, is

$$F_{as} = \frac{F_a}{1.6 - L/200r} \qquad (8.27)$$

This formula permits higher stresses than Eq. (8.26) and allows k to be taken as unity. The higher stresses are warranted by the relative unimportance of the members and the likelihood of restraint at their end connections.

Tables of allowable stresses are given in the AISC Steel Construction Manual.

Wood Column Formulas

For wood columns of structural lumber with rectangular cross sections, allowable column compression stress (F'_c) is determined by one of three formulas. Use of a particular formula, is based on the value of the slenderness ratio, which is determined as L/d for the rectangular section column, d being the dimension of the side of the section in the direction of buckling (the least dimension of the column if it is freestanding). The formulas and the limits for L/d are as follows:

Zone 1: $0 < L/d < 11$

$$F'_c = F_c$$

where F_c = allowable maximum compression parallel to grain.

Zone 2: $11 < L/d < K$

$$F'_c = F_c \left[1 - \frac{1}{3} \left(\frac{L/d}{K} \right)^4 \right] \qquad (8.28)$$

where $K = 0.671 \sqrt{E/F_c}$ and E = modulus of elasticity of the particular species and grade of wood

Zone 3: $K < L/d < 50$

$$F'_c = \frac{0.3\,E}{(L/d)^2} \qquad (8.29)$$

The value of 50 for L/d is the maximum permitted for a solid wood column.

Adjustments are made to the formulas for columns with other cross sections, such as round columns.

Reinforced Concrete Columns

Ultimate-strength design, with factored loads, is used for reinforced concrete columns. The material is treated as a composite. Axial load capacity is taken as the sum of the capacity of the concrete and the capacity of the reinforcing steel. The ACI Building Code, however, applies a capacity reduction factor $\phi = 0.75$ for columns with spiral reinforcement around the longitudinal bars and $\phi = 0.70$ for other types of columns.

The capacity of the longitudinal reinforcement of an axially loaded column can be taken as the steel area A'_s times the steel yield stress F_y. The capacity of the concrete can be taken as the concrete area times 85% of the 28-day compressive strength f'_c of the standard test cylinder. The ACI Building Code, however, requires that all columns be designed for a minimum eccentricity of load of 1 in. but not less than $0.10h$ for tied columns and $0.05h$ for spiral-reinforced columns, where h is the overall column thickness in the direction of bending. Design tables for columns are given in the ACI Design Handbook, SP 17.

8.9. TRUSSES

In structural frameworks, loads from roofs and upper floors are transmitted to the ground through columns. Structural members, therefore, must be provided to carry loads from the roofs and floors to the columns. When the column spacing is large, trusses often are an economical choice for those structural members. For economy, however, trusses usually have to be deep and consequently they can be used only if there will be sufficient space for them and adequate headroom under them. (Open-web joists, which are actually shallow trusses, are used on close spacing, about 24 in. center-to-center and for shorter spans or lighter loads than ordinary trusses. While open-web joists can be designed in the same way as ordinary trusses, the joists are excluded from the discussion in this section.) Because of the space requirements, the principal application of trusses in buildings is for supporting roofs.

Basically, a truss is a stable configuration of interconnected tension and compression members. The connections between members are assumed in truss design to be pinned, free to rotate, although actually the types of connections used may apply some restraint against rotation of truss joints.

To preclude bending of the truss members, location of loads applied to trusses should be restricted to the truss joints. Also, at each joint, the centroidal axes of all members at the joint and the load at the joint must pass through a single point, called the *panel point*.

Three members pinned together to form a triangle comprise the simplest type of truss. If small changes in the length of the members under load are ignored, the relative position of the joints cannot change. Hence, the configuration is stable. More complicated trusses are formed by adding members in a continuous sequence of triangles.

Ordinary trusses are coplanar; that is, all the component triangles lie in a single plane. Typical roof trusses shown in Fig. 8.27 are of this type. Trusses, however, also may be three dimensional. Such trusses usually are called *space trusses* or *space frames*.

As indicated in Fig. 8.27, the top members of a truss are known as the upper, or top, chord. Similarly, the bottom members are called the lower, or bottom, chord. Vertical members may simply be called verticals, or posts when they are undergoing compression, or hangers when they are undergoing tension. Inclined members incorporated between chords are called diagonals. Verticals and diagonals are collectively referred to as web members.

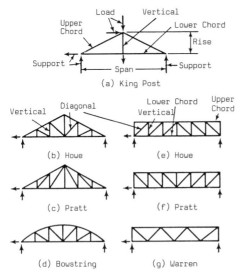

(a) King Post

(b) Howe

(c) Pratt

(d) Bowstring

(e) Howe

(f) Pratt

(g) Warren

Fig. 8.27. Types of roof trusses.

Truss Joints

Figure 8.28 shows how connections are often made in steel and wood trusses with fasteners.

Figure 8.28*a* shows a bolted joint at the top chord of a steel Warren truss in which the members are composed of pairs of angles. The connection is made through a *steel gusset plate* inserted between each pair of angles at the joint. In the case illustrated, the gage lines of the fasteners closest to the outstanding legs of the angles meet at the panel point. The applied load would be centered directly above the panel point.

Connections in steel trusses also may be made by welding. In that case, if the web members are single or double angles, the connections can be made without gusset plates by using for the

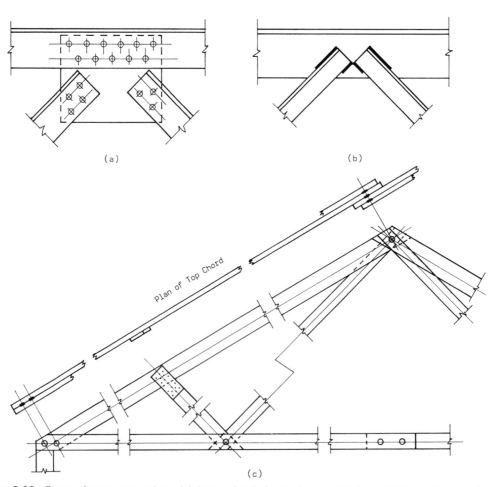

(a)

(b)

Plan of Top Chord

(c)

Fig. 8.28. Forms of truss connections. (*a*) At top chord of a steel truss with bolted joints and gusset plate. (*b*) At top chord of steel truss with welded joints. (*c*) In a light wood truss.

top and bottom chords WT shapes, made by
cutting a wide-flange shape longitudinally
along the web, and then inserting the WT web
between each pair of angles (see Fig. 8.28*b*).

Fig. 8.28*c* shows half of a symmetrical wood
truss with top chord inclined on a slope of 4 on
12. This truss is of a type that might be used to
support lightly loaded roofs on spans from 20
to 30 ft, for example, for houses. The connection at the support, or *heel*, and connections
where more than two members meet are made
with machine bolts and split rings, as is the bottom-chord splice at midspan. Elsewhere, the
connections are nailed and made through a
piece of lumber, called a *scab*. In more heavily
loaded trusses, members may be composed of
two or more pieces of lumber instead of the single component shown in Fig. 8.28*c*.

Stresses in Trusses

The reactions of a truss usually can be calculated from the laws of equilibrium [see Eqs.
(8.1) and (8.2)]. (For wind loads, the horizontal components of the reactions often are
assumed equally divided between the two supports, to simplify calculations.) After the reactions have been determined for specific loads,
the stresses in the truss members can be determined by vector analysis, either analytically or
graphically, except for unusual trusses. The
stresses can be found simply by applying the
laws of static equilibrium. These laws require
that, if a section is cut in any manner through
a truss, the vector sum of internal and external
forces must be zero on either side of the section, and the sum of the moments of the forces
about any point must be zero.

A commonly used section is one taken completely around a single joint, to isolate it from
the rest of the truss. The forces acting then are
the load at the panel point and the stresses at
the cut ends of the members of the joint. The
stress in each member is directed along the centroidal axis of the member. Because all the
forces intersect at the panel point, the moments
of the forces about that point is zero. To satisfy Eq. (8.1), the sum of the horizontal components and the sum of the vertical components
of all the forces also must be zero.

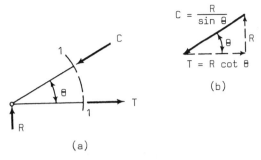

Fig. 8.29. Stresses in truss members. (*a*) Forces
acting at truss heel (support). (*b*) Force vector triangle indicating equilibrium at the joint.

As an example, consider the heel of the wood
truss in Fig. 8.28*c*. A circular section may be
taken around the joint, as indicated by section
1-1 in Fig. 8.29*a*. This section cuts through the
support and the top and bottom chords, which
are represented in Fig. 8.30*a* by their centroidal axes. The forces acting on the isolated
portion of the truss then are the reaction R, the
stress C in the top chord, assumed to be
compression, and the stress T in the bottom
chord, assumed to be tension. Note that R is
vertical, and the top chord is the only member
at the joint with a vertical component. Hence,
the vertical component of C, by Eq. (8.1), must
be equal to R but oppositely directed. As indicated in Fig. 8.29*b*, if θ is the angle the top
chord makes with the horizontal, C must be
equal to R/sin θ. This value of C can now be
used to determine the stresses in the next top-
chord joint. The value of T can also be found
by application of Eq. (8.1) to the isolated joint.
The law requires that the sum of the horizontal
forces be zero. As a result, T must be equal to
the horizontal component of C but oppositely
directed. As shown in Fig. 8.29*b*, $T = R \cot \theta$.
This value of T can now be used to determine
the stresses in the next bottom-chord joint.

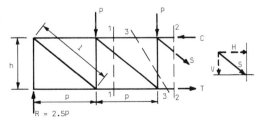

Fig. 8.30. Stresses in a truss with parallel chords.

A section commonly used for trusses with parallel chords is one taken vertically between panel points. Consider, for example, the part of the Pratt truss in Fig. 8.30 isolated by cutting the truss with section 2-2. The forces acting are the reaction R; loads P at two panel points; stress C in the cut top chord, assumed to be compression; stress T in the cut bottom chord, assumed to be tension, and stress S in the cut diagonal, assumed to be tension. All the stresses may be calculated by application of Eq. (8.2), which requires that the sum of the moments of all forces be zero. To find T, moments should be taken about the panel point where C and S intersect:

$$2Rp - Pp - Th = 0$$

where p is the panel length. In this case, with $R = 2.5P$, $T = 4Pp/h$, where h is the depth of the truss. To find C, moments should be taken about the panel point where T and S intersect:

$$3Rp - 2Pp - Pp - Ch = 0$$

In this case, $C = 4.5Pp/h$. (If the directions of C and T had been incorrectly assumed, the solutions would have appeared with negative signs, indicating that the directions should be reversed.)

The stress S in the diagonal can be determined in either of two ways.

One method is as follows: For the truss with parallel chords shown in Fig. 8.30, the diagonal is the only member cut by section 2-2 that has a vertical component. This vertical component is the only force available at the section for resisting the imbalance of the vertical components of the external forces. This *vertical shear* equals $R - 2P = 2.5P - 2P = 0.5P$. Hence, the vertical component of S must equal $0.5P$, and $S = 0.5P \sec \theta = 0.5Pl/h$, where l is the length of the diagonal.

The second method is useful when the truss chords are not parallel, as is the case with bowstring trusses (see Fig. 8.27d). In this method, S is first resolved into a vertical component V and a horizontal component H, both located at the intersection of the diagonal and the bottom chord. Then, moments are taken about the bottom-chord panel point just inside section 2-2 and set equal to zero:

$$2Rp - Pp - Ch + Vp = 0$$

from which, since C has already been computed, V can be determined.

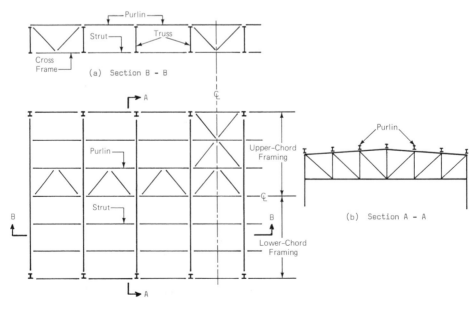

(a) Section B - B

(b) Section A - A

(c) Partial Plan of Roof Framing

Fig. 8.31. Roof framing for an industrial building, symmetrical about both centerlines.

Bracing of Trusses

Because the components of ordinary trusses are coplanar, such trusses offer little resistance to forces normal to their plane or to buckling of the compression chord unless adequate bracing is provided. Roof or floor framing can be used to brace the top chord. Usually, however, additional horizontal and vertical bracing are necessary, because the bottom chord, although in tension, is long and slender.

Figure 8.31b shows a Pratt truss for a roof with just enough slope in two directions to provide good drainage of rainwater. The roof is supported on eight such trusses, each truss supported at its two ends on columns. The roof framing is symmetrical in two directions. Hence, Fig. 8.31c shows only five of the eight trusses. Also, the upper half of the plan view shows only the framing for the upper chords, whereas the lower half of the drawing shows only the framing for the lower chords. Similarly, Fig. 8.31a shows only half of the cross section through the roof normal to the trusses. In all cases, the halves not shown are identical to the halves shown.

Figure 8.31a shows that three pairs of trusses are braced laterally by cross frames containing two diagonals in a vertical plane. Purlins supporting the roof and carried by the trusses brace all the top chords. Struts lying in the same plane as the purlins and the cross frames brace the bottom chord.

Figure 8.31c indicates that, in addition, horizontal diagonal bracing is placed in the plane of the top chords. This bracing should be designed to transmit wind loads on the building to bracing in vertical planes along the sides and ends of the building, for transmission to the foundations.

Economics of Trusses

Trusses offer the advantage of lighter weight for the long spans or heavy loads for which they are usually used than that for beams. Also, the openings between truss members often are useful for passage of pipes, ducts and wiring.

Costs of trusses, however, are not necessarily lower than costs of beams, despite the lesser amount of material in trusses. Fabrication and erection costs for trusses and their bracing must be taken into consideration. Also, use of trusses may be restricted to types of buildings in which fire protection of the trusses is not required or to locations in which costs of fire protection are sufficiently low.

SECTIONS 8.7-8.9

References

J. McCormac, *Structural Analysis*, 4th ed., Harper & Row, New York, 1984.

F. Merritt, *Building Design and Construction Handbook*, 4th ed., McGraw-Hill, New York, 1982.

J. Ambrose, *Building Structures*, Wiley, New York, 1988.

Words and Terms

Axial load
Buckling
Effective length (column)
Equilibrium: stable, unstable
Euler load
Moment of inertia
Net section
Radius of gyration
Slenderness: column, ratio
Truss parts: chords, heel, panel points, gusset plate

Significant Relations, Functions, and Issues

Axial load in linear members.
Net section in tension members.
Column actions: buckling, failure stress related to slenderness, end conditions and effective length.
Stresses (internal forces) in trusses.
Bracing of truss systems.

8.10. BEAMS

Like trusses, beams are used to support floors, roofs, walls, machinery and other loads over spaces below. Unlike trusses, however, beams are solid between their top and bottom surfaces. Consequently, they are subjected to both bending and shear stresses. Beams generally are more economical than trusses for short spans and light loads and are necessary where space for structural members is limited or headroom below is restricted.

The term *beam* is applied in general to structural members subjected principally to bending stresses. In specific applications, beams may be called by other names. For example, *joists* are light floorbeams; *stringers* support stairs; *headers* frame openings in floors and roofs; *purlins* are light, horizontal roof beams; *rafters* are light, inclined roof beams; *girts* are light members that span between columns to support curtain walls; *lintels* are light members that carry walls at floor levels in multistory buildings or over window or door openings; *girders* are heavily loaded beams or beams supporting other beams; *spandrels* support exterior walls and floor edges in multistory buildings; grade beams are shallow walls at ground level and extending slightly below to enclose the bottom of a building.

Types of Beams

Beams may be supported in various ways. One type of support is applied at the end of a beam and restricts translation but permits free rotation. This condition is represented by the symbol shown in Fig. 8.32*a* or by the forces in Fig. 8.32*b*. A support instead may restrict translation but permit a member that is continuous over it to rotate freely. This condition is represented by the symbol shown in Fig. 8.32*c* and by the forces, including moments represented by curved arrows, in Fig. 8.32*d*. The symbol shown in Fig. 8.32*e* indicates a support that restricts only vertical movement. Only a vertical reaction is present (see Fig. 8.32*f*). Alternatively, the end of a beam may be fixed, or com-

pletely restricted against movement. The symbol for this condition is shown in Fig. 8.32*g* and the forces are shown in Fig. 8.32*h*.

Beams may be classified in accordance with the methods of support. A beam free to rotate at both ends is called a simply supported, or *simple beam* (see Fig. 8.33*a*). The beam in Fig. 8.33*b* is a *cantilever*. The beam in Fig. 8.33*c* is a simple beam with overhangs. The overhangs are also called cantilevers. *Hung-span*, or *suspended-span, construction* results when a beam is connected between the overhangs of two other beams (see Fig. 8.33*d*). Figure 8.33*e* shows a *fixed-end beam*. Figure 8.33*f* shows a beam with one-end fixed and one end free to rotate and move horizontally. A three-span *continuous beam* is illustrated in Fig. 8.33*g*.

Reactions for the beams in Fig. 8.33*a* to *d* can be computed from the laws of equilibrium [Eqs. (8.1) and (8.2)]. When the reactions have been determined, internal forces in the beams also can be calculated from the laws of equilibrium, as will be demonstrated later. Such beams are called *statically determinate beams*.

The equations of equilibrium, however, are not sufficient for determination of the reactions or internal forces for the beams of Fig. 8.33*e* to *g*. Additional information is needed. This may be obtained from a knowledge of beam deformations, which permits development of additional equations. For example, the knowledge that no rotation or translation of a beam end can occur at a fixed end permits development of equations for obtaining the reactions of the beams in Fig. 8.33*e* and *f*. Beams such as those in Fig. 8.33*e* to *g* are called *statically indeterminate beams*.

Internal Forces

The external loads on a beam impose stresses within the beam. Consider, for example, the simple beam in Fig. 8.34*a*. Section 1-1 is cut vertically through the beam to the right of the first load *P*. The part of the beam to the left of section 1-1 nevertheless must remain in equilibrium. Consequently, as shown in Fig. 8.34*b*, the loads must be counterbalanced by a shearing force V_1 and a moment M_1, to satisfy the laws of equilibrium. If the left reaction R_L is

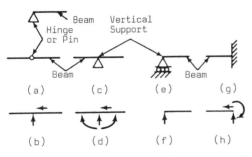

Fig. 8.32. Beam supports. (*a*) Hinge, or pin. (*b*) Forces acting on beam at hinge. (*c*) Continuous. (*d*) Forces acting in continuous beam at support. (*e*) Hinge support that permits horizontal movement. (*f*) Force acting on beam at support shown in (*e*). (*g*) Fixed (clamped) end. (*h*) Forces acting on beam at a fixed end.

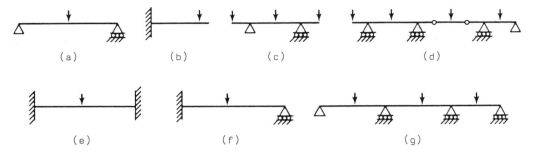

Fig. 8.33. Types of beams. (*a*) Simple beam. (*b*) Cantilever. (*c*) Simple beam with overhangs. (*d*) Hung-span, or suspended-span, construction. (*e*) Fixed-end beam. (*f*) Beam with one fixed end, one simply supported. (*g*) Three-span continuous beam.

known, V_1 and M_1 can be computed. From Eq. (8.1), $V_1 = -R_L + P$. From Eq. (8.2), $M_1 = -R_L x_1 + P(x_1 - a)$.

The moment M_1 that counterbalances the moment about section 1-1 of R_L and P is provided within the beam by a couple consisting of a force C acting on the top part of the beam and an equal but oppositely directed force T acting on the bottom part of the beam (see Fig. 8.34*c*). C at section 1-1 is the resultant of unit compressive stresses acting over the upper part of the beam. T at section 1-1 is the resultant of unit tensile stresses acting over the lower part of the beam.

The surface at which the unit stresses change from compression to tension and at which the stress is zero is called the *neutral surface*.

Shear

The unbalanced forces on either side of any section taken normal to the neutral surface of a beam is called the shear at the section. For the portion of the beam on the left of the sec-

tion, forces that act upward are considered positive and those that act downward are considered negative. For the portion of the beam on the right of the section, the signs should be reversed.

Bending Moment

The unbalanced moment of the forces on either side of any section taken normal to the neutral surface of a beam is called the bending moment at the section. For the portion of the beam on the left of the section, clockwise bending moments are considered positive and counterclockwise bending moments are considered negative. When the bending moment is positive, the bottom of the beam is in tension, the top in compression. For the portion of the beam on the right of the section, the signs should be reversed.

When the bending moment and shear are known at any section, the bending moment and shear at any other section through the beam can be computed from the laws of equilibrium. For example, for the beam in Fig. 8.34*a*, the bending moment at section 2-2 can be determined

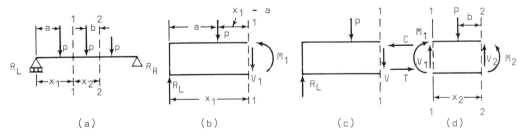

Fig. 8.34. Bending stresses in a beam. (*a*) Simple beam with concentrated loads. (*b*) Bending moment and shear at section 1-1. (*c*) Bending moment replaced by an equivalent force couple, *C* and *T*, at Section 1-1. (*d*) Bending moments and shears at sections 1-1 and 2-2.

from the bending moment and shear at section 1-1. Figure 8.34*d* shows the shear and bending moments acting at sections 1-1 and 2-2. From Eq. (8.2), the sum of the moments about section 2-2 for the portion of the beam between section 1-1 and 2-2 must be zero. Hence,

$$M_1 + V_1 x_2 - Pb - M_2 = 0$$

from which

$$M_2 = M_1 + V_1 x_2 - Pb$$

This result can be generalized as follows:

The bending moment at any section of a beam equals the bending moment at any other section on the left, plus the shear at that section times the distance between sections, minus the moments of intervening loads. If the section with known shear and moment is on the right, the moment of the shear and the moment of intervening loads should both be subtracted from the known bending moment because of the reversal of the sign convention.

Shear Variation along a Beam

A beam must be capable of resisting the design shear at every normal section along the neutral surface. The design shear at a section is the maximum that can be produced there by any possible combination of dead, live and other loads.

For a specific set of loads, it is convenient for design purposes to plot graphically the variation of shear along the span. Usually, the shear diagram can be speedily drawn by application of simple principles.

For instance, from the definition of shear, if a beam is horizontal and the loads are vertical, the shear at any section is the algebraic sum of the forces that lie on either side of the section. Consequently, if only concentrated loads are applied to the beam, the shear curve is a straight, horizontal line between the loads. Furthermore, at a concentrated load, the shear curve moves in the direction of the load abruptly vertically a distance, in accordance with the scale selected for the diagram, equal to the magnitude of that load.

For a uniformly distributed downward load

w, however, the shear curve is a straight line sloping downward from left to right. The slope of the line equals −*w*; that is, at any distance *x* from the start of the line, the vertical drop of the line equals *wx*, which is the total uniform load within the distance *x*.

Figure 8.35 shows some shear diagrams for simple beams with from one to three concentrated loads and a uniform load. Note, for example, that the shear diagram for the single concentrated load in Fig. 8.35*a* can be drawn by starting on the left by plotting to a selected scale the magnitude of the left reaction of the beam *bP/L* upward. From the left reaction over to the load *P*, the shear curve is a horizontal straight line, as drawn in Fig. 8.35*b*. At the location of *P*, the shear curve drops vertically a distance *P* (to the same scale at which the reaction is plotted). From *P* to the right reaction, the shear curve is again a horizontal straight line. At the location of the right reaction, its magnitude *aP/L* is plotted upward, to complete the diagram.

Shear diagrams for a combination of different

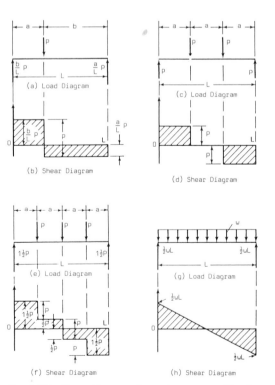

(a) Load Diagram

(b) Shear Diagram

(c) Load Diagram

(d) Shear Diagram

(e) Load Diagram

(f) Shear Diagram

(g) Load Diagram

(h) Shear Diagram

Fig. 8.35. Shear diagrams for various types of loads on a simple beam.

loading conditions can be plotted by superposition of the diagram for each condition. For example, the shear diagram for the loads in Fig. 8.35*e* and a uniformly distributed load *w* can be plotted by adding the ordinates of the shear diagram in Fig. 8.35*f* to the ordinates of the shear diagram in Fig. 8.35*h*, when these diagrams are drawn to the same scale.

Moment Variation along a Beam

A beam must also be capable of resisting the design bending moment at every normal section along the neutral axis. The design bending moment at a section is the maximum that can be produced there by any possible combination of dead, live and other loads.

For a specific set of loads, it is convenient for design purposes to plot graphically the variation of bending moment along the span. Usually, the bending-moment diagram can be speedily drawn by application of simple principles.

For instance, from the definition of bending moment, at any section the bending moment is the algebraic sum of the moments of the forces on either side of the section. Consequently, for a specific set of concentrated loads, bending moment varies linearly with distance from the loads. As a result, the bending-moment curve between any two concentrated loads, for a beam subjected only to concentrated loads, is a straight line. For a uniform load, however, the curve is parabolic. Bending moment for a uniform load varies as the square of distance, inasmuch as the total load increases with distance.

For any set of loads, the maximum bending moment occurs where the shear curve passes through zero shear.

Figure 8.36 shows some bending-moment diagrams for a simple beam with from one to three concentrated loads and for a uniformly distributed load. Note, for example, that the bending-moment diagram for the single concentrated load *P* in Fig. 8.36*a* can be started at either the left or the right beam end. In either case, the bending moment at a distance *x* from either reaction equals the reaction times *x*. The bending-moment curve, therefore, is a straight line on both sides of *P*, with a slope *bP/L* on the left of *P* and a slope $-aP/L$ on the right of

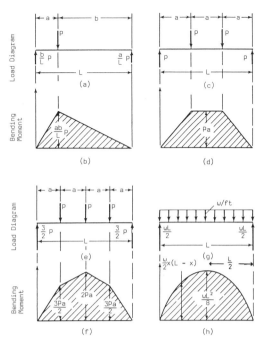

Fig. 8.36. Bending moment diagrams for various types of loads on a simple beam.

P (Fig. 8.36*b*). The maximum bending moment occurs at the location of *P*, where the two lines intersect, and is equal to *Pab/L*.

The bending-moment diagrams for the loading conditions in Fig. 8.36*c* and *e* can be drawn by computing the bending moments at each concentrated load, plotting the moments to scale at the location of the corresponding loads and then connecting the plotted points with straight lines.

The bending-moment diagram for the uniform load *w* in Fig. 8.36*g* is a parabola symmetrical about midspan. Its equation can be formulated by determining the bending moment at a point *x* from the left reaction. The reaction equals *wL/2*, and its moment is *wLx/2*. The load totals *wx*, its moment arm is *x/2*, and its moment therefore is $wx^2/2$. Consequently, the bending moment for a uniformly loaded simple beam is

$$M = \frac{w}{2}Lx - \frac{w}{2}x^2 = \frac{w}{2}x(L - x) \quad (8.30)$$

This is the equation for the parabola shown in Fig. 8.36*h*. The maximum bending moment occurs at midspan and equals $wL^2/8$.

Bending-moment diagrams for combinations of different loading conditions can be plotted by superposition of the diagram for each condition. For example, the bending-moment diagram for the loads in Fig. 8.36e and a uniform load w can be plotted by adding the ordinates of the bending-moment diagram in Fig. 8.36f to the ordinates of the bending-moment diagram in Fig. 8.36h, when these diagrams are drawn to the same scale.

Fixed-End Beams

A fixed-end beam has both ends completely fixed against translation and rotation. It is statically indeterminate, because its reactions, bending moments and shears cannot be determined from the laws of equilibrium. For vertical loads, each reaction consists of a vertical force and a moment, which prevent vertical movement and end rotation. These reactions can be computed by adding to Eqs. (8.1) and (8.2) equations that indicate that the end rotations are zero.

Figure 8.37 shows for a fixed-end beam of constant cross section the shear and bending-moment diagrams for a single concentrated load P and a uniform load w. The shear diagram for the uniform load (see Fig. 8.37e), because of the symmetry of loading and beam geometry,

is the same as for a simple beam (see Fig. 8.37h). The shear diagram for the single concentrated load (see Fig. 8.37b) is the shear diagram for the simple beam in Fig. 8.37a displaced vertically, because of the presence of equal but oppositely directed forces at each end of the beam, to counterbalance the fixed-end moments.

The bending-moment diagram for the single concentrated load (see Fig. 8.37c) is the bending-moment diagram for the simple beam in Fig. 8.37a displaced vertically, because of the occurrence of negative moments at each end of the beam. The maximum positive bending moment still occurs at the location of the concentrated load. Similarly, the bending-moment diagram for the uniform load (see Fig. 8.37f) is the bending-moment diagram for the simple beam in Fig. 8.37h displaced vertically, because of the occurrence of negative end moments. The maximum positive bending moment still occurs at midspan. For the uniform load, however, the maximum negative moment is larger than the maximum positive bending moment. For fixed-end beams, in general, maximum bending moments may occur at either beam end or at one of the loads on the span.

Note that in a simple beam (see Fig. 8.37) bending moments vary from zero at the beam ends to a maximum in the interior of the beam, usually near midspan. As a result, much of a simple beam is subjected to low bending stresses. In a fixed-end beam (see Fig. 8.37), in contrast, the ends as well as the center of the beam are subjected to large bending stresses. Consequently, a greater portion of a fixed-end beam than of a simple beam is useful in carrying loads.

With a uniformly loaded fixed-end beam, it may be economical to use a light section for the center portion and deeper sections, or haunches, at the beam ends, to resist the larger bending moments at the ends.

The bending moments and shears in a fixed-end beam are influenced by the cross-sectional dimensions throughout the beam. In contrast, the bending moments and shears in a simple beam are independent of the beam cross-sectional dimensions. The effects of the cross-

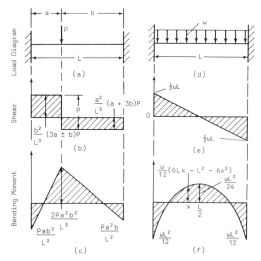

Fig. 8.37. Shear and bending moment diagrams for a single concentrated load and a uniformly distributed load on a fixed-end beam.

sectional dimensions of a fixed-end beam are introduced by the equations developed by setting the end rotations equal to zero. As a consequence of these equations, deepening the ends of an initially prismatic fixed-end beam increases the negative bending moments at the ends. At the same time, the positive bending moments decrease. Hence, the effect of haunches on the prismatic beam is similar to that of a decrease in span.

Continuous Beams

The variation of bending moments and shears in a loaded span of a continuous beam is similar to that of a fixed-end beam. If the loads on a horizontal continuous-beam span are vertical, the reaction at an interior support or at an end that is not simply supported consists of a vertical force and a bending moment. The beam is statically indeterminate, because the reactions, shears and bending moments cannot be determined from the laws of equilibrium. Additional equations are needed to account for the continuity of the beam over each interior support and for restraints, if any, on rotations at the ends of the beam. The equations can be developed from the continuity requirement that the rotation of the beam at a support must be the same on both sides of the support. As a consequence of these equations, vertical loads on a span induce negative end bending moments, and the bending-moment diagram resembles those in Fig. 8.37 for fixed-end beams.

Figure 8.38a shows one loaded span of a continuous beam. For a single concentrated load, the bending-moment diagram for that span (see Fig. 8.38e) looks very much like the bending-moment diagram in Fig. 8.37c for a fixed-end beam with the same loading. The continuous span also has negative bending moments at its supports, and the maximum positive bending moment is smaller than that in a simple beam.

The bending-moment diagrams in Figs. 8.37c and 8.38e can be obtained by superposition of the bending-moment diagrams for each of the span's external forces considered acting on a simple beam of the same span. For example, the loaded span in Fig. 8.38a may be taken as equivalent to the sum of the loaded simple beam in Fig. 8.38b, the simple beam with negative end moment M_L at the left support and the simple beam with negative end moment M_R at the right support. The bending-moment diagram in Fig. 8.38e then is equivalent to the sum of the corresponding bending-moment diagrams for those load conditions, as indicated in Fig. 8.38f to h.

As for a fixed-end beam, bending moments and shears are influenced by the cross-sectional dimensions throughout a continuous beam. The effects of the cross-sectional dimensions are introduced by the equations developed by equating the end rotations of spans continuous at a support.

Note also that for a continuous beam the laws of equilibrium require that at every support the algebraic sum of the end bending moments must be zero.

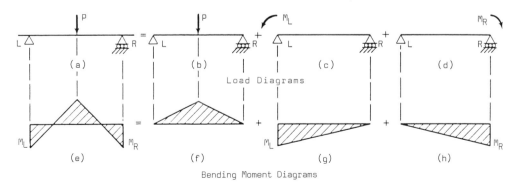

Fig. 8.38. **Bending moment diagram for a span of a continuous beam resolved into component moment diagrams for a simple beam with the same span.**

Bending Stresses in a Beam

The bending moments imposed on beams by loads are resisted at every section normal to the neutral surface by unit compressive and tensile stresses parallel to the neutral surface. The product of the average compressive stress above the neutral surface and the section area in compression equals the toal compressive force C at the section (see Fig. 8.34c). Similarly, the product of the average tensile stress below the neutral surface and the section area in tension equals the total tensile force T at the section (see Fig. 8.34c). Determination of the unit stresses requires that assumptions concerning beam geometry, loads and strain distribution be made.

To insure that bending is not accompanied by twisting, the beam cross-section should be sym-metrical about a plane perpendicular to the neutral surface. Also, the loads should lie in that plane. As a result, the line of action of the loads will pass through the centroidal axis of the beam.

Because many structural materials behave like the idealized structural material with a stress-strain curve as shown in Fig. 8.39b, beams will be assumed to be made of this ma-terial. Accordingly, within the proportional limit, unit stress equals the product of the mod-ulus of elasticity of the material E and the unit strain. Also, the modulus of elasticity in ten-sion is the same as that in compression.

Tests and experience indicate that it is reason-able to assume furthermore that, in a beam sub-jected to pure bending, cross sections that are plane before bending occurs remain plane dur-ing bending. As a result, both total and unit

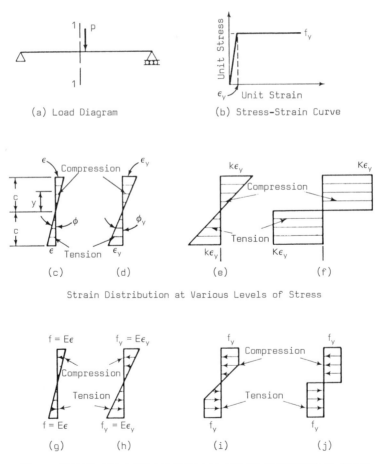

(a) Load Diagram

(b) Stress-Strain Curve

(c)   (d)   (e)   (f)

Strain Distribution at Various Levels of Stress

(g)   (h)   (i)   (j)

Stress Distributions Corresponding to Strain Distributions Above

Fig. 8.39. Bending stresses in the elastic and plastic ranges.

strains vary linearly with distance from the neutral surface.

Elastic Range. Consider, for example, a vertical section through a loaded horizontal beam, such as section 1-1 in Fig. 8.39a. The material of the beam is assumed to have the stress-strain curve shown in Fig. 8.39b, with stress proportional to strain up to the yield stress f_y and corresponding unit strain ϵ_y. For larger strains, stress remains constant at f_y.

Under service loads, unit strains at section 1-1 vary linearly with distance from the neutral surface, where unit stresses are zero. The maximum unit strain is ϵ, and the plane of the cross section rotates through an angle ϕ (see Fig. 8.39c). At a distance y from the neutral surface then, the unit strain equals $\epsilon y/c$, where c is the distance from the neutral surface to the outer surface at the section. From Hooke's law, the maximum unit stress is $f = E\epsilon$, and the unit stress at a distance y from the neutral surface equals fy/c. The force exerted on section 1-1 by the stress equals the product of the stress and the very small area ΔA of a strip parallel to the neutral surface. The moment of this force about the neutral surface then is $fy^2 \Delta A/c$.

The bending moment M on section 1-1 is equal in magnitude to the total resisting moment of the stresses. Thus,

$$M = \Sigma fy^2 \Delta A/c \tag{8.31}$$

where the summation is taken over the whole cross section of the beam. By definition, $\Sigma y^2 \Delta A$ is the moment of inertia I of the cross section about the neutral axis. Substitution in Eq. (8.31) yields

$$M = \frac{fI}{c} \tag{8.32a}$$

from which the maximum unit stress is

$$f = \frac{Mc}{I} \tag{8.32b}$$

Equation (8.32) is known as the *flexure formula*.

The ratio I/c in Eq. (8.32) is called the *section modulus*. For a rectangular section with depth d normal to the neutral surface and width b,

$$I = \frac{bd^3}{12} \tag{8.33}$$

and the section modulus is

$$S = \frac{I}{c} = \frac{bd^2}{6} \tag{8.34}$$

The AISC Steel Construction Manual lists moments of inertia and section moduli for structural-steel shapes.

Plastic Range. Suppose the load P on the simple beam in Fig. 8.39a were increased until the maximum unit stress at section 1-1 became the yield stress f_y and the corresponding unit stress became ϵ_y. The interior of the section would remain in the elastic range. The strain distribution would be the same as that shown in Fig. 8.39d and the stress distribution the same as that shown in Fig. 8.39h, both still linear.

Suppose now the load were increased until the maximum unit strain was several times larger than ϵ_y, say $k\epsilon_y$. The cross section would remain plane (see Fig. 8.39e). But the stress distribution would no longer be linear. The part of the cross section with strains greater than ϵ_y would be stressed to f_y. In the rest of the section, the unit stresses would decrease linearly to zero at the neutral surface (see Fig. 8.39i).

With further increases in load, the unit strains would increase rapidly (see Fig. 8.39f). The unit stresses would become a constant f_y throughout the whole section, which then would be totally plastic (see Fig. 8.39j).

During the process of load increase, the cross section would rotate from an angle ϕ at service loads (see Fig. 8.39c) to ϕ_y when the yield stress is reached initially (see Fig. 8.39d). Under larger loads, the section will rotate freely, like a hinge, without change in bending moment. The section is said to have become a *plastic hinge*.

Before a plastic hinge could form at section 1-1, however, the beam would fail, because a plastic hinge would form first at the section where maximum bending moment occurs, under the load. The beam would then have three hinges—one at each end and one under the load—and it would act like a mechanism, rotating freely with little or no increase in load.

The capacity, or ultimate strength, of a beam can be measured by the resisting moment for the stress distribution in Fig. 8.39*j*. The capacity can be expressed by the plastic moment:

$$M_p = Zf_y \qquad (8.35)$$

where

Z = plastic section modulus

From Fig. 8.39*j*, for a rectangular beam with depth *d* and width *b*, the plastic section modulus is

$$Z = \frac{bd^2}{4} \qquad (8.36)$$

(The AISC Steel Construction Manual lists plastic section moduli for structural-steel shapes.) Comparison of Eq. (8.36) with Eq. (8.34) and letting $f = f_y$ in Eq. (8.32*a*) indicates that a rectangular beam can resist 50% more moment after the yield stress is first reached at the outer surfaces.

Recall now that in a uniformly loaded fixed-end beam, the negative end moments are much larger than the positive moments. Suppose the uniform load were increased beyond the service load. Plastic hinges would form first at the beam ends, where the bending moment is maximum. At this stage, the beam would start to act as if it were simply supported. Load then could be further increased without failure occurring, until a third hinge formed at midspan, the point of maximum positive bending moment. At this stage, the bending moment at each hinge is the same and equals $w_p L^2/16$, where w_p is the uniform load at which the third hinge forms and *L* is the span. Assume now that the beam has a constant cross section with plastic section modulus *Z*. Substitution in Eq. (8.35) yields

$$w_p = 16f_y Z/L^2 \qquad (8.37)$$

From Fig. 8.37*f*, the bending moment at which the first hinges form is $w_y L^2/12$, where w_y is the load at that stage. Then, by Eq. (8.35),

$$w_y = 12f_y Z/L^2 \qquad (8.38)$$

Consequently, the capacity of the beam is 33% greater when failure occurs than at formation of the first hinge, as indicated by comparison with Eq. (8.37). For a simple beam, the moment at which the third hinge forms is $w_p L^2/8$, as indicated in Fig. 8.36*h*, and by Eq. (8.35),

$$w_p = 8f_y Z/L^2 \qquad (8.39)$$

Hence, by comparison with Eq. (8.37), a fixed-end beam with the same span and cross section can carry twice as much uniform load as a simple beam.

Behavior of a uniformly loaded continuous beam is similar to that of a fixed-end beam. Thus, a span of a continuous beam can carry much larger loads than a simple beam of the same length and cross section. This indicates that it is advantageous to use continuous beams instead of simple beams whenever conditions permit.

Combined Bending and Axial Stresses

Some structural members may be subjected to both bending and tensile axial loads. The resulting unit stress at any point is equal to the sum of the bending stress and the tensile stress at that point. Thus, if the assumptions of the flexure formula hold, the maximum stress at any vertical section is tensile and equal to

$$f = \frac{P}{A} + \frac{Mc}{I} \qquad (8.40)$$

where

P = axial load
A = cross-sectional area of the member
M = bending moment at the section
c = distance from neutral surface to surface where maximum unit stress occurs
I = moment of inertia of the cross section about the neutral surface

The bending moment need not be caused by transverse loads but instead by the eccentricity *e* of the load with respect to the centroidal axis. In that case, $M = Pe$ and the maximum stress is

$$f = \frac{P}{A} + \frac{Pec}{I} = \frac{P}{A}\left(1 + \frac{ec}{r^2}\right) \qquad (8.41)$$

where r = radius of gyration of the cross section [see Eq. (8.20)].

When the axial load is compressive, it has an eccentricity nevertheless, because of the beam deflection due to bending. The bending moment $P\Delta$ due to the deflection Δ should be added to the bending moment M. For a beam with relatively small bending deflections, however, the maximum stress is compressive and given closely by Eq. (8.40).

Unit Shear Stresses in Beams

The shear on a section of a beam normal to the neutral surface is resisted by nonuniformly distributed unit shear stresses. According to Eq. (8.10), the horizontal unit shear stress at any point in the section equals the vertical unit shear stress at that point. So when the horizontal unit shear stress has been determined, the vertical unit shear stress is also known.

Consider the loaded beam in Fig. 8.40a. Vertical sections 1-1 and 2-2 are taken very close together. The bending moment at section 2-2 is larger than that at section 1-1. Consequently, the unit compressive or tensile stress at any distance y from the neutral surface is greater on section 2-2 than on section 1-1 (see

Fig. 8.40c). For equilibrium, the unbalance must be resisted by a horizontal shear. This force, in turn, induces horizontal unit shear stresses f_v on the surface at the distance y from the neutral surface (see Fig. 8.40d). The corresponding vertical unit shear stresses on section 2-2, f_v in Fig. 8.41e, combine to resist the total shear on section 2-2. The relationship between this shear and the change in bending moment between sections 1-1 and 2-2, and use of the flexure formula [see Eq. (8.32)], leads to the following formula for unit shear stress at any point of section 2-2:

$$f_v = \frac{VQ}{Ib} \qquad (8.42)$$

where

V = shear on the section
I = moment of inertia of the beam cross section
b = width of the beam at the point
Q = moment, about the neutral surface, of the cross-sectional area of the beam included between the nearest surface free of shear (outer surface, for example) and a line, parallel to the neutral surface, drawn through the point

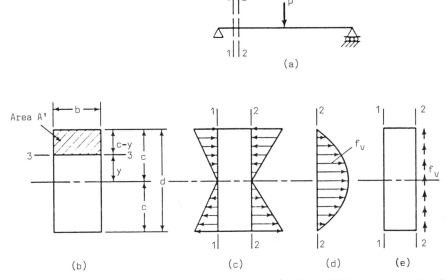

Fig. 8.40. Unit shear in a beam. (a) Simple beam with concentrated load. (b) Beam cross section. (c) Bending unit stresses at sections 1-1 and 2-2. (d) Distribution of unit shear stresses at section 2-2. (e) Vertical unit shear stresses at section 2-2.

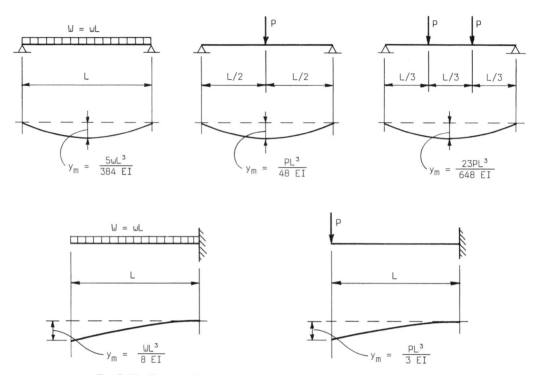

Fig. 8.41. Formulas for maximum deflection—common beam loadings.

For a rectangular cross section (see Fig. 8.40b), for determination of the unit shear at a distance y from the neutral axis,

$$Q = b(c - y)\left(y + \frac{c - y}{2}\right)$$

$$= \frac{b}{2}(c^2 - y^2)$$

$$= \frac{bc^2}{2}\left(1 - \frac{y^2}{c^2}\right)$$

$$= \frac{bd^2}{8}\left(1 - \frac{y^2}{c^2}\right) \qquad (8.43)$$

The moment of inertia I of the rectangular cross section equals $bd^3/12$. Hence, the unit shear stress at any point is, by Eq. (8.42),

$$f_v = \frac{12V}{8bd}\left(1 - \frac{y^2}{c^2}\right) = 1.5\frac{V}{A}\left(1 - \frac{y^2}{c^2}\right) \qquad (8.44)$$

where A = cross-sectional area of the beam.

This equation indicates that for a rectangular cross section the unit shear stress varies parabolically over the depth of the beam (see Fig. 8.40d). The maximum shear stress occurs at middepth, where $y = 0$. The maximum shear on a rectangular section then is

$$f_v = 1.5\frac{V}{A} \qquad (8.45)$$

Thus, the maximum shear is 50% larger than the average shear V/A.

In structural design, however, it is common practice to compute the average shear stress and compare it with an adjusted allowable unit stress.

Beam Deflections

When a horizontal beam is loaded by vertical downward-acting loads in its vertical plane of symmetry, the unsupported portion of the beam moves downward, that is, bends. The new, curved position of its originally straight, longitudinal centroidal axis is called the *elastic curve* of the beam. The vertical distance between the initial and final position of a point on the centroidal axis is called the deflection of the beam at that point.

Beam deflections are determined for several reasons. One of the most important reasons is that the effect of beam deflection on supported objects, especially those that may be damaged by large movements, needs to be known. Another important reason is that large deflections are unsightly. In addition, deflections may be needed in determination of reactions, shears and moments of statically indeterminate beams, as discussed previously.

Beam deflections may be calculated by any of several methods, all based on the same assumptions as the flexure formula [see Eq. (8.32)]. Of principal concern in most cases is the maximum deflection of a beam. For a simple span, symmetrically loaded beam, this will occur at the midspan point. For a cantilever beam, it will occur at the unsupported end. Derived by various theories, formulas for maximum deflection take the general form,

$$y_m = C \left(\frac{WL^3}{EI} \right) \qquad (8.46)$$

where

y_m = maximum vertical deflection
W = beam load (in pounds, kips, etc.)
E = elastic modulus of the beam material
I = moment of inertia of the beam section about the bending axis

Figure 8.41 shows several common loadings for beams and the corresponding formulas for maximum deflection. Note that for the uniformly loaded beams, with unit linear load of w, the load is expressed in total force units; thus $W = w(L)$. For beams with concentrated loads, the load is traditionally expressed as P, as shown in the formulas in Fig. 8.41.

Limits on Deflections. Design standards generally set limits on the maximum deflections of beams. This is done to prevent damage to objects supported on the beams and to control vibrations of beams. In addition, the deflection of roof beams supporting flat roofs must be restricted to prevent ponding of water on the roof. The weight of water trapped on a sagging roof causes additional deflections, which, in turn, traps more water, as a result of which the beams may fail.

One method of limiting beam deflections is to require the maximum deflection to be less than a specific fraction of the span. For example, for a beam supporting plaster ceilings, maximum deflection for live loads may be restricted to not more than $L/360$, where L is the beam span.

Another method is to set limits on the ratio of beam depth to span. For example, for a beam of A36 steel subjected to normal loading, the beam depth should not be less than $L/22$.

Camber. A sagging beam is not aesthetically appealing and may induce in observers a lack of confidence in the safety of the structure. Consequently, heavily loaded or long-span beams and trusses often are cambered, that is, given an upward deflection equal to the deflection anticipated under dead load. Then, when the structure is completed, the beam or truss becomes straight under the dead load.

Buckling of Beams

Because parts of a beam are in compression, the beam may buckle, just as a column may buckle under a critical load. For example, a narrow, rectangular beam may deflect normal to the plane of the loads as well as in the direction of the loads. This buckling can be prevented by use of a low allowable compressive stress, low ratio of span to beam width or short unsupported length of compression flange. The unsupported length can be decreased by placement of intermittent bracing at sufficiently close intervals or reduced to zero by a continuous diaphragm, such as plywood sheathing or a concrete floor or roof slab, firmly secured to the compression flange.

Local buckling of a compression flange or of a beam web under a concentrated load also may occur. This may be prevented by use of a low allowable compressive stress or low ratio of width or depth of beam components to their thickness. For cold-formed steel beams, it is common practice to consider only part of the cross-sectional area effective in resisting compression, depending on the width-thickness ratio of components. Thin webs of deep structural steel beams are often reinforced at concentrated loads with *stiffeners*. These may be plates set perpendicular to the web and welded to it or angles with one leg bolted or riveted to the web and the other leg set normal to the web. Compression flanges of cold-formed steel beams often are reinforced with a *lip*, or edge of flange bent parallel to the web (see Fig. 8.12).

Design standards for the various structural materials contain requirements aimed at prevention of beam buckling.

Reinforced Concrete Beams

While structural steel, cold-formed steel and wood behave structurally much like the idealized structural material for which the flexural formula was derived, especially in the elastic range, concrete does not. As pointed out in Sec. 8.5, concrete has a much lower tensile strength than compressive strength. If plain concrete were used for beams, they would fail due to tension well before the ultimate compressive strength is attained.

To make use of concrete more economical for beams and to avoid sudden, brittle failures, one method employs steel reinforcement, usually rebars, to resist the tensile bending stresses. The bars are most effective for this purpose when placed as close to the outer surface that is in tension as proper concrete cover for fire and corrosion protection permits.

Design Assumptions. Computation of bending strains, stresses and load-carrying capacity is based on the following assumption:

Unit strain in the concrete and reinforcing steel is the same. Hence,

$$f_s = nf_c \qquad (8.47)$$

where

f_s = unit stress in reinforcing steel
f_c = unit stress in concrete
n = *modular ratio* = E_s/E_c
E_s = modulus of elasticity of the steel
E_c = modulus of elasticity of the concrete (secant modulus)

Sections that are plane before bending remain plane during bending. Hence, total and unit strains are proportional to distance from the neutral surface.

The maximum usable unit strain at the outer concrete surface in compression equals 0.003 in. per in.

Steel unit stress less than the specified yield strength f_y equals $E_s \epsilon_s$, where ϵ_s is the steel unit strain. For unit strains larger than f_y/E_s, the steel unit stress is independent of unit strain and equals f_y.

Unless unit tensile stresses are very low, tensile strength of concrete is zero, because of cracks.

Behavior under Service Loads. Cast-in-place concrete beams are usually cast at the same time as floor or roof slabs. As a result, the beams and slabs are monolithic. While the part of the beam below the slab may be rectangular in cross section, the beam that is effective in resisting bending is actually T-shaped, because the slab works together with the concrete and steel protruding below. Precast concrete beams, however, often are rectangular in cross section, but other shapes, including tees and double tees, also are commonly used. In the following discussion of the behavior of concrete beams, however, a rectangular cross section is used for illustrative purposes. The principles presented nevertheless are also applicable to other shapes and to beams reinforced in both tension and compression.

Figure 8.42a shows a simply supported, reinforced concrete beam, with width b and depth h. Reinforcing steel is placed at a distance d below the outer surface in compression (see Fig. 8.42c and d). Total steel area provided is A_s. If f_s is the unit stress in the steel, then the force available for resisting tension is

$$T = A_s f_s \qquad (8.48)$$

Under service loads P, cracks are formed on the tension side of the beam (see Fig. 8.42a). (Because the reinforcing steel resists opening of the cracks, they may not be visible.) Section 1-1 is taken normal to the neutral surface between cracks and section 2-2 is taken at a crack. Figure 8.42b shows the portion of the beam be-

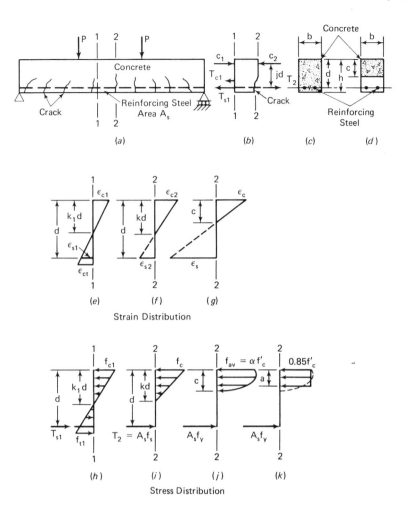

Fig. 8.42. Bending stresses in a reinforced concrete beam, as load increases from service load to ultimate load.

tween the two sections. A compressive force C_1 acts on the upper part of the beam at section 1-1, while tensile force T_{s1} acts on the steel and tensile force T_{c1} acts on the concrete bottom portion. For equilibrium, $C_1 = T_{s1} + T_{c1}$. Simultaneously, a compressive force C_2 acts on the upper part of the beam at section 2-2, while tensile force T_2 acts on the steel. Because of the crack, the concrete is assumed to offer no resistance to the tension (see Fig. 8.42d). For equilibrium, $C_2 = T_2$.

The unit strains at section 1-1 vary linearly (see Fig. 8.42e). Also, the unit stresses in the concrete vary linearly over the whole depth of the beam (see Fig. 8.42h). Conditions, however, are much severer at section 2-2.

Since cracks may occur at any place along the tension surface of a reinforced concrete beam, design must be predicated on the assumption that conditions similar to those at section 2-2 may develop at any section considered. The magnitude of the stresses and strains, however, will depend on the bending moment at the section.

At section 2-2, unit strains are proportional to the distance from the neutral surface (see Fig. 8.42f). Consequently, the steel unit strain is related to the concrete unit strain by

$$\epsilon_{s2} = \frac{1 - k}{k} \epsilon_{c2} \qquad (8.49)$$

where kd is the depth of the portion of the beam in compression. The compressive unit stresses also vary linearly (see Fig. 8.42i). Hence, the total compressive force is

$$C_2 = \tfrac{1}{2} f_c kbd \qquad (8.50)$$

where

f_c = maximum compressive unit stress in the concrete

With the use of Eq. (8.47), Eq. (8.49) can be converted to the useful relationship:

$$\frac{nf_c}{f_s} = \frac{k}{1 - k} \qquad (8.51)$$

where

f_s = unit stress in the steel

Equating C_2 as given by Eq. (8.50) and T_2 as given by Eq. (8.48) yields

$$\tfrac{1}{2} f_c kbd = A_s f_s = \rho bd f_s \qquad (8.52)$$

where

$\rho = A_s/bd$ or ratio of reinforcing steel area to concrete area

Simultaneous solution of Eqs. (8.51) and (8.52) for k yields

$$k = \sqrt{2n\rho + (n\rho)^2} - n\rho \qquad (8.53)$$

The bending moment at section 2-2 is resisted by the couple C_2 and T_2, which have a moment arm jd (see Fig. 8.42b). Because of the linear variation of concrete stress, C_2 acts at a distance $kd/3$ from the outer beam surface in compression. Hence, $jd = d - kd/3$ and

$$j = 1 - \frac{k}{3} \qquad (8.54)$$

The resisting moment of the beam in compression can be found by taking moments about the centroid of the reinforcing steel:

$$M_c = \tfrac{1}{2} f_c kjbd^2 \qquad (8.55)$$

Similarly, the resisting moment of the beam in tension can be found by taking moments about the centroid of the compression area:

$$M_s = f_s A_s jd = f_s \rho jbd^2 \qquad (8.56)$$

When allowable unit stresses are used in design, Eqs. (8.51) through (8.56) provide sufficient information for design of reinforced concrete beams for resistance to bending moments. In such cases, it is desirable that M_c be greater than M_s so that failure will occur in tension. Yielding of the steel, permitting wide cracks to form, will give warning that failure is imminent.

Bending Strength. If the loads P on the beam in Fig. 8.42a are gradually increased, the cracks will lengthen, the neutral surface will move upward and eventually the reinforcing steel will be stressed to its yield point f_y. With further increase in load, the steel will carry no greater tensile force than $A_s f_y$. (It is assumed here that the beam is so proportioned that failure by crushing of the concrete will not occur before the steel reaches its yield point.) At this stage, unit strains are still proportional to distance from the neutral surface (see Fig. 8.42g); but the compressive stress in the concrete no longer is proportional to the unit strain. The stress distribution is likely to resemble that shown in Fig. 8.42j. The ratio α of the average compressive stress to the 28-day compressive strength of the concrete f'_c is 0.72 for f'_c up to 4 ksi, but decreases by 0.04 for each ksi above 4 ksi. The resultant compressive force C_2 at this stage acts at a distance βc below the outer concrete surface in compression, where c is the depth of the part of the beam in compression (see Fig. 8.42j). $\beta = 0.425$ for f'_c up to 4 ksi, but decreases by 0.025 for every ksi above 4 ksi.

While the strength of the beam can be computed from the preceding information, it is common practice to assume a simpler but equivalent distribution of compressive stresses. Usually, the unit compressive stress when the steel reaches its yield point f_y is assumed constant at $0.85 f'_c$. The depth of this rectangular stress distribution (see Fig. 8.42k) is taken as

$a = \beta_1 c$, where $\beta_1 = 0.85$ for f'_c up to 4 ksi, but decreases by 0.05 for every ksi above 4 ksi. These assumptions make the rectangular stress distribution equivalent in computations of moment-resisting capacity to the stress distribution shown in Fig. 8.42*j*.

The depth c of the compression area in Fig. 8.42*j* can be obtained by equating the total compression $C_2 = \alpha f'_c bc$ to the tensile force $T_2 = A_s f_y = \rho bd f_y$. The result is

$$c = \frac{A_s f_y}{\alpha f'_c b} = \frac{\rho f_y}{\alpha f'_c} d \qquad (8.57)$$

The resisting moment in tension then equals

$$
\begin{aligned}
M_u &= A_s f_y jd = A_s f_y \left(d - \frac{a}{2} \right) \\
&= A_s f_y \left(d - \frac{\beta_1 c}{2} \right) = A_s f_y d \left(1 - \frac{\beta_1 f_y}{2\alpha f'_c} \rho \right)
\end{aligned}
$$

$$(8.58)$$

The steel ratio ρ should be chosen small enough that the unit strain in the concrete does not reach 0.003 in. per in. to avoid failure by crushing of the concrete.

The ACI Code requires that a capacity reduction factor $\phi = 0.90$ be applied to the strength of a beam [see Eq. (8.58)] to provide for small adverse variations in materials, workmanship and dimensions.

Shear Strength. Longitudinal reinforcement placed close to the tension surface of a reinforced concrete beam resists tension due to bending. There are also, however, tensile unit stresses associated with shear unit stresses. As pointed out previously, maximum unit shear stress occurs away from the beam outer surfaces and at or near the neutral surface. Consequently, tensile stresses due to shear are also at a maximum near the neutral surface. These stresses are inclined at an angle of 45° with the vertical section on which the shear acts. At some distance from the neutral surface, the tensile stresses due to shear, which decrease with distance from the neutral surface, occur at a section where there is appreciable horizontal, tensile bending stresses, which increase with

distance from the neutral surface. The vector sum of the shear and bending tensile stresses act at an angle to the horizontal and is known as *diagonal tension*. Usually, it is necessary to provide steel reinforcement in concrete beams to resist this diagonal tension, which can cause flexural cracks to curve toward midspan, as indicated in Fig. 8.43*a* and *c*.

If the shear strength at a vertical section of a horizontal beam is V_u, lb, the nominal shear stress, psi, at the section is given by

$$v_u = \frac{V_u}{\phi b_w d} \qquad (8.59)$$

where

b_w = width of a rectangular beam or of the web of a T beam, in.

d = depth from outer compression surface to centroid of reinforcement for tensile bending stresses

ϕ = capacity reduction factor = 0.85

Because the associated tension acts on a diagonal, the section for maximum shear V_u may be taken at a distance d from the face of each support in simple beams, unless the beams are very deep.

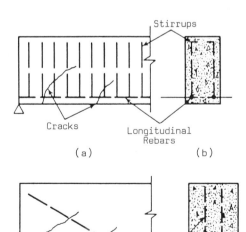

Fig. 8.43. Shear reinforcement for a reinforced concrete beam. (*a*) and (*b*) Stirrups. (*c*) and (*d*) Bent-up longitudinal rebars.

If the concrete will not crack at the stress computed from Eq. (8.59), no shear reinforcement is needed to resist the diagonal tension. Tests indicate that the shear unit stress v_c that causes cracking in normal-weight concrete lies between $2\sqrt{f'_c}$ and $3.5\sqrt{f'_c}$, where f'_c is the 28-day compressive strength of the concrete, psi. The actual value of v_c depends on the ratio of shear to moment at the section and the ratio of longitudinal steel area to the area of the concrete web. The lower value of v_c can be used for conservative calculations.

If $v_u > v_c$, steel reinforcement must be provided to resist the tensile stresses due to $v_u - v_c$, but the ACI Code requires that $v_u - v_c$ not exceed $8\sqrt{f'_c}$.

Stirrups are commonly used as shear reinforcement in reinforced concrete beams. A stirrup has bar legs lying in a vertical plane and close to the outer, vertical beam surfaces. The legs are hooked around the longitudinal steel reinforcement or have plain hooks of 90° or more (see Fig. 8.43b). The cross-sectional area A_v of the legs provides the required resistance to diagonal tension. Stirrups may be set vertically as in Fig. 8.43a or at an angle. In either case, they should be set closely enough so that at least one stirrup will pass through every possible location of a 45° crack. Also the ends, or hooks, of a stirrup should be anchored in a compression zone of the beam. The area, sq in., required in the legs of a vertical stirrup is

$$A_v = \frac{v_u - v_c}{f_y} b_w s \qquad (8.60)$$

where

f_y = yield strength of the shear reinforcement, psi

s = spacing, in., of the stirrups

For inclined stirrups, the right-hand side of Eq. (8.60) should be decreased by dividing by sin α + cos α, where α is the angle of inclination of the stirrups with the horizontal.

An alternative method of providing shear reinforcement is to bend up reinforcing bars at an angle of 30° or more, as indicated in Fig. 8.43c and d. Bending moments producing tension at

the bottom of a beam generally are maximum near midspan and decrease rapidly with distance toward the supports. Hence, fewer reinforcing bars are needed as the supports are approached. Those bars not needed may be bent up to serve as shear reinforcement. The area required for a single bar or a single group of parallel bars all bent at an angle α at the same distance from the support can be computed by dividing the right-hand side of Eq. (8.60) by sin α.

The ACI Code requires, however, that at least some shear reinforcement be incorporated in all beams with total depth greater than 10 in., or more than $2\frac{1}{2}$ times the flange (slab) thickness, or exceeding half the web thickness, except where v_u is less than $\frac{1}{2} v_c$. The area of this minimum shear reinforcement should be at least

$$A_v = \frac{50 b_w s}{f_y} \qquad (8.61)$$

Prestressed Concrete Beams

The concept of prestressed concrete is presented and the advantages of prestressing are discussed in Sec. 8.5. In the following, the application of prestressing to beams is discussed.

Consider as an example the prestressed concrete beam in Fig. 8.44a. Assume that it is simply supported and that it will carry uniformly distributed dead and live loads. The bending moment due to the total load then will vary parabolically from a maximum M_1 at section 1-1, at midspan, to M_2 at section 2-2, and then to zero at the supports. Suppose now that tendons are placed in the beam in a parabolic shape, sagging close to the tension surface at midspan and anchoring close to the centroidal axis of the beam cross section at the supports. A total force P_s is applied to tension the tendons. (The angle of inclination of this force at the supports is actually so small that the horizontal component of this prestress may also be taken equal to P_s.)

At any section of the beam, the prestress applies an axial compressive stress P_s/A_c, where A_c is the cross-sectional area of the beam. The prestress also imposes bending unit stresses whose magnitude depends on the eccentricity, at the section, of the tendons. For example, at

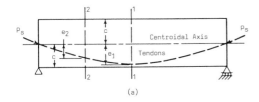

(a)

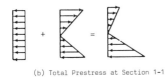

(b) Total Prestress at Section 1-1

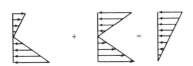

(c) Prestress Plus Load Stress at Section 1-1

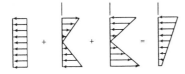

(d) Prestress Plus Load Stress at Section 2-2

Fig. 8.44. (a) Prestressed concrete beam with draped tendons. (b) Axial compression adds to bending stresses induced by tendons to impose linearly varying prestress at section 1-1. (c) Bending stresses due to load imposed on prestress at section 1-1 results in net stresses with no tension. (d) At section 2-2, bending stresses due to load imposed on prestress results in compression over the whole section.

section 1-1, the maximum bending stress is $P_s e_1 c / I_g$, where e_1 is the eccentricity of the tendons (see Fig. 8.44a), c is the distance from the outer beam surfaces to the centroidal axis and I_g is the moment of inertia of the beam cross section about the centroidal axis. Note that the prestress bending causes tension at the upper beam surface and compression at the lower beam surface. The total prestress is given by Eq. (8.41), as indicated in Fig. 8.44b.

At section 1-1, the vertical load on the beam causes maximum bending stresses $M_1 c / I_g$, with compression at the upper beam surface and tension at the lower beam surface. These stresses

are superimposed on the prestress already present in the beam. The result is a linear stress variation, which, for the case shown in Fig. 8.44c, varies from zero at the bottom of the beam to a compressive unit stress at the top. Hence, under service loads, the section is completely in compression.

At section 2-2, the eccentricity of the tendons is smaller and produces smaller bending stresses than those at section 1-1. The resulting prestress consequently exerts less compression at the bottom of the beam than that at section 1-1, but the bending moment at section 2-2 is also proportionately less than that at section 1-1. Hence, when the prestress is superimposed on the bending stress due to the loads, the net stresses are compressive (see Fig. 8.44d).

If a straight tendon with constant eccentricity had been used, sections near the supports would have net tensile stresses at the upper surface of the beam, because of the near-zero bending stresses imposed near the supports by the loads. An advantage of curving the tendons to decrease the eccentricity as they approach the supports is that the whole beam cross section can be kept in compression. Design of the beam can then be based on Eq. (8.41), with the whole cross section of the beam effective in resisting bending.

Specifications for the design of prestressed concrete beams are contained in the ACI Code.

Composite Steel-Concrete Beams

In the preceding discussions of reinforced concrete and prestressed concrete beams, methods are presented for use of steel to increase the load-carrying capacity of concrete in beams. In the following, methods are described for use of concrete to increase the load-carrying capacity of structural steel in beams. Specifications for the design of such steel-concrete composite beams are contained in the AISC Specification for the Design, Fabrication and Erection of Structural Steel for Buildings.

Structural steel beams are often used to support reinforced concrete floor or roof slabs. The slabs may be designed to support their own weight and other dead and live loads over the spaces between the steel beams. The beams, in turn, may be designed to support their own weight and the load from the slab. Because the slab has a relatively large cross-sectional area and the compressive strength of this area is not fully utilized, an opportunity exists for more economical design of the beams and slab. If the slab is secured to the compression flanges of the beams, the reserve compressive strength of the slab can be utilized to assist the beams in carrying their loads over the span between beam supports.

Interaction between the concrete slab and the steel beams by natural bond can be obtained by fully encasing the beams in reinforced concrete (see Fig. 8.45a). For this purpose, the top of the compression flange of each beam must be at least 2 in. above the slab bottom and, for fire protection, at least $1\frac{1}{2}$ in. below the slab top. Also, there must be at least 2 in. of concrete on the sides of the soffit. This method of construction, however, has the disadvantage that the encasement adds considerable weight to the loads the beams have to support and to the loads transmitted to the columns and foundations.

An alternative method of insuring interaction between slab and beams is to use shear con-nectors between them. The shear connectors must prevent separation of the beam flange and slab and must transfer horizontal shear between the steel and concrete. For this purpose, steel studs (see Fig. 8.45b), steel channels set with web vertical and perpendicular to the beam web, or steel wire spirals may be welded to the top of the compression flanges of the steel beams.

The beams should be designed to support their own weight and that of the concrete and formwork imposed on them during construc-tion. Shores, however, may be used until the concrete has hardened, to assist the beams if the steel would be overstressed.

After the concrete slab has attained sufficient strength to support loads, the composite section may be assumed to support the total load on the beam spans. The flexure formula [see Eq. (8.32)] may be used to design the composite beams. In this formula, the section modulus I/c for the transformed section should be used. The *transformed section* consists of the actual steel area plus the concrete area divided by the modular ratio $n = E_s/E_c$ (equivalent steel area). The concrete area equals the slab thickness times the *effective width* (see Fig. 8.45b).

Number of Connectors. Shear connectors may be spaced uniformly over the compression flanges of the steel beams. (Two or more welded studs may be placed in line at any cross section of a beam.) The total number of shear connectors required equals the total shear V_h to be transmitted between steel and concrete di-vided by the allowable shear load on a single connector q. This load q depends on the type and size of shear connector and the strength f'_c of the concrete. The shear V_h is the smaller of the values given by Eqs. (8.62) and (8.63).

$$V_h = \frac{0.85 f'_c A_c}{2} \tag{8.62}$$

$$V_h = \frac{A_s F_y}{2} \tag{8.63}$$

where

f'_c = 28-day compressive strength of the con-crete

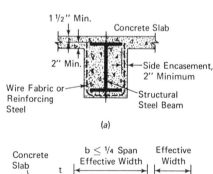

(a)

1½″ Min.

Concrete Slab

2″ Min.

Side Encasement, 2″ Minimum

Wire Fabric or Reinforcing Steel

Structural Steel Beam

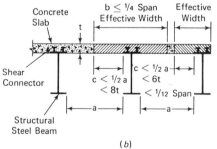

(b)

Concrete Slab

$b \le \frac{1}{4}$ Span Effective Width

Effective Width

t

Shear Connector

$c < \frac{1}{2} a$
$< 8t$

$c < \frac{1}{2} a$
$< 6t$
$< \frac{1}{12}$ Span

a

a

Structural Steel Beam

Fig. 8.45. Composite construction with steel beams and concrete slabs. (a) Steel beam completely en-cased in concrete. (b) Steel beams attached to con-crete slab with shear connectors.

A_c = actual area of effective concrete slab
A_s = cross-sectional area of steel beam
F_y = specified yield stress of the steel

The AISC Specification also sets a minimum for the number of shear connectors when a concentrated load occurs between the points of maximum and zero bending moment.

References

THEORY:

H. Parker and J. Ambrose, *Simplified Mechanics and Strength of Materials*, 4th ed., Wiley, New York, 1986.
H. Laursen, *Structural Analysis*, 3rd ed., McGraw-Hill, New York, 1988.
J. McCormac, *Structural Analysis*, 4th ed., Harper & Row, New York, 1984.

DESIGN:

F. Merritt, *Building Design and Construction Handbook*, 4th ed., McGraw-Hill, New York, 1982.
H. Parker and J. Ambrose, *Simplified Engineering for Architects and Builders*, 7th ed., Wiley, New York, 1989.
J. Ambrose, *Building Structures*, Wiley, New York, 1988.
S. Crawley and R. Dillon, *Steel Buildings: Analysis and Design*, 3rd ed., Wiley, New York, 1984.
Manual of Steel Construction, 8th ed., American Institute of Steel Construction, Chicago, 1980.
P. Rice, et al., *Structural Design Guide to the ACI Building Code*, 3rd ed., Van Nostrand Reinhold, New York, 1985.
Building Code Requirements for Reinforced Concrete, ACI 318–83, American Concrete Institute, Detroit, 1983.
CRSI Handbook, 4th ed., Concrete Reinforcing Steel Institute, Schaumburg, IL, 1982.
D. Breyer, *Design of Wood Structures*, 2nd ed., McGraw-Hill, New York, 1986.
Timber Construction Manual, 3rd ed., American Institute of Timber Construction, Wiley, New York, 1985.

Words and Terms

Beam
Beam buckling
Beam stresses: bending (flexure), shear
Bending moment
Composite beam
Deflection
Diagonal tension
Neutral surface
Prestressed beam
Shear (force)
Stirrups

Significant Relations, Functions, and Issues

Special purpose beams: joists, stringers, headers, purlins, rafters, girts, lintels, girders, spandrels.
Types of beams: simple, continuous, cantilever, fixed-end, statically indeterminate.
Beam loadings: distributed, concentrated.
Interrelation of shear and bending moment.
Plastic behavior of a steel beam.
Service load behavior versus ultimate-strength behavior of a reinforced concrete beam.

8.11. ARCHES AND RIGID FRAMES

An arch is a structural member with its centroidal axis lying in a plane and curved so that loads are carried between supports principally by compression (see Fig. 8.46).

Bending moments and shears can be kept small by proper shaping of the axis. Supports that are capable of resisting the reactions of the arch without translation, however, must be provided. Arch supports are subjected to forces inclined to the line of action of loads applied to the arch.

The distance between arch supports is called the *span*. The distance between the line between supports and a point on the centroidal axis farthest from the line, measured along a normal to the line, is called the *rise*. The smaller the ratio of rise to span, the flatter the arch and the larger the bending moments and shear are likely to be.

Usually, an arch axis is set in a vertical plane in buildings, and loads are applied in that plane. The arch supports may or may not be at the same level. The thickness, or depth, of arches is generally very small compared with the span and may be varied along the span for efficient resistance to stresses. In this section, however, only symmetrical arches of variable cross section, with supports at the same level—an arrangement generally used—are discussed.

Arches are more efficient than beams and can eliminate the need for columns by carrying loads directly to foundations; however, the arch shape is not always desirable in a building and the space required for the large ratio of rise to span for efficient arch action is not always available. Consequently, the most frequent ap-

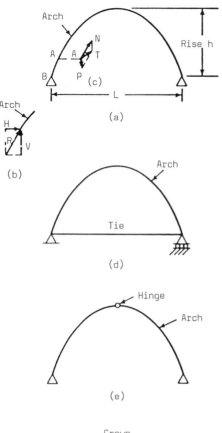

(a)

(b)

(c)

(d)

(e)

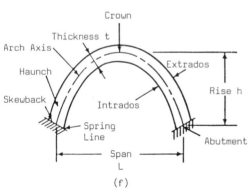

(f)

Fig. 8.46. Arches. (a) Two-hinged. (b) Arch reaction resolved into horizontal and vertcal components. (c) Resolution of thrust at point in arch above base. (d) Two-hinged arch with tie. (e) Three-hinged arch. (f) Fixed-end arch and names of arch components.

plication of arches in buildings is support of roofs, especially when long spans are desired. The added cost of fabricating curved members usually makes beams or trusses more economical for short spans.

Types of Arches

Arches are classified as two-hinged, three-hinged or fixed, in accordance with conditions provided at supports and in the interior.

Two-hinged arches are shown in Fig. 8.46a and d. Only rotation is permitted at the supports of the arch shown in Fig. 8.46a. The reactions have both horizontal and vertical components (see Fig. 8.46b). The arch must be supported on abutments that prevent both horizontal and vertical movements.

For the arch shown in Fig. 8.46d, the horizontal components of the reactions are resisted by a tie between supports. Small horizontal movements can occur at the supports because of elastic elongation of the tie on application of wind, snow or live loads.

Two-hinged arches are statically indeterminate. The laws of equilibrium do not provide sufficient equations for determination of the two horizontal and two vertical reaction components.

A three-hinged arch is shown in Fig. 8.46e. In practice, hinges are placed at the two supports and at the crown, or highest point of the arch. Conditions at supports are similar to those for two-hinged arches. Three-hinged arches, however, are statically determinate. Free rotation of the two halves of the arch at the crown hinge provides information for adding another equilibrium equation to those available for two-hinged arches.

A fixed arch is shown in Fig. 8.46f. No translation or rotation is permitted at either support. Consequently, the reaction at each support has a bending moment as well as horizontal and vertical components. The arch is statically indeterminate. Determination of the reactions requires two more equations than that needed for a two-hinged arch.

Fixed arches are not often used for buildings. Conditions for prevention of rotation at supports may not exist on a building site or may be expensive to provide. Furthermore, if an arch should be designed as fixed and the assumption of fixity should prove to be unwarranted by field conditions, portions of the arch may be dangerously overstressed.

Figure 8.46*f* also indicates the names commonly used for parts of an arch. The *extrados* is the upper surface, or back, of the arch. The *intrados* is the under surface, or soffit, of the arch. *Thickness*, or *depth*, of the arch is the distance, measured in a plane normal to the axis, between the intrados and extrados. A deepened portion of the arch near each support is called a *haunch*. The intersection of the abutment and a side of the arch is known as a *skewback*, and the intersection of the abutment and the arch soffit is called a *springing line*.

Arch Curvature

For any type of arch, minimization of bending, and hence more efficient arch action, depends on the shape of the arch and characteristics of the loads. If service loads did not change, it would be possible to shape an arch to eliminate bending moments completely. For example, consider conditions at point A in Fig. 8.46*a*. If there is no bending moment or shear, only a thrust N and a load P are present (see Fig. 8.46*c*). These combine vectorially to impose a thrust T on an adjoining point on the arch axis. If T does not cause bending moment at the adjoining point, T must pass through that point. Hence, to eliminate bending moment, the arch axis should be tangent to the line of action of T. For bending moment to be eliminated throughout the arch, the axis should be so curved that the line of action of the thrust at every point coincides with the axis.

For a vertical load uniformly distributed over the horizontal projection of an arch, the axis would have to be parabolic if the arch is to be subjected only to compression. Similarly, the axis would have to be shaped as an inverted catenary for a vertical load uniformly distributed along the arch axis, as would be the case for the weight alone of an arch with constant cross section.

These loading conditions rarely are attainable in building applications. Like other structural members, arches should be designed to resist at every section the maximum stresses that may occur. Such stresses may be imposed when live,

snow or wind loads are applied over only part of the arch. Also, arches may be subjected to concentrated loads. Consequently, at best it is practical to shape arches so that the thrust nearly coincides with the axis at every point. This, however, may not always be possible, because of limitations placed on the rise by other design requirements. Costs of roofing, for example, increase with increase in rise, as does the cost of heating the space under the arch, because volume increases with increase in rise.

Hundreds of years ago, when arches were made of stone, the matching of the arch profile to the loading was critical, due to the tension-stress limitation of the material. With modern construction, use of steel, reinforced concrete, or glued-laminated timber makes this less significant.

Arch Cross Sections

Thickness, or depth, of arches, often is varied along the axis as stresses change. In this way, with cross sections subjected mainly to uniform compression, arches use the materials of which they are made much more efficiently than beams do.

Arches may have solid cross sections of almost any shape. Wood or concrete arches usually have rectangular cross sections. Steel arches may be formed with wide-flange rolled shapes or built-up of plates in similar shapes. Steel box or circular tubes are also used.

Alternatively, wood or steel arches may be built-up like trusses. In such cases, if the loads are applied only at panel points, the chords and web members are subjected only to axial forces, usually compressive.

Reactions of Three-Hinged Arches

The horizontal and vertical components of the reactions of a three-hinged arch can be determined with the laws of equilibrium. Four equations are available for determination of the four unknowns:

1. The sum of the horizontal components of the loads and both reactions must be zero.
2. The sum of the moments about the left re-

action of all the forces acting on the arch must be zero.

3. The sum of the moments about the right reaction of all the forces acting on the arch must be zero.

4. The bending moment at the crown hinge must be zero. (The sum of the moments about that hinge of all forces acting on the arch also is zero, but this condition does not yield an independent equation.)

As a check, the sum of the vertical components of the loads and both reactions must be zero. This condition, too, does not yield an independent equation.

The reactions of a three-hinged arch can also be determined graphically. Consider, for example, the arch shown in Fig. 8.47a. It carries a load P, which may be a concentrated load or the resultant of a distributed load. There are no loads between the crown hinge and the right reaction. Since there is no bending moment at the crown, the line of action of the right reaction R_B must pass through the hinge. The line intersects the line of action of P at point X. Because P and the two reactions are in equilibrium, the line of action of the left reaction R_A must also pass through X. Then R_A and R_B can be determined graphically, as indicated in Fig. 8.47b. First, P is drawn as an arrow with length proportional to its magnitude. From both extremities of the arrow, lines are drawn parallel to AX and BX, the lines of actions of the reactions, to form a parallelogram of forces. If P is the equilibrant, the other two forces comprising

the sides of the parallelogram are, to scale, the reactions R_A and R_B.

Three-hinged arches do not use material as effectively as two-hinged arches, because of the large bending moments midway between crown and supports. Also, insertion of a hinge at the crown can in some cases add to construction costs. Maintenance costs also may be increased, because care should be taken that dirt or corrosion products do not cause the hinge to freeze and prevent rotation of the arch, thus making design assumptions invalid and causing over-stressing of parts of the arch.

One advantage of the three-hinged arch is the elimination of thermal stresses, inasmuch as rotation at the crown permits dimensional changes to occur freely. Another advantage rises for three-hinged arches of modest span and is the possibility for fabrication of the arch in two pieces that can be relatively simply erected and connected in the field. The latter advantage is often significant for buildings requiring short to moderate spans (50 to 150 ft).

Supports for Arches

Arches generally are supported on concrete abutments set in the ground. If the soil is sufficiently strong, for example, sound stable rock to resist the arch thrust, a tie between supports is not needed. When the soil is incapable of withstanding the arch thrust, the supports must be tied. In such cases, the tie is often placed under the ground floor. It is sometimes desirable, however, to install the tie at a higher level, for example, between two points on the arch. This tie may be a deep girder also used to support a floor. If the girder span is long, the girder may also be suspended at intervals from the arch.

Sometimes, it is desirable or necessary to support an arch above ground, on walls or columns. In such cases, provision for resisting the thrust, such as a tie between supports, must be made.

Arch Bracing

The maximum unit stress in an arch, the sum of the axial thrust and bending stresses, is compressive. Consequently, arches should be

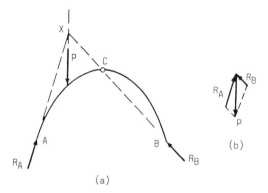

(a)

(b)

Fig. 8.47. Determinations of reactions for a three-hinged arch.

braced to prevent buckling. In addition, arches should also be braced to prevent overturning in the direction normal to the plane of the centroidal axis. Bracing may have to be put in place during erection of an arch.

Transition from Arch to Rigid Frame

As mentioned previously, arches with a suitable shape for the loads to be supported can be shaped so that little or no bending moments are imposed. Such arches, though, require a large ratio of rise to span. A high rise, however, may

be undesirable for many reasons, primarily economic, for a specific structure. A flat arch springing from the ground may be used instead, but in that case the space near the supports often cannot be used, because of the small clearance between floor and arch. One alternative to this undesirable condition is to make a portion of the arch in the vicinity of the supports vertical, like columns. The result is an *arched bent* (see Fig. 8.48a).

Such a bent has many of the characteristics of an arch. It may be two- or three-hinged or fixed. Reactions and stresses may be determined in the same way as for an arch. An arched

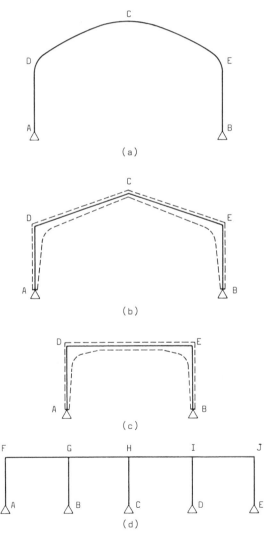

Fig. 8.48. Types of bents. (*a*) Arched bent. (*b*) Tudor arch, or gable frame. (*c*) Rigid frame. (*d*) Continuous rigid frame.

bent, however, is subject to much larger bending moments and shears than an arch with the same span and rise. Nevertheless, the arched bent is more efficient than a simple beam of the same span supported on two columns.

Curving structural members to a shape, such as DCE in Fig. 8.48a, adds to fabrication costs. Consequently, two straight rafters DC and CE often are substituted for the arched portion of the bent. The result is a *Tudor arch*, or *gable frame* (see Fig. 8.58b).

This bent, too, has many of the characteristics of an arch, but like an arched bent, a gable frame has much larger bending moments and shears than an arch of the same span and rise. In particular, large stresses occur at the *knees*, the intersections D and E of the rafters DC and EC with the columns AD and BE (see Fig. 8.48b). Consequently, a smooth, curved transition is usually provided at the knees. The rafters may be haunched near the knees or tapered from the depth at the knees to a lesser depth at the crown C (see dashed lines in Fig. 8.48b). Similarly, the columns may be tapered from the depth at the knees to a lesser depth at their supports. Also, a rigid connection is provided at the knees, to insure that the column and rafter ends at each knee have the same rotation under loads. As a result, the structural efficiency of a gable frame is lower than that of an arch, but the gable frame permits more efficient utilization of space and may be more economical to fabricate, especially for short spans.

When a flat roof is preferred or a floor has to be supported, the upper portion of the bent must be made straight. The result is the rigid frame shown in Fig. 8.48c, with horizontal beam DE and columns AD and BE. This bent has fewer characteristics of arches than the preceding bents. Nevertheless, reactions and stresses can be computed in the same way as for an arch.

As for a gable frame, rigid connections are provided at the knees at D and E. The beam therefore should be treated as a continuous beam. It is subjected to negative end moments at D and E, which make the positive moments near midspan much smaller than those in a simple beam. For greater economy, DE may be deepened where bending moments are largest, for example, at D and E. Because of continu-

ity, the columns AD and BE are subjected to axial compression and bending moments and may be haunched or tapered to vary the cross section with change in stresses, as indicated by the dashed lines in Fig. 8.48c.

Bents such as that in Fig. 8.48c can be placed in line in a vertical plane to form a continuous rigid frame (see Fig. 8.48d). Where columns are permitted at wide spacings within a building, this type of bent is advantageous for supporting floors or flat roofs of considerable length. Beam FGHIJ may be made continuous for all or part of its length, depending on the economics of fabrication and erection of long lengths and shipping limitations on length.

The spaces between columns in a bent are known as *bays*. Figure 8.48d shows a one-story rigid frame, four-bays wide.

Rigid frames may also be stacked one above the other in a vertical plane, to form a multistory, continuous rigid frame. Figure 8.49e shows a two-story, single-bay rigid frame.

Applications of Rigid Frames

Rigid frames are widely used for skeleton frames of buildings for many reasons.

One important reason is that the vertical columns and horizontal beams of such frames are compatible with the rectangular spaces generally desired in building interiors. (Rigid frames, however, can be constructed with inclined columns and beams, when it is desirable.) In buildings in which it is required that structural framing be hidden from view, it is simple to conceal the vertical columns in walls and the horizontal beams between floors and ceilings.

Another important reason for the popularity of rigid frames is that they often cost less than other types of construction for the spans frequently used in buildings. The economic advantage of rigid frames stems from the utilization of continuity of beams and columns. Continuity reduces bending moments from those in simple beams and permits more efficient utilization of the material in structural members. Also, the rigid connections between beams and columns enables rigid frames to resist lateral, or horizontal, forces, such as wind and earthquake loads. Often, such frames are capable of withstanding lateral loads without assistance from bracing or walls in their vertical planes. Elimination of bracing often is desirable for architectural reasons, because bracing with truss diagonals, X-bracing, or shear walls may interfere with door or window locations or other use of space.

Stresses in Rigid Frames

Continuous rigid frames are statically indeterminate to a high degree. Special techniques had to be developed to permit determination of the numerous reaction components and stresses. Tedious calculations, though, are often required for multibay, multistory buildings. Computer programs, however, are available for rapid analysis with electronic computers.

A complicating factor in analysis of continuous frames is that, because of continuity, the structural properties of each member affects the response of every other member to loads. To illustrate, Fig. 8.49a shows a single-bay, one-story rigid frame subjected to a vertical load P at midspan of beam DE. The deflected position of the centroidal axes of the beam and columns is indicated by a dashed line. Note that the sagging of the beam compels the columns AD and BE to bow outward, because the rigid connections at D and E make the beam and column ends rotate through the same angle. Note also that the reaction components for the bent under the vertical load P are the same in direction as those for an arch.

Stresses in the rigid frame, however, differ from those in an arch because of the presence

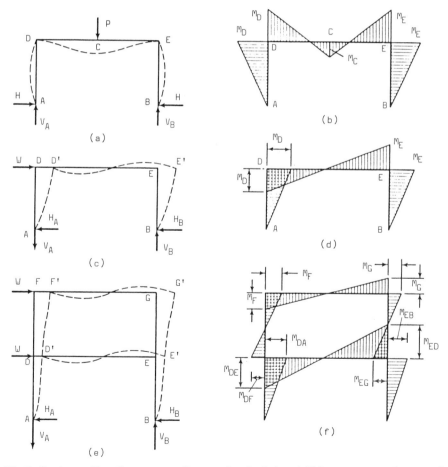

Fig. 8.49. Deflection and bending moment diagrams for single-bay rigid frames, one- and two-stories high, under vertical and horizontal loads.

of large bending moments and shears. The columns under load P are subjected to an axial compression, as in an arch, but are also subjected to a bending moment that increases linearly with distance from the support (see Fig. 8.49b). For example, the bending moment in column AD at any point at a distance y from the support equals $H_A y$, where H_A is the horizontal component of the reaction at A. For equilibrium, the bending moment in the beam at D equals the bending moment in the column at D. Consequently, as shown in Fig. 8.49b, the beam is subjected to negative end moments that reduce the positive moment at midspan from what it would be in a simple beam. (In Fig. 8.49b, bending moments are plotted on the tension side of the centroidal axis of each member.)

Consider now the same rigid frame subjected to a horizontal load W at the level of DE (see Fig. 8.49c). As indicated by the dashed line representing the centroidal axes of beam and columns, the bent sways, or drifts, in the direction of W, restrained by the rigid connections at D and E. This restraint curves the beam and columns. The beam, in particular, has an *inflection point*, or point where curvature reverses, at or near its midpoint.

The reaction components at A are reversed in direction from those at A in Fig. 8.49a. The horizontal component H_A at A must be directed to add to the horizontal component H_B at B to equal W. The vertical components V_A and V_B must form a couple to resist the moment of W about the supports. Hence, V_A must be directed downward, while V_B acts upward. Consequently, in tall, narrow buildings with light gravity loads, the net axial unit stress in windward columns may be tensile. Also, the support must be anchored against uplift to prevent overturning.

Bending moments in the columns vary linearly with distance from the support (see Fig. 8.49d). The bending moments in the beam at D and E are equal respectively to those in the columns at D and E. Because DE is not loaded in this case, the bending moment in the beam varies linearly, passing through zero at the inflection point.

Conditions in a two-story rigid frame (see

Fig. 8.49e) are not greatly different from that in a one-story bent. One important difference is the presence of a bending moment at the base of each second-story column (see Fig. 8.49f). This induces reverse curvature in these columns and reduces the moment at the top of the columns from that for a simple support at the base. For equilibrium, the sum of the column moments at D must equal the beam end moment at D. Similarly, the sum of the column moments at E must equal the beam end moment at E.

8.12. SHELLS AND FOLDED PLATES

Section 8.11 discusses the efficiency of arches, which is due to their curvature. Arches, in effect, are linear structural members, generally lying in a vertical plane. They transmit to the ground or other supports loads carried by other structural members. Greater structural efficiency, however, can be achieved by utilizing curvature of three-dimensional structural members, such as shells and folded plates.

A shell is a curved structure capable of transmitting loads in more than two directions to supports. A thin shell is a shell with thickness which is relatively small compared with its other dimensions (see Fig. 8.50). Such a shell is highly efficient when it is so shaped, proportioned and supported that it transmits loads on it to the supports without bending or twisting.

Shells are defined by the shapes of their *middle surfaces*, halfway between the extrados, or outer surface, and intrados, or inner surface. Thickness of a shell is the distance, normal to the middle surface, between extrados and intrados. Shapes commonly used for thin shells are the dome, often a hemisphere; barrel arch, often a circular cylindrical shell; and hyperbolic paraboloid, a saddle-shaped shell.

Because of their curvature, shells generally have a high rise relative to their spans, as do arches. Hence, like arches, shells are often used for roofs. They are especially efficient for long spans and light loads. They receive loads directly for transmission to the ground or other supports, thus eliminating the need for other structural members to carry loads to them. Be-

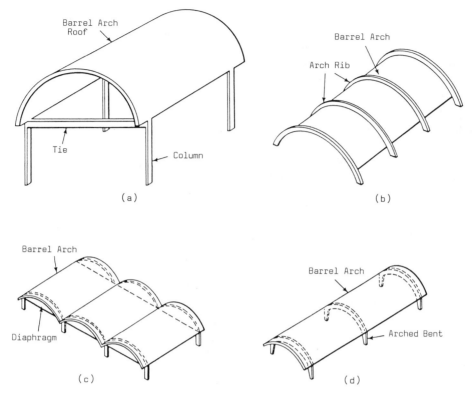

Fig. 8.50. Thin cylindrical shells. (*a*) Barrel arch on columns with ties. (*b*) Ribbed shell. (*c*) Multiple shells. (*d*) Continuous shells.

cause of the curvature of the middle surface, shells usually are subjected only to tensile, compressive and shear stresses, called *membrane stresses*. Consequently, the full cross section is effective in resisting internal forces.

Shell Construction

Because of the structural efficiency of shells, they may be built of almost any rigid material—cold-formed steel, wood, reinforced concrete, plastics. An egg is an appropriate example. It may have a ratio of radius of curvature to thickness of 50 or more.

Thin shells often have a solid cross section, like an egg. Because they are subjected to compression, however, precautions must be taken, in many cases, to prevent buckling or a failure due to punching shear, where concentrated loads have to be supported. Reinforced concrete shells, for example, are usually made just thick enough to provide the minimum cover required for steel reinforcement.

Buckling may be prevented by bracing a shell with deepened sections, called *ribs*. In such cases, the ability of ribs to carry loads to supports may also be utilized to supplement the load-carrying capacity of the shell. Sometimes, however, shells are made corrugated or sharply folded to increase their resistance to buckling.

An alternative method of constructing shells is with skeleton framing and a lightweight enclosure supported by that framing. The framing may lie in a curved surface or it may be a curved, three-dimensional truss or space frame.

Cylindrical Shells

One way to develop a shell conceptually is to start with a roof framed with arches spaced at equal intervals. For generality, assume that the arches are supported on columns, with ties between column tops to resist the arch thrusts. To eliminate the roof framing carrying loads to the arches, each arch may be made so wide that it extends to the adjoining arch on either side.

When this is done, the arches may as well be joined to each other to become a unit. While the arches are widened, they can also be made shallower, because the roof loads are now spread out over the whole surface instead of being concentrated on a linear arch. The resulting structure is a thin cylindrical shell spanning between the original columns (see Fig. 8.50a). It is also known as a *barrel arch*.

In the preceding concept of the development of a cylindrical shell from an arch, it is not necessary to consider all of each arch to be spread out. Part of the material may be left in place as an arch rib, to brace the shell against buckling. Figure 8.50b shows a ribbed cylindrical arch. In the case shown, the ribs support the shell and transmit roof loads to the ground.

Different arrangements of barrel arches may be used. Figure 8.50a shows a single shell. Figure 8.50c shows a roof composed of multiple shells, or barrel arches placed side by side with edges joined. With this type of roof, consideration must be given to drainage of the valleys between the shells and the probable load from snow collecting in the valleys. In the case shown, the shells are supported and tied by solid diaphragms, which are supported by columns. Shells also may be constructed to be continuous over one or more interior supports (see Fig. 8.50d). In the case shown, the supports are arched bents.

Figure 8.50 illustrates only some of the ways in which shells may be supported. Other methods also may be used. For example, shells may be carried to the ground and supported on spread footings.

Membrane Stresses in Cylindrical Shells

Shells are highly statically indeterminate. In the interior of thin shells, however, bending moments and shears normal to the middle surface usually are small and may be ignored. When this is done, the shell becomes statically determinate.

Cylindrical arches, such as the one shown in Fig. 8.50a, may be treated as a beam with a curved cross section. Longitudinal tensile and compressive stresses then may be computed from the flexure formula [see Eq. (8.32)].

Shear stresses may be determined with Eq. (8.42). Tangential compression may be computed by applying the laws of equilibrium to the other stresses.

For circular barrel arches, beam theory yields acceptable accuracy when the ratio of the radius of the shell to the span in the longitudinal direction is less than about 0.25. For larger ratios, more accurate stress analysis is necessary.

Folded Plates

Curved surfaces generally are more expensive to construct than flat surfaces. Consequently, it is sometimes economical to use planar surfaces to approximate the shape of a cylindrical shell. The result may be a folded-plate roof (Fig. 8.51a).

In the case shown in Fig. 8.51a, the roof is composed of five plates. They are shown supported near their ends on solid diaphragms, in turn seated on columns. As in the case of cylindrical shells, however, many arrangements are possible, such as multiple folded plates and continuous folded plates. Also, there may be more or less than the five plates in the example.

Behavior of folded plates resembles that of cylindrical shells. Folded plates, however, are subjected to significant bending moments in both the longitudinal and transverse directions. Consequently, stresses are generally determined by a different method from that used for thin shells.

A common procedure for folded plates is to analyze a transverse strip of unit width (see Fig. 8.51a). This strip is temporarily considered supported at each fold by vertical reactions supplied by the plates (see Fig. 8.51b). The strip is then treated as a continuous beam for determination of transverse bending moments and shears normal to the planes of the plates. The reactions for this condition may be resolved into components in the planes of the plates at each fold. Each plate so loaded now may be treated as a beam spanning between supports in the longitudinal direction, for determination of longitudinal bending moments and shears in the planes of the plates (see Fig. 8.51c). Some modification of the resulting

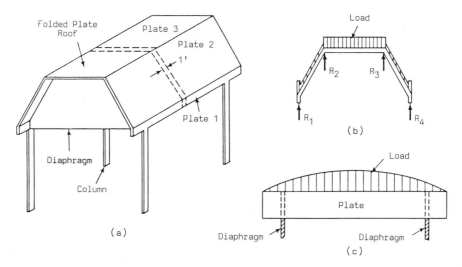

Fig. 8.51. (*a*) Folded plate roof. (*b*) Loaded transverse strip of unit width. (*c*) Loaded longitudinal girder.

stresses generally is necessary to adjust for the assumptions made in this simplified analysis.

Domes

As indicated previously, a shell may be generated by widening arches. Alternatively, a cylindrical shell may be considered generated by translation of the centroidal axis of an arch normal to the plane of the arch. With the use of this concept, a dome may be considered generated by rotation of the arch centroidal axis 180° about the axis of symmetry (see Fig. 8.52*a*). Such shells are also called *shells of revolution*.

Domes have double curvature; that is, they are curved in both horizontal and vertical planes. The double curvature improves the structural efficiency of a thin shell over that of the singly curved cylindrical shell. Use of domes, however, is limited to applications where circular spaces and high roofs are acceptable.

As indicated by the dashed lines in Fig. 8.52*a*, the intersection of the dome middle surface with a vertical plane is called a *meridian* and the intersection with a horizontal plane is called a *parallel*. The highest point on a dome is called the *crown*.

Membrane Stresses in Domes

To determine the membrane stresses in a dome, a tiny element at an interior point P is selected for analysis. For convenience, a set of three-dimensional, rectangular coordinate axes is established at P (see Fig. 8.52*b*). The z-axis is

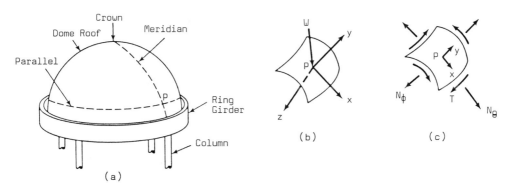

Fig. 8.52. (*a*) Thin-shell dome. (*b*) Coordinate system and load at a point. (*c*) Unit forces at a point.

oriented in the direction of the normal to the middle surface at P. The y-axis is tangent to the parallel through P. The x-axis is tangent to the meridian through P. A load w on the element then may be resolved into components in the x, y and z directions.

The internal forces *per unit of length* acting on the element at P are shown in Fig. 8.52c. For simplification, only the forces on two edges of the element are represented by symbols. These unit forces are shears T, meridional thrust N_θ and hoop stress N_ϕ. They may be resolved into components in the x, y and z directions.

All the forces on the element must be in equilibrium. This condition applied to components parallel to each coordinate axis yields three equations, from which T, N_θ and N_ϕ can be determined. As for cylindrical shells, one equation is algebraic and two equations are differential equations. They may be readily evaluated only for simple shapes and loads.

One simple case is that of a thin shell with constant curvature, or *spherical shell*, subjected to uniform vertical loading w per unit area over the horizontal projection of the shell. For this case:

$$T = 0 \qquad (8.64)$$

$$N_\theta = -\frac{wr}{2} \qquad (8.65)$$

$$N_\phi = -\frac{wr}{2} \cos 2\theta \qquad (8.66)$$

where

r = the radius of the shell

θ = angle between the normal to the surface at P and the shell axis

The solution given for N_θ, with a negative sign, indicates that there is a constant meridional thrust throughout the shell. The hoop forces N_ϕ are compressive in the upper portion of the shell, where θ is less than $45°$, and tensile in the lower portion.

At the base of the shell, reactions must be provided to counteract N_θ and N_ϕ. Usually, a dome is terminated at the base, along a parallel, in a wide, deep structural member, called a *ring girder*, which is designed to resist the hoop tension. The ring girder, in turn, may be seated on a circular wall or on columns capable of resisting the meridional thrust.

When a spherical dome is subjected to a uniform vertical load w over the area of the shell:

$$T = 0 \qquad (8.67)$$

$$N_\theta = -\frac{wr}{1 + \cos \theta} \qquad (8.68)$$

$$N_\phi = wr \left(\frac{1}{1 + \cos \theta} - \cos \theta\right) \qquad (8.69)$$

This type of loading may be imposed by the weight of a shell of constant thickness. In this case, the meridional compression N_θ increases with θ and therefore is larger at the crown than at the base of the dome. The hoop forces convert from compresion in the upper portion to tension in the lower portion at $\theta = 51°51'$. Again, a ring girder is desirable to resist the hoop tension at the base of the dome.

Design Considerations for Shells

Shells can readily accommodate openings for natural lighting and ventilation. For example, a round opening may be provided at the top of a dome. Similarly, openings may be left at the base of a cylindrical shell for use as entrances or windows. These openings, however, are discontinuities in the shell. Provision must be made for resisting the shell forces at the discontinuities or else the membrane theory will not be applicable in those regions of the shells.

The membrane theory does not apply where boundary or discontinuity conditions are incompatible with the requirements of equilibrium or shell geometry, including shell deformations under load, temperature change, shrinkage or creep. Generally, the effects on membrane action of geometric incompatibility at boundaries or discontinuities are significant only in a narrow region about each source of disturbance. Often, the resulting higher stresses can be taken care of by thickening the shell in the effected region or by adding a reinforcing beam. But much larger stresses result from incompatibility with the requirements of equilibrium. Consequently, it is important that reactions along boundaries and discontinuities are equal in magnitude and direction to the shell forces indicated by the membrane theory for those edges.

The ring girder used at the base of a dome to resist hoop tension there is a good example of the reinforcement of a shell to resist reactions at a boundary. Similarly, a stiffening beam is often desirable along the bottom edges of cylindrical shells to resist the shears and tangential thrusts there. If an opening is provided around the crown of a dome, a stiffening ring may be necessary to resist the hoop compression around the discontinuity.

In all these cases, the stiffening member would be much thicker than the shell. The deformations of the stiffening member then would be geometrically incompatible with those of the shell. To reduce the effects of this incompatibility, the shell should be gradually thickened to provide a transition from the typical shell thickness to the thick stiffening member. In general, abrupt changes in shape and dimensions should be avoided in shells.

Disturbances due to equilibrium or geometric incompatibility arise from the imposition of bending and torsional stresses. When shell design cannot eliminate these, provision must be made for the shell to resist them.

Bending theory may be employed for the calculation of such stresses, but the method is complex and tedious. The presence of bending and torsional stresses makes a thin shell highly statically indeterminate. A knowledge of shell deformations is needed to supplement equilibrium conditions in development of differential equations for determination of the unknown forces and moments. Even for simple cases, these equations usually are difficult and time-consuming to solve. Therefore, designers generally try to satisfy the assumptions of the membrane theory to the extent possible and minimize disturbances due to equilibrium or geometric incompatibility.

In computation of stresses in shells for the membrane theory, it is common practice to assume that forces are uniformly distributed over the thickness t of the shells. The unit forces derived from the membrane theory are forces per unit of length. Hence, the unit stresses in a shell are calculated by dividing the unit forces by the shell thickness t.

SECTIONS 8.11 AND 8.12

References

F. Merritt, *Building Design and Construction Handbook*, 4th ed., McGraw-Hill, New York, 1982.

E. Gaylord and C. Gaylord, *Structural Engineering Handbook*, 2nd ed., McGraw-Hill, New York, 1982.

R. White and C. Salmon, *Building Structural Design Handbook*, Wiley, New York, 1987.

C. Wilby and I. Khwaja, *Concrete Shell Roofs*, Wiley, New York, 1977.

Words and Terms

Arch: barrel, fixed, three-hinged, Tudor, two-hinged
Dome
Folded plate
Membrane stresses
Rigid frame
Ring girder
Shell

Significant Relations, Functions, and Issues

Rise-to-span ratio of arches related to stress in the arch and thrust at supports.
Hinged conditions in arches related to structural behavior.
Bracing of arch systems.
Rigid-frame behavior versus pinned-frame behavior.
Beam, arch and membrane actions in shells and folded plates.
Meridional thrust and hoop stresses in domes.

8.13. CABLE-SUPPORTED ROOFS

For structural purposes, a cable is a structural member with high resistance to tensile stresses and no resistance to compression or bending. Consequently, cables are used to sustain tensile loads. Because steel has high tensile strength, cables usually are made of steel, as discussed in Sec. 8.5.

Common applications of cables in buildings include use as hangers, to support gravity loads; ties, to prevent separation of building components, such as the bases of arches or rigid frames; and roof supports. This section deals only with the last type of application.

Types of Cable Structures

Cable-supported roofs may be classified basically as cable-supported, or cable-stayed, framing or cable-suspended roofs.

In a *cable-stayed structure*, the roof is supported directly on purlins and girders. One or more cables are used to assist each girder in carrying the roof loads.

Figure 8.53a shows one way to obtain a long span that has often been used for airplane hangars. The construction is much like that for a cantilever truss. The major difference is that in the cable-stayed roof the bottom chord is a girder subjected to bending moments and axial compression, whereas in the truss, the bottom chord carries only axial compression. The girder is supported at one end on a column and near the other end by a cable, which extends on an incline from a mast seated atop the column. The cable usually is continued past the mast to an anchor. In Fig. 8.53a, the cable extension, or cable stay, is anchored to a rear column. This column must be capable of resisting the uplift imposed by the cable. The tension in the cable makes it nearly straight. (There may be a slight sag due to the weight of the cable.) For long girder spans, or for heavy loads, another cable may be extended from the mast to support the girder at a second point.

The cable-stayed girders usually are placed parallel to each other at intervals along the length of the building. Roof loads are transmitted to them by purlins. In a hangar, a sliding door is installed under the end of the cantilever.

Figure 8.53b shows a cross section through a building with similar cable-stayed girders. In this case, however, the girders cantilever on both sides of a pair of columns. The cross section is typical of two cases. One case, like the construction in Fig. 8.53a, applies to girders set parallel to each other. The second case applies to a circular roof. The columns are set around an inner ring. The girders and their cables are placed radially.

In this type of construction, vertical loads impose bending moments in the girders. The

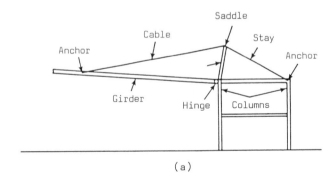

(a)

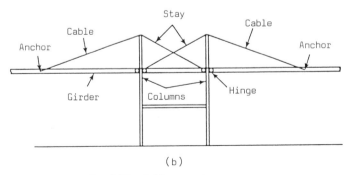

(b)

Fig. 8.53. Cable-stayed structures.

vertical reactions of the girders are supplied by the columns and the inclined cables. One vertical reaction therefore equals the vertical component of the cable stress. The girder, in turn, has to resist the horizontal component of the cable stress and transmit it to the column. The horizontal force may then be counteracted by the horizontal component of the cable stay and by bracing or rigid-frame action with the second column. Maximum unit stresses in the girder may be computed from Eq. (8.40).

In a *cable-suspended structure*, roof loads usually are supported directly on one or more cables. The cables may lie in a single or double surface, in either case with single or double curvature. The cables may be used singly, set in parallel or near parallel vertical planes or placed radially between concentric rings. Or the cables may form a net, crossing each other at intervals.

Figure 8.54a shows one way of supporting a roof with cables used singly. The cable extends on a curve between two masts. The roof may consist of panels set directly on the cables or suspended from them. If the cables are set in parallel vertical planes with supports at the same level, the resulting surface has single curvature.

In Fig. 8.54b, single cables extend radially between an inner tension ring and an outer compression ring. The latter is shown supported on columns. The tension ring is a device for interconnecting the radial cables at the center of the roof, to avoid a common intersection for numerous cables, which would be massive and clumsy. In effect, the tension ring floats in space. The compression ring provides the necessary reactions around the circumference of the roof. The compression is imposed by the inward pull of the cables under the roof gravity loads. The roof, being dish-shaped, has double curvature.

A pair of cables forming part of a double-surfaced roof is shown in Fig. 8.54c. If the cable pairs are set in parallel vertical planes with supports at the same level, the surface has single curvature. The upper, or primary cable, of each pair is the main load-supporting member. It is kept at fixed distances from the secondary cable by compression rods, or struts. These are

shown vertical in Fig. 8.54c, but they also may be inclined, as in a Warren truss. The purpose of this type of construction is to make the roof construction more resistant to vibrations than single-surface roofs.

A cable used singly is very flexible. Under dynamic loads, it can develop large or rapid vibrations, which may damage the roof. Such vibrations are unlikely to occur, however, if the cable were to be forced to move in unison with a second cable that has a different natural period of vibration. This is the principal reason for the use of double-surface cable roofs.

Figure 8.54d shows a double-surface roof with radial cables. This roof has double curvature.

Figure 8.54e shows a single-surface, cable-net roof with double curvature. This type of construction also has greater resistance to vibrations than cables used singly. In the structure illustrated, the primary cables are strung between arches at opposite ends of the roof. Cross cables are curved in the opposite direction to that of the primary cables. The resulting surface is saddle shaped. The secondary cables are anchored to boundary cables, also strung between the end arches.

Cable Reactions and Stresses

The basic principles of cable action may be illustrated with the simple example of a cable strung between two supports at the same level and subjected to vertical loads.

The cable offers no resistance to bending. Consequently, its shape is determined by the loads imposed. Figure 8.55a to c show the shape taken by a weightless cable as a concentrated load P is shifted in succession from one quarter point of the span to the next one on the right. Figure 8.55d shows the curve adopted by a cable under distributed vertical loading.

The reactions at supports A and B in Fig. 8.55d are inclined. For equilibrium, the horizontal component H at A must be equal in magnitude to the horizontal component at B but oppositely directed. The vertical component V_A at A can be determined by taking moments about B of all the loads and setting the moments equal to zero. The moment of H is zero, because its

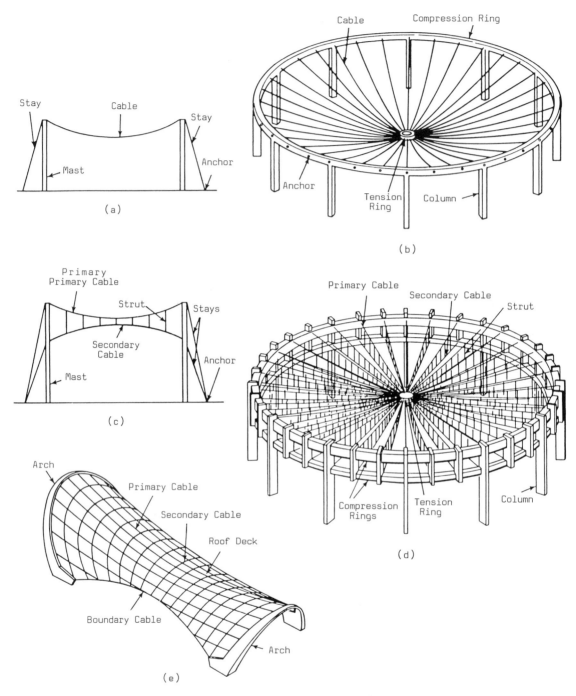

Fig. 8.54. Cable-suspended structures. (*a*) Single-surface cable roof with single curvature. (*b*) Single-surface roof with double curvature. (*c*) Cross section through double-surface cable roof. (*d*) Double-surface cable roof with double curvature. (*e*) Single-surface, saddle-shaped roof.

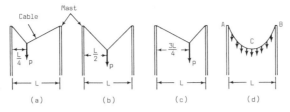

Fig. 8.55. Cable shape varies with load location and distribution.

line of action passes through B. The vertical component V_B can be obtained by setting moments about A equal to zero.

H can be determined from the condition that the bending moment in the cable is zero at every point. At the lowest point C of the cable, the cable stress equals H (see Fig. 8.56). This stress and the reaction component H form a couple Hf, where f is the sag of the cable below the supports. Since the bending moment is zero at C, $Hf = M_C$, where M_C is the bending moment due to the vertical forces. (M_C is the same bending moment that would be present at the corresponding location in a simple beam with the same span L as the cable and with the same loading.) Hence,

$$H = \frac{M_C}{f} \qquad (8.70)$$

For example, for a load w uniformly distributed over the horizontal projection of the cable, $M_C = wL^2/8$ and

$$H = \frac{wL^2}{8f} \qquad (8.71)$$

Because all the loads are downward-acting vertical forces, the maximum stress in the cable

occurs at the supports and equals the reaction:

$$R = \sqrt{H^2 + V_A^2} \qquad (8.72)$$

Cable Shapes. The shape of the cable can be determined by setting equal to zero the bending moment at any point D on the cable at a horizontal distance x and vertical distance y from the low point C. Because there are only vertical loads on the cable, the horizontal component of the cable stress equals H throughout. At D, this component forms with the horizontal component of the reaction a couple $H(f - y)$. For the bending moment at D in the cable to be zero, $H(f - y) = M_D$, where M_D is the bending moment due to the vertical forces. (M_D is the same bending moment that would be present at the corresponding location in a simple beam with the same span as the cable and with the same loading.) Solution for $f - y$ yields

$$f - y = \frac{M_D}{H} \qquad (8.73)$$

Since H is a constant, the vertical coordinates of the cable with respect to the line between supports is proportional to the simple-beam bending moments. Consequently, a cable subjected to loads assumes the same shape as the bending-moment diagram for the same loads acting on a simple beam of the same span as the cable. It follows from this conclusion that the low point of the cable occurs at the point of maximum bending moment and therefore also at the point where the shear is zero.

Substitution in Eq. (8.73) for H from Eq. (8.70) and solution for y gives

$$y = f - \frac{M_D}{H} = f - \frac{M_D}{M_C}f = f\left(1 - \frac{M_D}{M_C}\right)$$

$$(8.74)$$

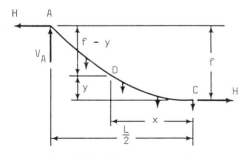

Fig. 8.56. Half-span of a cable.

Equations for cables with various types of loading can be determined from this equation. For example, for a load w uniformly distributed over the horizontal projection of the cable, $M_D = (L^2/4 - x^2) w/2$ and $M_C = wL^2/8$. The cable then is a parabola:

$$y = 4fx^2/L^2 \qquad (8.75)$$

For a load w uniformly distributed along the length of the cable, as is the case for cable weight, the shape is a *catenary*:

$$\begin{aligned} y &= \frac{H}{w} \left(\cosh \frac{wx}{H} - 1 \right) \\ &= \frac{wx^2}{2H} + \frac{w^3x^4}{24H^3} + \cdots \end{aligned} \qquad (8.76)$$

In Eq. (8.76), y is expressed in the alternative form obtained by substituting an infinite series of terms equivalent to the hyperbolic cosine. Terms after the first in the infinite series in Eq. (8.76) are usually very small. If those terms are ignored, the cable shape approximates a parabola. Because it is difficult to compute cable stresses for the load uniformly distributed along the cable length from a catenary equation, because of the hyperbolic functions, a parabola may be conveniently substituted for the catenary, generally with little error.

More complicated cable structures often require that changes in cable shape and sag with addition of live and other loads to dead loads be taken into account. Also, the possibility of damaging vibrations under variations in load must be investigated.

8.14. PNEUMATIC STRUCTURES

Section 8.12 points out that shells are highly efficient structurally and can be made thin because:

1. Loads are spread out over large areas.
2. Curvature enables the entire cross section to be effective in resisting stresses, in the absence of bending and twisting.

One way that the efficiency of shells can be improved, however, is to reduce the loads they have to carry. The weight of a shell is a high percentage of the total load imposed. It follows, therefore, that if a shell can be constructed of a very light material, such as fabric, the load to be supported would be reduced substantially so that the shell can be made very thin, virtually a true membrane.

Shells, however, must be capable of resisting compression. If they are made very thin, they would buckle and collapse. Hence, for a membrane to become structurally useful, compressive stresses must be eliminated from it. One way to achieve this objective is to let a membrane hang freely from supports around its edges, like a dish-shaped, single-surface cable net. Then, all the stresses would be tensile. This principle has been widely used for tents. Another method that appears to have greater potential for economical, long-span enclosures for buildings is use of prestress.

Prestressed Membranes

The application of prestress was discussed previously for concrete. For that material, prestress is used to overcome the weakness of concrete in tension. Prestress, however, can also be used to counteract the inability of a membrane to sustain compression. If a pretension were applied larger than any compressive stress that loads are likely to impose, then compressive forces would merely reduce the tensile stresses and buckling could not occur.

In practice, membrane roofs have been prestressed by pressurization. The construction is controlled and stabilized by application of pressure differentials achieved by the uniform loading actions of air or other gases, liquids or granular solids. An *air-stabilized roof*, in effect, is a membrane bag held in place by small pressure differentials applied by environmental energy.

Materials for Structural Membranes

A prime structural requisite for a structural membrane is that it should have high tensile and shear strength, so that it will be tear-resistant. It should have a high modulus of elasticity, to avoid excessive deformations; yet, it should be flexibile. It should have high fatigue resistance. In addition, for structural efficiency, it should have a large strength-to-weight ratio.

To prevent leakage of air that creates the pressure differentials acting on the membrane, the material should be airtight. Joints that may be necessary also should be airtight and at least as strong as the basic material.

In addition, a membrane should possess properties that are required of other roof materials. For example, a membrane should be durable, water and chemical-resistant and incombustible. It should provide good thermal insulation. Furthermore, its properties should be stable despite climatic changes and passage of time.

Four basic types of material have been used for membranes: coated fabrics, plastic films, woven metallic fabrics and metal foils. Coated fabrics have been used most frequently. They are generally made of synthetic fibers, such as nylon and Dacron. Coatings, which may be applied to one or two sides of the fabric, usually are relatively impervious materials, such as vinyl, butyl, neoprene and Hypalon. Weight generally ranges between 0.1 and 0.5 lb per sq ft.

When consideration is being given to use of an air-stabilized roof, the potential service life of membrane materials should be investigated. Some of the materials have a considerably shorter service life than many of the materials ordinarily used for roofs, and replacement costs, including installation and facility shutdown costs, may be high.

Types of Pneumatic Construction

Shapes of air-stabilized roofs often resemble those frequently used for thin-shell construction. Spherical and hemispherical domes (see Fig. 8.57a) are quite common. Semicircular cylinders with quarter-sphere ends also are often used (see Fig. 8.57b). But a wide variety of shapes have been and can be used. Figure 8.57c illustrates a more complex configuration.

Air-stabilized roofs may be classified as air-inflated, air-supported or hybrid structures.

In an *air-inflated structure*, the membrane completely encloses pressurized air. There are two main types.

Inflated-rib structures are one type. They consist of a framework of pressurized tubes that support a membrane in tension. The tubes serve much like arch ribs in thin-shell construction. The principle of their action is demonstrated by a water hose. When the hose is empty, it will collapse under its own weight on short spans or under light loads normal to its length; but when it is filled with water, the hose becomes stiff. The water pressure tensions the hose walls, significantly improving their ability to sustain compression.

Inflated dual-walled structures are the second type of air-inflated construction. These structures comprise two membranes tensioned by the air pressure between them (see Fig. 8.58a). The membranes are tied together by drop threads and diaphragms. Because of the large volume of air compressed into dual-walled structures, they can economically span larger distances than can inflated-rib structures.

Even though the membranes used for air-inflated structures are fairly impervious, provision must be made for replenishment of air. Some leakage is likely to occur, particularly at joints. Also, air pressure changes with variations in temperature inside and outside the building. Consequently, air must be vented to

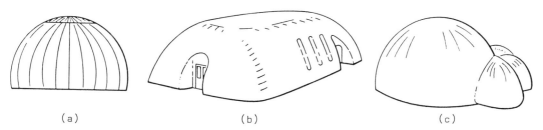

(a)   (b)   (c)

Fig. 8.57. Some shapes for air-supported structures.

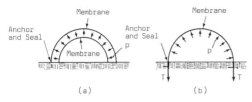

Fig. 8.58. (a) Inflated dual-walled structure. (b) Air-supported structure.

relieve excessive pressures, to prevent overtensioning of the membrane, and air must be added to restore pressure when reductions occur, to prevent collapse.

An air-supported structure consists of a single membrane that is supported by the excess of the internal pressure in a building over exterior atmospheric pressure (see Fig. 8.58b). The pressure differential produces tension in the membrane and pushes it upward. To resist the uplift, the air-supported structure must be securely anchored to the ground. Also, to prevent leakage of air, the membrane must be completely sealed all around its perimeter.

Note that if loads, including membrane weight, were uniform and completely balanced by the internal pressure, there would be no pressure differential. There would be no tension in the membrane and no uplift forces requiring anchorage. The membrane would just be a medium serving merely to separate the building interior from the outside environment. In that case, the air-supported structure theoretically could span over an enormous space. Actually, however, loads are never uniform over the whole area of a large membrane. Hence, pressure differentials large enough to prevent compressive stresses in the membrane must be applied to the membrane.

In practice, small pressure differentials are used. Often, they are in the range of 0.02 to 0.04 psi (3 to 5 psf) above atmospheric pressure.

While there may be some air leakage through the membrane, a more important source of air loss is the entrances and exits to the building. These losses can be minimized with the use of air locks and revolving doors. Nevertheless, provision must be made for continuous replenishment of the air supply with blowers to maintain the pressure differential.

Bubble Analogy

A soap film is a natural analogy to an air-supported structure. Surface-tension forces determine the shape of a bubble. The membrane is stressed equally in all directions at every point. As a result, the film forms shapes with minimum surface area, often spherical. Because of the stress pattern in soap films, any shape that can be obtained with them is feasible for an air-supported structure. Figure 8.57c shows a configuration formed by a group of bubbles as an illustration of a shape that can be adopted for an air-supported structure.

Membrane Stresses

When a spherical membrane with radius R, in., is subjected to constant radial internal pressure p, psi, the unit tensile force, lb per in., is given by

$$T = \frac{pR}{2} \qquad (8.77)$$

In a cylindrical membrane, the unit tensile force in the circumferential direction is

$$T = pR \qquad (8.78)$$

where R is the radius, in., of the cylinder. The longitudinal membrane stress depends on the conditions at the cylinder ends. With immovable end abutments, for example, the longitudinal stress would be small. If the abutments were not fixed to the ground, however, a tension about half that given by Eq. (8.78) would be imposed in the longitudinal direction in the membrane.

Hybrid Structures

The necessity of providing a continuous air supply and access to a building against a pressure differential is a significant disadvantage of pure pneumatic construction. An economical alternative is to use light metal framing, such as cables or cable nets, to support and tension a membrane under light loads. Pressurization may then be employed to supplement the framing under heavy wind and snow loads.

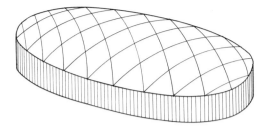

Fig. 8.59. Pneumatically stabilized membrane surface with restraining cables.

Cable-Restrained Membranes

Use of air-supported membranes often results in buildings with a high profile and a general bulbous look. A method for holding down the high arching profile of a pneumatically stabilized membrane is to employ a net of restraining cables, as shown in Fig. 8.59. In this case a perimeter structure—usually a form of compression ring—is used to anchor the edges of an inflated membrane, and is also used to anchor a set of cables that are tensioned by the inflation forces; thus restraining the membrane from developing its natural inflated profile. Major spans with a relatively low profile have been achieved with this system.

SECTIONS 8.13 AND 8.14

References

H. Irvine, *Cable Structures*, MIT Press, Cambridge, MA, 1981.

F. Merritt, *Building Design and Construction Handbook*, 4th ed., McGraw-Hill, New York, 1982.

P. Drew, *Tensile Architecture*, Granada, St. Albans, England, 1979.

R. Dent, *Principles of Pneumatic Architecture*, Architectural Press, London, 1971.

M. Salvadori, *Structure in Architecture*, Prentice-Hall, New York, 1986.

Words and Terms

Cable

Cable structures, types: cable-stayed, cable-suspended, cable net, cable-restrained

Pneumatic structures, types: air-inflated (ribbed, dual-walled), air-supported

Significant Relations, Functions and Issues

Structural use of tension for spanning structures: cables, cable nets, draped and stretched membranes, pneumatically stabilized membranes.

Support and anchoring systems for tensile structures.

8.15. HORIZONTAL FRAMING SYSTEMS

Structural roofs and floors are systems, but are subsystems of floor-ceiling or roof-ceiling systems, which, in turn, are part of the building system. Consequently, design of structural roofs and floors must take into account their effects on other systems and the effects of other systems on them.

An enormous number of different structural floor and roof systems have been and still are in use. It is not feasible, therefore, to describe and compare more than a few commonly used systems in this section which deals only with substantially horizontal, flat floor and roof systems. Steeply sloped and curved roof systems, however, are discussed in Sec. 8.9 and Secs. 8.11 to 8.14.

Decks

Prime components of a floor or roof are materials that serve as a wearing or weathering surface, or both, and a supporting, or structural, material. A structural material is necessary because the wearing or weathering material can be very thin. In some cases, the structural material may also be the wearing or weathering material. The structural material is often referred to as a deck or, in the case of concrete, a slab.

Being horizontal and flat, a deck is subjected to vertical dead and live loads, which tend to bend and twist it. (A deck may also serve as a horizontal diaphragm as part of the lateral-load-resisting structural system.) The bending and twisting limit the distance that the deck can span between supports. Spacing of supports depends mainly on the loads to be carried and the structural characteristics of the deck. Heavy loads or a weak or thin deck generally require closely spaced supports.

Decks may be made of any of a wide variety of materials. They are discussed in Chaps. 15

and 16. For this section, it is desirable to note that flat roof decks generally consist of a waterproof membrane fastened to a structural material, such as concrete, plywood, cold-formed steel or gypsum. Floor decks usually consist of a wearing surface, such as carpeting, concrete, wood, linoleum or asphalt, plastic or ceramic tile, and a structural material like those listed for roof decks.

Types of Horizontal Framing

In load-bearing construction, the prime vertical supports are walls. These must be spaced far enough apart to meet the architectural or functional requirements of the building. Sometimes, wall spacing falls within a range where it is economical to employ a deck that can span between the walls. Often, however, the most economical construction consists of a thin deck placed on beams that span between the walls. Spacing of the beams can be made appropriate to the loads to be carried and the structural characteristics of the deck.

Similarly, in skeleton framing, columns must be spaced far enough apart to meet the architectural or functional requirements of the building. This spacing may lie within a range where it is economical to employ a deck that can span, without additional supports, between the columns. But often it is more economical to use a thin deck and support it on beams. The beams, in turn, may span between columns, but usually only the edge beams of a structural bay do so, while the interior beams are supported on transverse girders that span between the columns.

In any case, it is convenient to classify decks as one way or two way. One-way decks develop, under vertical loads, curvature in only one direction due to bending. Loads are transmitted by such decks only in the plane of curvature to the deck supports. Two-way decks develop, under vertical loads, curvature in two perpendicular directions due to bending. Loads therefore are transmitted by such decks in more than one plane to the deck supports.

Factors in Selection of Horizontal Framing

Many things must be considered in selection of the structural floor or roof system for a build-

ing. Spacing of supports and magnitude of loads are major factors, as are architectural and functional considerations. Interaction of the structural system with other systems also must be taken into account. Cost of the system, erected, is important, too. Bear in mind, though, that the lowest-cost structural system may not always yield the lowest-cost building.

Other factors that should be considered include fire resistance, depth of structural system and weight. Often, the deeper the horizontal structural system, the higher the building must be. The greater the height, the more costly become the walls and, for high-rise buildings, the piping, air-conditioning ducts and electrical risers. Also, the heavier the floor framing, the more costly the columns and foundations become.

Two-Way Concrete Slabs

When conditions are propitious for their use, flat-plate floors and roofs (see Fig. 8.60a) offer many advantages over other types of framing. This type of construction generally employs cast-in-place, reinforced concrete. Sometimes, however, prestressed concrete or precast concrete is used to achieve greater economy. The concrete may be made with ordinary stone aggregate, to obtain high strength for long spans or heavy loads, or with lightweight aggregate, for lower weight.

The flat plate is usually supported on columns arranged to form rectangular bays, but other column layouts are feasible. Thickness of the plate is kept uniform throughout each bay (see Fig. 8.60a). This simplifies the formwork for the concrete and helps keep construction costs low. Without beams, lighting, from windows or ceiling fixtures, is unobstructed. Often, the underside of the flat plate can be plastered and painted, or simply painted, to serve as a ceiling. Piping and electrical conduit may be incorporated in the plate by installing them on the formwork before concrete is cast (see Fig. 8.61).

Under vertical loads, a flat plate, developing two-directional curvature, assumes a dish shape. When column spacing is not equal in the perpendicular directions or when loads are unsymmetrically placed in a bay, twisting occurs at

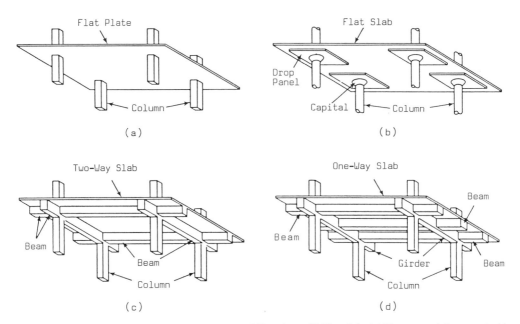

Fig. 8.60. Horizontal concrete framing systems. (*a*) Flat plate. (*b*) Flat slab. (*c*) Two-way slab supported by edge beams. (*d*) One-way slab supported by parallel beams; beams supported by column-line girders.

Fig. 8.61. Rebars and electrical conduit set on formwork, ready to receive concrete for a flat slab. (Courtesy Symons Manufacturing Co.)

the corners. Hence, the plate transmits loads to the columns through a combination of bending and torsion.

The structural action makes flat plates highly efficient for loads and spans such as those encountered in residential buildings. Consequently, for short spans and light loads, thickness of reinforced concrete flat plates may be as small as 5 in. Thus, because of the small depth of the horizontal framing, flat-plate construction is advantageous for high-rise buildings. Some of this advantage is lost, however, when lighting fixtures are set on the floor underside or air-conditioning ducts are run horizontally under the floor and especially when they have to be hidden above a ceiling.

Because of continuity, bending moments are larger at the columns than at the center of each bay. (The moments near columns are negative, producing tension at the top of the slab, thus requiring reinforcing steel to be placed close to the top.) Also, maximum shear occurs at the columns. To increase the capacity of flat plates, it becomes necessary in some cases to provide a *column capital*, an enlargement of a small length of column near its top (see Fig. 8.60*b*). For longer spans or heavier loads, flat plates, in addition, may be thickened in the region around the columns. This type of construction is called flat slab.

Flat slabs are flat plates with an abrupt thickening, called a *drop panel*, near columns (see Fig. 8.60*b*). The ACI Code requires that a drop panel be at least 25% thicker than the slab thickness elsewhere. Sides of the panel should have a length at least one-third of the column spacing. Design of flat slabs is similar to that for flat plates.

Note that in the center portions of flat plates and flat slabs, bending moments are positive, producing tension at the bottom of the slabs, thus requiring steel to be placed near the bottom. This steel is usually uniformly spaced in two perpendicular directions.

More efficient use may be made of the concrete, however, if larger but fewer bars are used at much larger spacing and the concrete in tension between the bars is omitted. The remaining concrete then forms a thin slab with ribs in two perpendicular directions incorporating the rebars. The underside of the slab has the appearance of a waffle.

Waffle flat slabs therefore consist, in the middle portion, of a thin, two-way top slab spanning between a grid of concrete joists in perpendicular directions. The joists terminate at drop panels, which may have a thickness equal to or greater than the depth of the joists. Waffle slabs are generally constructed by casting concrete over square dome forms with lips around the perimeter, so that concrete placed between the domes becomes the joists. Formwork may be more expensive for waffle slabs than for flat slabs, but for a given volume of concrete, waffle slabs can carry heavier loads or span longer distances than can flat slabs.

Another way to improve the capacity of flat slabs and flat plates is to make them thick but hollow. *Hollow*, or *cellular, slabs*, in effect, consist of a top and bottom slab connected by ribs.

Two-way slabs provide still another way of improving the capacity of flat plates. In this type of construction, a slab with constant thickness is supported on beams that span between the columns (see Fig. 8.60*c*). The design procedure is similar to that for flat slabs and flat plates, except that the structural characteristics of the beams are taken into account. As an alternative, walls may be substituted for the beams and columns.

One-Way Decks

The structural behavior of the framing changes significantly if one or more beams are placed within a bay in either or both directions normal to the beams in Fig. 8.60*c*.

A grid of beams placed in perpendicular directions is often economical for very large column spacing. (For roof construction, trusses often are used instead of beams in such cases.)

For more usual column spacings, it is often economical to place intermediate beams in only one direction (see Fig. 8.60*d*). Ends of these beams are supported on girders spanning between the columns or on load-bearing walls.

With the type of construction illustrated in Fig. 8.60d, the span of the deck is much shorter in the direction normal to the beams than the span in the direction parallel to the beams. Consequently, under vertical loads, the deck may be classified as one way, because there is very little curvature due to bending in the direction parallel to the beams. Load may be considered transmitted only in the direction normal to the beams. One-way slabs therefore may be designed by treating a unit-width strip as a beam spanning in the short direction.

Spacing of beams and hence the number of beams per bay depend on the width of bay and the type of deck selected. Often, for economic reasons, the spacing is chosen close to the maximum span with the minimum thickness permitted for the deck material. For example, if the deck is made of reinforced concrete and the local building code requires that a one-way concrete slab have a thickness of at least 4 in., the most economical beam spacing usually is the maximum distance that the 4-in. slab can span with the design loads. This spacing may be about 8 ft for lightweight concrete in a high-rise office building. If the deck is plywood, joists or rafters may be spaced 16 to 48 in. center-to-center, depending on the strength and thickness of the plywood (see Fig. 8.62). This type of construction often is used in one- and two-story houses.

The added depth of framing when beams are used is not a complete disadvantage. The space between beams can often be utilized for useful

Fig. 8.62. Plywood deck is nailed to rafters spaced 24-in. center-to-center for a roof. (Courtesy American Plywood Association)

purposes. For example, the space between rafters can be filled with insulation, for which space would be needed in any case. Often, air-conditioning ducts can be run in the spaces between beams and can be hidden behind a ceiling extending between the bottom flanges of the beams. Figure 8.63 illustrates horizontal framing with a cold-formed steel deck supported on steel beams, with a duct paralleling the beam on the right. (Note that a fire-resistant material has been sprayed on the steel beams and deck, whereas the steel columns have been fire protected with concrete enclosure.) If ducts have to be run transverse to the beams, however, holes have to be cut for passage of the ducts. The holes usually are made in the beam webs, because the web bending stresses are lower than those in the flanges. Also, the holes should be formed where bending moments are small, to minimize the loss of strength from removal of material. The perimeter of the hole often has to be reinforced to compensate for the loss of material. When open-web joists or trusses are used instead of beams, however, they offer the advantage that ducts and piping can be passed through the spaces between web members.

Connections in Horizontal Framing

A chain is no stronger than its weakest link. The weakest links in the chain of structural members transmitting loads from points of application to foundations are likely to be the joints, or connections. These must be capable of transmitting between connected members the largest loads the members may impose on each other. Also, the connections must be made without causing undesirable changes in the strength or stiffness of the members joined. In addition, at joints, deformations of the ends of members must be geometrically compatible and should conform closely to assumptions made in design of the structural framing. For example, connections between the ends of simple beams and supporting members should be capable of transmitting the reaction, resisting end shears and allowing the beam end to rotate freely. In contrast, where continuity of a beam is desired, a rigid connection should be provided, to insure that

Fig. 8.63. Cold-formed steel deck supported on steel beams, with sprayed-on fireproofing. (Courtesy H. H. Robertson Co.)

all connected members will rotate through the same angle.

Wood Connections. The type of connections used for wood framing depends, in addition to the preceding conditions, on the size of members to be joined. Thin pieces, for example, may be nailed, screwed or glued to other members. Thick or heavily loaded members should be joined with more substantial fasteners. A few of the many fastening methods for such joints, in addition to those described in Section 8.5, **Wood**, are illustrated in Fig. 8.64.

One method of supporting a wood beam on a wall is shown in Fig. 8.64a. The beam is held in position by a bolt through a pair of clip angles. Each angle in turn is secured to the wall

with an anchor bolt, embedded in the masonry. A steel bearing plate under the beam end distributes the beam reaction to the masonry and may be set in mortar, if necessary, to bring the beam end to the required level.

Figure 8.64b illustrates the use of a steel bent-strap hanger to support the end of a purlin at a wood beam. A steel tie strap is nailed to the tops of the purlin and beam and to a purlin on the opposite side of the beam.

For a heavily loaded wood beam framed into a wood girder, the connection may be made with a pair of steel angles bolted to both members (see Fig. 8.64c). A bolted connection with one or two steel angles also may be used for purlins or beams seated atop a girder (see Fig. 8.64d). Bolted angles also may be employed for connecting beams or girders below the tops of columns.

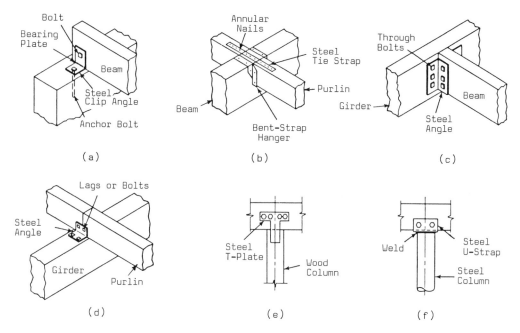

Fig. 8.64. Connections for wood beams.

Figure 8.64e and f show two methods for seating wood beams atop columns. In Fig. 8.64e, two beams are bolted to the wood-column top through a pair of steel T plates. In Fig. 8.64f, a steel U strap is welded to the top of a steel pipe column. The wood beams are seated in and bolted to the U strap.

Steel Connections. Steel members usually are connected to each other with bolts or welds. (In older buildings, rivets were generally used, but now riveted joints are more expensive.) For reasons of economy and quality control, steel parts are often welded in the fabricating shop and assembled with bolts in the field. Sometimes, however, both shop and field welding are economical, and field welding offers such advantages as savings in connection material and quieter operations. Bolts are tightened with impact wrenches, which are noisy.

Steel beams supported on walls usually are seated on a bearing plate, like the wood beam in Fig. 8.64a. The plate distributes the beam reaction to the masonry to prevent crushing.

Simple-beam connections between steel members may be classified as framed or seated.

Figure 8.65a illustrates one type of framed connection made with bolts. In the case shown, each beam is delivered to the building site with a pair of steel angles bolted to the web at the end to be connected to the girder. The outstanding legs of the angles are field bolted to the girder. Also, in the case shown, the beams frame into the girder with tops of all members at the same level. To prevent the top flanges of the beams and girders from interfering with the connection, the beams are notched, or coped, to remove enough of the flanges and webs to clear the girder flange. Framed connections generally require less steel than seated connections.

Examples of seated connections are shown in Fig. 8.65b and c. Seated connections often are used where there is insufficient clearance for framed connections. For example, seated connections are used for connections of girders to column webs to make erection of the girders between the column flanges easier. Seated connections, in general, however, are helpful in erection. They provide support for members to be connected while holes for field bolts are aligned and while bolts are installed or welds made.

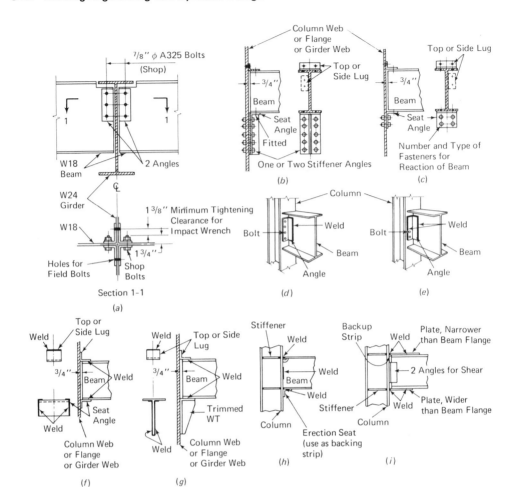

Fig. 8.65. Connections for steel beams.

A stiffened seat (see Fig. 8.65b) may be used when loads to be carried exceed the capacities of the outstanding legs of standard unstiffened seats (see Fig. 8.65c). An unstiffened seat consists only of an angle with one leg shop bolted to the supporting member and the outstanding leg horizontal, for field bolting to the seated member. The seat is designed to carry the full beam reaction. A top or side lug angle is also bolted to the connecting members to provide lateral support (see Fig. 8.65b and c). In a stiffened connection, additional support is provided to the outstanding leg of the seat angle. This may be done by bolting a pair of angles or a WT (W shape split by a cut through the web) to the supporting member so that the seat bears against the outstanding legs of the stiffening angles or WT web (see Fig. 8.65b).

Framed connections also may be welded. Figure 8.65d shows a case where the framing angles are shop welded to a beam and field bolted to a column flange. Figure 8.65e illustrates a framed connection that is both shop and field welded. A few temporary field bolts are used to secure the framing angles to the column flange until the field welds can be made between that flange and the angles.

Similarly, seated connections may be welded. Figure 8.65f illustrates an unstiffened welded seat connection and Fig. 8.65g, a stiffened welded seat connection. The supporting member is delivered to the site with the seat angle welded to it. The top or side lug is welded to the connecting members in the field to prevent interference with erection of the seated member.

Two types of **rigid connections,** often used when continuity of beams or rigid-frame action to resist lateral forces is desired, are shown in Fig. 8.65*h* and *i*. The connections illustrated are welded. In Fig. 8.65*h*, the beam is supported on an erection seat during welding. The seat also serves as a backup strip for the weld between the beam bottom flange and the column flange. Both beam flanges are welded directly to the column flange, to resist the end bending moment in the beam, while the beam web is welded to the column flange to resist the end shear. The type of connection shown in Fig. 8.65*i* may be used for larger bending moments and shears. Plates are welded to the top and bottom beam flanges and to the column flange, to resist the bending moment, while a pair of steel angles is welded to the beam web and column flange, to resist the end shear. In both types of connection, horizontal steel plates are welded between the column flanges at the level of the beam flanges to brace the column flanges against the forces imposed by beam bending.

Concrete Connections. In any consideration of concrete connections, a distinction must be made between joints between cast-in-place members and those between precast members. In cast-in-place construction, concrete for members at a joint can be cast continuously so that the members are monolithic, or integrated.

Reinforcement from each member is extended through a joint into the other members, to tie them together with steel (see Fig. 8.66). Often, the deck and beams for a floor or roof are cast simultaneously with the columns of the story below. Reinforcement for those columns are extended vertically above the deck to dowel the columns for the next story (see Fig. 8.66).

Precast-concrete joints require careful design. The members to be connected are discrete, as are wood and steel beams. Two techniques are in general use. One method extends reinforcement from the members to be connected and embeds the overlapping steel in cast-in-place concrete. The other method anchors steel plates to the concrete at the surfaces to be connected, and the plates of adjoining members are field welded. An alternative sometimes used is to connect members with steel tendons, which apply prestress.

8.16. VERTICAL STRUCTURAL SYSTEMS

A vertical structural system is that subsystem of the structural system of a building that transmits loads from the level at which they occur to the foundations. The vertical system may be hidden or exposed for aesthetic purposes. It may be used in addition to structural purposes, as air-conditioning ducts or to enclose piping, but its prime function is load transmission.

The loads on the vertical system may be re-

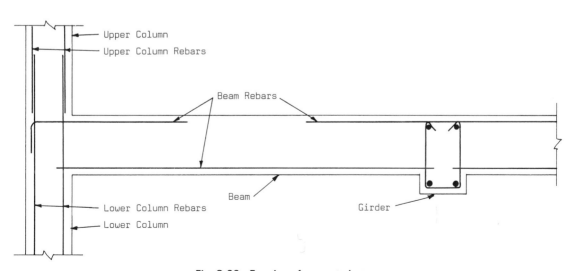

Fig. 8.66. Framing of concrete beams.

solved into vertical components, which usually are gravity loads, and horizontal components, which generally consist of such lateral forces as wind or earthquake loads. Causing sidesway or drift of the building, the horizontal components tend to rack and overturn the structure. The vertical structural system must prevent this and keep drift, including oscillations, within acceptable limits. Gravity loads often are helpful in preventing uplift under the overturning loads, but only dead loads can be relied on in design to offset uplift forces.

Lateral loads usually are of short duration. Furthermore, the probability of simultaneous occurrence of maximum lateral and maximum gravity loads is small. Hence, building codes permit a smaller safety factor to be used in the design of structural components for combinations of lateral and gravity loads than for either type of load alone. As a result, sometimes little or no increase in size of structural members over that required for gravity loads alone is needed for the combination loading. Systems design of structural framing should seek to take advantage of such conditions.

Vertical structural systems may consist of components designed to resist the combination loading or of components designed to sustain only gravity loads while assisted by other components in resisting lateral loads. Selection of either type depends on the height-width ratio of the building, loads, column spacing and structural materials.

Load-Bearing Walls

Floors, beams, girders and trusses may be seated conveniently on walls that transmit the loads directly to the foundations. Such walls are advantageous in that they may also serve as enclosures of rooms or the building exterior. They have the disadvantage, however, that they have to be made thicker in the lower portion as height is increased and occupy considerable space in very tall buildings. Also, walls usually take longer to build than skeleton framing.

For small, lightly loaded buildings, it is often economical to construct load-bearing walls of closely spaced columns, called *studs*, with non-load-bearing facings. In houses, studs usually

Fig. 8.67. Erection of a prefabricated wood-stud wall. (Courtesy Western Woods Products Association)

are spaced 16 or 24 in. center-to-center. They generally are about one story high. They are seated at the bottom on a *sole plate* and are capped at the top by a *top plate*, which supports

Fig. 8.68. Plywood sheathing is nailed to studs of a load-bearing wall. (Courtesy Western Woods Products Association)

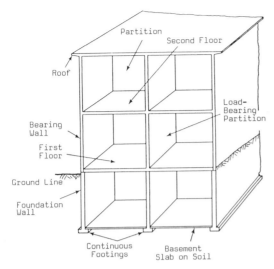

Fig. 8.69. Load-bearing construction for a two-story building with basement.

have a diaphragm securely attached to them to prevent racking. Figure 8.68 shows plywood sheathing nailed to wood studs for that purpose.

For taller buildings, load-bearing walls may be built of unit masonry, reinforced unit masonry or reinforced concrete. Figure 8.69 illustrates load-bearing construction for a two-story building with basement. Similar construction may be used for a high-rise building but with thicker walls at the base. Reinforced concrete walls may be cast in place, usually one story at a time, or precast. Figure 8.70 shows construction of a multistory building with one-story, precast wall panels and precast, hollow floor deck.

horizontal framing members or roof trusses at that level. Figure 8.67 shows a wood wall, with studs spaced 24 in. center-to-center, being erected. Studs may be diagonally braced or

Skeleton Framing

When skeleton framing is used, columns are the main components of the vertical structural system. (Their structural behavior is described in Sec. 8.8.) They may be required to carry only gravity loads or both lateral and gravity loads.

Fig. 8.70. Multistory building with precast load-bearing walls and floors.

In either case, columns need lateral support from other framing members to keep their slenderness ratios within acceptable limits, to prevent buckling. Stiff floors, roofs, beams, girders or trusses connected to shear walls, rigid frames or bracing may serve this purpose.

Wood and cast-in-place concrete columns usually are erected in lengths one story high. The concrete columns generally are cast at the same time as the floor or roof they will support. Precast concrete and structural steel columns often are erected in lengths two or three stories high in multistory buildings.

Steel columns usually are spliced 2 or 3 ft above a floor, for convenience, when extension upward is necessary. Stresses are transferred from column section to column section by bearing. Consequently, splice plates used to join the sections need be only of nominal size, sufficient for structural safety during erection. When the sizes of sections to be spliced are significantly different, stresses should be transmitted from the upper to the lower section with the aid of bearing flange plates on the upper section or by placing a horizontal butt plate between the two sections to distribute the load to the lower section.

A bearing plate is desirable under wood or steel columns seated on masonry or on concrete footings. The plate distributes the column load to prevent crushing of the masonry or concrete. Also, the plate may be set in mortar so that it will be level and the bottom of the column will be at the required elevation. In addition, column bases should be secured with anchor bolts embedded in the masonry or concrete.

Gravity and lateral loads are transmitted to columns by floors, beams, girders, walls or trusses. The columns require lateral bracing for both types of loads not only because horizontal components of the loads must be transmitted to the foundations but also because the horizontal framing may impose eccentric vertical loading and bending moments.

Even in low buildings, the vertical structural system must provide resistance to racking, sway and overturning. Figure 8.71a illustrates how trusses and X bracing may be employed in the framing for a one-story industrial building. Inclined trusses are incorporated in the planes of the roof to transmit wind loads to the ends of the building. X-braced bays at each end carry the wind loads to the foundations. Diagonal braces in the end walls resist wind loads acting on the sides of the building.

Multistory Framing

Resistance to lateral loads in high-rise buildings may be provided in a variety of different ways. Some commonly used methods are indicated in Fig. 8.71b through m. Sometimes, combinations of these methods are used.

Shear Walls. The primary function of a shear wall is to resist horizontal forces parallel to the plane of the wall. The lateral forces may be transmitted to the wall by other walls, floors or horizontal framing. Under these loads, the shear wall acts as a vertical cantilever. Shear walls also may be used as load-bearing walls for gravity loads and as enclosures for elevator shafts, stairways or closets or as large hollow columns. The walls usually are constructed of reinforced concrete or reinforced unit masonry.

Figure 8.71f illustrates the use of shear walls for resisting the component of lateral loads acting in the narrow direction of a building. The lateral loads in the perpendicular direction are carried by rigid-frame action of girders and columns.

Shear walls have little resistance to loads perpendicular to their plane. Consequently, shear walls often are constructed in perpendicular planes (see Fig. 8.71g) to resist lateral loads that may come from any direction. Figure 8.71h illustrates the use of shear walls as enclosures for elevator shafts and stairways.

Wind Bents. As indicated in Fig. 8.71c, lateral loads may be distributed to specific bents, combinations of girders, columns and often diagonal bracing lying in a vertical plane, which are designed to transmit the loads to the foundations. Because a bent is a planar structure, it has little resistance to loads perpendicular to its plane. Consequently, the special bents, often called wind bents, are placed in two perpendicular directions to resist lateral loads that may come from any direction.

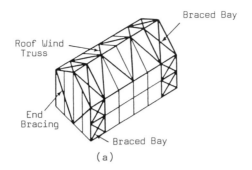

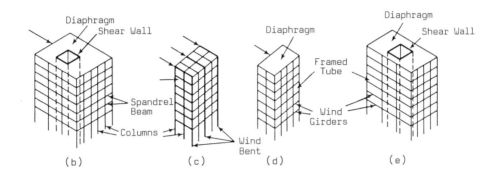

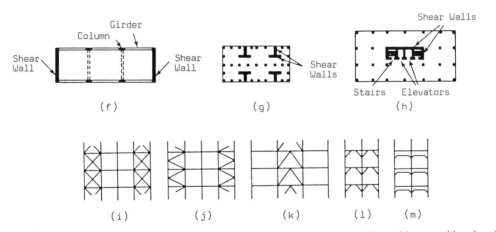

Fig. 8.71. Lateral bracing systems for buildings. (*a*) One-story industrial building with trussed bracing. (*b*) Core shear wall. (*c*) Selected vertical planar bents as rigid frames. (*d*) Framed tube. (*e*) Tube-in-tube. (*f*) Exterior (peripheral) shear walls. (*g*) interior shear walls. (*h*) Core shear walls. (*i*) X-braced bents. (*j*) K-bracing. (*k*) Inverted V-bracing. (*l*) Knee bracing. (*m*) Moment-resistive connections for beams to columns.

The proportion of the component of the lateral loads in the direction of a bent that is distributed to the bent depends on the relative stiffness of the bent compared with the stiffness of all the bents parallel to it. Stiffness is measured by the relationship of sidesway, or drift, of a bent to the load causing the drift. Load distribution also is dependent on stiffness of the floors and horizontal framing.

Figure 8.71*i* through *m* illustrates some of the commonly used types of construction used for wind bents. Figure 8.71*i* shows part of a wind bent with two X-braced panels per story in a building four-bays wide. The bracing forms a pair of vertical cantilever trusses. This type of bracing has the disadvantage of obstructing the center of the panel formed by the girders and columns, so that doors or windows cannot be placed in such panels. An alternative form of bracing, K bracing shown in Fig. 8.71*j*, has a similar disadvantage. The action of the two types of bracing differs in that one of the diagonals in each panel of K bracing must be capable of carrying compression, whereas the diagonals of X bracing may be designed only for tension.

Another alternative is the inverted V bracing shown in Fig. 8.71*k*. This type leaves the center of the panel open. Hence, windows or doors can be placed in the panel. The bracing has the disadvantage, however, that it is subjected to gravity loads from the girders.

Still another alternative method for developing resistance to lateral loads is through rigid-frame action. One way of doing this is to place knee bracing, short diagonals, near the intersection of girders and columns (see Fig. 8.71*l*). This type of bracing has the disadvantage that it must be placed in every panel and may not be architecturally desirable because the bracing may protrude beyond partitions and ceilings into rooms, or may interfere with window placement.

Rigid-frame action developed with rigid connections between girders and columns (see Fig. 8.71*m*) generally is more compatible with architectural objectives. Haunched girders, as shown in Fig. 8.71*m*, often are acceptable for wind bents along a building exterior. The haunches, however, are usually eliminated from interior bents so that structural framing can be hidden in partitions or above ceilings.

Framed Tubes. If the floors and roof of a building are made stiff enough to act as horizontal diaphragms, wind bents can be placed along the exterior only. Under the action of lateral loads, the bents will act together as a vertical cantilever tube.

Such a framed tube was used for the 100-story John Hancock Building in Chicago (see Fig. 8.72). The exterior bents are exposed for ar-

Fig. 8.72. John Hancock Building, Chicago. 100 stories high, with X-braced, framed-tube construction.

chitectural effect. In these bents, lateral loads are resisted by X braces placed across the full width of the bents.

A different type of framed tube was used for the 110-story twin towers of the World Trade Center in New York City (see Fig. 8.73). The bents along the exterior develop resistance to lateral loads through rigid-frame action. In this type of construction, the exterior columns can be placed closer together than in the building interior because they do not interfere with the use of interior space. Spacing may be 10 ft or less. The short spacing permits use of much smaller columns than would be required for the

Fig. 8.73. Twin, 110-story towers of World Trade Center, New York City, with perforated-tube construction. (Courtesy Port Authority of New York and New Jersey.)

larger spacing usually used. Under lateral loads, the structural system acts like a vertical cantilever perforated tube.

Similar construction was used for the 110-story Sears Tower in Chicago (see Fig. 8.74). In this case, however, several narrow tubes are combined, or bundled. More tubes are provided in the lower portion of the building, where greater lateral-load resistance is needed, than in the upper portion.

Another variation of tubular construction is the tube-in-tube illustrated in Fig. 8.71e. Wind bents are placed around the exterior of the building, but, in addition, a tube is also constructed around the core of the building. The interior tube is usually formed with shear walls, which may be load-bearing and which also may enclose elevator shafts and stairways. The exterior and interior tubes act together in resisting lateral loads.

SECTIONS 8.15 and 8.16

References

F. Merritt, *Building Design and Construction Handbook*, 4th ed., McGraw-Hill, New York, 1982.

W. Scheuller, *Horizontal-Span Building Structures*, Wiley, New York, 1983.

W. Scheuller, *High-Rise Building Structures*, Wiley, New York, 1977.

Fig. 8.74. Sears Tower, Chicago, 110 stories high, with bundled tube construction.

Words and Terms

Connections
Deck (structural)
Lateral bracing systems: shear wall, lateral bents, framed tubes
Load-bearing walls
Skeleton framing
Spanning systems: one-way, two-way, continuous

Significant Relations, Functions, and Issues

Horizontal system development: choice of type; relation to vertical supports; selection of deck; integration with roofing, flooring, ceilings, building services.
Vertical system development: planning; lateral bracing; coordination with horizontal system design; relations to architectural plan and spatial development; integration with doors, windows, stairs, elevators, building services.

8.17. SYSTEMS-DESIGN APPROACH TO STRUCTURAL SYSTEMS

Many of the innumerable variables in systems design of structural systems develop from the interaction of structural systems with other building systems. As a result, structural design generally is an iterative process. In each step of the process, potential structural systems are generated, evaluated and compared. As design proceeds, the framing being developed must be checked to insure compatibility with other systems as well as to verify that other systems will be properly supported. Often, structural systems have to be modified in all design stages, from preparation of schematics through final design, to meet architectural, mechanical, electrical or other building needs. A good building team, however, will insure that only minor changes will be necessary in the later design stages.

Because of the interaction of the various building systems with each other, selection of the lowest-cost structural system does not necessarily result in an optimum building. Such a system, for example, may be a dome with such a large volume of space enclosed that heating costs would be excessively high; or the system may require closely spaced columns that would interfere with activities planned for the building and make production costs unacceptably high; or the beams may obstruct passage of ducts and piping or increase the building height with consequent increases in costs of walls, ducts and piping. Instead, design of the structural system must be consistent with the objective of an optimum building.

Design of a structural system generally starts after a schematic architectural floor plan has been prepared. This floor plan should be based on conditions at the building site. At this stage also an estimate should be available of the number of stories the building is to contain. Little mechanical or electrical information may be available at this time because design of the mechanical and electrical systems too may be just starting. The structural engineer therefore can only rely on his judgment and experience and select potential structural systems for investigation. These systems must be fitted to the proposed floor plan and made compatible with design of an optimum site-foundation system. The systems-design steps illustrated in Fig. 3.4 should be applied to all structural systems considered.

Data Collection and Problem Formulation

Basic information for structural design comes from site studies, architectural floor plans, elevations and cross sections, and the owner's program of requirements. Additional information that will be needed should be obtained from local building and zoning codes, from design standards for various structural materials and from studies of local construction practices, including the type of work handled by the local construction trades or unions and union restrictions.

The floor plans and type of occupancy often determine the live loads that will be imposed on the structural system. The local building code, however, may set minimum loads on which design must be based, rather than the actual service loads. In addition, provision must be made for future changes in type of occupancy or floor plans.

The zoning code generally will set a limit on the height of the building. This limit, in turn, may restrict the depth of the horizontal framing, if the building is to contain the specified number of stories and also provide acceptable headroom or floor-to-ceiling heights. Also, the zoning code may require the building to have setbacks as height increases. The setbacks, in turn, alter the structural framing in upper and lower portions of the building.

In the later design stages, further information will come from electrical and mechanical engineers and other consultants. This information should provide details on equipment to be supported, ducts, pipes and wiring to be installed and access to be provided to all these building components. In addition, more details should become available on walls and partitions, elevator shafts and stairways. The new information may make necessary substantial modifications in the structural system or even a complete change.

The goal may be stated as: To design, as a component of an optimum building, a structural system that will enable the building to sustain all anticipated loads under ordinary and emergency conditions with no risk of injury to persons in or near the building, or of damage to any building components, or of motions causing human discomfort, except under very extreme conditions, such as cyclones or earthquakes, in which case the risk will be small.

Objectives

Most buildings have many objectives in common. Foremost among these objectives is the aim of selecting from among all the possible structural systems that can achieve the goal the one that will have the lowest life-cycle cost. This cost includes the cost of material, fire protection, fabrication, shipping, erection, maintenance and repairs.

Another objective is to provide a structural system that is compatible with architectural requirements and with mechanical, electrical and other systems needed for the building.

Still another objective is to design a structural system of materials that can be obtained quickly and then can be fabricated and erected speedily.

Specific buildings may require unusual objectives. For example, for various reasons, an objective may be to have far fewer columns or walls, or even no walls, in the first story than may be used at higher levels; or an objective may be to have a very long-span roof, for example, for an airplane hangar or a stadium; or sometimes, the objective may be to design a system that can be easily dismantled, to permit easy alterations or for movement to and reassembly on a different site.

Constraints

Numerous constraints may be imposed even for a simple building (see Sec. 8.6). Design live loads are among the most important. These depend on the purpose for which the building is to be used. The loads vary with the number of persons permitted to occupy a space, the equipment to be installed, and materials or vehicles to be stored. Geographic location and foundation conditions, however, may be equally important. The location of the building site and the shape of the building determine wind and seismic loads that the structural system must be capable of resisting. Foundation conditions may restrict the weight of the structural system and influence the spacing of columns or load-bearing walls.

Building and zoning codes, in addition, impose many constraints. Building codes may set limits on dimensions of materials, specify allowable unit stresses or loads and dictate design and construction methods, including safe working conditions and safe use of equipment. Building codes also incorporate requirements for fire protection of structural systems. Zoning codes generally constrain structural systems through limits on building height and requirements for setbacks from lot lines with increase in height.

Construction labor also imposes some constraints. Wages paid some trades, for example, may be higher than those paid others, especially when productivity is taken into account, so that structural systems requiring employment of the higher-paid trades may be uneconomic. In some regions, masons may be so expensive that skeleton framing should be chosen in preference to load-bearing walls, whereas in other regions, the reverse may be true. As another example, union work rules may prohibit some types of construction. In some regions, construction workers may refuse to handle components prefabricated off the site, whereas in other regions, prefabrication is an accepted practice. In some regions also, a union may require employment of a worker full time, although the worker may be needed only to start a machine in the morning and shut it off in the afternoon. These labor constraints may have a significant influence on selection of structural systems.

Synthesis and Analysis

In the schematics stage, the structural system should be laid out to be compatible with foundation conditions and the architectural floor plans, elevations and cross sections. Analysis of the proposed design should verify that the goal, objectives and constraints established for the structural system have been met.

In the preliminary design stage, with much more information available from the owner, architect, mechanical and electrical engineers and other consultants, preliminary designs of alternative structural systems should be pre-pared. These designs should be checked for compatibility with other systems of the building. Cost estimates for the systems should then be made.

Value Analysis and Appraisal

The benefits and costs of the alternative structural systems should be compared. In general, evaluation and selection of an optimum structural system will be difficult. Many criteria must be satisfied. Cost, as usual, is important, but there are likely to be several others about as important, such as those concerned with compatibility with other systems and speed of erection. The final decision on which system to use may have to be subjective, based on opinions of the members of the building team and how the benefits and costs are weighted in an evaluation.

In the final design stage, design of the chosen structural system may be refined. Value engineering may be helpful in reducing both weight and cost of the system and finding ways to speed construction.

During the early design stages, ways should be investigated to cut costs by making the structural system serve several purposes, for example, to serve also as walls or partitions or as ducts or as conduit for wires. Costs also can be cut through standardization of components and operations. Repetition usually reduces fabrication costs and speeds erection, because workers become familiar with the procedures. For this reason, in multistory buildings, the roof system often is made similar to the more heavily loaded floor system, thus avoiding a change from a repeated procedure at the lower levels to a new procedure at the top.

When repetition is feasible, prefabrication of large sections of the structural system often saves money and speeds construction. Figure 8.67 illustrates erection of a preassembled, load-bearing, wood stud wall. Figure 8.70 shows the construction of a multistory building with precast concrete, load-bearing walls and floors. Figure 8.75 shows placement of a one-story-high section of a precast concrete wall for an elevator shaft. Often, it is economical to prefabricate formwork for concrete in large sections, make it easily demountable and mov-

Fig. 8.75. Story-high, precast concrete unit for an elevator shaft. (Courtesy High Concrete Structures of New Jersey)

able, and reuse it many times during construction of a building (see Fig. 8.21c). (Prefabrication of small units, such as domes for waffle slabs or pans for ribbed floors, is common practice.)

A novel example of the use of repetition is illustrated in Fig. 8.76. The construction system shown utilizes one-story-high beams or trusses repeatedly. Extending the width of the building, these structural members may be uniformly spaced at each level, but their locations

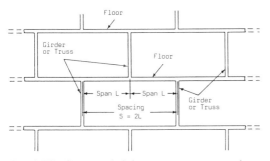

Fig. 8.76. Staggered girder, or truss, construction.

are staggered in adjoining stories. Each floor is supported at the base of one beam or truss and at the top of the adjoining beams or trusses. Consequently, the span of the floor is only half the spacing between the structural members. Hence, even with a thin floor deck, structural members can be widely spaced, with the result that fewer beams or trusses and columns are needed than for conventional construction. In addition, with the beams or trusses being as deep as one story, they can span long distances economically and hence it is feasible to eliminate interior columns in many cases. Openings for doors and corridors can be provided in the members.

GENERAL REFERENCES AND SOURCES FOR ADDITIONAL STUDY

These are books that deal comprehensively with several topics covered in this chapter. Topic-specific references relating to individual chapter sections are listed at the ends of the sections.

Structural Theory

American Standard Minimum Design Loads for Buildings and Other Structures, American National Standards Institute, 1982.

H. Parker and J. Ambrose, *Simplified Mechanics and Strength of Materials*, 4th ed., Wiley, New York, 1986.

R. Gutkowski, *Structures: Fundamental Theory and Behavior*, 2nd ed., Van Nostrand Reinhold, New York, 1987.

E. Gaylord and C. Gaylord, *Structural Engineering Handbook*, 2nd ed., McGraw-Hill, New York, 1979.

H. Laursen, *Structural Analysis*, 3rd. ed., McGraw-Hill, New York, 1988.

J. McCormac, *Structural Analysis*, 4th ed., Harper & Row, New York, 1984.

Structural Materials

H. Rosen, *Construction Materials for Architects*, Wiley, New York, 1985.

F. Wilson, *Building Materials Evaluation Handbook*, Van Nostrand Reinhold, New York, 1984.

F. Merritt, *Building Design and Construction Handbook*, 4th ed., McGraw-Hill, New York, 1982.

R. Smith, *Materials of Construction*, 3rd ed., McGraw-Hill, New York, 1988.

Structural Design

F. Merritt, *Building Design and Construction Handbook*, 4th ed., McGraw-Hill, New York, 1982.

H. Parker and J. Ambrose, *Simplified Engineering for Architects and Builders*, 7th ed., Wiley, New York, 1989.

J. Ambrose, *Building Structures*, Wiley, New York, 1988.

R. White and C. Salmon, *Building Structural Design Handbook*, Wiley, New York, 1987.

D. Breyer, *Design of Wood Structures*, 2nd ed., McGraw-Hill, New York, 1986.

Timber Construction Manual, 3rd ed., American Institute of Timber Construction, Wiley, New York, 1985.

S. Crawley and R. Dillon, *Steel Buildings: Analysis and Design*, 3rd ed., Wiley, New York, 1984.

B. Johnson and F. Lin, *Basic Steel Design*, 3rd ed., Prentice-Hall, New York, 1986.

Manual of Steel Construction, American Institute of Steel Construction, Chicago, 1986.

P. Rice, et al., *Structural Design Guide to the ACI Building Code*, Van Nostrand Reinhold, New York, 1985.

M. Fintel, *Handbook of Concrete Engineering*, 2nd ed., Van Nostrand Reinhold, New York, 1985.

Building Code Requirements for Reinforced Concrete, (ACI 318–83), American Concrete Institute, Detroit, 1983.

PCI Design Handbook—Precast and Prestressed Concrete, Prestressed Concrete Institute, Chicago, 1985.

J. Amrhein, *Masonry Design Manual*, 3rd ed., Masonry Institute of America, Los Angeles, 1979.

Construction Methods

F. Merritt, *Building Design and Construction Handbook*, 4th ed., McGraw-Hill, New York, 1982.

D. Watson, *Construction Materials and Practices*, 3rd ed., McGraw-Hill, New York, 1986.

E. Allen, *Fundamentals of Building Construction: Materials and Methods*, Wiley, New York, 1985.

American Plywood Association Design/Construction Guide: Residential and Commercial, American Plywood Association, Tacoma, WA.

ACI, *Manual of Concrete Practice*, American Concrete Institute, Detroit, MI, 1988.

EXERCISES

The following questions and problems are provided for review of the individual sections of the chapter.

Section 8.1

1. What is the prime function of the structural system of a building?

2. What is the relationship between applied loads and reactions?

3. How does a load differ, by definition, from stress?

4. What is the difference between dead loads and live loads?

5. What types of deformations do the following loads cause in a structural member?
   (a) Tensile forces?
   (b) Compressive forces?
   (c) Shearing forces?

6. How many unknowns can be determined with the laws of equilibrium for a set of nonconcurrent coplanar forces acting on a rigid body?

7. The 10-kip load on beam AB in Fig. 8.2 is replaced by a uniformly distributed load of 0.5 kip per ft. Determine the reactions for the beam.

8. The 10-kip load on beam AB in Fig. 8.2

is replaced by joists at 5-ft intervals. Each joist imposes a 2-kip load on the beam. Determine the reactions for the beam.

9. The 10-kip load on beam AB in Fig. 8.2 is replaced by a uniformly distributed load of 0.25 kip per ft and joists at 2.5-ft intervals. Each joist imposes an 0.8-kip load on the beam. Determine the beam reactions.

10. A simple beam with a span of 20 ft is loaded from one end to midspan with a uniformly distributed load of 0.5 kip per ft.

    (a) What are the magnitudes of the reactions?

    (b) What is the maximum shear?

    (c) Where does the maximum bending moment occur?

    (d) What is the value of the maximum moment?

11. A load of 30 kip is applied 2 in. from the centroidal axis of a thick column. What is the value of the bending moment imposed on the column?

Sections 8.2 and 8.3

12. A steel hanger 120-in. long with a cross-sectional area of 1.25 sq in. is subjected to an axial tension load of 20 kip.

    (a) What is the value of the unit tensile stress, ksi?

    (b) If the elongation is 0.006 in., what is the value of the unit tensile strain, in. per in.?

13. A steel hanger 84 in. long is subjected to a 16-kip axial tensile load.

    (a) If the steel is permitted to carry a unit stress of 24 ksi, what is the minimum cross-sectional area required for the hanger?

    (b) If the unit tensile strain is 0.00008 in. per in., how much will the hanger lengthen?

14. A steel hanger 96 in. long has a cross-sectional area of 2 sq in. If the steel is permitted to carry a unit stress of 24 ksi, what is the maximum tensile load allowed on the hanger?

15. A short concrete bracket, part of a concrete column, is subjected to a 14-kip vertical reaction from a beam. At the intersection with the column, the bracket is 12 in. deep and 6 in. wide. What is the value of the unit shearing stress at the face of the column?

16. One end of a beam rests on a steel bearing plate seated on a concrete pier. The purpose of the bearing plate is to distribute the beam reaction of 8 kip to the top of the pier. The concrete is allowed to carry a bearing stress of 0.4 ksi. What is the minimum area required for the bearing plate?

17. What distinguishes the structural behavior of a slender column from that of a short compression member?

Section 8.4

18. Why is a ductile structural material desirable?

19. What condition determines whether a material is elastic?

20. Define proportional limit.

21. What is the significance of yield point and yield strength?

22. A material is elongated 0.00004 in. per in. by a tensile load. If Poisson's ratio for the material is 0.25, what change takes place normal to the direction of the load?

23. What is Hooke's law?

24. A 300-in.-long steel bar hanger with a modulus of elasticity of 30,000 ksi and a Poisson's ratio of 0.3 is subjected to a 96-kip tensile load. The cross section of the bar is 2 sq in.

    (a) What is the unit tensile stress in the bar?

    (b) If the proportional limit is not exceeded, how much does the bar elongate?

    (c) How much does the width of the bar decrease?

25. What is the distinguishing characteristic of:

    (a) A homogeneous material?

    (b) An isotropic material?

(c) A tough material?

(d) A plastic (responsive) material?

26. A steel beam is subjected to a shearing unit stress of 11.5 ksi. The material has a modulus of elasticity of 30 ksi and Poisson's ratio of 0.3.

    (a) What is the value of the modulus of rigidity of the beam?

    (b) What is the magnitude of the shearing unit strain?

27. A structural steel member has a yield point of 36 ksi. If the safety factor is 1.67, what is the allowable unit stress in tension?

28. Why are yield point and yield strength important in structural design?

29. What characteristic of a material is determined by:

    (a) Modulus of elasticity?

    (b) Modulus of rigidity?

Section 8.5

30. What are the advantages in structural applications of:

    (a) Structural steels?

    (b) Wood?

31. What type of material is indicated by A572, Grade 45 steel?

32. What is designated by W 14 × 84?

33. What are the basic chemicals in structural steels?

34. How does the structural behavior of the following materials compare with that of the idealized material with the stress-strain curve shown in Fig. 8.7?

    (a) Steel.

    (b) Wood.

    (c) Concrete.

35. How do structural steels and cold-formed steels compare in thickness?

36. Describe at least three ways of protecting structural steels against corrosion.

37. How does bridge rope differ from bridge strand?

38. How does the strength of steel cables compare with that of carbon steels?

39. What are the advantages of connecting steel pieces with welds?

40. What is the purpose of lips on cold-formed shapes?

41. In what direction is wood strongest in tension and compression?

42. What is the purpose of the grading of lumber?

43. What is meant by dimension lumber?

44. A piece of lumber has a compressive strength parallel to the grain of 4 ksi, and cross grain of 0.8 ksi. What is the compressive strength at an angle of 45° with the grain?

45. Which has greater withdrawal resistance under load for the same length of embedment in wood, a nail or a wood screw?

46. Explain why fewer bolts are required in a connection between wood members if split-ring connectors are used.

47. Why does each ply in plywood have its grain perpendicular to the grain in adjoining plies?

48. What materials are generally used to make:

    (a) Portland cement?

    (b) Normal weight concrete?

    (c) Lightweight concrete?

    (d) Heavyweight concrete?

49. Why is reinforcing or prestressing needed for concrete?

50. What is the purpose of curing concrete?

51. Describe briefly the various types of composite materials and the advantages of each type.

52. Which of the commonly used structural materials is:

    (a) Strongest?

    (b) Lightest?

    (c) Most flexible?

    (d) Most fire resistant?

Section 8.6

53. How can a designer, to reduce construction costs, take advantage of the fact that part of the dead load is a controllable variable?

54. What can be done to reduce the snow load on a building?

55. What can be done to reduce the wind load on a building?

56. Three structural members are formed into a triangle with pinned connections where the members intersect. Discuss the stability of this arrangement.

57. A structural member is designed by the ultimate strength method. Is any part of the member likely to be stressed to or beyond the yield strength under service loads?

58. Why should a limit be placed on beam deflections, even if there is no danger of the beam being overstressed?

59. Why is a least-weight structural system not necessarily the lowest-cost alternative?

Sections 8.7 and 8.8

58. A round steel rod with allowable stress of 24 ksi supports a tension load of 24 kips.
    (a) What diameter is required if connections are welded?
    (b) If the rod is threaded for a nut, and the cross-section area is reduced by 25% at the root of the threads, what diameter is required?

59. A steel bar 0.5-in. thick supports a gravity load of 48 kips in tension. Allowable unit stress for the material is 24 ksi. If connections are made with a single row of $\frac{3}{4}$ in. bolts in the line of the load, what width is required for the bar?

60. Why is stress capacity of a material not important in determination of the load capacity of a slender column?

61. To what value does the critical unit stress converge, for a column made of a material with a well-defined yield point, as slenderness ratio becomes very small?

62. After a material and cross-sectional dimensions have been selected for a column, what two options are available to the designer for increasing the load-carrying capacity of the column?

63. A nominal wood 4 X 4 (actually 3.5 X 3.5) post 8-ft high is subjected to an axial compressive load. What is the slender-

ness ratio of the column if its ends are fixed against translation but free to rotate?

Section 8.9

64. What is the advantage of restricting loads on trusses to panel point locations?

65. What is the advantage of aligning truss members at a joint so that their centroidal axes pass through the panel point?

66. What is the purpose of a gusset plate?

67. What is the purpose of a scab?

68. Compute the stresses in the members of the truss in Fig. 8.30 that are cut at section 1-1.

69. From a consideration of the part of the truss in Fig. 8.30 on the left of section 3-3, show that the stresses in the top and bottom chords cut by the section are equal. Show also that the chord stresses form a couple that balances the moments of the external forces about the vertical cut by section 3-3.

70. What purpose is served by the cross frames and struts in Fig. 8.31?

71. In Fig. 8.31c, what is the purpose of the diagonals attached to the purlins?

72. The heel of a truss imposes a vertical reaction of 14 kips on its support. If the inclined top chord and the horizontal bottom chord intersect at an angle of 45°, what are the stresses in the chords?

Section 8.10

73. What are the advantages and disadvantages of beams, compared with trusses?

74. How does a fixed end of a beam differ from a simple support?

75. How are statically determinate beams distinguished from statically indeterminate beams?

76. In what way do bending unit stresses determine the neutral surface of a beam?

77. Define shear in a beam.

78. Define bending moment in a beam.

79. What type of bending unit stress does a

positive bending moment develop on the upper portion of a beam?

80. What type of bending unit stress does a negative bending moment develop on the upper portion of a beam?

81. What is the relationship between maximum bending moment and shear?

82. If the beam in Fig. 8.35*e* has a 20-ft span and each load *P* is 24 kip:
    (*a*) What is the value of the left reaction?
    (*b*) What is the shear at 1.5*a* from the left reaction?
    (*c*) Where does the maximum bending moment occur and what is its value?

83. If the beam in Fig. 8.35*g* has a span of 20 ft and w = 0.3 kip per ft:
    (*a*) What is the value of the left reaction?
    (*b*) What is the shear at the left quarter point?
    (*c*) What is the bending moment at the left quarter point?
    (*d*) Where does the maximum moment occur and what is its value?

84. A rectangular wood beam has a maximum bending moment of 750 ft-lb. Allowable bending unit stress is 1200 psi. If the beam is to have a nominal width of 2 in. (actually 1.5 in.), what should be the least nominal depth (full even number).

85. What are the characteristics of a plastic hinge?

86. A W 8 X 10 steel shape has a cross-sectional area of 3 sq in. and a section modulus of 7.8 cu in. It is subjected to an axial tension load of 18 kip and a bending moment of 6.5 ft-kip. What is the maximum tensile unit stress?

87. Where does the maximum unit shear stress occur in a simply supported 2 X 6 wood beam carrying a uniformly distributed load of 660 lb per ft on a 10-ft span and what is the magnitude of this stress?

88. A simply supported steel beam carries a load of 0.384 kip per ft on a 10-ft span. If the beam has a modulus of elasticity of 30,000 ksi and moment of inertia of 50 in.^{4}, where does the maximum deflection occur and what is its magnitude?

89. Explain the purpose of cambering a beam.

90. A 10-in. wide by 14-in. deep reinforced concrete beam is reinforced with 1.2 sq in. of rebars placed with centroid 12 in. below the top of the beam. If the allowable steel stress is 20 ksi, the allowable concrete stress is 1.2 ksi in compression and $n = 10$, what is the maximum allowable bending moment for the beam?

Section 8.11

91. What are the distinctive characteristics of an arch?

92. What are the advantages and disadvantages of an arch over a beam of the same span.

93. Under what conditions is a tie required for an arch or a rigid frame?

94. What enables a rigid frame with vertical columns and horizontal beams to resist lateral forces, such as wind and earthquakes?

95. What are the advantages and disadvantages of rigid frames over arches for the same span?

96. Copy Fig. 8.48*d* and draw on the diagram:
    (*a*) the shape of the elastic curve of the bent for a vertical, uniformly distributed load extending from G to I on the beam.
    (*b*) The shape of the bending moment diagrams for the beams and columns for the loading in (*a*).

97. Copy Fig. 8.48*d* and draw on the diagram:
    (*a*) the shape of the elastic curve of the bent for a horizontal load acting to the right at F.
    (*b*) the shape of the bending moment diagrams for the beams and columns for the loading in (*a*).

Section 8.12

98. What characteristics make thin shells more structurally efficient than framing composed of arches and beams?

99. What should be done to a thin shell to prevent buckling under service loads?

100. How should shell edges that are not at supports be treated?

101. In the membrane theory for thin shells, what types of forces act at an interior point:
    (a) Of a cylindrical shell?
    (b) Of a dome?

102. Summarize briefly the procedure for determining bending moments in folded plates

103. What are the advantages of a thin cylindrical shell compared to arch construction?

104. Why does a dome have high structural efficiency?

Section 8.13

105. Describe the structural characteristics of cables.

106. What is the role of cables in a cable-stayed structure?

107. What is the role of cables in suspended roofs?

108. A cable strung between two masts 50 ft apart carries a vertical load of 2 kips per ft uniformly distributed over the horizontal projection of the cable. Maximum sag is 5 ft below the cable supports.
    (a) What are the magnitudes of the vertical and horizontal components of the reactions?
    (b) What is the maximum stress in the cable?
    (c) What is the shape of the cable?

109. What is a major purpose of a double-surface, cable-suspended roof?

Section 8.14

110. Define an air-stabilized structure.

111. List criteria for selection of a structural membrane.

112. Describe briefly two types of air-inflated structures.

113. Define an air-supported structure.

114. Why are very long spans theoretically possible with an air-supported structure?

115. A spherical, air-supported membrane 0.125-in. thick and weighing 0.5 psf has a radius of 80 ft. If the pressure differential is 0.03 psi, what is the tensile unit stress in the membrane?

116. What are the advantages of hybrid pneumatic structures?

Section 8.15

117. How do one-way and two-way decks differ?

118. What are the economic advantages of:
    (a) A shallow, horizontal structural system?
    (b) A lightweight, horizontal structural system?

119. What are the advantages of flat-plate construction for a high-rise building?

120. Why should rebars be placed close to the top of a concrete flat plate or flat slab at columns?

121. How do flat slabs differ from flat plates?

122. Why do waffle slabs have greater load-carrying capacity than flat plates containing the same volume of concrete.

123. What are the advantages of two-way slabs compared with flat plates?

124. In which direction should rebars to resist bending moments be placed in a one-way slab?

125. What is the major difference between simple-beam and rigid connections?

126. What is the purpose of a bearing plate?

127. For steel beam connections, what are the advantages of a seated over a framed connection?

Section 8.16

128. Why are load-bearing walls often less expensive than skeleton framing with curtain walls for low buildings?

129. What are the purposes of a bearing plate at the base of a steel column?

130. Why is lateral bracing needed in structural framing even if only gravity loads have to be supported?

131. What is the prime purpose of a shear wall?

132. Why must shear walls or wind bents be erected in two perpendicular directions?

133. From an architectural viewpoint, why are rigid frames often preferred instead of braced frames?

134. From an architectural viewpoint, what is the advantage of:

(*a*) Inverted V-bracing over X-bracing?

(*b*) A rigid frame over a V-braced frames?

135. What are the advantages of framed tubes over a grid of lateral bents?

136. How does a framed tube differ from a tube-in-tube?

Chapter 9

Plumbing

Plumbing comprises major subsystems for conveyance of liquids and gases in pipes within a building. The pipes generally extend beyond the building walls to a supply source or a disposal means, such as a sewer. The plumbing subsystems have different objectives and must be kept independent of each other.

The subsystems generally include water supply, wastewater removal, heating gas and in some cases other liquids and gases. In this section, for simplification, the subsystems will be referred to as systems, because they generally consist of two or more subsystems. For example, the wastewater system may consist of subsystems for removal of domestic wastewater, industrial wastes and rainwater.

Systems design may be used in many ways to optimize the plumbing. To start, systems design may aim at conserving water and heating gas; that is, making the amount of these substances consumed in the building as small as possible. Next, systems design may seek to minimize the cost of removing wastes from the building. In some cases, this may require changes in the processes for which the building is being used, to eliminate the need for sewage treatment before the wastewater is discharged into a public sewer or a body of water. Also, systems design may try to minimize the sizes (diameters) of pipes and the length of pipe runs. In addition, an objective should be to minimize the life-cycle costs of plumbing. This objective generally requires selection of appropriate materials for the pipes, pipe accessories, plumbing

fixtures, such as sinks, lavatories, bathtubs and water closets, and other equipment associated with plumbing.

Comprehension of the role of plumbing in buildings requires a knowledge of:

1. Characteristics of materials to be conveyed
2. Piping materials, pipe accessories, fixtures and other plumbing equipment, for example, pumps
3. Types of plumbing systems and their characteristics
4. Hydraulics and pneumatics
5. Methods of installing plumbing

Plumbing analysis and design are the responsibility of mechanical engineers and plumbing designers. Plumbing generally is installed by plumbing contractors. They often also install heating and air-conditioning equipment, especially when piping is required for these installations. Building codes usually require that installation of plumbing be carried out under the supervision of licensed master plumbers. The work is done by plumbing journeymen and apprentices.

Only a few of the many types of plumbing systems are discussed in this section. These include water supply, disposal of domestic wastewater and rainwater, and supply of heating gas.

9.1. WATER SUPPLY

Building codes require that potable water be provided in every building intended for human occupancy or use: that is, the water must be fit to drink. Obtaining such water can be relatively easy if a building is located where sufficient public water supply may be tapped economically. If a public supply is not available or is too costly, development of an independent source may be feasible. For this purpose, surface water, such as lakes or rivers, or groundwater, such as well water, may be tapped, if available. If such supplies are not available or their use is prohibited by local authorities, as may be the case with wells in built-up areas, selection of a different site for the building is necessary.

Purity of Water

Regardless of source, water is almost never chemically pure. It contains various amounts of impurities. Acceptability of a water supply consequently depends on the amounts or types of impurities and on the use to which the water is to be put. For example, water that is potable, may not be suitable for use in industrial processes or in boilers for production of steam. Also, some potable waters may make washing or laundering difficult or costly.

In general, as a minimum requirement, water should meet the latest U.S. Public Health Service Drinking Water Standards if it is to be used for domestic purposes. If no source of potable water is readily available, however, water often can be economically treated to remove objectionable characteristics. Treatment generally is necessary for water for industrial processes or boiler feed. Water obtained from natural sources usually contains objectionable chemicals or is acidic and corrodes metals.

Public water supplies may be safely assumed to be potable. But water obtained from other sources and water from any source, if the water is to be used for industrial processes or boiler feed, should be tested to determine its physical, chemical and biological characteristics and suitability for the proposed end use.

Physical Factors. The tests should determine the temperature, turbidity, color, taste and odor of the water supply. Except for temperature, the tests should be made on samples of the water in a laboratory.

Turbidity is indicated by cloudiness of a water sample. This characteristic is caused by fine, visible material, usually colloidal soils, in suspension.

Often, these physical factors do not affect the potability of water, but they require correction if they are found objectionable.

Chemical Characteristics. Tests should be made to detect the presence of poisonous chemicals in the water supply. If any are present in amounts exceeding the maximum limits in the U.S. Public Health Service Drinking Water Standards, the water should be treated to remove the chemicals completely or at least the excess over the limits, or else the water supply should be rejected.

Tests also should evaluate the degree of hardness and alkalinity or acidity (pH) of the water supply. In addition, they should determine the amounts of iron, manganese and total dissolved solids present in the water.

Hardness is indicative of the difficulty of getting soap to lather and of the probability of deposition of scale in pipes. Scale is a hard layer of chemicals that builds up inside a pipe. This layer is undesirable because it reduces the flow of water and also can cause boiler-tube failures from overheating. Although any of several chemical compounds may be present in scale, the degree of hardness usually is expressed in terms of calcium carbonate, which is a common substance in scale. As little as 150 parts per million (milligrams per liter) of calcium carbonate may be objectionable as a soap waster, but even this quantity is generally unacceptable for boiler feedwater.

Water is acidic when the pH is less than seven. Such water is undesirable because it can cause corrosion.

Iron and manganese are objectionable because they can discolor laundry and plumbing fixtures.

To limit the mineral content of potable water, the U.S. Public Health Service recommends that total dissolved solids in potable water not exceed 500 ppm (mg per liter).

Biological Characteristics. Further tests should be made of a water supply to determine the types and amounts of pathogenic bacteria present. These organisms usually are present in water that is contaminated with sewage. The degree of contamination of water generally is measured by the amount of coliform bacteria present. The presence of such bacteria may be observed with a microscope.

A microscope also may be used to measure the amounts of other organic matter in the water. Such observations are desirable, because the presence of some organisms, such as free-floating plankton and algae, may have deleterious effects on water treatment.

Maintenance of Quality. It is not sufficient that potable water be delivered to a building. The quality of the water must be maintained while the water is being conveyed within the building to the point of use. Hence, the potable-water distribution system must be completely sealed to prevent contamination.

No cross connections may be made between this system and any portion of the wastewater-removal system. Furthermore, the potable-water distribution system should be completely isolated from parts of plumbing fixtures or other devices that might contaminate the water. Check valves or air gaps may be used to prevent backflow or back-siphonage.

Backflow is the flow into the distribution piping of any substance from any source other than the intended water supply source, such as a public water main.

Back-siphonage is the flowing back into the distribution piping of used or polluted water from a plumbing fixture because of a pressure differential.

Water Treatment

The main reason for treating water obtained from a public water supply usually is to reduce hardness. Sometimes also, the water may be treated to remove odors, color and turbidity and to improve the taste. Water obtained from other sources may, in addition, have to be disinfected to kill bacteria.

Accordingly, the most common water treatment methods include softening, filtering, sedimentation and disinfection.

The softening process may employ lime-soda precipitation or the base-exchange (zeolite) process to remove calcium carbonate, magnesium sulfate and other compounds. In lime-soda precipitation, lime (CaO), hydrated lime [$Ca(OH)_2$] and soda ash (Na_2CO_3) are added to the water. They create chemical reactions that form insoluble compounds of calcium and magnesium which are removable by precipitation from the water. In the base-exchange process, undesirable cations, such as calcium and magnesium, are replaced by soluble sodium and hydrogen cations as the water passes through a base-exchange material, such as zeolite. The calcium and magnesium are thus separated from the water supply.

There are many different filtering processes. An activated-carbon filter, for example, may be used to improve taste and remove odors. Screens are used to remove coarse particles. A rapid sand filter is used to remove small particles and some of the larger bacteria. This filter consists of an underdrained, watertight container in which is a thick layer of sand over a thin layer of gravel. The sand filters out the particles in the water as it passes downward through the filter. Water fed to the rapid sand filter usually is pretreated to produce flocculation and coagulation. The effluent from the filter generally is disinfected to kill remaining bacteria.

Particles also may be removed by passing water through sedimentation tanks, or settling basins. In such containers, water flow is slowed so that particles precipitate to the bottom and are removed from the water supply. The settling rate may be increased by addition of coagulants, such as alum (aluminum sulfate), to the water.

Disinfection of a water supply to kill bacteria is generally accomplished with chlorine. It may be injected into the water as a gas, liquid or hypochlorite. The chlorine dose used depends on the degree of contamination of the water and the desired residual amount of chlorine to be left in the water. A dose of 1 or 2 ppm (mg per liter) is usually sufficient to destroy bacteria present and leave an adequate residual amount of 0.1 ppm.

Although chlorine is widely used as a disinfectant for water, objections have been raised to its use because it reacts with some organic matter in water to form potential cancer-causing chemicals. As a result, other disinfectants, such as ozone, should be considered.

Water Flow

Not only must potable water be provided in a building but also the quantity of water supplied must be adequate for the needs of occupants and processes to be carried out in the building. The total water demand may be calculated by adding the maximum flows at all points of use and applying a factor less than unity to account for the probability that only some of the fixtures will be operated simultaneously. This is discussed in Sec. 9.5.

In addition, the pressure at which water is delivered to a building must lie within acceptable limits. Otherwise, low pressures may have to be increased by pumps and high pressures decreased with pressure-reducing valves. Table 9.1 lists minimum flow rates and pressures generally required at various water outlets.

The pressure in Table 9.1 is the pressure in the supply pipe near the water outlet while the outlet is wide open and water is flowing. At fixtures supplied with both hot and cold water, the flow rate given in Table 9.1 is for each of the two conditions. The wide range listed for flushometer valves for water closets is necessary because of the variety of designs of water closets and valves.

In the delivery of water to the outlets, there is a pressure drop in the distribution pipes because of friction. Therefore, water supplied at the entrance to the distribution system must exceed the minimum pressures required at the water outlets by the amount of the pressure loss in the system. But the entrance pressure should not exceed 85 psi, to prevent excessive flow and damage to fixtures. Velocity of water in the distribution system should not exceed 8 fps.

A separate supply of water must be reserved for fire fighting. This supply must be of the most reliable type obtainable. Usually, this requirement can be met with water from a reliable municipal water supply. Otherwise, pumps or storage in an elevated water tank must be provided to supply fire-fighting water. Generally, such water must be provided at a pressure of at least 15 psi at the highest level of fire sprinklers, while flow at the base of the supply riser is a minimum of 250 gpm for light-hazard occupancies and 500 gpm for ordinary-hazard occupancies.

If a building is so located that it cannot be reached by a fire department with 250 ft of hose, a private underground water system may have to be installed. Hydrants should be placed so that all sides of a building can be reached with hoses.

Table 9.1. Minimum Flow Rate and Pressure at Plumbing Fixtures[a] and Minimum Pipe Size[b]

Location	Pressure, psi	Flow, gpm	Branch-pipe diameter, in.
Ordinary lavatory faucet	8	2.0	$\frac{3}{8}$
Self-closing lavatory faucet	8	2.5	$\frac{3}{8}$
Sink faucet	8	4.5	$\frac{3c}{4}$ or $\frac{1}{2}$
Bathtub, shower or laundry-tub cock, home dishwasher or washing machine	8	5.0	$\frac{1}{2}$
Ball-cock for water closet	8	3.0	$\frac{3}{8}$
Flushometer valve for water closet	10–20	15–40	1
Flushometer valve for urinal	10	15.0	$\frac{3}{4}$
Drinking fountain		0.75	$\frac{3}{8}$

[a]Required by the Building Code of the City of New York.
[b]Required by "National Plumbing Code," American Society of Mechanical Engineers.
[c]For commercial installations.

Hydraulic Principles. Water flow and pressures can be calculated from principles of hydraulics.

For continuity, the flow Q is constant between any two points A and B of a section of pipe if no water is added or removed between A and B. Also, at any junction of pipes, the algebraic sum of the flows into and out of the junction must be zero. What goes into the junction must come out unchanged.

Water flow and velocity are related to pipe size by

$$Q = AV \tag{9.1}$$

where

Q = flow in pipe, cfs (1 cu ft = 7.5 gal)
A = cross-sectional area of flow, sq ft, or cross-sectional area of pipe interior, sq ft, if the pipe is full
V = velocity of water, fps

The specific weight of water w can be taken as 62.4 lb per cu ft. Hence, the hydrostatic pressure under 1 ft of water is $62.4/144 = 0.433$ psi.

The depth of water creating a hydrostatic pressure is called a *head*. The pressure due to a head h may be computed from

$$p = 0.433h \tag{9.2}$$

where

p = pressure, psi
h = pressure head, ft

The total head H, ft, on water at any point in a pipe is given by

$$H = Z + \frac{p}{w} + \frac{V^2}{2g} \tag{9.3}$$

where

Z = elevation, ft, of the point above some arbitrary datum
p/w = pressure head, ft
$V^2/2g$ = velocity head, ft
g = acceleration due to gravity, 32.2 ft per sec^2

When water flows in a pipe, the difference in total head between any two points in the pipe equals the friction loss h_f, ft, in the pipe between the points.

Any of several formulas may be used for estimating h_f. One often used for pipes flowing full is the Hazen-Williams formula:

$$h_f = \frac{4.727}{D^{4.87}} L \left(\frac{Q}{C_1}\right)^{1.85} \tag{9.4}$$

where

Q = discharge, cfs
D = pipe diameter, ft
L = length of pipe, ft
C_1 = coefficient

The value of C_1 depends on the roughness of the pipe, which, in turn, depends on pipe material and age. A new pipe has a larger C_1 than an older one of the same size and material. Hence, when pipe sizes are being determined for a new installation, a future value of C_1 should be assumed to insure adequate flows in the future. Design aids, such as charts (see Figs. 9.6 and 9.7) or nomograms, are often used to evaluate Eq. (9.4).

Meters are generally installed on the service pipe to a building to record the amount (volume) of water delivered. The meters may be installed inside the building, for protection against freezing, or outside, in a vault below the frost line. Meters should be easily accessible to meter readers. Meter size should be determined by the maximum probable water flow, gpm.

References

Water Quality and Treatment, 3rd ed., American Water Works Association, 1971.

B. Stein et al., *Mechanical and Electrical Equipment for Buildings*, 7th ed., Wiley, New York, 1986.

Words and Terms

Flow
Head
Potable water
Pressure (in piping)
Purity of water
Treatment of water
Water meter

Significant Relations, Functions and Issues

Treatment of water required to make it potable.
Hydraulic principles in piping systems: flow (velocity and amount), pressure, head.
Measurements in piping systems: flow velocity, flow amount (volume), flow rate (volume per time unit), pressure.

9.2. WASTEWATER DISPOSAL

For protection of the health of building occupants, wastes produced in a building must be removed swiftly in a sanitary manner from the building and then disposed of in a manner acceptable to the community or to state and federal health and environmental agencies. Liquid wastes or wastes carried in water usually are caught in basins and conveyed through pipes out of the building. The fixtures and pipes comprising this plumbing subsystem should meet the minimum requirements of the local building code and health department regulations, or in the absence of these, those of the National Plumbing Code, American National Standard A40.8.

Types of Wastewater

There are three main types of wastewater: domestic, storm and industrial. Separate plumbing systems are generally required for each type.

Domestic wastewater is primarily spent water from the building water supply, to which is added wastes from bathrooms, kitchens and laundries. The wastewater usually contains less than 0.1% of solid matter. The flow usually looks like bath or laundry effluent with garbage, paper, matches, rags, pieces of wood and feces floating on top. The wastewater can become septic in a short time and develop the odor of hydrogen sulfide, mercaptans and other sulfur compounds.

Storm water is primarily the water that runs off the roof of the building. The water usually is led to roof drains or gutters. These then feed the water to drainpipes, which convey it to a storm sewer. Special conditions at some building sites, such as large paved areas or steep slopes, may require capture of land drainage in inlets. From these, the storm water is conveyed to storm sewers. Discharge into sanitary sewers is objectionable because the large flows interfere with effective wastewater treatment and increase treatment costs.

Wastewater Treatment

Stormwater, if kept separate from other types of wastewater, usually can be safely discharged into a large body of water. Raw domestic wastewater and industrial wastes, on the other hand, have objectionable characteristics that make some degree of treatment necessary before they can be discharged.

Industrial wastes generally require treatments engineered to remove the specific elements injected by industrial processes that make the wastes objectionable. Often, these treatments cannot be carried out in public sewage treatment plants. Special treatment plants may have to be built for the purpose.

There are numerous treatment methods available for domestic sewage after its removal from a building. Some treatments are relatively simple, such as those afforded by septic tanks, Imhoff tanks, sewage lagoons or sedimentation tanks (clarifiers). Some treatments are complex, requiring many steps. These include filtration, activated-sludge and aeration methods. The degree of treatment necessary generally depends on the assimilation potential of the body of water to receive the effluent, primarily the ability of the body to dilute the impurities and to supply oxygen for decomposition of organic matter present in the wastewater. The amount of oxygen used for this purpose is called the *biological oxygen demand (BOD)*.

The degree of treatment of wastewater often is classified as preliminary, primary, secondary and tertiary, or complete.

Preliminary treatment may be the conditioning of industrial waste before discharge to remove or neutralize substances injurious to sewers or treatment processes, or it may be unit operations that prepare wastewater for major treatment.

Primary treatment is generally the first and sometimes the only treatment of sewage. This process removes floating and suspended solids.

The effluent, however, is considered only partly treated.

Secondary treatment applies biological methods to the effluent from primary treatment. Organic matter still present is stabilized by aerobic processes.

Tertiary treatment, sometimes also referred to as water renovation, removes a high percentage of suspended, colloidal and organic matter.

The wastewater also may be disinfected.

The solids and liquid removed during the treatment of wastewater are called *sludge*. Often, before sludge can be disposed of, it has to be treated to counteract objectionable characteristics. The treatment to be chosen depends on the quantity and characteristics of the sludge, means of disposal available and cost of treatment. Sludge often consists of 90% water and solids that are about 70% organic.

Treatment of sludge is generally done in digestion tanks. There, anaerobic decomposition of organic matter, carried out by bacteria in the absence of oxygen, takes place. The digested sludge may be used as a soil conditioner and weak fertilizer under certain conditions, after it has been dried. Otherwise, it may have to be disposed of by burning or in landfills.

Industrial wastes are primarily the effluent of industrial processes, but cooling water that is not recycled is also considered an industrial waste. Process wastes generally contain substances that, when discharged into a body of water, such as a lake, river or ocean, cause some biological, chemical or physical change in the water. Sometimes, the wastes cause discoloration or increase the water temperature. Chemicals present may be toxic to aquatic life, animals or humans using the water, or may produce unpleasant tastes or odors. Some wastes may contain large quantities of solids. As a result, some communities will not permit release of some industrial wastes into public sewers or bodies of water without prior treatment.

Types of Sewers

A sewer is a conduit, usually underground, through which wastewater flows to sewage treatment plants or other points of disposal.

Sanitary sewers carry primarily domestic wastewater. Storm sewers convey storm water to disposal points. Combined sewers carry both domestic and stormwater. They cost less than separate sanitary and storm systems, but large flows make sewage treatment difficult and costly.

A building drain, also called a house drain, is the discharge piping at the lowest level of the wastewater system of a building. The drain is generally considered to extend 2 to 10 ft beyond the exterior walls, depending on local building-code requirements.

A building sewer, also known as a house sewer, is a pipe that connects the end of the building drain to a public or private sewer or other point of disposal.

Reference

B. Stein et al., *Mechanical and Electrical Equipment for Buildings*, 7th ed., Wiley, New York, 1986.

Words and Terms

Plumbing
Biological oxygen demand (BOD)
Sewer: sanitary, building drain, building sewer, storm
Sludge
Wastewater: domestic, storm
Pressure head

Significant Relations, Functions and Issues

Water quality
Wastewater treatment: preliminary, primary, secondary, tertiary.
Water pressure: at fixtures, at entrance to building, elevated tanks

9.3. BASIC PRINCIPLES OF PLUMBING

Section 9.1 describes the main requirements for water supply to buildings and Sec. 9.2 discusses the prime requirements for disposal of wastewater from buildings. This section presents general requirements for distribution of water within buildings and collection and removal of wastewater from buildings. These requirements are based on those given in the National Plumbing Code and adopted by the City of New York for its building code.

All buildings must be provided with potable water in quantities adequate for the needs of their occupants. Plumbing fixtures, devices and appurtenances should be supplied with water in sufficient volume and at pressures adequate to enable them to function properly. The pipes conveying the water should be of sufficient size to provide the required water without undue pressure reduction and without undue noise under all normal conditions of use.

Plumbing should be designed and adjusted to use the minimum quantity of water consistent with proper performance and cleansing of fixtures and appurtenances.

Devices for heating and storing water should be designed, installed and maintained to guard against rupture of the containing vessel because of overheating.

The wastewater system should be designed, constructed and maintained to guard against fouling, deposit of solids and clogging. It should be provided with adequate cleanouts. A *cleanout* is an opening that provides access to a pipe, either directly or through a short branch, to permit cleaning the pipe. The opening is kept plugged, until the plug has to be removed for a cleansing operation.

Provision should be made in every building for conveying stormwater to a sewer if one is available.

Pipes, joints and connections in the plumbing system should be gastight and watertight for the pressure required by tests. Tests should be made to discover any leaks or defects in the system.

Plumbing fixtures should be located in ventilated enclosures and should be readily accessible to users, but they should not interfere with normal operation of windows or doors or other exits.

Plumbing fixtures should be made of smooth, nonabsorbent materials. They should not have concealed fouling surfaces. Plumbing fixtures, devices and appliances should be protected to prevent contamination of food, water, sterile goods and similar material by backflow of

wastewater. Indirect connections with the building wastewater system should be provided when necessary.

Every fixture directly connected to the wastewater system should be equipped with a *liquid-seal trap*. This is a fitting so constructed that passage of air or gas through a pipe is prevented while flow of liquid through the pipe is permitted. An example is a vertical J bend in a pipe. When water is not flowing in the pipe, water trapped in the bend of the J blocks passage of gases. When flow in the pipe resumes, the pressure moves the water through the bend.

Foul air in the wastewater system should be exhausted to the outside, through vent pipes. These should be located and installed to minimize the possibility of clogging and the return of foul air to the building.

If a wastewater system is subject to backflow of sewage from a sewer, suitable provision should be made to prevent overflow of the sewage into the building.

The structural safety of a building should not be impaired in any way as a result of the installation, alteration, renovation or replacement of a plumbing system or any of its parts. Pipes should be installed and supported to prevent stresses and strains that would cause malfunction of the system. Provision should be made for expansion and contraction of the pipes due to temperature changes and for structural settlements that might affect the pipes.

Where pipes pass through a construction that is required to have a fire-resistance rating, the space between the pipe and the opening or a pipe sleeve should not exceed $\frac{1}{2}$ in. The gap should be completely packed with mineral wool or equivalent noncombustible material and closed off with close-fitting metal escutcheons on both sides of the construction.

Pipes, especially those in exterior walls or underground outside a building, should be protected, with insulation or heat, against freezing. Underground pipes should be placed deep enough so as not to be damaged by heavy traffic. Pipes subject to external corrosion should be protected with coatings, wrappings or other means that will prevent corrosion. Dissimilar metals should not be connected to each other, because corrosion will result.

9.4. WATER-SUPPLY SYSTEMS

The water-supply system for a building consists of the water-service pipe, water-distribution pipes and the necessary connecting pipes, fittings, control valves and appurtenances used for conveying water. The water-service pipe is that portion of the water-supply system that extends from a source of water, such as a public water main, to the house control valve inside the building or to a point where the supply is fully metered. The water-distribution pipes convey water from the water-service pipe to the plumbing fixtures and other water outlets.

Pipe Materials

Local building codes generally specify materials that are acceptable for use for water pipes.

Generally, water-service pipe, which usually is placed underground, may be made of red brass or copper pipe, copper tube, cast-iron water pipe, galvanized wrought iron, or galvanized steel.

Water-distribution pipes may be made of red brass or copper pipe, copper tube, galvanized wrought iron, galvanized steel or plastics, such as polyvinyl chloride, polyethylene, and acrylonitrile-butadiene styrene, a synthetic rubber.

Steel, also called black iron, is the least expensive piping material. It, however, is subject to corrosion. Galvanized steel and wrought iron are preferable because of their greater corrosion resistance. Copper or red brass are often used, though, despite higher material costs, because these materials are lighter, easier to assemble, corrosion resistant and smoother, thus offering less resistance to water flow. Copper tubing is advantageous because connections can be made with solder and the tubing metal can be thin. Connections in brass pipe usually are threaded, as a result of which the pipe has to be thicker than copper tubing. Plastics are even lighter and also easy to assemble, but not so strong. They, however, are not subject to attack by electrolytic corrosion or acidic water as are the metals, but they may not be suitable for conveying hot water.

Fittings

Fittings are those parts of a pipeline that are used to change the direction of water flow, because it usually is not practical to bend pipe in the field; to make connections between pipes, and to plug openings in pipes or close off the terminal of a pipe. In a water-supply system, fittings and joints must be capable of containing pressurized water flow. Fittings should be of equal quality to the pipe connected.

A few of the commonly used fittings are shown in Fig. 9.1. Standard fittings are available and generally may be specified by reference to an American National Standard Institute or a federal specification. Fitting sizes indicate the diameters of the pipes to which they connect. For threaded fittings, the location of the thread should be specified: a thread on the outside of a pipe is called a *male thread*, whereas an internal thread is known as a *female thread*.

Cast-iron pipe is generally available with standard cast-iron or malleable-iron flanged or threaded fittings (see Fig. 9.1). Brass or bronze fittings for copper or brass pipe also may be flanged or threaded. Flanges are held together with bolts. In some cases, to make connections watertight, a gasket may be placed between flanges, whereas in other cases, the flanges may be machine-faced. Threaded fittings often are made watertight by coating the threads with a pipe compound, made with graphite or iron, or by wrapping the threads with lamp wick before the fittings are screwed onto the pipe.

Wrought copper or cast or wrought bronze solder fittings are used for soldered connections on copper tubing (see Fig. 9.1). Solders are alloys with low melting points, such as tin and antimony. (Lead, often used in the past, is not desirable for pipes carrying drinking water, because it may be leached from the solder and cause lead poisoning.) A solder joint relies on capillary action to draw molten solder into the annular space between a fitting and a copper tube. Strong, leakproof connections with copper tubing also can be obtained by brazing with filler metals, which melt in the temperature range between 1,100 and 1,500°F. In such cases, wrought fittings are generally preferred, because brazed connections with cast fittings must be made with greater care.

Changes in pipe direction may be made with elbows or bends (see Fig. 9.1a, b, h, i and o). A pipe and a 90° branch may be connected with

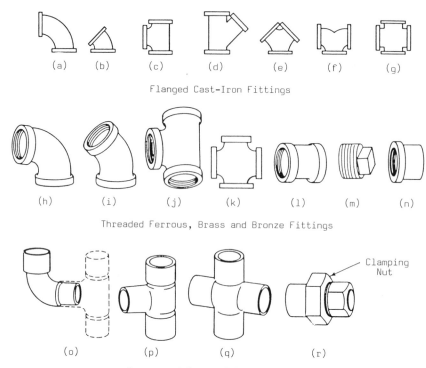

Fig. 9.1. Fittings used for water distribution piping. Flanged cast iron fittings: (*a*) elbow or bend. (*b*) 45° elbow or bend. (*c*) Tee. (*d*) Lateral. (*e*) Wye. (*f*) Double branch elbow or bend. (*g*) Cross. Threaded fittings: (*h*) Elbow or bend. (*i*) 45° elbow. (*j*) Tee. (*k*) Cross. (*l*) Coupling. (*m*) Plug. (*n*) Cap. Solder fittings: (*o*) 90° street ell. (*p*) Tee. (*q*) Cross. (*r*) Union.

a tee (see Fig. 9.1*c, j* and *p*). Fittings also are available for branches at other angles (see Fig. 9.1*d*). Other fittings permit multiple branching (see Fig. 9.1*e, f, g, k* and *q*).

Straight lengths of pipe may be connected end to end with a nipple, coupling or union. A *nipple* is a short length of pipe with male threads that joins two pipe ends with female threads. A *coupling* is a short length of pipe with female threads (see Fig. 9.1*l*) used to connect two pipe ends with male threads. A *union* is used for a similar connection when it may be dismantled and reassembled in the future. To uncouple the pipes, the clamping nut is disengaged by turning and then slid away. Other fittings often used include plugs (see Fig. 9.1*m*) and caps (see Fig. 9.1*n*).

Valves

Valves are devices incorporated in pipelines to control the flow into, through and from them.

Valves are also known as faucets, cocks, bibs, stoppers and plugs. The term *cock* is generally used with an adjective indicating its use; for example, a sill cock (also called a hose bib) is a faucet used on the outside of a building for connection with a garden hose. A *faucet* is a valve installed on the end of a pipe to permit or stop withdrawal of water from the pipe.

Valves usually are made of cast or malleable iron, brass or bronze. Faucets in bathrooms or kitchens are usually faced with nickel-plated brass or white alloy for aesthetic reasons.

Most water-supply systems contain numerous valves. Building codes generally require a house valve near the water meter inlet and usually also a valve at the meter outlet. In multistory buildings, a valve is needed at the bottom of each up-feed riser and at the top of each downfeed riser. Stop valves should be provided near each fixture on the hot- and cold-water pipes to it. In multifamily buildings, control valves should be inserted in the supply branch so that the water to any dwelling unit may be shut off without stopping the water flow to other units. In buildings with other types of occupancy, the supply branch to a group of adjacent fixtures or to each fixture should be provided with a valve to shut off the water to the fixtures. Supply lines to and from pressure or gravity tanks should be equipped with valves within the tank room. The cold-water branch to each hot-water storage tank or water heater should be provided with a valve near the equipment. Each tank or heater should be equipped with an automatic relief valve. All valves should be readily accessible for service and maintenance.

To completely seal off the flow of liquid or gases, valves are made tight by application of pressure between a fixed seat and a movable gasket or a ground-faced metal plug (see Fig. 9.2). The gasket is made of a soft material, such as rubber or plastic. It is pressed tightly against the seat when the valve stem is turned.

Valve stems that must be turned or lifted usually are enclosed in a soft packing to prevent leakage. The packing is compressed against the stem by tightening a threaded compression cup or gland.

The types of valves generally used in water-supply systems are gate, globe, angle and check valves. They are manufactured with different types of construction. One example of each type of valve is shown in Fig. 9.2.

Gate valves (see Fig. 9.2a) control flow by sliding a disk perpendicular to the water flow.

The disk has machined faces that fit tightly against the sloping faces of the seat rings when the hand wheel is turned to push the disk against the seat. This type of valve is usually used in locations where it can be left completely open or closed for long periods of time.

Globe valves (see Fig. 9.2b) control flow by changing the size of the passage through which water can flow past the valves. Turning a hand wheel moves a disk attached at the end of the valve stem to vary the passage area. When the valve is open, the water turns 90° to pass through an orifice enclosed by the seat and then turns 90° again past the disk, to continue in the original direction. Flow can be completely stopped by turning the hand wheel to compress the disk or a gasket on it against the seat. This type of valve usually is used in faucets.

Angle valves (see Fig. 9.2c) are similar to globe valves but eliminate one 90° turn of the water flow. Water is discharged from the valves perpendicular to the inflow direction.

Check valves (see Fig. 9.2d) are used to prevent reversal of flow in a pipe. In the valves, water must flow through an opening with which is associated a movable plug. When water flows in the desired direction, the plug automatically moves out of the way; however, a reverse flow forces the plug into the opening, to seal it.

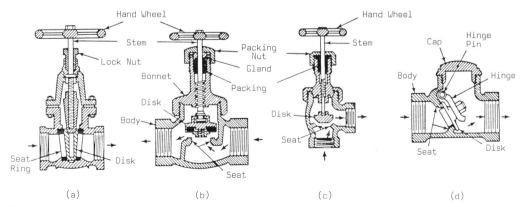

Fig. 9.2. Examples of valves used in water distribution systems. (a) Gate valve. (b) Globe valve. (c) Angle valve. (d) Check valve.

Pipe Supports

When standard pipe is used for water supply in a building, stresses due to ordinary water pressure are well within the capacity of the pipe material. Unless the pipe is supported at short enough intervals, however, the weight of the pipe and its contents may overstress the pipe material. Also, horizontal pipes may sag and vertical pipes may buckle. To prevent such undesirable conditions, pipes should be supported or braced at appropriate intervals.

Generally, it is sufficient to support vertical pipes at their base and at every floor. Maximum support spacing for horizontal pipes depends on pipe diameter and material. Building codes may set limits such as the following:

Threaded pipe, 1 in. or less
in diameter . . . .   8 ft
Threaded pipe, $1\frac{1}{4}$ in. or more
in diameter . . . .   12 ft
Copper tubing, $1\frac{1}{4}$ in. or less
in diameter . . . .   6 ft
Copper tubing, $1\frac{1}{2}$ in. or more
in diameter . . . .   10 ft

While the supports should be firmly attached to the building, they should permit movement of the pipe due to thermal dimensional changes or differences in settlement of building and pipe. Risers should pass through floors preferably through sleeves and transfer their load to the floors through tight fitting collars. Horizontal pipe runs may be carried on rings or hooks on metal hangers attached to the underside of floors.

Horizontal pipe runs should not be truly horizontal. They should have a minimum slope of about $\frac{1}{4}$ in. per ft toward the nearest drain valve.

Expansion and Contraction

Ferrous pipe 100 ft long changes in length about $\frac{3}{4}$ in. for every 100°F change in temperature. Copper changes in length about 50% more. Consequently, for long pipe runs, especially for hot-water pipe, special provision must be made for dimensional changes. Note that if water temperature is changed from 40 to 212°F, a copper pipe 100 ft long carrying the water will change 2 in. in length, some plastic pipe, 10 in.

To provide for expansion and contraction, expansion joints should be incorporated in pipelines. Such joints should be spaced not more than 50 ft apart in hot-water pipe. While special fittings are available for the purpose, flexible connections are a common means of providing for expansion. Frequently, such connections consist of a simple U bend or a spiral coil, which permits springlike absorption of pipe movements.

Pipe Insulation

The advantages of thermal insulation around hot-water pipes to prevent loss of heat are obvious. But thermal insulation is also advantageous for cold-water pipes. It can prevent the water from freezing when heat is shut off in a building in cold weather. Insulation can also prevent formation of condensation on cold pipes. Many different types of insulation are available for convenient, economical installation on pipes.

Pipe Noise

If proper precautions are not taken in pipe installation, pipes can produce annoying noises in buildings. Among the most frequent causes of pipe noise are water hammer, flow in undersize pipes, restraint against expansion and contraction of pipes and transmission of vibrations from pipes to building components.

Water hammer is caused by pressures developing during sudden changes in water velocity or sudden stoppage of flow. The result is a banging sound. It frequently results from rapid closing of valves, but it also may be produced by other means, such as displacing air from a closed tank or pipe from the top.

Water hammer can be prevented by filling a closed tank or pipe from the bottom while allowing the air to escape from the top. Water hammer also can be prevented by installing on pipelines air chambers or other types of water-hammer arresters. These generally act as a cushion to dissipate the pressures.

The roaring noise of flow in undersize pipes can be prevented by use of diameters more appropriate to the flow and pressures or by installation of pressure-reducing valves.

Noise from pipe expansion and contraction or

from vibrations can be prevented by avoiding direct contact between pipes or pipe supports and building components. Pipes and their supports should be isolated from the building with sound deadening materials.

Plumbing Fixtures

A prime objective of the water-supply system of a building is to deliver water to plumbing fixtures. These include kitchen sinks, water closets, urinals, bathtubs, showers, lavatories, drinking fountains, laundry trays and slop (service) sinks. Many building codes and the National Plumbing Code list the minimum number of each type of fixture that must be installed in buildings of various occupancies.

The plumbing fixtures are at the terminals of the water-supply system and the start of the wastewater system. To a large extent, the flow from the fixtures determines the quantities of wastewater to be drained from the building.

Materials used for plumbing fixtures must meet the sanitation requirement that they be smooth and nonabsorbent, so that wastes will be completely removed by flowing water. Consequently, building codes generally limit fixture construction to such materials as enameled cast-iron, vitreous-glazed earthenware, enameled steel, stainless steel, vitreous china and certain plastics. The fixtures are generally selected to satisfy federal specifications or product standards promulgated by the U.S. Department of Commerce.

Traps are required near the drains of most fixtures. Some fixtures, such as the water closet in Fig. 9.3, have integral traps. Others, such as lavatories and bathtubs, must be provided with separate traps. The traps should provide a water seal of at least 2 in., but not more than 4 in. Cleanouts should be provided on all traps, except those integral with the fixtures or those having a portion that can be removed easily for cleaning of the interior.

Water closets consist of a bowl and integral trap, which always contain water, and a tank or a flushometer valve, which supplies water for flushing the bowl. The passage through the trap to the discharge usually is large enough to pass a solid ball 2 to 3 in. in diameter. Flush-

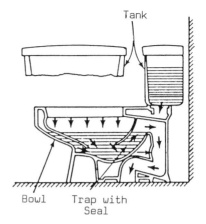

Fig. 9.3. Siphon-jet water closet.

ometer valves generally require a pressure of at least 15 psi for operation. Tanks, in contrast, are raised above the water level in the bowl so that gravity provides sufficient pressure for flushing.

The cleansing action of water flow in the bowl may be achieved in any of several different ways. One method is illustrated by the *siphon-jet* type pictured in Fig. 9.3. The tank discharges water around the rim and also jets water into the upleg of the trap. As a result, the contents of the bowl are siphoned out of the downleg of the discharge pipe. Other types of action include the *siphon-vortex*, in which water from the rim washes the bowl, creates a vortex, becomes a jet and discharges by siphonage; the *washdown*, in which pressure buildup causes the upleg to overflow and create a discharge siphon; and the *blowout*, used with a flushometer valve, which projects a strong jet into the upleg to produce the discharge.

Air Gaps. To prevent backflow of wastewater into the water supply at plumbing fixtures, an air gap must be provided between the fixture water-supply outlet and the flood-level rim of the receptacle. Building codes usually require a minimum gap of 1 to 2 in. for outlets not affected by a nearby wall and from $1\frac{1}{2}$ to 3 in. for outlets close to a wall.

In addition to the usual drain at the lowest point, receptacles generally are provided with a drain at the flood-level rim to prevent water from overflowing. The overflow should discharge into the wastewater system on the fixture side of the trap.

Methods of Distributing Water

Some plumbing fixtures are served only by cold water, and some by both hot and cold water. Hot water may be delivered at a temperature between 120 and 160°F.

For economy, pipe runs should be as short as possible. Theoretically, therefore, pipes should be laid in a straight line from water source to points of use. Generally, this is impractical, for many reasons. For one thing, in many buildings, for aesthetic reasons, pipes have to be concealed from view, within walls or between floors and ceilings. As a result, direct routes from water sources to plumbing fixtures are unavailable. Also, the paths of pipes may be blocked by structural members, requiring detours. In addition, it may be necessary to keep the pipes of different plumbing subsystems bunched together to avoid cutting numerous separate holes in the building structure for passage of pipes and for convenience of maintenance and repair.

Cold water may be conveyed to plumbing fixtures under the pressure of a water source, such as a public water main, by pumps, or by gravity flow from elevated storage tanks. Hot water may be distributed by pressure of a water source, by pumps, or by temperature differences. In the last case, the hot water rises above a heater or hot-water storage tank because it is lighter in weight than cold water. If the water is in a closed loop that is heated at the bottom, the water will rise in one leg and drop in the other leg on cooling, because of the difference in pressure in the two legs. As a result, there will be continuous circulation of water in the loop. Hot water can be withdrawn from either leg of the loop.

The water-distribution system should be so laid out that, at each plumbing fixture requiring both hot and cold water, the pressures at the outlets for both supplies should be nearly equal. This is especially desirable where mixing valves may be installed, to prevent the supply at a higher pressure from forcing its way into the lower-pressure supply when the valves are opened to mix hot and cold water. Pipe sizes and types should be selected to balance loss of pressure head due to friction in the hot- and cold-water pipes, despite differences in pipe lengths and sudden large demands for water from either supply.

Upfeed Water Distribution

To prevent rapid wear of valves, such as faucets, water is usually supplied to building distribution systems at pressures not more than about 50 psi. This pressure is large enough to raise water from four to six stories upward and still retain desired pressures at plumbing fixtures (see Table 9.1). Hence, in low buildings, cold water can be distributed by the upfeed method, in which at each story plumbing fixtures are served by branch pipes from risers that carry water upward under pressure from the water source.

Figure 9.4 illustrates a two-dimensional, schematic routing diagram for upfeed distribution in a small building. The diagram has been simplified by omission of many valves and controls that usually are necessary and the pipes do not follow the actual paths that would be used in walls and between floors and ceilings.

In Fig. 9.4, cold water is distributed under pressure from a public water main.

The hot-water distribution is a discontinuous system. Hot water rises from the water heater in the basement to the upper levels under pressure from the cold-water boiler fill. This type of distribution has the disadvantage that, when the system is not in use, hot water in the risers cools and, when a faucet is opened, cooled-off water flows until hot water reaches the faucet.

In long, low buildings, friction may reduce the flow of hot water to unacceptable levels. In such cases, pumps have to be used to maintain pressure.

Downfeed Water Distribution

For buildings more than about four stories high, it is conventional to pump water to one or more elevated storage tanks, from which pipes convey the water downward to plumbing fixtures and water heaters. Water in the lower portion of an elevated tank often is reserved for fire-fighting purposes (see Fig. 9.5).

Generally, also, the tank is partitioned to pro-

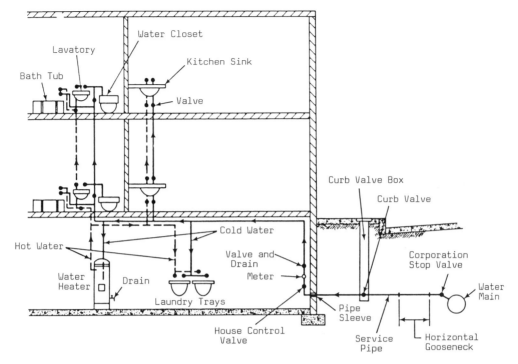

Fig. 9.4. Upfeed water distribution system for two-story building.

vide independent, side-by-side chambers, each with identical piping and controls. During hours of low demand, a chamber can be emptied, cleaned and repaired, if necessary, while the other chamber supplies water as needed. Float-operated electric switches in the chambers control the pumps supplying water to the tank. When the water level in the tank falls below a specific elevation, the switches start a pump, and when the water level becomes sufficiently high, the switches stop the pump.

Usually, at least two pumps are installed to supply each tank. One pump is used for normal operation. The other is a standby, for use when the first pump is inoperative.

When a pump operates to supply a tank, it may draw so much water from a public main that the pressure in the main would be considerably reduced. To avoid such a condition, water often is stored in a suction tank at the bottom of the building for use by the pumps. The tank is refilled automatically from the public main. Because refilling can take place even when the pumps are not operating, water can be drawn from the public main without much pressure drop.

Figure 9.5 illustrates a simplified, two-dimensional, schematic routing diagram for a downfeed distribution system of a type that might be used for buildings up to about 20 stories high. The service pipe supplies water to a suction tank in the basement. A pump draws water from that tank and pumps the water to a roof tank. Water in the lower part of that tank is reserved for fire fighting. Water from the upper part of the tank flows into a house header, which supplies two downfeed risers.

One riser conveys water to a water heater. The other feeds water to a cold-water header in the penthouse. That header, in turn, serves downfeed risers (only two of which are shown in Fig. 9.5) for supplying cold water to plumbing fixtures in each story.

For fire fighting, a pump must be capable of refilling the tank at a rate of at least 65 gpm.

Hot-water distribution in Fig. 9.5 is a continuous system. Hot water from the heater rises to a hot-water header in the penthouse. That header distributes the water to downfeed risers (only two of which are shown in Fig. 9.5) for supplying hot water to plumbing fixtures in each story. Unused hot water is recirculated to

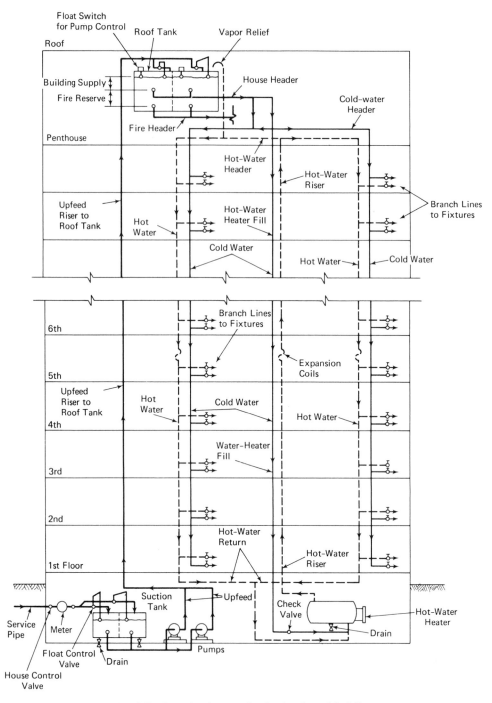

Fig. 9.5.   Downfeed water distribution for tall building.

the water heater. A check valve in the water heater fill line prevents backflow from the heater into the fill pipe.

Tall buildings may be divided into zones, each of which is served by a separate downfeed system. (The first few stories may be supplied by an upfeed system under pressure from a public main.) Each zone has at its top its own storage tank, supplied water by its own set of pumps in the basement. All the pumps draw on a common suction tank in the basement. Also, each zone has at its base its own water heater and a hot-water circulation system. In effect, the distribution in each zone is much like that shown in Fig. 9.5.

Standpipes for Fire Fighting

A standpipe is a vertical pipe in which water is stored under pressure or which can be rapidly supplied with water under pressure. Standpipes are usually required for fire fighting in buildings more than six stories or 75 ft in height or in buildings with a large floor area in each story; for example, over 7,500 sq ft per floor in buildings four or more stories high.

Standpipes may be supplied with water from a source separate from the regular water-distribution system of a building or from the fire reserve of an elevated tank (see Fig. 9.5). The downfeed risers should be so located that every point of every floor can be reached by a 20-ft stream from a nozzle attached to 125 ft or less of hose connected to a riser outlet valve.

All risers should be cross connected at, or below, the street entrance-floor level. In buildings that have zoned water-distribution systems, standpipes in each zone should be cross connected below, or in, the story with the lowest hose outlets from the water source in each zone. Cross connections should have a diameter at least as large as that of the largest riser supplied by the cross connection.

The cross connection between standpipe risers at the street entrance-floor level and at the bottom of each zone should extend to the exterior walls of the building and terminate in a siamese connection on the outside, 18 to 36 in. above grade. A siamese connection is a wye to which two fire hoses can be connected. The hoses, when connected by fire fighters to a street

hydrant or to a mobile pump, can supply the standpipes with water in an emergency. Building codes may require a siamese connection for each 300 ft of building wall facing on a street or public place.

Hose Stations. Standpipes should be placed within stairway enclosures, or as near such an enclosure as possible, for safe access to hose outlets by fire fighters. Risers up to 150 ft high, measured to the highest hose outlet, should be at least 4 in. in diameter. Taller risers should be at least 6 in. in diameter.

On each floor served by a standpipe, a $2\frac{1}{2}$-in.-diameter hose outlet should be provided for fire fighters. The valve should be readily accessible from a floor or a stairway landing and should be placed about 5 or 6 ft above that level.

Hose stations should be located at the standpipes. The hoses should be stored on hose racks. Local building codes generally specify the minimum diameter for hoses, depending on hose material.

The standpipe system should be zoned by the use of gravity tanks, automatic fire pumps, pressure tanks and street pressure so that the maximum gage pressure at the inlet of any hose valve does not exceed 160 psi. Building codes often require that the hose valve be equipped with an adjustable-type, pressure-reducing valve to limit the pressure on the downstream side of the hose valve. Usually, this pressure should be less than 50 psi for a discharge of 200 gpm from a 1-in. hose nozzle.

Fire Sprinklers

Automatic sprinklers have an excellent record for curtailing property damage and loss of life in buildings due to fires. Sprinklers are particularly effective, when they are properly installed and maintained, in early extinguishment of fires, preventing spread of fire and smoke, and cooling hot gases (see Sec. 6.4).

Water is supplied to sprinklers from sources such as those used for standpipes. The water is conveyed through upfeed or downfeed risers to a network of horizontal pipes installed at the roof and at ceilings in each story. On the horizontal pipes, at appropriate intervals, heat-sensitive sprinkler heads are installed. Any of these heads will discharge water when its temperature is raised to a predetermined level. Often, an alarm is connected to the system and set to give warning when water flows in the sprinkler pipes.

Building codes generally contain provisions for design and installation of automatic sprinkler systems, including specifications indicating the types of buildings and locations in which sprinklers must be installed. Often, these codes adopt by reference the National Fire Protection Association Standard, "Installation of Sprinkler Systems," NFPA 13. Insurance company requirements for sprinklers also should be checked.

Types of Sprinkler Systems. Several different types of sprinkler installations employ the basic system. Choice of type depends mainly on the temperature to be maintained in the building, the amount of damage a fire may cause, anticipated rate of spread of fires and total fire load.

In wet-pipe systems, the pipes always contain water. Hence, temperatures in the building must always be maintained above freezing. Water does not flow in the piping until a head starts discharging water. An alarm may be installed to signal when this happens.

Dry-pipe systems differ from wet-pipe systems in that the piping normally does not contain water. Dry-pipe systems are needed in locations in buildings where temperatures may drop below freezing. In such systems, when a head operates, air held under pressure in the piping is released to the atmosphere. This allows water to enter the piping and discharge through the open head. The valve that controls the water flow must be installed in a heated area or in a heated enclosure, to prevent water in piping at the valve from freezing.

Deluge systems permit all heads to open and discharge water simultaneously. A detection system installed throughout the area to be so protected actuates a deluge valve, which permits water to flow through the piping to the heads. The purposes of deluge systems are to isolate fires and cool exposed equipment not initially involved in the fires. Such systems are useful in chemical plants where flammable materials are handled.

Preaction systems are similar to deluge systems but the sprinkler heads are kept sealed until some time after the fire detection system sounds an alarm. The delay enables fire fighters to extinguish the fire before the heads open. As a result, total damage from fire and water may be less than when other systems are used. Preaction systems therefore are useful where valuable materials that may be damaged by water are stored.

Dry-pipe and preaction combined systems are suitable for large structures, such as piers, where piping would be needed in unheated areas and more than one dry-pipe system would be required. In such combined systems, the pipes contain compressed air and, instead of deluge valves, dry-pipe valves are installed in parallel. When the detection system discerns a fire, an alarm is sounded and water is fed into the sprinkler pipes.

Recycling systems are similar to other systems described but turn off the water when the fire is extinguished. If fire flares up again, heads will again discharge water. The systems are useful for protecting valuable materials that may be damaged by water.

Sprinkler-Pipe Materials. NFPA 13 requires that sprinkler pipes be made of wrought steel (ASTM A120), which preferably should be galvanized, or wrought iron (ASTM A72). The pipe should be capable of withstanding a pressure of at least 175 psi. (Standard-wall pipe may be used for pressures up to 300 psi.) NFPA 13 also contains tables that give the minimum diameters of pipe that may be used with specific numbers of sprinklers for various types of building occupancies.

Fittings on sprinkler pipes should be capable of resisting the working pressures in the pipes, but at least 175 psi. Fittings may be made of cast or malleable iron.

Connections should be made with screwed or flanged fittings. Joints in risers and feed mains, however, may be welded.

Spacing and Location of Sprinklers. In buildings with light-hazard occupancies, NFPA 13 allows a protection area per sprinkler of up to 200 sq ft for sprinklers under smooth ceilings or beams and girders. Under open wood-joist construction, the protection area is limited to 130 sq ft per sprinkler. For other types of construction, the protection area may be taken up to 168 sq ft.

In buildings with ordinary-hazard occupancies, NFPA 13 permits in general a protection area up to 130 sq ft per sprinkler. For high-piled storage, however, the protection area is limited to 100 sq ft. (High-piled storage applies to closely packed material more than 15 ft high or material on pallets or in racks piled more than 12 ft high.) For extra-hazard occupancies, the protection area per sprinkler may not exceed 100 sq ft.

Horizontal branch pipes with sprinklers are usually laid out in parallel lines in each story. In buildings with light- and ordinary-hazard occupancies, the maximum distance permitted between horizontal branch pipes and between sprinklers on those pipes is 15 ft. For high-piled storage, however, maximum spacing generally is limited to 12 ft. For extra-hazard occupancies, maximum spacing also is 12 ft. The distance from a wall to an end branch pipe or to an end sprinkler on a branch pipe should not exceed one-half the allowable spacing for the pipes or sprinklers, respectively.

Locations of sprinklers should be staggered on successive branch pipes in buildings with extra-hazard occupancies. Sprinklers also should be staggered in ordinary-hazard buildings when sprinkler spacing on branch pipes exceeds 12 ft or when the sprinklers are placed under solid beams spaced 3 to $7\frac{1}{2}$ ft on centers. In areas where aesthetics is important, however, the sprinklers need not be staggered.

Sprinklers may be installed above a branch pipe (upright) or below the pipe (pendent). Generally, the upright position is preferable, so that the heads are protected against damage by vehicles or other objects passing below the pipes.

Sprinklers should be located to prevent structural members from interfering with the discharge pattern. Where branch pipes are installed transverse to lines of beams, the deflectors of sprinklers, which direct the discharge, should preferably be located above the bottom of the beam. If the branch pipes are installed under beams, sprinkler deflectors should be placed 1 to 4 in. below the beams, but not more than 14 in. below combustible ceilings nor more than 16 in. below noncombustible ceilings. Under smooth-ceiling construction, deflectors of upright sprinklers should be at least 3 in. and of pendent sprinklers at least $2\frac{1}{2}$ in. below the ceiling; but they should be not more than 10-in. below combustible ceilings nor more than 12 in. below noncombustible ceilings.

Deflectors of sprinklers should be parallel to ceilings, roofs or the incline of stairs but horizontal when installed in the peak of a pitched roof.

Tests of the Water-Supply System

The water-supply system of a building may be tested when a section has been completed and when the whole system has been finished. In each test, the completed portion should be proven tight under a pressure at least equal to the working pressure and preferably 50% greater. Water used in the tests should be obtained from a potable source.

9.5. SIZING OF WATER-SUPPLY PIPES

Building codes require that the water-distribution system be designed and adjusted to supply fixtures and equipment with the amount of potable water required for proper use, cleansing and performance. Pipe diameters should be determined by application of the principles of hydraulics (see Sec. 9.1). While economy dictates use of the smallest sizes of pipes permitted by building-code requirements, other factors often make larger sizes advisable. These factors include:

1. Pressure at the water-supply source, usually the public main, psi.
2. Pressure required at the outlets of each fixture, psi.

3. Loss of pressure because of height of outlets above the source, psi.
4. Water demand (requirements) for the whole system and for each of its parts, gpm.
5. Pressure losses, psi, from friction of water flow in pipe.
6. Limitations on velocity of water flow, fps, to prevent noise and erosion.

Pipe Sizes for Small Buildings

Sizes of pipes for small buildings, such as single-family houses, can usually be determined from experience of the designer and applicable building-code requirements, without extensive calculations. For short branches to individual fixtures, for example, the minimum pipe diameters listed in Table 9.1 generally will be satisfactory. Usually also, the following diameters can be used for the mains supplying water to the fixture branches:

$\frac{1}{2}$-in. for mains with up to three $\frac{3}{8}$-in. branches
$\frac{3}{4}$-in. for mains with up to three $\frac{1}{2}$-in. or five $\frac{3}{8}$-in. branches
1-in. for mains with up to three $\frac{3}{4}$-in. or eight $\frac{1}{2}$-in. or 15 $\frac{3}{8}$-in. branches

The adequacy of these sizes, however, depends on the pressure available at the water source and the probability of simultaneous use of the plumbing fixtures. Determination of pipe sizes for more complex distribution systems requires detailed analysis with consideration given to the six design factors listed previously.

Water Demand

For each fixture in a building, a maximum requirement for water flow, gpm, can be anticipated. Table 9.1 indicates the minimum flow rate and pressure that building codes require must be provided. The maximum flow may be considerably larger. Branch pipes to each fixture should be sized to provide the maximum flow and minimum pressure the fixture will require.

Mains serving these branches, however, need not be sized to handle the sum of the maximum flows for all branches served. It is generally unlikely that all the fixtures would be supplying maximum flow simultaneously or even that all the fixtures would be operating at the same time. Consequently, the diameters of the mains need be sized only for the probable maximum water demand.

In practice, the probable flow is estimated by weighting the maximum flow in accordance with the probability of fixtures being in use. The estimate is based on the concept of *fixture units*.

A specific number of fixture units, as listed in Table 9.2, is assigned to each type of plumbing fixture. These values take into account:

1. The anticipated rate of water flow from the fixture outlet, gpm
2. The average duration of flow, minutes, when the fixture is used
3. The frequency with which the fixture is likely to be used

Table 9.2. Water Demand of Plumbing Fixtures, Fixture Units[a]

Plumbing Fixture	Private Use	Public Use
Bathtub (with or without shower)	2	4
Dishwasher (home)	2	...
Drinking fountain	1	2
Hospital sink (flushing rim)	...	10
Kitchen sink	2	4
Lavatory	1	2
Laundry tray	2	4
Service sink	2	4
Shower (separate, each head)	2	4
Urinal (tank)	...	3
Urinal (flush valve)	...	5
Washing machine	2	4
Washup sink (circular spray)	...	4
Water closet (tank)	3	5
Water closet (flush valve)	6	10
Bathroom group (tank)	6	...
Bathroom group (flush valve)	8	...

[a]Values are for total demand, hot and cold. For either hot or cold demand by itself, use 75% of the total fixture unit value. (Courtesy Copper Development Association, New York, N.Y.)

Table 9.3. Estimate of Water Demand, Gpm, for Total Fixture Units Served

Total fixture units served	With mostly tank-type w.c.	With mostly flush-valve w.c.	Total fixture units served	With any type of w.c.
5	4	22	1,000	208
10	8	27	1,500	270
20	14	35	2,000	327
30	20	42	3,000	435
40	24	46	4,000	560
50	28	51	5,000	670
100	43	68	6,000	770
200	65	91	7,000	870
300	85	110	8,000	965
400	106	127	9,000	1,060
500	124	142	10,000	1,160
600	143	157	15,000	1,640
700	161	170	20,000	2,110
800	178	183	25,000	2,575
900	194	197	30,000	3,040

The ratings in fixture units listed in Table 9.2 represent the relative loading of a water-distribution system by the different types of plumbing fixtures. The sum of the ratings for any part or all of a system is a measure of the load the combination of fixtures would impose if all were operating. The probable maximum water demand, gpm, can be determined from the total number of fixture units served by any part of a system by use of Table 9.3.

The demand obtained from Table 9.3 applies to fixtures that are used intermittently. If the system serves fixtures, such as air-conditioning units, lawn sprinklers or hose bibs, that are used continuously, the demand of these fixtures should be added to the intermittent demand.

Example. To illustrate the method of computing the maximum probable water demand of a distribution system, a building is assumed to contain the number and type of fixtures listed in the first two columns of Table 9.4. The water demand for each type of fixture is obtained from Table 9.2 and listed in the third column of Table 9.4. The number of fixtures of each type is then multiplied by the rating in fixture units of that type and entered in the last column. The resulting products are then totaled. For this sum, the water demand is found by interpolation from Table 9.3 to be 57 gpm, for a system with predominantly flush valves.

Table 9.4. Computation of Water Demand, Fixture Units

Number of fixtures	Type of fixtures	Fixture rating	Total fixture units
3	Flush-valve w.c. (public)	10	30
2	Flush-valve w.c. (private)	6	12
2	Flush-valve urinals (public)	5	10
4	Lavatories (public)	2	8
2	Lavatories (private)	1	2
2	Showers (private)	2	4
1	Kitchen sink	2	2
	Total		68

Effects of Pressure

The rate of flow, cfs, in a pipe is determined by Eq. (9.1), $Q = AV$, where A is the pipe area, sq ft, and V the water velocity, fps. In general, it is desirable to keep V less than about 8 fps to avoid noise and prevent erosion at valve seats. Consequently, the pipe area should be at least Q divided by 8.

For example, for the water demand in the preceding example of 57 gpm, the minimum pipe size can be computed as follows: Gallons per minute can be converted to cubic feet per minute by dividing by 7.5, and to cubic feet per second by dividing the quotient by 60. Thus, 57 gpm = 57/(7.5 × 60) = 0.1267 cfs. From Eq. (9.1), for a specified maximum velocity $V = 5$ fps, the pipe area should be at least 0.1267/5 = 0.0253 sq ft, or 3.648 sq in. This indicates that the pipe diameter should be at least 2.16 in. Hence, a $2\frac{1}{2}$-in. pipe, the nearest standard size should be used, to satisfy the requirement for maximum velocity. Required pressures at points of water use, however, may require a larger size of pipe.

The minimum pressures at plumbing fixtures generally required by building codes are listed in Table 9.1. These pressures are those that remain when the pressure drop due to height of outlet above the water source and the pressure lost by friction in pipes are deducted from the pressure at the water source. The pressure loss due to height can be computed from Eq. (9.2); that is, the pressure loss, psi, equals the height, ft, times 0.433. The pressure reduction caused by pipe friction depends, for a given length of pipe and rate of flow, on pipe diameter. Hence, a pipe size can be selected to create a pressure drop in the pipe to provide the required pressure at a plumbing fixture, when the pressure at the water source is known. If the pipe diameter is too large, the friction loss will be too small and the pressure at the fixture will be high. If the pipe size is too small, the friction loss will be too large and the pressure at the fixture will be too small.

Selection of pipe diameter can be made with the Hazen-Williams formula [see Eq. (9.4)] or more conveniently from either Fig. 9.6 or 9.7. Figure 9.6 was plotted for smooth pipe, such as copper tubing, which will not become rough

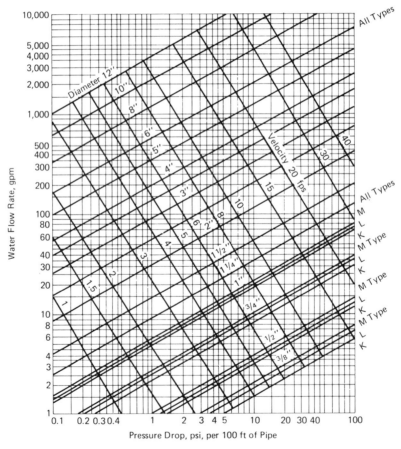

Fig. 9.6. Chart for determination of flow in copper tubing and other pipes that will be smooth after 15 to 20 years of use.

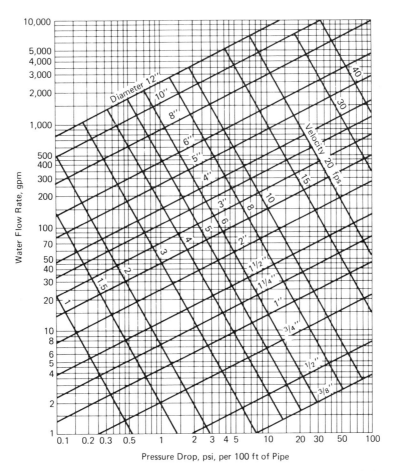

Fig. 9.7. Chart for determination of flow in pipes, such as galvanized steel and wrought iron, that will be fairly rough after 15 to 20 years of use.

over a period of 15 years or more. Figure 9.7 was plotted for fairly rough pipe. Cast iron and galvanized steel and wrought iron are considered to be in this category, although they may be smooth when initially installed.

Figures 9.6 and 9.7 are based on pressure drops due to friction in a 100-ft length of pipe of constant diameter. For lengths of pipe different from 100 ft, the pressure drop can be determined by proportion. Thus, the friction loss in a pipe 300 ft long is obtained by multiplying the value obtained from Fig. 9.6 or 9.7 by 3. Similarly, the pressure drop in a pipe 50-ft long is obtained by multiplying the chart reading by 50/100, or $\frac{1}{2}$.

In addition to friction loss in pipes, there also are friction losses in meters, valves and fittings. These pressure drops can be expressed for convenience as equivalent lengths of pipe of a spe-

cific diameter. Table 9.5 indicates typical allowances for friction loss for several sizes and types of fittings and valves.

In practice, pipe diameter is selected from Fig. 9.6 or 9.7 as the nearest larger size to the point of intersection of a horizontal line at the specified rate of flow, gpm, and a vertical line at the desired pressure drop in 100 ft of pipe due to friction. For example, suppose a 500-ft length of pipe carrying 100 gpm is required to lose 10 psi due to friction. The loss per 100 ft of pipe is 10/5 = 2 psi. If the pipe may be considered fairly rough, the pipe size can be determined from Fig. 9.7 as 3 in., at the intersection of the horizontal line through 100 gpm and the vertical line through 2 psi. Also, from Fig. 9.7, the velocity can be estimated as 4.7 fps, as determined by a velocity line through the point of intersection.

Table 9.5. Allowances for Friction Losses in Valves and Fittings, Expressed as Equivalent Length of Pipe, Ft[a]

Fitting size, in.	Standard ells		90° tee		Coupling	Gate valve	Globe valve
	90°	45°	Side branch	Straight run			
$\frac{3}{8}$	0.5	0.3	0.75	0.15	0.15	0.1	4
$\frac{1}{2}$	1	0.6	1.5	0.3	0.3	0.2	7.5
$\frac{3}{4}$	1.25	0.75	2	0.4	0.4	0.25	10
1	1.5	1.0	2.5	0.45	0.45	0.3	12.5
$1\frac{1}{4}$	2	1.2	3	0.6	0.6	0.4	18
$1\frac{1}{2}$	2.5	1.5	3.5	0.8	0.8	0.5	23
2	3.5	2	5	1	1	0.7	28
$2\frac{1}{2}$	4	2.5	6	1.3	1.3	0.8	33
3	5	3	7.5	1.5	1.5	1	40
$3\frac{1}{2}$	6	3.5	9	1.8	1.8	1.2	50
4	7	4	10.5	2	2	1.4	63
5	9	5	13	2.5	2.5	1.7	70
6	10	6	15	3	3	2	84

[a]Allowances are for streamlined soldered fittings and recessed threaded fittings. For threaded fittings, double the allowances shown in the table. (Courtesy Copper Development Association.)

SECTIONS 9.3, 9.4 AND 9.5

References

National Plumbing Code, American National Standards Institute.

Uniform Plumbing Code, International Association of Plumbing and Mechanical Officials.

B. Stein et al., *Mechanical and Electrical Equipment for Buildings*, 7th ed., Wiley, New York, 1986.

F. Merritt, *Building Design and Construction Handbook*, 4th ed., McGraw-Hill, New York, 1982.

Words and Terms

Cleanout
Demand
Fittings
Fixture
Fixture units
Head
Liquid-seal trap
Piping
Pressure
Sprinkler
Standpipe
Valves: cock, faucet, gate, globe, angle, check
Vent
Water closet

Significant Relations, Functions and Issues

Elements of a plumbing system: fixtures, piping, valves, cleanouts, traps, vents, meter.
Special plumbing requirements: structural support; provisions for expansion and contraction, insulation, noise.
Methods of water distribution: upfeed, downfeed.
Pipe sizing by fixture units.
Determination of flow in piping.

9.6. WASTEWATER-REMOVAL SYSTEMS

Water discharged from plumbing fixtures and other equipment and not used, waterborne wastes, storm water, water removed from drains or drain valves and any water that may be polluted must be removed speedily from buildings and piped to a treatment plant or an acceptable disposal point. Otherwise, obnoxious odors may develop or humans may be exposed to substances deleterious to their health.

A wastewater-removal system consists of a drainage system and a venting system. The drainage system includes all the pipes within the building that are installed to convey wastewater, rainwater or other liquid wastes to a

legal disposal point. The system does not include the mains of a public sewer system,
sewage-treatment plant or disposal area. The
venting system consists of pipes that provide
an air flow to or from the drainage system or
that circulate air within the drainage system to
protect trap seals from siphonage and back
pressure.

Many of the principles for designing wastewater-removal are given in Sec. 9.3. Many of
the design considerations for water-supply
systems discussed in Sec. 9.4 apply also to
wastewater-removal systems. Note, however,
that flow in the removal systems is generally
either horizontal or downward. Horizontal
branch pipes convey wastewater from fixture
drains in each story of a building to a vertical
pipe, or stack. This conducts the flow to a
main horizontal drain at the base of the building for removal to a sewer. The plumbing design, which should conform to the requirements
of the National Plumbing Code or the local
building code, should take into account the
space available for pipes, fittings and other appurtenances; settlement of the building; supports needed for the pipes; provision for expansion and contraction; offsets or bends in
stacks and maintenance requirements.

Elements of Wastewater Systems

Collection of wastewater for a drainage system
starts at a drain. It may be located in a plumbing fixture, in equipment that discharges water
or in a floor. A short length of pipe conducts
the wastewater from the drain to a trap. Next, a
branch pipe conveys the wastewater to a stack.
The branch pipe also may collect wastewater
from several drains.

The drainage pipes are classified as soil pipes
or waste pipes. *Soil pipes* carry the discharge
of water closets, urinals or fixtures having similar functions, with or without the discharge
from other fixtures, to the building drain. *Waste
pipes* convey only water free of fecal matter.
Pipes from kitchen sinks, lavatories, laundry
trays, bathtubs, showers, washing machines
and food-waste grinders are considered waste
pipes. In Fig. 9.8, the stack carrying only the
wastewater from a sink in each story is a waste
stack. The stack in Fig. 9.8 carrying waste-

water from water closets, bathtubs and lavatories is a soil stack. A building may have several
stacks. All may be soil stacks, or some may be
waste stacks.

The waste and soil stacks feed the wastewater
to the building, or house, drain, which must be
below the level of all other drains. The building
drain conducts the wastewater to the building
sewer outside the building, for delivery to a
main sewer (see Fig. 9.8). Cleanouts, accessible from a building floor, should be provided
for the building drain to permit removal of any
blockage that may occur.

A few building codes require that a trap, called
a building, or house, trap be inserted near the
end of the building drain to prevent circulation
of air between the drainage system of the building and the building sewer. If such a trap is
installed, it should be provided with a cleanout
and a relieving vent or fresh-air intake (see Fig.
9.8). Many engineers, however, believe that
such a trap is undesirable because it prevents
escape from the sewer of air and gases, and a
buildup of pressure from the gases may cause
an explosion.

If the building drain is below the level of the
main sewer, provision must be made to pump
the wastewater up to the sewer. For this purpose, the house drain may discharge into a
sump pit, from which the wastewater collected
may be pumped to the sewer by a pneumatic
ejector or a motor-driven pump.

Vents

If the waste and soil stacks were not supplied
with a continuous flow of air, discharge of water into them could develop air pressures in the
pipes that would siphon the liquid seal from
traps. Also, there could be a buildup of gases in
the stacks and branches. Consequently, vent
pipes must be connected to the waste and soil
stacks and branches to supply air from outdoors. Such pipes are represented in Fig. 9.8
by dashed lines.

A fixture may be connected by an individual
vent to a vent stack, as indicated for the sinks
in Fig. 9.8, or several fixtures may be served
by a branch vent, which extends to a vent stack.
The individual vents are connected to the waste

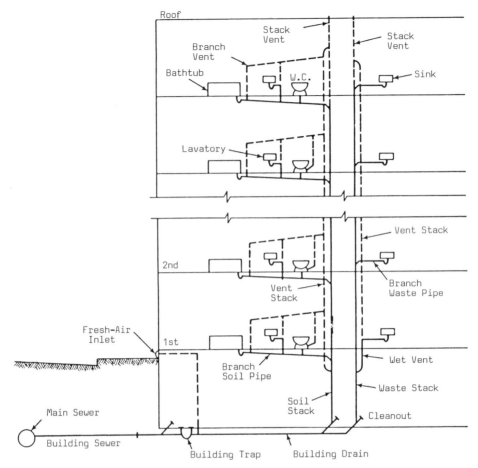

Fig. 9.8. Wastewater-removal system for a multistory building.

and soil branches downstream from the fixture traps. The venting system is discussed later in greater detail.

Pipe Materials

Above ground in buildings, soil, waste, vent and stormwater or leader pipes may be brass; copper; uncoated, extra-heavy, cast-iron soil pipe; threaded cast-iron pipe; galvanized wrought iron; galvanized steel; lead or plastics.

All underground building drains may be uncoated, extra-heavy, cast-iron soil pipe, brass or copper. Building sewers may be uncoated, extra-heavy, cast-iron soil pipe, vitreous pipe or asbestos-cement. Other materials may be used for drains and sewers if approved by the local building official.

Roof drains may be made of cast iron, bronze, copper, brass, stainless steel, lead or other equivalent corrosion-resistant material. Exterior stormwater leaders and gutters installed above ground may be made of sheet metal—copper, aluminum, galvanized steel, stainless steel or other equivalent corrosion-resistant material. Pipe made of galvanized steel, galvanized wrought iron, cast iron or brass may be used for the first 15 ft of vertical pipe, or leader, extending up from grade. Subsoil drains may be clay tile that is open jointed, horizontally split or perforated; open-jointed cast-iron soil pipe; porous concrete pipe; or open-jointed, horizontally split or perforated asbestos-cement pipe.

Fittings and Joints

Fittings such as the commonly used types shown in Fig. 9.9 are required wherever there

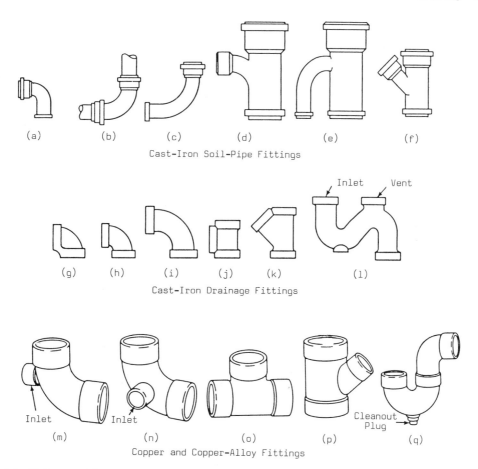

Fig. 9.9. Fittings used for sanitary drainage systems. Cast-iron soil-pipe fittings: (*a*) Quarter bend. (*b*) Short sweep. (*c*) Long sweep. (*d*) Tee. (*e*) Vent branch. (*f*) Wye branch. Cast-iron drainage fittings: (*g*) Short turn. (*h*) Long turn. (*i*) Extra-long turn. (*j*) Tee. (*k*) 45° wye branch. (*l*) S trap. Copper fittings: (*m*) Quarter bend with high-heel inlet. (*n*) Quarter bend with side inlet. (*o*) Tee. (*p*) Wye branch. (*q*) P trap.

is a change in direction in drainage piping. These fittings may be 45° wyes, long- or short-sweep quarter bends (90°), or sixth (60°), eighth (45°) or sixteenth (22.5°) bends, or a combination of these. Single and double sanitary tees and quarter bends may be used in drainage lines only where the direction of flow is from the horizontal to the vertical. Short sweeps not less than 3 in. in diameter may be used in soil and waste lines where the change in direction of flow is from the horizontal to the vertical, or vice versa. Short sweeps, however, also may be used for making necessary offsets between a ceiling and the next floor above.

Drainage-pipe fittings should have a smooth, continuous interior surface and no abrupt

changes in direction. In brief, fittings should not obstruct or retard flow of air or wastewater in pipes.

Sizes of fittings are designated by the diameters of the pipes with which they are to be used.

Fittings for cast-iron soil pipe come with bell-and-spigot ends (see Fig. 9.10), threaded ends

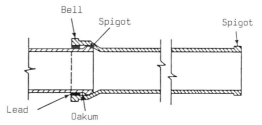

Fig. 9.10. Bell-and-spigot joint for cast-iron pipe.

or a combination of such ends. A bell-and-spigot joint is made by inserting a spigot within a bell, with the spigot end resting against the bottom of the bell hub. Then, an oakum gasket is forced into the bell and around the spigot with a calking tool and hammer. Next, the remainder of the space is filled with molten lead, which is calked after it cools, to compensate for shrinkage of the lead and insure that the space between spigot and bell is completely filled. More skill is required for making such joints than for threaded joints, but less precise cutting and fitting is needed. A few of the many different types of bell-and-spigot fittings for cast-iron soil pipe are shown in Fig. 9.9.

Threaded cast-iron fittings are generally referred to as drainage fittings, but they are sometimes called recessed fittings or Durham fittings. Made with a shoulder and recess, they are threaded so as to permit a joint to be made that is smooth and has an interior surface of the same diameter as the pipe when the pipe end is screwed up tightly against the shoulder. Examples of drainage fittings are shown in Fig. 9.9.

Soldered fittings usually are used for copper pipe and tubing. A few of the many types of fittings are shown in Fig. 9.9.

Pipe Supports

Pipes of wastewater-removal systems should be supported and braced in the same way as pipes of water-supply systems (see Sec. 9.4). Vertical pipes generally should be supported at every floor. Horizontal pipes should be supported at intervals not exceeding the following: cast-iron soil pipe, 5 ft and behind every hub; threaded pipe, 12 ft; copper tubing, 10 ft. Supports also should be provided at the bases of stacks.

Consideration should be given to the possibility of building settlement and its effects on vertical pipes and also to thermal expansion and contraction of pipes, especially when the pipes are made of copper.

Layout of Soil, Waste and Vent Piping

Wastewater-removal systems should be sized and laid out to permit use of the smallest-diameter pipes capable of rapidly carrying away the wastewater from fixtures without clogging the pipes, without creating annoying noises and without producing excessive pressure fluctuations at points where fixture drains connect to soil or waste stacks. Such pressure changes might siphon off the liquid seals in traps and force sewer gases back through the fixtures into the building. Positive or negative air pressure at the trap seal of a fixture should never be permitted to exceed 1 in. of water.

The drainage system is considered a nonpressure system. The pipes generally do not flow full. The discharge from a fixture drain is introduced to a stack through a stack fitting, which may be a long-turn tee-wye or a short-turn or sanitary tee. The fitting gives the flow a downward, vertical component. As the water accelerates downward under the action of gravity, it soon forms a sheet around the stack wall. If no flows enter the stack at lower levels to disrupt the sheet, it will remain unchanged in thickness and will flow at a terminal velocity, limited by friction, to the bottom of the stack. A core of air at the center of the stack is dragged downward with the wastewater by friction. This air should be supplied from outdoors through a stack vent (see Fig. 9.8), to prevent creation of a suction that would empty trap seals.

When the sheet of wastewater reaches the bottom of the stack, a bend turns the flow 90° into the building drain. Within a short distance, the wastewater drops from the upper part of the drain and flows along the lower part of the drain.

Slope of Horizontal Drainage Pipes. Building codes generally require that horizontal pipes have a uniform slope sufficient to insure a flow with a minimum velocity of 2 fps. The objective is to maintain a scouring action to prevent fouling of the pipes. Codes therefore often specify a minimum slope of $\frac{1}{4}$ in. per ft for horizontal piping 3 in. in diameter or less and $\frac{1}{8}$ in. per ft for larger pipe.

Because flow velocity increases with slope, greater slopes increase pipe carrying capacity. In branch pipes, however, high velocities can cause siphonage of trap seals. Therefore, use of larger-size pipes is preferable to steeper slopes for attaining required capacity of branch pipes.

Venting Systems. Various types of vents may be used to protect trap seals from siphonage or back pressure.

The *main vent* is the principal artery of the venting system. It supplies air to *branch vents*, which, in turn, convey it to individual vents and wastewater pipes.

Every building should have at least one main *vent stack*. It should extend undiminished in size and as directly as possible from outdoor air at least 6 in. above the roof to the building drain. The main vent should be so located as to provide a complete loop for circulation of air through the wastewater-removal system. As an alternative to direct extension through the roof, a vent stack may be connected with a stack vent, if the connection is made at least 6 in. above the flood-level rim of the highest fixture.

A *stack vent* is the extension of a soil or waste stack above the highest horizontal drain connected to the stack. This vent terminates above the roof.

Generally, a group of fixtures, consisting of one bathroom group and a kitchen sink or combination fixture, may be installed without individual fixture vents in a one-story building or on the top floor of a building—but only if each fixture drain connects independently to the soil stack and the water-closet and bathtub drains enter the stack at the same level. In that case, all the fixtures are vented through the stack extension, or stack vent.

An *individual*, or *back*, *vent* is a pipe installed to vent a fixture trap and connected with the venting system above the fixture served or terminated outdoors. To insure that the vent will adequately protect the trap, building codes limit the distance downstream that the vent opening may be placed from the trap. This distance generally ranges from $2\frac{1}{2}$ ft for a $1\frac{1}{4}$-in fixture drain to 10 ft for a 4-in. fixture drain, but not less than two pipe diameters. The vent opening should be located above the bottom of the discharge end of the trap.

To reduce the amount of piping required, two fixtures may be set back to back, on opposite sides of a wall, and supplied air by a single vent (*common vent*). In that case, however, the fixtures should discharge wastewater separately into a double fitting with inlets at the same level.

A *branch vent* is a pipe used to connect one or more individual vents to a vent stack or to a stack vent.

A *wet vent* is a pipe that serves both as a vent and as a drainage pipe for wastes other than those from water closets. This type of vent reduces the amount of piping from that required with individual vents. For example, a bathroom group of fixtures may be vented through the drain from a lavatory, kitchen sink or combination fixture if such a fixture has an individual vent (see Fig. 9.11).

A battery of fixtures is any group of similar fixtures that discharges into a common horizontal waste or soil branch. A battery of fixtures should be vented by a circuit or loop vent. (Building codes usually set a limit on the num-

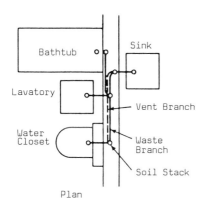
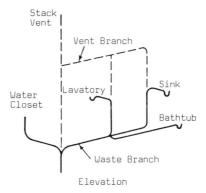

Fig. 9.11. Wet venting of bathtub drainage pipe.

ber of fixtures that may be included in a battery.)

A *circuit vent* is a branch vent that serves two or more traps and extends from the vent stack to a connection to the horizontal soil or waste branch just downstream from the farthest upstream connection to the branch (see Fig. 9.12*a*).

A *loop vent* is like a circuit vent but connects with a stack vent instead of a vent stack (see Fig. 9.12*b*). Thus, air can circulate around a loop.

Soil and waste stacks with more than ten branch intervals should be provided with a relief vent at each tenth interval installed, starting with the top floor. A *branch interval* is a section of stack at least 8 ft high between connections of horizontal branches. A *relief vent* provides circulation of air between drainage and venting systems. The lower end of a relief vent should connect to the soil or waste stack, through a wye, below the horizontal branch serving a floor where the vent is required. The upper end of the relief vent should connect to the vent stack, through a wye, at least 3 ft above that floor. Such vents help to balance the pressures that are continuously changing within a plumbing system.

Combined Drainage and Vent Systems

The possibility of considerable cost savings over the separate drainage and venting systems previously described has been created by development of combined drainage and venting systems.

A combination waste and vent system, intro-duced by the Western Plumbing Officials' Association, is one such system. It employs horizontal wet venting of one or more sinks or floor drains by means of a common waste and vent pipe adequately sized to provide free movement of air above the flow line of the pipe. Relief vents are connected at the beginning and end of the horizontal pipe. The traps of the fixtures are not individually vented. Some building codes permit such a system only where structural conditions preclude installation of a conventional system and, where it can be used, require larger than normal waste pipes and traps in the system.

The Sovent system, invented by Fritz Sommer, a Swiss, and promoted in the United States by the Copper Development Association, Inc., is another type of combination system. It requires drainage branches and soil stacks, with special fittings, but no individual and branch vents and no vent stacks. This system is discussed in Sec. 9.9.

Slopes and Connections for Vent Pipes

While the venting system is intended generally to convey only air to and from the drainage system, moisture may condense from the air onto the vent pipe walls. To remove the condensation from the venting system, all individual and branch vent pipes should be so sloped and connected as to conduct the moisture back to soil or waste pipes by gravity.

At horizontal drainage pipes, a vent pipe should be connected above middepth of those pipes. From the connection, the vent pipe

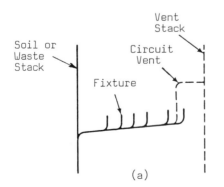
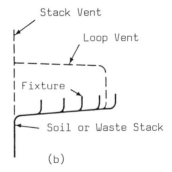

Fig. 9.12. (*a*) Circuit venting for a group of plumbing fixtures. (*b*) Loop venting for a group of plumbing fixtures.

should rise at an angle of at least 45° with the horizontal to an elevation at least 6 in. above the flood-level rim of the fixture it is venting. Above that elevation, the vent pipe may be offset horizontally or connected to a branch vent. The objective is to prevent flow of wastewater into the venting system.

For the same reason, the connection between a vent pipe and a vent stack or a stack vent should be at least 6 in. higher than the flood-level rim of the highest fixture served by the vent. Also, horizontal portions of branch, relief or loop vents should be at least 6 in. above the flood-level rim of the highest fixture served.

Indirect Waste Piping

To protect food, sterile materials and the potable water supply from contamination by backing up of wastes, most building codes require wastewater from food handling, laundry washers, commercial refrigerators, hospital sterilizers and sinks, and other possible sources of nondomestic pollution to be discharged through an indirect waste pipe. Such a pipe must not be connected to the building drainage pipes. Instead, the indirect waste pipe discharges wastewater into a plumbing fixture or receptacle, in such a way that an air gap is always present between the pipe outlet and the flood-level rim of the basin. Building codes generally specify the minimum size of air gap allowed. The plumbing fixture or receptacle is permitted to drain into the building drainage system.

Interceptors

An interceptor is a device for separating deleterious, hazardous or other undesirable materials from normal wastewater and for retaining those materials. After they are removed, the wastewater flows into the building drainage system. The retained materials should be removed periodically from the interceptor.

To prevent obstruction of pipes or interference with wastewater treatment, grease interceptors are installed in the drains serving kitchens, cafeterias, restaurants and other establishments where large amounts of grease may be discharged. Interceptors also are used to separate

out oil or other flammable liquids, sand or other solids, and chemicals.

Storm Drainage Pipes

Building codes require, in general, that all roofs, terraces, outside balconies, paved areas, yards, courts and courtyards be individually drained into a storm sewer. If, however, only a combined storm and sanitary sewer is available, stormwater may be discharged into such a sewer. Other exceptions are runoffs from one- and two-family dwellings. Building codes may permit such flows to be discharged on flat areas, such as driveways and lawns, if the stormwater will flow away from the buildings toward an unpaved area on the same lot and that area is capable of absorbing the discharge.

In any case, stormwater may not be discharged into sanitary sewers.

Drains carrying only clear water; for example, air-conditioning drips, pump drips, or cooling water, may discharge into the stormwater system, but only via an indirect waste pipe. That pipe should discharge into a trapped funnel or drain before the connection to the storm drain.

Roofs should be sloped so that rainwater flows by gravity to drains or to gutters, or troughs. Gutters are placed along the edges of a roof, or eaves, and sloped to a drain. A strainer usually is set over the drain to prevent solid materials from entering and clogging the drain pipe below (see Fig. 9.13a). A vertical drain pipe, called a leader or a conductor, carries the rainwater from the gutter drain down to an outlet above grade or to a building storm drain (see Fig. 9.13c).

Roof areas not draining to gutters should direct the flow of rainwater to roof drains. In general, these drains should be protected by strainers extending at least 4 in. above the roof. A strainer should have an inlet area above the roof at least 50% larger than the area of the leader to which the drain is connected (see Fig. 9.13b). For flat roofs used for sun decks, parking decks and similar areas normally serviced and maintained, roof-drain strainers may be of the flat-surface type and may be set flush with the deck. In that case, a strainer should have an inlet area at least twice the area of the leader to which the drain is connected.

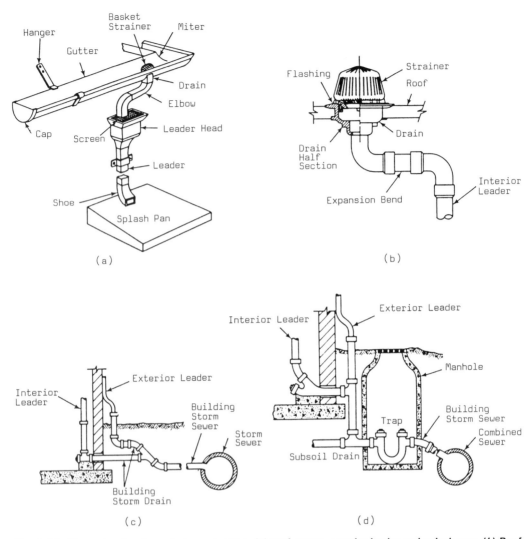

Fig. 9.13. Elements of a storm-drainage system. (a) roof gutter, exterior leader and splash pan. (b) Roof drain and top portion of interior leader. (c) Piping to a storm sewer. (d) Piping to a combined sewer.

Provision should be made for dimensional changes of leaders and drains because of temperature variations and for building settlement. Pipe movements may be accommodated through the use of expansion joints, sleeves, or bends in the piping. In Fig. 9.13a and b, for example, the S bend below the drain will permit expansion and contraction of the leader.

Leaders may be placed on the exterior of a building along a wall or within a building. Interior leaders should be connected to a horizontal building storm drain, which extends below grade to a point at least 3 ft outside the building wall (see Fig. 9.13c). There, the drain should transfer the flow to a building storm sewer,

which extends to a main sewer or other disposal point.

If the main sewer is a combined sewer, leaders and storm drains must be provided with a trap. (A fresh-air inlet is not required, although needed with building traps for building sanitary drains.) Individual stormwater traps should be installed on the stormwater drain branch serving each leader or a single trap should be installed in the main storm drain just before its connection with the building storm sewer (see Fig. 9.13d).

Some building codes permit a building drain to carry both sanitary wastes and rainwater for discharge to a combined sewer. In such cases, the leader should be trapped and connected to the combined drain through a single wye fitting. The connection should be at least 40 pipe diameters downstream from the connection of the nearest soil stack or branch to the building drain.

Leaders of one- or two-family dwellings may discharge onto a graded splash pan (see Fig. 9.13*a*), from which the rainwater may be distributed over a lawn or other absorbent area; or a leader may discharge into a seepage pit or a dry well. The latter, consisting of a 3-ft length of 18-in.-diameter vitrified-clay or concrete pipe, filled with crushed stone, is suitable for disposal of small quantities of stormwater.

Note that exterior sheet-metal gutters and leaders are usually considered the work of the sheet-metal trade and not included in plumbing contracts. Interior leaders and drains, however, are considered plumbing.

Tests of Wastewater-Removal Systems

Drainage and venting systems, except for outside leaders and subsoil drains, should be inspected to insure compliance with building-code requirements and project specifications and tested on completion of the rough piping installation to insure watertightness.

Rough plumbing comprises the parts of the plumbing system, including fixture supports, that can be completed before installation of plumbing fixtures.

The usual test of rough plumbing employs water to create pressures in the entire system or any section of it. When the entire system is to be tested, all openings except the highest one in the plumbing should be tightly closed. Then, the pipes are filled with water to the point of overflow. When a section is tested, every opening except the highest should be tightly plugged and the section filled with water. No section, pipe, fitting or joint should be subjected to a head of less than 10 ft of water. In all tests, the water should be retained for at least 15 minutes before inspection starts. The system should be tight throughout.

An alternative test employs air. All openings except one are closed tightly. An air-compressor testing apparatus is attached to the unclosed opening to force air into the system, to create a uniform gage pressure of 5 psi or sufficient pressure to balance a column of mercury 10 in. high. The system should be capable of holding this pressure without introduction of additional air for at least 15 minutes.

After plumbing fixtures have been installed and their traps sealed with water, the entire drainage system should be subjected to a smoke or peppermint test, to insure gastightness. In these tests, smoke under a pressure of 1 in. of water, or 2 oz of oil of peppermint washed into each stack with 10 quarts of hot water, are used to locate leaks.

9.7. SIZING OF WASTEWATER AND VENT PIPES

Determination of the sizes of drainage and vent pipes is difficult because the flow of wastewater and air to be carried involves complex phenomena. Consequently, in practice, design of wastewater-removal systems is based on experiments, experience and judgment. Requirements in the National Plumbing Code and local building codes for minimum diameters of pipes reflect this.

Usually, the prime source of wastewater in sanitary drainage pipes consists of plumbing fixtures that are operated intermittently. In a large building, it is likely that, at any given time, only a few of the fixtures will be discharging water into the drainage system. Hence, design of drainage systems should be based on the probable maximum flow. For this reason, the concept of fixture units, as presented in Sec. 9.5 for water-supply pipes, is also applied in determination of sizes of drainage pipes.

Drainage pipes should be neither too large nor too small for the flows to be accommodated. If the pipes are too small, they may become clogged or noisy. If they are too large, the flow velocity may be too small to scour the pipes and keep them clean. Generally, a velocity of at least 2 fps is desirable for branch pipes and preferably at least 3 fps for building drains and sewers.

Drainage Branch Pipes

The minimum diameter permitted for horizontal sanitary drainage pipes leading from a plumbing fixture is determined by the diameter of the trap pipes. The drainage pipes must be at least as large as the traps to which they connect. Many building codes specify the minimum size of trap required for various plumbing fixtures (see Table 9.6).

When a soil or waste branch receives flow from two or more fixture branches, the minimum size of the branch may be determined from the total number of fixture units contrib-

uting flow to it. Fixture-unit ratings for various plumbing fixtures are given in Table 9.6. The maximum number of fixture units that may be connected to specific sizes of pipes is listed in Table 9.7. Hence, the minimum pipe diameter permitted for a horizontal fixture branch can be obtained from Table 9.7. For example, suppose that a branch is connected to plumbing fixtures with a total rating of 100 fixture units. Table 9.7 indicates that a 3-in. pipe is permitted

Table 9.6. Fixture Unit Ratings and Minimum Trap Diameters for Plumbing Fixtures[a]

Fixtures	Fixture units	Trap size, in.	Fixtures	Fixture units	Trap size in.
Automatic clothes washer (2-in. standpipe)	3	2	Shower stall	2	2
			Showers, group, per head	2	3
Bathroom group: lavatory, bathtub or shower stall and a flushometer water closet	8	$\cdots$	Sink, surgeon's	3	$1\frac{1}{2}$
			Sink, flushing-rim type, flushometer valve	6	3
Bathroom group: lavatory, bathtub or shower stall and a flush-tank water closet	6	$\cdots$	Sink, service type, with standard trap	3	3
Bathtub with or without overhead shower	2	$1\frac{1}{2}$	Sink, service type, with P trap	2	2
Bidet	3	$1\frac{1}{2}$	Sink, pot, scullery or similar type	4	2
Combination sink and laundry tray	3	2	Urinal, pedestal, with 1-in. flush valve	6	3
Dental unit or cuspidor	1	$1\frac{1}{2}$	Urinal, stall or wall-hung, $\frac{3}{4}$-in. flush valve	4	2
Dental lavatory	1	$1\frac{1}{4}$	Urinal, flush tank,	4	2
Drinking fountain	$\frac{1}{2}$	$1\frac{1}{2}$	Water closet, flushometer	6	3
Dishwasher, domestic type	2	$1\frac{1}{2}$	Water closet, flush tank	4	3
Floor drain	2	3	Unlisted fixture with $1\frac{1}{4}$-in. fixture drain	1	$1\frac{1}{2}$
Kitchen sink, domestic type	2	2			
Laboratory cup sink	1	$1\frac{1}{2}$	Unlisted fixture with $1\frac{1}{2}$-in. fixture drain	2	$1\frac{1}{2}$
Laboratory sink	2	2			
Laundry tray (1 or 2 compartments)	2	$1\frac{1}{2}$	Unlisted fixture with 2-in. fixture drain	3	2
Lavatory	1	$1\frac{1}{2}$			
Lavatory (barber shop, beauty parlor, surgeon)	2	$1\frac{1}{2}$	Unlisted fixture with $2\frac{1}{2}$-in. fixture drain	4	$2\frac{1}{2}$
Lavatory, multiple type (wash fountain or wash sink), per each equivalent lavatory unit or set of faucets	2	$1\frac{1}{2}$	Unlisted fixture with 3-in. fixture drain	5	3
			Unlisted fixture with 4-in. fixture drain	6	4

[a]Based on the Building Code of the City of New York.

Table 9.7. Maximum Permissible Loads for Sanitary Drainage Pipes, Fixture Units[a]

| Pipe diameter, in. | Any horizontal fixture branch or at one story of stack | Total for stack | Building sewer, building drain and building branches from stacks | | | |
| | | | Slope, in. per ft | | | |
			$\frac{1}{16}$	$\frac{1}{8}$	$\frac{1}{4}$	$\frac{1}{2}$
1[b]	3	4	d	d	d	d
2[b]	6	8	d	d	21	26
$2\frac{1}{2}$[b]	12	30	d	d	24	31
3[c]	20	97	d	20	27	36
4	160	507	d	180	216	250
5	360	1,445	d	390	480	575
6		2,918	d	700	840	1,000
8		6,992	1,440	1,600	1,920	2,300
10			2,500	2,900	3,500	4,200
12			3,900	4,600	5,600	6,700

[a] From the Building Code of the City of New York.
[b] No water closets permitted.
[c] Not over two water closets permitted.
[d] Not permitted.

20 units and a 4-in. pipe, 160 units. The 3-in. pipe is too small for 100 units. Therefore, the 4-in. pipe should be specified.

Note, however, that the footnotes limit the number of water closets permitted for the smaller pipe sizes. The smallest size permitted for a soil pipe is 3 in.

Soil and Waste Stacks

In a one-story building or at the top of a multistory building, where the fixture branches connect to a drainage stack, the size of the stack is determined in the same way from Table 9.7 as for the horizontal branches.

The minimum size of a multistory stack is determined also by the total number of fixture units contributing flow to it. This size can be obtained from the fixture-unit ratings given in the third column of Table 9.7.

Building Drains and Sewers

The minimum size of a building drain or sewer may be determined in a similar manner from Table 9.7 as for multistory soil or waste stacks, except that the size also depends on the slope of the pipe. For example, if plumbing fixtures with a total rating of 200 fixture units contribute flow to a building drain with a slope of

$\frac{1}{8}$ in. per ft, the pipe should be at least 5 in. in diameter. (A 4-in. pipe is permitted 180 units and a 5-in. pipe, 390 units, according to Table 9.7). Portions of a drainage system installed underground, however, should be at least 2 in. in diameter. Copper or brass drip pipes, though, may be 1 in. in diameter.

Allowances for Continuous Flow

Table 9.7 may also be used for determining the size of drainage pipes receiving continuous or semicontinuous flows, such as those from pumps, ejectors, air conditioners and similar equipment. For such discharges, the rate of flow may be converted into an equivalent fixture-unit rating. The National Plumbing Code specifies two fixture units for each gpm.

Sizing of Vents

Diameters of vent pipes are determined by the total number of fixture units to be vented and by the developed lengths of the pipes. The developed length of a pipe is its length along its centerline, including length of fittings. Building codes usually establish minimum diameters for different types of vents:

Table 9.8. Minimum Sizes of Vent Stacks and Branch Vents[a]

Diameter of soil or waste stack, in.	Max. fixture units connected	Diameter of vent required, in.								
		$1\frac{1}{2}$	2	$2\frac{1}{2}$	3	4	5	6	8	10
		Maximum developed length of vent, ft[b]								
$1\frac{1}{2}$	4	100	c	...	...	...	...	...	...	...
2	8	30	170	c	...	...	...	...	...	...
$2\frac{1}{2}$	30	15	70	175	c	...	...	...	...	...
3	97	6	24	89	250	c	...	...	...	...
4	507	d	d	11	78	310	c	...	...	...
5	1,445	d	d	d	16	110	380	c	...	...
6	2,918	d	d	d	d	34	143	380	c	...
8	6,992	d	d	d	d	d	14	73	340	c
10	...	d	d	d	d	d	d	d	d	c

[a] From Building Code of the City of New York.
[b] A $1\frac{1}{2}$ in. vent may be used for 6 or less fixture units for a developed length of 15 ft from the fixture to header regardless of development length limiting the header size.
[c] Unlimited.
[d] Not permitted.

Fresh-air inlet: one-half the diameter of the building drain at the point of connection but not less than 3 in.

Individual vent: one-half the diameter of the drainage pipe at the point of connection but not less than $1\frac{1}{2}$ in.

Relief vent: one-half the diameter of the soil or waste branch at the point of connection but not less than $1\frac{1}{2}$ in.

Branch vent: as indicated by Table 9.8. The developed length of the branch vent should be measured from its connection to a vent stack or stack vent to the farthest fixture drain connection served by the branch vent. For example, suppose a branch vent with a developed length of 50 ft is to vent plumbing fixtures with a total rating of 50 fixture units. According to either Table 9.7 or 9.8, the stack diameter should be at least 3 in. Table 9.8 indicates that a branch vent connected to a 3-in. soil or waste stack should be at least 2 in. in diameter if the vent has a developed length of 24 ft and at least $2\frac{1}{2}$ in. in diameter if the developed length does not exceed 89 ft. Hence, for a developed length of 50 ft, the branch vent should be $2\frac{1}{2}$ in. in diameter.

Vent stacks: as indicated by Table 9.8. The developed length of a vent stack should be measured from the base or from the point of connection to a soil or waste stack to a connection with a vent header or its termination above the roof.

Sizing of Storm Drainage Pipes

Diameters of pipes in the storm-drainage subsystem may be determined empirically or by hydraulic principles. If the latter method is used, controlled flow and storage of some stormwater on the roof may be assumed. In that case, building codes may require also assumption of a probable maximum rate of rain intensity, in. per hr, for example, the maximum intensity of a 10-year storm. In addition, the codes may limit the depth of water on the roof, for instance, to 3 in. Flashing should extend above this limiting depth, and scuppers should be placed in parapet walls, to insure that greater storage of water on the roof cannot occur because the water will flow through the scuppers. Codes may also relate the minimum required number of roof drains to the roof area. The Building Code of the City of New York, for example, requires two drains for roof areas up to 10,000 sq ft and four drains in roof areas exceeding 10,000 sq ft. Minimum sizes of pipes may be determined from Table 9.10 for the calculated flow, gpm.

If the empirical method is used, sizes may be determined as follows:

Table 9.9. Minimum Sizes of Roof Gutters[a]

Diameter of semicircular gutters, in.	Maximum projected roof area for gutters of various slopes, sq ft			
	$\frac{1}{16}$ in. per ft	$\frac{1}{8}$ in. per ft	$\frac{1}{4}$ in. per ft	$\frac{1}{2}$ in. per ft
4	144	192	272	385
4	288	409	575	815
5	500	705	1,000	1,420
6	770	1,090	1,540	2,220
7	1,150	1,560	2,220	3,120
8	1,590	2,250	3,180	4,490
10	3,600	4,080	5,780	8,000

[a]Gutters other than semicircular may be used if they have the same cross-sectional area as that required for semicircular gutters. From Building Code of the City of New York.

Table 9.10. Minimum Sizes of Stormwater Leaders[a]

Leader diameter, in.	Flow, gpm	Maximum projected roof area, sq ft
2	22.6	433
$2\frac{1}{2}$	39.6	779
3	66.6	1,278
4	143	2,745
5	261	4,992
6	423	8,121
8	911	17,491
10	1,652	31,718

[a]From Building Code of the City of New York.

Roof gutters: as given in Table 9.9. For example, suppose a semicircular gutter with a slope of $\frac{1}{8}$ in. per ft is required to carry the runoff from a roof with an area, when projected on a horizontal plane, of 400 sq ft. Table 9.9 indicates that a 4-in.-diameter gutter may be used. (It is permitted to carry the runoff from 409 sq ft on the given slope.)

Leaders: as given in Table 9.10. For example, for a roof with a projected area of 800 sq ft, a 3-in.-diameter leader should be used. The table indicates that this size of pipe is permitted to drain the runoff from 1,278 sq ft. The equivalent diameter of a square leader may be taken as the diameter of the circle that can be inscribed within the cross section of the leader.

Building storm drains: as given in Table 9.11, except if the drain is a branch serving a single roof drain and having a slope of $\frac{1}{2}$ in. per ft or less. The size of such a branch may be obtained from Table 9.10. Suppose a storm drain with a slope of $\frac{1}{4}$ in. per ft is required to receive the runoff from a roof with a projected area of 2,000 sq ft. Table 9.11 indicates that a 4-in.-

Table 9.11. Minimum Sizes of Horizontal Storm Drains[a]

Drain diameter, in.	Maximum projected roof area, sq ft, for various slopes		
	$\frac{1}{8}$ in. per ft	$\frac{1}{4}$ in. per ft	$\frac{1}{2}$ in. per ft
2	250	350	500
$2\frac{1}{2}$	357	505	714
3	690	930	1,320
4	1,500	2,120	3,000
5	2,700	3,800	5,320
6	4,300	6,100	8,700
8	9,300	13,000	18,400
10	16,600	23,500	33,000
12	26,700	37,500	53,000
15	47,600	67,000	95,000

[a]From Building Code of the City of New York.

Table 9.12. Fixture-Unit Equivalents of Roof Drainage Areas[a]

Drainage area, sq ft	Fixture-unit equivalent
180	6
260	10
400	20
490	30
1,000	105
2,000	271
3,000	437
4,000	604
5,000	771
7,500	1,188
10,000	1,500
15,000	2,500
20,000	3,500
28,000	5,500
Each additional 3 sq ft	1

[a]From Building Code of the City of New York.

diameter drain should be used. That size of pipe is permitted to drain 2,120 sq ft.

Combined sanitary and storm drains and sewers: Since the flow from soil or waste stacks is measured in fixture units, it is convenient to convert the drainage area to an equivalent fixture-unit rating. Table 9.12 may be used for the purpose. The minimize size of building drain or sewer may then be obtained from Table 9.7.

If equipment, such as a pump, ejector or air conditioner, discharges a continuous flow into a building storm drain or sewer, the size of pipe may be determined by converting each gpm of discharge into 19 sq ft of drainage area and using Table 9.11.

SECTIONS 9.6 and 9.7

References

National Plumbing Code (ANSI A40.8), American National Standards Institute.

Uniform Plumbing Code, International Association of Plumbing and Mechanical Officials.

B. Stein et al., *Mechanical and Electrical Equipment for Buildings*, 7th ed. Wiley, New York, 1986.

F. Merritt, *Building Design and Construction Handbook*, 4th ed., McGraw-Hill, New York, 1982.

Words and Terms

Interceptor
Rough plumbing
Soil pipe
Storm drain
Vents
Waste pipe

Significant Relations, Functions and Issues

Components of wastewater systems: horizontal branch, stack, vents, traps, sewer.
Need for adequate slope of horizontal drainage piping.
Sizing of piping based on fixture units and drainage slopes.

9.8. PIPING FOR HEATING GAS

Some buildings require a supply of natural or manufactured gas for heating. The heat may be used for cooking on domestic or commercial ranges, space heating, industrial heating, or production of hot water or steam. The gas may be delivered to the buildings in pipes from an underground public utility main, a pressurized container (bottled gas) or other source. Inside a building, the heating gas may be delivered to points of use through distribution pipes meeting requirements much like those for water-supply pipes. Special precautions are necessary, however, because of hazardous characteristics of the gas.

Design and installation of gas pipes are governed by requirements in local building codes and often also by regulations of state agencies, such as the Public Service Commission. Building codes may adopt by reference appropriate standards, such as the American National Standards Institute "Installation of Gas Appliances and Gas Piping," Z21.30. If the gas is supplied by a public utility, the supplier will require installation of a meter to measure the quantity of gas supplied and will have requirements for the service installation up to the outlet side of the meter. In addition, the gas piping system may have to meet requirements of insurance companies.

Characteristics of Heating Gases

Natural gas consists mostly of methane (CH_4) with some ethane (C_2H_6). Manufactured gases usually are a mixture of hydrogen, carbon monoxide, methane, ethylene (C_2H_4) or other fuel gases. Either type of gas burns when ignited in the presence of oxygen. Unless the rate of combustion is properly controlled, an explosion can occur. Hence, equipment using heating gas must be provided with devices for controlling the rate of combustion, and distribution pipes must be airtight. Also, stopcocks used must be reliable in completely halting flow of gas.

The heat content of heating gas depends on the constituents. The rating may be obtained from the supplier or from tests. It usually ranges between 500 and 3,000 Btu per cu ft. The required rate of flow of gas, cu ft per hr, may be calculated by dividing heat needed by equipment, Btu per hr, by the gas heat content, Btu per cu ft.

Heating gas is colorless. It also may normally be odorless, in which case the supplier may add to the gas a gaseous substance with a strong odor, such as ethyl mercaptan, or a nose and throat irritant, such as crotonaldehyde, to warn of gas leakage. Weight of heating gas is about 40% less than that of air.

Heating gas may contain some moisture. This must be taken into account in layout of gas piping.

Products of combustion of heating gas, as well as the gas itself, are hazardous to humans. Hence, enclosed spaces containing gas appliances should be well ventilated. Local building codes generally contain requirements for minimum sizes of vents and for vent materials.

Materials for Gas Pipe

Distribution gas pipe within a building may be made of copper or brass, with threaded connections, or of wrought iron or steel meeting the requirements of ANSI B36.10, with fittings of malleable iron or steel, except for stopcocks, valves, pressure regulators and controls. The iron or steel fittings may be threaded, welded or flanged type with gaskets.

Installation of Gas Piping

Building codes usually require the gas meter to be installed in a basement or cellar, or if such spaces are not available, on the ground floor. The meter location should be clean, dry and well ventilated. The meter should not be exposed to steam or chemical fumes, nor to extreme heat or cold. The meter should be readily accessible for reading and inspection. In addition, it should not be located in a boiler room or other space containing a heater, in a stair hall or in a public hall above the cellar, or above the ground floor if there is no cellar.

Generally, if gas is supplied by a public utility, the supplier will install the service pipe and the meter. The service pipe and distribution piping should be enclosed in pipe sleeves where they pass through walls and floors. A shut-off stopcock or valve should be installed within the building in an accessible location within 2 ft of the point where the service pipe enters the building but upstream of the meter and pressure regulator. This entrance point should be more than 10 ft from the cellar termination of a stairway, for safety reasons. This requirement applies also to meters and pressure regulators. The requirement may be waived, however, if the service entrance point, meters and pressure regulators are separated from the stairway termination by a wall with a fire-resistance rating of at least 1 hr.

Building codes usually limit the pressure at which heating gas may be distributed within buildings. For example, the Building Code of the City of New York permits pressures in distribution piping only up to 3 psi for commercial and industrial use and for other large volume use in which fuel requirements for boiler-room equipment exceed 2,400 cu ft per hr and the large flow is supplied through separate piping to the boiler room. For other applications, the code does not permit pressures in distribution piping to exceed 0.5 psi. If gas is supplied to a building at greater pressures, a pressure regulator must be installed. This device should reduce the pressures to that permitted by the code or to that appropriate for equipment that uses the gas, whichever is less. Each regulator should be provided with a vent pipe that leads directly to outdoors. But the vent outlet should not be located under a window or an overhang nor near any opening through which leaking gas may enter the building.

Because heating gas may contain some moisture, provision should be made to drain away water if any should collect in the gas pipes. For the purpose, service pipes should be sloped downward from the building toward the supply main. Also, piping inside the building should be sloped downward toward accessible drips. (A drip is a nipple, or short length of vertical pipe, that extends downward from a connection with a distribution pipe and that is installed to drain that pipe. When the drip is not being used for draining, its bottom end is closed off with a threaded cap.) Drips should be located at the bottom of every riser and at every low point of a horizontal pipe.

Where branches are connected to a gas pipe, they should be joined at the top or sides, to prevent moisture from accumulating.

Because of the hazard of escaping gas from leaks, gas pipes should be installed so that access to them is easy for stopping the flow of gas and repairing the leaks. It is especially important that valves and stopcocks, and also unions and bushings, be accessible. (A *union* is a pipe fitting for joining the ends of two pipes that cannot be turned. A *bushing* is a plug that threads into the end of a pipe and has a threaded borehole to receive a smaller pipe). Despite the desirability of having gas pipes accessible, architectural requirements generally make it necessary that gas pipes, like other plumbing, be concealed from view in many types of buildings.

Outlets for fixtures should extend at least 1 in. beyond finished walls and ceilings. Until the fixtures are installed, the outlets should be closed off with threaded caps. Outlets should have at least the following diameters: $\frac{3}{4}$ in. for cooking appliances, 1 in. for water heaters and $\frac{1}{2}$ in. for room space heaters.

Diameters of distribution piping should be sufficiently large that gas in required quantities will be supplied to points of use without excessive pressure losses between those points and the meter. Generally, the pressure loss should not exceed 3 in. of water pressure during periods of maximum gas demand. Gas pipes, in general, should be at least $\frac{3}{4}$ in. in diameter. Pipes embedded in concrete, however, should be at least 1 in. in diameter.

Tests of Gas Piping

After a section of a gas system or the entire system has been completed, but before appliances are connected to it, the installation should be inspected to insure that appropriate materials have been used and properly installed. Tests should be made to verify that the system is gastight under a pressure at least 50% greater than the proposed operating pressure. The test pressure, however, should be at least 3 psi gage and retained for at least 10 min. For the test, the pipes may be filled with fuel gas, air or an inert gas (but not oxygen), and the source of pressure should be isolated before pressure readings are made. Before appliances are connected to the pipes, they should be purged with the heating gas to be used.

Reference

Installation of Gas Appliances and Gas Piping (ANSI Z21.30), American National Standards Institute.

Words and Terms

Bushing
Drip
Shutoff
Union

Significant Relations, Functions, and Issues

Hazardous character of gas: noxious fumes and explosions.
Need for access to piping and fittings for rapid control and repair of leaks.

9.9. SYSTEMS DESIGN OF PLUMBING

Plumbing systems consist of major subsystems—water supply, wastewater removal, heating gas and sometimes other liquids and gases—that must be kept physically independent of each other. For architectural purposes, however, for minimum interference with or disruption of the layout of other systems and for convenience in installation of piping, it generally is desirable that the pipes for the different subsystems be located close to each other and share plenums and shafts intended to house piping. Consequently, the layout of piping for each subsystem is often dependent on the layout of piping for other subsystems.

Also, plumbing systems, in general, are affected by other systems, principally architectural, structural and HVAC. Usually, the designs of those systems are given higher priority than the design of the plumbing subsystems, because changes in the other systems to accommodate plumbing frequently are more costly than changes in the plumbing to accommodate those systems. Systems design, however, may be useful in coordinating the various building systems to produce an optimum overall design. For example, in some cases, systems design may indicate that use of open-web structural members will reduce both the cost of the structural system and the cost of the plumbing.

Data Collection and Problem Formulation

Basic information for plumbing design comes from examination of the owner's program of requirements, type of occupancy of the building and anticipated number of occupants. This information is useful in determining the number and type of plumbing fixtures, equipment and appliances that will be needed. From these data, estimates can be made of supply needs, such as those for water and heating gas, and disposal requirements, such as those for wastewater and products of combustion. Then, additional information must be obtained on possible sources of supply and permissible means of disposal of wastes.

The architect and the mechanical engineer or plumbing designer should collaborate in location of plumbing fixtures and of equipment and appliances to be served by the supply piping. Additional collaboration also is desirable in determination of spaces in which piping may be placed within the building. The results of the cooperative efforts should be displayed on architectural plans, elevations and cross sections and on schematic drawings of the plumbing and submitted to all the members of the building team for review.

Further information for design of the plumbing should be obtained from local building codes, government agencies that also may have jurisdiction over plumbing installations, national standards, requirements of utilities that may supply water, gas or steam to the building, and restrictions imposed by insurance companies.

In later stages of design, architectural, structural and HVAC drawings should be studied to insure compatibility of the plumbing installation with the other systems.

The goal of plumbing design may be stated as: To design, as a component of an optimum building, plumbing that will supply needed water, heating gas and other liquids and fuels and that will remove wastewater, products of combustion and other liquid and gaseous wastes before conditions that may be injurious to the health, safety or welfare of building occupants or the community can develop.

Objectives

The foremost objective of systems design of plumbing should be safety. Plumbing should be designed to protect building occupants against the hazards of pressurized water supply; pollution from wastewater; poisonous, asphyxiating, explosive heating gas and any other dangerous substances that plumbing may be installed to carry.

Another highly important objective is reliability. Occupants of a building and often processes to be carried out in a building require continuous availability of potable water and of heating gas, if equipment or appliances have been installed to use it. Thus, a dependable source of supply must be selected for each of the substances needed, and the systems that deliver them to and within the building should operate automatically to provide quantities needed. With little maintenance, plumbing should serve continuously for long periods of time with no breakdowns or shutdowns for repairs. If trouble should develop in a portion of a system, it should be easy to isolate that portion and continue to serve other parts of the system. Similar reliability is required of wastewater-removal systems, venting systems for removal of products of combustion and other systems that may be needed for disposal of liquid or gaseous wastes.

Another important objective is the aim of choosing from among all the possible plumbing systems that can achieve the goal the one that will have the lowest life-cycle cost. This cost includes the cost in place of plumbing fixtures, piping, fittings, supports, valves and other controls, meters, pumps, tanks, water treatment, wastewater treatment and disposal and other plumbing appurtenances and the cost of maintenance and repairs.

Other objectives of systems design of plumbing include conservation of water and heating gas, protection of the environment in disposal of wastes, compatibility of plumbing with other systems, prevention of annoying noises from plumbing and fast, easy installation of plumbing systems.

Constraints

The major constraints on plumbing design are discussed in the preceding sections. These constraints result from the nature of services that must be provided for a building and that can be supplied by plumbing and from the need for protection against hazards that arise in delivery of those services.

Each of the major plumbing subsystems are subjected to numerous constraints applicable only to itself.

Treatment of water supply before distribution of water within a building, for example, depends on the quality of water available and treatment costs. Treatment of wastewater after removal from the building depends on whether a public sewer is available, whether it is permissible to dispose of the wastes in the sewer and whether it is economical and feasible to treat the wastewater and dispose of the effluent.

When a public water supply is to be used, design of the water-supply system has to be based on the water pressure at the main. Operating pressures at plumbing fixtures place additional constraints on design. For some buildings, the supply pressure may have to be reduced for distribution within the building; for other buildings, the supply pressure may have to be increased for distribution purposes.

Building codes, regulations of state agencies and requirements of insurance companies impose numerous constraints on the plumbing subsystems in the interests of public safety, health and welfare and prevention of property damage. These constraints place limitations on materials that may be used in plumbing, govern methods of installation of plumbing elements, affect the types, number and location of controls on flow and influence the location of many plumbing elements. The constraints determine the type and minimum number of plumbing fixtures that should be installed. Also, the constraints set maximum or minimum requirements, or both, for pressures and rate of flow in pipes and at fixtures, for pipe sizes and for spacing of pipe supports. In addition, venting constraints are imposed on wastewater-removal systems.

Other constraints generally are imposed on plumbing design by other building systems. A common constraint is the architectural requirement that pipes be concealed from view. This makes it necessary to install pipes in assigned shafts, in walls and in the spaces between floors and ceilings. Often, as a result, a preferred path for pipes is obstructed by structural members or HVAC ducts.

Furthermore, key points in the subsystems, such as the start and termination of the subsystems, usually are constraints. A public utility may determine where its water main may be tapped or its sewer entered. An architect may determine where plumbing fixtures are to be installed.

Synthesis and Analysis

In the schematics stage, a source of water supply and of heating gas, if required, should tentatively be selected. Also, a means of disposing of wastewater should tentatively be chosen. Then, schematic drawings should be prepared for the plumbing subsystems to show how plumbing fixtures, equipment and appliances will be supplied and wastes will be removed. Analysis of the proposed design should verify that the goal, objectives and constraints established for the plumbing system have been met.

When more information becomes available from the owner, public utilities, architect and structural, mechanical and electrical engineers, a preliminary design for each of the plumbing subsystems should be prepared. These designs should also be checked to verify that the goal, objectives and constraints have been met and that the plumbing subsystems will be compatible with the other building sytems. Cost estimates for the subsystems should then be made. Alternative subsystem designs also should be investigated.

Value Analysis and Appraisal

The benefits and costs of the alternative plumbing subsystems should be compared. Selection of a subsystem is complicated by the necessity of achieving an optimum building, which may not result from installation of an optimum subsystem. Installation of the lowest-cost water-supply or wastewater-removal subsystem, for example, may increase the cost of walls, floors or structural components more than the amount of savings over the cost of an alternative installation.

In the final design stage, design of the chosen plumbing subsystems may be refined. Value engineering should be applied to reduce the subsystem costs and to find ways to speed installation; but efforts to save money by reducing pipe sizes should be carefully studied, because smaller pipe may be noisy, result in flow blockages and require more frequent supports.

Reduction of the length of pipe runs or elimination of piping, in contrast, offers considerable potential for cost reduction. Designers and value engineers should seek out alternatives with shorter pipe runs and fewer pipes. For this purpose, it may often be necessary to obtain the cooperation of other members of the building team, especially architects, to change the location of plumbing fixtures, equipment and appliances and of walls and partitions.

For example, shorter pipe runs and fewer pipes may be needed for plumbing fixtures in the same story of a building if the fixtures can be closely grouped. Two bathrooms, for instance, may be located so that corresponding fixtures are back-to-back, close to or against a partition that separates the bathrooms. The two water closets may be placed to drain directly into a double-wye fitting on a soil stack in the partition. A single waste branch can drain the two bathtubs and two lavatories, and a single branch vent can vent those fixtures.

Similarly, shorter pipe runs and fewer pipes may be required in multistory buildings if plumbing fixtures in the various stories are grouped above each other. Thus, in an apartment building, kitchens may be placed above kitchens and bathrooms above bathrooms.

Another potential for cost reduction for plumbing lies in use of fittings that simplify and speed pipe installation. One example is a sleeve connection for hubless cast-iron soil, waste or vent pipe. Installed in accordance with Cast Iron Soil Pipe Institute Standard 301, a joint is made by slipping the spigoted ends of two pipes into a cylindrical neoprene gasket. A cylindrical, stainless-steel shield is then slid over the gasket and screw clamps are tightened with a calibrated wrench to compress the gasket and prevent the pipes from separating (see Fig. 9.14). Such joints are much more compact and can be made faster than bell-and-spigot joints. Special fittings designed for the gasketed connections considerably reduce the amount of labor required in pipe installation.

Piping sometimes can be eliminated economically by use of unconventional plumbing subsystems. One example is installation of upfeed water distribution for a multistory building. Section 9.4 points out that, because of the pressures at which water usually is supplied to buildings, upfeed distribution generally is limited to buildings up to about four stories high. For taller buildings, water often is pumped to an elevated tank and distributed by a downfeed system. It sometimes is more economical, however, to install pumps in a building and supply water through an upfeed system. This type of installation has the advantages of eliminating downfeed risers and the cost and weight of elevated tanks. To meet varying water demands, pumps may be operated or shut down in sequence automatically by controls activated by pressure sensors.

Another example is the copper Sovent system mentioned in Sec. 9.6, which eliminates vent stacks. The system has four basic parts: a soil or waste stack, a Sovent aerator fitting on the stack at each floor, horizontal branches,

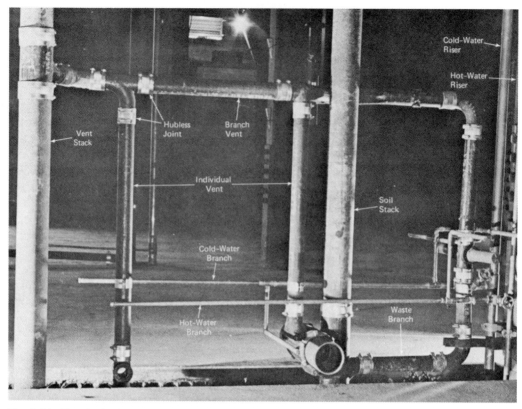

Fig. 9.14. Rough plumbing near several plumbing fixtures, with copper water-supply tubing and hubless cast-iron drainage pipe. (Courtesy Tyler Pipe.)

and a Sovent deaerator fitting on the stack at its base and at horizontal offsets (see Fig. 9.15). The aerator and deaerator provide means for self venting the stack.

In the discussion of flow in a stack under Layout of Soil, Waste and Vent Piping, Sec. 9.6 points out that the flow draws air with it as it drops. If a continuous supply of air is not furnished, a suction would be created in the drainage branches and the liquid in the trap seals at fixtures would be sucked out. In a conventional drainage system, a vent stack is installed to supply air to vent pipes connected to the drainage branches and to the soil or waste stack to prevent destruction of the trap seals. In the Sovent system, however, the vent stack is not needed because the aerator, deaerator and stack vent avoid creation of a strong suction.

The aerator does the following: It reduces the velocity of both liquid and air in the stack. It prevents the cross section of the stack from fill-

ing with a plug of water. And the fitting mixes the wastewater from the drainage branches with the air in the stack.

The deaerator separates the air flow in the stack from the wastewater. As a result, the wastewater flows smoothly into a horizontal offset or the building drain. Also, air pressure preceding the flow at 90° turns is prevented from rising excessively. This is achieved by installation of a pressure relief line between a deaerator and a stack offset or the building drain to allow air to escape from the deaerator.

As with conventional systems, the size of the stack is determined by the number of fixture units of the plumbing fixtures contributing flow. An aerator is required on the stack at each level where a horizontal soil branch or a waste branch the same size as or one tube size smaller than the stack discharges to it. Smaller waste branches may drain directly into the stack. At any floor where an aerator fitting is

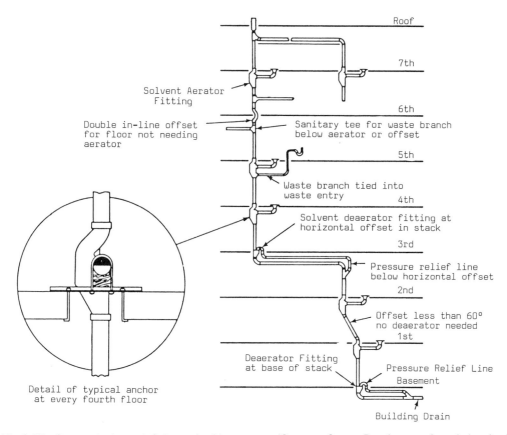

Roof

7th

Solvent Aerator
Fitting

6th

Double in-line offset
for floor not needing
aerator

Sanitary tee for waste branch
below aerator or offset

5th

Waste branch tied into
waste entry   4th

Solvent deaerator fitting at
horizontal offset in stack

3rd

Pressure relief line
below horizontal offset

2nd

Offset less than 60°
no deaerator needed
1st

Deaerator Fitting
at base of stack

Pressure Relief Line
Basement

Detail of typical anchor
at every fourth floor

Building Drain

Fig. 9.15. Copper single-stack Solvent plumbing system. (Courtesy Copper Development Association, Inc.)

not required, the stack should have a double in-line offset, to decelerate the flow (see Fig. 9.15). No deaerators are required at stack offsets of less than 60°.

While designers and value engineers will find it worthwhile to investigate unconventional plumbing subsystems, such as the preceding, they should also seek ways to reduce the amount of water consumed in the building. Reduction in water use is not only desirable because it conserves a scarce resource but also because it reduces life-cycle costs for the owner. The less water used, the less the owner pays for water. Often, the savings in water costs will offset in a short time the higher costs, if any, of plumbing fixtures that make possible such savings. Also, reduction in water demand may make feasible use of smaller-diameter pipes and thus additional costs savings.

One way of reducing water costs is by installation of sprays on lavatory faucets. Sprays discharge less water for a washing task than the flow from an open orifice. (Showers require less water for bathing than bathtubs.) Another simple way to reduce water use is to install water-conserving water closets. These require less water than conventional tank-type units and are available from several manufacturers.

The alternatives discussed in this section are only a few of the many possibilities that can result in better and less costly buildings.

Systems Integration

Installation of elements of the plumbing systems requires considerable coordination with the architectural and structural designs. Many of the plumbing fixtures are built into walls or floors, often requiring more than the normally available space in ordinary construction. For masonry and concrete construction, built-in elements offer special problems, requiring pre-established void spaces in the ordinarily solid construction. Hollow construction of walls and ceiling spaces in most wood and steel construction allows for easier installation with minimal advance provision in many cases.

Intelligent planning of the building to minimize piping runs and to generally group together the building elements that require plumbing will usually result in cost savings as well as simpler construction—with fewer vent stacks and long, pitched, horizontal piping runs. The latter—required for piping such as that for waste lines—is a particularly troublesome issue. Pressurized piping can be ducked over, under, or around structural elements, but lines requiring gravity flow have major form constraints.

In large buildings a considerable amount of the piping consists of lines that service the HVAC system. This situation requires close coordination of the architectural, structural, plumbing, and HVAC designs. Plumbing lines, plumbing fixtures, and HVAC equipment must be given adequate space and passage through the building without major disturbance of architectural spaces or unreasonable modifications of structural elements. In the best of coordinated design work, these considerations may affect design choices in any of the areas involved.

GENERAL REFERENCES AND SOURCES FOR ADDITIONAL STUDY

These are books that deal comprehensively with several topics covered in this chapter. Topic-specific references relating to individual chapter sections are listed at the ends of the sections.

National Plumbing Code (ANSI A40.8), American National Standards Institute.

Uniform Plumbing Code, International Association of Plumbing and Mechanical Officials.

B. Stein et al., *Mechanical and Electrical Equipment of Buildings*, 7th ed. Wiley, New York, 1986.

F. Merritt, *Building Design and Construction Handbook*, 4th ed., McGraw-Hill, New York, 1982.

EXERCISES:

The following questions and problems are provided for review of the individual sections of the chapter.

Section 9.1

1. Name some possible sources of water for use in buildings.
2. What is the minimum requirement for water quality for use by humans in buildings?
3. What are the disadvantages of hard water? What can be done to soften water?
4. What is the disadvantage of water that is nearly pure?
5. If a potential source of water does not meet USPHS drinking water standards, what can be done to permit use of the water in a building?
6. What can be done to reduce turbidity of water?
7. What can be done to remove bacteria from water?
8. Why are cross connections between different plumbing systems undesirable?
9. At a specific instant, several plumbing fixtures in a building are in use. If the water meter at the service inlet pipe registers 30 gpm, what is the total discharge from the fixtures at the time?
10. A branch pipe draws 15 gpm of water from a riser in which the flow is 25 gpm. What is the flow past the branch connection?
11. Water flow in a 1-in. pipe with a roughness coefficient $C_1 = 100$ is 24 gpm. What is the:
    (a) water velocity in fps?
    (b) friction loss in ft in a 100-ft length of pipe?
12. What is the water pressure at a point in a riser 100 ft above a water meter if the pressure at the meter is 64.5 psi and the friction loss is 0.5?

Section 9.2

13. What is the purpose of sewage treatment?
14. Why should flows of the three main types of wastewater from buildings be kept separated?
15. What is the relationship of building drains to building sewers?
16. Why is secondary treatment generally desirable for domestic wastewater?

17. What is the significance of BOD in wastewater?
18. What is sludge?

Section 9.3

19. In what part of a building should a building drain be installed?
20. What characteristics are necessary for materials used in plumbing fixtures?
21. Why must plumbing pipes, joints, and connections be gastight and watertight?
22. Where should wastewater system vent pipes be terminated?
23. What is the purpose of a liquid-seal trap on a plumbing fixture?
24. What should be done to prevent damage because of the effects of temperature changes on pipes?
25. Why is it necessary to close off the space around a pipe where it passes through a floor?

Section 9.4

26. Where is a water-service pipe placed?
27. What is solder made of and where is it used in plumbing?
28. How is brazing accomplished in plumbing installations?
29. What is the purpose of valves in plumbing installations?
30. What type of valve is usually used in faucets?
31. What is the purpose of a check valve?
32. What provision should be made for draining horizontal pipes?
33. Describe a common means of providing for thermal changes in length of hot water pipes.
34. What is the purpose of a trap in the discharge pipe of a plumbing fixture?
35. Why is an air gap required at plumbing fixtures and where should it be located?
36. What are the common methods of providing cold water under required pressures to plumbing fixtures?
37. Why should the pressures of hot and cold water be nearly equal at plumbing fixture outlets?

38. What are the major differences between upfeed and downfeed water distribution systems?
39. Name the main components of a fire sprinkler system.
40. How are sprinkler heads actuated?
41. How does a wet-pipe sprinkler system differ from a dry pipe system? Where should a dry-pipe sprinkler system be used?
42. What are the common causes of noisy water-distribution pipes?

Section 9.5

43. A single-family house contains a kitchen sink, dishwasher, and washing machine, the branches to which are fed by a main. What is the smallest size of copper tubing that can be used for the main?
44. A single-family house contains two flush-tank water closets, two lavatories, two bathtubs, and two showers, the branches to which are fed by a main. What is the smallest size of copper tubing that can be used for the main?
45. Suppose the water closets and urinals in Table 9.4 were tank type.
    (a) What would the total water demand be in fixture units?
    (b) What would the probable maximum flow rate be in gpm?
    (c) If the maximum velocity permitted is 6 fps, what is the minimum diameter that may be used with copper tubing for the probable maximum flow rate?

Section 9.6

46. Name the two major subsystems of a wastewater removal system.
47. What plumbing elements are part of the:
    (a) drainage system?
    (b) venting system?
48. Why are so-called horizontal drainage and vent pipes not truly horizontal?
49. What is the distinction between soil and waste pipes?
50. Where should a vent pipe be located relative to a fixture trap?
51. What characteristics are desirable for drainage-pipe fittings?

52. What are the three common methods of connecting fittings to drainage pipes?
53. What is the maximum positive or negative air pressure permitted at a trap seal?
54. How does a stack vent differ from a vent stack?
55. What distinguishes a loop vent from a circuit vent?
56. What is the purpose of a relief vent? What pipes does it connect?
57. A 2-in. diameter horizontal drainage branch slopes downward away from a trap at $\frac{1}{4}$ in. per ft.
    (a) What is the maximum distance that a vent may be placed away from the trap and satisfy the requirement that the vent opening be higher than the bottom of the discharge end of the trap?
    (b) If the vent pipe is connected 12 in. from the trap, what is the least distance that the vent opening may be set above the bottom of the 2-in. branch?
58. What is the purpose of indirect waste piping?
59. In what ways may storm water runoff from a building roof be legally disposed of?
60. Why do building codes set a minimum velocity for wastewater flows?
61. How does indirect waste piping prevent backflow of wastewater?
62. What are the advantages of a combined drainage and vent system over separate drainage and vent systems?
63. Describe a method for draining rainwater from:
    (a) A steeply sloping roof.
    (b) A flat roof.

Section 9.7

64. What is the minimum diameter of pipe that should be used for:
    (a) A horizontal waste branch draining a lavatory?
    (b) A horizontal waste branch receiving discharges from two lavatories and two bathtubs?
    (c) A horizontal soil branch connected to a water closet?
    (d) A soil stack to which plumbing fixtures with a total rating of 95 fixture units contribute flow?

(e) A building drain with a slope of $\frac{1}{4}$ in. per ft receiving discharge from plumbing fixtures with a total rating of 25 fixture units?

65. What is the minimum diameter pipe permitted for:
    (a) A branch vent connected to a waste pipe for a combination sink and laundry tray?
    (b) A 50-ft-long branch vent connected to a 3-in. diameter soil stack?
    (c) A 100-ft-long vent stack connected to a 3-in. diameter soil stack?
66. What is the minimum diameter pipe permitted for:
    (a) A building storm drain to carry 100 gpm?
    (b) A semicircular roof gutter with a slope of $\frac{1}{4}$ in. per ft for draining a roof with a projected area of 800 sq ft?
    (c) A storm water leader for a 1500 sq ft roof?
    (d) A building storm drain with a slope of $\frac{1}{4}$ in. per ft for a roof with a projected area of 10,000 sq ft?
67. A combined building drain with a slope of $\frac{1}{4}$ in. per ft receives intermittent flow from plumbing fixtures with a total rating of 200 fixture units. The drain also receives runoff from a 10,000-sq-ft roof. What is the smallest diameter that should be used for the drain?

Section 9.8

68. What characteristics of heating gases make special precautions necessary for piping installations compared with water supply piping?
69. A domestic range consumes a maximum of 50,000 Btu per hr. If gas supplied has a heat content of 1000 Btu per cu ft, at what rate should gas be supplied to the range?
70. Heating gas supplied to equipment consuming 48,000 Btu per hr has a heat content of 800 Btu per cu ft. At what rate must gas be supplied to the equipment?
71. What is the purpose of a pressure regulator in a gas system?
72. What is the objective of tests of a gas piping system after completion of the installation of the system?

Chapter 10

Heating, Ventilating, and Air Conditioning

This chapter presents the basic principles employed in design of HVAC systems. Because of the many different methods that may be used for environmental control, however, only a few of the more common control concepts are discussed.

10.1. DESIGN CONSIDERATIONS

Heating, ventilation and air conditioning (HVAC), along with lighting and sound control, are part of the environmental control systems within buildings. HVAC is desirable for the health and comfort of building occupants. In some cases, though, HVAC may be necessary in manufacturing processes, storage of products or operation of equipment, such as computers. The systems-design approach to HVAC can be useful in attaining a system that meets functional objectives with low initial and long-term costs.

The physical environment within a building or any enclosed portion, or zone, of it that HVAC is intended to control is influenced by many variables. The most important of these are temperature, humidity, air movement and air quality.

Different persons have different abilities to adapt to their environment. Adaptability depends on a person's age, activity, state of health and degree of acclimatization to the environmental conditons. With regard to the last factor, it is noteworthy that a person's initial reaction to an environment is likely to be considerably different from the reaction after exposure for several hours. Determination of what is a comfortable environment, therefore, requires a subjective decision, which may be different for different persons or even for the same person at different times.

It is possible, however, to establish a bracketed range for any single factor—or, more significantly—for combinations of the interactive factors that affect comfort, such that, within the bracketed range, a majority of persons will feel reasonably comfortable. Taking this idealized range (or the average value of the range as a more specific value) as representative of an ideal building interior environment condition, various design criteria can be developed. With regard to temperature alone, the single value of 75 degrees Fahrenheit is close to an ideal for interiors, and as shown in Fig. 10.1, may be used to establish design maximum limits with regard to exterior conditions derived from local climate histories. Requirements for heating and cooling equipment and for development of need for thermal insulation may use this criteria.

Temperature, humidity, and air movement can all be considered separately for determination of what is too hot or too cold, too dry or too humid, or too still or too breezy. However, in the end, the interrelations of these fac-

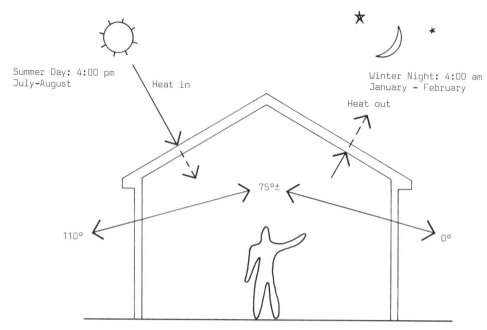

Fig. 10.1. Design extremes: summer afternoon and winter night.

tors must be considered. Figure 10.2 represents in graph form the plot of the climate range of outdoor temperature and humidity for a specific locale and approximate bracketed comfort ranges for both still air and a mild breeze situation.

Single conditions—relating to the extremes of climate—are important for establishing the maximum limits for design, which are signifi-

cant, for example, to choose equipment size and other construction details. However, variations in conditions are also significant to the activities that occur in buildings over periods of time—days, weeks, seasons and years. Figure 10.3 indicates the form of temperature variations that may occur on a single 24-hour day for both the interior of a building and the outdoors. Relating this to the period of actual use

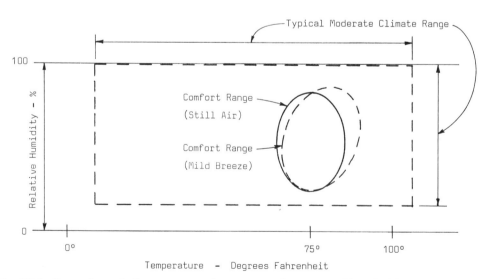

Fig. 10.2. Comparison of climate range and average comfort range, with and without air movement.

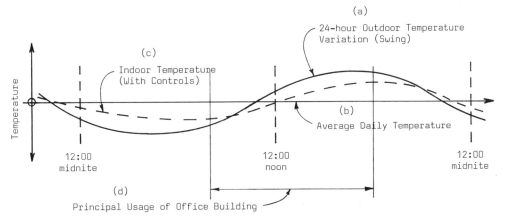

Fig. 10.3. Effects of daily temperature variation. (*a*) Form of typical daily variation. (*b*) Average 24-hour outdoor temperature. (*c*) Indoor temperature swing with limited control. (*d*) Peak usage period—based on building occupancy.

of the building, it is possible to evaluate the real nature of significant environmental control that must be considered for the comfort of the building occupants (an office building in the illustration).

Evaluations of interior needs and exterior events produce some data that are used for the development of the various components of a building's HVAC system. These components are provided for the major tasks, described as follows:

Ventilation is the process of supplying air to any space within a building, without noticeable odors and without an objectionable level of contaminants, such as dust or obnoxious or harmful gases, and removing vitiated or polluted air. Outdoor, or fresh, air is a generally acceptable source for ventilation air. Often, however, a mixture of fresh air and air present in a building is also acceptable. In either case, the air may or may not be conditioned; that is, filtered and adjusted in temperature and humidity.

Heating is a process of transferring heat from a heat source, such as a furnace, heat exchanger, radiant panel or electric coil, to any space within a building.

Air conditioning is the process of treating air to control simultaneously its temperature, humidity, cleanliness and distribution to meet de-

sign criteria for any space within a building. Air conditioning may be applied to raise the temperature of the space by heating or to lower the temperature by cooling.

Cooling is a heat-removal process and may also be accomplished without air conditioning; for example, with chilled radiant panels.

For maintenance of a desired temperature within a space, heat usually must be added to or removed from the space. This heat is considered a load on the heating, ventilation and air-conditioning systems. The component of this heat supplied by any source is considered the *heat load* on the space from that source.

Basically, the design process for an HVAC system consists of establishment of criteria for maintenance of an optimum indoor environment, determination of heat loads, selection of environmental control methods and equipment for achieving the objectives, and estimates of initial and life-cycle costs for the system. The designer's responsibility, however, extends beyond mere design of a system of adequate capacity to satisfy assumed criteria. The design goal also must include achievement of an optimum building.

To achieve this goal, the HVAC system must be integrated with other building systems. Its design is affected by and, in turn, influences the design of other building systems, such as construction of walls and roofs and shape and orientation of the building. These systems, and especially thermal insulation in exterior components and amount and type of glass in exterior walls, have a considerable effect on indoor temperatures. Consequently, for a thorough evaluation of the indoor environment, the effects of all building systems must be considered. This evaluation should start in the schematic stage of design. Otherwise, design objectives for the HVAC system may not be achieved when the building is completed or life-cycle costs may be excessively high.

Comprehension of the role of HVAC in buildings requires a knowledge of:

1. Human physiology, psychology and local socioeconomic and climatological conditions
2. Aerodynamics, thermodynamics—an extension of mechanics to include temperature—and sources of heat gains and losses in buildings
3. Properties of air and water vapor in air
4. Thermal insulation and vapor barriers
5. Design concepts for HVAC and available environmental control equipment
6. Methods of fire prevention and control

HVAC systems are designed by mechanical engineers and installed by mechanical contractors. Ducts may be fabricated and placed by sheet-metal workers and pipe may be installed by pipefitters. Necessary electrical equipment and wiring may be placed by electricians.

Reference

B. Stein et al., *Mechanical and Electrical Equipment for Buildings*, 7th ed., Wiley, New York, 1986.

Words and Terms

Air conditioning
Comfort
Heat load
Humidity
HVAC
Swing (temperature)
Temperature
Ventilation

Significant Relations, Functions, and Issues

Relations of indoor and outdoor conditions: quantified differences, daily swing, type and degree of interior modification and control, comfort criteria.

Components of interior control: ventilation, heating, cooling, air conditioning.

HVAC system design factors: design criteria, choice of system type, size (capacity, rating, etc.) of equipment, operation and control of systems.

10.2 MEASUREMENT OF HEAT

Heat is a form of energy that is transferred from one substance to another because of a temperature difference between them. Heat applied to a substance increases the average velocity of its molecules or electrons and hence increases their kinetic energy. Similarly, heat removed decreases the average molecular velocity and therefore also electron or molecular kinetic energy.

A thermometer is a device used to measure the heat in a substance, but it does not do so directly. The thermometer scale indicates the temperature of the substance, and change in temperature, as read on a thermometer, is a measure of heat transferred to or from the substance. A unit of temperature is called a *degree*.

By convention, the scale used on thermometers is an interval scale (see Sec. 3.2). In the *Celsius system* of measuring temperatures, 0°C is assigned to the temperature at which water freezes and 100°C to the temperature at which water boils at normal atmospheric pressure. Therefore, on a Celsius thermometer, 1°C is represented by a distance on the scale equal to 0.01 of the distance between the points representing the freezing and boiling temperatures of water. In the *Fahrenheit system*, 32°F is the temperature at which water freezes and 212°F the temperature at which water boils at normal atmospheric pressure. Therefore, on a Fahrenheit thermometer, 1°F is represented by a distance on the scale equal to $\frac{1}{180}$ of the distance between the points representing the freezing and boiling temperatures of water.

$$°F = 1.8 \times °C + 32$$

$$°C = \frac{5}{9}(°F - 32)$$

British Thermal Units

In HVAC calculations, the unit heat quantity usually used is the British thermal unit (Btu). This unit, by definition, is the amount of heat required to raise the temperature of 1 lb of water 1°F at or near 39.2°F, which is its temperature of maximum density.

If 1 Btu was added to or removed from 1 lb of a substance other than water, the temperature of the substance usually will change more or less than 1°F. The temperature change will depend on a property of the substance called thermal capacity.

The *thermal capacity* of a substance is the number of Btu required to raise the temperature of 1 lb of the substance 1°F.

Specific heat is the ratio of the thermal capacity of the substance to the thermal capacity of water. By definition then, the specific heat of water is unity. In HVAC calculations it is often more convenient to use specific heat rather than thermal capacity, because Btu may then be used as a unit of heat quantity without conversions. Specific heats of air and some common building materials are listed in Table 10.1. (Data for other substances may be obtained from tables in the *Handbook of Fundamentals*, American Society of Heating, Refrigerating and Air-Conditioning Engineers.) Note that the specific heat of common materials is less than unity. This indicates that of all common substances water has the largest thermal capacity and specific heat.

Table 10.1. Specific Heats of Some Common Substances

Substance	Specific heat
Air at 80°F	0.24
Water vapor	0.49
Water	1.00
Aluminum	0.23
Brick	0.20
Brass	0.09
Bronze	0.10
Gypsum	0.26
Ice	0.48
Limestone	0.22
Marble	0.21
Sand	0.19
Steel	0.12
Wood	0.45–0.65

Sensible Heat

When heat is added to or removed from a substance, the resulting temperature changes can be detected by the sense of touch. For this reason, the heat is called sensible heat. There is another type of heat, called latent heat, which is not associated with temperature change. It will be discussed later.

Sensible heat is the heat associated with a change in temperature. The quantity of sensible heat transferred in a heat exchange can be calculated from the basic heat equation:

$$H = mc(t_2 - t_1) \qquad (10.1)$$

where

H = sensible heat, Btu, absorbed or removed

m = mass, lb, of substance undergoing a temperature change

c = specific heat of the substance

t_2 = temperature, °F, after the heat exchange

t_1 = temperature, °F, at the start of the heat exchange

Laws of Thermodynamics

HVAC calculations often are based on two laws of thermodynamics. These laws may be expressed in different but equivalent ways. The following representations of the laws have been selected for simplicity.

The first law of thermodynamics states that when work produces heat, the quantity of heat generated is proportional to the work done. Conversely, when heat is employed to do work, the quantity of heat dissipated is proportional to the work done. (*Work*, ft-lb, is equal to the product of the force, lb, acting on a body by the distance, ft, the body moves in the direction of the application of the force.)

The first law may be expressed mathematically by

$$W = JH \qquad (10.2)$$

where

W = work, ft-lb

H = heat generated by the work, Btu

J = Joule's constant = mechanical equivalent of heat

Experiments have shown that $J = 778$ ft-lb per Btu.

The second law of thermodynamics states that it is impossible for a machine unaided by an external agency to transfer heat from any substance to another substance at a higher temperature.

The law implies that the supply of energy available in the universe for doing useful work is continuously decreasing. The law also indicates that it is impossible to devise an engine that can convert a specific quantity of heat into an equivalent amount of work.

The second law of thermodynamics can be expressed mathematically with the use of the concept of entropy.

Entropy is the ratio of the heat added to a substance to the absolute temperature at which the heat is added.

$$S = \frac{dH}{T} \qquad (10.3)$$

where

S = entropy
dH = differential (very small change) of heat
T = absolute temperature

Suppose an engine, which transforms heat into mechanical work, receives heat H_1 from a heat source at temperature T_1. The engine gives off heat H_2 at temperature T_2 to a heat sink after doing the work. By the law of conservation of energy, or the first law of thermodynamics, H_2 is less than H_1 by the amount of work done. By the second law of thermodynamics, T_2 is less than T_1. The entropy of the heat source decreases by $\Delta S_1 = H_1/T_1$. The entropy of the heat sink increases by $\Delta S_2 = H_2/T_2$. The net change in the entropy of the universe because of this process then is $\Delta S_2 - \Delta S_1$. The second law of thermodynamics requires that this net change be greater than zero; that is, that entropy increase is an irreversible thermodynamic process:

$$\Delta S_2 - \Delta S_1 > 0 \qquad (10.4)$$

Furthermore, the energy that has become unavailable for doing work because of the irreversibility of the process is

$$H_u = T_2(\Delta S_2 - \Delta S_1) \qquad (10.5)$$

Absolute Temperature

The definition of entropy involves the concept of absolute temperature measured on a ratio scale (see Sec. 3.2). The unit of absolute temperature is measured in degrees Kelvin (°K) in the Celsius system and in degrees Rankine (°R) in the Fahrenheit system. Zero degrees in either system is determined from the theory of behavior of an ideal gas. For such a gas,

$$PV = mRT \quad \text{or} \quad Pv = RT \qquad (10.6)$$

where

P = absolute pressure on the gas, psf
V = volume of the gas, cu ft
v = specific volume of the gas, cu ft per lb = reciprocal of density of gas
m = mass of the gas, lb
T = absolute temperature
R = universal gas constant divided by molecular weight of the gas

For a gas under constant volume, the absolute temperature should be zero when the pressure is zero and molecular motion ceases. This temperature has been determined to be $-273°C$ and $-460°F$. Hence,

Kelvin temperature
  = Celsius temperature + 273   (10.7a)

Rankine temperature
  = Fahrenheit temperature + 460 (10.7b)

In the Rankine temperature system, gas constant R equals $1,545.3$ divided by the molecular weight of the gas. For air, $R = 53.4$, and for water vapor $R = 85.8$.

Latent Heat

Previously, sensible heat was defined as heat associated with a change in temperature. In contrast, latent heat is heat associated with a change in state of a substance, for example, from solid to liquid or from liquid to gas. Latent heat must be taken into account in calculations of the total heat content in air, which almost always contains some water in the form of vapor. Because of this concern with water vapor, the concept of latent heat will be illustrated by consideration of the changes in state of water.

When heat is added to ice (solid water) the

temperature of ice rises until the melting point is reached. During melting, the water absorbs heat without change in temperature as the solid state changes to the liquid state. Similarly, if liquid water is cooled to the freezing point, heat is removed without change in temperature as the liquid is converted to ice. This heat is called the latent heat of fusion of water. Experiments show that 144 Btu are required to convert 1 lb of water at 32°F to 1 lb of ice at the same temperature. Thus,

Latent heat of fusion of water = 144 Btu per lb
(10.8)

If the pound of water is now heated by addition of 180 Btu, it will reach its boiling point, 212°F, under normal atmospheric pressure. Further addition of heat causes the water to vaporize, or turn into steam, a gas. During vaporization, the water absorbs heat without change in temperature as the liquid changes to gas. This heat is called the latent heat of vaporization of water. Experiments show that 970 Btu are required to convert 1 lb of water at 212°F to 1 lb of steam at the same temperature. Hence, at 212°F,

Latent heat of vaporization of water
= 970 Btu per lb   (10.9)

When steam is cooled, 970 Btu per lb must be removed to convert it to liquid water; that is, to condense the vapor.

Under normal atmospheric pressure, 29.92 in. of mercury, some water will evaporate from a water surface at temperatures below the boiling point. The amount of vapor that will be absorbed by the air above the water surface depends on the capacity of the air to retain water at the existing temperature and the amount of water vapor already in the air. The latent heat of vaporization in such cases is supplied by the water and the air. Both thereby become cooler.

Table 10.2 lists the latent heat of vaporization of water for various air temperatures and normal atmospheric pressure. More extensive tables of thermodynamic properties of air, water and steam are given in the *Handbook of Fundamentals*, American Society of Heating, Refrigerating and Air-conditioning Engineers (ASHRAE).

Enthalpy

Enthalpy is a measure of the total heat in a substance and equals the sum of its internal energy plus its capacity to do work, or PV/J, where P is the pressure of the substance, V its volume and J the mechanical equivalent of heat. *Specific enthalpy* is the heat per unit of weight, Btu per lb.

For purposes of HVAC calculations, the specific enthalpy of dry air is taken as zero at 0°F. The enthalpy at higher temperatures h_a equals the product of the specific heat, about 0.24, by the temperature, °F (see Table 10.2).

The specific enthalpy of saturated air h_s, which includes the latent heat of vaporization of the moisture, also is listed in Table 10.2. The specific enthalpy of the water vapor at the air temperature may be obtained from Table 10.2 by subtracting h_a from h_s.

Table 10.2 also lists, for various temperatures, the humidity ratio of the air at saturation, or weight, lb, of water vapor in saturated air per lb of dry air. In addition, Table 10.2 gives h_g, the specific enthalpy, Btu per lb, of saturation water vapor, the sum of the latent heat of vaporization and the specific enthalpy of water at various temperatures.

The specific enthalpy of unsaturated air equals the sensible heat of dry air at the existing temperature, with the sensible heat at 0°F taken as zero, plus the product of the humidity ratio of the unsaturated air and h_g for the existing temperature.

Cooling by Evaporation

When water evaporates without heat being supplied by an outside source, the water provides the heat of vaporization. Thus, sensible heat of the liquid is converted to latent heat in the vapor. As a result, the temperature of the remaining liquid drops. Since no heat is added or removed during this process of evaporation, it is called *adiabatic cooling*.

Humans are cooled by evaporation of perspiration from their skin. Similarly, relatively dry air may be cooled by evaporation of water into the air. Also, refrigeration may be accomplished by evaporation of a liquid refrigerant.

Table 10.2. Some Thermodynamic Properties of Air at Atmospheric Pressure

Air temperature, dry bulb (or dew point) (°F)	Specific volume, cu ft per lb		Lb of water in saturated air per lb of dry air (Humidity ratio)	Latent heat of vaporization of water (Btu per lb)	Specific enthalpy of dry air h_a,[a] (Btu per lb)	Specific enthalpy of saturated air h_s,[b] (Btu per lb)	Specific enthalpy of saturation vapor h_g (Btu per lb)
	Dry Air	Saturated Air					
0	11.58	11.59	0.0008	· · ·	0	0.84	· · ·
32	12.39	12.46	0.0038	1,075.2	7.69	11.76	1,075.2
35	12.46	12.55	0.0043	1,073.5	8.41	13.01	1,076.5
40	12.59	12.70	0.0052	1,070.6	9.61	15.23	1,078.7
45	12.72	12.85	0.0063	1,067.8	10.81	17.65	1,080.9
50	12.84	13.00	0.0077	1,065.0	12.01	20.30	1,083.1
55	12.97	13.16	0.0092	1,062.2	13.21	23.22	1,085.2
60	13.10	13.33	0.0111	1,059.3	14.41	26.46	1,087.4
65	13.22	13.50	0.0133	1,056.5	15.61	30.06	1,089.6
70	13.35	13.69	0.0158	1,053.7	16.82	34.09	1,091.8
75	13.47	13.88	0.0188	1,050.9	18.02	38.61	1,094.0
80	13.60	14.09	0.0223	1,048.1	19.22	43.69	1,096.1
85	13.73	14.31	0.0264	1,045.2	20.42	49.43	1,098.3
90	13.85	14.55	0.0312	1,042.4	21.62	55.93	1,100.4
95	13.98	14.80	0.0367	1,039.6	22.83	63.32	1,102.6
100	14.11	15.08	0.0432	1,036.7	24.03	71.73	1,104.7
150	15.37	20.58	0.2125	1,007.8	36.1	275.3	1,125.8
200	16.63	77.14	2.295	977.7	48.1	2,677	1,145.8
212	· · ·	· · ·	· · ·	970.2	· · ·	· · ·	1,150.4

[a]Specific enthalpy of dry air h_a is taken as zero for dry air at 0°F.
[b]Specific enthalpy of water vapor in saturated air = $h_s - h_a$, including sensible heat above 32°F.

Heating by Condensation

An *adiabatic process* is a process that takes place without addition or removal of heat. *Adiabatic heating* takes place when air is cooled to a temperature at which some moisture can no longer be retained. The excess water vapor condenses, becomes liquid. If no heat is removed during condensation, the latent heat of vaporization of the water vapor is converted to sensible heat in the air. As a result, the air temperature rises.

Thus, formation of a fog is often accompanied by an increase in air temperature. Similarly, air temperature generally rises when rain or snow begins to fall.

Psychrometry

Psychrometry is the branch of physics that deals with measurement or determination of atmospheric conditions, particularly regarding the moisture mixed with air. Some psychrometric principles have been presented in the preceding discussions. Additional principles will now be discussed.

The mixture of dry air and water vapor at ordinary temperatures and pressures behaves nearly like an ideal gas. For example, temperatures, volumes and pressures may be computed with Eq. (10.6), $Pv = RT$. Also, *Dalton's law* applies:

When several gases occupy a common space, each gas fills the volume and behaves as if the other gases were not present.

From this it may be concluded, that each gas in a mixture not only occupies the same space but also is at the same temperature as every other gas in the mixture. The total weight of the mixture, however, equals the sum of the weights of the gases. Also, the enthalpy of the mixture equals the sum of the enthalpies of the gases.

Dalton's law of partial pressures states that the pressure of a mixture of gases equals the sum of the pressures each gas would exert if it alone occupied the volume enclosing the mixture. Thus, *barometric pressure*, the atmospheric pressure registered by a barometer, equals the sum of the partial pressure of dry air and the partial pressure of the water vapor in the air.

Partial pressures of water vapor and dry air in the atmosphere can be used to calculate the degree of saturation of the air, or relative humidity, at a specific temperature.

Relative humidity is the ratio of the mole fraction of water vapor present in the air to the mole fraction of water vapor present in saturated air at the same temperature and pressure. (A *mole* is the weight of a substance numerically equal to its molecular weight.) This definition is awkward to apply in practice. Instead, relative humidity is approximated by the ratio of the partial pressure of the water vapor in the air to the saturation pressure of water vapor at the same temperature and is usually expressed as a percentage. Thus, 100% relative humidity indicates saturated air and 0% relative humidity, dry air. Relative humidity may also be computed from humidity ratios but with somewhat less accuracy.

Humidity ratio, or *specific humidity*, W_a, at a specific temperature, is the weight, lb, of water vapor in air per lb of dry air. Let W_s be the humidity ratio of saturated air at the same temperature (see Table 10.2). Then, relative humidity can be obtained approximately from

$$RH = \frac{W_a}{W_s} \times 100 \qquad (10.10)$$

Dalton's law of partial pressures and Eq. (10.6) can be used to evaluate the humidity ratios:

$$W_a = 0.622 \frac{p_w}{P - p_w} \qquad (10.11)$$

where

P = barometric pressure, psi

p_w = partial pressure of water vapor, psi

Measurement of partial pressures, however, requires use of scientific instruments that are generally available only in research laboratories. In practice, therefore, humidity measurements are made in the field with simpler instruments that give readings that can be converted readily into humidity ratios or relative humidity.

One method utilizes the wet-bulb temperature of the unsaturated air. This temperature is determined with a thermometer that has its bulb covered with a wick that is saturated with water. The thermometer is slung, or rotated, to evaporate water quickly from the wick. The evaporation lowers the temperature of the water and the thermometer. After a while, the temperature registered by the thermometer remains constant. This temperature is called the *wet-bulb temperature*.

To differentiate ordinary temperatures from wet-bulb temperatures, the readings on an ordinary thermometer are called *dry-bulb temperatures*.

The difference between a dry-bulb temperature for unsaturated air and the wet-bulb temperature for the air is related to the relative humidity of the air. Various books on psychrometry and HVAC contain tables or charts giving the relative humidity for specific wet-bulb and dry-bulb temperatures.

Dew-Point Temperatures

During warm weather, dew often is found on grass, trees and other outdoor things in early morning. Dew is the condensation of water vapor from the atmosphere as the temperature drops during the night.

As unsaturated air cools, it may reach a temperature at which it becomes saturated. With a further drop in temperature, the air is unable to retain as much water vapor as before. Moisture in excess of the amount of vapor the air can hold at saturation at the lowered temperature condenses and forms drops of water, or dew.

Dew-point temperature is the name given to the temperature at which the condensation of water vapor in a space begins for a given state of humidity and pressure as the temperature of the vapor is reduced. At the dew-point temperature, the air is saturated; that is, it has 100% relative humidity, with the same weight of water per cu ft of air and at the same pressure as the unsaturated air before it was cooled.

When relative humidity is known, the dew-point temperature can be obtained from Eq. (10.10) and Table 10.2. For the temperature of the unsaturated air, the humidity ratio at saturation can be read from Table 10.2. Multiplication of this ratio by the relative humidity gives the humidity ratio for the dew-point temperature, which can be determined by examination of Table 10.2. For example, to determine the dew point for air at 85°F and 80% relative humidity, note that Table 10.2 indicates a humidity ratio at saturation of 0.0264 at 85°F. Multiplication by 80% yields a humidity ratio of 0.0211. By interpolation in Table 10.2 between humidity ratios at saturation for temperatures at 75 and 80°F, the dew point is found to be 78.3°F.

A convenient way to obtain many of the properties of mixtures of dry air and water vapor is to use a *psychrometric chart*. A typical chart relates dry-bulb and wet-bulb temperatures to relative humidity, humidity ratio, dew point and specific volume of air. A psychrometric chart usually is provided with books on psychrometry and HVAC.

Refrigeration Ton

For convenience, a special unit, called a ton, is used for measuring the amount of cooling accomplished in air conditioning. A ton of refrigeration equals 12,000 Btu per hr, or 200 Btu per min, of cooling.

10.3. HEAT FLOW AND HUMAN COMFORT

This section deals with methods of heat transfer and its effects on human comfort. The concepts may be easily adapted to environmental control of building interiors for industrial processes or computer installations.

Indoor temperatures are determined over a period of time by a heat-exchange balance. In cold weather, heat flows out of the building. A comfortable temperature can be maintained only by adding enough heat to counterbalance the heat loss. In contrast, in hot weather, heat

flows into the building. In that case, a comfortable temperature can be maintained only by removing from the interior enough heat to offset the heat gain.

In both cases, the rate of heat flow is a function of the difference between indoor and outdoor temperatures. Hence, the heat that must be added to the building interior in cold weather or removed in hot weather depends on the indoor temperature that should be maintained for human comfort and the outdoor temperatures that may cause discomfort. Consequently, before HVAC equipment can be selected for environmental control, design criteria must be established for indoor and outdoor conditions. In addition, the factors affecting heat flow from indoors to outdoors must be determined and evaluated.

In practice, it is not necessary to establish design criteria with great exactness nor to maintain design conditions with extreme accuracy. For one thing, there are too many variables involved. For another, it is usually impossible to maintain indoor conditions that will please every occupant. In addition, building owners often are unwilling to pay for the cost of equipment and controls that will maintain optimum conditions. Therefore, in practical design of HVAC, some dependence should be placed on the adaptability of humans to small variations from optimum comfort conditions. Also, designers should bear in mind that human comfort depends not only on environmental conditions but also on the type of clothing occupants wear, the activities being performed, the difference in sensation between immediately entering or leaving an environmentally controlled space and being exposed to the environment for a long time, and the density, or closeness, of occupants within the space.

In addition, the performance expected by the owner from the HVAC system should be taken into account. To judge whether expected performance can be attained and maintained, designers should learn how the capacities of all the system components have been determined, what are the limits of accuracy of published performance data and what is the capability of available equipment. In summary, designers should exercise engineering judgment in establishing design criteria and selecting a system for satisfying these criteria.

Methods of Heat Transfer

According to the second law of thermodynamics, heat flows in the direction of decreasing temperature. The flow may occur in any of three ways—conduction, convection or radiation—or in any combination of them.

Thermal conduction is the process of heat transfer through matter by transmission of kinetic energy from particle to particle without large displacement of particles. When one end of a short metal bar is heated, conduction is the method by which heat flows from the heated end to the cooler end.

Thermal convection is the process of heat transfer by movement of a liquid or gas. Natural convection generally occurs because of differences in density in parts of a liquid or gas and is common occurrence where density changes are caused by changes in temperature. Forced convection results from pressures, such as those imposed by a blower, pump or jet.

Thermal radiation is the process of heat transfer in which the internal energy of one substance is converted to electromagnetic energy of very long wavelengths, which is absorbed by another substance and changed back to heat. Because electromagnetic energy, like light from the sun, can travel through a vacuum, thermal radiation can be transmitted through empty space. In fact, the presence of matter in the path of the radiation can curtail or block its passage.

Thermal Conduction

The rate at which heat is transferred from one face of a flat sheet of material to the opposite face by conduction is proportional to the difference in temperature between the faces. If the temperature difference is held constant so that the rate of heat flow is constant, steady-state heat conduction occurs. In that case, the heat-flow rate through a sheet with unit thick-

ness may be calculated from

$$q = kA(t_2 - t_1) \qquad (10.12)$$

where

q = heat-flow rate, Btu per hr

A = cross-sectional area of the sheet, sq ft, normal to the heat flow

t_2 = temperature, °F, on the warmer side of the sheet

t_1 = temperature, °F, on the cooler side of the sheet

k = coefficient of thermal conductivity for a unit thickness of the material

The coefficient depends on the characteristics of the sheet material. Also, the numerical value of k depends on the units chosen for the other variables in Eq. (10.12). Hence, when values of k are obtained from tables, the units with which those values are to be used should be ascertained.

In practice, thicknesses of materials used are usually more or less than unit thickness. Consequently, it is convenient to define and use a coefficient of conductivity applicable to the actual material thickness. This coefficient, called thermal conductance, can be obtained by dividing k by L, the thickness, or length of path along which the heat flows.

Thermal conductance C is the rate of heat flow across a unit area of a homogeneous material from one face to the opposite face for a unit temperature difference between the two faces, under steady-state conditions. The heat-flow rate, Btu per hr, then may be computed from

$$q = CA(t_2 - t_1) \qquad (10.13)$$

Values of k and C for various building materials may be obtained from tables in the ASHRAE *Handbook of Fundamentals* and other publications.

Thermal resistance is a measure of the obstruction a material offers to flow of heat through it. Thermal resistance R is defined by

$$R = \frac{1}{C} \qquad (10.14)$$

This coefficient is useful in computing the heat-flow rate through a barrier consisting of a sequence of different materials, as explained later.

The thermal resistance to radial flow across the thickness of a long cylinder, such as a pipe, is given by

$$R = \frac{\log_e r_o/r_i}{2\pi kl} \qquad (10.15)$$

where

r_o = outer diameter of the cylinder, in.
r_i = inner diameter of the cylinder, in.
k = coefficient of thermal conductivity
l = length of the cylinder, ft

Air Films. In HVAC calculations, the heat flow that is of concern is that which passes from the air on one side of the enclosure of a space to the air on the opposite side. Before external heat can reach the enclosure face that is at the higher temperature or internal heat can leave the face that is at the lower temperature, the heat in each case must pass through a thin air film that clings to the face. The air film resists passage of heat.

The amount of this resistance depends on the physical characteristics of the enclosure face, the rate of air movement along the face and the inclination of the direction of heat flow with respect to the horizontal. Calculation of the heat-flow rate through an enclosure from the air on one side to the air on the other side, therefore, must include the resistance of two air films.

The thermal conductance of air films, denoted by f_i for an indoor film and by f_o for an outdoor film, are given approximately in Table 10.3. For convenience also, the thermal conductance of an enclosed air space is also given in the table. (These coefficients include the effects of the air films along both enclosing surfaces.) More extensive data are presented in tables in the ASHRAE *Handbook of Fundamentals*.

Metals have a high coefficient of thermal conductivity k; for example, for steel k is about 300 Btu per hr per sq ft per °F per in. thick, and for aluminum k is about 1,400. Comparison of these values with those in Table 10.3 indicates that still air is a poor heat conductor.

Table 10.3. Thermal Conductance of Air[a]

f_i for indoor air film (still air)	
Vertical surface, horizontal heat flow	1.5
Horizontal surface	
Upward heat flow	1.6
Downward heat flow	1.1
f_o for outdoor air film, 15-mph wind (winter)	6.0
f_o for outdoor air film, 7.5-mph wind (summer)	4.0
C for vertical air gap, $\frac{3}{4}$ in. or more wide	1.1
C for horizontal air gap, $\frac{3}{4}$ in. or more wide	
Upward heat flow	1.2
Downward heat flow	1.0

[a]Btu per hr per sq ft per °F.

Air-to-Air Heat Transfer. For steady-state conduction of heat through several different materials in sequence, the rate of heat flow, Btu per hr, can be computed from

$$q = UA(t_2 - t_1) \qquad (10.16)$$

where

U = coefficient of thermal transmittance

The *coefficient of thermal transmittance U*, also known as the *overall coefficient of heat transfer*, is the rate of heat flow per unit area, under steady-state conditions, from the air on the warm side of an enclosure to the air on the cooler side when there is a unit temperature difference between the air on opposite sides. When the heat flow passes through several materials in sequence, the total resistance to heat flow equals the sum of the resistances of the materials, and U is the reciprocal of the total resistance.

Suppose, for example, that an enclosure is a composite of several materials with thermal resistances $R_1, R_2, \ldots R_n$. Then, the total thermal resistance is

$$R = \frac{1}{f_i} + R_1 + R_2 + \cdots + R_n + \frac{1}{f_o}$$
$$(10.17)$$

with the resistances of the indoor air film $1/f_i$ and outdoor air film $1/f_o$ included. Consequently, the coefficient of thermal transmittance is

$$U = \frac{1}{R} \qquad (10.18)$$

This coefficient may be substituted in Eq. (10.16) for computation of the heat-flow rate through the enclosure.

As a simple example, consider an exterior wall made of 8-in.-thick, lightweight concrete block with a thermal conductance of 0.5. Exterior area of the wall is 100 sq ft. Temperature indoors is to be maintained at 75°F when the outdoor temperature is 0°F with a 15-mph wind blowing. What is the heat-flow rate through the wall?

From Table 10.3, the indoor air film has a resistance of $1/1.5 = 0.67$ and the outdoor air film, a resistance of $1/6.0 = 0.17$. The concrete block has a resistance of $1/0.5 = 2.0$. Total resistance of the wall then is

$$R = 0.67 + 2.0 + 0.17 = 2.84$$

Hence, the coefficient of thermal transmittance is

$$U = \frac{1}{2.84} = 0.35$$

and the heat-flow rate through the wall is

$$q = 0.35 \times 100(75 - 0) = 2,625 \text{ Btu per hr}$$

The ASHRAE *Handbook of Fundamentals* contains tables with U values for various types of construction. If the type of construction to be used for a building is not included in a table of U values, the preceding procedure can be used to calculate U from the thermal coefficients of the components.

Convection

Air readily transfers heat by natural convection. Air contacting a warmer wall surface receives heat by conduction. Becoming warmer, the air also becomes less dense, hence, it rises (see Fig. 10.4a). As it moves away from the warm surface, other air moves to the surface to replace the rising air. If the moving air should now come in contact with a cooler wall surface, this air would become denser. It would then fall due to gravity (see Fig. 10.4a). As it falls other air moves to the cool surface to replace the falling air. As a result, heat is transferred from the warm wall to the cool one. Convection will continue as long as there is a temperature difference between the two surfaces.

Convection can be used to heat a building interior. For this purpose, a heater consisting of a pipe containing steam or hot water is placed along one or more walls of a room near the floor. Closely spaced metal fins may be attached to the pipe perpendicular to its axis to increase the heating surface. This type of heater is called a *convector*. Cool air along the floor of the room contacts the hot pipe and fins. The air consequently becomes warmer and its density increases, causing it to rise. As it does, more cool air is drawn into the convector, becomes warm and rises (see Fig. 10.4b). In this manner, the room is heated by convection. It also receives from the convector some heat by radiation.

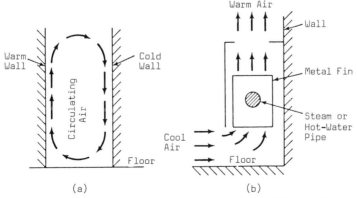

Fig. 10.4. Convection air currents caused by temperature differences. (a) Air circulation between two walls (vertical section). (b) Air flow induced by convector.

Convection also is used by the human body to lower the skin temperature. When the skin temperature is higher than the ambient air temperature, the skin warms the air in contact with it. The air rises and is replaced by cooler air. In this way, convection currents are established, cooling the skin. If the air is moving, for example, because of wind or pressure from a fan, the rate of heat flow from the skin increases. That is why people feel cooler in a breeze without a drop in air temperature. People who are perspiring will be additionally cooled by convection, because evaporation of perspiration will be speeded by the air movement.

Radiation

Radiation is the most common method of heat transfer. All substances radiate energy and absorb energy radiated by other substances. The earth is heated by radiation from the sun. Solar energy also heats the exteriors of buildings. Walls and windows radiate the solar energy they receive to occupants and furnishings in the buildings. When walls and windows are cold, however, the occupants and furnishings lose heat by radiation to the cold surfaces. At night, the earth and buildings on it lose heat by radiation to the cold sky.

Materials differ in their ability to radiate energy or absorb radiant energy. Those with black, rough surfaces are both good radiators and good absorbers. Those with smooth, shiny surfaces are good reflectors of radiation and consequently poor absorbers of radiant energy. They also are poor radiators.

Glass is transparent to short-wave radiation and therefore transmits much of the radiation from the sun that strikes the surface. After this radiation enters a building and strikes interior objects, some of the energy is radiated outward by those objects, but at longer wavelengths. Glass, however, is an inefficient transmitter of the longer wavelengths. It prevents most of this energy from escaping from the building. The glass thus acts much like a one-way valve for solar radiation. In the winter, the trapped solar energy is useful for heating building interiors adjacent to windows. But in the summer, the trapped heat adds to the cooling load for the building.

According to the *Stefan-Boltzmann law of radiation*, the energy radiated from a perfect radiator, or black body, is proportional to the fourth power of the absolute temperature of the body. Ordinary materials, however, are not perfect radiators, nor are they perfect absorbers. Hence, in practice, the Stefan-Boltzmann proportionality constant has to be modified by application of a factor called *hemispherical emittance*. Techniques for estimating heat-transfer rates by radiation are presented in the ASHRAE *Handbook of Fundamentals*.

Occupants of a building receive radiated heat from and lose heat by radiation to surrounding walls, floor and ceiling and to furniture and other indoor objects. The amount of heat transferred depends on the temperatures of the various radiating bodies. For convenience in HVAC calculations, the differing temperatures are combined into a mean radiant temperature.

Mean radiant temperature is the temperature of a black enclosure in which a solid body or occupant would exchange the same amount of radiant heat as in the existing nonuniform environment.

Thermal Criteria for Building Interiors

Research has indicated that the main conditions to be controlled for human comfort indoors are the dry-bulb temperature, relative humidity and rate of air movement. The first two conditions are discussed in the following paragraphs.

The indoor conditions to be maintained should be evaluated at the breathing line, 3 to 5 ft above the floor. Measurements should be made where average conditions for the building or for a zone exist and where the measurements would not be affected by abnormal heat gains or losses. Conditions, however, may be permitted to vary from design criteria within acceptable limits in the occupied zone.

The occupied zone of an indoor space is the region between levels 3 in. above the floor and 72 in. above the floor and more than 2 ft from walls or fixed heating or cooling equipment.

Tests have shown that for inactive persons,

lightly clothed and immersed in relatively still air indoors, the design dry-bulb temperature may be determined from

$$t = 180 - 1.4t_r \qquad (10.19)$$

where

t = dry-bulb temperature, °F

t_r = mean radiant temperature of the space, °F

Note that when the dry-bulb temperature and the temperatures of walls, floor, ceiling, furnishings and other objects in the space are the same, $t = t_r = 75$°F. Hence, for each 1°F the mean radiant temperature lies below 75°F, the design dry-bulb temperature should be increased 1.4°F above 75°F. Thus, in cold weather, when the exterior walls of a building are cold, the temperature in outer rooms should be maintained above 75°F, to compensate for radiant heat lost to the cold walls. In hot weather, when the exterior walls are warm, the temperature in outer rooms should be held below 75°F, to offset heat gain from the warm walls. [Equation (10.19), however, applies only for mean radiant temperatures in the range between 70 and 80°F.]

Criteria for Humidity. For temperatures close to 75°F, relative humidity has little effect on the comfort of occupants who are not active. Relative humidity, however, should be controlled for reasons other than human comfort. It is not desirable that relative humidity should exceed 60%, because some materials deteriorate when too moist. Also, it is not desirable that relative humidity be less than 20%, because human nostrils become too dry and some items, such as furniture, may be damaged by drying out.

In winter, low relative humidity is desirable indoors to prevent condensation on windows and in exterior walls and roofs. Often, in such cases, a relative humidity of 30 to 35% is desirable.

In summer, a relative humidity of 50% normally is acceptable, but 40% may be more practical in arid regions.

Permissible Variations in Criteria. Design indoor conditions may have to be varied from the preceding criteria in accordance with the activities of the building occupants and the intended use of the building. Appropriate criteria for a variety of conditions are presented in the "Applications" volume of the ASHRAE *Guide and Data Book*.

When the design dry-bulb temperature is about 75°F, the actual temperature may be permitted to vary from about 73 to 77°F within the occupied zone. The air temperature at the ceiling may be somewhat higher, normally about 0.75°F more per foot of height above the breathing level. Additional details on indoor thermal design criteria are given in ASHRAE Standard 55, *Thermal Comfort Conditions*.

Outdoor Design Conditions

Outdoor dry-bulb temperatures, relative humidity and winds that may be expected to occur at a building site are of importance in building design for two reasons. One reason is that these conditions influence determination of the capacities, or sizing, of the heating and cooling plants to be installed. The second reason is that these conditions affect estimates of the cost of operation of HVAC.

Capacities of the heating and cooling plants should be based on the heating and cooling loads for the building. A large proportion of these loads, in turn, generally depend on the difference between outdoor and indoor temperatures. For a given indoor design temperature, sizes of heating and cooling plants, therefore, often will be determined predominantly by assumed outdoor conditions.

These plants may be sized for extreme outdoor conditions; for example, for temperatures that may occur in winter or summer only once in every 50 years; but unless this type of design is essential for the intended use of the building, initial cost of the plants and cost of operation will be unnecessarily large.

Extreme outdoor conditions are likely to occur only occasionally and may last for only a few hours. Plants sized for such conditions will be oversized for operation under less extreme conditions. The plants accordingly will not operate at design capacity or will operate at short intervals (cycle frequently). In either

Table 10.4. Outdoor Design Conditions for Some Cities

City	Winter dry-bulb temperatures, °F	Winter wind velocity, mph	Summer dry-bulb temperatures, °F	Summer wet-bulb temperatures, °F
Atlanta, Ga.	14	15	95	78
Baltimore, Md.	12	10	94	79
Boise, Idaho	0	10	96	68
Boston, Mass.	-1	15	91	76
Chicago, Ill.	-5	10	94	78
Cincinnati, Ohio	2	10	94	78
Dallas, Texas	14	15	99	79
Denver, Colo.	-9	10	92	65
Detroit, Mich.	0	10	92	76
Houston, Texas	24	10	96	80
Jacksonville, Fla.	24	10	96	80
Las Vegas, Nev.	18	10	108	72
Lincoln, Neb.	-10	10	96	78
Little Rock, Ark.	13	10	99	80
Los Angeles, Cal.	38	10	94	72
Miami, Fla.	39	10	92	80
Minneapolis, Minn.	-19	10	92	77
New Orleans, La.	29	10	93	81
New York, N.Y.	6	15	94	77
Phoenix, Ariz.	25	10	108	77
Portland, Maine	-15	10	88	75
Portland, Ore.	21	10	89	69
Richmond, Va.	10	10	96	79
St. Louis, Mo.	1	10	96	79
San Francisco, Cal.	38	10	80	64
Washington, D.C.	9	10	94	79

case, operation will be inefficient. Furthermore, when the equipment is cycling, building occupants are likely to feel uncomfortable during the periods when the equipment is not operating.

In contrast, if heating and cooling plants are designed for conditions that occur frequently, the plants will operate more often at design capacity, or cycles of operation will be longer. When extreme outdoor conditions occur, such plants may not be able to maintain design indoor temperatures, but such occurrences are likely to be infrequent and the duration of uncomfortable indoor conditions is likely to be short.

One other factor should be considered. Capacities of heating and cooling plants should be larger than the sizes needed merely to maintain design indoor conditions. The plants must be large enough to change indoor temperatures from some abnormal value to the design value

with reasonable rapidity. For example, in cold weather, the temperature of a building may be allowed to fall considerably below design indoor temperature at night when the building is unoccupied. The heating plant must be capable of raising the temperature rapidly to the design temperature in the morning when occupants arrive. To do this, the plant must have a larger capacity than that needed to maintain the design indoor temperature after it has been attained.

For the preceding reasons, outdoor conditions for design purposes should not be extremes for the building site but should be chosen in accordance with the intended use of the building, comfort of building occupants and acceptability to the owner of installation and operating costs. Outdoor design conditions that may be assumed for several cities are given in Table 10.4. More detailed data are presented in the ASHRAE *Handbook of Fundamentals*.

The design conditions in Table 10.4, however, are not likely to yield accurate results in estimates of operation costs. Temperatures at any given location vary considerably seasonally, daily and hourly. These variations should be taken into account when operating costs are estimated.

Also, the assumption of steady-state heat flow usually used for computing thermal loads in sizing heating and cooling plants may not hold for computations of operating costs. Because of the ability of some materials used for exterior walls and roofs to retain heat, indoor temperatures may not fluctuate as rapidly or over as large a range as outdoor temperatures. This condition is especially likely to occur in buildings with massive walls and roofs and relatively small window areas.

Hence, cost estimates should be based on average daily or hourly differences in temperature between outdoors and indoors. The necessary computations, however, are complex and tedious. They are best executed with the aid of high-speed electronic computers. Programs for this purpose are available.

10.4. THERMAL INSULATION

The method for computing thermal transmittance of various types of construction described in Sec. 10.3 indicates that design of HVAC systems should be integrated with design of the building exterior. For the heat-flow rate through the exterior determines a major portion of the heating and cooling loads for the building, and the heat-flow rate is proportional to the thermal transmittance of the exterior. A significant reduction in heating and cooling loads, therefore, can be achieved by decreasing the thermal transmittance, or overall coefficient of heat transfer U, of the exterior construction. By Eq. (10.18), this would be equivalent to increasing the thermal resistance of the exterior.

By Eq. (10.17), thermal resistance of a sequence of materials equals the sum of the resistances of the component materials. To increase the thermal resistance of any portion of a building exterior, therefore, it is necessary to increase the thermal resistance of at least one component or to add to the construction a material with high thermal resistance.

Materials with high thermal resistance are called thermal insulation. They are purposely incorporated in construction to reduce the heat-flow rate. Note that insulation does not completely stop heat flow but reduces it considerably.

Types of Thermal Insulation

Generally, materials with a coefficient of thermal conductivity less than unity are considered thermal insulation. Good insulation has a coefficient less than 0.30.

Insulation may be divided into two types. One type depends on a porous structure and irregular molecular structure for thermal resistance. The second type relies on an ability to reflect heat and on poor absorption of radiant energy.

Gases, such as air, having an irregular molecular structure with spaces between molecules that are large relative to molecular size, are good insulators. Hence, porous materials or materials separated by air spaces generally are good insulators, but only if the air is trapped, prevented from moving and transferring heat by convection.

Insulation such as cork, cellular glass and foamed plastics derive their thermal resistance from tiny volumes of air enclosed in separated cells. Granular insulation such as pumice, vermiculite and perlite entrap air in relatively large pores. Glass fiber and other fibrous materials owe their high thermal resistance to poor conductivity of the base materials and thin films of air that cling to the surfaces of the materials. In general, thermal conductance of these porous materials is inversely proportional to their thickness.

Insulation that depends for thermal resistance on reflectivity usually consists of a sheet of shiny metal, such as aluminum foil. Such materials also are poor radiators of heat. They are incorporated in a construction with an air gap on at least one side (see Fig. 10.5a). If, for example, bright aluminum foil receives heat by radiation across the air gap, it will reflect back about 95% of the radiation. If the foil receives

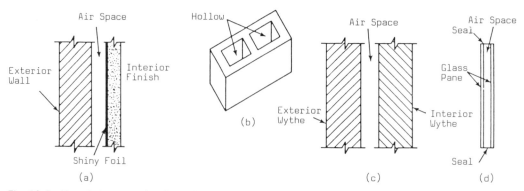

Fig. 10.5. Use of air spaces for thermal insulation. (*a*) Section through an exterior wall with a gap between outer facing and interior finish. (*b*) Hollow concrete block. (*c*) Section through a masonry cavity wall. (*d*) Double glazing for a window.

heat by conduction from the side opposite the air gap, the reflective insulation will lose only about 5% of the energy by radiation from the surface next to the air gap.

The air gap used with reflective insulation should be from $\frac{3}{4}$ to 2-in. wide. If greater thermal resistance is desired with a single foil, an air gap should be placed on both sides of it. In any case, supports for the foil should be poor conductors of heat, to prevent passage of heat past the air spaces.

Some precautions to be taken with aluminum-foil installations are as follows: Contact with other metals should be avoided or else galvanic corrosion may occur. Exposure to concrete or plaster should be avoided to prevent attack by alkalies in those materials. Also, installation should be on the warm side of a construction, because the foil acts as a vapor barrier and would cause condensation of the water vapor in the building air if the foil were cold.

Information on thermal conductance of the various types of insulation may be obtained from the manufacturers.

Air spaces often are used in building construction in lieu of solid insulation. For example, brick, block or concrete slabs frequently are made hollow to improve thermal resistance (see Fig. 10.5*b*). Cavity walls are built by placing two parallel, vertical layers, or wythes, of brick or block 2 or more in. apart (see Fig. 10.5*c*). Similarly, insulated glass windows are formed with two or three panes of glass separated by air spaces (see Fig. 10.5*d*).

Heat Flow Through Insulated Construction

The heat-flow rate through construction in which thermal insulation is incorporated can be calculated in the same way as for any assembly of materials through which heat flows in sequence, as described in Sec. 10.3. The total thermal resistance is computed with Eq. (10.17). The overall coefficient of heat transfer U is obtained from Eq. (10.18). And the heat-flow rate through the assembly is calculated with Eq. (10.16).

If the conductance of a wall without insulation is known, the effect of adding insulation can be readily computed by adding the resistance of the insulation to the resistance of the wall. The new U then is the reciprocal of the sum. The effect of adding still more insulation can be calculated in the same way.

Example. Consider an exterior wall of a type that might be used for a house. Such a wall would have an exterior facing of wood siding, backed by waterproofing building paper and plywood sheathing. These materials would be attached to 2 × 4-in. wood studs. On the inner face of the studs would be the interior finish, gypsum lath and plaster. Between the studs, which would be spaced 16 in. center to center, would be an air space. Up to 3 in. of insulation could be placed in the air space, if desired (see Fig. 10.6*a*). In the calculation of heat flow through the wall, the flow through the wood studs may be ignored because of the relatively low conductance of the wood compared with the air space.

The computations for the overall coefficient U for the uninsulated wall are shown in the upper part of Table 10.5. Each component of the wall is listed with its thermal conductance and resistance. In addition, the thermal conductance and resistance of interior and exterior air films along the wall are given for winter conditions (15-mph wind outdoors). The total resistance of the assembly is 3.62, the sum of the resistances of the components. The U value of the assembly is the reciprocal, 0.28.

Suppose now 1 in. of insulation with a thermal conductance of 0.27 and resistance of 3.70 was inserted in the air space between the studs. As indicated in Table 10.5, the sum of the resistance of the uninsulated wall and the resistance of the insulation is 7.32. Then, U for the assembly with 1 in. of insulation is the reciprocal, 0.14. Comparison with U for the uninsulated wall indicates that the addition of 1 in. of insulation reduces the heat flow 50%, as shown in the last column of Table 10.5.

The remainder of Table 10.5 shows the effects on U for the wall of adding 2 or 3 in. of insulation. Note that doubling the thickness of the insulation, from 1 in. to 2 in., decreases the U value of the wall 35%, compared with the 50% reduction achieved with the initial 1 in. of insulation. Also, increasing the insulation thickness from 2 in. to 3 in. produces only a 25.3% reduction in U for the wall. Thus, the percent savings in reduction of heat loss decreases with additional increments in thickness of insulation.

Installation of Thermal Insulation

Thermal insulation comes in a variety of forms. These include metal foils previously described, loose fill, cement, and flexible, semi-rigid, rigid and foamed-in-place materials.

Loose-fill insulation consists of powders, granules or nodules. Perlite and vermiculite, for example, are often supplied in this form. They are usually poured or blown into air spaces between facings and between structural components of exterior construction.

Insulating cement is supplied as a powdery or

Table 10.5. Effect of Additions of Insulation to a Wall (Winter Conditions)

Construction	C or U	R	Incremental decrease in heat flow, percent
Exterior air film, 15-mph wind	6.0	0.17	
Wood siding	1.2	0.83	
Building paper	16.7	0.06	
Plywood sheathing	2.1	0.48	
Air space ($3\frac{1}{2}$ in. for studs)	1.0	1.00	
Gypsum lath	3.1	0.32	
Plaster	11.1	0.09	
Inside air film	1.5	0.67	
Assembly	0.28	3.62	
1 in. of insulation	0.27	3.70	
Assembly with 1 in. of insulation	0.14	7.32	50.0
1 in. more of insulation	0.27	3.70	
Assembly with 2 in. of insulation	0.091	11.02	35.0
1 in. more of insulation	0.27	3.70	
Assembly with 3 in. of insulation	0.068	14.72	25.3

granular mixture. After being mixed with water, it may be troweled or blown onto a surface and then dried in place to serve as insulation.

Flexible and semi-rigid insulation may be supplied as blankets, batts or felts. They come with various degrees of compressibility and flexibility. Often, a vapor barrier or reflective surface is applied to one face of this form of insulation. Figure 10.6a shows glass-fiber batts, supplied in long rolls, being installed between wood studs and rafters in a house. These batts come in widths that enable them to fit between wood 2 × 4's spaced 16 in. center to center. The batts are applied with the reflective, vapor-barrier surface facing the interior (warm) side of the construction. Flaps on the edges of the batts may be nailed or stapled to the studs. An interior finish, such as wallboard or lath and plaster, may be attached to the inner face of the studs, covering the insulation. Figure 10.6b shows an air-conditioning duct being wrapped with a glass-fiber blanket.

Rigid insulation may be supplied in the form of blocks, sheets, boards, or slabs. Figure 10.6c shows insulation board being placed atop metal-deck roofing. The board will be held in place by a waterproof, bituminous adhesive mopped over the decking.

Rigid insulation for pipes comes curved to appropriate radii and in half sections that may be strapped around the pipes or in full sections slit longitudinally that may be snapped around the pipes. Figure 10.6d shows calcium-silicate insulation, suitable for high-temperature applications, being applied.

Savings by Use of Thermal Insulation

The purchase price of insulation increases with the amount of insulation installed. The cost of installing insulation, however, increases very little with thickness of insulation. Hence, the cost of insulation in place is not proportional to insulation thickness but increases less rapidly than increases in thickness.

Although insulation adds to the cost of exterior construction or piping installations, use of insulation is desirable because it conserves energy, reduces HVAC operating costs, makes possible use of smaller-size HVAC plants and makes it easier to maintain comfort conditions for building occupants. For the purposes of systems design of a building, therefore, a major objective of designers is to determine what type of insulation to specify and how much of it should be installed.

In general, the type of insulation with the lowest cost in place should be selected. Cost comparisons for different insulation materials can be readily made by computing the ratio of the installed cost per sq ft per inch of thickness to the thermal resistance per inch of thickness for each material. The one with the lowest ratio would be the best buy for the money.

The thickness of insulation to be used depends on the space available for its installation and, when ample space is available, on the savings that can be achieved relative to the investment in the insulation. Table 10.5 shows that smaller reductions in heat transfer, and hence less savings in operating costs, result with each increment in insulation thickness. Hence, with costs of insulation, including interest and amortization, fuel and power known, it is possible to determine an optimum thickness for the insulation.

Weather Stripping

Although not strictly thermal insulation, weather stripping serves the same purpose as insulation: decreasing heat transfer between the interior of a building and outdoors. The economics of weather stripping installations, therefore, is the same as that for insulation. Also, weather stripping helps conserve energy.

The major objective of weather stripping is to prevent passage of air between the building interior and outdoors around joints between doors and their stops, or portions of the building against which doors close, between separate components of window sash, and between window sash and its stops or guides. Generally also, weather stripping serves the additional purpose of preventing rain from passing through such joints.

For the purpose of preventing leakage, weather stripping is placed along the edges of doors and windows at the joints where air or water leakage may occur. There are several

(a)

(b)

(c)

(d)

Fig. 10.6. Installation of building thermal insulation. (a) Glass-fiber batts between wall studs. (b) Glass-fiber wrapping for air-conditioning ducts. (c) Formed rigid insulation boards on top of steel deck. (d) Enclosure of pipes with high-temperature, calcium silicate insulation.

types of weather stripping available. It may consist of a narrow, curved strip of sheet metal attached on opposite sides of a joint. When a door or window is closed, the curved strips hook into each other. Or weather stripping may consist of a plastic gasket or a narrow strip of resilient, compressible material against which a door or window presses when it is closed.

By cutting off the flow of air around windows and doors, weather stripping thus prevents infiltration of cold air into a building in cold weather and exfiltration of cool air in warm weather. Consequently, savings in cost of HVAC operation should pay for the cost of the weather-stripping installation in a short time.

SECTIONS 10.2, 10.3 AND 10.4

References

ASHRAE Handbook of Fundamentals, American Society of Heating, Refrigerating, and Air Conditioning Engineers, Atlanta, GA, 1981.
Uniform Mechanical Code, International Association of Plumbing and Mechanical Officials.
J. Threlkeld, Thermal Environmental Engineering, 2nd ed., Prentice-Hall, Englewood Cliffs, NJ, 1970.
B. Stein et al., Mechanical and Electrical Equipment for Buildings, 7th ed., Wiley, New York, 1986.
W. Turner and J. Malloy, Thermal Insulation Handbook, McGraw-Hill, New York, 1981.

Words and Terms

Absolute temperature
British thermal unit (Btu)
Coefficient of thermal transmittance (U)
Comfort (human)
Condensation
Enthalpy
Entropy
Evaporation
Heat transfer: conduction, convection, radiation
Humidity: specific, relative
Latent heat
Sensible heat
Specific heat
Temperature: Celcius, Fahrenheit, dry-bulb, wet-bulb
Thermal capacity
Thermal insulation
Thermal resistance (R)

Significant Relations, Functions, and Issues

Heat: amount (temperature), quantity (Btu), capacity, resistance.

Sensible heat versus latent heat.
Laws of thermodynamics: first and second.
Means of heat transfer: by conduction, convection or radiation.
Cooling by evaporation; related to human comfort.
Interrelations of temperature, humidity and air movement; related to human comfort.
Air-to-air heat transfer related to control of building thermal conditions (indoor-outdoor heat loss or gain).
Criteria for HVAC system design: comfort standards, outdoor conditions assumed for design.
Usage factors for thermal insulation: type, amount, construction installation, cost-effectiveness.

10.5. PREVENTION OF DAMAGE FROM CONDENSATION

In many buildings, during cold weather, condensation may appear on windows. This occurs because the temperature of the windows is below the dew point of the air inside the building. The air in contact with the windows is cooled and no longer can hold the same amount of water vapor as it did when it was warmer. Hence, some of the vapor condenses as liquid water on the windows. If the window temperature is below the freezing point, the condensation will form frost.

A small amount of condensation may be merely annoying; however, air convection currents within the building will move the saturated air away from the windows and replace it with warmer moist air. The replacement air will become chilled below its dew point and release additional condensation on the windows. The condensation process will continue until the amount of water vapor in the building air at room temperature no longer exceeds the amount of vapor the air can retain when saturated at the window temperature. If the indoor relative humidity is high or if humidity is repeatedly added to the building air, considerable amounts of condensation can occur on the windows and run down over the window stools, down the walls and onto the floor. Damage may result.

This type of condensation trouble may be prevented by maintaining low relative humidity within the building and by use of insulated windows. This class of windows includes those with two or more panes of glass separated by air spaces and those protected by storm windows. The latter are windows placed outside the ordinary windows and separated from them by an air space. Note that whenever an air space is used, the space should be sealed so that heat cannot be transferred to or from it by convection.

Visible condensation may also occur on the inside of buildings in warm weather when the relative humidity is high. Such condensation may appear on basement walls kept cool by the earth around them and on uninsulated cold-water pipes. Water vapor in the air condenses on the walls and pipes because their temperatures are below the dew point of the building air. Often, the only convenient way to prevent this from becoming troublesome is to reduce the relative humidity of the building air by use of dehumidifiers. This type of equipment usually has a cooling coil on which moisture in the air condenses, drops into a receptacle and is piped out of the building.

Invisible Condensation

Condensation can be especially troublesome when it occurs where it cannot be seen—within walls, roofs or floors over crawl spaces (floors placed several feet above exposed earth). This invisible condensation occurs when the dew-point temperature exists within the construction. The materials exposed to the moisture may rot, rust or become stained. If insulation becomes saturated, it loses much of its thermal resistance. If the condensation should freeze, the construction may be damaged by the expansion when the ice thaws. The danger from invisible condensation is particularly acute because its occurrence may not be discovered until damage has been observed.

This danger exists because ordinary building materials are porous. Air and the water vapor in it can to some extent pass through such materials. Even when vapor barriers are incorporated in the construction, some vapor gets by, because vapor barriers are not likely to be perfect. As a result, if the dew-point temperature exists within the construction, condensation will occur.

Trouble from invisible condensation is more likely to occur in cold weather than in warm weather. For example, suppose a room adjoining an exterior wall contains air at 75°F dry bulb and 50% relative humidity when the outside temperature is 40°F. The dew-point temperature for the room air is 55°F. With the inside face of the wall at 75°F and the outside face at 40°F, the temperature somewhere between the two faces is 55°F or less. Condensation, therefore, will occur in the portion of the wall below 55°F.

If the wall contains thermal insulation, the sharpest temperature drop within the wall will occur in the insulation. Consequently, the dew point is likely to occur in the insulation, but condensation will not necessarily form in the insulation. The air motion will carry the moisture toward the cold exterior face of the wall until a surface with greater resistance to air movement is encountered. The moisture then will be deposited on that surface. This surface is likely to be the inner face of the exterior facing of the wall or sheathing, if used. Depending on the degree of humidity in the room, the condensation may soak the barrier and run down it and collect at the base of the wall. Damage to insulation, sheathing and interior and exterior facings may result.

For some types of building occupancy, condensation may occur only for a short time or intermittently. In such cases, the moisture may evaporate before damage results. In warm weather, when the inside of a building is kept cooler than outdoors, condensation that occurs because the dew-point temperature of the outdoor air exists within a construction often evaporates at night when the outdoor air becomes cooler.

Control of Invisible Condensation

The most effective measure for limiting the amount of condensation that can occur within a construction is installation of a vapor barrier in the construction to reduce to a minimum the amount of vapor that can reach the dew-point zone. Since condensation that forms in cold weather is likely to be more troublesome than condensation created in warm weather, the vapor barrier should be placed close to the room face of the construction. If the construction contains thermal insulation, the vapor barrier should be installed on the side of the insulation that would be warmer under winter conditions. Some types of thermal insulation are supplied with a vapor barrier on one face. (See, for example, Fig. 10.6a). Care must be taken that vapor barriers are not torn during or after installation and that all edges are sealed.

Aluminum foil and laminates containing asphalt often are used as vapor barriers, but other materials such as plastic films or sheets and aluminum, plastic, asphalt and rubber-base paints are satisfactory alternatives. They are not, however, equally effective.

Vapor barriers can be compared on the basis of their cost in place and their resistance to vapor transmission. This resistance is the reciprocal of a property of a material called *permeance*. Permeance is measured in units called *perms*. A *perm* represents a vapor-transmission rate of 1 grain (7,000 grains equals 1 lb) of water vapor per hour through 1 sq ft of material when the vapor-pressure difference equals 1 in. of mercury. For a 1-in. thickness of material, the permeance is measured in *perm-inches*.

Whether or not vapor barriers are installed within a construction, it is advisable that spaces where condensation may occur be ventilated and drained to the outside of the building. Provision of ventilation openings on the cold-side of a construction will permit water vapor that penetrates the porous inner face of the construction to escape to the outdoors without condensing. In walls, these openings may be small holes covered with porous, water-repellent material. With clapboard exterior facings, openings may be provided by insertion of thin wedges under the lower edge of each clapboard. Where insulated ceilings are placed below an attic or where an air space is provided under a roof, louvers should be installed for venting the attic or air space to outdoors. In such spaces, at least one opening should be high and at least one other low, so that convection will assist the movement of air to the outdoors. In general, vent area should be at least 1/300 the area of the horizontal projection of the roof.

Ventilation is also advisable in roof construction where insulation is placed above a roof deck and sandwiched between a vapor barrier and waterproof roofing. Unless the insulation has low permeance and is rigid, solar heat will cause the air in the insulation to expand and form bubbles under the waterproofing. At night, the air will cool and contract. If this process is repeated many times, the bending back and forth of the roofing will cause it to fail because of fatigue. This situation can be prevented by side-venting the insulation, allowing the entrapped air to escape to the outdoors when heated. The side vents should be louvered to prevent entrance of rain.

To drain spaces within walls, small holes, called *weep holes*, should be formed at the base of the walls and over wall openings. In masonry walls, the weep holes can be formed in the mortar joints by insertion of $\frac{3}{8}$-in.-diameter rubber or plastic tubing between masonry units before placement of the mortar. The tubing may be withdrawn after the mortar has hardened. Metal and precast-concrete wall panels should be designed to drain to the outdoors any moisture that may collect at the inside face.

Ventilation is also desirable for occupied spaces of a building, especially when the exterior construction contains a vapor barrier. Because a vapor barrier impedes escape of moisture from the building interior to outdoors, humidity will build up inside the building unless some means of removing water vapor is provided. Ventilation of the interior is effective for this purpose. In winter weather, normal infiltration of outdoor air usually provides a sufficient number of air changes per hour to control the indoor humidity within a desirable range. But additional ventilation may be necessary in spaces where large amounts of moisture may be evaporated, such as in kitchens and laundries. In some cases, dehumidifiers may have to be installed.

10.6. VENTILATION

Section 10.5 discusses uses of ventilation for humidity and condensation control within buildings. Ventilation is also necessary for many other purposes. It replaces vitiated, or stale, air with fresh air. In addition, ventilation can exhaust to the outdoors excessive heat, odors, smoke and undesirable gases. It can improve indoor temperature conditions when outdoor conditions are more comfortable than those indoors. And it can produce air movements that make building occupants feel more comfortable. Building codes generally contain requirements specifying the minimum amount of fresh air that must be supplied to occupiable areas.

Ventilation Methods

Ventilation may be achieved by natural or mechanical means. In either case, the basic source of fresh air must be outdoor air. For this purpose, openings, or intakes, must be provided in the exterior walls or roof of the building. Also, openings, or outlets, must be provided for exhaust air. An opening, however, may serve as both an inlet and an outlet. When mechanical ventilation is used, supply air may be conveyed to interior spaces and exhaust air removed from those spaces in ducts.

Natural ventilation may be produced by wind acting through openings in the building exterior, by differences in temperature that exist between outdoor air and the building air, or by both means. The amount of ventilation that can be achieved by either method varies with time. The wind can change considerably in velocity and direction in hours, or even minutes, and therefore can vary commensurately the amount of air that wind forces drive through an opening. Temperature-caused air movements can be substantial in cold weather, when there are large differences between indoor and outdoor temperatures, but in warm weather, when indoor and outdoor temperatures are nearly equal, air movement produced by temperature differences may become negligible. Air flow, however, may be augmented by a stack effect, achieved by locating openings in opposite walls at different levels.

Natural ventilation incurs no operating costs, whereas mechanical ventilation consumes energy for which the building owner must pay; however, mechanical ventilation may be necessary to serve indoor spaces without adequate ventilation openings. In addition, at times, natural ventilation may be impractical for the following reasons:

1. Outdoor air may be undesirably hot or cold.
2. Outdoor air may carry large amounts of dust, smoke, gases or odors.
3. The amount of fresh air admitted is difficult to control.
4. The fresh air may not penetrate far enough into an interior space to ventilate remote areas.

Mechanical ventilation employs fans to create a pressure difference that causes movement of air. A *fan* is an assembly of rotatable blades or runners which may or may not be enclosed in a casing. A fan may be a blower or an exhauster. A *blower* is a fan used to force air under pressure into ducts, plenums, rooms or other spaces. An *exhauster* is a fan used to withdraw air by suction from a space.

Ventilation systems should be installed to satisfy the provisions of the National Fire Protection Association "Standard for Installation of Air-Conditioning and Ventilating Systems," NFPA 90-A. The purposes of this standard are to restrict the spread of smoke, heat and fire through ducts, to maintain the fire resistance of building components affected by duct installations and to minimize ignition sources and combustibility of elements of the duct system. The standard also contains provisions governing installation of air intakes and outlets.

Minimum Ventilation

The amount of fresh air required for ventilation depends on the type of occupancy of a building, the number of persons occupying a space, type of activity, volume of the space and amount of heat, moisture and odors generated in the space. The minimum amount of ventilation may be specified in terms of the number of air changes required per hour in the space or the airflow rate, cfm, per sq ft of floor area or per occupant.

Some building codes may recommend a minimum flow rate, cfm per occupant, in accordance with the type of activity as follows: Inactive, such as in theaters, 5; light activity, such as in offices, 10, or in bars and restaurants, 15-20; active work, such as in shipping rooms and

404	Building Engineering and Systems Design

gymnasiums, 30–50. Multiplication of these flow rates by the maximum number of occupants yields the air supply required, cfm.

The number of air changes per hour in the space may be computed from:

$$N = \frac{60Q}{V} \qquad (10.20)$$

where

Q = air supplied, cfm
V = volume, cu ft, of the ventilated space

When the number of air changes per hour is too small; for example, less than one change per hour, ventilation will initially be inadequate, because it will take too long to produce a noticeable effect. Usually, five changes per hour is a practical minimum. Large changes also are undesirable because of high air velocities. A desirable maximum is 60 changes per hour.

Ventilation for Pollution Control

For removal of dust, odors, gases, smoke and excessive heat, ventilation is, at best, a dilution process. Fresh air is mixed with indoor air and a portion of the mixture is then exhausted to the outdoors. Ventilation should be capable of removing undesirable airborne substances at the same rate as that at which they are generated in the space.

For example, suppose that C is the allowable concentration, lb per cu ft, of gas in a room. If fresh air is blown into the room at a rate Q, cfm, and room air is exhausted at the same rate, the exhaust will remove QC, lb per min, or $60QC$ lb per hr, of the gas. Hence, if the gas is being generated at the rate of X lb per min, $QC = X$. The amount of ventilation air needed to maintain the allowable concentration is then

$$Q = \frac{X}{C} \qquad (10.21)$$

If the substance generated is water vapor, the ventilation air, cfm, needed to remove it is also given by Eq. (10.21), but with X taken as the lb of moisture vaporized per minute and C taken as the allowable concentration of moisture, lb per cu ft, in excess of the moisture content of the outdoor air.

Ventilation for Heat Control

When there is a net increase in heat that must be removed by ventilation air, Eq. (10.21) also can be used to determine the amount of ventilation air. In this case, X should be taken as $q_v/60$, where q_v is the heat, Btu per hr, to be removed. From Eq. (10.1), C can be obtained with the density of air, 0.075 lb per cu ft, substituted for m and the specific heat of air under constant pressure, 0.24, substituted for c. Thus, Eq. (10.21) transforms into

$$Q = \frac{q_v}{1.08(t_i - t_o)} \qquad (10.22)$$

where

Q = required flow rate of ventilation air, cfm, for removal of q_v
t_i = indoor temperature maintained, °F
t_o = outdoor temperature, °F

Where large quantities of heat or gases are produced at a concentrated source, such as kitchen ranges or laboratory work benches, a canopy hood placed over the source and connected to an exhaust duct is a much more efficient means of removing the heat or gases from the premises than general ventilation.

Air Replacement

Air exhausted from a space must be replaced at the same rate by air from outdoors. This air may enter the building either by infiltration through doors and windows or via a fresh-air make-up system. During cold weather, the fresh-air supply usually has to be heated before injection into an indoor space. During hot weather, the supply air may be cooled. These temperature adjustments are called *tempering*.

Natural Ventilation

Natural ventilation may be provided through windows, skylights, monitors, roof ventilators, doors, louvers, jalousies or similar ventilating openings. Intakes should be so located as to admit fresh air from outdoors without also drawing in polluted air from nearby sources. Outlets should be able to exhaust building air without discharging into air intakes of the same or other buildings. Air also may be supplied through or exhausted by way of special vent shafts.

Minimum Areas of Openings. Building codes usually set a minimum size for ventilating openings for all habitable rooms or spaces. The Building Code of the City of New York, for example, requires a free openable area of at least 5% of the floor area of the room or space ventilated. To qualify as an opening for natural ventilation, an opening should have an area of at least 1% of the floor area.

In crawl spaces where the floor above is constructed of wood or metal, ventilating openings and vapor barriers equivalent to either of the following should be provided:

1. At least four widely distributed ventilating openings with a total net free area of at least 1/800 of the crawl-space ground area should be placed in the foundation walls. The ground within the crawl space should be covered with a vapor barrier.
2. At least two ventilating openings with a total net free area of at least 1/1,500 of the area of the crawl space should be placed in the foundation walls. Also, a vapor barrier should be installed to cover the entire under side of the first-floor construction and overlap the foundation walls.

The crawl space should be at least 18 in. high, measured from the ground to the lowest point of the floor construction, such as beams and sills. If one side of a crawl space is completely open to a ventilated basement with a floor area at least equal to that of the crawl space, foundation-wall vents are not required for the crawl space.

Air Flow Due to Temperature Differences. As mentioned previously, movement of ventilation air can be produced by temperature differences between indoor and outdoor air. Because warm air within a building tends to rise to a ceiling or roof whereas cold air tends to drop to the floor, convection currents can be used to aid the flow of ventilating air if one ventilating opening is placed near the floor and another is placed near the ceiling or roof. The resulting air motion is called the *stack effect*.

The air flow due to the stack effect can be computed from

$$Q = 9.4A \sqrt{h(t_i - t_o)} \qquad (10.23)$$

where

Q = flow of ventilating air, cfm
A = free area of intakes or of outlets, sq ft
h = difference in height between intakes and outlets, ft
t_i = average indoor air temperature within the distance h, °F
t_o = outdoor air temperature, °F

The derivation of Eq. (10.23) assumes that the areas of intakes and outlets are equal. The largest flow per unit area of opening occurs when this condition holds. Increasing the relative area of intakes or of outlets increases the air flow but at a much smaller rate than the increase in area. Equation (10.23) also assumes little resistance to air flow within the building between intakes and outlets.

Several conclusions can be drawn from Eq. (10.23) and the principle of stack effect:

1. Window openings must be large enough for the required air flow. Openings larger than calculated areas are often desirable, in case of unusual or emergency conditions.
2. Windows should be opened at top and bottom (see Fig. 10.7a or b) or opened windows should be provided in different walls (see Fig. 10.7c), to create pressure differences sufficient for ventilation purposes. For the arrangement in Fig. 10.7c, but with all windows at a high level, however, the air flows through the space only at that level. The flow may ventilate along its route but will not produce appreciable ventilation at the level of occupancy if it lies below the windows. For a temperature difference to work to maximum advantage, the vertical distance between intakes and outlets should be as large as possible.

Mechanical Ventilation

Mechanical ventilation generally is an acceptable alternative to natural ventilation, but mechani-

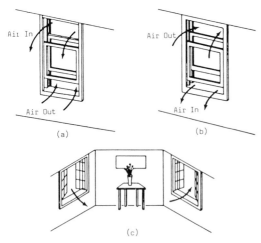

Fig. 10.7. Ventilation of a room with open windows. (a) Air flow past a double-hung window when outdoor temperature is higher than indoor temperature. (b) Air flow past a double-hung window when indoor temperature is higher than outdoor. (c) Cross ventilation.

cal ventilation usually is required by building codes for the following locations:

1. All occupiable rooms or spaces where the requirements for natural ventilation cannot be met.
2. All rooms or spaces where dust, fumes, gases, vapors, noxious or injurious impurities or substances that can create a fire hazard are generated.

The minimum flow rate of ventilation air with mechanical ventilation should be at least that required for natural ventilation. The air may be supplied from an air-conditioning system.

Air that has been exhausted from an air-conditioned space may be reconditioned by air-conditioning equipment and recirculated as equivalent outdoor air. Building codes may require, however, that the amount of fresh air be at least one-third of the required total, but not less than 5 cfm per occupant. These requirements, though, may be reduced 50% if the recirculated air is passed through adsorption devices capable of removing odors and gases. But air drawn from any of the following spaces may not be recirculated: bathrooms or toilet rooms, mortuary rooms, hospital operating rooms, rooms that must be isolated to prevent spread

of infection, or rooms where an objectionable quantity of flammable vapors, dust, odors or noxious gases is generated.

Exhaust may be accomplished by raising the pressure within a space and thus causing leakage of air past doors and windows; or vitiated air may be exhausted by suction into return-air ducts of air-conditioning equipment or into an exhaust duct discharging directly to the outdoors.

For an exhaust system to be effective, a sufficient quantity of make-up air should be provided to the ventilated space by one or more of the following methods:

1. Injection of make-up air into the space by a blower.
2. Infiltration of air through cracks around window sash and doors.
3. Infiltration of air through louvers, registers or other permanent openings in walls, doors or partitions that are next to spaces to which air is supplied by either of the preceding methods.

Fans

Fans create the pressure differences that cause air to flow in an air-distribution system. A fan consists basically of an impeller, or set of rotatable blades that push against the air, an electric motor that rotates the impeller, and often also a housing, or casing that encloses the impeller.

Fans are classified as centrifugal or axial, depending on the direction in which air flows through the impeller. A *centrifugal fan* employs centrifugal force to accelerate air to high velocity and then converts part of the velocity head to pressure head as the air leaves the impeller. In this type of fan, the air usually enters the impeller near and parallel to the axis of rotation, flows radially through the impeller and discharges at the outside edge. In *axial fans*, in contrast, air flows through the impeller generally parallel to its axis of rotation.

The impeller of a centrifugal fan is often referred to as a wheel, because of its cylindrical shape. The impeller blades are spaced uniformly around the circumference of the wheel.

In accordance with the configuration of the blades, centrifugal fans are classified as airfoil, backward-inclined or curved, forward-curved, and radial. Airfoil or backward-inclined blades are preferable when large volumes of air have to be moved, because of higher efficiency. Forward-curved blades, however, are satisfactory for small volumes of air. Centrifugal fans are available with capacities up to 500,000 cfm and can operate against pressures up to 30 in. water gage.

Air pressures in air-distribution systems are usually measured in inches, water gage, rather than in feet as in water-distribution systems. One inch water gage is equivalent to 5.2 psf.

Axial fans are classified as propeller, tube-axial and vaneaxial, depending on the configuration of the fan housing relative to the impeller. Propeller fans are often used for small air-moving chores, such as exhausters for kitchens and bathrooms and air cooling of condensers in air-conditioning systems; but this type of fan can be used for moving more than 200,000 cfm of air. It can operate against pressures only up to $\frac{1}{2}$ in. water gage, however, and it is usually noisier than centrifugal fans of similar capacity. Vaneaxial fans come with capacities of nearly 175,000 cfm and can operate against pressures up to 12 in. water gage. Tubeaxial fans have somewhat lower maximum capacities and can operate against pressures of only 1 in. water gage.

There are two special classifications for fans: tubular centrifugal fans and power roof ventilators.

A tubular centrifugal fan discharges air from the impeller in the same way as standard centrifugal fans, but the air then is directed through straightening vanes to flow parallel to the axis of rotation of the impeller. Thus, the air flow through the fan housing is radial, but through the impeller, centrifugal. Tubular centrifugal fans can move more than 250,000 cfm of air and can operate against pressures up to 12 in. water gage.

Power roof ventilators may employ centrifugal or axial fans and are used to exhaust air through roofs. The fans are protected by a cover, called weatherhood. Power roof ventilators generally have capacities up to about 30,000 cfm and can operate against pressures only up to about $\frac{1}{2}$ in. water gage.

Fan performance characteristics, including variation of static pressure and horsepower with air-flow rate, can be obtained from fan manufacturers. Space requirements for fans also can be secured from the manufacturers. These requirements should be known by the HVAC designers and the architect because adequate space, often in special fan or machine rooms, must be allotted for the fans within buildings.

Filters

Filters are installed in air-distribution systems to cleanse air of pollen, dust, smoke, odors, gases, and bacteria, spores and other living organisms. Screening or impingement-type filters should be capable of removing particles ranging in size from 150 to 10 μ. (1 μ equals 0.001 mm.) Electrostatic precipitators should be able to remove particles down to about 0.25 μ in diameter. Activated charcoal or chlorophyll are often used in devices for removing gases and odors. Ultraviolet or germicidal lamps may be employed to kill bacteria.

Dry-screen filters depend on a fine mesh to prevent passage of particles while air passes through the mesh. In viscous-impingement filters. the air passes through oil- or grease-coated meshes, to which particles in the air adhere. Both types of filters are available with renewable, cleanable or throwaway meshes.

Fibrous filters are a throwaway type that are available with a bag or pocket design with a large filter surface area.

Electrostatic precipitators have a relatively high initial and operating cost but can remove much finer particles than other types of air-cleaning devices and with much less resistance to air flow. In these devices, the air supply passes between electrically charged plates, which give particles in the air a strong electric charge. Then, the air flows between collector plates, which have an electric charge of opposite sign. The particles in the air are attracted to the collector plates and trapped. The plates should be cleaned periodically.

Ductwork

Mechanical ventilation cannot always serve all of a building interior by simply blowing air from or exhausting air to one or more openings in the building exterior. In many buildings, some spaces may be enclosed or so located that they can be ventilated effectively only with air conducted through special sealed passageways. Ducts are usually used for the purpose, especially in large buildings with central air-conditioning plants.

A duct is a tube used to convey air for ventilation, heating or air conditioning. Ducts may carry supply, or fresh, air, make-up air, exhaust air or air returned to air-conditioning equipment for reconditioning and recirculation. Ducts generally are placed horizontally or vertically (risers) to distribute air. Horizontal ducts often are run under floors or roofs, and where, for aesthetic reasons, they must be concealed from view, between floors and ceilings. In tall buildings, risers are often located in special shafts. The layout of ductwork must be carefully coordinated with structural and architectural plans to prevent obstruction of the paths of ducts by other building components, excessive cutting away of structural members, undesirable intrusions into spaces or aesthetically objectionable arrangements.

Ductwork must also be considered part of the fire-protection system of a building. Although ducts should not be combustible, they are a potential fire hazard nevertheless. If a fire should occur, they can spread smoke and hot gases to spaces far from the source. Precautions should be taken to prevent this. In addition, ducts can be used to assist in fire fighting. For example, they can deliver air to stairways, refuge areas and escape routes to pressurize those spaces and prevent entrance of smoke and hot gases. Also, the supply of air to areas involved in a fire can be cut off and the ducts used only for exhaust, to remove smoke and heat.

Ducts usually are made of galvanized-iron or aluminum sheet metal or of glass fiber (see Fig. 10.8). When sheet metal is used, it may be lined with sound-absorbing material, to reduce noise from movement of air. Air-conditioning metal ducts passing through nonconditioned

Fig. 10.8. Installation of glass-fiber air-conditioning duct. (courtesy Johns-Manville Corp.)

spaces should be wrapped with thermal insulation enclosed in a vapor barrier. This will prevent heat transfer between air within and carrying cold air. Glass-fiber ducts, however, are inherently sound absorbing and resistant to passage of heat. Regardless of materials used for ducts, the interiors should be smooth to keep resistance to air flow small.

Sizing of Ducts. Resistance offered by a duct to a specific flow rate of air depends on the ratio of the perimeter of the duct interior to the cross-sectional area of the interior. The smaller this ratio, the lower the resistance will be. For a given cross-sectional area, therefore, round ducts offer less resistance than square ducts, and square ducts offer less resistance than rectangular ducts.

Rectangular ducts, however, often have to be used instead of round or square ones to fit available space. Shallow ducts, for example, are essential when large volumes of air are required and ducts have to be placed between a ceiling and the floor above or where vertical clearance is tight.

The mean velocity of air and the pressure drop in a duct depend on the flow rate, cross-sectional duct area and the roughness of the duct interior surface. The relationships between these variables have been determined empirically. Information on them can be obtained from manufacturers of duct materials or from references, such as the ASHRAE *Guide and Data Book.*

In general, the smaller the area of the duct for conveying a specific flow rate of air, the larger will be the air velocity and the friction loss per unit length of duct. Good practice requires that the friction loss not exceed 0.15 in. water gage per 100 ft of duct, to limit power requirements for air circulation. Also, it is advisable to limit the air velocity in ducts. If too low a velocity is selected, uneconomical, bulky ducts will be required. Too high a velocity can cause noisy and panting ductwork. Building codes may set maximum velocities to prevent noise that depend on whether a duct is a main, submain or branch.

(A duct that connects directly to any terminal device, such as a grille, diffuser or register, is considered a *branch duct* for a distance of at least 4 ft from the terminal device. A *submain* is a duct that connects a branch duct to a main duct or to a fan. A *main* is a duct connected to a fan or to two or more submains.)

It may be advisable, however, to choose air velocities much less than the maximums set by a building code for noise control. For the higher the velocity, the larger will be the power requirements for air circulation. It is good practice to keep the starting velocity in main ducts below 900 fpm in residences, 1,300 fpm in schools, theaters and public buildings, and 1,800 fpm in industrial buildings. Maximum velocity in submains should be about two-thirds of the preceding velocities and in branches and risers about one-half.

Sizes of ducts for conveying specific flow rates of air are usually determined in practice by the equal-friction method. In this method, areas of sections of the ductwork are chosen so that the pressure loss due to friction per unit length of duct will be the same throughout the ductwork. Calculations start with determination of the flow rates required in mains, submains and branches. Then, a starting velocity is selected for the mains. The air-flow rate and starting velocity determine the area of the main and establish the friction loss per unit length for the main. The same friction loss is used to deter-

mine the sizes of submains and branches. As air is distributed from the main to submains and branches, the sizes of the ducts and the air velocities in those ducts are reduced by this method.

Bends in a duct cause sharp increases in pressure losses. To reduce the losses, turning vanes (see Fig. 10.10a) may be installed at bends. These are thin, curved partitions, placed at intervals at a bend, to split the air flow and guide it around the curve.

Dampers

Dampers are rotatable partitions used to reduce the flow rate of air in ducts. These devices intentionally introduce a pressure drop in a duct by sharply reducing the cross-sectional area of the duct over a short length. The parallel-leaf and opposed-leaf dampers shown in Fig. 10.9a and b are the most commonly used types. The angle at which the dampers are turned with respect to the direction of air flow determines the flow rate downstream.

After an air-distribution system has been installed, measurements should be made of actual air-flow rates at terminal devices. The dampers should then be adjusted to make the air-flow rates correspond to those specified by the designers. This operation is called *balancing the system.*

In some systems, dampers may rarely be readjusted after the initial balancing. In other systems, the position of dampers may be changed automatically by electric or pneumatic controls to meet changing air requirements in building spaces. The controls may be activated by a measuring device, such as a thermostat.

For efficient operation of a damper, its resistance when wide open should be between 3 and 6% of the total system resistance. A damper meeting this requirement, however, may have to be smaller than the duct. In that case, a barrier called a safing plate (see Fig. 10.9c) should be installed in the duct. A small damper also has the advantage of better sealing when it is fully closed rather than a damper the same size as the duct, because the total length of crack is less for a small damper.

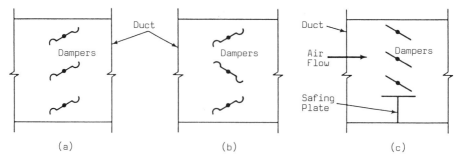

Fig. 10.9.   Types of dampers for ductwork. (a) Parallel leaf. (b) Opposed leaf. (c) Dampers with safing plate.

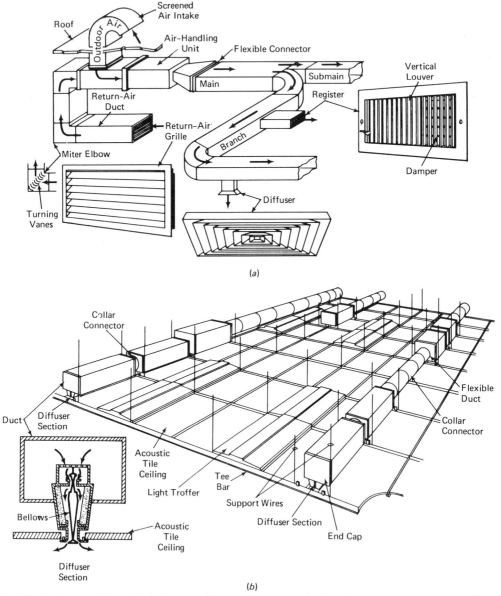

Fig. 10.10.   Ductwork for air distribution. (a) Ducts discharging air through square or rectangular grilles. (b) Ducts discharging air through linear diffusers.

Terminal Devices

At the ends of ducts distributing ventilation or conditioned air, special terminal devices should be installed, depending on the function of the terminal. Where ducts simply end without taking in or discharging air, the end should be capped (see Fig. 10.10*b*). For air intakes and outlets, duct ends should be left open and protective air-control devices inserted in them.

Intakes for outdoor air and exhaust outlets to the outdoors should be protected with screening, to prevent entrance of large particles, birds, animals and unauthorized humans, and shielded against rain and snow. For this purpose, openings in exterior walls may be protected with louvers. To control air intake or exhaust, the louvers may be equipped with adjustable blades, to vary the opening area, or a damper may be installed in the duct. Vertical ducts extending above a horizontal surface, such as a roof or the ground, may be bent into a gooseneck before the screened opening (see Fig. 10.10*a*) to prevent rain or snow from entering.

Like intakes and outlets on the building exterior, terminal devices on ducts within the building should protect duct openings from entrance of undesired objects. Usually, louvers or a grillwork are used for this purpose. Such terminal devices accordingly are called *grilles*. Those equipped in addition with a damper or control valve are called *registers*. A special type of register, called a *diffuser*, is used to discharge air into an interior space in many directions (see Fig. 10.10*a* and *b*). Selection of a suitable register and its location is important in achieving the desired distribution of air in the space to be served.

In selection of a register for a supply-air duct, consideration should be given to the dispersion of the supply air that will result and the ability of the device to control direction and rate of air flow.

The supply-air register in Fig. 10.10*a*, for example, is equipped with a damper for controlling the air-flow rate and with vertical louvers that can be adjusted to discharge air to the left or the right, or both. This register is a type that might be chosen because of its ability to throw the air outward a desired distance.

The diffusers in Fig. 10.10*a* and *b*, in con-

trast, are designed to spread out the discharged air and are aimed generally downward. The diffuser shown in Fig. 10.10*a* is square, but round ones are also available. The diffuser illustrated in Fig. 10.10*b* is a linear type. Viewed from below, the opening appears as a slot in the ceiling. The diffuser, however, may instead be installed above a partition, where the opening may be almost unnoticeable, and may discharge on either or both sides of the partition. The linear diffuser shown is a variable-volume type made by the Carrier Corp. It employs two internal bellows to control the air flow in accordance with the air requirements in the space to be served.

The return-air grille in Fig. 10.10*a* has fixed blades, which screen the opening. Because it is an air intake, there is no need for movable blades to control air direction. A damper behind the blades controls the air-flow rate.

Sections 10.5 and 10.6

References

ASHRAE Handbook of Fundamentals, American Society of Heating, Refrigerating, and Air Conditioning Engineers, Atlanta, GA, 1981.

B. Stein et al., *Mechanical and Electrical Equipment for Buildings*, 7th ed., Wiley, New York, 1986.

F. Merritt, *Building Design and Construction Handbook*, 4th ed., McGraw-Hill, New York, 1982.

Words and Terms

Air change; replacement
Blower
Condensation
Damper
Duct: main, submain, branch
Exhaust
Exhauster
Fan
Filter
Pollution
Stack effect
Terminal devices: grilles, registers, diffusers
Vapor barrier
Ventilation: natural, mechanical

Significant Relations, Functions and Issues

Problems and design considerations for condensation: use of vapor barriers, relation to air movement and dew point.

Ventilation methods: natural or mechanical, use of stack effect, equipment, control factors (air velocity, change per hour, quality of air).

Components of ventilation systems: openings, fans, filters, intake and exhaust, ducts, terminal devices, control equipment.

10.7. HEAT LOSSES

Heat loads for a building include losses due to low temperature of outdoor air and gains due to hot outdoor air, solar radiation and heat generated within the building. These loads must be calculated for use as a guide in selection of heating and cooling plants. Principles for calculation of the heat loads were presented in preceding sections of this chapter. This section describes methods often used for computing heat losses. It assumes that design indoor and outdoor temperatures have been established, as discussed in Sec. 10.3.

Walls and Roofs

For most buildings a high percentage of the total heat load in cold weather consists of the heat losses through walls and roofs. These losses may be computed with Eq. (10.16), with the temperature gradient through the exterior taken as the difference between design indoor and outdoor temperatures.

For the calculations, the exposed area and heat-transmission coefficient U for each type of construction used for the building exterior must also be determined. The value of U may be obtained from tables or computed by the method described in Sec. 10.3. The areas of exposed walls, roofs, windows and other exterior construction should be secured from the architectural roof plans and exterior elevations. The total heat loss through walls and roofs is the sum of the products of the area, temperature gradient and U for each component of the exterior.

Basement Floors and Walls

Heat losses through basement floors and walls also may be computed with Eq. (10.16). The temperature gradient, however, may be uncertain, because the temperature of the ground under and around the basement is difficult to determine. As a result, in practice, the heat losses through basement walls and floors often are estimated. The effect of errors in such estimates are rarely likely to have a significant effect on the calculated total heat load for ordinary buildings.

The estimates may be based on local groundwater temperatures. These normally range from 40 to 60°F in the northern sections of the United States and from 60 to 75°F in the southern sections. The heat loss through basement floors, Btu per hr per sq ft of floor area, may be taken as 3, 2 or 1 for groundwater temperatures of 40, 50 or 60°F, respectively. The heat loss through basement walls per sq ft of wall area may be assumed to be twice that through the floors.

Floors on Grade

As for basement walls and floors, the heat loss through floors on grade are difficult to calculate with Eq. (10.16) because of the uncertainty of the temperature gradient. In practice, therefore, the loss directly downward is ignored, and the loss around the exposed edges of the floors is estimated. Experience indicates that a high percentage of the heat loss occurs through those edges. Hence, it is advisable to place insulation around the edges to reduce the heat losses. The estimate of heat losses should be based on the thermal resistance of the insulation to be installed. Typically, with 2 in. of insulation, the heat loss, Btu per hr per lin ft of exposed edge, may be taken as 50 in the cold, northern sections of the United States, 45 in the temperate zones, and 40 in the warm south. Without insulation, the corresponding losses are 75, 65 and 60.

Unheated Attics and Other Unheated Spaces

Heat loss from the top story of a building through the roof can be computed from Eq. (10.16), with the temperature gradient taken as the difference between design indoor and outdoor temperatures. The heat loss when the top story underlies an unheated attic, however, requires special treatment, although Eq. (10.16) still is applicable.

Computation of this heat loss requires the assumption that the attic temperature is established by equilibrium between indoor and outdoor heat transfers. For equilibrium in cold weather, the heat gain to the attic from the heated top story must be equal to the heat loss through the roof to the outdoors:

$$U_c A_c(t_i - t_a) = U_r A_r(t_a - t_o) \qquad (10.24)$$

where

U_c = thermal transmittance of the top-story ceiling
U_r = thermal transmittance of the roof
A_c = ceiling area, sq ft
A_r = roof area, sq ft
t_i = design indoor temperature, °F
t_a = attic temperature, °F
t_o = design outdoor temperature, °F

Solution of Eq. (10.24) for the attic temperature yields:

$$t_a = \frac{U_c A_c t_i + U_r A_r t_o}{U_c A_c + U_r A_r} \qquad (10.25)$$

With t_a known, the heat loss from the top story can be calculated from the left-hand side of Eq. 10.24.

The same method can be used to compute the heat loss from other heated spaces to adjoining unheated spaces.

Ventilation and Infiltration Air

When ventilation air is supplied to an indoor space during cold weather, the cold outdoor air must be heated if design indoor temperature is to be maintained, because the warm indoor air mixes with and loses heat to the air introduced. The amount of this heat loss can be computed by solving Eq. (10.22) for q_v:

$$q_v = 1.08Q(t_i - t_o) \qquad (10.26)$$

Even if no ventilation air is introduced into the building, however, there will be a heat loss to outdoor air. Buildings seldom are completely airtight. Even if windows and doors are weather stripped, outside air will infiltrate through cracks and indoor air will leak out. The consequent rate of heat loss depends on crack areas, wind velocity, the number of ex-

posures and many other factors. As a result, an exact computation of the heat loss often is impractical.

Instead, calculations may be based on the assumption that cold outdoor air will be heated and then distributed in sufficient quantities throughout the building to create a static pressure large enough to prevent air infiltration.

If ventilation air supplied to a space produces N air changes per hour, the rate of air flow Q, cfm, is given by

$$Q = \frac{VN}{60} \qquad (10.27)$$

where V is the volume of the space. The heat loss then can be computed from Eq. (10.26), with Q as given by Eq. (10.27).

10.8. HEAT GAINS

Section 10.7 presents methods for computing heat losses for buildings in cold weather. Methods for computing heat gains are described in the following paragraphs. The discussion assumes that design indoor and outdoor dry-bulb and wet-bulb temperatures have been established, as described in Sec. 10.3.

Some heat gains may occur throughout the year. A few examples of these are heat generated by solar radiation in daytime, by occupants, by electric lighting and by operation of machines. These heat gains should be deducted from heat losses in cold weather in calculations of room temperatures or in determination of the heat load for sizing a heating plant, if the heat gains can be utilized.

Heat gain from ventilation or air infiltration occurs in warm weather. This heat gain should be added to that from other sources in determination of the heat load for sizing a cooling plant or in calculations of room temperatures.

Latent Heat

Part of a cooling load for a building is due to sensible heat and part is due to latent heat. The sensible-heat portion would increase indoor temperatures if it were not taken care of by cooling. The latent-heat load results from removal of water vapor from the air, either as part of the cooling process or for control of humidity. Thus, sizing of a cooling plant requires determination of the sum of the sensible heat gain in the building, the latent heat gain in the building and the total (sensible plus latent) heat gain from outdoor air processed by the cooling plant.

The latent heat gain in a building results from moisture given off by baths, showers, washing, cooking, building occupants, humid outdoor air brought indoors, combustion products, industrial processes and any surface of stored water. A cooling unit must remove about 1,000 Btu for every pound of water vapor it condenses from the air.

Walls and Roofs

Heat gains through the building exterior result either from the difference in temperature between outdoor and indoor air or from solar radiation striking the building enclosure.

Heat gain from the temperature gradient through walls and roofs can be calculated in the same way as for heat losses through the exterior in cold weather, as described in Sec. 10.6. The temperature gradient for use in Eq. (10.16) is the difference between design outdoor dry-bulb temperature and design indoor dry-bulb temperature. The total heat gain through walls and roof is the sum of the products of the area, temperature gradient and U for each component of the exterior.

For most buildings, the effect of solar radiation on the walls can be ignored. Because of the thermal capacity of walls, it takes considerable time for them to absorb the radiated energy and then reradiate it to the interior. As a result, the peak cooling load is not likely to be significantly augmented by the heat from solar radiation on the walls.

Solar radiation through glass and through roofs, however, is likely to produce large heat gains in the interior. Heat gain from solar radiation on windows can amount to about 200 Btu per hr per sq ft for a single sheet of unshaded, common window glass facing east or west. The heat gain might be about three-fourths as much for such windows facing northeast or northwest, and about half as much for

such windows facing south. Shading windows with overhangs, vertical fins or louvers on the building exterior or with trees outdoors can reduce the solar heat load considerably.

Solar radiation on roofs also can impose a large heat load. For many roofs, the heat gain from this source is equivalent to that for a temperature difference of 50°F. This heat gain can be reduced by spraying a roof with water. This also has the advantage of increasing roof life by preventing swelling, blistering and vaporization of volatile roofing components and of avoiding thermal shock from sudden rainstorms in hot weather. The heat gain from solar radiation on a sprayed roof is equivalent to that for a temperature difference of 18°F.

Interior Sources of Heat Gain

Among the more significant contributors to heat gain generated within a building are lights, motors, appliances and people. Electric lights and most electric appliances convert the electric energy supplied into an equivalent amount of heat. The heat gain from such sources may be computed from

$$q = 3.41W \qquad (10.28)$$

where

q = heat generated, Btu per hr
W = rated capacity of lamps or appliances, watts

For fluorescent lighting, add 25% of the lamp rating to account for heat generated in the ballast.

Some appliances, such as coffee makers, also evaporate water. An estimate of sensible-heat and latent-heat gain from such appliances may be obtained from tables in the ASHRAE *Handbook of Fundamentals*.

Heat gain from electric motors depends on the horsepower rating, load factor, or fraction of the horsepower being delivered when cooling of the building is required, and motor efficiency. This heat gain, Btu per hr, is approximately 2,500 times the rated horsepower for motors rated at 1 hp or more and about 5,000 times the rated horsepower for motors of $\frac{1}{4}$ hp or less.

For some buildings, such as schools, theaters and churches, the heat given off by occupants may be a large percentage of the total heat load on the cooling system. The sensible and latent heat gains from people depend on age and sex of each person, type of activity, environmental influences and duration of occupancy. The ASHRAE *Handbook of Fundamentals* provides data for estimating these heat gains. Sensible-heat gain ranges from about 250 Btu per hr for moderately active persons to 580 Btu per hr for persons doing heavy work. Latent-heat gain ranges from 105 to 870 for the corresponding activities. Persons walking at 3 mph or doing moderately heavy work generate about 400 Btu per hr of sensible heat and 600 Btu per hr of latent heat.

Ventilation and Infiltration Air

Warm humid air that comes into a building as ventilation air or by infiltration imposes both a sensible-heat and latent-heat gain on the cooling system. The sensible-heat gain may be computed from

$$q_s = 1.08Q(t_o - t_i) \qquad (10.29)$$

where

q_s = sensible heat from outdoor air, Btu per hr
Q = flow rate of outdoor air into the interior, cfm
t_o = design outdoor dry-bulb temperature, °F
t_i = design indoor dry-bulb temperature, °F

The latent-heat gain from outdoor air may be computed from

$$q_l = 4,840Q(W_o - W_i) \qquad (10.30)$$

where

q_l = latent heat from outdoor air, Btu per hr
W_o = humidity ratio of outdoor air, lb of water vapor per lb of dry air
W_i = humidity ratio of indoor air, lb of water vapor per lb of dry air

The humidity ratio of air at various dry-bulb and wet-bulb temperatures can be obtained from a psychrometric chart.

Miscellaneous Sources of Heat Gain

A minor amount of the total heat load for a building often is generated within the building from such items as exposed pipes, air-conditioning ducts, pumps, fans and compressors. For example, the fan used in an air-conditioning unit develops some horsepower, the amount depending on the air-flow rate and the total resistance of the ducts and other components of the distribution system. This horsepower will be converted to sensible heat, which causes a rise in air temperature. For most low-velocity duct systems, the heat gain from this fan may range from 5% of the total sensible load for small systems to $3\frac{1}{2}\%$ of the total sensible load for large systems. The heat gain from such miscellaneous sources should be included in the computation of the total heat load.

Sections 10.7 and 10.8

References

ASHRAE Handbook of Fundamentals, American Society of Heating, Refrigerating, and Air Conditioning Engineers, Atlanta, GA, 1981.

B. Stein et al., *Mechanical and Electrical Equipment for Buildings,* 7th ed., Wiley, New York, 1986.

J. Threlkeld, *Thermal Environmental Engineering,* 2nd ed., Prentice-Hall, Englewood Cliffs, NJ, 1970.

Words and Terms

Edge loss
Heat transmission
Infiltration
Latent heat
Solar heat gain

Significant Relations, Functions and Issues

Nature and significance of heat loss or gain as related to: (1) building indoor temperature, (2) comfort of occupants.

Relative significance of different components of heat loss or gain.

Methods of computation of significant factors for heat loss or gain.

10.9. METHODS OF HEATING BUILDINGS

Preceding sections of this chapter present principles and methods for calculation of heat loads for use in determining the capacity required for a heating plant. After the heating requirements have been established, appropriate equipment for meeting the requirements can be selected. This section describes some of the equipment and methods used for heating buildings.

Sizing a Heating Plant

The total heat load for a space to be heated equals the heat lost from the space through conduction, radiation and infiltration. If, however, the space contains equipment that produces heat, such as lights, motors and steam-heated vessels, and this equipment is in constant use during periods in which heating of the building is required, the equipment may be considered a supplementary heating device. Its heating capacity may be deducted from the total heat load. Similarly, the sensible-heat gain due to occupants of the space may be subtracted from the total heat load, to obtain the net heat load.

Equipment with a heating capacity equal to the net heat load will be able to maintain the design indoor temperature in the space when the design outdoor temperature prevails. In most buildings, despite cold weather, the temperature is allowed to fall (perhaps $20°F$) below the design indoor temperature at night or when the buildings are not in use. To raise the temperature to the design indoor temperature, the heating equipment requires a larger heating capacity than that for maintaining equilibrium conditions. For this purpose, it is common practice to select equipment with a heating capacity about 20% larger than the net heat load.

Available heating equipment, however, may not have a rated capacity equal to the computed capacity. In that case, equipment with the nearest larger capacity should be selected. It is not advisable, however, to select too large a unit. Not only is a larger plant likely to cost more, but also operating efficiency will be less and fuel consumption will be higher than for a plant with a capacity closer to that required.

Suppose, for example, that equipment with 50% larger capacity than the net heat load were installed. After the morning pickup, this equipment, when design conditions prevail, will operate 100/150, or two-thirds, of the time, to maintain the design indoor temperature. The design outdoor temperature, though, will occur only infrequently during the heating season. Consequently, the over-sized equipment will usually operate less than two-thirds of the time. It will cycle rapidly, supplying heat for short periods of time and then shutting down. This repeated on-off operation is inefficient. Furthermore, during the off periods, occupants are likely to feel uncomfortable, because of heat losses by radiation and drafts. Therefore, it is good practice to select a heating plant no smaller than that required for the net heat load and the morning pickup but not much larger.

A heating plant may also be required to provide heat to produce domestic hot water. In that case, the additional heating requirement should be included in the required heating capacity of the heating equipment.

Heating Equipment

Heat for a building may be obtained from electricity, solar radiation or combustion of coal, gas or oil. Generally, the relative cost of operating a heating plant determines the heat source that will be chosen. For example, the cost of solar energy may be zero, but the costs of entrapping it, converting it to heat as needed and supplementing it with conventional heating when there are several days of cloudy weather may make conventional heating less costly. As another example, the cost of electricity may be greater than that of coal, gas or oil but electrical equipment may cost less, requires no pumps, pipes, tanks, combustion-air supply or chimneys and may have lower maintenance and repair costs. Selection of the heat source, therefore, should be based on a comparison of life-cycle costs for the heating plant.

Heat is delivered to building occupants by radiation or convection, or a combination of these. The radiant energy or the warm air currents may be produced in any of several ways.

For example, air may be heated directly and then distributed throughout a building or a space that is to be warmed. Or water may be heated or steam formed, piped to spaces to be warmed and then passed through equipment that either transfers the heat to the air by conduction or radiates heat to the spaces.

A central plant that employs combustion to heat air directly is usually referred to as a *furnace*. A central plant that heats water is called a *boiler*. Equipment installed in a space or room for the purpose of heating only that space or room is called a *unit heater*. A *heat exchanger* is equipment whose primary purpose is not to supply heat to a space but to transfer heat from one medium, such as water or steam, to another, such as air.

Building codes generally contain requirements governing the installation and operation of heating equipment. The codes may adopt by reference appropriate standards of the American National Standards Institute, American Society of Mechanical Engineers and the Underwriters' Laboratories.

Heating equipment should be installed with adequate clearances from combustible construction to prevent it from catching fire. Either the heating equipment should be insulated or the building construction should be fire protected. During operation of the equipment, the surface of combustible construction should not be raised to a temperature higher than $170°F$. The equipment should be located, arranged and protected so that the means of access to it for ordinary operation or maintenance will not be hazardous. Floor-mounted heating equipment may be seated on noncombustible construction or on fire-protected combustible construction if flames or gaseous products of combustion do not impinge on the base of the equipment.

Accessible pipes carrying steam, water or other fluids at temperatures exceeding $165°F$ in occupiable spaces should be insulated with noncombustible material. The insulation should prevent the temperature at its outer surface from being more than $60°F$ above the indoor air temperature. Piping carrying a fluid at a temperature below $250°F$, however, need not be insulated if the insulation would interfere with the functioning of the system; but the pipes should be far enough from combustible construction not to raise its surface temperature to more than $170°F$, but at least $\frac{1}{2}$ in. away.

Air Supply and Exhaust for Heaters

Equipment that generates heat by combustion must be connected to the outdoors so that combustion products are removed from the building. The combustion products are conducted to the outdoors through an enclosed passageway, called a *flue*. This should be made of an incombustible material capable of withstanding high temperatures without loss of strength and of resisting corrosive products of combustion.

Where a flue passes through building construction, the flue should be housed in an incombustible *chimney* for support and protection against damage. A chimney may contain more than one flue. If so, the flues should be insulated from each other without loss of support. For heating equipment for which combustion products do not have temperatures exceeding $600°F$, chimneys should extend vertically to a point at least 3 ft higher than the roof, including roof ridges within 10 ft of the chimney. For higher-temperature combustion products, building codes may require a chimney to terminate 10 to 20 ft above the nearest construction.

The vent connection between heating equipment and the chimney also should be made of heat-resistant material and should have a cross-sectional area at least equal to that of the heater flue. The length of this connection should be made as short as possible. Also, the conduit should pitch upward toward the chimney at least $\frac{1}{4}$ in. per ft. Long connections should be insulated to prevent combustion products from cooling and condensing.

Sufficient air for combustion should be provided for the heating equipment. For the purpose, outdoor air should be brought into the space housing the equipment. The air may be supplied through an open window or a duct with an area at least equal to twice that of the vent connection between the heater and the chimney.

Warm-Air Heating

A warm-air heating system heats a room by injecting air at a temperature higher than the ambient room temperature. Systems with a central heating plant may be classified as gravity or forced types. In either case, the warm air is distributed through ducts from a furnace to the spaces to be heated. In gravity systems, the furnace is installed below the spaces to be heated and the warm air rises through the ducts because it is less dense than the cooler room air. In forced systems, pressures created by a blower push the air through the ducts. Gravity systems are seldom used, because forced systems offer many advantages. A forced warm-air system, for example, can provide higher air velocities through the ducts. Hence, smaller ducts can be used. In addition, this type of system provides much more sensitive control.

The heat supplied by warm air to a space may be computed from

$$q = 1.08Q(t_h - t_i) \qquad (10.31)$$

where

q = heat added by warm air, Btu per hr
Q = flow rate of warm air into the space, cfm
t_h = temperature of warm air as it leaves the grille, °F
t_i = ambient room temperature, °F

This equation indicates that, for a specific amount of heat q provided, the higher the temperature t_h of the warm air, the smaller Q need be; that is, the smaller the amount of air that the system has to heat and distribute. With less air to be handled, operating costs of the system should be less.

There are practical limits, however, to the range of temperatures t_h that can be used. Discharge temperatures as high as 170°F, for example, may produce uneven temperature distribution within a room and cause persons close to a discharge point to feel uncomfortably warm. Temperatures much below 135°F may produce drafts. Consequently, discharge temperatures are generally kept between 135 and 150°F.

Grilles supplying warm air to a room should be arranged to blow a curtain of warm air across cold, or exterior, walls and windows. The supply-air grilles should be placed in or near the floor. The warm air, being less dense than the room air, will rise to the ceiling.

To prevent buildup of air pressure in the room and loss of heat by air exfiltration to the outdoors under the pressure, ducts should be provided to return room air to the heater. The return-air grilles should be placed in interior locations, so that air is drawn across the room to produce a uniform temperature distribution. Preferably also, these grilles should be placed in or near the ceiling. This location has the advantage of reducing the difference in temperatures between air at the floor and air at the ceiling. Ceiling air may be as much as 10°F warmer than floor air. Furthermore, return of the warmer ceiling air places less of a heat load on the heating plant than would return of floor air and thus results in lower operating costs.

The heating of outdoor air in cold weather to warm an indoor space may make the indoor air too dry. To prevent this, moisture may be added to the air by incorporation of a humidifier in the furnace discharge bonnet.

Sequence of Operation. A thermostat in the space to be heated is set to signal for heat at temperatures below a specific room temperature. The signal turns on the heater. When the air chamber in the heater reaches about 120°F, a sensor in the chamber starts the blower. (An earlier starting of the blower is inadvisable, because cool air would be discharged from the grilles and cause drafts.) For safety reasons, another sensor will turn the heater off if the temperature in the air chamber exceeds about 180°F, because of a malfunction in the system. In ordinary operation, temperature in the air chamber will be maintained at the design value, which might be about 150°F.

When the temperature at the thermostat reaches that for which the thermostat has been set, a signal turns off the heater. The blower continues to operate. It will be turned off by the sensor in the heater air chamber when the temperature there drops below 120°F. Because the blower operates after the heater is shut off, the heat given off while the temperature in the air chamber drops is utilized for space heating instead of escaping up the chimney.

Hot-Water Heating

For lower initial and operating costs than with warm-air heating, but with some sacrifice in occupant comfort, a building may be heated with a hot-water system. In this type of system, heat is distributed in water to the spaces to be heated. There, devices are provided for transferring the heat in the water to the room air.

A hot-water heating system requires a boiler for heating the water, a circulator (circulating pump) for delivering the water, pipes for conducting the water, an expansion tank to accommodate volume changes in the system water when it is heated, and terminal heat-transfer equipment. Sometimes, the heat-transfer equipment consists of *radiators*, a series of bent pipes or coils, which convert much of the heat to radiation and transfer some heat by conduction to the air. The air distributes the heat throughout a room by convection. In other cases, the heat-transfer equipment consists of an arrangement of closely spaced, vertical fins on a pipe and is called a *convector*. Much of the heat is transferred to the air in a room by conduction and distributed by convection. Some of the heat is also radiated. When the finned pipe extends along the base of the wall, the heat-transfer device is referred to as *baseboard heating*.

Water usually is supplied to the radiators or convectors at about 180°F. Thus, with hot-water heating, higher temperatures can be used for distribution of heat than with warm-air heating. In addition, 1 lb of water can carry more heat than 1 lb of air. Hence, the hot-water heating system need handle much smaller quantities than a warm-air system delivering the same amount of heat. Pipes used for hot-water heating can be much smaller than ducts used for warm-air heating.

Piping. Any of several different water-distribution systems may be used for hot-water heating. Generally, they may be classified as one-pipe or two-pipe systems. In a *one-pipe system*, a single pipe supplies water from a boiler to the radiators or convectors in sequence. After the water passes through each heat-transfer device, it is returned to the pipe and eventually recirculated to the boiler. In a *two-pipe system*, one pipe supplies water from a boiler to the radiators or convectors and a second pipe returns the water from each heat-transfer device to the boiler. In a one-pipe system, the temperature of supply water to each heat-transfer device is lower than the supply-water temperature for the preceding device, because of the heat transferred by that device to the room air and by radiation to the room. Consequently, such systems are difficult to design and to keep in adjustment after a period of use. Therefore, two-pipe systems are usually preferred.

A two-pipe hot-water heating system with reversed return is illustrated in Fig. 10.11. It is called reversed return because the radiators or convectors with the shorter supply pipes from the boiler have the longer return pipes back to the boiler. Because the hot water is supplied to each heat-transfer device without passing through any other radiator or convector, supply-water temperatures for these devices are nearly equal. Also, because the lengths of supply plus return pipes for each radiator or convector are nearly equal, the loss of pressure due to friction for the devices is about the same. For these reasons, the two-pipe reversed-return system is preferred to the direct-return type, in which the return pipe from each device takes the shortest route back to the boiler, as a result of which the device farthest from the boiler suffers the greatest pressure loss due to friction.

Radiators or convectors should be selected in accordance with the heat each is required to supply to a space. The water flow required for each heat-transfer device can be calculated from Eq. (10.1), taking into account that 1 gpm of water flow is equivalent to 500 lb per hr and

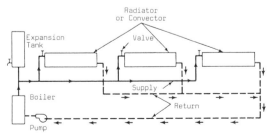

Fig. 10.11. Two-pipe hot-water system with reversed return.

that the specific gravity of water is unity. The drop in temperature across each radiator or convector may be taken as 20°F. Substitution of these values in Eq. (10.1) and solving for Q, the water flow, gpm, yields

$$Q = 0.0001q \qquad (10.32)$$

where

q = heat supplied by radiator or convector, Btu per hr

The hot-water pipes should be sized so that water velocity in them will not exceed 4 fps. At higher velocities, flow may be noisy. If the velocity is too low, however, large, costly pipes will be needed. Loss in pressure head due to friction should be held between 0.25 and 0.60 in. per ft of pipe. The circulator (pump) should be capable of supplying the required water flow at the required pressure head.

Also, provision should be made in the piping for expansion and contraction due to temperature changes. For this purpose, the piping layout should employ arrangements with bends that permit the pipes to flex, swing joints at branch take-offs that allow the threads in elbows to screw in and out slightly, or expansion joints.

An *expansion tank* should be connected to the piping at a level at least 3 ft higher than the highest radiator or convector and in a location where the tank water cannot freeze. Volume of the tank should be at least 6% of the total volume of the water in the hot-water system. The tank may be either of the open or closed type. The open type has a vent at the top to let air escape as the tank fills. The closed type is sealed and kept only partly filled with water. The air above the water surface in the tank acts as a cushion to allow for expansion and contraction of the volume of water in the system.

It is important that entrapment of air in hot-water systems be prevented. One precaution is to pitch all pipes so that vented air will collect at points that can be vented manually or automatically. Vents should be provided at all heat-transfer devices.

Valves should be placed at appropriate locations on pipes, especially at the beginning of a zone, when parts, or zones, of a building are served by different distribution systems. The objective should be the ability to shut down parts of a system or a zone in emergencies or for repairs without shutting down the whole system or zones not affected.

Tall buildings should be divided into zones, each consisting of several stories served by a separate distribution system. Otherwise, the static water pressure at the boiler may be undesirably high. In a zoned building, with the boiler at the base of the building, the bottom zone can be supplied directly from the boiler as in a conventional two-pipe system. The circulator serving that zone can also supply water from the boiler to a heat exchanger at the bottom of the next zone above. The heat exchanger transfers heat to the supply pipe serving the upper zone. A circulator in the zone circulates water through the zone piping and back to the heat exchanger. The pump also supplies water to a heat exchanger at the bottom of the next zone above. The distribution system for this and succeeding upper zones are the same as for the second lowest zone. Each zone has to be designed for a somewhat lower supply-water temperature than the one below, because of heat lost in distribution and heat transfer.

Sequence of Operation. When the hot-water system is in use, the water in the boiler is maintained at the design temperature, usually about 180°F. To attain and hold this temperature, an immersion thermostat in the boiler water controls the heat source, such as an oil burner or gas solenoid valve. Thus, an immediate supply of hot water is available for space heating and also for domestic use, if desired.

A thermostat in the space to be heated is set to signal for heat at temperatures below a specific room temperature. The signal starts the circulator, which forces hot water from the boiler to flow to the radiators or convectors. As an economy measure, if there is a system malfunction that leaves the supply-water temperature too low, for example, below 70°F, the supply water is shut off, because it would be ineffective for space heating. To accomplish this, a low-limit immersion stat, located in the boiler water and wired in series with the room thermostat and the circulator, will shut down the circulator. In normal operation, a signal turns off the circulator when the temperature at the room thermostat reaches that for which the thermostat has been set.

Steam Heating

An economical alternative to hot-water heating is steam heating. There can be considerable savings in heating costs with steam heating when steam is available as a by-product of an industrial process or of power generation. In a steam-heating system, heat is distributed in steam piped to the spaces to be heated or to a heat exchanger that transfers heat to water or air. In the spaces to be heated, radiators or convectors are usually used to transfer the heat from the steam to the room air.

In this type of system, the heat used for space heating is primarily the heat given off when steam condenses to water. This heat is equal to the heat of evaporation of water at the boiling point and is much larger than the heat contained in water at the boiling point. In condensing, 1 lb of steam generates about 1,000 Btu of latent heat. Hence, a steam-heating system need handle much smaller quantities of the heat-supplying medium than hot-water or warm-air heating systems for delivering a specific amount of heat.

Steam-heating systems can be designed to operate at low pressures—from below atmospheric pressure to 5 to 10 psi above. Total pressure drop due to friction may be kept as low as $\frac{1}{4}$ to $\frac{1}{2}$ psi. The steam flows through the pipes to radiators or convectors under boiler pressure or suction. The hot condensate is returned to the boiler by gravity flow or suction.

Steam-heating systems may be classified as one pipe or two pipe. In both systems, supply steam is fed from a heat source to a loop that returns to the heat source. Risers to convectors or radiators take off from the upper part of the loop.

In a one-pipe system, the condensate is returned to the loop through the same risers that supply the steam.

In a two-pipe system, one pipe supplies steam to the radiators or convectors and a second pipe returns the condensate. Steam is prevented from entering the return pipe by a thermostatic drip trap at the condensate end of each radiator or convector. The trap stays closed when its temperature exceeds 180°F, keeping steam from escaping into the return pipe. When enough condensate collects to drop the trap temperature below 180°F, the trap opens and allows the condensate to flow into the return pipe and return to the heat source. There are many variations of the one-pipe and two-pipe systems.

One type often used is a *vacuum heating system*. This is a two-pipe system with a pump for returning the condensate to the boiler. The system is usually operated at a boiler pressure below atmospheric. The vacuum pump therefore must draw noncondensables from the piping and heat-transfer devices for discharge to atmosphere. Steam pressure eliminates air from the piping and heat-transfer devices, when the system starts, by opening thermostatic vents to the atmosphere.

Capacities for components of steam-heating systems are given in the ASHRAE *Guide and Data Book* and in manufacturers' catalogs. Capacities may be expressed in sq ft of equivalent direct radiation (*EDR*).

$$1 \text{ sq ft } EDR = 240 \text{ Btu per hr} \quad (10.33)$$

Capacities also may be given in lb per hr, with 1 lb per hr of steam equivalent to 970 Btu per hr.

In many buildings with steam heating, temperature control for indoor spaces is crude. Valves at radiators or convectors at best may be manually adjusted, or in some cases, either kept fully open or fully closed. Or windows may be partly opened to admit cold outdoor air for temperature adjustment. Better temperature control may be achieved, but with an increase in initial and operating costs. For example, steam may be supplied to heat exchangers, which transfer heat to air for distribution to indoor spaces. In that case, thermostats can be used to control the air flow and thus also room temperature.

Unit Heaters

A unit heater is a heating device usually employed to heat only the space in which it is located. It is suitable for heating large, open areas, such as garages, showrooms, stores and factory production areas. Best location for a unit heater is near the ceiling. The heater contains a fan that blows air past the heat source, and guide vanes direct the heated air downward and outward.

Heat may be supplied by electricity, steam, hot water or combustion of gas. For gas-fired unit heaters, a flue to the outdoors must be provided for removal of combustion products, and adequate air for combustion must be supplied.

Radiant Heating

Radiant heating, or panel heating, employs heated floors, ceilings or walls, or a combination of these, to heat indoor spaces. Much of the heat from the warm panels is conveyed by radiation to occupants or contents of those spaces. Some heat is transferred to the room air by conduction. If the warm panels are in or near the floor, the heated air will rise and distribute the heat by convection. The amount of heat transmitted by either method depends on the difference in temperature between the panel and the air, occupants or room contents.

Heat transmitted by conduction is proportional to the temperature difference. Heat transmitted by radiation, in contrast, is proportional approximately to the difference between the fourth power of the absolute temperatures. Hence, the larger the panel temperature, the larger will be the amount of heat transmitted by radiation. High panel temperatures, however, would make occupants of the space feel uncomfortable. Floor temperatures, in particular, cannot be too high, because the occupants' feet are in contact with the floor and would become too warm. In practice, floors used for radiant heating are maintained at about 85°F or lower. Ceilings may be slightly higher, about 100°F, because of the air space interposed above the occupants' heads. For a panel at about 85°F, the percentage of radiant heat is about 56% of the total given off, whereas at about 100°F, the percentage of radiant heat is about 70%.

Because of the heat supplied by radiation, it is feasible to maintain indoor temperatures lower than the usual for other types of heating, without discomfort to the occupants. But this holds true only when the panel design temperature is maintained. If the premises reach the design indoor temperature and the heat source is shut off, allowing the panel temperature to decrease, the occupants will feel uncomfortable at the low air temperature. Consequently, it is advisable to design radiant heating to maintain customary design indoor temperatures.

Panels may be warmed with hot water, steam, warm air or electricity. Most often, hot water is used. It flows through coils of pipe or tubing embedded in floors, walls or ceilings. Figure 10.12a shows tubing embedded in a concrete slab on the ground. Rigid insulation placed below the tubing prevents loss of heat to the ground. The concrete slab transfers the heat to the space above. Fig. 10.12b shows tubing embedded in a plaster ceiling, which radiates heat to the space below.

Tubing should be laid out to balance pressure losses due to friction in the various piping circuits serving the building. For this purpose, the tubing can be arranged in coils over large areas. Balancing valves should be provided for each coil. In addition, coil geometry can be employed to achieve different degrees of resistance. High resistance can be achieved with the continuous coil shown in Fig. 10.12c, low resistance with the grid in Fig. 10.12d and medium resistance with a combination of the two arrangements (see Fig. 10.12e).

Radiant heating may be combined with other types of heating. Sometimes, this is necessary because a floor or ceiling provides insufficient area for maintaining design conditions. Sometimes, a combination is desirable for economy. For example, a *perimeter warm-air heating system* distributes air through hollows in a floor or through underfloor ducts from a central location to the floor perimeter and heats the floor. The warm air may also be circulated through ducts around the perimeter. Then, the air is discharged through grilles at the perimeter to supply additional heat by convection.

Heat Pumps

A heat pump supplies heat to a building through a sequence of operations similar to the direct-

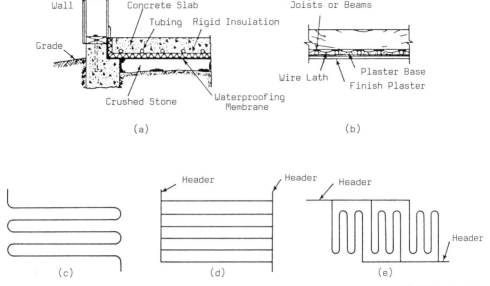

Fig. 10.12. Radiant heating with hot water. (*a*) Tubing in concrete slab on grade. (*b*) Tubing in plaster ceiling. (*c*) Continuous coil. (*d*) Grid. (*e*) Combination of continuous coil and grid.

expansion refrigeration cycle described in Sec. 10.10, which is used for cooling buildings. The difference between the heat-pump cycle and the refrigeration cycle is that the heat pump utilizes the heat of condensation of a refrigerant for space heating, whereas the refrigeration cycle discharges the heat outdoors.

To maintain design indoor temperature with a heat pump, indoor air is passed over coils containing a hot refrigerant gas under pressure (at right in Fig. 10.13). The air is thus heated, whereas the pressurized gas is cooled and becomes liquid. The liquid refrigerant then passes through an expansion valve (at left in Fig.

10.13) where the pressure and temperature of the liquid are reduced.

At the low pressure and temperature, the refrigerant then flows through coils, called an *evaporator*, over which warm water or outdoor air is passed (also at left in Fig. 10.13). The refrigerant absorbs heat from the air or water and vaporizes. Finally, the gas is fed to a *compressor* (center of Fig. 10.13), which compresses it to produce the hot gas that is used to heat the indoor air. This completes the heat-pump cycle, which can now be repeated. The heart of the system is the compressor, which performs work on a low-pressure gas to raise it

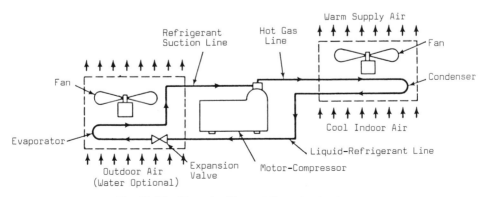

Fig. 10.13. Flow of refrigerant through a heat pump.

to a high pressure and a correspondingly high temperature for space heating.

The economy of a heat pump arises from the ability of the system to produce several times as much heat for space heating as the heat equivalent of the mechanical work performed by the compressor. The reason for this is that a large percentage of the heat used for space heating is supplied by the outdoor air or by warm water. (The water may be provided from a well and later returned to the ground.) When outdoor air is used as the heat source, a heat pump has the disadvantage that its efficiency decreases as outdoor temperatures drop, while heating requirements increase.

Solar Heating

Radiation from the sun can provide sufficient energy to supply at least the domestic hot water for a building and also often space heating and cooling. The big advantages of solar heat are that it is free and will be available for a long, long time. Thus, its use eliminates the cost of fuels and conserves fuels in limited supply. The disadvantages are that solar heat is available only when the sun is shining; the intensity varies hourly, daily and with the weather; and the energy received per square foot of radiated surface is small, generally considerably under 400 Btu per hr per sq ft. Because of the low energy flux, it is necessary to collect solar energy over a wide area to obtain sufficient energy for heating purposes.

Solar heating or cooling of buildings is advantageous when the cost of the heat produced is less than the cost of heat produced by conventional heat sources. In general, the cost of solar heating can be kept small by low capital charges and amortization rates for the equipment required, by favorable climatic conditions, by continuous heat loads and by efficient design of the heating system.

The efficiency of solar energy applications depends on the efficiency of energy collection and the efficiency of the conversion of radiation to a more useful form of energy.

One way to collect the energy is to use lenses or curved mirrors to focus the radiation on a point, in order to develop high temperatures at the point. Another way, and one that is more frequently used for heating buildings, is to entrap the radiation that strikes a flat surface held normal, or nearly so, to the sun's rays.

For conversion of the radiation to other forms of energy, there are two principal technologies. One is photovoltaic conversion, and the second is solar-thermal conversion. The former process converts radiation directly to electricity. At present, the materials available are very inefficient and the cost per kilowatt is very high. Research under way, however, is gradually reducing the cost. Solar-thermal conversion, which produces heat by absorption of the radiation, has been used for some buildings. Its efficiency is acceptable. The deterrents to its use have been relatively high capital charges and amortization rates for equipment required. In addition, investment costs are kept high because generally conventional equipment must also be installed to supplement solar heating during periods when adequate sunshine is not available.

A typical solar-thermal conversion system (see Fig. 10.14a) includes the following:

Solar collector—a means for collecting and absorbing solar radiation and converting it to heat.

Heat-storage materials and equipment—a means for storing solar heat as received and discharging it as needed.

Source of supplemental energy—a means of supplementing the solar-heating output when it is inadequate.

Auxiliary equipment—a means, as in conventional systems, of distributing heat and transferring it to water or air, as required.

Cooling equipment—a means of utilizing solar heat for cooling a building, if desired.

Solar Collector. The commonest means of collecting and absorbing solar energy for heating of buildings employs a flat, black surface in an enclosure with a transparent cover, usually glass or plastic. The cover is transparent to the incident solar radiation, opaque to re-radiated energy and resistant to transmission of heat by conduction. Thus, solar energy is trapped within the enclosure, absorbed by the black surface and converted to heat.

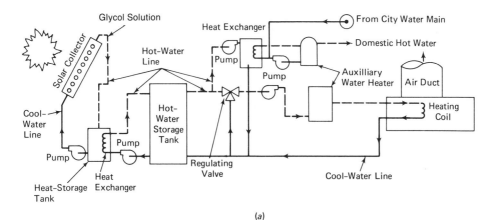

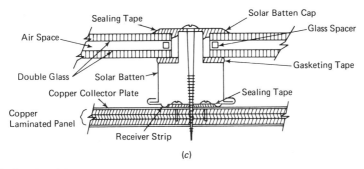

Fig. 10.14.   Solar heating. (a) Heat absorbed from solar radiation by a glycol solution is transferred to water for heating a building and producing domestic hot water. (b) Solar collector for absorbing solar radiation. (c) Detail of batten assembly of the solar collector.

Figure 10.14b shows a solar collector developed by the Copper Development Association. The panel is about 2 × 8 ft. It is designed to be interlocked with several similar panels installed on the roof of a building on a slope nearly normal to the rays of the winter sun. The cover consists of twin panes of glass, to reduce loss of heat by conduction. The radiation absorber is a blackened copper sheet, 0.016-in. thick. It is bonded to $\frac{3}{8}$-in.-thick plywood, backed with thermal insulation. For removal of the absorbed heat, blackened, rectangular copper tubes are clipped to the absorber sheet. A mixture of water with 10% propylene glycol antifreeze flows through the tubes in a closed circuit and absorbs heat from the tubes and the solar absorber. A detail of the solar-collector enclosure is shown in Fig. 10.14c.

Heat Storage. As indicated in Fig. 10.14a, for a typical solar-heating system, the hot glycol solution flows from the solar collector to a heat exchanger, which transfers heat to water. The cool glycol solution then is pumped back to the solar collector. The heated water flows in a closed circuit to a hot-water storage tank and back to the heat exchanger. Hot water is withdrawn from the storage tank as needed.

The hot water, if desired, may be pumped to a heat exchanger, where heat is transferred to a hot-water supply for domestic use. Also, the hot water from the storage tank may be pumped to a heat exchanger that transfers heat from the water to air for space heating. In both cases, after heat transfer, the cooled water is returned to the hot-water storage tank or recirculated to the heat exchanger served by the glycol solution.

Supplemental Heating. Figure 10.14a indicates that an auxiliary water heater is provided for the domestic hot-water supply and for space heating, for use when insufficient solar heat is available. Heat for the auxiliary heater may be furnished by electricity or combustion of coal, gas or oil.

Cooling. If desired, the system in Fig. 10.14a can be adapted to employ solar heat for cooling the air supplied to the air duct. A three-way valve placed in the hot-water line before the heat exchanger for space heating can be operated to divert the hot-water to an absorption-type cooler, which is described in Sec. 10.10. The output of this cooler then may be used to cool air supplied to the building.

References

B. Stein et al., *Mechanical and Electrical Equipment for Buildings*, 7th ed., Wiley, New York, 1986.

J. Threlkold, *Thermal Environmental Engineering*, 2nd ed., Prentice-Hall, Englewood Cliffs, NJ, 1970.

Uniform Mechanical Code, International Association of Plumbing and Mechanical Officials.

ASHRAE Handbook of Fundamentals, American Society of Heating, Refrigerating, and Air Conditioning Engineers, Atlanta, GA, 1981.

Words and Terms

Chimney
Flue
Heat exchanger (pump)
Radiator
Solar heater (collector)
Supplemental heating
Unit heater

Significant Relations, Functions, and Issues

Types of heating systems: warm-air, radiant (hot-water, steam), mixed.

Components of heating systems: source of heat, method of heat transfer, equipment for distribution, controls.

Special systems: heat pumps, solar heaters, heat storage.

10.10. METHODS OF COOLING AND AIR CONDITIONING BUILDINGS

Sections 10.2 to 10.8 present principles and methods for calculation of cooling loads for use in determining the capacity required for a cooling plant. After the cooling requirements have been established, appropriate equipment for meeting the requirements can be selected. This section describes some of the equipment and methods used for cooling buildings.

There are several possible ways for cooling an indoor space. Among them are, injection of cool air, evaporative (cooling of room air by evaporation of water sprayed into it) and radiant cooling, which is similar to radiant heating. The most commonly used method, however, is mixing of cool air with room air.

In warm weather, relative humidity in excess of about 60% has an adverse effect on human comfort. Hence, control of humidity indoors as well as control of temperature is desirable. As a result, if cooling is to be used for a building, or spaces in it, air conditioning is a logical choice, because complete environmental control can be secured with little additional cost over plain cooling.

For air conditioning, fresh air for ventilation is mixed with return air from indoor spaces, after which the air is filtered, cooled, dehumidified, distributed through ducts and discharged into rooms to mix with room air. The same distribution system can be used for warm-air heating in winter, or for ventilation when neither heating nor cooling is required.

In most buildings, air is cooled and dehumidified either by absorption or by direct expansion of a refrigerant (basic refrigeration cycle). The former employs heat to cool and the latter extracts heat with the aid of a compressor and an external heat sink.

Sizing an Air-Conditioning Plant

The total cooling load for a space to be cooled equals the sum of the sensible heat and the latent heat to be removed. The refrigeration equipment selected should have a capacity, Btu per hr or refrigeration tons, nearly equal to the cooling load. The most critical part of an air-conditioning system, however, usually is the air-distribution system. Proper cooling cannot be achieved if insufficient air is circulated; however, circulation of large quantities of air is expensive, because both initial investment and operation costs are high. Consequently, selection of air-conditioning equipment with appropriate air-handling capacity is also important.

To determine the size of the air-conditioning plant needed for a building, the heat load and air-supply requirements for each space to be air conditioned should be determined. The total heat load is the sum of the loads for each space, and the total air that must be handled is the sum of the air quantities required for each space.

The air required for a space may be computed

from

$$Q = \frac{q_s}{1.08(t_i - t_c)} \quad (10.34)$$

where

Q = flow rate of cool air into the space, cfm
q_s = heat absorbed by the cool air, Btu per hr
t_i = ambient room temperature, °F
t_c = temperature of cool air as it leaves the grille, °F

Equation (10.34) indicates that the closer t_c is to t_i, the more air that has to be handled. On the other hand, if t_c is much smaller than t_i, other undesirable conditions may result. For example, occupants near a grille may feel a draft. Also, if t_c is equal to or less than the dew-point temperature for indoor conditions, the grilles may condense moisture from the supply air and drip, and uninsulated ducts will sweat. It is good practice, therefore, to set the discharge-air temperature at about 3°F higher than the room dew point.

For design indoor dry-bulb temperatures of 75 to 80°F and relative humidity about 50%, t_c should be about 18°F less than t_i, to prevent t_c from dropping below the dew point.

The air supply usually is a mixture of fresh air taken from outdoors for ventilation and indoor air returned from conditioned spaces. This supply air is filtered and passed over cold tubing. Moisture in the air condenses on the tubing and drips into a receptacle, the drainage from which is generally removed from the building. Thus, the latent heat load on the air-conditioning plant is absorbed at the cold tubing, whereas the sensible heat loads are absorbed in the conditioned spaces.

If, when the air leaves the cold tubing, the temperature is too low for discharge into a room, the air has to be reheated. The air in that case may be heated directly or mixed with air warmed for that purpose.

Basic Refrigeration Cycle

A frequently used means of cooling and dehumidifying air is the direct-expansion refrigeration cycle (see Fig. 10.15). In this process, supply air is passed over cooling coils, tubing containing a cold liquid. The cooling coils at the left-hand side of Fig. 10.15a contain a cold refrigerant, which evaporates when it absorbs heat from the air. The cooled air, at a lower relative humidity than the entering air, is distributed through ducts to condition indoor spaces.

The refrigerant gas flows from the cooling coils, or *evaporator*, to a *compressor* (center of Fig. 10.15a) which compresses the gas. Hot and under pressure, the gas then flows through coils, called a *condenser* (at right in Fig. 10.15a), over which air or water is passed, to absorb heat from the refrigerant.

If air is the condenser coolant, outdoor air is used for the purpose. If water is the coolant, it may be reused after cooling in a cooling tower. (Well water may be returned to the ground.)

In the condenser, the pressurized gas is cooled and becomes liquid. The liquid refrigerant then passes through an expansion valve (at left in Fig. 10.15a) where the pressure and temperature of the liquid are decreased. Finally, the cooled liquid refrigerant flows to the cooling coils, to cool the supply air.

This completes the basic refrigeration cycle, which can now be repeated. The heart of the system is the compressor, which performs work on a low-pressure gas to raise it to a high pres-

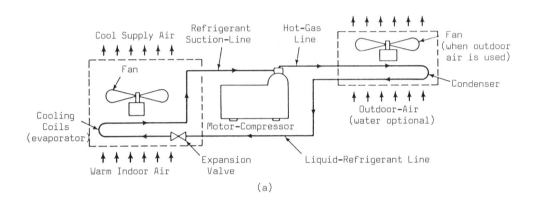

(a)

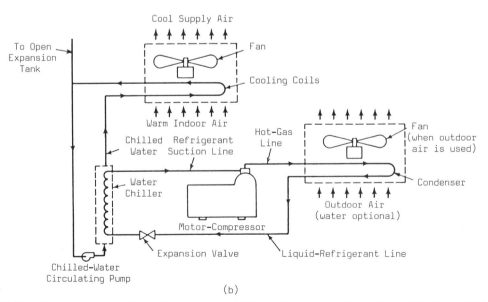

(b)

Fig. 10.15. Direct-expansion refrigeration cycle used (*a*) for cooling air directly for air conditioning and (*b*) for chilling water, which then cools supply air.

sure and a correspondingly high temperature for efficient transfer of heat in the condenser to an air or water heat sink.

Chilled-Water Refrigeration Cycle. In the direct-expansion cycle shown in Fig. 10.15a, the evaporator cools supply air directly. When large refrigeration capacity is required, however, it may be more economical to chill water instead (see Fig. 10.15b). The chilled water then is distributed to cooling coils either at central air handlers or in spaces to be conditioned, where supply air is passed over the coils and cooled. Chilled-water refrigeration systems often are used when 100 tons or more of refrigeration are needed.

From Eq. (10.1) and the fact that 1 ton of refrigeration equals 12,000 Btu per hr, the flow of chilled water through the cooling coils may be computed:

$$Q = \frac{24q}{t_l - t_e} \qquad (10.35)$$

where

Q = flow rate of chilled water, gpm

q = cooling required, tons

t_e = temperature of water entering the coils, °F

t_l = temperature of water leaving the coils, °F

The temperature rise of the chilled water through the cooling coils usually is 8 to 10°F.

Controls. Operation of the refrigeration system usually is controlled by a thermostat in the conditioned space or a manually operated switch. In either case, a signal that starts or stops motors and opens or closes valves is transmitted. Basically, the thermostat or switch controls the compressor motor and a fan in the air-handling unit. If the condenser is air cooled, a condenser fan is also controlled. If the condenser is water cooled, a pump or valves may be controlled, and in addition, if the water is cooled in a cooling tower, the tower fan is also controlled by the thermostat signal.

Basic Absorption Cycle

The basic absorption cycle differs from the direct-expansion cycle in two ways: 1. Heat is used instead of electricity. 2. An absorber, pump and generator replace the motor-compressor. Expansion valves, evaporators and condensers are used in both types of cycles and serve the same functions.

In the absorption cycle, the refrigerant usually is water. It is used to form a solution with a chemical, frequently lithium bromide, in both the absorber and the generator that substitute for the motor-compressor. Heat is applied to the generator to evaporate water from the solution. The strong, or more concentrated solution thus formed is pumped from the generator to the absorber, where the solution is cooled. Cooling is accomplished by passing the solution over cooling coils containing flowing water. In the absorber, the refrigerant, as cold vapor, is mixed with the strong solution. The resulting weak, or less concentrated solution is then returned to the generator, to complete the absorber-generator cycle, which can then be repeated.

The flow of the refrigerant through the absorption cycle is similar to that through the direct-expansion cycle. The refrigerant vapor leaves the generator at a high pressure and temperature. The vapor next is cooled and condensed in the condenser, then passed through an expansion valve or spray nozzle. The resulting evaporation process produces vapor at a low pressure and temperature. Then, the cold vapor flows past cooling coils, which are used to chill water for air conditioning. Finally, the vapor flows to the absorber, goes into the lithium bromide solution, and returns with the weak solution to the generator. This completes the absorption cycle, which can then be repeated.

This cycle requires more cooling water than does the direct-expansion system, but it has the advantage that very little mechanical work is required for its operation, only that required for the pump circulating the solution between the generator and the absorber.

Absorption units have been used for air-conditioning requiring cooling capacities of 100 tons or more. Such systems may be more economical than direct-expansion systems when low-cost steam is available; for example, as a by-product of an industrial process or power generator.

Cooling Equipment

An air-conditioning plant can range from the relatively simple and inexpensive installation of a window cooling unit in one room to the extensive and often expensive installation of complete and automatic environmental control all year round for a whole building. The choice depends on how much money the owner is willing to spend.

At a minimum, a cooling system requires a refrigeration cycle, as a source for heat removal from the building; an air handler to circulate air, and air distribution to bring cool air to the spaces to be conditioned.

These items may be part of a central plant or they may be unit coolers. The latter usually are *package units*, in which all the items are delivered for installation factory preassembled. Often, however, only the refrigeration equipment and air handler are preassembled and ductwork, if needed, is installed separately.

In addition to the minimum items, an air-conditioning plant usually includes filters for cleaning the air; fans for moving the air; grilles for discharging the air; louvers for fresh-air intake; auxiliary heat exchangers to transfer heat from chilled water, when used, to air; heaters and reheaters for adjusting supply-air temperatures, and also often cooling towers when water is used to cool condensers. Pumps, dampers, valves and controls also are often essential.

Depending on the type of installation, supply air, after passing over the cooling coils of a refrigeration machine, may be blown directly into a room or distributed through an apartment, zone or whole building by ductwork. Ducts may be constructed as described for ventilation in Sec. 10.6. Dampers often are necessary in the ducts to control air flow, balance the system and adjust the air flow to the desires of the building's occupants.

From the ducts, supply air is delivered through one or more registers to each room, and return air is removed from the rooms through grilles. Because cool air is dense, supply air tends to drop toward the floor after discharge from a register. Consequently, the best location for a supply-air grille for cooling is in or near the ceiling. Generally, also, these registers should be arranged to blow the supply air across sources of heat, such as windows, to neutralize their effect on room temperatures before they cause discomfort. A good location for a return-air grille is near the ceiling, to remove warmer air, and near an indoor wall, to stimulate air distribution by convection.

Zoning. In the design of an air-conditioning plant, it may be necessary to provide more than one system to secure nearly equal comfort conditions throughout a building, despite variations in heat loads for various parts. A building containing spaces with considerably different cooling requirements should be zoned so that each zone with about the same requirements can be served by a separate system.

For example, a two-story building should be air conditioned with two separate systems, one for the first story and one for the second story. The second, or top story will have a higher heat gain than the bottom story at midafternoon on a mild sunny day, because of the sun heating up the roof. Despite roof insulation, the top story will need to be cooled, but the first story will be comfortable without cooling, since outdoor temperatures are mild. If the building were served by only a single cooling system, operating the system to cool the top floor will make the story below too cold. If, however, the system were designed to maintain comfort conditions in the bottom story, the system would not operate, because of the mild temperature, and the top story would be too warm. Hence, installation of separate systems for each story is advisable.

Package Air-Conditioning Units

Factory preassembled air-conditioning plants, called package units, are available with cooling capacities up to 100 tons for cooling supply air directly. The smaller units, up to about 10 tons, are often used for conditioning a room. They may supply air without ductwork. The larger-size units are generally used for a central plant, with ductwork.

The simplest and least expensive air conditioner is a package unit that is set in a window opening. It has the disadvantage of obstructing the window. This disadvantage may be eliminated by use of a through-the-wall package unit, installed in an opening in an exterior wall.

Both window and through-the-wall units are installed to take care of cooling requirements of a room and offer the advantage of separate temperature control for that room. They employ the basic refrigeration cycle (see Fig. 10.15a). Some operate as a heat pump in cold weather. These units, however, have the disadvantages of producing noise in the room and of an appearance that may be objectionable from an architectural or interior-decoration viewpoint.

The disadvantages of room units can be overcome by installation of a central plant from which air is distributed through ducts to the spaces to be conditioned. A package unit also may serve as a central refrigeration machine and air handler. The central unit may be floor-type, ceiling-type or roof-type. In any case, these package units, like the window and through-the-wall units, usually have air-cooled condensers. The condensers may be projected on the outside of the building, while the cooling coils extend inside, or the condensers may be located near a window or other exterior-air intake to reduce the duct runs needed for cooling air. Equipment for the refrigeration cycle and control wiring are factory installed. Other wiring, ductwork and gas piping are installed at the building site.

Most package units are designed to supply about 400 cfm of air per ton of refrigeration. For spaces with high latent-heat loads, the latent-heat capacity of a package unit may be less than that desired, but if the sensible-heat capacity is higher than that needed, the deficiency in latent-heat capacity can be offset by setting the thermostat below design indoor dry-bulb temperature. At the lower temperature, occupants should feel comfortable, despite a higher than usually desired relative humidity. For spaces with high-sensible heat loads, air quantities supplied by a package unit may be too small for the installed tonnage. But if the latent-heat capacity is higher than that needed, occupants will feel comfortable, because of low relative humidity, even though the dry-bulb temperature is higher than the design temperature.

Split Systems. For a quieter air-conditioning plant for buildings with low cooling requirements, such as apartments or houses, the noisier portion of the plant—the motor-compressor and the air-cooled condenser—can be placed outdoors. The air handler, with the cooling coils, can be installed inside the building. This layout eliminates the need for a wall opening for the condenser or for air intakes and exhausts for condenser cooling air. The layout, however, requires that refrigerant piping and electric wiring be extended between the indoor and outdoor components.

Package Chillers. Package units with water chillers, employing the refrigeration cycle in Fig. 10.15b, also are available for central-plant installations. These units may come with air-cooled condensers for complete outdoor installation. As an alternative, a package chiller may consist of a condenserless portion that is installed indoors and an air-cooled condenser that is placed outdoors. Very large units may be shipped knocked down and assembled by the manufacturer's representative at the building site. There, the units are sealed, pressure-tested, evacuated, dehydrated and charged with the proper amount of refrigerant.

Built-up Air-Conditioning Units

Large central plants, with capacities exceeding about 50 tons, usually are assembled on site, to provide cooling in hot weather, heating in cold weather and ventilation in intermediate seasons. Such plants may be classified as all-air or air-water systems.

In an all-air system, refrigeration machines and air-treating and handling equipment are usually located in a central area, remote from the spaces to be conditioned. Cooling and heating loads are taken care of entirely by air.

An air-water central plant is similarly located but circulates hot or cool water to heat-transfer devices in the spaces to be conditioned. There, room air is heated or cooled by the water. Air-water systems consequently have to handle much smaller air quantities than all-air systems, can take care of large differences in loads in different parts of a building and have lower operating costs.

All-Air Systems. There are five commonly used types of all-air systems: double duct, dual conduit, multizone, single duct with terminal reheat, and variable-volume single duct. In all cases, the central plant contains a refrigeration machine and a heater, or boiler. The air handler has a connection to a damper-controlled, outdoor-air intake, with a preheater supplied heat by the boiler, to temper the outdoor air. Return air from conditioned spaces is mixed with the fresh air downstream of the preheater. The mixture then is filtered. After the filter, the air-handler and air-distribution equipment differs in the different types of all-air systems.

The double-duct system employs two sets of ducts. One set carries warm air and the other set, cold air. In the air handler, a fan blows filtered air over cooling coils into the cold-air duct and over heating coils into the warm-air duct. Near spaces to be conditioned, the warm and cold air are combined in mixing boxes, then discharged into the spaces. A thermostat in the conditioned spaces controls the discharge temperature by increasing or decreasing the volumes of warm and cold air.

The double-duct system costs more than other types because of the need for two sets of ducts, mixing boxes and controls. Also, the dual ducts occupy considerable space. This system is advantageous for buildings with extremes in loads, such as laboratories and commercial buildings.

The dual-conduit system also employs two sets of ducts, but both carry air at high velocity. Consequently, the system has to incorporate sound traps to absorb noise created by air movements. One set of ducts carries primary supply air and the other set, secondary supply air.

In the air handler, filtered air is passed through equipment that adds or absorbs moisture and cools the air. The air is then blown by a fan into a separator equipped with a sound trap, where the air stream is divided into the primary and secondary supply.

The secondary air stream, at 50 to 55°F, is fed directly to spaces to be conditioned, to handle the internal loads from lights, people and the sun. A thermostat controls the flow rate.

The primary air stream goes first to a reheater, where its temperature is adjusted to between 55 and 115°F in accordance with the heat load at the building perimeter. A constant volume of this tempered air is supplied to the spaces to be conditioned.

Dual-conduit systems have lower initial and operating costs than double-duct systems. Dual conduits are suitable for use in buildings where water piping is undesirable for air conditioning and in office buildings for which overhead heating is acceptable, for example, in mild climates.

The multizone system conditions several zones in a building without the use of a separate air handler for each zone. Each zone is served from a central air handler by a single supply-air duct. In the air handler, filtered air is blown by the fan in two streams, one over hot and one over cold coils. Motor-operated dampers on the downstream side of each of these coils and assigned to a specific zone, mix the proper quantities of warm and cold air in response to the zone thermostat. The mixture is then distributed through the zone duct to the rooms in the zone. Multizone systems have higher initial and operating costs than some of the other types of all-air systems because of the many duct runs and controls.

The single-duct, terminal-reheat system uses a single duct carrying cold air and employs heating coils in the duct or in room units to reheat the air, as necessary, to take care of room heat loads. For this purpose, the room terminal units contain small steam, hot water or electric reheat coils.

In the air handler, filtered air is passed through equipment that adds or absorbs moisture and cools the air. Then, a fan blows the air through the supply duct to terminal units in each room, where a thermostat controls the discharge temperature.

The room terminal unit may be of the straight reheat, induction reheat or reheat induction type.

The straight reheat unit (see Fig. 10.16a)

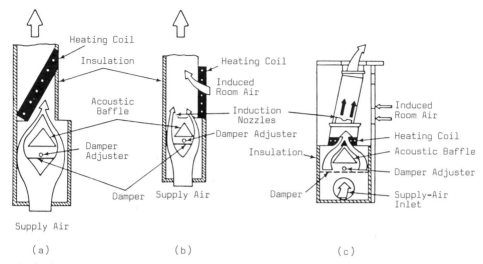

Fig. 10.16. Room terminal units for all-air central-plant systems. (a) Straight reheat. (b) Induction reheat. (c) Reheat induction.

takes the air from the supply duct, passes it over a reheat coil, then discharges the air into the room. No heat is used when full cooling is required.

The induction reheat unit (see Fig. 10.16b) sends supply air through nozzles, called induction nozzles. The discharge from the nozzles induces a nearly equal quantity of room air to flow through a reheat coil into the terminal unit and mix with the supply air. The mixture is then discharged into the room.

In a reheat induction unit (see Fig. 10.16c), the reheat coil is upstream of the induction nozzles. The supply air is heated, then mixed with induced room air for discharge into the room. Because of lower resistance to induction of room air into the unit, the reheat induction unit provides greater air circulation than the other types. This type of unit is especially suitable for hospitals because it does not contain wet surfaces where bacteria can collect and grow.

Of all the various types of air-conditioning systems, the terminal reheat types, when properly engineered, provide the best temperature and humidity control; however, operating costs may be high. Like the three other types of central-plant air-conditioning systems previously described, terminal-reheat systems waste energy, because they first cool the air to some low temperature, usually 55°F, and then reheat it again to maintain desired space temperatures.

The variable-volume single-duct system conserves energy by eliminating reheat at outlets. Instead, thermostats in individual zones or rooms control the volume of cool air being distributed and discharged into the spaces to be conditioned.

In the air handler, filtered air is passed through equipment that adds or absorbs moisture and cools the air. A fan blows the air via a sound trap into the single supply-air duct. A static pressure regulator controls the air flow at the discharge of the fan.

Air flow from the supply duct into a room is controlled by a diffuser like that shown in Fig. 10.10b. This type of diffuser not only throttles air flow into the room but also reduces the total system air-flow rate. Changes in system pressure are compensated by use of constant-volume regulators and aerodynamically designed outlets that can throttle air down to 10% of design air-flow rate without excessive loss of air velocity that could result in poor air circulation. Also, sound levels are low at small loads.

Variable-volume single-duct systems generally have the lowest initial and operating costs of the all-air types of systems. Yet, temperature and humidity control are excellent.

Air-Water Systems. An air-water system distributes chilled water to room terminal units that absorb heat from room air. Temperature of room air is controlled by varying the quantity of water flowing through the coils in the terminal units as the load varies. In addition, conditioned air for ventilation is supplied to the rooms.

Chilled water is usually distributed through two-pipe or four-pipe systems. In a two-pipe system, one pipe supplies the water, which may be chilled for cooling or hot for heating, and the second pipe returns the water from the room terminal units to the central plant. A four-pipe system provides two pipes for hot-water supply and return and two pipes for chilled-water supply and return. Initial cost of the two-pipe system is less, but it lacks flexibility. There are days when both heating and cooling are required, whereas the single supply pipe cannot supply both hot and cold water. The four-pipe system, however, also has temperature-control drawbacks, especially when changeover from cooling to heating is required. Consequently, the system is usually equipped for manual changeover by adjustment of thermostats that permit selection of either heating or cooling as desired by occupants.

Room terminal units for air-water systems generally are fan-coil or induction types.

A fan-coil terminal consists basically of water coils, a fan for blowing room air over the coils, an enclosure for the equipment and controls for fan speed or water flow. For two-pipe systems, the terminal may incorporate an electric resistance heater, to provide four-pipe performance without the added cost of the extra pipes. Fan-coil units may be floor mounted and placed under windows, to neutralize perimeter heat loads; or the units may be recessed in a ceiling and connected to ducts. Conditioned ventilation air in either case may be supplied from a central plant through ducts that discharge into the rooms and bring back return air. In some cases, fan-coil units placed at exterior walls may take in outdoor air through wall openings directly, but such units generally are noisy, and the coils may freeze in cold weather when outdoor air requires heating.

Induction terminal units are supplied with a constant volume of high-velocity, high-pressure conditioned air from a central plant. Flow of this air through induction nozzles draws room air into the units, where the air is either heated or cooled, as required. Usually, 65 to 80% of the heat load is taken care of by the water flow through the units and the remainder of the heat load by the conditioned air. For two-pipe systems, the unit may incorporate an electric heater. In that case, the pipes may carry only chilled water, yet heating or cooling can be achieved at any time.

The air-water systems have lower operating costs than all-air systems but do not have as good control of temperature, humidity, air quality, noise and air movement. Of the air-water systems, fan-coil units with supplementary conditioned air from a central plant provides the best control.

References

B. Stein et al., *Mechanical and Electrical Equipment for Buildings*, 7th ed., Wiley, New York, 1986.

Carrier Air Conditioning Company, *Handbook of Air-Conditioning System Design*, McGraw-Hill, New York, 1965.

J. Gladstone, *Air-Conditioning Testing—Adjusting—Balancing: A Field Practice Manual*, 2nd ed., Van Nostrand Reinholds, New York, 1981.

N. Harris, *Modern Air-Conditioning Practice*, 3rd ed., McGraw-Hill, New York, 1983.

ASHRAE Handbook of Fundamentals, American Society of Heating, Refrigerating, and Air Conditioning Engineers, Atlanta, GA, 1981.

Words and Terms

Air-water systems
Basic absorption cycle
Basic refrigeration cycle
Chilled-water refrigeration cycle
Compressor
Condenser
Double duct system
Dual conduit system
Evaporator
Package unit
Single duct, terminal reheat system
Variable-volume, single duct system

Significant Relations, Functions, and Issues

Basic components of cooling systems: cooling source (chiller), heat-exchange systems: interior (piped, ducted, air or water), heat-exhausting systems: exterior.

Forms of multizoned systems and their design factors: equipment type, zone planning, control systems, delivery systems, individual zone reconditioning systems and controls.

10.11. PASSIVE DESIGN

In recent years, increasing attention has been given to what are described as passive aspects of building design with general regard to energy use. The use of the word passive is derived by comparison of non-energy-utilizing techniques with what are considered to be active responses—notably, energy-consuming mechanical equipment for the building. In general, anything is considered to contribute to passive response if it does not involve consumption of electricity or fossil fuels (coal, gas, and oil). Thus, use of natural ventilation, stack effects for air movement, and daylight for illumination are all contributions to the enhancement of passive response of the building to thermal and lighting needs.

The general thrust of efforts in passive design is to affect the architectural design decisions that impact on energy use. Principal components of this are sun orientation, window placement and detail, building form related to wind and sun, and choice of materials and systems of construction in terms of their thermal properties. Use of building or landscaping elements for sun shading may also be significant in some situations. Precisely defining the goals for passive design begins with consideration of the building use and the significant issues based on local climate, occupant requirements and hours of critical usage. Cost effectiveness of various passive-design measures must relate to comparisons of their additional cost (if any) to long-term gains in reduction of energy use and equipment costs for maintenance and replacement.

Many of the passive design elements have been parts of traditional HVAC design. Natural ventilation, natural light (daylight), insulation in and high thermal capacity of ordinary building construction, the shading effects of roof overhangs and trees, and placement of windows to avoid excessive solar heat gain, have—all along—been used by good designers of any HVAC system. What is currently different is

the degree of attention in the design effort to a more deliberate optimization of these natural factors, rather than to rely so heavily on motorized-equipment-design solutions as the principal response to thermal and lighting needs. Also generally innovative is the increased involvement of architects in HVAC design. As a result, design decisions regarding building form, window placement, building material choices and other basically architectural design concerns are being made with more emphasis on HVAC considerations.

Nevertheless, unusual, elaborate and expensive passive-design solutions must be carefully judged as to their real cost effectiveness, especially when they seriously compromise the general building design solution in terms of occupant needs or desires. When passive design can be effectively achieved with little compromise of building usage and little or no increase in total construction cost, any benefits derived will be positive, regardless of energy or equipment costs. For example, if general planning of a building indicates the need for large expanses of windows on two opposite sides of a building and it is relatively arbitrary which direction these windows face, placing them on the north and south sides will clearly reduce sun glare and solar heat-gain problems. Thus with no change in building construction cost, a major passive response gain is achieved.

Before HVAC systems achieved the level of mechanical sophistication and complexity—and major cost—that they currently have in most buildings, passive-design measures evolved through more instinctive pragmatic responses to the building's user needs. Much of the naturally logical character of passive design was squeezed out of common construction practices as the availability of total motorized-equipment control of interior air, heat and lighting became possible. Mechanical ventilation, electric lighting and air conditioning made natural ventilation and natural lighting in buildings seem obsolete. With diminished fuel resources, higher energy costs and the expense of elaborate equipment clearly in view, application of passive design is not merely desirable but also necessary now and can be profitably employed for HVAC.

References

R. Knowles, *Sun, Rhythm, Form*, MIT Press, Cambridge, MA, 1981.

G. Brown, et al., *Inside Out: Design Procedures for Passive Environmental Technologies*, Wiley, New York, 1982.

G. Brown, *Sun, Wind, and Light*, Wiley, New York, 1985.

B. Stein, et al., *Mechanical and Electrical Equipment for Buildings*, 7th ed., Wiley, New York, 1986.

10.12. SYSTEMS-DESIGN APPROACH TO HVAC

Usually, HVAC systems affect and are affected by other building systems, principally architectural and structural. Those systems often are given higher priority than the HVAC systems because changes in architectural or structural elements to accommodate HVAC systems cost more than changes in HVAC to accommodate the other systems. If, however, life-cycle costs are taken into consideration, there are many conditions under which building designers should give higher priority to HVAC. These conditions generally prevail when designs of other systems adversely affect operating costs of HVAC. Note that architectural and structural elements do not have operating costs, whereas operating costs for HVAC continue over the life of the system.

For HVAC, systems design should be employed in two ways. One way is to provide adequate environmental control for the owner's purposes at minimum life-cycle cost. The second way is to coordinate HVAC and the other building systems to produce an optimum overall design. These two objectives are not impossible to attain. There is a wide range of conditions that can meet the requirements of adequate environmental control for an owner's purposes. This leeway makes feasible selection of an HVAC system with low life-cycle cost while design of the building as a whole is optimized.

Data Collection and Problem Formulation

Basic information for HVAC design comes from examination of the owner's program of requirements, type of occupancy of the building, anticipated number of occupants and type of activity of the occupants. Additional infor-

mation is supplied by design drawings for the other building systems. Of major importance are the climate in which the building site is located, shape of building, orientation of building, roof area, amount of glass on the exterior, provisions for shading walls and windows from the summer sun, mass of walls and roof, availability of space for thermal insulation of walls and roof, availability of space for HVAC equipment, ducts, pipes and terminal units, and heat injected into the building by lights, motors and other heat sources.

From these data, estimates can be made of heating and cooling loads. Also, estimates can be made of the capacities needed for heating and cooling equipment, air handlers and air distribution. With this information, strategies can be developed for providing the desired environmental control.

The architect, structural engineer and the mechanical engineer or HVAC designer should collaborate in determining location of heating and cooling equipment, air handlers, ductwork, piping and room terminal units. The results of the cooperative efforts should be displayed on the architectural and structural drawings and on schematic drawings of the HVAC system and submitted to all the members of the building team for review. In later design stages, all design drawings should be studied to insure compatibility of the HVAC installation with the other systems.

Further information for HVAC design should be obtained from local building codes and government agencies, such as the departments of health and labor, and from national standards. Also, information should be obtained on the requirements of utilities that may supply water, electricity, gas or steam to the building and on costs and availability of fuels.

The goal of HVAC design may be stated as: to design, as a component of an optimum building, the lowest-cost (life-cycle) HVAC system that will provide acceptable environmental control for the owner's purposes.

Objectives

A primary objective of HVAC is provision of air quality within a building that equals or exceeds the minimum essential for maintenance of the health of building occupants. This objective requires that clean, odor-free outdoor air be supplied to indoor spaces in quantities necessary for breathing and dilution of odors. In addition, tobacco smoke, products of combustion, excessive moisture, gases and other airborne pollutants generated indoors must be removed from the building.

Another important objective of HVAC is provision of comfort conditions within a building insofar as these conditions can be attained by control of radiant heat and air movements. This objective requires that air temperature indoors be closely controlled, in accordance with the physical condition and activity of occupants. Humidity also must be controlled but not as closely. Relative humidity between 20 and 60% generally is acceptable. In addition, air motion should be controlled so that air velocity is between 25 and 40 fpm. Too slow a velocity would make an indoor space feel stuffy. Too high a velocity when the air is cool would cause drafts. Finally, air cleanliness and freedom from odors are as essential to comfort as they are to health.

Safety of building occupants is another important objective. Heating requires heating equipment, which is a potential fire hazard. Precautions should be taken to prevent building fires from these heat sources. Also, if a fire should occur from any cause, air distribution equipment could spread the fire. Again, precautions should be taken to prevent spread of fire and also, if feasible, to employ the HVAC system to assist in fire extinguishment.

Still another important objective is reliability of HVAC operation. Unhealthy or uncomfortable conditions can occur within a building when HVAC equipment malfunctions or ceases to operate. Consequently, HVAC equipment selected should be capable of operating continuously for long periods without breakdowns and of being easily repaired or replaced when breakdowns occur. From a cost viewpoint, little maintenance should be necessary, and the cost of maintenance and repairs should be low.

In addition, energy conservation should be a design objective. Energy used in HVAC operation consumes precious natural resources. Hence, it is necessary to keep fuel consumption

for HVAC as small as possible. This objective requires therefore a highly efficient HVAC system. But other conservation measures should also be taken, such as appropriate use of thermal insulation, solar radiation, building shape and orientation, and heat-resistant walls and roof.

Furthermore, lowest life-cycle cost should be a major HVAC design objective. Operating costs for HVAC often can be of greater importance to an owner than initial costs.

Constraints

One of the most important constraints on HVAC design is the nature of the climate in which the building is located. The climate determines design outdoor dry-bulb and wet-bulb temperatures, and wind velocity. In addition, the geographic location of the site determines the angle of incidence of solar radiation, which is of importance in employing solar heat for heating buildings and in determining cooling loads.

Building codes and regulations of local departments of health and labor impose numerous constraints on HVAC systems in the interests of public safety, health and welfare. These constraints place restrictions on equipment and materials that may be used for HVAC and govern methods of installation of HVAC equipment.

Other constraints usually are imposed on HVAC design by other building systems. Architectural requirements are among the most important of these. For example, shape and orientation of a building have a considerable effect on heating and cooling loads, as does the amount of glass on the building exterior. Even the color of the building exterior can have a significant effect. Also, architectural considerations govern the locations of spaces in which HVAC equipment may be installed and the locations in which ducts or piping may be placed. In addition, architectural considerations influence the type of room terminal units that may be used for supplying and exhausting air and the location of these units.

A common constraint is the architectural requirement that HVAC ducts and piping be concealed from view. This makes it necessary to install these elements in assigned shafts, in walls and in the spaces between floors and ceilings. Often, as a result, a preferred path for ducts and piping is obstructed by structural members.

Availability and cost of water needed for heating or cooling are additional constraints on HVAC design. These constraints, for example, may indicate the need for air-cooled condensers or for recycling of water for water-cooled condensers.

Other major constraints are availability and cost of electricity and fuels. These may have a considerable effect on selection of a strategy for heating and cooling a building. For example, if steam is available at low cost, it not only can be used to heat the building but also to cool it by use of absorption-type refrigeration. If electricity is available at low cost, a heat pump may be economical for space heating. If fuels are expensive or scarce, solar radiation may be the only economic alternative for heating or cooling a building.

Synthesis and Analysis

The design process for an HVAC system basically examines three major factors: criteria, analysis and control. These require answers to three questions:

1. What is the optimum indoor environment, not only for occupants but also for processes carried out in the building?
2. What factors affect that environment? Principally, what are the dynamic nature, source and magnitude of heat and air flows?
3. What types of system strategies, equipment or devices can be used to control the factors influencing the environment in order that desired criteria can be achieved?

It is not sufficient, however, merely to design an HVAC system of adequate capacity to satisfy given criteria. It is necessary, in addition, to design, as closely as possible, an optimum building. This generally will require the most efficient utilization of energy and investment capital and low operating costs. To achieve these objectives, it is essential that HVAC be treated as part of an integrated building system.

Criteria. Design criteria established for HVAC influence the size and type of the HVAC system required for a building and the extent of interaction and interdependence of the HVAC system with other building systems, such as walls, roof, windows and ceilings.

Temperature criteria have the largest and most widespread effect on building design. Design indoor and outdoor temperatures determine the temperature gradient across the building exterior, which, in turn, determines heat loads on the HVAC system. Consequently, thermal criteria should be selected with care so as to achieve design objectives for comfort or building processes and yet conserve energy and minimize system costs. It is not economical to design for outdoor conditions that will occur infrequently and last for only a few hours; nor is it economical to design for indoor conditions that will make every occupant of the building feel comfortable.

Criteria for air humidity, quality and velocity influence the design of the HVAC system but usually have only a small effect on other building systems. Generation of humidity or pollutants within a building, however, will impose loads on the HVAC system, for which it must be designed. Also, criteria for air temperature, quality and velocity in emergency conditions, such as fire, may influence HVAC design without much effect on other systems.

Analysis. Ventilation requirements and heat loads for a building are not static but vary almost continuously. Consequently, energy consumed and operating costs for HVAC also vary with time, but system capacity required and initial cost are determined by peak loads. In the early design stages, peak loads, system capacity required and initial and operating costs can be estimated with manual computations, but in the later design stages, the variations in load with time should be taken into account and, as a result, computations become very complicated. Use of a computer to make the computations would be very helpful. Programs are available for the purpose.

In every design stage, the building team should investigate the consequences of all design decisions on environmental conditions. In doing so, the team should be able to identify significant opportunities to optimize construction and operating costs. To capitalize on these opportunities, the team should consider alternative designs for those elements for which alternatives are available. For an efficient design process, decisions on alternatives should be made in the early design stages insofar as possible. Usually, such decisions have a considerable effect on the indoor environment and the cost of its control.

Control. After design criteria have been selected and heat loads determined and evaluated, strategies can be formulated to satisfy the criteria. A strategy comprises one or more concepts for achieving desired objectives for a system. For example, a strategy for satisfying temperature criteria might include mechanical ventilation and heating or cooling, or it might include natural ventilation and heating. Several strategies can and should be used to satisfy criteria for different conditions and for different parts of a building.

Design strategies for satisfying temperature and humidity criteria generally include use of ventilation, heating and humidification or air conditioning. But consideration also should be given to removal of heat at its source, exclusion from the building of unwanted heat and utilization of waste heat. In addition, provision should be made for incorporation in building construction of optimum thermal insulation.

Strategies for controlling air velocity usually include use of fans, appropriate sizing of ducts and suitable choice of room terminal units. Air-quality control generally concerns choice of air-intake and air-exhaust locations and use of filters. Environmental control in fire emergencies involves control of air flow in critical portions of a building, methods for limiting smoke concentration and processes for restricting rises in air temperature.

Basically, in design of an HVAC system, a strategy will be selected for each criterion. Design decisions will interact and conflicts will result. The main concern of the HVAC designer is to resolve these conflicts in such a way that an optimum building will be produced. This requires substitution of alternative strategies. In particular, consideration should be given to

techniques for reducing heat loads and to changes in other components of the building that influence the environment. The latter requires the cooperation of other members of the building team and preferably should be done in the early design stages.

Value Analysis and Appraisal

Because HVAC design interacts with the output of other members of the building team and because several environmental criteria must be satisfied simultaneously, design of an HVAC system is complex and iterative and involves many trade-offs and recalculations. Trade-off studies should examine the performance and cost impact of all affected building systems. Resulting changes in any of the systems to optimize the building design should be made in the early design stages. Later design changes are costly and will delay completion of design and construction. Consequently, value engineering should be applied to the maximum extent in the early design stages for environmental control. Value analysis should consider alternatives not only for the HVAC system and its components but also for other building elements that are affected by or affect the HVAC system. Evaluation of the alternatives should take into account not only initial construction costs but also life-cycle costs and energy conservation, especially operating costs for HVAC.

The efficiency of energy employment in a building can be determined by preparing an energy budget and comparing the energy consumed with a generally accepted standard. The budget should include energy required for HVAC, interior and exterior lighting, electric power distribution, solid-wastes disposal and elevator and escalator operations. Energy consumption may be expressed in Btu per yr per sq ft of gross floor area or per sq ft of exposed area. In 1975, the U.S. General Services Administration recommended a maximum energy consumption of 55,000 Btu per yr per sq ft of gross floor area for new buildings. Since that time, many states and local building regulatory agencies have developed requirements for energy consumption, generally resulting in the need for an annual energy-consumption analysis for part of the process of obtaining a building permit. Designing the HVAC and lighting systems to satisfy these codes is a major constraint, generally forcing efficiency standards more stringent than those used for many systems of the past.

Energy Conservation. Value engineers should investigate the following potentials for reduction in energy consumption:

1. Reduction of heat loads. This would make possible a smaller capacity HVAC plant and lower energy consumption and operating costs. Heat loads can be reduced with added thermal insulation. Peak loads may be decreased by increasing the mass of exterior walls and roof so that heat is absorbed instead of being immediately transmitted. Heat loads can be decreased also by reducing the amount of window area. In particular, in the northern hemisphere, little or no window area should be incorporated on northern exposures. Where windows are used, they should be an insulated type: double-glazed, tinted, reflective or heat-absorbing. Canopies or vertical fins should shade windows from the summer sun. Windows of low buildings may also be shaded by trees. In addition, exterior walls should be reflective or light colored, to reflect solar radiation. Roofs may be sprayed with water to mitigate the effects of solar heat in warm weather.

2. Reduction of ventilation and air infiltration and exfiltration. Excessive ventilation imposes unnecessary heat loads on the HVAC system. So does air not needed for ventilation that leaks into the building. Windows and doors should be weather stripped and calked to reduce air leakage.

3. Reduction of heating and cooling of unoccupied spaces. For spaces used only occasionally, for example, for storage or maintenance, manual controls may be installed for supplying heat or conditioned air only when required. For spaces occupied only during the daytime and only on weekdays, time-clock controls may be advantageous.

4. Selection of equipment optimally suited for the task to be performed. Capacities of oversized equipment should be decreased. In a multiboiler installation, one boiler can often be eliminated by more efficient use of the others. Consider use of room fans for cooling.

5. Reduction of electric lighting. Time-clock controls or photoelectric switches may be installed to turn on and off exterior and decorative lighting. Provisions should be made for turning off most or all lights in premises unoccupied at night or weekends. Also, provision should be made for turning off indoor lighting close to the perimeter of the building when adequate daylight is available. To reduce cooling loads without impairing the quantity of illumination provided, fluorescent or other low-heat-emitting lamps should be used for indoor lighting.

6. Confinement of heat to points where needed. In sparsely occupied buildings with high ceilings, unit heaters blowing warm air directly on occupants may be more economical than other types of installations. Infrared heat is an alternative. It requires less fuel than other kinds of heating, because it heats solid objects without heating all the air in the space.

7. Utilization of waste heat. Heat from people, lights and motors not only reduces the heating load at the center of a large building but also can be distributed to offset heat loads around the perimeter of the building. Also, waste heat from burner stacks and industrial processes can often be recovered and used, for example, for preheating boiler feedwater or for absorption-cycle cooling. In addition, recuperators can be installed on combustion equipment to preheat incoming combustion air. Furthermore, heat is recovered when steam condensate is returned to boilers.

8. Recycling of waste materials, such as sawdust and oil, as supplemental fuels.

9. Establishment of maintenance programs for use after equipment is placed in operation, to insure efficient performance. Heating surfaces of boilers, both on the

water and the fire side, should be kept clean. Also, the proper air-fuel ratio should be maintained for the burner. Stack temperatures and concentrations of carbon dioxide, carbon monoxide and oxygen in flue gases should be monitored to insure that heat is not being wasted and that complete combustion is occurring. Incomplete combustion is indicated by the presence of carbon monoxide in the flue gas. Oxygen should be less than 2% and carbon dioxide should be between 10 to 15% of the flue gas, depending on the fuel used. Flames in the flue represent unnecessary heat loss. In addition, fans should be set to supply only the amount of air that is actually required for HVAC.

10. Employment of controls to monitor HVAC operation, to insure efficient performance of the system. Outdoor dry-bulb and wet-bulb temperatures (or relative humidity) should be monitored and dampers adjusted to mix the proper quantities of fresh and return air so as not to waste energy on unnecessary heating or cooling and dehumidification of outdoor air. On cool days, when the interior of a building nevertheless requires cooling, outdoor air may be used instead of refrigerated air to cool the interior.

Integrated Systems. Initial construction costs for a building often can be reduced by integrating HVAC in other building systems.

Dead space in a building, for example, can be used for wireways, pipe runs and air supply and return. Ventilation or conditioned air can be distributed through the space between ceiling and floors; through tubular structural members, such as columns and beams; and through hollow metal decking or hollow concrete slabs.

As another example, collector panels for solar heating may be used as the exterior facade for walls and roofs. In multistory buildings, vertical exterior walls facing the sun usually provide enough area for collector panels to heat the building, even though the panels are not inclined at the most efficient angle for absorbing solar radiation.

As still another example, high-pressure steam may be produced to generate electricity for a large building or group of buildings. The low-pressure steam exhausted from the turbines then can be used to heat or cool the buildings.

GENERAL REFERENCES AND SOURCES FOR ADDITIONAL STUDY

These are books that deal comprehensively with several topics covered in this chapter. Topic-specific references relating to individual chapter sections are listed at the ends of the sections.

B. Stein, et al., *Mechanical and Electrical Equipment for Buildings*, 7th ed., Wiley, New York, 1986.

ASHRAE Handbook of Fundamentals, American Society of Heating, Refrigerating, and Air Conditioning Engineers, Atlanta, GA, 1981.

G. Brown et al., *Inside Out: Design Procedures for Passive Environmental Technologies*, Wiley, New York, 1982.

J. Threlkeld, *Thermal Environmental Engineering*, 2nd ed., Prentice-Hall, Englewood Cliffs, NJ, 1970.

F. Merritt, *Building Design and Construction Handbook*, 4th ed., McGraw-Hill, New York, 1982.

EXERCISES

The following questions and problems are provided for review of the individual sections of the chapter.

Section 10.1

1. What are the four primary factors of concern for human comfort in most buildings?
2. Why is it not possible to establish a single set of comfort criteria for indoor temperature and humidity?

Section 10.2

3. What distinguishes heat from other forms of energy? What distinguishes sensible heat from latent heat?
4. A thermometer reads 77°F.
   (a) What is the corresponding reading in the Celsius system?
   (b) What is the absolute temperature, °Rankine?
   (c) What is the absolute temperature, °Kelvin?

5. A blower is sending air through a duct at the rate of 1000 cfm. The air has a specific volume of 12.5 cu ft per lb. How many pounds of air per hour is the blower handling?

6. Standard atmospheric pressure is 29.92 in. of mercury. This metal has a specific gravity of 13.6.
   (a) What is the standard atmospheric pressure, ft of water?
   (b) What is the standard atmospheric pressure, psi?

7. A blower is handling 2400 cfm of dry air with a specific volume of 13.5 cu ft per lb. Temperature of the air is 35°F.
   (a) How many pounds of air per hour is the blower handling?
   (b) How much heat, Btu per hr, is required to warm this air to 120°F? (See Table 10.1, Section 10.2, for the specific heat of air.)

8. A machine room contains electric motors operating at 50 horsepower (1 hp = 33,000 ft-lb per min) How much heat, Btu per hr, can the motors add to the room?

9. Outside, dry air at 20°F and with a specific volume of 12.0 is heated by a furnace to 180°F at the rate of 1200 cfm. To humidify the air for the comfort of occupants, 0.014 lb of water per lb of air is added per hour. Water supplied to the humidifier is at 50°F. (Latent heat of vaporization of water at 180° is 990 Btu per hr per lb.)
   (a) How many lb per hr of dry air is the furnace blower handling?
   (b) How much sensible heat, Btu per hr, is being added to the dry air?
   (c) How much water, lb per hr, is evaporated from the humidifier?
   (d) How much sensible heat, Btu per hr, is being added to the water vapor?
   (e) How much latent heat, Btu per hr, is supplied to the water?
   (f) How much heat, Btu per hr, does the furnace supply?

10. Using Eq. (10.6), compute the volume, specific volume and density of 30 lb of dry air at 80°F and standard atmospheric pressure.

11. The dry-bulb temperature of air at standard atmospheric pressure with a relative humidity of 50% is 95°F.
   (a) What would the enthalpy of 1 lb of the air be if it were dry?
   (b) How much water vapor, lb, would 1 lb of the air hold if it were saturated?
   (c) What is the humidity ratio of the 95°F air at 50% relative humidity?
   (d) What is the enthalpy of the water vapor in 1 lb of the dry air?
   (e) What is the enthalpy of 1 lb of the air at 95°F and 50% relative humidity?
   (f) What is the dew-point temperature of the air?

12. A compressor is operated to produce 2 tons of cooling continuously. How much heat, Btu, does it remove in one day in air conditioning a building?

Section 10.3.

13. What are the three basic methods of heat transfer in a building?

14. What is the major difference between heat transfer by conduction and by convection?

15. Which method of heat transfer can take place in a vacuum?

16. Under what conditions does steady-state heat transfer occur?

17. A building has an exterior wall of aluminum panels $\frac{1}{8}$-in. thick. The coefficient of thermal conductivity of aluminum is 1460. Indoor temperature is 62°F when outdoor temperature is 32°F and a 15-mph wind is blowing.
   (a) What is the thermal conductance of the metal?
   (b) What is the thermal resistance of the metal?
   (c) What is the thermal resistance air-to-air of the wall?
   (d) What is the coefficient of thermal transmittance of the wall?
   (e) What is the heat-flow rate, Btu per hr, through 200 sq ft of the wall?

18. A wall is constructed of two $\frac{1}{8}$-in. thick, parallel aluminum panels with $\frac{3}{4}$-in. space between them. The coefficient of thermal conductivity of the aluminum is 1460.

What is the coefficient of thermal transmittance of the wall?

19. What causes heat transfer by convection?
20. Why do glass walls exposed to the sun trap heat within a building?
21. The mean radiant temperature in a room is 73°F. What should the design dry-bulb temperature for the room be?
22. What are the disadvantages of oversized heating and cooling plants?
23. On what temperature conditions should estimates of HVAC operating costs be based?
24. In cold weather, the mean radiant temperature of a room is 74°F. What should the design indoor dry-bulb temperature be for lightly clothed, inactive occupants?

Section 10.4

25. Why is thermal insulation advantageous in building construction?
26. Why should a vapor barrier be placed on the warmer side of construction?
27. Does the thermal conductance of reflective insulation change much with changes in insulative thickness?
28. A cavity wall has an exterior of 4-in. brick with a thermal resistance of 0.76 and an inner wythe of 4-in. concrete block with a resistance of 0.71. The wythes are separated by a 2-in.-wide air space. An interior finish of plaster with a resistance of 0.32 is applied directly to the concrete block.
    (a) What is the U value of the wall?
    (b) What would the U value become if the 2-in. cavity was filled with expanded perlite with a coefficient of thermal conductivity of 0.36?
29. What are the advantages of weather stripping?

Section 10.5

30. Under what conditions will water vapor be condensed from air?
31. What makes condensation a continuing process within a building?

32. How can condensation on windows be prevented or reduced?
33. What is the most effective means of limiting condensation within a roof or exterior wall?
34. Where should a vapor barrier be placed within a roof or wall of an ordinary building (one not used for cold storage)?
35. What provision should be made for removal of condensation that may occur within:
    (a) A wall?
    (b) An attic?
    (c) A basement?
36. An attic has a floor area of 1200 sq ft. What is the minimum area of openings to outdoors that should be provided for attic ventilation?
37. What is the purpose of weep holes?

Section 10.6

38. What are the purposes of ventilation?
39. How is air moved in a room by natural ventilation?
40. How is air moved within a building by mechanical ventilation?
41. What is the basic source of fresh air?
42. What is the purpose of ducts?
43. If mechanical ventilation costs more than natural ventilation, why is mechanical ventilation used?
44. What is a blower?
45. A theater has a maximum capacity of 480 persons. It has a volume of 180,000 cu ft. What is the maximum air-flow rate, cfm, of fresh air that should be supplied to the theater if the building code requires:
    (a) 5 cfm per occupant?
    (b) 5 air changes per hour, with at least one third the total being outdoor air?
46. Ozone leaks into the air of a chemical plant at the rate of 0.0001 lb per hr. The threshold limit for ozone is 0.2 ppm (lb per 1,000,000 lb). Air density may be assumed as 0.075 lb per cu ft. At what rate, cfm, should outdoor air be supplied to the plant interior to maintain the threshold limit for ozone?

47. Heat is generated in a room at a rate of 20,000 Btu per hr. Room temperature to be maintained is 75°F. Outdoor temperature is 55°F. At what rate, cfm, should outdoor air be supplied to maintain the specified room temperature?

48. Windows and monitors, each with a total openable area of 100 sq ft, are provided on opposite sides of a factory. The monitors are 32 ft higher than the windows. When the indoor temperature is 65°F and the outdoor temperature is 57°F, at what rate, cfm, will ventilation air enter and leave the factory when the windows and monitors are wide open?

49. Air pressure maintained in a duct is 10 in. water gage. What is the pressure, psf?

50. Which type of fan should be selected to distribute air through a duct with high total resistance?

Section 10.7

51. A building has the following wall and roof components:
    Walls, area 6000 sq ft, $U = 0.30$
    Windows, area 1200 sq ft, $U = 1.13$
    Roof, area 5000 sq ft, $U = 0.25$
    The building also has a first story floor on grade with a perimeter of 300 ft and 2-in.-thick edge insulation. Groundwater temperature is 60°F. If design indoor temperature is 75°F and design outdoor temperature is 15°F, what is the building heat load, Btu per hr?

52. Design indoor temperature of a two-story dwelling is 75°F and design outdoor temperature is 15°F. The roof over an unheated attic in the building has an area of 3000 sq ft; $U = 0.50$. The ceiling below the attic has an area of 2000 sq ft; $U = 0.14$. What is the rate of heat loss, Btu per hr, from the top story to the attic?

53. Design indoor temperature of a building is 75°F and design outdoor temperature is 15°F. The building has a volume of 48,000 cu ft. For ventilation and to prevent air infiltration from outdoors, five changes per hour of heated air are provided. What is

the heat load, Btu per hr, caused by the ventilation air?

Section 10.8

54. A building has the following wall and roof components:
    Walls, area 6000 sq ft, $U = 0.30$
    Windows (shaded), area 1200 sq ft, $U = 1.13$
    Roof, area 5000 sq ft, $U = 0.25$
    (a) What is the sensible-heat gain, Btu per hr, resulting from a design indoor dry-bulb temperature of 75°F and a design outdoor dry-bulb temperature of 95°F?
    (b) What is the sensible-heat gain, Btu per hr, from solar radiation impacting the roof?

55. Fluorescent lamps in an office building have a total rating of 5000 watts. How much heat, Btu per hr, do the lamps generate?

56. In an office building with 50 employees, 90% may be doing moderately active work and the rest may be walking.
    (a) What is the sensible-heat gain, Btu per hr, due to these occupants?
    (b) What is the latent-heat load, Btu per hr, due to these occupants?

57. A room with a volume of 30,000 cu ft is supplied $7\frac{1}{2}$ air changes per hour, of which one-third of the air is fresh air. Design dry-bulb temperatures are 95°F outdoors and 75°F indoors. For the design conditions, the humidity ratio of the outdoor air is 0.025 and the humidity ratio of the indoor air is 0.010.
    (a) What is the sensible-heat gain, Btu per hr, due to the ventilation air?
    (b) What is the latent-heat load, Btu per hr, due to the ventilation air?

Section 10.9

58. The net heat load for a building is 150,000 Btu per hr. Catalogs show that heaters are available with capacities of 145,000, 165,000, 185,000 and 205,000 Btu per hr. Costs of the heaters are nearly proportional

to the capacities. Which heater size should be selected? Explain your answer.

59. What precautions should be taken to prevent a heater from causing a fire?

60. What provisions should be made to remove from a building combustion products from a heater?

61. What provisions should be made to supply combustion air to a heater?

62. A room heated by a warm-air system requires 180,000 Btu per hr.
    (a) How much air, cfm, should be supplied to the room at 150°F to maintain a room temperature of 70°F?
    (b) If the heater is oversized by 25%, during what percentage of the time will the heater operate to maintain the room temperature?

63. With a warm-air heating system, what is the best location in a room for a supply-air grille:
    (a) Relative to exterior and interior walls?
    (b) Relative to floors and ceilings?

64. What is the best location in a room with hot-water or steam heating for radiators or convectors:
    (a) Relative to exterior and interior walls?
    (b) Relative to floors and ceilings?

65. A room heated by a hot-water system requires 10,800 Btu per hr. The temperature drop across each convector in the room is 20°F. At what rate, gpm, should hot water flow through the convectors?

66. In a two-pipe steam-heating system, what device should be used and where should it be located in the system to prevent steam from escaping into the condensate return pipe?

67. A room heated with steam-heated radiators requires heat at the rate of 4800 Btu per hr.
    (a) How many sq ft EDR should the radiators have?
    (b) How much steam, lb per hr, should be supplied to the radiators?

68. What are the advantages of a perimeter warm-air heating system?

69. What produces the heat for space heating in a heat pump?

70. When is solar heating more economical than other types of heating?

Section 10.10

71. A building to be air conditioned has a total sensible-heat load of 180,000 Btu per hr and a latent-heat load of 60,000 Btu per hr.
    (a) What is the capacity, tons, required of the refrigeration machine?
    (b) If supply air is 18°F cooler than room air, what is the flow rate, cfm, of air required for cooling?

72. What is the function in the basic refrigeration cycle of:
    (a) The refrigerant?
    (b) The compressor?
    (c) The evaporator?
    (d) The condenser?

73. A building to be air conditioned requires 24 tons of refrigeration. Cooling is to be accomplished with chilled water. If the water temperature increases 8°F in the cooling coils, what is the flow rate, gpm, of water required?

74. What functions do the compressor in the direct-expansion cycle and the absorber-generator on the absorption cycle serve in common?

75. Describe briefly the minimum equipment needed for a cooling system for a building.

76. What are the advantages and disadvantages of a package air conditioner?

77. How do all-air air conditioning systems differ from air-water systems?

78. Which type of central-plant air conditioning system can give the best temperature and humidity control?

79. Which type of central-plant all-air system has the lowest initial and operating costs?

80. Which type of central-plant air-water system can provide the best environmental control?

Chapter 11

Lighting

Good lighting within a building is desirable for many reasons, the most important of which include the following:

1. Better visibility in work areas in home, office and factory for comfort, efficiency and safety, and accident prevention.
2. Better visibility in recreation areas for better performance, enjoyment in observing play and accident prevention.
3. Development of color effects for pleasure and accident prevention.
4. Better visibility for security reasons.
5. Exploitation of light effects to influence human moods, accent select spaces or objects, and decorate.

As indicated in Sec. 2.9, good lighting requires an adequate quantity of light, good quality of illumination and proper colors in the light. The effects of lighting, however, depend on the characteristics of the floor, walls and ceiling around the light source and of the objects to be illuminated.

In the design of a lighting system, it is not sufficient merely to provide good lighting. It is necessary to take into account also the life-cycle costs of providing the lighting and the need for energy conservation.

11.1. ACCIDENT PREVENTION

More accidents happen in buildings than in autos, trains and airplanes. These accidents usually result from carelessness or neglect. Often, they can be avoided by use of adequate lighting.

Outdoor Lighting

Although building interiors and the areas around buildings are often illuminated at night for the sake of appearance or for security reasons, for example, to discourage intruders, nighttime lighting also is highly desirable for accident prevention. Accidents, for example, can happen right at the front door or on the walk leading to it. Someone might trip and fall in the dark over an unseen bump or stair. Good lighting reduces the chances of such accidents.

Work and Play Areas

Many human activities are potentially hazardous in home, office and factory. In particular, operation of machines with exposed moving parts is especially dangerous. Good lighting, however, reduces the possibilities of accidents. In addition, switches to operate lights should be placed at the entrance and exit of every room to permit a person, before entering a dark room, to put on a light.

Stairway Lighting

Besides machine areas, stairways are among the most hazardous areas in a building. To prevent accidents, lights should be placed at the top and bottom of all flights of stairs. Preferably, the lights should be shielded to prevent glare from interfering with the visibility of persons on the stairs.

Bathrooms

Good lighting at mirrors in bathrooms is highly important, but because bathrooms usually contain medicine cabinets, good general lighting may be even more important, to eliminate the possibility of labels on medicine bottles from being misread in dim light. Also, lighting should be provided over bathtubs for safety reasons. Fixtures placed over bathtubs and stall showers preferably should be moistureproof.

Switches for operation of lights in bathrooms should be located where persons in a bathtub or shower cannot reach them. Such locations minimize the possibility of wet persons receiving an electric shock when touching a faulty switch.

Emergency Exits

In many types of buildings, such as theaters and hotels, building codes usually require that lights be used to indicate the location of emergency exits and the direction in which to travel to reach them. Often, illuminated signs are used for this purpose.

11.2. QUANTITY OF LIGHT

Quantity, or intensity, of light depends on the brightness and concentration of the light source and its distance from objects to be viewed. Quantity of light that should be provided in a room generally depends on the activity to be performed in the room.

Dim lighting is sometimes desirable for mood effects, but when tasks have to be carried out within a building, sufficient light must be provided on the work if it is to be executed without eyestrain. The amount of illumination needed depends on the type of work. (The Illuminating Engineering Society publishes recommended minimum lighting levels for various tasks.) In general, more light is needed with small details, with dark colors or when there is poor contrast with the background against which

the task is being performed. Also, more light is required for tasks lasting several hours than for the same tasks when completed in a few minutes. Furthermore, more light may be needed for subnormal eyes and for older persons.

Higher than the minimum recommended levels of illumination usually improve visibility, but often with greater energy consumption and increased life-cycle costs. In addition, when electric energy is consumed, heat is released, as indicated by Eq. (10.28). Unless means can be found to utilize this heat, it must be removed from the building, at additional cost, to prevent discomfort of the occupants, especially in warm weather.

In addition to its effects on visibility, lighting intensity can be used to influence the mood of persons in a room. High levels of lighting produce cheerful and stimulating effects. Low levels instead tend to create an atmosphere of relaxation, intimacy and restfulness.

Measurement of Light Quantity

The unit of light quantity is called a *lumen*. A more useful unit in practical illumination applications, however, is the *foot-candle*, the unit of light intensity, or light quantity per unit area. For a uniformly distributed light,

$$ft\text{-}c = \frac{L}{A} \tag{11.1}$$

where

$ft\text{-}c$ = number of foot-candles
L = lumens striking a uniformly lit area
A = area, sq ft

The unit used for measuring the intensity of a light source is *candlepower*. A light source equal to one candlepower produces a light intensity of 1 ft-c on a surface 1-ft away from the light source.

The intensity, ft-c, of light striking a surface varies inversely as the square of the distance of the surface from the light source.

Lighting Distribution

A theoretical point source radiates equal quantities of light in all directions. Lamps used in buildings, however, are not true point sources. If curves are plotted showing for a specific lamp the lumens measured on spherical zones of $10°$ width and 1-ft radius, centered on the light source, these candlepower distribution curves will be irregularly shaped. Lighting intensity usually is a maximum in a specific direction and diminishes as the angle with that direction is increased.

11.3. QUALITY OF LIGHT

Quantity of light alone is not an adequate measure of good lighting. Quality of lighting also is important. Quality is determined by the contrast of light and dark areas. Contrast is created by shadows or by relatively bright areas. The differences in lighting intensity that result where contrast occurs affect visibility, mood, comfort and eyestrain.

When shadows are indistinct, the lighting is said to be *soft* or *diffused*. This effect is produced by lighting that is uniform or nearly so. It provides an atmosphere more relaxing and visually less compelling than one with contrast.

When deep shadows are present, the lighting is called *hard*. Such lighting, when used artfully, provides highlights and contrasts. It can emphasize texture, delineate form and create a sense of life and gaiety. Presence of some hard lighting in a room makes the space more interesting than when soft lighting alone is used. Also, use of both hard and soft lighting in a room can produce distinct changes in visibility, atmosphere and mood to meet specific objectives.

Contrast between light and dark spaces or areas, however, can also be a source of glare, which can cause severe discomfort. Glare, for example, often occurs when a bright light source is reflected from a shiny surface. For good lighting then, contrast must be strictly controlled.

While visual tasks are mostly grouped in terms of their light-quantity requirements, the quality of the light may also be critical. Diffused light is generally more pleasant for general viewing purposes, but reading, drafting, and other visual work of a fine nature requires some degree of hard quality for heightened contrast visibility of the work. Providing such light without accompanying glare is a critical requirement for designers of lighting systems.

11.4. COLOR

In addition to light of appropriate quantity and quality, to provide good lighting both the color of light and the color of objects to be illuminated should be chosen to produce desired effects. Color of light affects the colors of objects observed.

Usually, the color of light should be chosen to enhance color identification of an object or surface. For example, proper use of light in a restaurant can make food served appear more appetizing, whereas selection of an inappropriate color would have the reverse effect. As another example, use of colored lights can alter the appearance of human skin, and under some circumstances the change may be objectionable. Often, bright colors may not be acceptable for a light source, because they can adversely affect the accepted appearance of people and things.

Perceived color of objects also is influenced by quantity of light. Low illumination levels tend to make colors appear grayer. High levels intensify color.

SECTIONS 11.1 TO 11.4

References

B. Stein, et al., *Mechanical and Electrical Equipment for Buildings*, 7th ed., Wiley, New York, 1986.
J. Nyckolls, *Interior Lighting for Environmental Design*, Wiley, New York, 1983.
IES Lighting Handbook, Illuminating Engineering Society of North America.

Words and Terms

Candlepower
Contrast
Foot-candle
Glare
Intensity of light
Light quality: diffused, hard
Lumen

Significant Relations, Functions and Issues

Good lighting as accident prevention.
Light quantity (intensity) related to visual tasks.

Measurement of light: lumens, foot-candles, candlepower. Diffusion or hardness of light related to contrast.

11.5. LIGHTING STRATEGIES

Lighting effects in a room depend not only on the light source but also on texture, color and reflectivity of objects illuminated. To meet specific illumination objectives, the following lighting strategies may be used alone or in combination:

General Lighting. This provides a low level of illumination throughout a space. This type of lighting is useful for performing ordinary activities and for making local lighting more pleasing.

Local or Functional Lighting. This is the application of a relatively high level of illumination to the relatively small area in which a task is to be performed, such as reading or operation of tools.

Accent or Decorative Lighting. This also requires application of a higher level of illumination than that used for general lighting; however, this type of lighting has the objective of creating focal points for observers, to emphasize objects on display or to form decorative patterns.

Any of these types of lighting can be attained with natural or artificial illumination, or both. Natural illumination is provided by daylight. Artificial illumination usually is provided by consumption of electric power in various types of lamps and sometimes by burning candles or oil or gas lamps. Lamps may be portable or permanently set in or on ceilings or walls and enclosed in luminaires, or lighting fixtures. Provision for lighting fixtures usually is made in the design of a building, and usually they are installed during the finishing operations in building construction.

Direct and Indirect Lighting

Illumination may be classified as direct or indirect lighting. Direct lighting travels in a straight line to observed objects from a light source, such as a window, skylight or lamp. Indirect lighting is reflected from room enclosures, such as walls, floor and ceiling, in traveling from a light source to observed objects. Unless the room enclosures are mirrored or act like mirrors, the reflected light is diffused and produces soft lighting.

A combination of direct and indirect lighting may be provided by windows or some types of luminaires. For example, a lamp may throw some light directly on objects below it, while it projects some light upward to be reflected downward by the ceiling onto those objects. Such lamps may provide a combination of general and local lighting in a room.

Lighting Distribution

Luminaires can be designed for a specific type of lamp to distribute light in a way that will meet desired objectives. For this purpose, use is made of various shapes of reflectors, various types of lenses and control of size and shape of openings through which light is emitted.

Generally, luminaires provide illumination that is symmetrically or asymmetrically distributed. With symmetrical distribution, the intensity of the light projected from a luminaire on a work plane (flat surface on which a task is to be performed) is nearly the same at equal distances from the point of maximum light intensity on the plane. With asymmetrical distribution, light is concentrated in a specific direction.

Luminaires used for general lighting usually have a symmetrical distribution. For accent lighting, however, a luminaire with an assymmetrical distribution sometimes is advantageous. For example, in a museum, to illuminate a picture on a wall, use could be made of a luminaire called a wall washer, which is recessed in the ceiling. This type of luminaire directs a broad beam of light at the picture to produce nearly uniform lighting on it, while surrounding wall areas receive much less light.

Lighting Control

An important aspect of the design of a lighting system is the type of control that is required. The simplest form of control is a simple on/off switch; more complex controls involve switching of banks or sets of lights, dimming of light intensity, directional variations, and so on. Control of daylight may involve manipulation of shading devices as the sun moves across the sky. The controls appropriate to a given situation will depend on the visual tasks or desired mood and the form of lighting equipment. Design of control systems are discussed more fully in Sec. 11.7.

Integrated Lighting Systems

Design of building lighting systems currently consists of developing an integrated combination of electrical illumination and daylighting. Daylight is used where possible and is augmented by some electrical light during daylight hours. At night, of course, any required light must be totally derived from artificial (notably electric) sources. Design of such a system calls for close coordination of the work of the architectural designer and the illumination engineer.

11.6. DAYLIGHT

Daylight itself is free, but there are costs involved in bringing it into a building and delivering adequate intensities to points where needed. To begin with, openings must be provided in the building exterior to admit the light. Structural supports are required over or around the openings. Also, to keep out the weather, but in addition, to transmit light, the openings must be glazed with a transparent or translucent material, such as glass or plastic, that is durable, weather resistant and incombustible. For economic reasons and to give building occupants an undistorted view of the outdoors, the glazing usually is thin. As a result, its thermal resistance is poor compared with that of the exterior wall or the roof. Hence, openings for daylighting increase heating and cooling costs. (Heating costs may sometimes be reduced when the windows are used as a sun trap, as mentioned in Sec. 10.3.) The life-cycle costs of daylighting therefore should be considered in cost comparisons of alternative lighting systems.

Daylight has the disadvantages of not being available between sunset and sunrise and of providing weak light on cloudy days and around twilight and dawn. When lighting is needed within a building at those times, it must be provided by artificial illumination. In such cases, the cost of luminaires and lamps is not saved by use of daylighting, although costs of power for lighting can be reduced by turning off lamps not needed when daylight is available.

Elements can be incorporated in building construction to control daylight to some extent to provide good lighting within short distances from fenestration. For example, to prevent excessive contrast or glare, windows should be shielded against direct sunlight. Preferably, use should be made of light reflected from the ground instead of skybrightness for indoor illumination. As another example, to illuminate large rooms, daylight should be admitted through more than one wall; also, through skylights or other roof openings, if possible; and ceiling, floor and walls should have good reflectance so that the light bounces off them into all parts of the room. In rooms that extend a long distance from windows, the top of the windows should be placed as close to the ceiling as possible, to permit deep penetration of the daylight. The windows also may be glazed with light-directing glass or glass block to throw daylight toward the ceiling, from which it will be reflected down to work stations away from the windows.

Design for effective and practical use of daylight requires a great deal of cooperation between the architect, structural designer, and illumination engineer. Issues regarding building form and orientation, window form, size, placement, and glazing, and any shading created by building elements—all affect the control and availability of daylight. Windows and skylights constitute structural openings and their size and location may affect the ability to use walls as bearing or shear walls or roofs as effective horizontal diaphragms. This truly represents a critical situation for the application of systems-design methods.

SECTIONS 11.5 AND 11.6
References

P. Sorcar, *Architectural Lighting for Commercial Interiors*, Wiley, New York, 1987.

J. Nyckolls, *Interior Lighting for Environmental Design*, Wiley, New York, 1983.

G. Brown, *Sun, Wind, and Light: Architectural Design Strategies*, Wiley, New York, 1984.

M. Egan, *Concepts in Lighting for Architecture*, McGraw-Hill, New York, 1983.

IES Recommended Practice of Daylighting, Illumination Engineering Society of North America.

Words and Terms

Accent lighting	General lighting
Daylighting	Indirect lighting
Direct lighting	Local lighting

Significant Relations, Functions, and Issues

Forms of lighting: (1) general, local, accent, or decorative; (2) direct or indirect; (3) artificial (electric) or natural (daylight).

Aspects of light distribution: form and placement of luminaires, use of interior surfaces and architectural features.

11.7. LIGHTING EQUIPMENT

Illumination should be provided in every space in a building in accordance with the activities to be carried out there. If zones in a room have different lighting requirements, illumination should be provided in each zone in accordance with its requirements.

After lighting requirements for a space have been determined, a strategy should be selected for meeting those requirements. Then, equipment can be selected to achieve the objectives of providing good lighting at lowest life-cycle cost.

Choice of the most suitable lamp as a light source for each space is critical to the performance and cost of the lighting system. This decision should be carefully made before a luminaire is selected.

In general, the higher the wattage rating, or power consumption, of a specific type of lamp, the greater will be the lumen output.

Greatest economy will be attained through use of the highest-wattage or highest-lumen output lamp that can be properly and comfortably applied to the space to be illuminated. In addition to costs, however, lighting quality and color of emitted light should also be considered in lamp selection.

Information on lamp characteristics and performance can be obtained from lamp manufacturers. The following information usually is useful:

Minimum mounting height of light source. This generally depends on the lamp wattage and the initial foot-candles to be applied on the work plane.

Lamp life. This is given as the probable number of hours of operation before failure.

Lamp efficiency. This is measured by the lamp output in lumens per watt of power consumed.

Lamp lumen depreciation. From tests, curves are plotted showing the gradual decrease in light output with length of time of lamp operation. Different types of lamps depreciate at different rates. This should be taken into account in lamp selection. In addition to the decrease in light output due to aging, light output is also reduced because of dirt accumulation on luminaires. This can be corrected with proper maintenance, but nevertheless should be considered in lighting design.

Lamp warm-up time. Some lamps, such as fluorescent and high-intensity discharge lamps, take some time to develop full light output. If instant illumination is needed, alternative light sources should be chosen.

Lamp re-start time. Some lamps require time to relight after they have been extinguished momentarily. The lamps may go out because of low voltage or power interruption. This characteristic, too, should be considered in lamp selection.

Light color. Different types of lamps radiate a combination of colors with different intensities. (Color is related to the frequency or wavelength of light.) Unless the lamp is designed for predominance of a specific color, such as red or blue, the resulting light usually is nearly white; for example, yellow-white or blue-white. For some applications, the tint of the light from a lamp that might otherwise be satisfactory may be objectionable. In such cases, an alternative light source should be chosen.

Electric voltages and frequency. In most buildings in the United States, electricity is distributed for light and power as an alternating current at 60 cycles per second. Voltage often is 120, but sometimes, especially for industrial buildings, voltages of 208, 240, 277 or 480 are used, because of lower transmission losses and more efficient operation of electrical equipment. In a few cases, in particular when batteries are used as a power source, direct current may be distributed. (The characteristics of alternating and direct currents are discussed in Chap. 13).

Lamps should be selected in accordance with the voltage at which they are to be operated. If a lamp is operated at a voltage somewhat larger than that for which it was designed, it usually will be brighter but its life will be considerably shortened. If a lamp is operated at a voltage slightly less than that for which it was designed, it will last longer, but it will be dimmer.

In normal practice, incoming power for lamp operation is carried by wires first to a switch and then to one or more lamps. The switch turns the lights on or off by closing or opening an electrical circuit including the power source and the lamps.

There is, however, an alternative method of operating lamps that sometimes is more economical. In this method, the circuit incorporating the lamps and the power source is closed or opened by a relay, to turn the lamps on or off. The relay, in turn, is activated by low-voltage power controlled by a switch operated by a building occupant, as desired. Control-system voltages generally range between 6 and 24 volts. The low voltage permits use of smaller wires, consumes less power and presents less of a fire or shock hazard than the higher voltages usually used. Transformers can be used to step down from the normal power-distribution voltage to the control voltage.

Noise. Some types of lamps, such as fluorescent and high-intensity discharge lamps, depend for operation on a device, called a ballast, which is housed in the luminaire. Some ballasts may hum when the lamp is operating. Whether the hum will be inaudible, audible or annoying under ordinary circumstances depends on the ambient noise level in the room in which the luminaire is installed. The ambient noise level is produced by the total of such sounds as those produced by operation of ventilating and air-conditioning equipment, motors, occupants talking or performing tasks, music or operation of equipment such as typewriters and cash registers. If the hum is louder than the ambient noise level of the room, the noise will be audible and possibly annoying. If the hum is not as loud, the ambient sounds will mask it and it will not be noticeable. Before a combination of lamp, ballast and fixture is selected, its noise rating should be obtained from the manufacturer and compared with the anticipated ambient sound level of the room in which the luminaire will be installed.

Ambient temperature. This is the temperature around a luminaire when the lamps are producing light. The temperature is higher than the ambient room temperature because of the heat generated from consumption of electrical energy in the luminaire. If the ambient luminaire temperature exceeds the rated maximum temperature for the luminaire, lamp life may be considerably reduced. For instance, operation of an electrical device at $18°F$ over its rated temperature may reduce its life about 50%, but operation at temperatures below the rated maximum will increase its life. Consequently, provision should be made in design of a lighting system for dissipation of heat produced by lamps. Also, lamps placed in a luminaire should not be rated at a wattage greater than that recommended by the luminaire manufacturer.

Lamp and Luminaire Output

Light output from a light source usually decreases with time, as mentioned previously. The initial lumen output of a lamp or of a luminaire with one or more lamps can be obtained from the manufacturer. This output should be multiplied by a lamp lumen depreciation factor to correct for effects of aging. The product equals the output after a period of time, usually at 70% of lamp life.

In addition, the output should be multiplied by a dirt-depreciation factor, to correct for the dimming effects of dirt accumulation on lamps and luminaires. Both depreciation factors are less than unity.

For convenience in lighting calculations, the two factors can be combined into a single maintenance factor M.

$$M = LLD \times LDD \qquad (11.2)$$

where

LLD = lamp lumen depreciation factor
LDD = luminaire dirt depreciation factor

Lamp output then after a period of operation is given by

$$L = L_i M \qquad (11.3)$$

where

L_i = initial lamp output, lumens

Minimum Number of Luminaires for General Lighting

From a knowledge of the use to which a space is to be put, the foot-candles of light that should be applied to the work plane can be chosen. (Recommendations of the Illuminating Engineering Society can be used as a guide in making this decision.) The lumens required on the work plane can then be calculated as the product of the area of the work plane by the foot-candles required, on the assumption of uniform lighting over the work plane [see Eq. (11.1)].

The number of lumens thus computed, however, is not necessarily equal to the required output of lamps or luminaires to be installed. Some light from lamps is lost by absorption in the luminaires. The output of light from luminaires therefore should be obtained from the luminaire manufacturer. Also, light intensity at any point on a work plane is not just the intensity of the light received directly from the luminaires. Light emitted by the luminaires bounces off ceiling, floor, walls and other objects in the space to be illuminated and then impinges on the work plane. The intensity of light at a point in the work plane is equal to the sum of the intensitites of both direct and reflected light.

The intensity of direct light received varies inversely as the square of the distance of the luminaire from the work plane. The intensity of the reflected light received depends on the shape and dimensions of the room and on the reflectance of floor, ceiling and walls. The *reflectance* of such surfaces is the percentage of the light striking the surfaces that is reflected. The combined effects of direct and reflected light in a room are expressed in a factor called coefficient of utilization.

The coefficient of utilization U is the ratio of the lumens received on a work plane to the lumen output of luminaires at a specific mounting height above the work plane and arranged to impose uniform lighting on the work plane.

$$U = \frac{L_r}{LN} \qquad (11.4)$$

where

L_r = lumens required on the work plane
L = lumen output of one luminaire
N = number of luminaires in the room

From Eq. (11.1), $L_r = A \cdot ft\text{-}c$, where A is the area of the room. From Eq. (11.3), $L = L_i M$. Substitution in Eq. (11.4) and rearrangement of terms yields

$$\frac{A}{N} = \frac{L_i U M}{ft\text{-}c} \qquad (11.5)$$

This equation gives the area per luminaire for provision of the required foot-candles on the work plane with a given initial lumen output per luminaire.

For design of a lighting system, the coefficient of utilization usually is obtained with the aid of tables and charts, which are found in catalogs of lamps and luminaires or in books on lighting. The table and charts are usually computed by the *zonal cavity method*. In this method, a room is divided into three horizontal zones, called floor, ceiling and room cavities.

The effective reflectance of each cavity is then determined from the reflectances of the floor and ceiling and from a factor called *cavity ratio*. This ratio depends on the size and shape of the room, the mounting height of the luminaires above the floor and the height of the work plane above the floor. For a specific room cavity ratio and luminaire, U can be obtained from a table, with the previously determined effective reflectances.

Suppose, for example, that 100 ft-c are to be provided on desks in a 1,000-sq ft office. The type of luminaire selected will provide initially 8,000 lumens. From the shape of the room, its dimensions and the mounting height of the fixtures, the floor, ceiling and room cavity ratios are determined. These are then used with assumed reflectances of the floor, ceiling and walls to obtain the effective reflectances of the cavities. Assume that for this case, a table gives the coefficient of utilization, based on the room cavity ratio and the effective reflectances, as 0.50. From the lamp lumen depreciation factor and the luminaire dirt depreciation factor, the maintenance factor is estimated as 0.7. Substitution in Eq. (11.5) gives

$$\frac{A}{N} = \frac{8,000 \times 0.50 \times 0.7}{100} = 28 \text{ sq ft per luminaire}$$

This indicates that the 1,000-sq-ft room will require 36 luminaires. If the room is square, the fixtures may be arranged in a square array, with six luminaires in each row and a spacing center-to-center of 5 ft.

In general, it is good practice to make the fixture spacing about equal to the mounting height of the fixture above the floor. The distance between a wall and the first line of fixtures should be about half the spacing of the fixtures.

Luminaires should be selected in accordance with the type and number of lamps and accessory equipment, such as ballasts, if needed, to be housed. For example, luminaires that are designed for incandescent lamps differ from those designed for fluorescent lamps, and the fixtures are not interchangeable.

Incandescent Lamps

These generate light by heating thin wires, or filaments, until they glow. Because the heated

metal evaporates gradually, the life of the filament is prolonged by enclosing the heated wire within a sealed glass bulb filled with an inert gas. In a tungsten-halogen incandescent lamp, the gas contains some halogens (iodine, chlorine, fluorine and bromine), which restore to the tungsten filament any metal that evaporates. This enables this type of lamp to last about twice as long as those with ordinary inert gases, with about the same efficiency in light output but less lumen depreciation.

Most of the light produced by an incandescent lamp is in the yellow to red portion of the spectrum. The color depends on the wattage at which the lamp is operated. Generally, the higher the wattage rating of the lamp, the whiter is the color. A reduction in wattage or voltage results in a yellower light.

Normally, lamps should be operated at the voltage for which they were designed. Although they are brighter at higher voltages, lamp life decreases. At lower voltages, lamp life increases but brightness decreases. For some illumination purposes, for example, to create mood effects, it may be desirable to decrease the brightness from normal. In such cases, a dimmer switch may be installed to control brightness by varying the voltage, as desired, from normal to zero.

In selection of luminaires, care should be taken that the size, shape, wattage and voltage of lamps and type of lamp base are compatible with the luminaires. Note that if lamps with wattage rating larger than the maximum specified by the manufacturer for a luminaire are installed, the excessive heat produced by the lamps may cause socket and wiring damage.

Also, in selection of lamps and luminaires, bear in mind that high-wattage incandescent lamps are more efficient in light output than low-wattage ones. For example, at 120 volts, a 100-watt lamp may generate 1,750 lumens, whereas two 50-watt lamps generate a total of 1,280 lumens. Hence, operating costs can be reduced by use of fewer but higher-wattage lamps.

The glass bulbs of incandescent lamps are often treated to obtain specific effects. The most common effect desired is diffusion of the emitted light, to reduce glare from the lamps. If a luminaire emits light with harsh shadows and veiled reflections, the condition can be corrected by use of a bulb treated to produce diffusion. The amount of diffusion depends on the treatment of the bulb. Bulbs with an inside frosting produce softer illumination than clear bulbs, but silica-coated (white) bulbs produce softer lighting than the inside-frosted type. Diffused light and other effects can also be obtained with control devices incorporated in luminaires.

Fluorescent Lamps

These generate visible light by bombarding phosphors with ultraviolet rays. The phosphors are coated on the inside surface of a sealed glass tube, which is filled with an inert gas, such as argon, and low-pressure mercury vapor. When an electric arc is passed through the mercury gas, it emits the ultraviolet rays that activate the phosphors to radiate visible light. The electric arc, in turn, is started and maintained by cathodes at the ends of the glass tube.

A device called a ballast initially provides the high voltage needed to form the arc and, after the arc has been formed, limits the current in the arc to that needed for its maintenance. Ballasts also may be designed to decrease the stroboscopic effect of the light output caused by the alternating current and to keep the variation in current nearly in phase with the variation in voltage.

Fluorescent lamps may be classified as rapid start or instant start. They differ in the method employed to decrease the delay in lighting up after a switch has been thrown to close the electric circuit. For a rapid-start lamp, the cathodes are electrically preheated rapidly. For an instant-start lamp, high voltage from a transformer forms the arc without preheating the cathodes.

Instant-start lamps may be of the hot-cathode or cold-cathode type, depending on cathode shape and voltage used. Both types are more expensive than rapid-start lamps and are less efficient in light output, but the instant-start lamps come in sizes that are not available for rapid-start lamps and can operate at currents that are not feasible for rapid-start lamps. Also, the instant-start lamps can start at lower temperatures, for instance, below $50°F$, than the rapid start.

The life of most types of fluorescent lamps is adversely affected to a considerable extent by the number of times they are started. Cold-cathode, instant-start lamps, however, have a long life, which is not affected significantly by the number of starts. They have the disadvantage of lower efficiency of light output than the hot-cathode type.

Fluorescent lamps last about ten times longer and provide more light for power consumed than incandescent lamps. Consequently, fluorescent lamps cost less to operate, but initial cost of luminaires may be larger. Fluorescent lamps require larger fixtures, because of tube length, and special equipment, such as ballasts and transformers. Consideration also has to be given to the possibility of ballast hum and interference by the electrical discharges during lamp operation with radio reception. On the favorable side, lamp brightness is relatively low, so that there is less likelihood of glare, even when the lamps are unshielded, than with incandescent lamps with the same lumen output.

Color of light produced by a fluorescent lamp depends on the coating material used in the tube. The efficiency of light output and general color effects of some commonly used types of lamps are as follows:

Deluxe cool white (CWX) lamps have medium efficiency, lumens per watt. Creating a neutral to moderately cool atmosphere in rooms illuminated, the light simulates natural daylight. This type of lamp provides the best overall color rendition of all the fluorescent types. Under its light, human complexions appear natural.

Deluxe warm white (WWX) lamps also have medium efficiency. Creating a warm atmosphere in a room, the light strengthens red, orange, yellow and green, but blue is grayed. White skin complexions appear ruddy. The light simulates incandescent light and provides good color rendition.

Cool white (CW) lamps have high efficiency. Like that from CWX lamps, the light creates a neutral to moderately cool atmosphere in rooms. The light strengthens orange, yellow and blue, but tends to gray red. White skin complexions appear pale pink. Blending with natural daylight, the light has good color acceptance.

In selection of luminaires for fluorescent lamps, care should be taken that the size, shape, wattage and voltage of lamps and type of lamp base are compatible with the luminaires. Depending on the wattage, the lamps vary in length and tube diameter. Hence, a luminaire designed for one wattage will not accept lamps of a different wattage. Also, different types of ballasts are required for rapid-start and instant-start lamps, and different transformers are needed for hot-cathode and cold-cathode lamps. For economy, a luminaire may house two fluorescent lamps served by a single ballast. Two fluorescent lamps often produce more lumens than a single lamp with the same total wattage.

Neon Lamps

These generate light when an electric current is passed through ionized neon gas contained in a sealed glass tube. A common use of such tubes, which may be bent to produce writing or designs, is for indoor or outdoor signs. These lamps require a high voltage, between 6,000 and 10,000 volts, which is obtained from normal electrical distribution voltages with a transformer. Color of the light usually ranges from pink to dark red, depending on the gas pressure, but other colors can be obtained by use of colored tubing or of mixtures of helium and neon.

High-Intensity-Discharge Lamps

These generate light by passing an electric arc through a metallic vapor contained in a sealed glass bulb. The lamps operate at pressures and current densities sufficient to produce desired quantities of light within their arc. Three types of high-intensity discharge lamps are in wide use: mercury vapor, metal halide and high-pressure sodium. They differ mainly in the material and type of construction used for the tube in which the arc is maintained and in the type of metallic vapor contained in the arc tube. The lamps require different luminaires and ballasts. In performance, the lamps differ in efficiency, starting characteristics, color rendition, lumen depreciation, price and life. The three types of high-intensity-discharge lamps are available with lumen outputs considerably greater than those of the highest-wattage fluorescent lamps available.

In general, high-pressure sodium lamps will have the lowest initial investment cost and operating cost of the three types of lamps. Metal halide lamps usually will cost somewhat more, but less than mercury vapor lamps. On the other hand, mercury vapor lamps generally will last two to three times as long as metal halide lamps, which, in turn, have a somewhat longer life than high-pressure sodium lamps. Fluorescent lamps usually have a slightly longer life than metal halide and sodium lamps.

Warm-up time for metal halide and sodium lamps may be 3 to 5 min. and for mercury lamps 5 to 7 min. Restart time for metal halide lamps, however, may be 10 to 15 min., compared to less than 1 min. for sodium lamps and 3 to 6 min. for mercury lamps.

The color of mercury lamps tends to be greenish. Deluxe white mercury lamps, however, contain additives that correct the color but with some loss in efficiency of light output. The light strengthens red, blue and yellow, while creating a warm, purplish atmosphere, but green is grayed. White skin complexions appear ruddy. Generally, color rendition is acceptable.

Acceptability of the color rendition of metal halide lamps is about the same, but the light strengthens green, blue and yellow, creating a moderately cool, green atmosphere, and grays red. White skin complexions appear gray.

Light from high-pressure sodium lamps blends with incandescent lamps but has poor color acceptance. Creating a warm, yellowish atmosphere, the light strengthens yellow, green and orange, but grays red and blue, and turns white skin complexions yellow.

Special luminaires are available for the high-intensity-discharge lamps. Some will accept either of the three types of lamps in specific wattages. Others will accept mercury or metal halides or only mercury.

Luminaires

A wide variety of luminaires are commercially available with designs capable of meeting all usual lighting objectives. Costs also vary widely, depending on luminaire size, materials and accessory equipment required.

Some luminaires are simple housings, providing support for one or more lamps, electric wires to the lamps, ballasts when needed and a surface for reflecting to the work plane light emitted by the lamps. Other luminaires may be sealed to keep out dirt. One surface of the enclosure may be plain glass for passage of emitted light, or a translucent material to diffuse the emitted light, or a lens to control distribution of the emitted light. Sometimes, louvers are installed in the luminaire opening to shield the eyes of building occupants from the glare of the lamps. Some luminaires are vented and filtered to dissipate heat and prevent accumulation of dust.

Luminaires may be classified in accordance with type and location of mounting and type of light distribution, as described in the following.

Recessed Luminaires. To illuminate a room without table or floor lamps and without projections from ceilings or walls, luminaires may be recessed in the ceiling. They send all their light downward. Thus, recessed luminaires provide only direct lighting. Figure 11.1 shows several types of recessed luminaires.

Figure 11.1a shows a wide-profile type, with an incandescent lamp. Similar luminaires may house fluorescent lamps. The lamps may be shielded with a flat diffuser or a lens that refracts the emitted light. Both devices spread out the beam for wide coverage of the work plane and to reduce brightness of the luminaire. General lighting of a room can be accomplished by installation of several such luminaires in the ceiling. A similar recessed luminaire with smaller opening, called a medium-profile luminaire, is useful for local lighting.

Figure 11.1b shows a narrow-profile type,

also known as a high hat. It is suitable for accent and decorative applications. The inside of the housing may have a dark matte finish to absorb light, so that light is directed straight down in a narrow beam, or an insert may be placed in the housing to reflect light, depending on the lighting objective. Often, a louver is placed in the opening as a shield against glare.

Figure 11.1c shows a recessed type with asymmetrical light distribution. It may be used for local or display lighting.

Ceiling-Mounted Luminaires. These are lighting fixtures attached to ceilings. This type of luminaire is useful for illuminating large rooms and when other than direct lighting is desired. As illustrated in Fig. 11.2, ceiling-mounted luminaires may be of the general-diffusing, diffuse-downlighting or downlighting type.

Figures 11.2a and b illustrate small general-diffusing types. The lamps are enclosed in a translucent material to diffuse the emitted light or in a clear glass with prisms cut into the outer surface to refract the light to reduce brightness. Figure 11.2c illustrates a large general-diffusing type. It is often used for one or more fluorescent lamps.

General-diffusing luminaires direct light in a wide pattern. They usually are used for general lighting. In selection of such luminaires, care should be taken to choose only a type that will not create glare in a specific application. Lamps should not be visible at eye level. They should be enclosed in a diffusing material. Lamps should not be too widely spaced nor too close to the diffusing element, so that no part of the diffuser will be more than twice as bright as the average brightness of the diffuser.

Figure 11.2d illustrates the diffuse-downlighting type of luminaire. The bottom surface of this type is a diffuser and the sides are

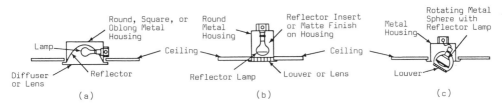

Fig. 11.1. Luminaires recessed in ceilings. (a) Wide-profile fixture with incandescent lamp. (b) Narrow-profile fixture with reflector lamp. (c) Asymmetrical luminaire.

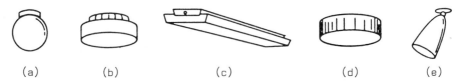

Fig. 11.2. Ceiling-surface-mounted luminaires. (*a*) Small, general-diffusing, globe type. (*b*) Small, general-diffusing, drum type. (*c*) Large, general-diffusing, fluorescent type. (*d*) Opaque-side drum, diffuse-down-lighting type. (*e*) Direct downlighting type.

opaque, or nearly so. This type does not throw light on the ceiling and therefore provides only downlighting, unless the fixture is placed close enough to walls to bounce light off them. For general lighting, there should be at least one of these luminaires for each 100 sq ft of floor area.

Figure 11.2*e* shows a downlighting type of luminaire. It is often used for accent lighting or for supplementary task lighting. Care must be taken to prevent glare, because the lighting is hard and produces sharp shadows.

Pendant Luminaires. These are lighting fixtures hung from ceilings. They generally are chosen for decorative effect. A few of the many designs used for this type of luminaire are shown in Fig. 11.3. They may be classified as general diffusing (see Fig. 11.3*a*), direct-indirect (see Fig. 11.3*b*), downlighting (see Fig. 11.3*c*) and exposed-lamp (see Fig. 11.3*d*), in accordance with softness or hardness of light emitted and with the directions in which light is thrown.

Wall-Mounted Luminaires. These are lighting fixtures attached to walls. Indoors, they may be used for accent or decorative lighting or they may be used for task lighting, for example, in bathrooms to provide light for shaving or for application of cosmetics. Outdoors, luminaires may be mounted on exterior walls to illuminate entrances and exits or for security lighting. A few of the many designs used for wall-mounted luminaires are shown in Fig. 11.4. They may be classified as exposed-lamp (see Fig. 11.4*a*), small diffuser (see Fig. 11.4*b*), linear, which may house several incandescent lamps or a fluorescent lamp (see Fig. 11.4*c*), and directional, such as the bullet (see Fig. 11.4*d*) or the night light (see Fig. 11.4*e*). The last is useful for providing enough light in a dark room for an occupant to see objects below eye level, yet dim so as not to be disturbing.

Structural Lighting. In addition to the luminaires previously described, lighting fixtures may be built into the structure of the building or built to use structural elements, such as the spaces between joists, as parts of luminaires. Structural lighting offers the advantage of having the lighting system conform to the architecture or interior decoration of a room. Basic considerations of lamp concealment, diffused brightness, light distribution and light-source color apply equally as well to structural lighting as to the other types of luminaires.

Some types of structural lighting are widely used in residences. For accent or decorative lighting, cornices, valences, coves or brackets generally are built on walls to shield fluorescent lamps. For task lighting, lamps may be

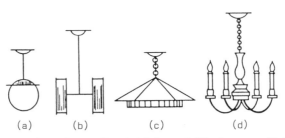

Fig. 11.3. Pendant luminaires (hung from ceilings). (*a*) General-diffusing type. (*b*) Direct-indirect type. (*c*) Downlighting type. (*d*) Exposed-lamp type.

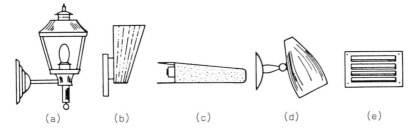

Fig. 11.4. Wall-mounted luminaires. (*a*) Exposed-lamp type. (*b*) Small diffuser type. (*c*) Linear type, for example, four-lamp incandescent or a single fluorescent. (*d*) Bullet-type directional luminaire. (*e*) Directional-type night light—behind downward-aimed, louvered grille.

housed in soffits or canopies. For general lighting, luminous ceilings may be installed.

Luminous ceilings often are used for general lighting in other types of occupancies, such as offices, as well as in residences. In such cases, the luminaire may be large enough to cover all or a substantial part of a room. The lamps are shielded by a continuous diffuser below them. The diffuser is supported on narrow bars, in turn hung from a ceiling or the underside of the floor above. Soft lighting is produced.

Portable Lamps. Lighting objectives can be partly or completely met with portable lamps in some types of building occupancies. For the purpose, a wide variety of table and floor lamps are commercially available. The lamps may be chosen for decorative or functional purposes, or both. Lighting may be direct, indirect or direct-indirect.

Since the light sources are usually mounted at a relatively low level, care should be taken to avoid glare, by appropriate placement of lamps and by selection of suitable lamp shades. Deep, narrow shades should not be used for lamps intended to provide useful illumination. The inner surface of a shade should be highly reflective. The shade material should transmit adequate light but also should be a good diffuser. (If an opaque shade is used, the room should also have supplementary lighting from other sources, to illuminate areas, such as walls, not lighted by the portable lamp.) For reading, the bottom of the shade should be at about eye level. For tasks such as the application of cosmetics, the center of the shade should be at

task level, and useful light should be transmitted through the shade onto the task.

References

B. Stein, et al., *Mechanical and Electrical Equipment for Buildings*, 7th ed., Wiley, New York, 1986.

J. Nyckolls, *Interior Lighting for Environmental Design*, Wiley, New York, 1983.

M. Egan, *Concepts in Lighting for Architecture*, McGraw-Hill, New York, 1983.

IES Lighting Handbook, Illuminating Engineering Society of North America.

Words and Terms:

Fluorescent lamp
High-intensity-discharge lamp
Illumination
Incandescent lamp
Lamp
Luminaire
Luminous ceiling
Neon lamp

Significant Relations, Functions and Issues

Lamp performance: life, efficiency, lumen depreciation, warm-up time, re-start time, color.

Luminaire design: output, number, coefficient of utilization, lamp type and size, installation, control, wiring.

Luminaire placement: recessed, ceiling-mounted, wall-mounted, pendant, portable.

11.8. SYSTEMS DESIGN APPROACH TO LIGHTING

Lighting, along with HVAC and sound control, is part of the environmental control systems within buildings. Adequate lighting is essential for the safety of building occupants at all times and in all parts of a building. Good lighting is desirable not only in the interests of safety but also to foster efficiency and to enhance the comfort of building occupants in carrying out their tasks. Also, lighting is useful in carrying out other design objectives; for example, to illuminate displays, to accent, or call attention to, objects or spaces in a room or to create desired atmospheres to influence human moods.

No matter how provided, lighting requires an initial investment. Daylighting requires glazed openings, such as windows or skylights. Artificial lighting requires lamps, luminaires, mountings, electric wiring and controls. There are no operating costs for daylight, but artificial lighting consumes energy, usually electrical, and therefore has operating costs associated with its use. In addition, artificial lighting requires maintenance, at least for lamp replacement and usually also for cleaning of lamps and luminaires periodically. Therefore, in selection of a lighting system, life-cycle costs are an important concern.

Lighting systems affect and are affected by other building systems. The interaction of the various systems requires close cooperation in building design between members of the building team.

For lighting, systems design should be employed in two ways. One way is to provide good lighting at minimum life-cycle cost. The second way is to coordinate lighting and the other building systems to produce an optimum overall design. These objectives are not incompatible. There is a wide range of conditions that can meet the objective of good lighting. This leeway makes feasible selection of a lighting system with low life-cycle cost while design of the building as a whole is optimized.

Data Collection and Problem Formulation

Basic information for lighting design comes from examination of the owner's program of requirements, type of occupancy of the building and tasks to be performed by the occupants. Additional information should be obtained from design drawings for the other building systems. Some of the information may be the result of decisions made jointly by the lighting designer and other members of the building team. This information most likely would include size, shape and location of window and roof openings for admission of daylight; glazing of the openings and other devices for control of daylight distribution; and size, shape and finishes to be used for walls, floors and ceilings. In addition, information should be obtained from the owner and the architect on lighting objectives and aesthetic requirements, as well as limitations on initial investment in lighting equipment.

Minimum illumination levels for various parts of the building may be established by the owner, required in some cases by building codes or local departments of health or labor, or adopted from national standards, such as the recommendations of the Illuminating Engineering Society. From such sources, information should be obtained for lighting levels for both normal and emergency conditions and for signs locating entrances, exits and stairs for certain types of buildings. For emergency purposes, power for lighting may have to be provided from batteries or auxiliary generators.

Because buildings usually require artificial illumination and the source of power usually is electricity, the availability and cost of a suitable supply of electric power for lighting constitutes essential information for lighting design. In some cases, it may be necessary as a result of such information to rely heavily on daylighting. For this purpose, large windows or skylights may be required, and rooms may be shaped to make effective use of the light thus admitted. Information on the type of electric power, such as alternating or direct current, frequency, voltages and phases, also is important. Often, the electricity used elsewhere in the building is also used for the lighting.

The goal of lighting design may be stated as: To design, as a component of an optimum building, the lowest-cost (life-cycle) lighting system that will provide good lighting throughout the building for the purposes for which the building is to be used.

Objectives

A primary objective of a lighting system is provision of adequate lighting throughout a building for the safety of occupants. Usually, to achieve this objective all that is required is the minimum quantity of light for visibility and color of light that permits identification of objects. This minimum lighting, however, usually is not satisfactory for performance of tasks by building occupants or for their seeing comfort; nor can such lighting meet other design objectives, such as decoration, accent and mood creation.

Another important objective therefore is provision of good lighting. This requires that:

1. The quantity of light provided be commensurate with the use to which the space is to be put and the tasks to be performed by occupants.
2. The quality of light enable occupants to observe their surroundings and their tasks comfortably.
3. The color of the light not only permit occupants to readily identify objects but also enhance the appearance of the environment and its contents.

In addition, an important objective is to provide convenient, easy control of lighting. Location of control points, such as switches, should be readily identified by building occupants. The controls that turn lights on and off or dim and brighten lights should also be safe and simple to operate. Furthermore, the controls should be placed near entrances and exits to rooms so that lights can be turned on by occupants before entering a dark room.

Still another important objective is lowest life-cycle cost for the lighting system. Operating costs for lights can be considerably larger than the initial investment for lighting equipment. Also, maintenance of luminaires and replacement of lamps can be costly over the life of a building. Hence, the aim should be optimization of the sum of the lighting costs—initial, operating, maintenance and repair. Often, optimization of operating costs provides an additional bonus by contributing to optimization of air-conditioning costs, because of decreases in energy consumption for lighting.

Consistent with this objective is the equally important objective of energy conservation. Because artificial illumination consumes electric energy and the supply of this is not inexhaustible, it is desirable that the lighting system be efficient in its use of electricity, that artificial lighting levels be no higher than necessary and that lamps be turned off when not needed.

No less important is the objective that the system be reliable. If lights should fail and daylight is not available to meet minimum requirements, the activities of occupants cannot be carried out and their safety can be endangered. For many types of occupancy, such as residences and offices, it may be reasonable to assume that the electrical supply is reliable and that if one lamp should fail some light will be available from other lamps or luminaires in the room. For other types of occupancy, such as industrial buildings and theaters, standby power should be available for emergency lighting. Lamps that fail should be replaced promptly. It may also be desirable to replace lamps on a schedule based on expected life before they burn out. In some cases, for example, with fluorescent lamps, this procedure is advisable to prevent ballast damage.

For some types of buildings, such as residences and offices, it may be important that the location of light sources be changeable as needs change or as use to which a space is put changes. In such cases, flexibility in relocating lighting equipment becomes an objective.

Constraints

Building codes or regulations of local health and labor departments may specify minimum illumination levels that must be provided for the safety of building occupants. These requirements usually are considerably less restrictive than those for good lighting and therefore can generally be easily satisfied.

For example, a building code may require that habitable rooms, with some exceptions, be provided with both natural and artificial illumination. The exceptions are likely to be rooms or spaces occupied exclusively during daylight hours and provided with daylight; also small storage rooms and small spaces for mechanical facilities containing no moving parts, no combustion equipment or no other hazardous equipment. The code may also specify that natural light should be provided by windows, skylights, monitors, glazed doors, transoms, fixed lights, jalousies or other light-transmitting media. These light sources should face or open on the sky, or on a public street, space, alley, park, highway, or right of way, or on a yard, court, plaza or space above a setback, if these elements are on the same lot as the building.

Sources of natural light may be required by the code to have an aggregate light-transmitting area of at least 10% of the floor area of the room or space served. Also, only that area of the light source above 30 in. from the floor may be considered as providing the natural illumination required in any space.

For certain critical spaces in buildings, a building code may indicate specifically what is considered minimum illumination. For example, bathrooms and toilet rooms may be required to have at least 10 ft-c at a level 30 in. above the floor. Also, illumination of at least 5 ft-c measured at the floor level should be maintained continuously in the following places when a building is occupied: in exits and their access facilities for their entire length; at changes in direction in and at intersections of corridors, and on balconies, exit passageways, stairs, ramps, escalators, bridges, aisles in theaters, tunnels, landings and platforms. (One- and two-story dwellings may be exempt from this requirement.) Illumination should be so arranged that the failure of any one light will not leave any area in darkness.

The building code may further require that when several lights, often five or more, are required for exit illumination, the electric circuits should be separate from the general lighting and power circuits. The exit-lighting circuits should either be taken off ahead of the main switch or connected to an emergency-lighting power source, if available. Phosphorescent materials may not be used to provide exit lighting. Also, battery-operated electric lights, portable lamps or lanterns may not be used as primary illumination for exits.

In addition, a building code may require that, except in one- or two-story dwellings, the location of every exit on every floor be clearly indicated by exit signs. Signs in places of assembly should be internally lighted when illumination at floor level is less than 5 ft-c during a performance. Signs in other types of occupancies may be internally or externally lighted or an electroluminescent type. The artificial light source on externally lighted signs should provide a red light by the use of a colored lamp or other coloring means that applies at least 25 ft-c on the exposed face of the sign. For internally lighted signs, the average initial brightness should be such that the letters will emit at least 25 lumens per sq ft, and where an illuminated background is used, it should emit initially an average of at least 250 lumens per sq ft. Letters of exit signs should be red. Power-source requirements for signs generally are the same as those for exit lighting.

Aesthetic Constraints. Selection of luminaires and lamps may be governed by aesthetic requirements of the owner and the architect or an interior decorator, depending on the type of occupancy and the use to which the space to be illuminated is to be put. For example, the lighting designer may not be able to specify the most efficient lamp because the color of its light may not be acceptable. The most effective luminaire may be ruled out because its appearance is not appealing, or it may be too noisy, or the lighting distribution does not create a desired atmosphere.

Aesthetics also may govern the mounting of luminaires. The owner, architect or interior decorator may, for example, want obscure light sources. In that case, recessed ceiling fixtures may be the only option. On the other hand, decorative lighting fixtures may be required. Then, pendant fixtures or wall-mounted luminaires may be the choice. In addition, the arrangement of luminaires in a room must be aesthetically appealing. Furthermore, in some cases, interesting lighting patterns or specific mood atmospheres may be required.

Power Constraints. Cost and availability of electricity for artificial illumination may be major constraints on lighting design. High cost of electricity may compel greater reliance on daylighting than when low-cost power is available. Type of electric power being supplied to other parts of the building may determine the type of lamps and luminaires that can be used for lighting.

Energy Budget. When an energy budget has been established for a building, as discussed in Sec. 10.12, the power needed for lighting becomes a component of that budget. If the budget is to be met, energy for lighting must be limited, as well as energy needed for HVAC, elevators and other energy-consuming devices. If a portion of the budget is allotted for lighting, this limitation is a constraint on design of the lighting system. To satisfy the budget requirement, not only will the lighting system have to be designed for optimum efficiency of operation but also lower illumination levels than those the lighting designer would ordinarily prefer may have to be provided. Also, special measures, such as automatic extinguishment of lights when not needed, may have to be taken to reduce energy consumption.

Interaction with Other Systems. Architectural and decoration requirements often impose major constraints on lighting design. The location, size and arrangements of such openings for daylighting as windows, skylights, monitors and clerestories and the type of glazing used may be only partly determined by lighting requirements. The appearance of the building exterior and the room interior may also be important influences. Also, HVAC requirements may be important. Hence, the lighting system may have to be designed to accommodate daylight sources and HVAC grilles already fixed in nature and location.

Provision of good lighting depends on room geometry and the reflectances of floor, ceiling and walls, as explained in Sec. 11.7. Room geometry usually is chosen by the architect. Reflectances depend on room finishes, which may be selected by the owner, architect or interior decorator, preferably in consultation with the lighting designer. Generally, however, room finishes will be a constraint on the lighting design. The effect of this constraint can be considerable. For example, a room with dark finishes may appear dimly lit with 80 ft-c on a task, whereas if the room had a white finish, it might appear overbright with 50 ft-c on the task.

Also, it is axiomatic that good lighting should be provided on tasks. But if the locations of work places are uncertain when lighting is being designed or if the locations are subject to change after the building is in use, the lighting system must either be designed to meet all possible contingencies or to be flexible, permitting movement of luminaires.

Another important constraint is that imposed by the structural system. For example, if ceiling fixtures are to be used, beams and girders that project below the mounting height of the fixtures will cut off some of the light. Similarly, columns will obstruct some of the light. Often, cooperation between lighting designer, architect and structural engineer can reduce or eliminate the adverse effects of this constraint. The structural engineer, for example, may be able to substitute a flat-plate concrete floor for the beam-and-girder construction and thus eliminate the obstructions. Or the architect may specify a flat ceiling under the beams, and the luminaires may be recessed in or mounted on the ceiling, thus reducing light cutoff.

Synthesis and Analysis

Design of a lighting system requires initially the determination of what constitutes good lighting in every space in a building. With these design criteria established, strategies can be selected to satisfy those criteria, while achieving the objectives for lighting within the limitations of imposed constraints. The objectives require both optimization of lighting costs and optimization of the building. To achieve these, it is essential that lighting be treated as part of an integrated building system.

Selection and location of lamps and luminaires are only one part of lighting design. The enclosures of spaces to be illuminated are also part of the lighting system, as are openings for admission of daylight.

Quantity, quality and color of light provided are interrelated. For good lighting, these elements must be properly balanced. Hence, lighting design must consider the effects of reflected light from floor, ceiling and walls on lighting quantity and quality.

The shape, texture and reflectance of those surfaces influence both illumination intensity on the work plane and seeing comfort. When illumination intensity is large enough to prevent eyestrain, seeing discomfort still may be caused by glare, or sharp contrast between brightness of an object to be viewed and its background. Glare may be caused by direct or reflected light. Direct glare comes from a light source and depends on the contrast of the brightness of the light source and the brightness of its background. Reflected glare comes from the reflection of a light source in a polished surface. In addition, seeing discomfort may be caused by veiled reflections, the reflection of a light source in dull or semimatte·finishes. In all cases, the result of the contrast in brightness is that an object is difficult to see. Since the reflections come from the room enclosure and the work surface, those elements should be considered part of the lighting system and designed to contribute to good lighting at lowest cost.

In the schematic-design stage of a building, aesthetics is likely to be a predominant factor in choice of window sizes and locations, but the building team should also consider the effects of this decision on lighting, HVAC, the energy budget and maintenance costs. In this early stage also, the types of lamps to be used and luminaire locations may be tentatively selected.

In the preliminary-design stage, as more information becomes available on room sizes and shapes; on ceilings, walls and floors, and on activities or tasks to be performed in the rooms, the lighting system can be designed in more detail, including the electrical supply and controls. Lighting design decisions will interact with decisions made in design of other building systems and conflicts will result. The building team should resolve these conflicts in such a way that an optimum building will be produced.

Value Analysis and Appraisal

Because lighting design interacts with the output of other members of the building team, many trade-offs between team members may be necessary for optimization of the building design. Trade-off studies should examine the performance and cost impact of all affected systems. Resulting changes in any of the systems to optimize the building preferably should be made before completion of the preliminary design stage.

Value analysis should compare alternatives not only for the lighting system but also for the other building components that affect or are affected by the lighting system. Evaluation of the alternatives should consider not only initial construction costs but also life-cycle costs and energy conservation, especially operating costs for lighting and HVAC.

Criteria for lighting intensity should be examined critically. A sufficient quantity of light is essential, but more light does not necessarily result in better lighting. Furthermore, it is the quantity of light applied to a task requiring illumination that is important. Intensity of general lighting usually is not critical except when it is necessary because of uncertainty in location of work stations or to provide for future changes in work places. Therefore, value analysts should ascertain that lighting-intensity criteria will not result in excessive lighting in any part of a building. This is important because the quantity of light provided has significant bearing on costs of lamps and luminaires, lamp operation, HVAC installation and operation, and the electrical installation.

Alternative lamp and luminaire combinations that will provide good lighting for each space in the building should be investigated. Generally, the combination providing the most lumens per watt of power consumed will have the lowest operating cost, but initial, maintenance and lamp replacement costs also must be taken into account in evaluations of alternatives.

The greatest opportunity for cost savings lies in integration of the lighting system into other building systems. For example, light sources may be made a part of the ceiling or wall systems. Figure 11.5 shows an instance where soft, general lighting is provided in an office by fluorescent fixtures mounted on modular ceiling panels. The panels are called *modular* because they are prefabricated in standard sizes and are coordinated to fit together on the build-

Fig. 11.5. Modular ceiling panels equipped with fluorescent luminaires and integrated with linear air-conditioning diffusers.

ing site without the need for cutting. In Fig. 11.5, the ceiling panels also are coordinated for assembly with linear air-conditioning diffusers. In other buildings, similar panels are coordinated for assembly with panels containing other types of air-conditioning grilles. Other types of lighting may be similarly integrated in ceilings.

Value analysts should also investigate methods of prevention of dirt depreciation of luminaires. In some cases, selection of self-cleaning luminaires that cost more than other types initially may prove more economical in the long run by reducing initial lumens required and hence initial, operating and maintenance costs of lighting.

Lighting and Building Design

Of all the areas of building engineering, it is hard to imagine one that is more crucial to the architectural design than lighting. The building form, detail, and general spatial disposition can not be experienced and appreciated if it cannot be *seen*. This applies to both the inside and outside of the building. Good lighting for the building interior—and, where applicable for night viewing or maneuverability, the building exterior—is essential as an architectural design component. Experienced architectural designers realize this and many of them take very active parts in the development of building lighting systems. Whenever the entire lighting design effort is assigned to illumination engineers, whatever their level of sensitivity to architectural problems, it is difficult to realize the architect's design intentions fully.

If major use of daylighting is a significant factor in the design of a lighting system, the architectural designer must place the requirements for logical placement of natural light-gathering components near the top of the ordered set of values for the design work. Choice of window type, size and materials, sun orientation, and use of building features for sun shading or reflection must be done to optimize the effective use of daylight. As in any design work, no single value can totally override all others, but effectiveness of a daylight-utilizing system is as unyielding as the laws of gravity.

GENERAL REFERENCES AND SOURCES FOR ADDITIONAL STUDY

These are books that deal comprehensively with several topics covered in this chapter. Topic-specific references relating to individual chapter section are listed at the ends of the sections.

B. Stein, et al., *Mechanical and Electrical Equipment for Buildings*, 7th ed., Wiley, New York, 1986.

J. Nyckolls, *Interior Lighting for Environmental Design*, Wiley, New York, 1983.

M. Egan, *Concepts in Lighting for Architecture*, McGraw-Hill, New York, 1983.

IES Lighting Handbook, Illuminating Engineering Society of North America.

G. Brown, et al., *Inside Out: Design Procedures for Passive Environmental Technologies*, Wiley, New York, 1982.

W. Turner, *Energy Management Handbook*, Wiley, New York, 1982.

EXERCISES

The following questions and problems are provided for review of the individual sections of the chapter.

Section 11.1

1. Why should areas around the exterior of a building be lit up at night?
2. Why should switches for lights be placed at the entrance and exit of every room?
3. How should the locations of emergency exits be indicated?

Section 11.2

4. What is the advantage of increasing the intensity of lighting in a room?
5. What are the disadvantages of increasing the intensity of lighting in a room above the minimum level needed for the activi-

ties to be comfortably and easily performed?
6. How does lighting intensity usually affect human moods?
7. An average of 80 ft-c are measured on the top of a 30×60 in. desk. How many lumens are required to produce that intensity?
8. A lamp produces 100 ft-c on a surface 10 ft away. What is the candlepower of the light source?

Section 11.3

9. How is quality of light determined?
10. How is soft lighting produced?
11. How is hard lighting produced?
12. What is the difference between soft and hard lighting in the effects produced on humans?
13. How is glare produced and why is it considered poor lighting?

Section 11.4

14. Why is color of light important for visibility?
15. What is the relationship between color and light intensity?

Section 11.5

16. What are the purposes of:
    (a) General lighting?
    (b) Local lighting?
    (c) Accent lighting?
17. How does indirect lighting differ from direct lighting?
18. How may indirect lighting be achieved with a luminaire?
19. How do luminaires control light distribution?
20. Under what conditions is an asymmetrical distribution of light by a luminaire advantageous?

Section 11.6

21. In a comparison of the costs of natural and artificial illumination, what costs should be included for use of daylight?

22. How can glare from windows be controlled?

23. How can daylight be obtained for reading at places in a room not close to a window?

Section 11.7

24. In selection of components of a lighting system, which component should be selected first?

25. In choosing a lamp wattage, would it be more economical to specify two 30-watt fluorescent lamps or one 60-watt lamp of the same type?

26. A 100-watt incandescent lamp may be supplied inside-frosted, white or silver bowl. They are rated respectively at 17.5, 17.1, and 14.5 lumens per watt initially. Which type will have the greatest efficiency of light output?

27. What are the effects on a lamp of operating it at a slightly lower voltage than the design voltage?

28. Why is dissipation of heat generated by a lamp important?

29. A lamp chosen for a lighting installation has a lumen depreciation factor of 0.85. The luminaire housing the lamp has a dirt depreciation factor of 0.80.

(a) What is the luminaire maintenance factor?

(b) If the luminaire has an initial output of 4000 lumens, what will the light output be at about 70% of lamp life?

(c) If the coefficient of utilization for the room is 0.25 and 80 ft-c are required on the work plane, how many square feet of work-plane area can be assumed in design to be illuminated by the luminaire?

30. How is light produced in an incandescent lamp?

31. What determines the color of light in:
(a) An incandescent lamp?
(b) A fluorescent lamp?

32. Under what conditions would a high-intensity discharge lamp be chosen instead of a fluorescent lamp?

33. What type of luminaire should be used when the light source in a room is required to be obscure?

34. What is the difference in lighting distribution between that produced by a diffusing luminaire and that produced by a downlighting luminaire?

Chapter 12

Sound and Vibration Control

Sound and vibration control, with HVAC and lighting, are parts of the environmental control systems of a building. Vibrations are undesirable. Some sounds also are objectionable, because they interfere with communication or the enjoyment of quiet or of music, or because they are meaningless or distracting. Such unwanted sounds are called *noise*. In contrast, certain sounds are necessary for communication or are desirable, because, as in the case of music, they produce pleasing sensations. Sounds useful for communication or for music are called *signals*.

Vibration control is necessary in a building to prevent annoyance of occupants, to eliminate noise due to rattling of loose objects, or to avoid cracking or breaking of the parts or contents of a building because of the cyclic movements.

Sound control is necessary to insure distinct hearing of conversation and music and to enhance musical sounds. Also, sound control is essential to eliminate noise or reduce it to an acceptable level. Noise can be an annoyance, block out signals, injure human hearing, cause psychological disorders or otherwise impair human health. For these reasons, building codes and the regulations of several government agencies set limits on the intensity of noise permitted in buildings. These limits affect such installations as HVAC, lighting, plumbing, elevators, escalators, motors and factory machinery.

Design for vibration and sound control requires initially establishment of design criteria limiting the intensities and frequencies of vibrations and sounds that will be permitted. Next, sources of sound and vibration and the nature of these are determined. Then, control methods can be selected and controls installed. The installation should contribute to optimization of the building design.

Effective, low-cost control of sound and vibrations requires a knowledge of:

1. Sound and vibration sources
2. Methods of controlling production of sounds and vibrations
3. Acoustic principles
4. Acoustic properties of materials
5. Human physiology and psychology
6. Methods of controlling transmission of sounds and vibrations

Except possibly for very simple buildings in quiet neighborhoods and with quiet activities, sound control usually requires knowledge of an expert. Similarly, vibration control often requires expert advice. For this purpose, the building team should call on the services of consultants who specialize in those fields.

12.1. NATURE OF SOUNDS AND VIBRATIONS

Vibrations originate with expenditure of power that produces a local molecular disturbance. In such a disturbance, a molecule moves until it collides with another molecule, reverses direction, moves until it collides with another molecule and continues to repeat the process. The original molecule need not move much in either direction from its initial position, but because the molecules involved in the collisions disturb other molecules and all move back and forth about their initial positions, the disturbance is propagated rapidly away from the point of application of power.

In solid materials, vibrations may travel in the form of a stress wave in which compression alternates with tension. In gases, vibrations may propagate in the form of a pressure wave in which compression alternates with rarefaction.

The velocity of spread of the disturbance, or wave, is distinct from the motion of the molecules producing it. The wave velocity is proportional to the square root of the ratio of the modulus of elasticity to the density of the media through which the vibration is transmitted.

Sounds are vibrations that are audible. Sounds, therefore, like vibrations, have the characteristics of waves. They travel through air at a velocity of about 1,100 fps at sea level, through water at about 4,500 fps, and through steel or aluminum at 16,000 fps.

Wave Characteristics

For a better understanding of the nature of sounds and vibrations, the characteristics of waves are briefly reviewed in the following.

A complete vibration, that is, a compression followed by tension or rarefaction, is called a *cycle*.

Frequency of a vibration equals the number of cycles that occur per unit of time. The unit of frequency usually is cycles per second (cps) or hertz (Hz).

Wavelength of a vibration is the distance between two successive compressions or rarefactions.

From these definitions,

$$\lambda = \frac{v}{\omega} \qquad (12.1)$$

where

λ = wavelength, ft
v = velocity of wave, fps
ω = frequency, Hz

Amplitude of a vibration equals the maximum compression, tension or rarefaction. It is often called the *intensity* of vibration.

If, at any instant, a wave can be represented by a sine curve, the wavelength would be twice the distance between points where the sine becomes zero, and the amplitude would be the maximum height of the curve above, or below, the zero points.

Wave Propagation

If the source of a vibration is a point in an open space, the vibration would propagate in all directions. As a result, the intensity of the vibration varies inversely as the square of the distance from the source. Velocity, wavelength and frequency, however, do not change with distance from the source.

A vibration may be initiated in and transmitted through a solid, liquid or gas. When a vibration transmitted through one medium encounters a different medium, some of the energy in the vibration may be transmitted through the new medium as a vibration and some of the energy may be absorbed, or lost, as heat in that medium. Also, some of the energy may be reflected back to the original medium as a vibration with the original wavelength and frequency.

Waves can readily cross each other or travel along with each other. Where two waves are combined, either at an intersection or by movement along the same path, the amplitude of the resultant at any time is the algebraic sum of the amplitudes of the individual waves at the time at the point of combination. (When this addition is performed, compression may be assigned a positive sign, and tension or rarefaction, a negative sign.) Accordingly, when two waves of different wavelength or frequency move along the same path, they combine into a resultant wave with wavelength or frequency different from those of the original waves.

In general, sounds and vibrations are propagated as a combination of waves with different intensities and wavelengths or frequencies. As a result, at any point away from the source of the disturbance, the intensity of the resultant wave changes abruptly with time, often from compression to tension or rarefaction and back to compression. Also, the frequency of the resultant wave is irregular. Computations of the effects of such a wave would be very complicated. Hence, in practice, simplifying assumptions concerning the variation of intensity and frequency with time are often made.

Waves have another property that is of importance in sound control. As mentioned previously, when a wave is transmitted from one medium to another, part of the wave energy is reflected as a similar wave of lesser intensity. If nearly all of the energy is reflected, the new medium may be considered a *barrier*. For example, when sound being transmitted through air strikes a massive wall, the wall acts as a barrier off which the sound bounces. Now, if there should be a small opening in the wall, sound will pass through the opening. On the other side of the wall, the sound will spread out in all directions. This property is called *diffraction*.

12.2. MEASUREMENT OF SOUNDS

The elements of importance in sound transmission are the source of the sound, the path the sound wave travels and the receiver. The last is usually the human ear. Consequently, the reactions of humans to sounds reaching their ears, as well as the characteristics of sound waves, must be a basis for measurements of sounds.

Experiments have shown that human sensitivity to stimuli usually is a function of the ratio of characteristics of the stimuli. For example, response to arithmetic differences in intensity of a stimulus is not as significant as the response to doubling or halving the intensity. Consequently, measurements of sound usually are based on the ratios of quantities being measured. In measuring frequencies, for instance, a common unit is the octave.

An **octave** is the difference in frequency between a sound wave at a specific frequency and another sound wave at a frequency twice, or half, as large.

Frequency Bands

Sound sources generally emit energy in many frequencies. The effects of these frequencies on construction materials and the human ear depend on the frequency. Hence, measurement of the total energy emitted by a sound source in all the frequencies cannot give an accurate indication of the effects that will be produced by the sound wave. On the other hand, it is impractical to measure the energy in every frequency emitted so that the effects of the sound wave can be predicted. Therefore, it has become common practice to make measurements of sound in the audible range in nine frequency bands, each an octave wide. Measurements are made at the center of each band.

Frequencies at which measurements are usually made have been standardized at 31.5, 63, 125, 250, 500, 1,000, 2,000, 4,000 and 8,000 Hz. Frequencies less than about 400 Hz are considered low frequencies, and those above about 1,500 Hz, high frequencies, for acoustical purposes. Frequencies between 400 and 1,500 Hz are considered middle frequencies. (For more detailed acoustical analyses, the audible range may be divided into half-octave, third-octave or smaller bands.)

Intensity and Power Levels

The ear recognizes sounds because of pressure changes in the air. The response varies nearly logarithmically with ratios of sound intensity.

Intensity is the power per unit of area applied by a sound wave and varies as the square of the sound pressure.

Intensity level is a measure of human response to intensity.

Intensity levels are measured in *bels*, or more commonly in *decibels* (dB), one-tenth of a bel. The intensity level of a sound at a given point and given time may be computed from

$$IL = 10 \log_{10} \frac{p^2}{p_o^2} = 20 \log_{10} \frac{p}{p_o} \quad (12.2)$$

where

IL = intensity level, dB
p = sound pressure measured, Pascals (Newtons per sq m)
p_o = reference pressure = 2×10^{-5} Pa for airborne sound

Thus, intensity level is measured by the ratio of sound pressure to the reference pressure. This reference pressure has been established at the average threshold of human hearing of sound at about 1,000 Hz and is equal to one-millionth of standard atmospheric pressure (one microbar).

When $p = p_o$, the logarithm of p/p_o equals zero. Therefore, the intensity level is zero. Hence, zero intensity level occurs at the threshold of human hearing.

Because sound intensity varies as the square of sound pressure, Eq. (12.2) can also be given as

$$IL = 10 \log_{10} \frac{I}{I_o} \qquad (12.3)$$

where

I = intensity measured, watts per sq cm
I_o = reference intensity = 10^{-16} watts per sq cm

When intensity is known, sound power at a source can be computed from the inverse-square law (see Sec. 12.1). At a distance d, ft, from the source, the power is uniformly distributed over a spherical surface with radius d. The sound intensity is the power per unit of area on the surface. Since the area of the surface is $4\pi d^2$, sq ft, and intensity is given in units of watts per sq cm, with 1 sq ft = 929 sq cm, sound power, watts, may be computed from

$$W = 929 \times 4\pi d^2 I = 11{,}674 d^2 I \qquad (12.4)$$

Suppose, for example, that sound intensity 10 ft away from a source of sound is 10^{-6} watts per sq cm. Then, by substitution in Eq. (12.4), the power of the source is

$$W = 11{,}674 \times 10^2 \times 10^{-6} = 1.17 \text{ watts}$$

Sound power level is defined in a manner similar to intensity level.

$$SPL = 10 \log_{10} \frac{W}{W_o} = 20 \log_{10} \frac{p}{p_o} \qquad (12.5)$$

where

W = power measured, watts
W_o = reference power = 10^{-12} watts

Equation (12.5) gives SPL in decibels. Note that SPL applies to human perception of a sound source whereas IL applies to human perception at a distance from the source.

For sound control, components of a building may be modified or combined with other materials to reduce the intensity of sound reaching a receiver. Usually, the objective is to reduce noise.

Noise reduction, dB, can be computed from

$$NR = 10 \log_{10} \frac{I_2}{I_1} = IL_2 - IL_1 \qquad (12.6)$$

where

I_1 = sound intensity, watts per sq cm, before sound treatment
IL_1 = intensity level, dB, before sound treatment
I_2 = sound intensity, watts per sq cm, after sound treatment
IL_2 = intensity level, dB, after sound treatment

Addition of Decibels

When sound from two sound sources and at the same frequency are combined, the result is a louder sound, but the intensity level is not the sum of the intensity level for each source. For example, if two musical instruments simultaneously play the same note at 50 dB, the intensity level is not 50 + 50 = 100 dB, but only 53 dB. This can be seen from Eq. (12.3).

Suppose n identical instruments produce simultaneously at a given frequency a sound intensity of nI. From Eq. (12.3), the intensity level is

$$IL_n = 10 \log_{10} \frac{nI}{I_o} = 10 \log_{10} n$$

$$+ 10 \log_{10} \frac{I}{I_o} = 10 \log_{10} n + IL$$

$$(12.7)$$

Thus, for the case where sound intensity is doubled, $n = 2$, $\log_{10} 2 = 0.3$, and $IL_2 = 10 \times 0.3 + IL = 3 + IL$.

Consequently, when the intensity or power level for a combination of two sounds is to be determined, the intensity for each sound should be substituted in Eq. (12.3). For example, suppose an intensity level of 60 dB is imposed on an intensity level of 65 dB. From Eq. (12.3), the intensity corresponding to 60 dB is $I_1 = 10^6 I_o$, and the intensity corresponding to 65 dB is $I_2 = 10^{6.5} I_o$. The intensity level for the combination of the sounds then is

$$IL_c = 10 \log \frac{I_1 + I_2}{I_o}$$

$$= 10 \log (10^6 + 10^{6.5}) = 66.2 \text{ dB}$$

Sound-Level Meters

Measurements of sound intensity or power levels are usually made with an electronic instrument that records sound pressures, but is calibrated in decibels. A meter can be used to indicate the intensity level over a frequency range from 20 to 20,000 Hz or over a narrow frequency band, such as one with a width of an octave or less. For different purposes, the electric circuits can weight the recordings for different frequencies and combine the results in a single intensity or power-level reading.

A commonly used weighting is registered on a scale called the A scale. The weighting is intended to make the meter respond to sounds like the human ear does. The ear is relatively insensitive to low frequencies and most sensitive to the middle frequencies. Hence, the meter applies greater weights to intensities of sounds in the middle frequencies than to those in the other frequencies. Intensity and sound power levels read on the A scale are indicated in dBA units.

SECTIONS 12.1 AND 12.2

References

B. Stein et al., *Mechanical and Electrical Equipment for Buildings*, 7th ed. Wiley, New York, 1986.

L. Beranek, *Noise and Vibration Control*, McGraw-Hill, New York, 1971.

L. Beranek, *Acoustic Measurement*, Wiley, New York, 1949.

Words and Terms

A scale
Amplitude
Cycle
Decibel (dB)
Diffraction

Frequency
Frequency band
Hertz (Hz)
Intensity level (*IL*)
Noise
Octave
Signal
Sound barrier
Sound power level (*SPL*)
Vibration
Wave
Wavelength

Significant Relations, Functions, and Issues

Wave characteristics: cycle, frequency, wavelength, amplitude.

Aspects of wave propagation: decay with distance from source, overlapping of separate waves, reflection, barriers, diffraction.

Measurement considerations: frequency (octave, bands, Hz), intensity level (bel, decibel), sound power (watts).

12.3. ACOUSTIC PROPERTIES OF MATERIALS

Sound may travel simultaneously over several paths from a sound source to a receiver. Because of this, the same sound wave arrives at the receiver at different times. As a result, signals may become indistinct, or annoying sounds may be received. Materials with desirable acoustic properties are therefore used to prevent these disturbances.

Sound may travel directly from source to receiver or may be reflected from barriers, such as walls, floor and ceiling. In either case, the intensity of the sound varies inversely as the square of the length of path the sound wave traverses. In a large room, distance from a source can be large enough to make the reduction of sound level significant. In a small room, the decrease in sound intensity with distance from the source is not significant for waves traveling directly to the receiver.

Sound waves that bounce off the room enclosures before reaching the receiver, however, may reinforce each other where they overlap. Thus, faint sounds may become distinctly audible. Consequently, an effective means for reducing the intensity of reflected sounds is to cover one or more of the room enclosures with a sound-absorbing material.

For noise reduction in general, an effective measure is erection of a barrier between the sound source and the receiver. If sound cannot get over, under or around the barrier, noise can reach the receiver only by going through the barrier. Different materials and types of construction may be used as barriers, and they differ in ability to block noise.

Sound Transmission across Barriers

When a sound wave strikes a room enclosure, the wave energy causes the enclosure to vibrate. The vibrating enclosure, in turn, causes the air on the other side of the enclosure to vibrate, producing sounds. The intensity of the far-side sound produced depends on the frequencies of the initial impinging wave, the mass of the barrier set into vibration and the stiffness of the barrier.

Sound-transmission loss in a barrier, dB, is the difference between the intensity level of the sound incident on the barrier and the intensity level of sound reradiated on the far side of the barrier.

The more massive and solid a barrier the greater the sound-transmission loss. At certain frequencies, however, sound-transmission loss is very sensitive to the stiffness of the barrier and decreases with increase in stiffness.

Thus, for a single, solid panel, the sound-transmission loss for low-frequency, or bass, sounds depends on the mass of the barrier primarily. In this range, the transmission loss in the panel is roughly proportional to the sound frequency and increases at about 6 dB per octave increase in frequency. For a single, solid panel, the sound-transmission loss for high-frequency, or hissing or whistling, sounds also depends greatly on the mass of the barrier. Again, the transmission loss is roughly proportional to the frequency of the sound. The sound-transmission loss for middle-frequency sounds, however, is affected by the panel stiffness and may change very little with change in sound frequency. For a double-layer panel, es-pecially one with sound-absorbing material between layers, sound-transmission loss will still vary with frequency but usually in a different way than for a single-layer barrier.

As a result of these characteristics of barriers, sound-transmission losses in actual construction components are determined experimentally. Usually, components are evaluated and rated by methods given in standards of the ASTM and the Acoustical and Insulating Materials Association. The tests are usually performed with sounds at several specific frequencies, but it would be difficult to compare different types of construction or to specify required noise-reduction performance if the ratings for construction components were given at several frequencies. Instead, a single rating is generally given for each component to represent its performance over the whole spectrum of sound frequencies. One method of representing the ability of a barrier to isolate sounds places components in a *sound-transmission class* (STC).

The *STC* rating is a single number that represents the ability of a barrier to isolate sounds; that is, to reduce the intensity level of airborne sounds that impinge on it and pass through. The higher the *STC* to which a construction component is assigned, the greater is its ability to isolate sounds.

When two types of components are combined in a barrier, such as a door or a window in a wall, the sound-transmission loss for the combination depends on the acoustic properties of each component. In general, however, the overall transmission loss for the combination will not differ much from that of the component with the smallest loss. Consequently, where sound isolation is important, care should be taken that the performance of a barrier is not impaired by incorporation of openings or of materials with small transmission losses.

Typically, $\frac{1}{4}$-in. plate glass is in *STC* 26; $\frac{3}{4}$-in. plywood in *STC* 28; a 2 × 4 wood stud wall faced on two sides with $\frac{1}{2}$-in. gypsumboard in *STC* 33; an 8-in. reinforced concrete wall in *STC* 51; and a cavity wall with two 6-in. concrete-block wythes separated by a 2-in. air space in *STC* 56. Data on acoustical performance of various building materials and assemblies can be obtained from their manufacturers or from the Acoustical and Insulating Materials Association. See also "Sound Insulation of Wall, Floor and Door Construction," National Bureau of Standards Monograph 77, Nov. 30, 1964, U.S. Government Printing Office, Washington, D.C. 20402.

Vibration Transmission through a Structure

Sound and other vibrations may be transmitted from their source to distance parts of a building through the structure. The vibrations may be carried by floors, walls, ceilings, beams, columns, ducts and pipes, if precautions are not taken to prevent passage.

Unless isolated properly from the building structure, any moving, rotating, oscillating or vibrating equipment will transmit some energy to the structure. This energy will be transmitted as vibration or sound from the source along any continuous, rigid building component and similar components connected to it. In distant parts of the building, the energy may be sensed as annoying vibrations or as noise.

To prevent these disturbing effects, the source should be isolated to the greatest extent possible from the building structure and potential paths for vibration transmission should be interrupted. Isolators and interrupters should be energy-absorbing materials or devices, such as air spaces, viscous materials, springs, rubber, glass fiber or cork. Springs are the most effective resilient isolators, because they can accommodate large movements.

Noise also may be transmitted across a barrier by an impact. Typical producers of impact noises are dropped objects, hammer blows, walking with hard-heeled shoes and furniture being shuffled along a floor. The noises may be carried across the building component struck and through the building structure. Different types of construction have different abilities to isolate these noises. One method of representing such capabilities places building materials or assemblies in an *impact insulation class* (*IIC*).

The *IIC* is a single number that represents the ability of a barrier to isolate impact sounds. The higher the *IIC* to which a construction com-

ponent is assigned, the greater is its ability to isolate impact sounds. Tests for establishment of *IIC* usually are made on floor-ceiling assemblies. In these tests, the top surface of the floor is tapped, and the sound pressure level is measured below the ceiling.

An alternative rating system sometimes used yields an impact noise rating (*INR*). The rating is a single number nearly equal to *IIC* − 51. Thus, if a floor-ceiling assembly is in *IIC* 61, its *INR* is 61 − 51 = 10.

Typically, an assembly consisting of oak flooring, $\frac{1}{2}$-in. plywood subfloor, 2 × 10 wood joists and $\frac{1}{2}$-in. gypsumboard ceiling is in *IIC* 23. If the floor is carpeted, *IIC* rises to 48. Similarly, an 8-in. concrete floor slab is in *IIC* 35, but when it is carpeted, *IIC* increases to 57. These results indicate that a thick carpet over a good pad is an effective isolator of impact sounds. Information on acoustic performance of damping materials, sound isolators, and vibration interrupters can be obtained from manufacturers.

Sound Absorption

To reduce the intensity level of sounds reflected from room enclosures, the surfaces may be covered with acoustical absorbents, materials that have a large capacity for absorbing sound and converting the sound energy to heat. These materials usually are lightweight and porous or fuzzy and are available as tile, boards, blankets and panels. They convert sound to heat either by pumping of air contained within the porous structure of the material or by flexing of thin sheets.

Porous sound absorbents that are frequently used for acoustical purposes contain a random matrix of fibers or particles and interconnected pores or capillaries. For sound absorption, the air within the matrix must be able to move to permit sound energy to be dissipated by friction between air molecules and the matrix fibers and particles.

Good sound absorbents, because of their porous structure, may be also good transmitters of sound. Sound absorbents, therefore, should not be used mistakenly to improve the airborne-sound isolation capacity of a barrier. The purpose of sound absorbents is to reduce the intensity level of reflected sounds.

Sound absorbents may be made with vegetable or mineral fibers, porous or granular aggregates, foamed elastomers or other materials with the proper internal construction. (Tuned chambers, with small openings and a restricted neck leading into the chambers, also are sometimes used for sound absorption.) The exposed surface of sound absorbents may be smooth or textured, fissured (see Fig. 12.1) or perforated, or decorated or etched in many ways.

Because of their fragile structure, sound absorbents may be covered with sturdy perforated or porous facings for protection. Thin perforated facings with closely spaced openings do not appreciably impair the acoustical performance of most absorbents.

The capacity of a material for absorbing sound, or *absorptivity*, depends on thickness, density, porosity and other characteristics. Absorptivity also varies with frequency of the impinging sound. A thin layer of sound absorbent may have high absorptivity for high-frequency sounds, but a thick layer may be needed to absorb low-frequency sounds. Absorptivity usually may be increased at all frequencies by increasing the thickness of the sound absorbent.

Absorptivity can be represented by a *sound-absorption coefficient*, usually derived from tests. The coefficient indicates the fraction of the incident sound energy at a specific frequency that a material absorbs. A material with a coefficient of zero absorbs no energy, whereas one with a coefficient of unity is a perfect absorber at that frequency. Most sound absorbents have a coefficient larger than about 0.2. (In comparisons of sound absorbents, differences in coefficients of less than about 0.1 are not significant).

Tests of absorptivity are usually performed with sounds at several specific frequencies; however, it would be difficult to compare different sound absorbents or to specify required noise reduction if the ratings for different materials were given at different frequencies. Instead, it is convenient to give a single number as a rating for each material to represent its performance over the whole spectrum of sound frequencies. One such rating is the *noise-reduction coefficient (NRC)*.

NRC is computed as the average of the sound absorption coefficients of a material, computed to the nearest 0.05, at frequencies of 250, 500, 1,000 and 2,000 Hz.

Because *NRC* is an average, it may not be representative of a material's acoustical performance at low and high frequencies. Also, two materials with the same *NRC* may have different absorptivity at the same frequency.

Sound-absorption and noise-reduction coefficients may be obtained from manufacturers' catalogs of acoustical materials or from "Performance Data Bulletins" of the Acoustical and Insulating Manufacturers Association.

Some common decorative materials also may serve as sound absorbents. For example, carpets and draperies may have noise-reduction coefficients ranging from about 0.2 to 0.5. (These materials, however, may have low absorptivity at low frequencies.) In theaters and auditoriums, also, seats and audiences may provide good absorption. Upholstered seats may have sound-absorption coefficients ranging from 0.5 to 0.7, depending on the frequency. Seated audiences may have sound-absorption coefficients ranging from 0.6 to 0.95, depending on the frequency.

Sound-absorption and noise-reduction coefficients measure the absorptivity of a 1-sq ft exposed surface of a sound absorbent, in units called sabins. Hence, for a sound-absorbing surface of any area,

$$a = A\alpha \quad \text{or} \quad a = A\,NRC \qquad (12.8)$$

where

   a = sound absorption, sabins
   A = area against which sound impinges
   α = sound-absorption coefficient for a specific frequency
NRC = noise-reduction coefficient

Consequently, to find the sound absorption for a room, Eq. (12.8) should be applied to each surface or parts of surfaces of different materials exposed to sound and the products of coefficients and surface areas should be totaled.

Because reflected sounds add to sounds traveling directly from source to receiver, the change in sound intensity level in a room is proportional to the change in total room absorption, sabins. From Eq. (12.6), therefore,

Fig. 12.1.   Installation of a fissured acoustical tile in a coffered ceiling.

noise reduction after treatment of a room with sound absorbents may be computed from

$$NR = 10 \log_{10} \frac{a_2}{a_1} \qquad (12.9)$$

where

NR = reduction in sound intensity level, dB, in the room

a_1 = total room sound absorption, sabins, before treatment

a_2 = total room sound absorption, sabins, after treatment

12.4. SOUND AND VIBRATION DESIGN CRITERIA

Acoustical design criteria should be established for all buildings to insure that an acoustical environment satisfactory for most occupants will be maintained. For some conditions, design criteria have been established by local building codes or government agencies. Such criteria generally limit the noise or vibrations that will be legally tolerated. In addition, criteria are necessary to insure distinct hearing and for enhancement of music.

Preferably, criteria should establish objective standards of performance. Human responses to sounds and vibrations, however, are mostly subjective. Different persons react differently to different frequencies and intensities. As a result, acoustical criteria are generally subjective. They usually establish broad classes of standards or ranges of values that will satisfy most persons under conditions that will probably prevail. In general, values that deviate 2 to 3 points or dB from design criteria may be considered acceptable.

Human Responses to Sounds and Vibrations

The principal response of humans to ordinary sounds is the sensation of hearing. (Additional sensations, ranging from pressure in the chest to pain in the ears, may be felt at very low frequencies or at very high sound intensities.) The ear relates objective, or physical, characteristics of sound to subjective judgments. As a result, humans relate:

Sound amplitude, pressure or intensity to loudness
Sound frequency to pitch
Spectral distribution of acoustic energy to quality

Loudness is the ear's interpretation of sound intensity level. The larger the intensity level, the louder a sound is considered to be. Table 12.1 indicates how humans generally relate various common sounds to intensity and power levels, dBA. Judgment of loudness, however, also depends on frequencies of sounds.

Table 12.1. Human Judgment of Sound Intensity and Sound Power Level *SPL*

Type of Sound	Relative intensity	*SPL*, dBA	Loudness
Very faint	1	0	Threshold of hearing
	10	10	Rustle of leaves in wind
	10^2	20	Whisper
Faint	10^3	30	Inside a bedroom
Moderate	10^4	40	Inside a private office
	10^5	50	Inside an office with several occupants
	10^6	60	Near an express highway
Loud	10^7	70	Face-to face conversation
	10^8	80	Restaurant with nonabsorbent walls
Very loud	10^9	90	Orchestra or noisy factory
	10^{10}	100	Loud auto horn at 10-ft distance
Deafening	10^{11}	110	Near an accelerating motorcycle
	10^{12}	120	Threshold of feeling or noisy band
	10^{13}	130	Threshold of pain
	10^{14}	140	Artillery fire

Pitch is the ear's interpretation of sound frequencies. Humans identify low frequencies as low pitch, or bass, and high frequencies as high pitch, or tenor, soprano or whistle. For example, middle C on the piano has a frequency of 261 Hz. The note an octave below is at 130 Hz and at an octave above, 532 Hz. The audible range of frequencies is from about 16 Hz to less than 20,000 Hz. Thus, for practical purposes, humans are deaf to very low and very high frequencies. Also, the ear interprets high-pitch sounds as louder than low-pitch sounds of the same intensity level. Thus, acoustical design criteria should take into account frequencies of sounds as well as their intensity level.

Humans consider that they hear distinctly when they can detect signals, or wanted sounds, clearly, despite the presence of noise. Weak sounds can be heard if the level of the background noise is low. If the noise level is increased, signals can still be heard if their intensity is increased to a level above that of the noise. Nevertheless, there is a limiting noise level above which even shouting or loud music becomes unintelligible. Such a noise level, if maintained for a period of time, can damage the human auditory system. At still higher noise levels, humans may feel pain and even suffer physical damage. Hence, acoustical design criteria should set limits on noise levels to insure that background noise will not be uncomfortably loud.

Criteria for Background Noise

Human response to background noise depends on its frequency. Most noises cover a wide range of frequencies. (Noise from machinery may extend over the whole audible range of frequencies.) For convenience in specifying limits on background noise, therefore, a single rating, called *noise criteria* (*NC*), has been established for specific sound intensity levels at specific frequencies. The higher the number following *NC*, the louder will be the sound at every frequency. Table 12.2 indicates the human response to various *NC*.

Sound intensity levels, dBA, measured with meters for background noise are only roughly related to *NC*. Identical dBA readings can be

Table 12.2. Noise Criteria Corresponding to Various Sounds

Type of sound	Equivalent dBA	Noise Criteria
Very quiet	25–35	*NC*-15 to *NC*-25
Quiet	40–45	*NC*-30 to *NC*-35
Noisy	50–65	*NC*-40 to *NC*-55
Very noisy	70–80	*NC*-60 to *NC*-70

Table 12.3. Typical Acceptable Noise Backgrounds

Listening condition	Type of space	*IL*, dBA
Excellent	Concert, opera, recording studio	25–30
Quiet	Bedrooms, libraries	30–40
Very good	Theaters, churches, meeting rooms	30–40
Good	Private offices, classrooms	40–45
Fair	Large offices, stores restaurants	45–55
Poor	Shops, garages, machine rooms	55–65

obtained for sounds with completely different spectrums. Nevertheless, it is often convenient to measure noise levels in dBA. *NC* can be estimated by subtracting 7 to 10 points from dBA values; for example, 40 dBA may lie between *NC*-30 and *NC*-33. Table 12.2 lists approximately equivalent values of *NC* and dBA.

Table 12.3 lists background noise levels that are acceptable under ordinary conditions. If the noise is steady and reliable, it is useful for masking intruding unwanted sounds with lower intensity levels. As an example, Fig. 12.2 illustrates a case where a wall reduces sound transmission from 60 dBA on one side to 20 dBA on the other side, where the noise background level is 30 dBA. Because of the background noise, the 20-dBA transmitted sound will be inaudible.

Sound Isolation in Residences

For limiting airborne noise, local building codes may require that spaces containing noise sources be isolated by enclosure with barriers with a minimum *STC* rating. For example, the Build-

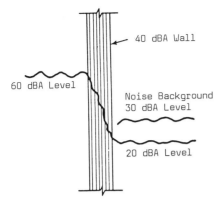

Fig. 12.2. Wall decreases sound level from 60 dBA on one side to 20 dBA on the other side. The transmitted sound is masked by the 30-dBA noise-background level in the receiving room.

ing Code of the City of New York contains the following requirements:

Walls, partitions and floor-ceiling construction separating dwelling units from each other or from public halls, corridors, stairs, boiler rooms or spaces or shafts containing air-conditioning, refrigeration or ventilating equipment, elevator machinery or other mechanical equipment should belong to an *STC* of 50 or higher. Dwelling entrance doors should be in *STC*

35 or higher. Ventilating openings into boiler rooms and other mechanical-equipment spaces should be so located that noise transmitted from them through windows into dwelling units should not exceed *NC*-35 in the dwellings. Noise output from mechanical equipment or from room air-conditioning terminal units, such as grilles, registers, diffusers, induction units and coil units, should not exceed the *SPL* listed in Table 12.4 by more than 2 dB in any octave band.

For limiting structure-borne noise, the Building Code of the City of New York requires that floor-ceiling construction separating dwelling units from each other or from public halls or corridors have a minimum impact-noise rating *INR* of zero (*IIC* = 51). Boilers supported on floors above a story with dwelling units should be supported on resilient isolators with a minimum static deflection of 1 in. When boilers are equipped with mechanical draft fans, the boiler breeching and piping that are supported on or from slabs, floors or walls that are contiguous to a dwelling unit should be supported for a distance of 50 pipe diameters on or from resilient isolators. Each isolator should have a mini-

Table 12.4. Maximum Permissible Sound Power Levels, dB, in Buildings[a]

Noise sources	Octave band mid-frequencies, Hz							
	63	125	250	500	1,000	2,000	4,000	8,000
Mechanical spaces or shafts with equipment rated at more than 75 hp adjoining dwelling spaces[b]	88	87	90	89	87	86	83	81
Exterior mechanical equipment at a distance from buildings of:								
12 ft[b]	87	80	73	68	65	63	62	61
25 ft[b]	94	86	79	74	71	69	68	67
50 ft[b]	100	92	85	80	77	75	74	73
100 ft[b]	106	98	91	86	83	81	80	79
unless the intensity level in the building of the noise of the exterior mechanical equipment is less than:	61	53	46	40	36	34	33	32
Room air-conditioning terminal units under design operating conditions	66	59	52	47	44	42	39	38

[a]Based on the Building Code of the City of New York noise-control requirements. Field measurements of *SPL* should not exceed the values in the table by more than 2 dB.
[b]The maximum permissible *SPL* in the table should be reduced 5 dB in any octave band where the equipment generates a pure tone.

mum static deflection of 1 in. The New York building code also sets minimum standards for isolators for piping, pumps, elevator machinery, metal chutes, ducts and HVAC equipment.

Sound Isolation between Rooms

For limiting transmission of airborne noise between rooms, criteria should require that walls and floor-ceiling construction have a sufficiently high *STC* rating. Table 12.5 contains minimum *STC* ratings that can be used as a guide for ordinary conditions for the types of occupancy listed. In addition, exterior walls should have a sufficiently high *STC* rating to limit intrusion of noise from outside the building.

For limiting transmission of structure-borne noise, floor-ceiling construction should have a sufficiently high *IIC* or *INR* rating. Table 12.6 contains minimum *IIC* ratings that can be used as a guide for ordinary conditions for the types of occupancy listed.

In buildings where conversations in one room, the source room, can be annoying to occupants of an adjoining, or receiving, room, the common barrier between the rooms should have an *STC* rating high enough that the speech will be unintelligible in the receiving room. The *STC* rating required for this purpose depends on how loudly the persons in the source room may speak; the sound absorption of walls, floor and ceiling in the source and receiving rooms; the surface area of the common barrier; the degree of privacy required in the receiving room, and the background noise level in the receiving room. If the background noise level is high enough, it will mask the conversations intruding from the source room.

As a rough guide, if the *STC* rating of a common barrier is equal to or greater than the value computed from Eq. (12.10), most persons will find the level of intruding conversations from an adjoining room acceptable.

$$STC = IL_1 + \frac{1,250}{A_1} + P - \frac{A_2}{S} + 3 - IL_2$$

(12.10)

where

IL_1 = intensity level of conversation in source room, dBA (60 for conventional speech, 66 for raised voices, 72 for loud talking)
A_1 = area, sq ft, of floor of source room
P = degree of privacy desired in receiving room (9 for normal conditions, 15 for confidential conversations)
A_2 = area, sq ft, of floor of receiving room
S = surface area, sq ft, of common barrier between the rooms
IL_2 = background noise level, dBA, for masking in receiving room

Reverberation Criteria

In rooms with reflective enclosures, sounds bounce off them. If the same sound is heard

Table 12.5. Minimum *STC* Ratings for Barriers between Rooms

Receiving room	Noise source	*STC*
General office	General office	33
Classroom	Corridor	33
Classroom	Classroom	37
Private office	Private office	42
Hotel bedroom	Hotel bedroom	47
Hotel bedroom	Corridor	47
Bedroom	Mechanical room	50
Theater	Classroom	52
Theater	Music rehearsal room	57

Table 12.6. Minimum *IIC* Ratings for Floor-Ceiling Construction

Upper room or noise source	Lower, or receiving, room	*IIC*
Office	Office	47
Classroom	Classroom	47
Hotel bedroom	Hotel bedroom	55
Music room	Classroom	55
Public spaces	Hotel bedroom	60
Music room	Theater	62

repeated distinctly with a time interval between, the later sound received is called an *echo* of the initial sound. There are different types of echoes. For example, *flutter* is a type of echo that is rapidly repeated as at least a partly distinguishable sound. It may be produced by sounds bouncing between parallel walls of a corridor. Both flutter and ordinary echoes are undesirable because they tend to confuse hearers.

Reverberation is another type of echo. It is produced by very rapid, repeated, jumbled echoes, which blend into an indistinct sound that continues for some time after the hearer receives the initial sound. Under some conditions, reverberation is undesirable because it makes speech within a room unintelligible. Under other conditions, reverberation may be desired to make speech or music seem more vibrant and alive. Consequently, criteria should be established for control of reverberation.

The usual basic criterion establishes limits on the time it takes for a pulse of sound to decay within a room. This period of time depends on the frequency of the sound and on size, shape and sound absorptivity of the room. A standard definition has been adopted for reverberation time as follows:

Reverberation time for an enclosed space is the time required for a sound pulse to decay 60 dB within the space after the sound source has ceased. This represents a decrease in sound intensity to one-millionth of the initial intensity.

In practice, reverberation time for a room is calculated from the Sabine formula:

$$T = \frac{0.049V}{a} \qquad (12.11)$$

where

T = reverberation time, sec
V = volume of room, cu ft
a = total sound absorption, sabins, of the room surfaces

As pointed out in Sec. 12.3, sound absorption depends on the frequency of the sound. Hence, reverberation time also depends on frequency. In practice, however, for ordinary rooms, reverberation time is determined for sound at a

frequency of 500 Hz. For concert halls and other spaces where sound quality is critical, reverberation times usually are determined for two octaves above and below 500 Hz.

Reverberation time should be controlled so that it is large enough to attain the desired fullness and vibrancy of music or speech, but not so large as to interfere with intelligibility of speech. The optimum reverberation time lies between these two limits and is a subjective determination. For normal conditions and ordinary spaces, reverberation time should not exceed

$$T = 0.4 \log_{10} V - 0.4 \qquad (12.12a)$$

but should not be less than

$$T = 0.4 \log_{10} V - 0.9 \qquad (12.12b)$$

Note that these criteria are based only on the room volume, whereas reverberation time actually is influenced by many factors. Hence, for critical spaces, such as concert halls, more detailed acoustical analysis by experts is desirable.

Vibration Criteria

To prevent physical discomfort of building occupants, fatigue failures of building components, noise from rattling of building components and contents of buildings and damage to delicate apparatus, vibrations transmitted through the structure should be limited. Effects of vibrations depend on both amplitude and frequency.

Human response to vibrations is plotted in Fig. 12.3. The spaces between curves indicate the ranges in amplitude and frequency that elicited specific sensations varying from imperceptible to painful in tests.

Damage to building components from vibrations is plotted in Fig. 12.4. The spaces between curves indicate the ranges in amplitude and frequency that caused specific degrees of damage varying from none to destruction.

Comparison of Figs. 12.3 and 12.4 indicates that if the amplitudes and frequencies are kept within the ranges where vibrations are just perceptible or imperceptible to building occupants, there will be no damage to building components. Therefore, criteria for control of vibrations should be based on human responses.

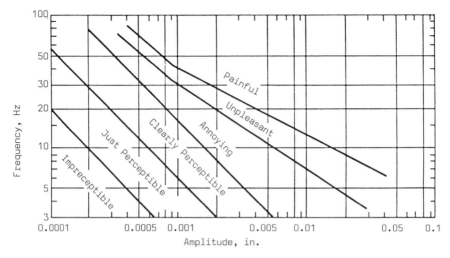

Fig. 12.3. Human sensitivity to vibration. (Courtesy British Research Station Digest No. 78.)

The most effective means of controlling vibrations is to prevent their transmission from the source to the building structure. This is usually done by mounting machinery and other potential sources of vibration on resilient supports. Design criteria often require these supports to have a specific minimum static deflection (under steady load), depending on the driving frequency of the vibration source. This is a characteristic of equipment that should be obtained from the manufacturer. As a guide, for ordinary conditions, the resilient support should deflect under the weight of the equipment at least an amount δ, in., computed from

$$\delta = \frac{200}{f^2} \qquad (12.13)$$

where

f = driving frequency of equipment, Hz

For example, if a machine has a driving frequency of 15 Hz, the static deflection of the support under the weight of the machine should be at least $200/(15)^2 = 0.89$ in., say 1 in. The resilient support should be set on a rigid base.

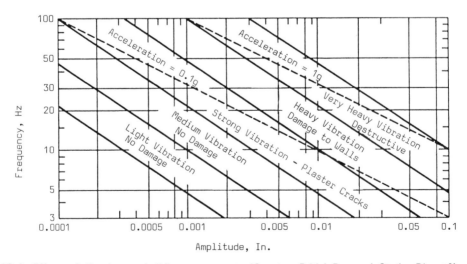

Fig. 12.4. Effects of vibrations on building components. (Courtesy British Research Station Digest No. 78.)

Equation (12.13) indicates that the lower the frequency, the more the support should be capable of deflecting and hence, the more flexible it should be.

SECTIONS 12.3 AND 12.4

References

L. Beranek, *Noise and Vibration Control*, McGraw-Hill New York, 1971.

R. Jones, *Noise and Vibration Control in Buildings*, McGraw-Hill, New York, 1984.

HUD, *A Guide to Airborne, Impact, and Structure-Borne Noise Control in Multi-Family Dwellings*, U. S. Department of Housing and Urban Development, Washington, D.C., 1968.

NBS, *Sound Insulation of Wall, Floor and Door Construction*, National Bureau of Standards Monograph 77, U.S. Government Printing Office, Washington, D.C., 1964.

Words and Terms

Absorptivity
Background noise
Decibel level (dBA)
Impact insulation class (*IIC*)
Impact noise rating (*INR*)
Loudness
Noise-reduction coefficient (*NRC*)
Noise criteria (*NC*)
Pitch
Reverberation
sabin (unit)
Sabine formula
Sound-absorption coefficient
Sound isolation
Sound transmission class (*STC*)
Structure-borne vibration and sound

Significant Relations, Functions, and Issues

Sound transmission across barriers: measurement (by decibel difference), related to mass and stiffness of barriers, for different frequencies, rating (by *STC*).

Structure-borne sound: impact insulation class (*IIC*), minimum ratings.

Sound absorption: at single frequencies, over a spectrum of frequencies (*NRC*).

Loudness as measured by sound power level (SPL); background noise rated by noise criteria (*NC*).

Minimum *STC* ratings for barriers (by building code).

Echos, flutter and reverberation in enclosed spaces.

12.5. SOUND AND VIBRATION CONTROL

Sound and vibration control is achieved by treatment or modification of the source or the potential paths that sound or vibration may traverse, or by protection of the receiver. Generally, the most effective method is treatment or modification of the source when that is possible and economical.

Noise Reduction at the Source

The most annoying or harmful sounds in buildings usually originate outside the buildings or with machines operating inside.

The sources of sounds outside a building often cannot be controlled, but the intensity of sounds reaching the building from the outside sources sometimes can be. Usually, the best method of reducing outside noise is to block it with fences, walls or trees. Also, exterior walls and doors facing noisy sources should have a high *STC* rating. Windows should be kept to a minimum size, kept closed and preferably should be double or triple glazed. If the noise source may be aircraft, the roof also should have a high *STC* rating.

If the source of noise is machinery, the quieting method depends on whether the noise is inherent in the process or caused by defects. When the latter is the case, noise can be eliminated by such measures as replacement of defective parts, balancing of moving parts or lubrication of bearings. If noise is inherent in the process, such as a forging machine, consideration should be given to either modification of the machine or a change to a different, less noisy process.

Sound Isolation with Barriers

Building designers often cannot influence the activities to be performed in a building or the industrial equipment to be installed or the maintenance that will be done on that equipment. Hence, they usually cannot quiet the noise source. Instead, to shield building occupants from excessive noise, designers have to employ noise-reduction techniques. One such control measure commonly used is acoustic isolation of the sound source.

Barriers or enclosures, such as walls, floors and ceilings, with a sufficiently high *STC* rating, and floor-ceiling construction with a sufficiently high *IIC* rating can be the most effective means for reducing noise that intrudes into a space from outside the building or from another space within the building. Such barriers are useful in increasing the privacy and comfort of occupants and improving communication between occupants. Barriers with high sound-transmission loss are especially desirable for protection of building occupants against damaging noise levels. In some cases, it may be necessary to enclose noise sources or persons who might be exposed to damaging noise levels in a closed booth or small room.

Sound, however, will travel along any path along which its passage is not blocked. Hence, if there is any opening in or around a barrier, sound will bypass it and defeat the purpose of the barrier. The opening need not be large. A hole or crack with an area of only 1 sq in. will transmit as much noise as a 6-in.-thick concrete block wall with a 100-sq ft area.

Sound can readily leak around the perimeter of pipes, ducts, doors and windows. It can penetrate through cracks at heads and sills of partitions or through joints between partitions and other space enclosures. Sound also can bypass partitions by traveling through the floor or ceiling (see Fig. 12.5a) and can bypass floors and ceilings by traveling through walls (see Fig. 12.6a). Hence, design of a barrier should provide means of blocking bypasses. Openings should be avoided when possible. Cracks and

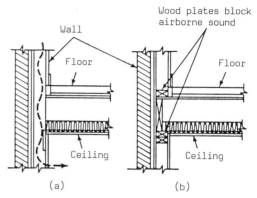

Fig. 12.6. (*a*) Sound bypasses a barrier–ceiling edge at vertically continuous wall. (*b*) Sound path in (*a*) blocked.

joints should be calked and sealed. If possible, floors should not be continuous at partitions (see Fig. 12.5b). Instead, a gap should be formed and filled with a damping material. Where a flanking path may exist around the end of a floor, the opening in the wall should be blocked with a material possessing a high sound-transmission loss (see Fig. 12.6b).

Sound Absorption

Another generally feasible control measure is modification or treatment of the surfaces of space enclosures to achieve desired acoustic effects. For example, if amplification of sounds at specific locations is desired, enclosures can be shaped and made reflective to build up sound at those locations. Similarly, enclosures can be designed to direct sound away from specific locations and to prevent echoes and flutter.

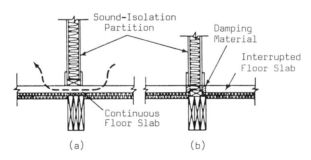

Fig. 12.5. (*a*) Sound bypasses (flanks) a barrier–partition on top of finish floor. (*b*) Sound path in (*a*) blocked by interrupting with construction break.

Note that concave surfaces, such as dome interiors, tend to focus sounds, whereas convex surfaces tend to spread sounds.

Treatment of the surfaces of enclosures with sound-absorbing material often is the most effective means, other than quieting the noise source, of reducing the general noise level. Such treatment can improve communication between building occupants, increase their comfort and reduce risk of hearing damage.

Use of sound absorbents often is the most effective means of controlling reverberation. This measure also may be necessary to eliminate echoes and flutter when the configuration of reflecting enclosure surfaces cannot be changed for the purpose.

Sound absorbents should not be used as a substitute for sound barriers. Sound absorbents are like blotters, in that they are intended to soak up sounds; however, absorbents cannot stop sound from traveling through them.

Absorbents should not be placed on surfaces that are needed for useful reflection of sound. In some rooms, carpets, drapery and heavily upholstered furniture provide sufficient sound absorption. In such rooms, some reflected sounds are necessary to provide desired reverberation. This is often the case in auditoriums. When more noise reduction or less reverberation is desired, ceilings should be covered with sound-absorbent materials. In long, narrow rooms, however, treatment of the walls may be more effective.

The acoustic effects of sound absorbents are significantly influenced not only by the amount and location of the material but also by the method of installation of the material. Typical methods of mounting sound absorbents include cementing to a smooth, solid surface; spraying; troweling; nailing or stapling to furring strips; or seating on small, light structural members suspended from the underside of a floor.

The surface formed by sound absorbents need not be flat. Coffers (see Fig. 12.1), grids of hanging panels and inclined surfaces often provide desired acoustic effects.

Structural behavior and fire resistance of sound absorbents as well as acoustic performance should be considered in selection of an installation method.

Background Noise

In some cases, a continuous background noise may be desirable. If the background noise level is too low, occupants of a room will be distracted by low-intensity sound, annoyed by talking of persons in distant parts of the room, hear conversations in adjacent offices or find the space oppressive. Some background noise may be provided naturally by general activity in a space or by injection of air from air-conditioning grilles. In some cases, however, it may be necessary to create background sounds to mask the objectionable sounds (see Fig. 12.2). One method of doing this is installation of a sound system to broadcast soft music.

In rooms in which the background noise level interferes with communication between occupants, either noise-reduction techniques must be employed or desired sounds must be amplified. In auditoriums, for example, where background noises from outside or from the audience may be too high, loudspeakers should be installed for amplification of the signal so that it can readily be distinguished from background noises.

Personal Protection Against Noise

In spaces where sound levels may be annoying or harmful and cannot be reduced, the most effective control measure may be personal protection of occupants. For example, they may be sheltered in sound-isolating enclosures. If such enclosures would have to be too large because of the number of occupants to be sheltered or because the occupants have to perform activities that require them to move about, they may be equipped with an ear-protection device, such as earmuffs or plugs.

Vibration Isolation or Damping

Usually, the most effective means of preventing annoying vibrations is correction of the source. If the source is machinery, vibrations may be reduced or eliminated by replacement of defective parts or by balancing moving parts. Generally, also, the machine should be isolated from the building structure. This can be done by mounting the machine on resilient supports that absorb the energy of vibration in deflecting.

Also, potential paths for transmission of vibrations should be interrupted with carefully designed discontinuities (see Fig. 12.5*b*). In assembly of building components, damping of vibrations can be accomplished by insertion in joints of energy-absorbing materials, use of viscoelastic adhesives, such as asphaltic compounds, that are neither completely elastic nor entirely viscous, and installation of connections that resist transmission of vibrations.

Difficulty of Complete Control

The best of design efforts for sound and vibration control are often thwarted by simple circumstances. It is very difficult to anticipate and prevent all the possible sound leaks and unintended sound paths. During construction, changes in various systems are frequently made and the implications of rerouting wiring, piping, and ducts with regard to sound leaks or vibrations are not always considered in approving the changes.

Changes in building equipment may be made before, during or after completion of construction, with resulting sound and vibration effects of a nature not anticipated in the original acoustic design. Remodeling or building tenant changes often result in changes of nonstructural partitions, ceilings, lighting systems, sound systems, and locations of noise-producing occupant-installed equipment. Retrofitting for adjustment of sound and vibration control is often required for such changes and does not necessarily reflect on the adequacy of the original design work.

References

R. Jones, *Noise and Vibration Control in Buildings*, McGraw-Hill, New York, 1984.

B. Stein et al., *Mechanical and Electrical Equipment for Buildings*, 7th ed., Wiley, New York, 1986.

Significant Relations, Functions, and Issues

Basic means of achieving sound and vibration control: modifications at source, treatment of paths, protection of receiver.

Controls at source: adjustments and maintenance of noisy equipment, shielding source with barriers, structural isolation and damping.

Control of paths: enhancement of barriers (better STC, ICC), closing of leaks and bypasses.

Controls in enclosures: form of surfaces, absorptivity of surfaces, background noise, protection of individual receivers.

Maintaining control: monitoring of design and construction changes, need for post-construction adjustments or retrofit.

12.6. SYSTEMS DESIGN APPROACH TO SOUND AND VIBRATION CONTROL

Sound and vibration control affects and is affected by nearly all building systems, especially those with moving parts. Unless constrained by legal requirements, such as building codes, to prevent excessive noise, other building systems usually are given higher priority in design than acoustics. The reason for this is that changes in those systems cost more than changes in acoustical treatment to accommodate the other systems. Sometimes, as a result, it is necessary to apply corrective acoustical measures after a building has been completed, at higher cost than for satisfactory installation during building construction. Also, the corrective work interferes with or halts normal use of the space requiring treatment.

Systems design should be applied to sound and vibration control to secure satisfactory environmental conditions within buildings at lowest life-cycle cost for the buildings. Cost of such control can be relatively low if buildings are designed from the start to integrate it in the overall design.

Two of the most critical stages of building design for control of sound and vibration are the beginning and end of the design process. At the beginning it should be noted which are the potential, significant sources of sound and vibration problems and which are the areas of the building where receivers need the most protection or sound privacy. If it is possible to make simple, logical design decisions at the start and that result in reduction of the magnitude of sound and vibration control problems, all the necessary design efforts for control done later will be made easier. Simply locating major problem sources as far away as possible from the most sensitive receivers is a critical first step, and easiest to achieve when it is part of the early design value system.

During the general design development period, sound and vibration implications should be included in all design choices that impinge on their control. Choice of construction materials and details, type and locations of noisy equipment, location and details of any expansion or movement-control joints in the building structure, means of attachment or support of ducts and piping, and the forms of interior spaces should all be reviewed for sound and vibration issues as the design work progresses.

As the design work proceeds to completion, a careful final review should be made, reaffirming the original design objectives and making sure that the various design adjustments have not caused changes in original assumptions or intentions. A carefully managed systems design process that involves continuous exchange and interaction between the various members of the design team is significantly beneficial to the design for control of sound and vibration (which is highly susceptible to being forgotten as the design work proceeds).

Data Collection and Problem Formulation

Basic information for sound and vibration control comes from examination of the owner's program of requirements, type of occupancy, type of activity of the occupants and use to which building spaces are to be put. Also, information may be obtained from manufacturers on potential noise- or vibration-producing equipment, such as motors, lighting fixtures, fans and other machinery.

Additional information is supplied by design drawings for the other building systems. Of particular importance are the architectural drawings, which provide information on size, shape and purpose of building spaces and on materials and types of construction used for enclosing the spaces. Also, inspection of the environment surrounding the building should disclose possible sources of disturbing noise and vibration from outside the building, such as highways, railroads, airplanes, industry and school playgrounds. Further information may be provided by requirements of government agencies and the local building code.

From all the information obtained, estimates can be made of the acoustic environment required in the various spaces, of sound and noise levels that may be generated in the performance of activities in the building and of potentially troublesome conditions. Then, strategies can be developed for providing the desired environmental control.

The goal of sound and vibration control may be stated as: to design, as a component of an optimum building, spaces and their enclosures that will insure a satisfactory environment for occupant communication, comfort, privacy and good health.

Objectives

A primary objective of sound and vibration control is prevention of annoying noises and vibrations that can be influenced by building design.

Another primary objective is reduction of the intensity level of noise and vibration that cannot be prevented. This objective requires that measures be taken to isolate, absorb and mask noise and to isolate and damp vibrations. When it is not feasible to meet this objective because of processes essential to the purpose of the building, an alternative objective that should be met is personal protection of occupants, to prevent injury to hearing.

Still another objective is sound control for distinct hearing, enhancement of music and masking of annoying sounds.

Constraints

Many factors that affect acoustic performance and vibration generation and transmission are beyond the control of the acoustics designer and sometimes also beyond the control of the other building designers.

A common constraint is the process to be carried out in a space. An office requires typewriting and conversation; a factory requires machinery; a school requires an auditorium and perhaps also a music practice room, spaces in which loud sounds are generated. In such cases, noise cannot be prevented. The best that can be expected of acoustics design is isolation of the noise and vibration, or in extreme cases in industry, provision of personal protection for occupants.

Another common constraint is noise and vibration generated outside the building. These disturbances cannot be controlled by the building designers.

Building codes and government agencies also impose constraints in the interests of public health and welfare. These constraints place restrictions on equipment and materials that may be used for HVAC and plumbing and on the type of mounting that can be used for equipment with moving parts.

Other constraints are usually imposed on sound and vibration control by other building systems. Architectural requirements are among the most important of these. For example, size and shape of spaces, and materials and type of construction of enclosures of the spaces have considerable effects on acoustics in the spaces. The construction adopted for roofs and exterior walls significantly influences the levels of noise transmitted into the building from outside. Also, the location of areas requiring quiet relative to noise-generating areas may determine whether adverse acoustic conditions will develop. In addition, interior decoration after the building has been completed may change acoustical performance in the spaces.

Still another important constraint is the lack of control over maintenance of equipment with moving parts. Improper maintenance can permit noise and vibration to develop and continue, because of defective and unbalanced equipment components.

Synthesis and Analysis

The design process for sound and vibration control basically examines three major factors: criteria, analysis and control. These in turn require answers to three questions:

1. What is the optimum acoustic environment for each space?
2. What factors affect that environment?
3. What strategies can be employed to influence the environment so that desired criteria can be achieved?

It is not sufficient, however, merely to design a satisfactory acoustical environment. In addition, it is necessary to design, as closely as possible, an optimum building. This generally requires close collaboration between the acoustics designer and all other members of the building team.

In the schematic-design stage, the acoustics designer and the architect should confer on location of spaces so that placement of noisy areas next to areas requiring quiet will be avoided, when practical. The designers should also discuss size and location of windows to minimize entry of noises generated outside the building. Consideration also should be given to shapes of spaces, inasmuch as some shapes, such as cubes and long, narrow rectangles, tend to make sound control difficult.

In the preliminary-design stage, as more information becomes available on other building systems and on activities to be performed in the various spaces, sound and vibration control strategies can be selected. The building team should collaborate to integrate required sound barriers in space enclosures; to quiet noise sources, such as lighting, plumbing and HVAC ducts and grilles; and to prevent generation and transmission of vibrations. Where sound absorbents are required, the acoustics designer and the architect should cooperate to insure that aesthetics and fireproofing requirements are met. If the ceiling is to be covered with a sound absorbent, it should be integrated with lighting fixtures and HVAC grilles.

It may not be possible to work out the details of sound and vibration control until the final-design stage, when all the needed information on the other building systems becomes available. As design proceeds, decisions made by the members of the building team will interact and conflicts may arise. As with other systems, the building team should resolve these conflicts in such a way that a satisfactory acoustical environment will be obtained in an optimum building.

Value Analysis and Appraisal

Because sound and vibration control interacts with the output of other members of the building team, many trade-offs between team members may be necessary for optimization of the building design. Trade-off studies should investigate the performance and cost impact of all affected systems. Changes needed should be made as early as possible in design.

Value analysis should compare alternatives not only for sound and vibration control but also for the other building components that affect or are affected by it. Alternatives investigated should include changes in room locations, sizes and shapes as well as in sound and vibration control techniques.

The greatest opportunity for cost savings lies in integration of sound control into other building systems. For example, sound barriers should be incorporated in or actually serve as walls, floors and roofs. Sound-absorbent materials should serve as ceilings or wall surfaces. Also, advantage should be taken of the capacity of carpets, drapery, upholstered furniture or seats and the bodies of building occupants to absorb sound, thereby reducing the need for other acoustic surface treatments. Particular attention should be paid to quieting noise sources, preventing vibrations from developing and isolating potential sound and vibration sources.

GENERAL REFERENCES AND SOURCES FOR ADDITIONAL STUDY

These are books that deal with several topics covered in this chapter. Topic-specific references relating to individual chapter sections are listed at the ends of the sections.

B. Stein et al., *Mechanical and Electrical Equipment for Buildings*, 7th ed., Wiley, New York, 1986.

L. Beranek, *Noise and Vibration Control*, McGraw-Hill, New York, 1971.

R. Jones, *Noise and Vibration Control in Buildings*, McGraw-Hill, New York, 1984.

EXERCISES

The following questions and problems are provided for review of the individual sections of the chapter.

Section 12.1

1. After seeing a bolt of lightning, a person hears thunder three seconds later. How far away was the lightning?

2. What is the wavelength of a sound in air at sea level if the frequency of the sound wave is 550 Hz?

3. What is the result at any point in space of imposing a sound wave on another sound wave of equal intensity and frequency when the maximum compressions in one wave coincide with the maximum rarefactions in the other wave?

4. If a person moves from 10 ft away from a speaker to 20 ft away, what is the ratio of the intensity of the speaker's voice at the new location to that at the former location?

5. A partition in an office does not extend to the ceiling. Explain how reflection permits a person on one side of the partition to hear a conversation on the other side.

6. A partition extending from floor to ceiling separates a room from a corridor. A door in the partition is separated from the floor by a narrow space. What property of waves permits a person any place in the room to hear a conversation held in the corridor near the closed door?

Section 12.2

7. How many octaves are there between sounds at frequencies of 200 and 800 Hz?

8. What do the following measure:
   (a) Sound intensity?
   (b) Sound intensity level?
   (c) Sound pressure level?

9. The intensity of music from an orchestra at a seat 31.6 ft away from the musicians is measured at 10^{-13} watts per sq cm.
   (a) What is the intensity level at the seat?
   (b) What is the sound power of the orchestra, watts?
   (c) What is the sound power level of the orchestra, dB?

10. Acoustic treatment reduces sound intensity at a point in a room from 80 to 8 watts per sq cm. What is the reduction in intensity level?

11. Sound power level of a musical instrument is measured at 50 dB. What is the combined sound power level of ten such instruments playing simultaneously?

12. A sound with an intensity level of 40 dB

is imposed on a sound with an intensity level of 50 dB. What is the resulting intensity level?

13. What is the acoustic significance of dBA units?

Section 12.3

14. What is the purpose of a sound barrier?
15. What is the purpose of a sound absorbent?
16. What acoustic property is indicative of the ability of a barrier to isolate sounds?
17. How are *STC* and sound transmission losses of a building component related?
18. A $\frac{1}{4}$-in. steel plate is in *STC* 36 whereas $\frac{3}{4}$-in. plywood is in *STC* 28. Which material would be better, in general, for sound isolation?
19. On what type of material should a machine be mounted to prevent it from causing vibrations in a building?
20. A floor-ceiling construction has an *IIC* of 65. What is its *INR*?
21. How is sound dissipated in a porous sound absorbent?
22. Why are sound absorbents usually useless for increasing the *STC* of a ceiling?
23. How are *NRC* and sound-absorption coefficients of a sound absorbent related?
24. A room 10 ft × 20 ft by 10 ft high has concrete walls, floor and ceiling. The *NRC* for the concrete is 0.02.
    (a) What is the total room sound absorption, sabins?
    (b) If the whole ceiling is covered with acoustic tile with *NRC* = 0.7, what does the total room absorption become?
    (c) What will the noise reduction due to the ceiling treatment be?

Section 12.4

25. To what physical characteristic of sound is pitch related?
26. What would a person be likely to call a sound with an *SPL* of 80 dBA?
27. What range of frequencies can the human ear normally detect?

28. What would a person be likely to call a continuous sound with a rating of *NRC*-40?
29. What is a major advantage and a major disadvantage of background noise?
30. What is the minimum *STC* rating that a wall between a classroom and a corridor ordinarily should have?
31. What is the minimum *IIC* rating that the floor-ceiling construction should have in the bedroom section of a hotel?
32. A conference room with a floor area of 300 sq ft is separated by a 100-sq ft wall from a private office with a floor area of 150 sq ft. (Occupants of the conference room may be expected to talk with raised voices.) Background noise in the private office, where confidential conversations are often held, is 40 dBA. What is the minimum *STC* rating the common wall should have to prevent speech from the conference room from being annoying to occupants of the private office?
33. A classroom has a floor area of 1,000 sq ft and a sound-absorbent ceiling 10 ft above the floor. The total sound absorption of the room is 500 sabins.
    (a) What is the reverberation time for the room?
    (b) Is the reverberation in the room likely to be considered acceptable by the students?
34. Vibrations transmitted through a building structure into a room have an amplitude of 0.001 in. and a frequency of 20 Hz.
    (a) Are occupants of the room likely to complain about the vibrations?
    (b) Are the vibrations likely to damage building components?
    (c) If the vibrations are caused by a fan and the vibrations are to be isolated at the source by mounting the fan on resilient supports, what is the minimum static deflection the supports should have?

Section 12.5

35. What is the most effective method of:
    (a) Reducing noise or vibration?

(b) Preventing transmission of sound between rooms?

(c) Reducing noise in a room if the sources cannot be quieted?

(d) Controlling excessive reverberation?

36. How can passage of airborne sound through walls past floors and ceilings be prevented?

37. What can be done when the occupant of a private office complains of being distracted by low sounds?

38. A worker in a factory has to tend personally to a very noisy machine. What can be done to prevent hearing damage from the machine?

39. How can transmission of structure-borne vibrations be prevented?

Chapter 13

Electrical Systems

Buildings usually contain several major electrical subsystems. All have in common the main purpose of controlled transmission of electricity. They differ, however, in results to be achieved. Hence, the major electrical subsystems may be classified in accordance with their objectives.

In general, these subsystems fall into either of two classes. One class has transmission of power as its main objective. The other class has communication, or the transmission of information, as its objective. In this section, for simplification, the subsystems will be referred to as systems, because they usually consist of two or more subsystems.

Power systems are used mainly for lighting, heating, refrigeration and operation of motors, pumps and other equipment. Communication systems include telephone; electronic devices, such as computers, radio and television; surveillance; alarms; clocks; programmers, or timers; intercommunication; paging and music. Electrical conductors for lightning protection may be considered a special type of power system. Conductors for the different classes of building electrical systems should be separated from each other.

These applications of electricity indicate that many functions of a building depend on availability of an adequate, reliable and appropriate supply of electricity. Systems design of the electrical systems, therefore, should aim at attainment of an optimum building. In such a building, low operating and maintenance costs

and high reliability of the electrical systems as well as continuous, efficient and frequently also automatic operation of equipment essential to building functions are often more important than low initial costs of electrical systems.

Comprehension of the roles of electrical systems in buildings requires a knowledge of:

1. Characteristics of electricity
2. Transmission of electricity
3. Electrical conductors and insulation
4. Control and protective devices for electrical systems
5. Types of electrical systems

Design of electrical systems is the responsibility of electrical engineers. The power systems are usually installed by electrical contractors. Building codes often require that the work be done by qualified journeymen and apprentices under the supervision of licensed master electricians.

13.1 CHARACTERISTICS OF DIRECT CURRENT

In many ways, transmission of electricity in buildings is analogous to water-supply distribution. In a water-supply system, water flows through pipes under pressure to points of use. In electrical systems, electrons flow through conductors, such as wires and cables, under a difference in potential energy, or voltage, to points of use. As water flows, there is a drop in pressure because of friction between the water and the pipes. As electrons flow (current), there is a drop in voltage because of electrical friction between the electrons and the conductors. In studying the characteristics of electricity, it may be helpful to keep this analogy in mind.

Electric Charges

Electrons and protons attract each other. Electrons, however, repel other electrons, and protons repel other protons. The forces causing this attraction and repulsion are attributed to electrostatic charges carried by electrons and protons. All electrons carry the same charge, which is negative. Protons carry a charge that is positive. Particles with opposite signs attract each other; those with the same sign repel each other. The force of attraction or repulsion, which is applied through empty space, without physical contact between the particles, is proportional to the magnitude of the electrostatic charge and inversely proportional to the square of the distance between particles.

A *coulomb* is the basic unit used in practical measurements of magnitude of electrostatic charges.

Voltages and Currents

Force is needed to make two particles with like charge approach each other. Hence, work is done on a particle when it moves toward another particle with the same charge. As a result, there is a change in the potential energy of the moving particle.

Voltage, or potential difference, is the term usually applied to the difference in potential energy between two points in the path of a moving charged particle. (Electromotive force is another term sometimes used.)

One volt is required to do one joule of work on a charge of one coulomb. Voltage is analogous to pressure in water flow.

Current is the name given to a flow of charges. A current that always flows in the same direction is called a *direct current*.

An *ampere* (amp) is the basic unit for measurement of flow of current. One amp is equivalent to a flow of one coulomb per second.

Amperage is analogous to rate of water flow, gpm, in water supply.

Voltages and currents for practical applications are usually produced by conversion of chemical or mechanical energy to electric energy. Batteries develop a potential difference between two points, or terminals, by chemical means. Generators can create much larger potential differences through use of mechanical energy to rotate electrical conductors in a magnetic field. The relationship between magnetism and electric currents will be discussed later.

One terminal of a battery or generator is assigned a positive sign and the other terminal, a negative sign. The terminal with the positive sign is conventionally considered to be at a higher potential than the terminal with the negative sign. If the two terminals are connected with a conductor of electricity, an electric current will flow between them. By convention, direct current will flow from a positive terminal of a battery or generator to the negative terminal; that is, from any point on a conductor to any other point on the conductor at a lower potential.

Electric equipment usually is designed to operate efficiently at a specific voltage. Voltage supplied to the equipment, however, may fluctuate as other electric equipment also being served varies the demand for electricity. The deviation from rated voltage must be limited to prevent damage to the equipment. Permissible deviation usually is specified in the form of voltage regulation VR, percent.

$$VR = \frac{E_o - E_f}{E_f} \times 100 \qquad (13.1)$$

where

E_o = voltage supplied to the building with no load on the power line

E_f = voltage supplied to the building with full load on the line

For example, if voltage drops from 120 volts at no load to 115 volts at full load, the voltage regulation is $100(120 - 115)/115 = 4.35\%$.

In a specific case, voltage regulation required depends on the type of electric equipment. For some motors, for instance, variations from rated voltage may be limited to 5% for undervoltage and 9% for overvoltage. Other equipment may require smaller limits on the low side and permit higher limits for overvoltage. Still other equipment may need tighter limits for both undervoltage and overvoltage.

Design of electrical systems must also take into account voltage drops that occur within the building from the point of entry of electric service to the point of use. Voltage regulation at the point of entry and voltage drop to the point of use are analogous to fluctuations in supply pressure at the water main and the pressure drop in building pipes in water-supply distribution.

Electric Resistance

All materials resist the flow of electrons, or electric current, that occurs when a voltage is applied. Resistance causes a voltage decrease. Electric energy also decreases as some energy is converted to heat.

The amount of resistance depends on the characteristics of the materials. Materials, such as copper and aluminum, with low resistance are called *conductors*. Materials, such as some plastics and glass, with high resistance are called *insulation* or *insulators*. Insulation is used to cover conductors to prevent other materials from contacting them. Insulators are used to separate conductors from supporting materials that might conduct electricity.

Resistance not only depends on the characteristics of a material but also on the cross-sectional area of the conductor and the length of the path over which current-flow occurs. The unit of measurement of resistance is the *ohm*.

A resistance of one ohm creates a 1-volt potential difference when a 1-amp current flows.

Ohm's law gives the relationship between resistance, voltage drop across the resistance and current as

$$E = IR \qquad (13.2)$$

where

E = potential difference between terminals of the resistance, volts

I = current through the resistance, amp

R = resistance, ohms

Figure 13.1a shows symbolically a resistance R placed between the terminals of a battery. The resistance is often referred to as the load on the battery. If the resistance R equals 20 ohms and the current I is 0.25 amp, Eq. (13.2) indicates that the battery applies a voltage $E = IR = 0.25 \times 20 = 5$ volts. In Fig. 13.1a, the zigzag line is the symbol for a resistance and the straight lines represent conductors with zero resistance. In actuality, conductors have some resistance, which also would be represented by a zigzag line.

The battery, the resistance and the conductors between them form a closed electrical circuit. Current flows from the battery through the conductors and the resistance and back to the battery. Direct current will not flow in the conductors of an open circuit.

Figure 13.1b shows symbolically a closed circuit with current flowing from a generator to a lamp and back to the generator. The generator is represented by a circle with a G in it, and the lamp, by a circle with an L in it. This cir-

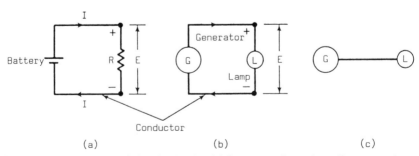

(a)   (b)   (c)

Fig. 13.1 Symbolic representation of electric circuits. (a) Battery applies voltage *E* across resistance *R*, and current *I* flows around the circuit. (b) Circuit similar to (a), except that the power source is a generator and the resistance (or load) is a lamp. (c) Single-line representation of the circuit in (b).

cuit also may be represented by the alternative diagram in Fig. 13.1c. In that diagram, for simplicity of showing interconnected components and their relative positions, the conductors to and from the generator are indicated by a single line. Diagrams such as Fig. 13.1b, however, are more suitable for analysis of circuits to determine currents and voltages.

In some cases, it is more convenient in analysis to use the capacity of a material to conduct current instead of using its resistance. This capacity is called conductance.

Conductance G is the reciprocal of resistance; that is

$$G = \frac{1}{R} \qquad (13.3)$$

The unit of measurement of conductance is called a *mho*. Substitution of Eq. (13.3) in Eq. (13.2) yields an alternative form of Ohm's law:

$$I = GE \qquad (13.4)$$

Power from Direct Current

From the definition of volt, potential difference $E = W/Q$, where W is the work done, joules, in moving electric charges Q. From the definition of ampere, current $I = Q/T$, where Q/T is the rate of flow, coulombs per second, of electric charge. These results can be used to determine the power expended when a current I flows between two points with a potential difference E.

Power is the rate of doing work, joules per second. Electric power P is measured in *watts*. One watt is equivalent to one joule per second. Thus, power, watts, can be computed from

$$P = \frac{W}{T} = \frac{W}{T} \cdot \frac{Q}{Q} = \frac{W}{Q} \cdot \frac{Q}{T} = EI \qquad (13.5)$$

For direct current, then, power, watts, is given by the product of potential difference, volts, and current, amperes.

With the use of Ohm's law, Eq. (13.5) can be written in several alternative forms:

$$P = EI = I^2 R = \frac{I^2}{G} = E^2 G = \frac{E^2}{R} \qquad (13.6)$$

Power Transmission. Suppose an electric power line has to deliver a specific amount of power. According to Eq. (13.5), this can be done either by delivering a high current at low voltage or a small current at high voltage; however, for economy, the power P that will be lost due to the resistance of the transmission lines should be kept as small as practical.

According to Eq. (13.6), this resistance R is determined by P/I^2 or E^2/P. Hence, for a given power loss P, R varies inversely as the square of the current and directly with the square of the voltage. The higher the current, the smaller the resistance must be, and therefore the larger the area of the conductor must be. In contrast, the higher the voltage, the greater the resistance may be for the same power loss, and consequently smaller conductors may be used than for transmission of high currents. Thus, it is advantageous to transmit electricity at high voltage. In practice, electricity to be transmitted long distances is carried at very high voltage.

High voltages, however, can be hazardous to persons and contents of buildings. For safety reasons, therefore, low voltages are used for electrical transmission in buildings.

For large amounts of power, the units used are the kilowatt (kw), which equals 1,000 watts, and the megawatt (Mw), which equals 1,000,000 watts or 1,000 kw. Electric energy consumed is usually measured in kilowatt-hours (kwhr).

Charges for use of electricity may be based on a combination of two or more different rates. One type of rate is applied to total energy usage, kwhr, in a specific period of time, for example, one month. A second type, the demand rate, is applied to the maximum kw required.

Direct-Current Circuits

Figure 13.1 illustrates the simple case of a direct-current (dc) circuit with a single load, or resistance. Usually, however, a generator is employed to serve several loads. These may be connected to each other and to the generator in any of several different ways. Two common types of connection are illustrated in Figs. 13.2 and 13.3.

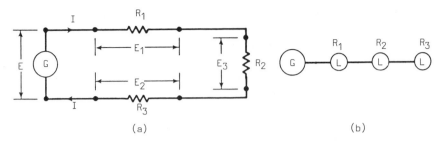

Fig. 13.2. Symbolic representation of a series circuit. (a) Generator applies voltage E across three resistance in series. (b) Single-line representation of circuit in (a).

Kirchhoff's Laws. Currents and voltages in dc circuits generally can be determined with the aid of Ohm's law and Kirchhoff's laws.

Kirchhoff's current law requires that the sum of the currents entering any junction of conductors be equal to the sum of the currents leaving the junction. If the flow toward the junction is given a positive sign and the flow away, a negative sign, the algebraic sum of the currents at the junction must be equal to zero.

Kirchhoff's voltage law requires that in any closed circuit the sum of all applied voltages be equal to the sum of all voltage drops. If increases in potential are given a positive sign and decreases, a negative sign, the algebraic sum of the potential differences in a closed circuit must be equal to zero.

Series Circuits. In a series circuit (see Fig. 13.2), loads are connected to each other in sequence. The current I at each load therefore is the same, as indicated in Fig. 13.2a. Figure 13.2b shows a single-line diagram for the series circuit in Fig. 13.2a.

According to Kirchhoff's voltage law, the sum of the voltage drops E_1, E_2 and E_3 in the circuit in Fig. 13.2a equals the voltage E applied by the generator:

$$E = E_1 + E_2 + E_3 \qquad (13.7)$$

From Ohm's law, $E_1 = IR_1$, $E_2 = IR_2$ and $E_3 = IR_3$, where R_1, R_2 and R_3 are the resistances connected in series and across which the voltage drops occur. Substitution in Eq. (13.7) yields

$$E = IR_1 + IR_2 + IR_3 = I(R_1 + R_2 + R_3) \qquad (13.8)$$

In a series circuit, therefore, the product of the current by the sum of the resistances in series equals the applied voltage.

Also, by Ohm's law, $E = IR$, where R is a resistance equivalent to the total resistance in the circuit. Comparison with Eq. (13.8) indicates that

$$R = R_1 + R_2 + R_3 \qquad (13.9)$$

Hence, in determination of currents and voltages in a circuit, resistances in series can be conveniently replaced by a single resistance equal to the sum of the resistances.

Series circuits are advantageous for applications in which high voltage can be applied to several low-voltage loads, for example, to a

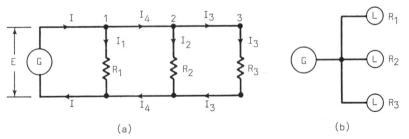

Fig. 13.3 Symbolic representation of a parallel circuit. (a) Generator applies voltage E across three resistances in parallel. (b) Single-line representation of circuit in (a).

group of lamps, and thus achieve efficient electrical transmission. The disadvantage is that if one of the resistances should fail and break the circuit, for instance, if a lamp should burn out, no current will flow anywhere in the circuit and all loads will be disconnected.

Parallel Circuits. In a parallel circuit (see Fig. 13.3), loads are connected across a common pair of terminals, so that the same voltage is applied to each load. Figure 13.3a shows a generator providing a voltage E across each of three resistances R_1, R_2 and R_3. The current supplied by the generator is indicated by I. At the first connection, a current I_1 is sent through R_1 and a current I_4 to the second connection. After the second connection, a current I_2 flows through R_2 and a current I_3 flows through R_3. Figure 13.3b shows a single-line diagram for the parallel circuit in Fig. 13.3a.

Application of Ohm's law in the form of Eq. (13.4) gives the currents through the resistances as

$$I_1 = G_1E, \ I_2 = G_2E, \ I_3 = G_3E \quad (13.10)$$

where $G_1 = 1/R_1$, $G_2 = 1/R_2$ and $G_3 = 1/R_3$. According to Kirchhoff's current law, $I_4 = I_2 + I_3$. Substitution of I_2 and I_3 from Eq. (13.10) yields

$$I_4 = G_2E + G_3E = (G_2 + G_3)E \quad (13.11)$$

From Kirchhoff's current law also, $I = I_1 + I_4 = I_1 + I_2 + I_3$. Use of Eq. (13.10) then gives

$$I = (G_1 + G_2 + G_3)E \quad (13.12)$$

In a parallel circuit, therefore, the product of the applied voltage and the sum of the conductances in parallel equals the current supplied by the power source.

Also, by Ohm's law, $I = GE$, where G is a conductance equivalent to the total conductance of the circuit. Comparison with Eq. (13.12) indicates that

$$G = G_1 + G_2 + G_3 \quad (13.13)$$

Hence, in determination of currents and voltages in a circuit, resistances in parallel can be conveniently replaced by a single resistance with a conductance equal to the sum of the conductances. Equation (13.13) can also be written in the form:

$$\frac{1}{R} = \frac{1}{R_1} + \frac{1}{R_2} + \frac{1}{R_3} \quad (13.14)$$

Hence, the reciprocal of the equivalent resistance for resistances connected in parallel equals the sum of the reciprocals of the resistances. For the special case of only two resistances in parallel,

$$R = \frac{R_1 R_2}{R_1 + R_2} \quad (13.15)$$

Parallel circuits are advantageous where it is important that the same voltage be applied to several loads. This type of circuit also has the advantage that if one load should fail, for example, if a lamp should burn out, current will still flow to the other loads connected in parallel and at the same voltage.

Still another advantage can be seen from the following: Suppose n equal resistances R_1 are connected in parallel. Then, from a generalization of Eq. (13.14), $1/R = n/R_1$ and $R = R_1/n$. Hence, the equivalent resistance is $1/n$-th of an individual resistance. In comparison, if n resistances R_1 are connected in series, the equivalent resistance equals nR_1.

Networks. In the distribution of electricity throughout a building, combinations of circuits usually are used. These combinations are called networks. They generally consist of elements interconnected in series and in parallel, rather than pure series and pure parallel circuits.

Networks can be analyzed with Ohm's and Kirchhoff's laws, as was done in the preceding discussion of series and parallel circuits. Usually, however, the networks are complex and analysis requires lengthy, tedious calculations. Solutions can often be obtained speedily with the aid of electronic computers.

Magnetic Fields

Every electric current flowing through a conductor generates a magnetic field. This field, which envelops the conductor and diminishes with distance from it, is invisible. Presence of the field, however, can be detected by observing the forces it exerts on certain materials. These materials are called magnets, because they react to a magnetic field. Soft iron, steel, nickel, cobalt and alloys of these metals exhibit magnetic properties.

If a rectangular piece of such a metal is magnetized and suspended at its center of gravity so that it is free to rotate in a horizontal plane, one end of the piece will turn toward the north or the south pole of the earth. (Needles of simple compasses are magnets.) The end that turns toward the north pole of the earth is called the north pole of the magnet. The other end is called the south pole of the magnet.

If two magnets are brought close together, like poles will repel each other; unlike poles will attract each other. In either case, the force exerted is proportional to the magnetic strength of the poles and inversely proportional to the square of the distance between the poles.

If a magnet, such as a compass needle, is brought close to a straight conductor carrying a direct current, the needle will turn crosswise to the conductor. This reaction indicates that the magnetic forces associated with the current encircle the conductor.

Right-Hand Rules. The direction of magnetic forces around a straight conductor can be determined in the following way: Curl the fingers of the right-hand around the conductor with the thumb pointing in the direction of the current. The fingers will then be pointing in the direction of the magnetic forces.

A magnetic field with different characteristics develops around a coiled conductor, or *solenoid* (conductor that looks like it is wrapped around a circular cylinder). When current flows through the solenoid, the current creates a magnetic field that envelops the coil. The magnetic forces are directed along the central axis of the coil in one direction and along the exterior of the coil in the opposite direction.

If the fingers of the right hand are curled around a solenoid with the fingers pointing in the same direction as the current in the coil, the thumb will point in the direction of the magnetic field along the central axis of the coil.

In effect, the solenoid becomes a magnet. The end of the coil toward which the thumb

points acts like a north pole and the opposite end, like a south pole.

Permanent magnets have a natural magnetic field. The magnetic field generated by an electric current is called an induced field. In either case, the characteristics of the field are the same.

Generators and Motors. If a coil is moved across a magnetic field, a voltage will be created across the coil. If a conductor connects the two ends of the coil, current will flow through the conductor and the coil. Thus, a magnetic field causes electric current to flow in a moving conductor, whereas an electric current creates a magnetic field. These relations between electricity and magnetism are used in practice to convert mechanical energy to electric energy and electric energy to mechanical energy.

Electric generators produce electricity by rotation of coiled conductors in a magnetic field. For this purpose, rotational energy usually is obtained from hydraulic or steam turbines.

Electric motors provide rotational energy when electric currents create a magnetic field that causes magnets to rotate.

References

B. Stein et al., *Mechanical and Electrical Equipment for Buildings*, 7th ed., Wiley, New York, 1986.

D. Leach, *Basic Electric Circuits*, 3rd ed., Wiley, New York, 1984.

Words and Terms

Alternating current	Insulator
Ampere (amp)	Magnetic field
Charge	Motor
Circuits: series, parallel	Network
Conductance	Ohm
Conductor	Resistance
Direct current	Right-hand rule
Generator	Voltage

Significant Relations, Functions and Issues

Resistance and conductivity: Ohm's law.

Currents and voltage drop in closed circuits: Kirchhoff's law.

Voltage and resistance in series and parallel circuits.

Relation of magnetic fields and rotations to electric current in motors and generators.

13.2. CHARACTERISTICS OF ALTERNATING CURRENT

Direct-current circuits, as described in Sec. 13.1, can be applied in many ways in buildings. Such circuits, however, usually are used in buildings only for special purposes, because electricity supplied by utility companies to buildings is in the form of an alternating current, a current that reverses direction at frequent and constant intervals of time. Alternating current is used because of the following conditions:

As indicated in Sec. 13.1, it is advantageous to transmit electricity at high voltages, because smaller conductors can be used than those required for low voltages and high currents. But use of high voltages for transmission of electricity in buildings is hazardous. For safety, therefore, high voltages used for transmission of electricity from generating plants are stepped down in substations to much lower voltages in areas where the electricity is to be utilized. These lower voltages may be reduced even more, for safety reasons, to service voltages before being furnished to buildings.

The most economical means available for voltage conversions are transformers, devices that accomplish the desired results through use of changing magnetic fields associated with changing currents. Such currents can be readily generated by alternators.

These generators supply an alternating current that varies in magnitude with time in the same way as a sine or cosine function. The current rises from zero to a maximum, drops to zero, then reverses direction and repeats the changes in magnitude. At the same time, voltages change in the same manner. Also, the magnetic field associated with the current develops and collapses as the current changes in magnitude.

These changes in the magnetic field are employed in transformers to step down high voltages to low voltages. Consequently, utility companies generate alternating current at high voltage for transmission over long distances, then use transformers for conversion to low

voltages. Alternating current can be efficiently used for almost all purposes for which electricity is needed in buildings.

Sine-Type Currents and Voltages

The variation of magnitudes of alternating currents and voltages is cyclic. The pattern and magnitudes are repeated at constant intervals.

A *cycle* is a complete set of positive and negative values through which an alternating current or voltage passes.

Frequency f is the number of cycles per second of alternating current or voltage.

Magnitudes of alternating currents and voltages, as generated in practice, generally can be plotted as sine or cosine curves. At any instant, the magnitude of the current can be computed from

$$i = I_m \sin (360ft + \alpha_I) \qquad (13.16)$$

where

I_m = maximum value of the current, amp
f = frequency, Hz (cycles per seond)
t = time, seconds
α_I = constant, or angle, degrees at time $t = 0$

In the U.S.A., alternating current usually is generated at 60 Hz. A single cycle, with $\alpha_I = 0$, is plotted in Fig. 13.4a.

Also, at any instant, the magnitude of the voltage can be computed from

$$e = E_m \sin (360ft + \alpha_E) \qquad (13.17)$$

where

E_m = maximum voltage
α_E = constant, or angle, degrees, at time zero

Figure 13.4a also represents a single cycle of alternating voltage, with $\alpha_E = 0$.

Phase is the time relationship between two currents, or two voltages, or a voltage and a current with the same frequency. If α_I is the same for two currents, or α_E is the same for two voltages, or α_I for a current equals α_E for a voltage, these pairs of currents and voltages are in phase with each other. They pass through zero, maximum or minimum values simultaneously.

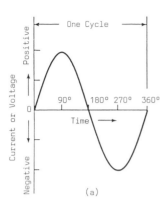 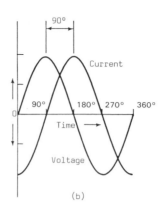 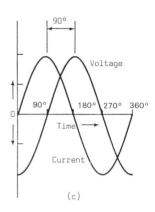

Fig. 13.4. Variations in current and voltage of alternating current are represented by sine curves. (a) A single cycle of current or voltage. (b) Voltage-current phase relationship when voltage leads current by 90°. (c) Voltage-current phase relationship when current leads voltage by 90°.

Two currents, or two voltages, or a current and a voltage may differ in phase. At any instant, one may lead or lag the other by up to 360°. For example, if α_E is 90° larger than α_I, the current passes through zero when the voltage is at its maximum positive or negative value. The voltage is said to lead the current by 90°, or to lag the current by 270° (see Fig. 13.4b). But if α_I is 90° larger than α_E, the voltage passes through zero when the current is at its maximum positive or negative value. The current is said to lead the voltage by 90°, or to lag the voltage by 270° (see Fig. 13.4c).

Phase angle is the smallest difference, measured in degrees, between the occurrences of maximum values of two currents, or two voltages, or a voltage and a current.

Effective Currents and Voltages. Components of alternating-current (ac) circuits can be arranged in series or in parallel in the same manner as dc circuits. Ac circuits can be similarly analyzed, with Eqs. (13.16) and (13.17) for the instantaneous magnitudes of current and voltage. It usually is more convenient, however, to use an effective current I and an effective voltage E in analyses of ac circuits.

An effective current I is numerically equal to the steady value, amp, of a direct current that will produce the same average heating (expenditure of power) in the same resistance. For ac in the form of a sine or cosine function, $I = I_m/\sqrt{2} = 0.707I_m$. Similarly, the effective volt-

age $E = E_m/\sqrt{2} = 0.707E_m$. When an alternating current or voltage is specified, it should be taken as the effective value unless some other value is specifically indicated.

Ohm's law applies to resistances in ac circuits, just as it does to resistances in dc circuits. Thus, $e = iR$ and $E = IR$, where R is the resistance, ohms.

Inductive Reactance and Susceptance

When ac passes through a coiled conductor, or solenoid, the magnetic field enveloping the coil changes as the current changes. The effect on the coil is similar to that which would be produced if the coil were moved through a steady magnetic field. A potential difference, or voltage, is generated in the coil. This voltage opposes the change of current that produces the changing magnetic field. The effect is called *self-inductance.*

The effective voltage of self-inductance E is proportional to the effective current I. The relationship can be expressed as:

$$E = IX_L \qquad (13.18)$$

where the constant of proportionality is known as the *inductive reactance* and, like resistance, is measured in ohms. For parallel circuits, it is convenient to use

$$I = B_L E \qquad (13.19)$$

where

$B_L = 1/X_L$ is called the *inductive susceptance* and is measured in mhos.

If the current through the coil varies as indicated by Eq. (13.16), the voltage across the coil due to self-inductance varies as indicated by Eq. (13.17). But because the voltage opposes the changes in current, the current lags 90° behind the voltage (see Fig. 13.4b). The current is zero when the voltage attains its maximum value.

Because of this 90° phase angle between voltage and current, the average power for self-inductance is zero. As the magnetic field about the coil develops, energy extracted from the electrical circuit is stored in the field. Then, as the magnetic field collapses, this energy is restored to the electrical circuit.

Capacitive Reactance and Susceptance

An electrostatic condenser, or capacitor, consists of two or more plates, capable of conducting electricity, with electric insulation between them. If a potential difference is applied across the terminals of a condenser, electric charges are stored on the plates.

A capacity circuit is a circuit that contains a capacitor. If dc is applied to such a circuit, no current will flow, because the circuit is broken by the insulation in the capacitor, but if ac is applied, the effect is like that of a closed circuit. Each time the direction of the voltage applied to the capacitor reverses, the capacitor gives up the charge stored in it and then stores up a charge of the opposite sign. As a result, current flows in the circuit with the same frequency as that of the applied voltage. The effect is called *capacitance*.

The effective current I is proportional to the effective voltage E. The relationship can be expressed as:

$$I = B_C E \qquad (13.20)$$

where the constant of proportionality B_C is known as the *capacitive susceptance* and, like conductance, is measured in mhos. For series circuits, it is convenient to use:

$$E = I X_C \qquad (13.21)$$

where

$X_C = 1/B_C$ is called the *capacitive reactance* and is measured in ohms.

If the voltage applied to a capacitor varies as indicated by Eq. (13.17), the current due to capacitance, or storage ability of the capacitor, varies as indicated by Eq. (13.16). Charge on the capacitor varies in phase with the voltage. Current, however, is the rate of flow of charges. It is zero when the applied voltage is maximum, and there is no flow of charges because they are being completely stored in the capacitor. The current is a maximum when the applied voltage is zero and the charges are flowing at the maximum rate to and from the capacitor. Consequently, in a capacity circuit, the current leads the voltage by 90° (see Fig. 13.4c).

Because of this 90° phase angle between voltage and current, the average power for capacitance is zero. Energy obtained from the circuit is stored in the capacitor as voltage increases. This energy is restored to the circuit as voltage decreases.

Impedance and Admittance

As in dc circuits, voltage E in ac circuits is proportional to the current I. In general, this relationship can be written as:

$$E = IZ \qquad (13.22)$$

where the constant of proportionality Z is called *impedance* and is measured in ohms.

From Ohm's law, in an ac circuit containing only resistances, Z equals the equivalent resistance, computed as for resistances in dc circuits. In such a circuit, current and voltage are always in phase.

In an ac circuit containing only pure inductance, Z equals the equivalent inductive reactance, computed in the same way as equivalent resistance. In such a circuit, the voltage leads the current by 90°.

In an ac circuit containing only pure capacitance, Z equals the equivalent capacitive reactance, computed in the same way as equivalent resistance. In such a circuit, the current leads the voltage by 90°.

In an ac circuit containing combinations of resistances, inductances and capacitances, the phase angle may lie between 0° and ± 90°. The positive sign applies when the voltage leads the current, and the negative sign when the current leads the voltage.

The relationship between current and voltage in an ac circuit may also be written as:

$$I = YE \qquad (13.23)$$

where

$Y = 1/Z$ is called the admittance and is measured in mhos.

In an ac circuit with only resistances, Y equals the equivalent conductance. In an ac circuit with only pure inductance, Y equals the equivalent inductive susceptance. In an ac circuit with only pure capacitance, Y equals the equivalent capacitive susceptance.

Series Circuits. Figure 13.5a shows an ac circuit with a resistance and a coil connected in series. The current I is the same through both the resistance R and the coil, which has an in-

ductive reactance X_L. By Kirchhoff's voltage law, the sum of the voltages across the resistance and the coil must be equal to the applied voltage E. The voltage and the current through the resistance are in phase, however, whereas the voltage across the coil leads the current by 90°, and therefore also leads the voltage through the resistance by 90°. Consequently, the voltages across the resistance and the coil must be added vectorially to obtain the applied voltage E, as indicated in Fig. 13.5b.

Now, the voltage across the resistance $E_R = IR$, and that across the coil $E_L = IX_L$. The applied voltage E equals IZ, where Z is the impedance of the resistance and coil in series. Since the current is the same for each vector, the vector diagram in Fig. 13.5b to an appropriate scale also indicates the relationship between impedance, resistance and inductive reactance. Thus, Z is the vector sum of R and X_L; or from a different viewpoint, R and X_L are the components of Z. The phase angle θ, or angle between E and I, equals the angle between Z and R.

Figure 13.6a shows an ac circuit with a resistance R and a capacitor with capacitive reactance X_C connected in series. The current I is the same through both the resistance and the

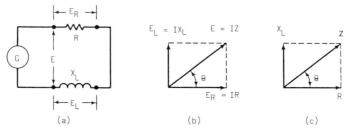

Fig. 13.5. (a) Alternating current circuit with resistance R and inductive reactance X_L in series. (b) Vector addition of voltages in (a). (c) Vector addition of R of X_L, which yields impedance Z of circuit.

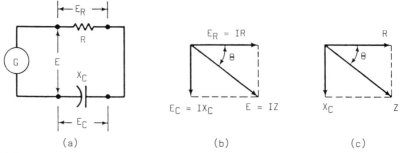

Fig. 13.6. (a) Alternating current circuit with resistance R and capacitive reactance X_C in series. (b) Vector addition of voltages in (a). (c) Vector addition of R and X_C, which yields impedance Z of circuit.

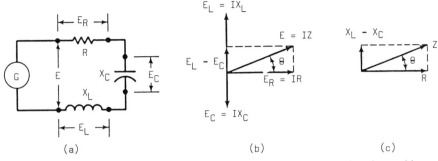

Fig. 13.7. (a) Alternating current circuit with resistance R, inductive reactance X_L and capacitive reactance X_C in series. (b) Vector addition of voltages in (a). (c) Vector addition of R, X_L and X_C, which yields impedance Z of circuit.

capacitor. By Kirchhoff's voltage law, the sum of the voltages across the resistance and the capacitor must be equal to the applied voltage E. Determination of impedance and phase angle is completed in the same way as for inductive reactance, except that X_C is considered to be negative, as shown in Fig. 13.6b and c.

Figure 13.7a shows an ac circuit with a resistance, coil and capacitor in series. The applied voltage E equals the vector sum of $E_R = IR$, $E_L = IX_L$ and $E_C = -IX_C$. (IX_C is taken with a minus sign because the phase angle for capacitance is negative. IX_L, in contrast, is taken as positive because the phase angle for inductance is positive.) Figure 13.7b shows the vector addition. Since the current is the same in the resistance, coil and capacitor, the impedance Z equals the vector sum of the resistance and the reactances (see Fig. 13.7c). The phase angle θ equals the angle between Z and R. This angle may be positive or negative depending on whether the inductive reactance of the circuit is larger than the capacitive reactance.

Parallel Circuits. Determination of phase angle θ is computed in the same way as for series circuits, except that θ is taken as the angle between the current supplied, $I = YE$, and the current I_R through the resistance (Fig. 13.8). Y is the vector sum of the inductive susceptance B_L and the capacitive susceptance B_C, with B_C considered as negative.

Power in AC Circuits

A pure inductance stores energy in a magnetic field, then releases the energy back to the circuit from which it was extracted. No power is expended. A pure capacitance stores energy in an electric field, then releases the energy back to the circuit from which it was taken. Again, no power is expended. Power in an electrical circuit, therefore, is consumed only in resistances. This power $P = E_R I_R$, where E_R is the effective voltage drop across the resistance and I_R is the effective current flowing through the resistance. In ac circuits, E_R and I_R are in phase.

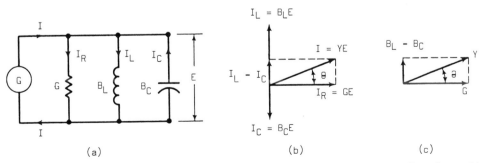

Fig. 13.8. (a) Alternating current curcuit with conductance G, inductive susceptance B_L and capacitive susceptance B_C in parallel. (b) Vector addition of currents in (a). (c) Vector addition of G, B_L and B_C, which yields admittance Y of circuit.

In an ac circuit where resistances and reactances are in series, $I_R = I$, the current supplied by the generator. $E_R = E \cos \theta$, as can be seen in Figs. 13.5b, 13.6b and 13.7b. Hence, in a circuit with impedance, the power, watts, is given by

$$P = E_R I_R = EI \cos \theta \qquad (13.24)$$

where θ is the phase angle between the current I and applied voltage E.

In an ac circuit where resistances and reactances are in parallel, $E_R = E$, the voltage applied by the generator. $I_R = I \cos \theta$ as shown in Fig. 13.8b. Therefore, in a circuit with impedance, the power, watts, is also given by Eq. (13.24).

Power Factor. Cos θ is called the power factor of the circuit. It can range from zero to unity. Therefore, if a circuit contains a reactance, the actual power, watts, expended is always less than the product of applied voltage and current, volt-amperes.

In practice, power factor is usually given as a percentage. Thus, the maximum power factor, which develops for a circuit containing only resistances, such as incandescent lamps, is 100%. For a factory with several large but lightly loaded induction motors, in contrast, the power factor may be as low as 40%.

Low power factor is usually undesirable for many reasons. One important reason, however, is that many more volt-amperes have to be supplied than the watts expended. According to Eq. (13.24), for a specific amount of power, current required is inversely proportional to the power factor. Thus, for an electric load with 50% power factor, twice as much current must be supplied as for an electric load with 100% power factor. To carry the current for the smaller power factor, conductors with double the cross-sectional area are required. Hence, the cost of conductors is at least twice as high as that for conductors to a load with 100% power factor. In addition, the I^2R power loss in transmission through the conductors is much larger for the 50% power factor.

Costs of supplying power for a specific kw load, therefore, are much higher for a low power factor than for a high power factor. To recover the larger costs, utilities charge extra when power factors are low.

Power factors are often low because of the use in a building of inductive apparatus, such as induction motors, transformers, electromagnets or fluorescent or vapor lamps. To avoid the extra power charges and to improve power transmission, the power factor should be corrected by incorporation in the circuit of capacitive equipment. This equipment should be connected in parallel with the inductive apparatus at or close to it.

Inductive apparatus has a positive phase angle, whereas capacitive equipment has a negative phase angle. When the two types of equipment are connected in parallel, the positive phase angle is reduced, resulting in an increase in the power factor. Generally, it is economical to correct the power factor to 90 to 95%.

For improving the power factor of inductive equipment, any of three types of capacitive equipment may be used: capacitors, or condensers; synchronous motors, and synchronous condensers (synchronous motors operating without a mechanical load, solely for the purpose of power correction). Synchronous motors may be substituted for some induction motors, where several motors are in use, to correct a low power factor. Usually, however, where induction motors are required, capacitors are more economical, unless the motors are large. Synchronous condensers are economical where correction is needed for very large amounts of power. They are also useful for voltage regulation.

Conversion of AC to DC

Generally, dc motors offer the advantage of more accurate speed control compared with ac motors. Hence, dc motors may be specified for installation in buildings for elevators, machine tools, conveyors and process equipment. The dc power required for these motors is often obtained by converting ac to dc. For this purpose, motor-generator sets or rectifiers are usually used.

Motor-generator sets consist of an ac motor and a dc generator. The motor, which powers the generator, is driven by ac.

Rectifiers convert ac taken from the building power supply to dc by combining elements that permit the flow of current in only one direction. Thus, each element cuts off half a cycle of ac, but also each passes half a cycle of ac in the same direction. The half-cycle supplied by one element follows or overlaps the half-cycle supplied by another element.

Rectifiers have the advantages that they may be located close to the points where the dc power is to be used and power distribution is the same as for ac. Some types also have the advantage that they have no moving parts to wear out.

Two-Phase and Three-Phase Systems

The ac circuits previously discussed are called single-phase systems. Under an alternating voltage at a specific frequency, current flows, at that frequency, from one terminal of an alternator to electric loads through one conductor and returns to a second alternator terminal through a second conductor.

It is often more economical, however, to transmit to the loads voltages that differ in phase but are at the same frequency. For this purpose, a single phase may be carried by each of several conductors from the alternator to the loads, and a combination of the phases may be carried by a conductor back to the alternator.

Circuits applying voltages at the same frequency but with different phases are known as multiphase systems. Two-phase systems apply two voltages differing in phase. Three-phase systems apply three voltages differing in phase.

One advantage of multiphase systems is that multiphase alternators and motors usually cost less than single-phase equipment with the same power rating. Another advantage is illustrated in Fig. 13.9.

Buildings usually have installed equipment that operate efficiently at different voltages. For example, for safety reasons, low voltage is supplied for portable tools and household appliances, but voltage about twice as large is provided for heaters and large motors that operate more efficiently at high voltage. If separate circuits are used to furnish low and high voltage, four wires are needed, as shown in Fig. 13.9a.

If, however, a two-phase system is used, one conductor can be eliminated. Figure 13.9b shows a two-phase system in which conductors 1 and 2 carry equal voltages that differ 180° in phase. The return wire N is called a neutral.

In Fig. 13.9b, the potential difference between either phase wire 1 or phase wire 2 and wire N is shown as 115 volts. (Other voltages, though, can be used.) Hence, equipment rated at 115 volts may be connected between either wire 1 and wire N or between wire 2 and wire N. The current returning through wire N equals the difference between the currents through wires 1 and 2. If the loads for each phase are

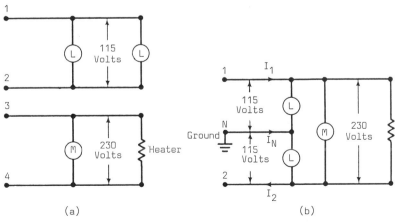

(a)   (b)

Fig. 13.9. (a) Four conductors are used to apply 115 volts across a lamp in one circuit and 230 volts across a motor and a heater in a second circuit. (b) Only three conductors are used in a two-phase circuit to apply 115 volts across each of two lamps and 230 volts across a heater and a two-phase motor.

nearly equal, or balanced—and good design arranges loads on the system with this objective—the neutral wire N carries no current.

The potential difference between phase wires 1 and 2 equals $2 \times 115 = 230$ volts, because the voltages in those wires differ by $180°$ in phase. Hence, equipment rated at 230 volts may be connected across the two phase wires. In this way, a two-phase system supplies power through only three conductors to the same loads for which four conductors are required for single-phase transmission.

The neutral wire N is usually connected to the ground. (The voltage at the connection is considered to be zero.) Then, the voltage between either phase wire and the ground is the same as the voltage between either phase wire and wire N. This is desirable for safety reasons.

A three-phase system offers the same advantages over a single-phase system as does a two-phase system. Also, a three-phase system offers better speed and torque control for motors than does a two-phase system. Hence, where such control is needed, for example, for industrial plants, three-phase power is usually furnished.

Figure 13.10a shows how lighting, heaters and motors may be served by a three-phase system. As indicated in Fig. 13.10a, four conductors are often used, with one serving as a neutral, which is grounded. (Basically, only the three phase wires are needed.) Conductors 1 and 2 carry voltages that differ by 120°. These voltages are represented in Fig. 13.10b by vectors E_{N1} and E_{N2} drawn at an angle of 120° with each other. Also, conductor 3 carries a voltage that differs in phase by 120° from the voltages in both wires 1 and 2. This voltage is represented by vector E_{N3} in Fig. 13.10b. The frequency is the same for all three voltages, and the voltages relative to the neutral wire are all equal in magnitude.

The potential difference between any two phase wires is obtained by vector subtraction, as shown in Fig. 13.10b for the vectors E_{N1} and E_{N2}. Because the phase angle is 120° and the voltages are equal in magnitude, the potential difference between phase wires equals $\sqrt{3}$ times that between any phase wire and the neutral wire or the ground. In Fig. 13.10a, the effective voltage between the neutral wire N and each phase wire is shown as 120 volts, but other voltages are often used. For this case, the effective voltage between any two phase wires is $120\sqrt{3} = 208$ volts.

In Fig. 13.10a, loads requiring 120 volts, single phase, are shown connected across the neutral wire N and one of the phase wires. A heater requiring 208 volts is supplied by two of the phase wires. A three-phase motor is connected to each of the three phase wires.

Good design of a three-phase system requires

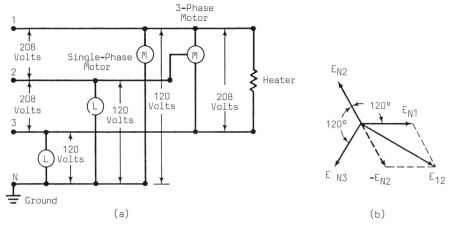

Fig. 13.10. Four-wire, three-phase circuit. (a) Lamps and single-phase motor operate on 120 volts between ungrounded conductors and a grounded, neutral conductor. A heater connected between two ungrounded conductors gets 208 volts. A three-phase motor receives power from all three ungrounded conductors. (b) Voltage E_{12} between two ungrounded conductors equals the vector difference of the voltages E_{N1} and E_{N2} between the conductors and the neutral.

that the loads for each of the phases be nearly equal, or balanced. When this happens, the neutral wire N carries no current. When the system is unbalanced, the wire carries the vector sum of the currents in the phase wires. Loads should be arranged so that this sum is small.

References

B. Stein et al., *Mechanical and Electrical Equipment for Buildings*, 7th ed., Wiley, New York, 1986.

D. Leach, *Basic Electric Circuits*, 3rd ed., Wiley, New York, 1984.

Words and Terms

ac circuits: single-phase, two-phase, three-phase
Alternating current (ac)
Capacitive reactance
Conversion of current: ac/dc
Cycle
Frequency
Impedance
Inductive reactance
Phase
Phase angle
Power factor
Susceptance

Significant Relations, Functions and Issues

Aspects of alternating currents: cycle, frequency, phase, phase angle, voltage.

Constants of proportionality in ac currents: inductive reactance, inductive susceptance, capacitive susceptance, capacitive reactance, impedance, equivalent conductance.

Circuits: series and parallel; power; conversion of ac/dc; single-phase, two-phase and three-phase systems.

13.3. ELECTRICAL LOADS

An electrical load is any electrical service that must be provided by an electrical system. The load may be specified by its requirements for power, amperes, voltage, frequency, number of phases, dc or ac, temperature conditions, type of insulation and time in service.

The electrical distribution system in a building often has to be designed before all loads are known. In the schematic stage of design, locations of utilization equipment may be only generally indicated on floor plans and the electrical requirements can only be roughly estimated. In later design stages, equipment requirements may be revealed gradually. Sometimes, during this design process, equipment and its location may be completely changed. As a result, electrical loads may have to be considerably revised.

Thus, it is important that preliminary estimates of electrical loads be carefully made so that the electrical design can be easily modified to accommodate load changes. Designers should bear in mind that such changes are not only likely to be made during building design but also after the building has been occupied. The cooperation of all the members of the building team usually is necessary for development of a satisfactory load estimate.

Some electrical loads, such as lighting, are continuous; that is, they may be applied for 3 hr or longer without interruption. Other loads, such as motors, may be applied intermittently. Hence, it is not likely that an electrical distribution system will have to serve the sum of all the loads simultaneously. To account for the probability that at most only a portion of the total load may be applied at any time to an electrical system, some or all of certain loads are multiplied by a demand factor, as permitted by the local building code or the "National Electrical Code," sponsored by the National Fire Protection Association (see Table 13.1).

Demand factor is the ratio of the maximum demand of part or all of an electrical system to the total connected load on the part or on the total system, respectively.

In computation of loads, it usually is convenient to treat separately lighting and other loads, such as those from motors, which are called power loads. The reason for this is that in buildings with large lighting and large power loads each type may be served from different load centers.

Lighting Loads

Local building codes or the National Electrical Code usually specify the minimum power for lighting for which a building electrical system

Table 13.1. Minimum Lighting Loads[a]

Type of occupancy	Load, watts, per sq ft	Correction for probability of demand	
		Total load, kw	Demand factor
Dwellings (except hotels)	3	First 3 or less	1.00
		Next 117	0.35
		Portion over 120	0.25
Hospitals	2	First 50 or less	0.40[b]
		Portion over 50	0.20[b]
Hotels and motels	2	First 20 or less	0.50[b]
		Next 80	0.40[b]
Office buildings	5	Total kw	1.00
Warehouses	$\frac{1}{4}$	First 12.5 or less	1.00
		Portion over 12.5	0.50
All others	. . .	Total kw	1.00

[a]Based on the "National Electrical Code," National Fire Protection Association.
[b]For areas in which all lights will be on at the same time, use 100% demand factor.

should be designed. This minimum is usually proportional to the floor area. This area is computed from the outside dimensions of the building or area being considered and the number of floors, but not including open porches, garages for dwellings, or unfinished spaces or unused spaces in dwellings unless adaptable for future occupancy. Table 13.1 lists minimum lighting power, watts per sq ft of floor area, based on requirements in the National Electrical Code.

The values in Table 13.1 may not be sufficient for a specific installation. For more accurate estimates of lighting loads during design, the intensity of illumination to be provided, type of lamps to be installed, mounting heights and fixture layout should be known.

Power Loads

Power loads are much more difficult to estimate than lighting loads. In the early design stages of a building, only rough estimates can be made, based on information obtained on power loads in similar buildings with the same type of occupancy. For example, combinations of lighting and power demands in industrial buildings may range from 3.5 volt-amp per sq ft for small-device manufacturing to 10 volt-amp per sq ft for general manufacturing and to as large as 25 volt-amp per sq ft for airplane factories.

Demand factors for power loads also vary over a considerable range. For example, for domestic electric ranges, the National Electrical Code lists demand factors that depend not only on the number of such appliances installed in a building but also on the kw rating of the ranges. In an industrial building, the demand factor for a group of motors driving a conveyor belt may be 100%, whereas the demand factor for a group of portable tools may be only 30%.

Load estimates also may use as a guide the approximate energy, kwhr, required to produce a unit quantity of product. For example, aluminum may take about 9 kwhr per lb, electric-process steel about 0.33 kwhr per lb and portland cement about 0.05 kwhr per lb. For an automobile, 1,050 kwhr may be required, and for 1,000 pairs of shoes, 470 kwhr may be consumed.

13.4. ELECTRICAL CONDUCTORS AND RACEWAYS

Metals such as copper, aluminum and steel conduct electricity. Because copper has lower resistance, however, it is normally used for distribution of electricity in buildings. Depending on the current to be carried, the conductor may be in the form of a solid round wire, strands or bars. All of these should have adequate strength, insulation and *ampacity*, or current-carrying capacity, for the conditions under which they are to serve. Usually, the conductors are wrapped in insulation, except where they are connected to the terminals of electrical devices or equipment.

In general, electrical distribution lines should be protected against damage from normal activities in a building, from work done in making building repairs or maintenance, from exposure to moisture or chemicals and from the environment. When a conductor is damaged, the metal may be exposed or broken. In either case, occupants may be exposed to electric shocks, electric arcs may be formed that can cause a fire, or large currents may be short-circuited; that is, transmitted through other conductors or metal building components not intended to carry such currents, thus creating hazardous conditions. To reduce the chances of damage or personal injury, conductors are usually enclosed in a protective cover and also in a housing, called a *raceway*.

Raceways may be made of metal or insulating material. Many types of raceways are in use, including rigid metallic and nonmetallic conduit, metallic tubing, flexible metal conduit and underfloor channels.

Also for safety, overcurrent protective devices, such as *fuses* and *circuit breakers*, are incorporated in circuits to break the circuit, or halt the flow of current from a supply source, when excessive currents flow. Other power-line controllers, such as surge protectors, filters, and isolation transformers, may be inserted in a circuit to protect sensitive electric and electronic equipment, such as motors and computers. Ground-fault interrupters are used to prevent injury to people from faulty grounding. All these devices operate automatically when protection is needed. Their purpose differs from that of switches. A *switch* is a device that operators of electric equipment use to close a circuit to permit flow of current or to open a circuit to interrupt the flow of current.

Safety Regulations

Local building codes usually specify safety measures that must be taken in installation and operation of electrical distribution systems and electric equipment. These codes are usually based on the National Electrical Code, obtainable from the National Fire Protection Association, Quincy, Mass. This code contains requirements for wiring design and protection,

wiring methods and materials, general and special equipment, installations in various types of occupancies and communication systems.

Underwriters Laboratories, Inc., tests and inspects electric materials, devices and equipment. If the agency approves the performance and safety of an item, it is permitted to carry the UL label of approval. The agency also periodically publishes lists of approved items.

Major Distribution Conductors

The electrical supply for a building may be obtained directly from a street main or from transformers. The supply conductors from the street main or the transformers to the service equipment of the building are called *service conductors*.

Service equipment comprises the main control and means of cutoff of the electrical supply to the building. The equipment usually consists of a circuit breaker or switch and fuses, and their accessories, and is installed near the point of entrance of service conductors to the building.

Conductors extending from the service equipment to various distribution points within a building are called *feeders*. Conductors between the feeders and *outlets*, or places where current is taken from the system to supply utilization equipment, are called *branch circuits*.

At an outlet, a junction box (see Fig. 13.11) may be installed, and within it protected connections may be made between branch-circuit wires and wires from the equipment. Or one or more receptacles may be installed at the outlet (see Fig. 13.12). A *receptacle* is a contact de-

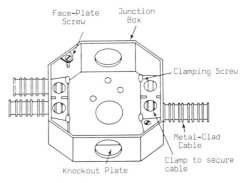

Fig. 13.11. Junction box.

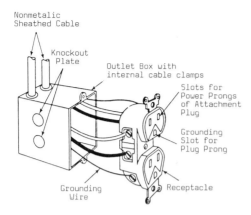

Fig. 13.12. Outlet box and duplex receptacle.

vice for the insertion of prongs of a single *attachment plug*, to establish electrical connection between the branch-circuit conductors and the conductors of a flexible cord attached to the plug.

Grounding of Conductors. One conductor of every interior wiring system should be grounded. A *ground* is a conducting connection, whether intentional or accidental, between an electrical circuit or electric equipment and the earth or some metal building component that serves in place of the earth.

The *grounded conductor* of a wiring system is intentionally grounded. It also is required to be continuously identified as a grounded conductor throughout the system. Furthermore, the grounded conductor of the interior wiring system must be electrically connected to the grounded conductor of the service wiring.

The purpose of a grounded conductor in interior circuits is to keep the potential difference between a power wire and the ground as small as possible, to provide protection against dangerous shocks from high voltages and against fire. For example, as indicated in Sec. 13.2, the neutral wire of a three-phase system usually is grounded. Hence, while the potential difference between the phase conductors may be 208 volts, the potential difference between a phase wire and the ground is only 120 volts.

A neutral wire, however, often carries current, because a multiphase system cannot be maintained in balance continuously. Consequently, if a fault should develop in a circuit

and any of the conductors should come in contact with a metal building component or metal equipment housing, occupants touching these would be subjected to an electric shock. To prevent such hazards, equipment housing should be grounded with a separate conductor, called a grounding conductor.

A **grounding conductor** is installed to connect either equipment or the grounded circuit of a wiring system to a grounding electrode or electrodes. (Often, metal cold-water pipes are used as grounding electrodes.) The installation should be such that there will be no objectionable flow of current over a grounding conductor. (Temporary currents set up under accidental conditions are not considered objectionable.) Sometimes, because a grounded neutral carries current, there will be a continuous flow of current to a nearby grounding conductor. Such a flow of current is objectionable. To prevent it, the location of the grounds should be changed or other measures should be taken to interrupt the flow of current.

Each service circuit, whether power alone, lighting alone or a combination of power and lighting service, should be grounded (see Fig. 13.13). The grounding electrode conductor should be connected to the circuit on the supply side of the service disconnecting means, preferably within the housing of the service disconnecting means. Also, the grounded conductor should be run to the electric equipment. Hence, any fault currents from the wiring to the ground that develop should return to the grounded conductor at the grounding electrode for the service instead of to the supply-source ground through the earth.

Wire Sizes

Copper wire is generally specified by a gage number or by area in circular mils. In the U.S.A., the standard gage is American Wire Gage (AWG), which is the same as Browne & Sharpe (B & S) gage. The larger the gage number, the smaller is the area of the wire.

Usually, the smallest wire used for distribution of electricity in buildings is No. 14, although No. 12 is a preferred minimum. The wire area doubles with every increase of three

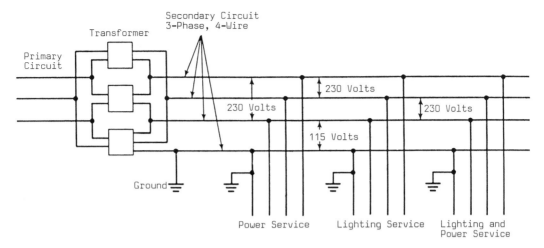

Fig. 13.13. Grounding of neutral wires in a three-phase system.

gage numbers. The largest gage size is No. 0000 (No. 4/0). Larger wires and cables are specified in circular mils.

A circular mil (CM) is the area of a circle 1 mil ($\frac{1}{1,000}$-in.) in diameter. The area of any circle can be obtained in CM by squaring the diameter, mils. For example, the area of a circle 3 mils in diameter is $3^2 = 9$ CM. To convert CM to square inches, multiply the number of CM by 0.7854 and divide by 1,000,000.

For building distribution systems, conductors up to No. 10 are usually solid wire. No. 8 may be solid or strands. Strands are used for No. 8 in raceways and for all larger-size conductors.

Cables up to 1,000,000 CM (1 in. in diameter) are made up of 7, 19, 37 or 61 strands.

Ampacity of Conductors

Current-carrying capacity, or ampacity, of bare conductors is determined by the safe metal temperature. Ampacity of insulated conductors depends on the temperature rating of the insulation, or maximum permissible operating temperature of the insulation. The conductor temperature for a specific current depends on the air temperature around the conductor and the heat generated in the conductor.

A conductor becomes heated because of its resistance to the flow of current. One kilowatt is equivalent to 3,415 Btu per hr.

The heat may be dissipated to the environ-

ment by radiation, convection of air currents or conduction via a raceway. When several conductors are placed in a raceway, heat is not dissipated as rapidly as for fewer conductors or for a single conductor suspended in air. Hence, the current rating is larger for such a single conductor, and the National Electrical Code limits the number of conductors that can be placed in the various sizes of raceways.

The National Electrical Code also contains tables listing the allowable ampacities of standard-size conductors, with and without insulation.

Protective Coverings and Insulation for Conductors

Conductors usually are wrapped in insulation. In turn, the insulation generally is wrapped in a protective covering. Nearly always, this outer covering is colored to identify the conductor.

Color Coding. The National Electrical Code requires that insulated grounded conductors should be colored white or natural gray. For feeders larger than No. 6, however, the grounded conductors may be identified instead by distinctive white marking at terminals during installation. In certain three-phase, four-wire systems where the potential difference between one phase wire and the neutral wire is about double that between the other phase wires and the neutral wire, the wire with the higher potential difference should be colored orange. The code contains no other requirements for color coding of feeders.

When several branch circuits are included in the same raceway, one system should incorporate a grounded conductor with white or natural gray identification. The grounded conductor of each of the other systems should have a colored stripe, other than green, running along the white outer covering, the stripe color being different in each of the systems.

In a branch circuit, the grounding conductor, if insulated, should be colored green or green with one or more yellow stripes.

Ungrounded conductors should be identified by colors other than white, gray, or green. Usually, for 3-wire systems, the wires are colored black, red and white. For 4-wire systems, the wires generally are blue, black, red and white. All ungrounded conductors of the same color should be connected to the same ungrounded feeder conductor.

Conductors for systems of different voltages should be assigned different colors.

Insulation Materials. The insulation around a conductor not only must have a high resistance to passage of electric current but also a high resistance to damage from heat and aggressive chemicals and to passage of water. Most conductors used for distribution of electricity in buildings are insulated with rubber, plastic or mineral fiber. The National Electrical Code assigns identifying letters to different types of insulation and presents tables listing the maximum operating temperature and permissible applications for each type of insulation.

The code also specifies materials with which each type of insulation should be covered. Some insulations require no outer covering. Others may be required to be covered with a moisture-resistant, flame-retardant, nonmetallic covering, a nylon jacket, copper, glass, lead or flame-retardant cotton braid.

Wiring Assemblies

Generally, the wires of a system—two-wire, three-wire or four-wire—are grouped together in runs to outlets or taps. For open wiring or con-

cealed knob-and-tube wiring, however, the National Electrical Code requires that only single wires be used.

Open Wiring. Insulated conductors not concealed by the building structure are considered open wiring. This type of wiring is often used for temporary installations, for example, in buildings under construction. It may not be used for permanent installations in commercial garages, theaters, assembly halls, motion-picture studios, hoistways or hazardous locations.

Open wiring may be secured to and supported by porcelain, glass or composition insulators. These supports, called knobs or cleats, should hold the conductors from $\frac{1}{2}$ to 1 in. away from building surfaces, depending on the voltage. In general, insulator supports should be placed not more than $4\frac{1}{2}$ ft apart, not more than 6 in. from a tap and not more than 12 in. from the last lampholder or receptacle served. In dry locations, however, when the wiring is not exposed to severe physical damage, a conductor may be encased in flexible nonmetallic tubing, which may be supported by straps up to $4\frac{1}{2}$ ft apart.

For open wiring with up to 300 volts between conductors, the wires should be spaced at least $2\frac{1}{2}$ in. apart. For potential differences between 300 and 600 volts, the conductors should be at least 4 in. apart. Also, conductors in open wiring should be separated from walls, floors, timbers or partitions through which they pass. Contact should be prevented by insertion of the wires in tubes or bushings of noncombustible, nonabsorptive insulating material. In addition, open wiring should be kept at least 2 in. away from metallic conduit, piping or other conducting material and from any exposed lighting, power or signal conductor. Alternatively, contact may be prevented by insertion of additional insulation between an open conductor and other conducting bodies.

Concealed Knob-and-Tube Wiring. Single, insulated conductors supported in a manner similar to that for open wiring may also be concealed in hollow spaces of walls and ceilings. Such installations, however, may not be used in occupancies for which use of open wiring is prohibited.

Concealed conductors should be separated at least 3 in. from each other, or encased in flexible metal tubing where space is limited. Also, the conductors should be kept at least 1 in. away from building surfaces. Where practicable, conductors should be extended singly along separate joists, rafters or studs.

Mineral-Insulated, Metal-Sheathed Cable (Type MI). Type MI cable contains one or more conductors insulated with a highly compressed refractory mineral, such as magnesium oxide, which has a fusion temperature of about 5,000°F and is stable at high temperatures. The conductors and insulation are sheathed in a watertight and gastight metallic tube. As a result, the cable is completely noncombustible.

This type of cable may be used in many hazardous locations. It may be embedded in plaster, masonry or concrete. Also, the cable may be laid underground. In buildings, the cable may be supported by staples, straps, hangers or similar fittings. Spacing of supports generally should not exceed 6 ft.

Metal-Clad Cables. One or more insulated conductors may be factory assembled in a flexible metallic sheath. Such cables are classified as Type MC or Type AC.

Type MC is a power cable with copper conductors No. 4 gage or larger. The sheath is either an interlocking metal tape or an impervious, close-fitting, corrugated tube. An uninsulated grounding conductor is also housed within the metal armor. This type of cable may be used in partly protected areas, such as in continuous, rigid cable supports. It may be used in wet locations if the metallic covering is impervious to moisture, or if a lead sheath or other impervious jacket is provided under the metallic covering, or if the insulated conductors are approved for use in wet locations. Spacing of supports should not exceed 6 ft.

Type AC, also known as *BX* cable, has an armor of flexible metal tape (see Fig. 13.11). Type ACL cable, for use in places exposed to water, or embedded in masonry or concrete, or in ducts or raceways underground, contains lead-covered conductors. Type AC cables, except ACL cables, have an internal copper bonding strip in intimate contact with the metal armor for its full length. Both the bonding strip and the armor provide a means of grounding outlet boxes, fixtures and other such equipment. The conductors have an overall moisture-resistant, fire-retardant fibrous covering around the insulation. (In Type ACT, only the individual conductors have a moisture-resistant fibrous covering.)

Type AC cable may be used only in dry locations, for underplaster extensions, embedded in plaster finish on brick or other masonry, or in the hollows of block or tile walls. The flexibility of the cable makes its use advantageous for concealed wiring. Its appearance, however, is disadvantageous for applications where the cable will be exposed. Also, the National Electrical Code prohibits its use, in general, in theaters and assembly halls, motion picture studios, hazardous locations, storage battery rooms or hoistways or on cranes, hoists or elevators.

Spacing of supports should not exceed $4\frac{1}{2}$ ft. Also, the cable should be securely supported within 12 in. of each outlet box or fitting, to prevent the cable from pulling away from the box connector. The supports do not have to be insulators.

Nonmetallic-Sheathed Cables. One or more insulated conductors may be factory assembled in a sheath of moisture-resistant, flame-retardant, nonmetallic material. The assembly may also include an uninsulated grounding conductor. This type of cable is also known as *Romex*. It is classified as Type NMC or Type NM.

Type NMC has a fungus-resistant and corrosion-resistant sheath. It may be used in damp locations and in block or tile interior or exterior walls. If the cable is embedded in plaster within 2 in. of the finished surface, the cable should be protected against damage from nails by a cover of corrosion-resistant-coated steel, at least $\frac{1}{16}$-in. thick and $\frac{3}{4}$-in. wide, placed under the finished surface.

Type NM cable may be used only in dry locations. It may be exposed or concealed in hollows of masonry or tile walls. It may not be embedded in masonry, concrete, fill or plaster.

Neither Type NM nor Type NMC may be

used in buildings with occupancies for which use of metal-clad cable is prohibited.

Supports for the cables should be placed not more than $4\frac{1}{2}$ ft apart or more than 12 in. from a box or fitting. Where the cable passes through a floor, the cable should be enclosed in rigid metal conduit or metal pipe extending at least 6 in. above the floor. Supports for the cable need not be insulators.

Switch, outlet and tap devices of insulating material may be used without boxes in exposed wiring. Outlet boxes, when used, may be nonmetallic. With such boxes, a completely nonmetallic enclosure system should be provided for the wiring. This is advantageous for locations where corrosive vapors are present.

Shielded Nonmetallic-Sheathed Cable (Type SNM). One or more insulated conductors may be factory assembled in an extruded core of moisture-resistant, flame-resistant, nonmetallic material. The core is covered with an overlapping metal tape and wire shield and sheathed in an extruded, nonmetallic material that is resistant to moisture, flames, oil, corrosion, fungus and sunlight. The metal tape is spiraled with a long lay.

Type SNM is intended for use in rigid cable supports or in raceways in hazardous locations.

Service-Entrance Cable. One or more insulated conductors may be factory assembled with a suitable overall covering for use as service-entrance wiring. The assembly may include an uninsulated grounding conductor. The cable may have a steel-tape armor, with a fire- and moisture-resistant braid over the armor for protection against fire and atmospheric corrosion. The cable may be classified as Type SE or Type USE.

Type USE may be used underground. Its covering should be moisture resistant but need not be fire retardant.

Type SE has a moisture-resistant, fire-retardant covering. Type SE may also be used in interior wiring systems; that is, as feeders or branch-circuit conductors, if all the circuit conductors are covered with rubber or thermoplastic.

Underground Feeder and Branch-Circuit Cable (Type UF). One or more insulated conduc-

tors may be factory assembled in a sheath that is resistant to flames, moisture, fungus and corrosion. The sheath also should be suitable for direct burial in the earth. The assembly may include an uninsulated grounding conductor.

If cables with a single conductor in each sheath are installed, all cables of the same circuit, including the neutral conductor, should be grouped in the same trench or raceway.

For buried cables, a minimum of 18 in. of earth cover should be provided. The depth of cover may be reduced to 12 in. if the cable is protected on the upper side by a 2-in.-thick concrete pad, a metal raceway, a pipe or similar guard.

This type of cable also may be used for interior wiring, including wet or corrosive locations. It may not, however, be used as service-entrance cable or in buildings with occupancies where metal-clad cable is prohibited.

Nonmetallic Extensions. Two insulated conductors may be factory assembled within a nonmetallic jacket or an extruded thermoplastic cover for either of the following purposes:

1. Surface extensions mounted directly on the surface of walls or ceilings.
2. Aerial, or overhead cable, containing a supporting messenger cable as an integral part of the cable assembly.

Nonmetallic extensions may be used for 15- and 20-amp branch circuits as an extension from an existing outlet. The wiring must be exposed, and the location must be dry.

Surface extensions may be used only in residences and offices, but not on or near the floor. Support spacing should not exceed 8 in.

Aerial cables may be used only for industrial purposes where a highly flexible means for connecting equipment is required. Supports for the messenger cable should be spaced not more than 20 ft apart. Clearance under the cable should be at least 10 ft above the floor for pedestrian traffic, 14 ft for vehicular traffic and 8 ft for work benches. The messenger cable may also serve as a grounding conductor, but not as a branch-circuit conductor.

Busways. For transmitting very large currents in a compact system, copper conductors in the form of bars, called busbars are often used. Busbars generally can carry about 1,000 amp per sq in. of bar cross section.

A busway or busduct consists of a factory assembled group of busbars in a sheet-metal trough. The busbars are insulated from each other and from the enclosure. Plug-in devices are available to permit switches, circuit breakers and other control equipment to be connected to the busbars in a manner similar to that for receptacles and attachment plugs. Also, trolley connections can be used for taps for portable tools and lamps. Branches, however, are usually made with busways, or metal raceways, conduit or tubing, or metal-clad cables.

Busways must be exposed, for accessibility and for dissipation of heat generated in the conductors. They may not ordinarily be used in wet or hazardous locations or in hoistways. Busways are used mainly in industrial buildings as an alternative to conduit or in high-rise buildings as feeders. Supports for horizontal busways usually may be spaced up to 5 ft apart.

Cablebus. To carry large currents, several insulated conductors may be mounted, at least one diameter apart, in a ventilated, metal structure that provides both protection and support. Fittings and conductor terminations are also housed in the framework.

Cablebus can carry large currents at less cost than busduct with conductors of the same cross-sectional area. Cablebus, however, is bulkier, because of the requirement for separation of the cables. In cablebus, the insulated conductors are supported on special insulating blocks that separate the conductors from each other and from the framework. Cablebus is usually assembled on the building site at point of installation, whereas busways generally are factory assembled. Like busways, cablebus may be used only exposed, to insure accessibility and heat dissipation.

Cablebus may be used as service-entrance conductors, feeders and branch circuits, but not ordinarily in wet or hazardous locations or in hoistways. The framework may be used as a grounding conductor. Supports for cablebus may be spaced up to 12 ft apart.

Electrical Connections

For various purposes, conductors have to be joined together to provide a connection with little electrical resistance. There are many methods and devices for doing this. Selection of the appropriate one depends on the type and size of conductor and importance of speed and ease in making the connection. Connections may be made by means of switches or circuit breakers (see Sec. 13.5), splices, taps, receptacles and attachment plugs or electrically interconnected terminals on electric fixtures or equipment.

Splices or taps may be made by soldering, welding or squeezing together the conductors to be joined. For small conductors, pressure connectors are frequently used. For example, screw-on pressure connectors, or wire nuts, may be screwed over the ends of two or three conductors that have been tightly twisted around each other. Bared ends of insulated conductors may be curled under the head of a screw, which is then tightened to press the conductors against an electrical contact; or conductors may be crimped together and covered with a sleeve by a special tool. Sometimes, the end of a conductor may be equipped with a lug for bolting to a terminal or to another conductor. Large conductors often are spliced by soldering and bolting.

Electrical connections should be covered with insulation and protective wrapping. The insulation should be electrically equivalent to that around the conductors.

For protection of connections from damage, a box should be installed at each conductor splice connection point, outlet, switch point, junction point or pull point for connection of cables, conduit, tubing or raceway (see Figs. 13.11 and 13.12). Weatherproof boxes should be used in damp locations.

Boxes are usually made of metal. Nonmetallic boxes, however, may be used with open wiring on insulators, concealed knob-and-tube work, nonmetallic-sheathed cable and nonmetallic conduit.

The size of a box should be such as to provide adequate free space for all conductors enclosed in the box. The National Electrical Code contains tables listing the maximum number of conductors permitted in a box.

Boxes should be supported on a structural member or securely embedded in concrete or masonry. Every box should be provided with a cover. At fixtures, however, a fixture canopy may serve as the cover for an outlet box.

Raceways

Raceways serve the following purposes:

1. Support for conductors
2. Protection of enclosed conductors from damage or corrosion
3. Grounded enclosure to prevent electric shocks
4. Wiring-system ground path
5. Protection of other building components against fire from overheating of or arcing between enclosed conductors

To meet these objectives, raceways usually are made rigid to provide continuous support. Often, they are made of metal, which can be used as a grounding conductor for equipment served by the wiring. If the metal is not corrosion resistant, it must be galvanized, painted or otherwise coated for protection against corrosion. In hazardous or corrosive locations, nonmetallic raceways may be used.

A continuous rigid structure often is used in industrial installations, for economy, to carry both power and signal wiring. For the purpose, the structure may be in the form of a trough, ladder, run, or a channel. The Underwriters Laboratories limits such applications to specific cables.

Metal sections of continuous rigid cable supports should be bonded and grounded. The purpose of this is to provide a continuous path to the ground for fault currents. The support nevertheless may not be used as a grounded circuit conductor or as an equipment grounding conductor, as may Type AC metal-clad cable and metal raceways.

Rigid Conduit. Special pipes, called conduit or tubing, are often used as raceways. Tubing has much thinner walls than conduit. Both types come in sections that are usually joined with threaded couplings. The National Electri-

cal Code contains requirements for spacing of supports and minimum radius of bends.

The number of conductors that can be housed depends on the ratio of the cross-sectional area of the conductors to that of the conduit or tubing. Except for lead-sheathed conductors, one conductor may occupy 53%, two conductors 31% and three or more conductors 40% of the conduit or tubing interior area. (Percentages are slightly different for lead-sheathed conductors.)

Rigid metal conduit may be used under all atmospheric conditions and in buildings with any type of occupancy, with some exceptions. Galvanized conduit often is used for branch circuits incorporated in concrete slabs of multistory buildings.

Flexible Metal Conduit. This resembles the armor of Type AC metal-clad cables. It is also known as *Greenfield*. Conductors may be inserted on the building site, within the area limitations previously given for rigid conduit.

Flexible conduit often is used in short lengths for distributing electricity to electric motors or other equipment where vibration may occur, movement takes place or where many bends around obstructions are required. A watertight type is available for use in wet locations.

Surface-Mounted Raceways. These may be classified as surface raceways and multioutlet assemblies. Oval or flat rectangles in cross section, either type may be metallic or nonmetallic. They are intended for exposed installations in dry locations, where raceways will not be subject to severe physical damage.

A multioutlet assembly is a raceway equipped at intervals with receptacles to which attachment plugs may be connected. The assembly may be installed on an interior building surface or may be flush mounted; that is, the assembly may be recessed if the front is not covered. Nonmetallic assemblies may be recessed in baseboards, at the intersection of floors and walls.

Underfloor Raceways. Also known as underfloor ducts, underfloor raceways are concealed sheet metal enclosures, generally rectangular in cross section. They were developed to make power, lighting and signal outlets readily accessible in buildings where the locations of utilization equipment are uncertain during building design or may be changed frequently after the buildings are in use.

Underfloor ducts are often used in office buildings. Placed in parallel runs just below the floor surface, about 6 to 8 ft apart, the ducts permit placement of power, lighting and signal outlets directly under desks and other furniture. Underfloor ducts also are often used in large retail stores to provide outlets for display-case lighting and electronic equipment, such as cash registers, at any desired location.

Underfloor raceways have the advantage that they are installed below the floor surfaces, usually in a lightweight, nonstructural fill above the structural deck, during building construction. They are equipped with inserts on top at frequent intervals, often about 4 to 6 ft, to provide access to the conductors. The inserts are brought level with the top of the fill where outlets may be attached to them. Inserts also may be placed after a building is in use, by drilling through the nonstructural fill and tapping the underfloor ducts.

Alternatives to raceways installed under the floor during building construction are often inconvenient and expensive. For floors on the ground, a channel must be cut in the floor to the nearest outlet for installation of a raceway, or a surface raceway must be laid on the floor. In multistory buildings, poke-through construction (see Fig. 13.14) may have to be used. Where an outlet is needed, a hole must be made through the floor and through the ceiling below, for insertion of a conduit extending to a ceiling fixture or back through the floor again to an outlet above.

Underfloor-duct installations usually are single level or double level.

In a single-level installation, feeder and distribution ducts are crisscrossed at the same level (see Fig. 13.15a). At least 1 in. of concrete fill is usually placed atop the ducts. Where ducts intersect, junction boxes are installed to connect them. The boxes are brought level with the floor surface and sealed against entrance of water. Metal junction boxes and metal inserts used with metal raceways should be electrically continuous with the raceways.

The advantage of a single-level installation is that the ducts can be kept close to the floor surface, thus permitting use of a shallow concrete fill. When there are numerous wires in ducts, however, large junction boxes become necessary. Their size may require lowering of the ducts and deeper concrete fill.

In a double-level installation, a header duct with feeders may be placed above or below the distribution ducts. With this arrangement, junction boxes may be smaller and feeder capacity of the ducts may be increased. Depth of fill, however, is greater. When the header is placed above the distribution ducts, the top of the header may be laid flush with the floor surface (see Fig. 13.15b). The header in that case should be provided with a rigid cover plate capable of supporting floor loads.

For all types of underfloor installation, the floor surface may be covered with usual flooring materials, such as carpet, linoleum or tile.

Cellular Metal Floor Raceways. In steel-frame buildings with cold-formed steel cellular decking, the cells of the structural deck may conveniently and economically be used as raceways (see Fig. 13.15c). Each cell may house

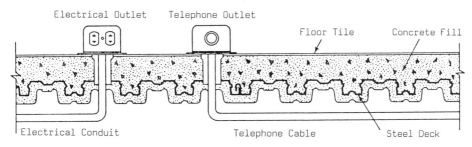

Fig. 13.14. Outlets on a floor are served by conduit or cable through holes cut in the floor.

(a)

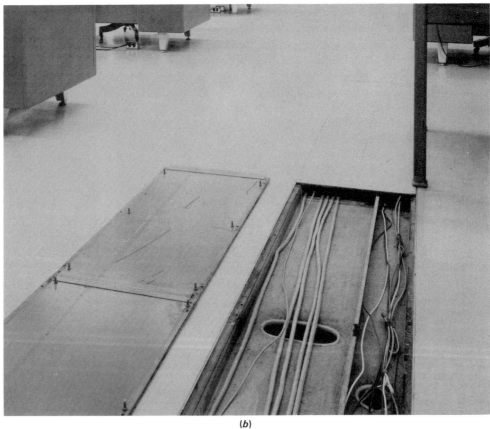

(b)

Fig. 13.15. Underfloor raceways. (a) Single-level ducts. (b) Header with feeders. (c) Closed cells in steel floor deck used as ducts with power and communication conductors in separate cells.

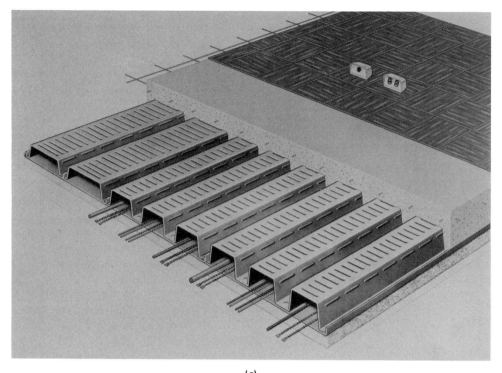

(c)

Fig. 13.15. (Continued)

the same area of conductors as an underfloor raceway with the same area.

Feeders are brought to the conductors in the cells through header raceways set atop the cellular decking in a concrete slab, which may be structural or a fill. The top of the header may be flush with the floor surface (see Fig. 13.15b). Each header may contain wiring for a specific purpose and connect only to those cells housing the distribution wiring for that purpose. A junction box should be installed where each header crosses a cell to which it connects. Splices and taps may be made only in junction boxes and header access units.

Inserts for access to the cells should be brought level with the floor surface. Metal inserts may be placed wherever desired over a cell containing the desired conductors. For this purpose, a hole with a diameter of $1\frac{5}{8}$ in. is cut in the top of the cell with a special tool. The lower end of the insert is then screwed into the cell top to form a secure mechanical and electrical connection.

Cellular Concrete Floor Raceways. In the same manner as the cells of steel decking are used as raceways, the cells of precast-concrete decks may serve as raceways. Transverse metal header ducts may be placed above or below the structural concrete deck for connection of feeders to the distribution cells. Junction boxes are used at connections, as with other types of underfloor ducts. Similarly, inserts for access to the cells may be provided.

A marker, usually a special, flat-head brass screw should be extended through the floor covering to indicate the location of each hidden access point between a header and a cell intended for future use. Different head designs can be selected to identify the type of service in the cell.

SECTIONS 13.3 AND 13.4

References

NFPA, *National Electric Code*, National Fire Protection Association, Quincy, MA, 1984.

J. McPartland, editor, *National Electric Code Handbook*, McGraw-Hill, New York, 1983.

B. Stein et al., *Mechanical and Electrical Equipment for Buildings*, 7th ed, Wiley, New York, 1986.

D. Leach, *Basic Electric Circuits*, 3rd ed., Wiley, New York, 1984.

Words and Terms

Attachment plug
Branch circuit
Circuit breaker
Current-carrying capacity of conductors (ampacity)
Demand factor
Electrical conductor
Electrical load
Feeder
Fuse
Grounded conductor
Lighting load
Power load
Raceway
Receptacle
Service conductor

Significant Relations, Functions and Issues

Conductor size, length and type related to current-carrying capacity.

Color coding to identify conductors.

Wiring assemblies: number of conductors, kinds of circuits (power, communication, etc.), protection and insulation required, installation form (conduit, cable, etc.), connections, grounds, construction details, access for hookup and modification.

13.5. POWER-SYSTEMS APPARATUS

In addition to a power source and electrical transmission lines, the electrical system for a building requires various types of apparatus for providing desired voltages; activating and deactivating circuits as needed for operation of utilization equipment; overvoltage, undervoltage and overcurrent protection; lightning protection, and power-factor correction. Apparatus for power-factor correction is discussed in Sec. 13.2. This section describes the other types of apparatus.

In all cases, installations should meet the requirements of the local building code or the National Electrical Code. Generally also, the electric apparatus should bear the UL label of approval. Much of the apparatus is manufactured to meet standards of the National Electrical Manufacturers Association (NEMA) or the American National Standards Institute.

Transformers

High voltages are used to transmit electricity economically over long distances. For safety at points of utilization of the electricity, however, power is distributed at low voltages. Transformers are used to convert high voltages to lower voltages.

Voltages of 600 volts or less are considered low, those between 601 and 15,000 volts are considered medium, and those above 15,000 volts are considered high.

Low voltages are generally required for utilization equipment, such as lighting and motors. Medium voltages, often 2,400, 4,160, 6,800 and 13,000 volts, are usually used for distribution to buildings or within large buildings. High voltages are used by utility companies to transmit power to substations equipped with transformers for stepping down the voltages to lower levels.

A transformer functions only with ac. Basically, the apparatus consists of an iron core with conductors from two separate circuits wrapped around it (see Fig. 13.16a). The circuit supplying energy is called the primary. The circuit receiving and distributing the energy is called the secondary. In a step-down transformer, used to convert a high voltage to a lower one, the high-voltage side of the transformer is the primary. In a booster transformer, often used to correct a low voltage caused by energy losses in distribution, the low-voltage side of the transformer is the primary.

Voltage conversion in a transformer occurs because the changing magnetic field associated with the primary ac imposes voltages on the coil of the secondary. The voltages in the primary and secondary circuits at the transformer are approximately proportional to the number of turns around the core in the primary and secondary coils.

$$\frac{E_s}{E_p} = \frac{n_s}{n_p} \tag{13.25}$$

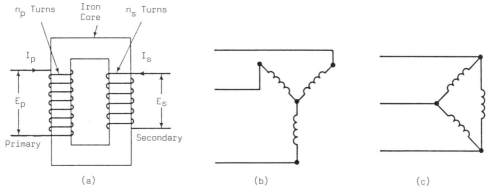

Fig. 13.16. (*a*) Transformer. (*b*) Y connection of coils in three-phase circuit. (*c*) Delta connection of coils in three-phase circuit.

where

E_s = voltage across the secondary coil
E_p = voltage across the primary coil
n_s = number of turns in the secondary coil
n_p = number of turns in the primary coil

Currents in the primary and secondary circuits at the transformer are nearly inversely proportional to the number of coil turns around the core.

$$\frac{I_s}{I_p} = \frac{n_p}{n_s} \qquad (13.26)$$

where

I_s = current in the secondary coil
I_p = current in the primary coil

Energy Losses. Power delivered to the secondary coil is slightly less than that supplied by the primary coil because of the impedance of the transformer. Also, because of the impedance, current and voltage are out of phase. Hence, instead of the power capacity of transformers being given in kilowatts, transformers are rated in kilovolt-amperes (kva), the product of rated voltage and maximum current.

Because of the energy losses, transformers become heated. This heat must be dissipated to keep the temperature of the conductors below the temperature rating of their insulation. For the purpose, the core and windings may be cooled by air or by immersion in an insulating liquid, such as oil or askarel. Askarel is a non-flammable fluid, which, when decomposed by an electric current, generates only nonexplosive gases.

Some of the energy lost in transformers produces noise. Transformers can be designed to limit the amount of noise generated. Manufacturers can supply sound-power ratings, dB, for their equipment, measured in accordance with NEMA standards.

Multiphase Transformers. Coil windings for single-phase circuits at transformers are basically as shown in Fig. 13.16*a*. For three-phase circuits, either of two types of windings may be used on either the primary or secondary side of the transformer. The coil arrangement shown in Fig. 13.16*b* is called a three-phase Y and that shown in Fig. 13.16*c*, a three-phase delta (because of the resemblance to the Greek letter delta). The power delivered by either type is the same, but the voltages between ungrounded conductors are different in the two arrangements. Hence, equipment designed for a three-phase Y cannot be used with a three-phase delta.

Transformer Installation Requirements. Transformers and the rooms or spaces in which they are installed should, in general, be readily accessible to qualified personnel for inspection and maintenance. These spaces should be provided with ventilation sufficient to dispose of the heat from transformer full-load energy losses without creating excessive ambient temperature. As for other electric equipment, exposed non-current-carrying metal parts of transformer installations, including fences and guards, should be grounded, to reduce the risk of shocks to inspection and maintenance personnel.

Meters

Utilities charge customers in accordance with electrical energy used. Consumption of electrical energy in a building is registered on the dials of watthour meters. For a single-purpose building, such as a one-family house, an office building occupied by one company, or a school or other public building, only one meter need be installed. For an apartment building or an office building with multiple tenants, a meter may have to be installed for each tenant as well as a meter for the owner.

Utilities also may include a charge for demand, maximum power used. A demand register may be placed on a watthour meter to indicate the maximum load served within a specific time interval. The device adds and averages the energy supplied within the interval, usually 15 to 30 min.

When electricity is purchased from a utility company, that company usually supplies the meters. A meter should be placed on the supply side of the service equipment for a building, to prevent the meter from being disconnected. Located near the entrance to the building of the service, or supply, conductors, the service equipment provides the main control and means of cutoff of the electric supply.

Three-wire meters usually are used for residences and provide 115/230 volt, two-phase service. For industrial and commercial installations with large motors, three-phase, four-wire meters often are used.

Switches

A switch is a disconnecting means inserted in a circuit to permit or halt the flow of electric current. The desired result is usually achieved by interrupting only ungrounded conductors. A switch, however, may also interrupt grounded conductors if they are disconnected at the same time as or after the ungrounded conductors. If

a switch disconnected only the grounded wire of a two-wire circuit, all devices on the circuit would have a voltage to ground equal to that between the ungrounded, or power, wires to ground. Then, if the grounded wire should be accidentally grounded between the switch and the devices, the circuit would not be controlled by the switch.

A circuit breaker, which is an overcurrent protection device, may also serve as a switch. The National Electrical Code requires that the disconnecting means for ungrounded conductors consist of either a manually operable switch or circuit breaker equipped with a handle or other suitable operating means, or a power-operated switch or circuit breaker if the device can be opened by hand when a power failure occurs.

Switches should be selected with the appropriate ampere and voltage rating for the loads to be controlled. The rating indicates the maximum current that can be transmitted continuously and interrupted repeatedly without damaging the switch or causing overheating. Some switches for control of motors may, in addition, have a horsepower rating.

Classification of Switches. Switches may be classified as general use, isolating or motor circuit, in accordance with their application in a circuit.

A general-use switch is intended for circuit control in general distribution and branch circuits.

An isolating switch is used to keep an electric circuit separated from its source of power. The switch carries no interrupting rating, as does a general-use switch. An isolating switch is not intended for disconnection of a circuit. The switch should be opened only after the circuit has been disconnected by some other means.

A motor-circuit switch is intended for interruption of the circuit for a motor. The switch, rated in horsepower, should be capable of interrupting the maximum operating overload current of a motor with the same horsepower rating as the switch at the rated voltage.

Types of Switches. Switches may be designed for power and manual operation or only for manual operation. Either type may also be designed as quick-make, quick-break type; that is, for very rapid closing and opening of circuits. The purpose of this type is to reduce the amount of arcing that may occur as the contact points being closed or opened approach each other. Rapid opening and closing is often achieved by use of springs in which energy is stored in advance of the desired action.

Snap and knife switches are commonly used manual types. Snap switches are usually operated by a toggle (see Fig. 13.17a), but they also may be operated by a pushbutton or a rotary snap. This type of switch is often used for control of lighting circuits and small motors.

Knife switches consist of copper blades and two contacts, one with projecting prongs and the other with a hinge, for each blade (see Fig. 13.17b). Hinged at one end to a contact and with an insulated handle at the other end, the blades are rotated into the prongs to close a circuit, and away from the prongs to open the circuit. The hinged end of each blade should be connected on the load side of the circuit, so that the blades are dead when the switch is open. When single-throw switches (see Fig. 13.18a to c) are installed vertically, the contact on the load side should be below the contact on the supply side. Then, the switch cannot be accidentally closed by action of gravity alone. When double-throw switches (see Fig. 13.18e) are installed vertically, they should be equipped with a locking device that will hold them open after they have been opened.

Switches may also be used in the form of electromagnets. In these devices, the contacts are closed or opened by magnetic forces. Electromagnetic switches for service below 10 amp are usually referred to as relays. Large switches, called contactors, often are equipped with a magnetic blow-out coil to disrupt an arc that might form between the contacts as they are opened or closed.

Switches also may be classified, in accordance with the number of positions they may take in making contact, as single throw, double throw, etc. Figure 13.18a to c shows single-throw switches. Figure 13.18e illustrates use of a double-throw switch for control of a lighting switch from either of two locations. As shown, the circuit is interrupted and the lights are out. If the switch at either location is turned to the alternate position, the circuit will be closed and the lights will operate. Figure 13.18d shows a switch with two alternate *on* positions and a central *off* position (a hand-off automatic switch).

In addition, switches may be classified, in accordance with the number of contacts that can be made simultaneously, as single pole, double pole, etc. Figure 13.18a, d and e show a single pole switch. The one in Fig. 13.18a, for example, is described as a single-pole, single-throw switch, and the one in Fig. 13.18e as a single pole, double-throw switch. Figure 13.18b shows use of a double-pole, single-throw switch in a three-wire, single-phase circuit. Figure 13.18c illustrates use of a three-pole, single-throw switch in a four-wire, three-phase circuit. The circuits in Fig. 13.18a to c

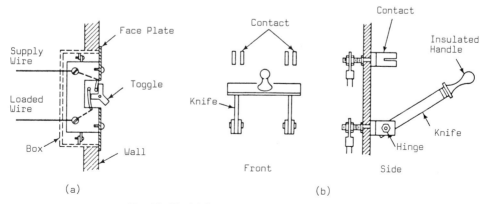

Fig. 13.17. (*a*) Snap switch. (*b*) Blade switch.

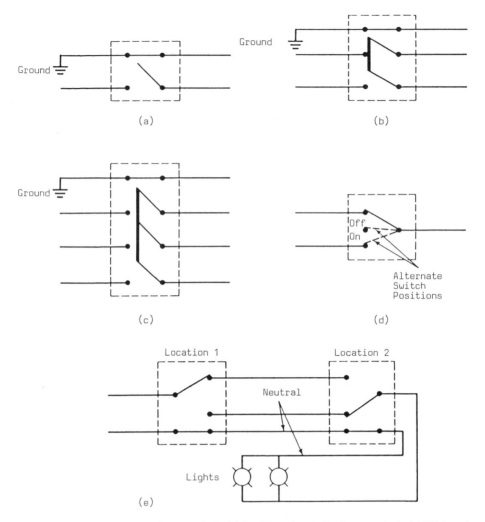

Fig. 13.18. (a) Single-pole, single-throw switch. (b) Double-pole, single-throw switch. (c) Triple-pole, single-throw switch. (d) Single-pole, three-position switch. (e) Single-pole, double-throw switches used for remote control of lights from two locations.

are shown with a solid (continuous) grounded neutral.

Except for snap and knife switches mounted on panelboards or open-faced switchboards, switches should be enclosed in boxes or cabinets and should be capable of operation by handle or other device on the exterior of the enclosure. The enclosure should be appropriate to the location of the installation. Cabinets are classified as general purpose and dust resistant, for dry indoor locations; drip-proof, for indoor use; raintight, weatherproof or watertight, for outdoor use; dust tight; submersible; hazardous; and industrial, for use indoors where resistance

to dust, lint, oil and moisture is required. Cabinets may be designed so that the door cannot be opened when the switch is in the closed position.

Protective Devices for Circuits

In design and installation of electrical systems, provision should be made for immediate detection of abnormal conditions and speedy disconnection of the faulted part of the systems. The objectives are to prevent damage to equipment and injury to personnel and to minimize interruption of electric service. Protection usually is provided for such abnormal conditions as overload, or currents larger than rated currents; short circuits, which generally produce very high currents; arcing ground faults, which usually produce lower overcurrents, and low voltage. Sometimes, protection is provided for overvoltage, abnormal frequencies, phase reversal or single phasing.

Overcurrent protection is required for all circuits. Devices installed for this purpose must be capable of interrupting the maximum available short-circuit current, but they should also be automatically actuated by moderately abnormal currents. Yet, it usually is not desirable that the protective devices automatically interrupt the circuit when such currents are of very short duration. For example, some equipment, such as motors, compressors and transformers, draw currents when starting that are considerably higher than the current for normal operation. The high currents last for only a short time and are not likely to cause damage. (These currents are much smaller than short-circuit currents.) If protective devices are actuated by the starting currents, circuits would be interrupted whenever the equipment started. To avoid this condition, the devices usually have a built-in time delay that permits the moderately abnormal, short-time currents to pass without interrupting the circuit.

For maximum reliability of electric service only the protective devices directly actuated by overloads or faults should open the circuit. Such selectivity is achieved by installation of overcurrent devices in key parts of the electrical system. Also, backup overcurrent protection should be provided as a safeguard against failure of the primary protective system.

In addition to opening circuits when overload currents occur, overcurrent devices should have interrupting-capacity (IC) ratings not less than the short-circuit current available at the line terminals of the devices. Otherwise, a short circuit between two or more ungrounded conductors or between any of them and grounded metal might cause damage or injury. When the short-circuit current may exceed the interrupting ratings of ordinary fuses or circuit breakers, a current-limiting fuse may be installed.

A **current-limiting overcurrent protective device,** when interrupting currents in its current-limiting range, also reduces the current in the faulted circuit. The resulting current is substantially less than that which would flow if the device were replaced with a solid conductor having comparable impedance.

A fuse or circuit breaker should be connected in series with every ungrounded conductor, in general at the point where the conductor receives its supply. When circuit breakers are used, they should open all ungrounded conductors of the circuit. (The National Electrical Code permits, under certain conditions, installation of individual single-pole circuit breakers on each ungrounded conductor of a multiphase system.) Overcurrent protective devices, however, are not permitted to be connected in series with intentionally grounded conductors.

So that a malfunctioning overcurrent device does not pose a hazard in its vicinity, it should be enclosed in a cabinet, as are switches, or a cutout box, unless the device is part of an assembly that provides equivalent protection. An enclosure is not required, however, for devices mounted on open-type switchboards, panelboards or control boards in dry rooms accessible only to qualified personnel and not containing easily ignited material. Also, the operating handle of a circuit breaker may be placed on the outside of an enclosure.

Fuses. To open circuits when an overcurrent occurs, a fuse incorporates a meltable part that is heated and severed by the overcurrent. The meltable part is connected in series with an ungrounded circuit. Melting of the link opens the circuit. Because functioning of a fuse depends on the heat generated by a flow of current, the larger the current, the shorter will be the time it takes a fuse to open the circuit.

Plug fuses house the meltable link in a porcelain cup. They must be replaced with new fuses after they function. They are used in circuits, such as those in residences, with not more than 150 volts to ground and with ratings up to 30 amp.

Cartridge fuses enclose the meltable link in an insulated fiber tube. They may be renewable or nonrenewable, or one time. With the renewable type, a new link can be inserted. With a one-time fuse, a new fuse must be installed. Cartridge fuses are available for 250, 300 and 600-volt circuits and with ratings up to 6,000 amp.

Fuses similar to cartridge fuses are available for medium-voltage and high-voltage circuits.

Current-limiting cartridge fuses are used in circuits where the short-circuit current may exceed the interrupting ratings of ordinary fuses or circuit breakers. After a link melts, such fuses introduce a high arc voltage that reduces the magnitude and duration of the overcurrent. This limits the mechanical and thermal stresses that otherwise would be imposed on system components.

High-interrupting-capacity fuses are a current-limiting type that have the ability to interrupt high values (usually over 10,000 amp) of current. Operating time, however, is slow compared with that of ordinary current-limiting fuses. As a result, high-interrupting-capacity fuses do not reduce mechanical and thermal stresses on system components when high short-circuit currents flow. Nevertheless, such fuses are needed with service equipment and main distribution panels, which are connected to thick cables or large buses that have small resistance.

Although all fuses take some time, but very little, to function, some fuses have a built-in delay and are classified as time-delay fuses. These include plug fuses that require at least 12 sec to open when carrying 200% of rated current and low-voltage cartridge fuses that require at least 10 sec to operate when carrying 500% of rated current. Such fuses are used to pass high short-time currents, such as starting currents of motors.

A switch should be placed on the supply side of every fuse in circuits with more than 150 volts to ground. The purpose is to deenergize the parts of the circuit containing a fuse that has functioned so that the fuse or the destroyed link can be replaced safely.

Circuit Breakers. Fuses have the disadvantages that they are destroyed when they function to protect a circuit and that an individual fuse is required for every ungrounded conductor of a multiphase circuit. In contrast, circuit breakers are not normally damaged when they function. They can be reset, if the fault condition or overload does not persist, by operation of a toggle or a handle. They can be opened automatically or manually. Hence, they can be used instead of a switch, and a single multiple circuit breaker can be installed in a multiphase circuit to disconnect all conductors simultaneously. Fuses, however, are available with higher interrupting capacity than circuit breakers.

Circuit breakers are like quick-make, quick-break switches. Tripping of springs opens or closes contacts rapidly. It may be accomplished thermally, magnetically or by a combination of both. A magnetic trip provides short-circuit protection by opening the circuit instantaneously. A thermal bimetal gives the necessary time delay for overload protection. When used together, the trip units operate a common trip bar for all poles.

In any case, a circuit breaker should be *trip free*; that is, the contacts cannot be held closed if a short-circuit current is flowing.

Circuit breakers for low-voltage circuits may be classified as molded case or power circuit breakers. The latter are available with much higher interrupting capacity.

Molded-case circuit breakers consist of an operating mechanism with contacts, arc interrupter and trip elements for each pole—all enclosed in a molded plastic case. This type of breaker may be operated manually by a handle or a toggle.

Low-voltage power circuit breakers consist of an electrically and mechanically trip-free circuit-breaker element and an enclosure, if one is required by code regulations. The circuit-breaker element comprises direct-acting, dual magnetic, overcurrent tripping devices for each pole; a handle or a solenoid for resetting the trip, and a multiple finger-type contact structure equipped with arc quenchers. The tripping devices may provide a long time delay, short time delay or instantaneous operation.

Protective Relays. These should be used in conjunction with circuit breakers to provide overload and fault protection against excessive voltage variations and phase reversals. A separate control power source must be provided for operation of a relay, which requires low current at low voltage.

Current-Limiting Reactors. A current-limiting reactor is a coil with large inductive reactance placed in series with the service. Its purpose is to protect circuits and equipment by storing energy and thus limiting the magnitude of short-circuit currents that may occur. In this way, reactors assist other protective devices, permitting use of lower-rated and less costly units.

Switchgear and Switchboards

Switchgear and switchboards have two major functions: protection of circuits against short-circuit currents and means for switching circuits and equipment into and out of service. Also, when necessary, switchgear and switchboards provide specialized control for electric equipment.

The terms switchgear and switchboard are often used interchangeably. Generally, however, switchgear applies to protection of high- and medium-voltage circuits (see Fig. 13.19).

The National Electrical Code defines a switchboard as a large single panel, frame or assembly of panels on which are mounted, on the face or back, or both, switches, overcurrent and other protective devices, buses and usually instruments. Switchboards are generally accessible from the rear as well as from the front. Usually, switchboards are completely enclosed in metal. The enclosure, the framing supporting switching equipment, and instruments, relays, meters, and instrument transformers installed on switchboards should all be grounded. Metal-enclosed switchboards are often referred to as *metalclad switchgear*.

Metal-clad structures are generally factory assembled and shipped in sections, ready for installation (see Fig. 13.19b). Main parts of the primary circuit, such as circuit breakers, transformers and buses, are isolated by grounded metal barriers (see Fig. 13.19a). For high- and medium-voltage circuits, power circuit breakers usually are used and are removable for maintenance. Control-power sources for closing and tripping the circuit breakers are also incorporated in the enclosure. Circuit instruments, protective relays and control switches are mounted on the hinged control panel.

Auxiliary sections with the same dimensions as and general appearance of power-circuit-breaker sections are available for various purposes. Such sections can house transformers, bus entrances for termination on the primary bus and specialized control devices for generators and motor-starting circuits (see Fig. 13.19b).

Low-voltage metal-clad switchgear performs similar functions for main, tie and feeder cir-

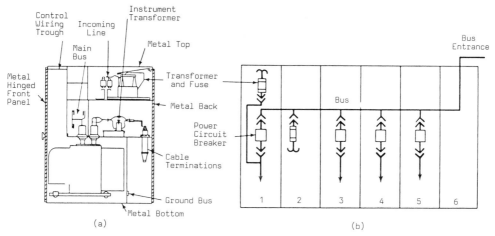

Fig. 13.19. (*a*) Transverse section through medium-voltage switchgear. (*b*) An assembly of switchgear sections: 1. Incoming-line compartment. 2. Auxiliary compartment with transformer. 3 and 4. Motor-starting circuits. 5. Outgoing feeder. 6. Bus-entrance compartment.

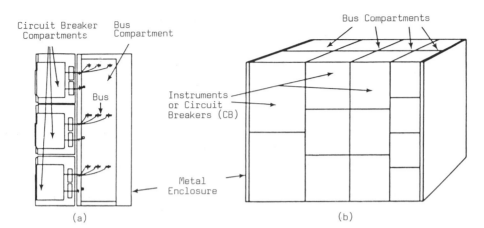

Fig. 13.20. (*a*) Transverse section through low-voltage switchgear. (*b*) Arrangements of stacked compartments.

cuits and for electric equipment rated 600 volts or less. Compartments, however, usually are smaller than those for high-voltage switchgear. They may be stacked one above the other and alongside each other (see Fig. 13.20). Front compartments contain drawout, power or molded-case circuit breakers, other protective devices and control components, such as instruments, meters and relays (see Fig. 13.20*a*). The rear, or bus, compartment houses the primary bus, cable or bus terminations, metal raceways, instrument transformers and control fuses. Figure 13.20*b* shows various arrangements of stacked compartments.

Switchgear may be designed for indoor or outdoor installation. Main switchgear for large buildings and installed indoors generally is used to house service equipment and is placed at the location where service conductors enter the buildings. In buildings with a basement, the switchgear may be placed in a well ventilated room in the basement. Adequate space should be allowed for access to the equipment and for maintenance and replacement of parts. Lifting equipment may be required for handling of the parts. In addition, doorways, shafts and hallways should be wide enough for transportation of parts in and out of the building.

Unit Substations

High voltages transmitted to a building are stepped down to medium voltages for distribu-

tion purposes, usually in an outdoor substation. A unit substation consists of well coordinated, mechanically and electrically connected, power transformers and metal-clad switchgear. In addition to stepping down incoming voltage, a unit substation also provides switching and short-circuit protection for feeders.

A master unit substation (see Fig. 13.21) is often used to transform high voltages to between 2,400 and 13,800 volts. It consists of the following components: a section for connection of incoming high-voltage circuits, a transformer section and metal-clad switchgear with connections for one or more feeders.

A load-center substation is used to convert medium-voltage power to low voltage (600 volts or less). Savings in the cost of an electrical

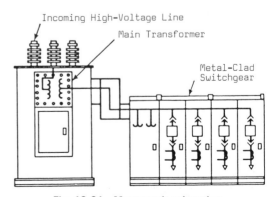

Fig. 13.21. Master unit substation.

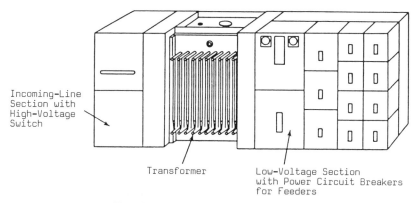

Incoming-Line Section with High-Voltage Switch

Transformer

Low-Voltage Section with Power Circuit Breakers for Feeders

Fig. 13.22. Load-center unit substation.

system result from installation of such a substation close to the location of major electrical loads. The substation permits use of relatively small primary feeders to carry medium-voltage power to the load center instead of large busways or cable transmitting low-voltage power. With load centers located close to the load, shorter low-voltage branch circuits are needed. Not only are transmission losses thus kept small, but better voltage regulation results.

A load-center unit substation (see Fig. 13.22) is a coordinated, factory-assembled apparatus that is easily bolted together in a building. The unit substation usually consists of a metal-enclosed incoming-line section with high-voltage switch and primary-cable termination facilities, a transformer section, and a section for connection of low-voltage feeders, which are protected and switched by power circuit breakers.

Panelboards

Feeders transmit power from substations or switchboards to panelboards. There, connections are made to branch circuits for distribution of the power to utilization equipment. A feeder connected to a panelboard may supply power to up to 42 branch circuits. For this purpose, a panelboard contains bus bars for supplying power to the branch circuits, circuit breakers or fuses and switches for overcurrent protection and switching circuits on and off, and a neutral bus bar for connection of the return wires from the loads (see Fig. 13.23).

The overcurrent devices, buses and wiring

gutters are assembled in the form of a single panel, which is housed in a cabinet or cutout box. Mounted on a wall or partition, the enclosure is accessible only from the front. Generally, panelboards are of dead-front constructions; that is, when the door of the enclosure is opened, no live (electrified) parts are exposed.

Power panelboards are used to supply power to lighting panelboards, motors and other loads, including other panelboards. Lighting panelboards are used to supply power to branch circuits containing lighting fixtures, convenience receptacles, small electric tools and miscellaneous appliances. At least one lighting panelboard should be installed to serve each story of a building.

Service Equipment. Switchboards and panelboards may be used as service equipment, but the National Electrical Code specifies that it must be possible to disconnect a set of service conductors to a building with only six circuit breakers or switches, in a single enclosure. Services for such purposes as fire pumps and emergency lighting are considered separate services and do not have to be grouped with the regular service equipment.

Motor Control Centers

Motors are provided with controllers for starting and stopping and for overload protection. A controller is equipped with a set of contacts for on-off control. For a manual starter, the contacts are operated by hand.

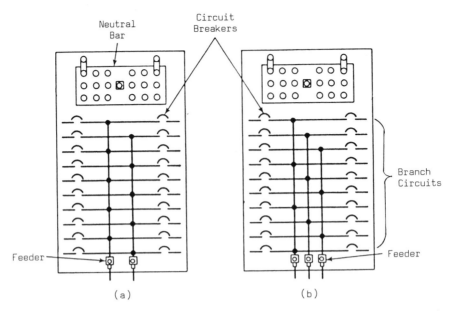

Fig. 13.23. Panelboards. (*a*) **Wiring for three-wire, single-phase power supply.** (*b*) **Wiring for four-wire, three-phase power supply.**

For a magnetic starter, the contacts are operated by a magnetic coil actuated by a pushbutton, thermostat or other type of control. The type of overload protection incorporated in a controller depends on the size of motor.

For motors larger than 1 hp, overload protection of the motors is accomplished with thermal relays. They open the contacts and cut off power to the motors when excessive current persists. For motors 1 hp or less, protection may be provided by a fused switch and a circuit breaker.

The National Electrical Code requires that a disconnecting means be provided within sight of each controller. The disconnecting means may be a motor-circuit switch, general-use switch, isolating switch, circuit breaker, attachment plug and receptacle, or the branch-circuit overcurrent protective device, depending on the motor size and other conditions. The disconnecting means should be capable of disconnecting both the motor and the controller from all ungrounded supply conductors. Therefore, although mounted on the same panel as the controller or enclosed in the same box, the disconnecting means must be an independent device, not part of the controller. The objective is to permit safe periodic inspection and servicing of motors and their controllers.

If the motor controller should fail to open the circuit when the motor stalls or when other heavy-overload conditions prevail, the disconnecting means can be used to cut off power to the motor. Usually, therefore, if a switch is used as the disconnecting means, it must be capable of interrupting very large currents.

While a controller is required for each motor, it is advantageous to assemble controllers for a group of motors in a single location and to supply them from a continuous power bus. Such an assembly is called a motor control center.

For a motor control center, complete units can be assembled and wired in a factory, thus reducing installation costs. For safety, the units may be constructed as dead front, housed in a metal enclosure.

The National Electrical Manufacturers Association (NEMA) classifies motor control centers in accordance with function. NEMA Class I control centers are basically a mechanical grouping of motor starters or control assemblies, or both, which can be applied without systems analysis or systems engineering. Class II centers are designed as complete control systems requiring systems analysis and engineering and including electrical interlocking between units or with outside devices. NEMA also classifies motor control centers in accordance with the type of wiring.

Lightning Protection

A special electrical system can be installed to protect a building against damage from lightning, which is, in essence, a high-voltage surge of electricity of short duration. A lightning-protection system for a building should be designed and installed by experts. For an incomplete or poor installation may be worse than no lightning protection at all. Lightning-protection systems should conform to the standards of the American National Standards Institute, National Fire Protection Association (Bulletin No. 78, "Lightning Code") and Underwriters Laboratories (Standard UL 96A, "Master Labeled Lightning-Protection Systems").

Lightning protection is generally accomplished by installation of conductors that extend from points above the roof of a building to the ground, with the objective of conducting to the ground lightning that may strike the building. Such an installation has the potential hazard that, if it is imperfect, lightning may leap from the lightning conductors to other metal components of the building that offer a path of lower resistance to the ground. Hence, the lightning-protection system must insure that lightning is conducted to the ground without causing damage to the building or exposing occupants of the building to injury.

Electrical systems for buildings, in particular, should be protected against lightning, which can damage conductors and their insulation and electrical equipment. Lightning may cause overvoltages in electrical circuits either by direct stroke on an outdoor transmission line or by electrostatic induction from a stroke to earth near the line. Either of such strokes can send large quantities of electric charges along the line in both directions from the stricken point to ultimate ground. The rate of rise of voltage along the line can be very rapid.

Direct-stroke overvoltages may be on the order of millions of volts and several hundred-thousand amp. Induced strokes, which occur more frequently, may be several hundred-thousand volts with currents ranging from 50 to 2,000 amp.

To prevent electrical-systems damage from lightning, lightning arresters connecting ungrounded conductors to the ground may be installed. Two types of arresters are in general use, valve type and expulsion type.

Primarily, the valve type is used. It provides an air gap in series with a valve element. The valve element offers very high resistance to currents at normal system voltage, but it readily passes currents at voltages of lightning magnitudes. The arrester must be positively grounded so that overvoltage currents are conducted to the ground. The valve element then becomes connected to the ground when an overvoltage of sufficient magnitude occurs to arc over the gap.

An expulsion arrester is designed so that when an arc is produced across a gap, the arc must pass over the surface of gas-evolving material. The overvoltage current is transmitted to the ground, but the emitted gas quickly deionizes the arc and interrupts the flow of following power-system current.

Ratings of arresters give the maximum permissible line-to-ground power-system voltage to which they can be subjected. An arrester should not be installed if its rating is less than the system line-to-ground voltage.

Arresters offer dependable lightning protection only near their locations. Hence, arresters should be placed principally at locations of such apparatus as transformers, switchgear and rotating machines. Each arrester should be connected so as to protect the specific apparatus to which it is assigned.

References

B. Stein et al., *Mechanical and Electrical Equipment for Buildings,* 7th ed., Wiley, New York, 1986.

D. Leach, Basic Electric Circuits, 3rd ed., Wiley, New York, 1984.

NFPA, *National Electric Code,* National Fire Protection Association, Quincy, MA, 1984.

Words and Terms

Circuit protection	Meter
Lightning protection	Motor controller

Panelboard
Switch
Switchboard

Transformer
Unit substation

13.6. ELECTRICAL DISTRIBUTION IN BUILDINGS

Many factors must be considered in planning electrical distribution in a building. Among the most important, however, are safety, reliability, simplicity of operation, life-cycle costs and especially operating costs, energy conservation, voltage quality, ease of maintenance and flexibility. The last requires that provision be made for future expansion of and changes in the electrical system to meet changes in requirements after completion of the building.

Characteristics of a distribution system for a building depend on its power requirements. Because power requirements for different buildings vary widely, distribution systems also vary in detail, although design may be based on established principles and may specify standard components. Planning and design of a distribution system, in general, may follow the following sequence. (Some of the steps listed, however, may be changed or omitted for specific buildings, depending on size, complexity, reliability and flexibility of power required.)

Step 1. Location and size of power source and present and future loads are shown on floor plans.

Step 2. The required capability of the power system is determined.

Step 3. A decision is made as to whether electrical power is to be purchased or generated on site and what voltage is to be supplied.

Step 4. Voltage levels to be provided throughout the building are established.

Step 5. Size, number and locations of substations are determined. In so doing, adequate provision should be made for future expansion of the system.

Step 6. The reliability of electrical service to be provided is determined in accordance with the importance the owner attaches to continuity of service. Then, electrical circuits are arranged and designed to provide the desired reliability.

Step 7. Provision is made for systems grounding and overcurrent, overvoltage and lightning protection.

Step 8. Voltage spread and voltage-regulation limits are checked.

Step 9. Equipment specifications are written so that equipment with adequate ratings and essential safety will be installed and operated at minimum life-cycle cost.

Required Capability of Power System

A close estimate of the sizes and locations of the electrical loads in a building is essential for proper design of the power system. But in the initial design stage, the loads are rarely known with exactness. Hence, it is necessary for the designer to estimate the loads from what information on the proposed building is known and then supplementing this usually vague information with data on loads in similar existing buildings.

More accurate early estimates of the demand for electric power generally can be made by estimating lighting and power loads separately, as recommended in Sec. 13.3, and then combining the estimates to determine the demand for the various areas of the building. The load for each area is usually expressed as load density, volt-amp (va) or kilovolt-amp (kva) per sq ft of floor area. Load density usually will be different for the various areas, depending on their function.

The required system capability can be estimated by adding the products of the individual loads and appropriate factors. The factors most frequently used are the demand factors (defined in Sec. 1.3), diversity factors and load factors.

Demand factor is the ratio of the maximum demand on a subsystem (or on a system) to the total connected loads of the subsystem (or the system). Branch-circuit maximum demand is equal to the sum of the loads connected to the branch circuit multiplied by the demand factor of the loads.

Diversity factor is the ratio of the individual maximum demands of the various parts of a subsystem (or of a system) to the maximum demand of the subsystem (or the whole system). Total maximum demand of a system or of a load center equals the sum of the maximum demands to be met divided by the diversity factor of the load circuits.

Load factor is the ratio of the average load, kwhr, over a specific period of time, usually 30 days = 720 hours, to the peak load occurring in that period, kw. For example, suppose a building consumes 800,000 kwhr in 30 days with a maximum demand of 2,000 kw. The average load was 800,000/720 = 1,100 kw. The load factor then is 1,100/2000 = 0.55. Utility companies generally have higher charges for low load factors.

As an example of the use of demand and diversity factors for load estimation, the electrical distribution system represented by the single-line diagram in Fig. 13.24 will be analyzed. The maximum demand on Branch Circuit A equals the demand factor for that circuit multiplied by the sum of the loads connected to the circuit. The maximum demand on Branch Circuits B, C and D are computed in a similar manner. The maximum demand on Feeder 1 equals the sum of the maximum demands on Branch Circuits A and B divided by their diversity factor. The maximum demand on Feeder 2 is similarly computed from the demands on Branch Circuits C and D. The sum of the maximum demands on Feeder 1 and Feeder 2 divided by their diversity factor yields the maximum demand on the main feeder or on the whole system.

In this way, the maximum demand on a system and on each of its subsystems can be computed, beginning at the loads and continuing toward the power source.

Selection of Voltage Levels

Voltage levels for a distribution system should be carefully chosen. Once a voltage level is selected, changing it can be difficult and costly.

Selection of a voltage level for small buildings, such as one-family houses, may be based mostly on common practice. Utilities generally supply three-wire service. This consists of a neutral wire, connected to a transformer midpoint, and two power lines differing in phase by 180°. This service provides for:

Single-phase, two-wire 230 volt, by tapping across both power lines.
Single-phase, two-wire 115 volt, by tapping one power line and the neutral.
Single-phase, three-wire 155/230 volt, by using both phase wires and the neutral.

For larger buildings, selection of distribution voltages depends on the following factors:

1. Magnitude of load.
2. Distance over which power is to be transmitted.
3. Rated voltages of utilization devices that are available. For example, single-phase,

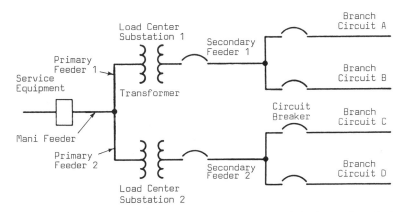

Fig. 13.24. Single-line diagram illustrating flow of current through feeders to branch circuits.

120-volt power must be supplied in buildings with incandescent lamps, appliances, hand tools, business machines and small, fractional-horsepower motors. For large, multiphase motors, however, the horsepower rating may determine the voltage to be supplied.

4. Safety. (With this objective, building codes and the National Electrical Code limit the voltages that may be used in equipment and distribution circuits within buildings.)
5. Life-cycle costs. (For high-voltage distribution systems, some components may cost more than for low-voltage systems, but operation costs may be lower because of lower transmission losses and more efficient operation of equipment.)

The service frequently used for large buildings is three-phase, four-wire, 120/208 volt. This consists of a neutral wire and three power wires with current differing in phase by 120°. It provides:

Single-phase, two-wire 208 volt, by tapping across two power wires.
Single-phase, two-wire 120 volt, by tapping across one power wire and the neutral.
Two-phase, three-wire 120/208 volt, by using the two power lines and the neutral.
Three-phase, three-wire 208 volt, by using the three power wires.
Three-phase, four-wire 120/208 volt, by using all four wires.

Selection of Substation Sizes

For a building supplied power at a voltage larger than utilization levels, the supply voltage must be stepped down to the utilization levels in one or more substations within the building. Required substation sizes, kva, depend on the number of substations to be installed. Sizes and number of substations should be selected to achieve the lowest life-cycle cost for the whole electrical system.

This cost is primarily determined, except for costs of the utilization equipment, by the costs of three major system components—primary cable, substation transformers and switchgear, and secondary cable. Decisions affecting selection of any of these components affect the other two. Hence, design of the electrical system must take into account the interrelationship of these components.

Section 13.5, in the discussion of substations, points out that it is usually economical to place a substation close to the load to be served. If a large number of substations are installed for the purpose, long lengths of primary feeder cable are required. But the lengths of secondary feeder cable needed will be small. Also, with a large number of small substations, initial cost per kva of the substations will be high. In contrast, if only one substation were to be installed in a large building with high power requirements, lengths of primary feeder cable will be smaller, whereas lengths of secondary feeder cable will be larger. Transformer cost will be smaller, but costs of the secondary switchgear may be higher. Hence, it is possible to determine a substation size that will yield the lowest system life-cycle cost.

In addition to costs, other factors also may be of importance in selection of substation sizes. For example, sometimes it is necessary to select large substations because of high primary voltages, to avoid complications in substation overcurrent protection. Also, sometimes it is desirable to use a large substation to serve a large spot load, such as a furnace. As another example, substation size may be determined by available space within the building. Sometimes, space can be found for one large substation but not for two small substations in different locations, or vice versa. In commercial buildings, installation of substations in other than basement or penthouse spaces may require use of valuable rentable floor area. Hence, in such buildings, it is more economical to use large substations.

Circuit Arrangements for Industrial Buildings

Some industrial buildings house processes that require continuous electrical service. Other plants can tolerate short-time outages without severe economic losses. The designer must provide an electrical system with the required degree of reliability at the lowest life-cycle cost.

Service reliability depends on use of high-quality equipment with a low failure rate and an appropriate power-system circuit arrangement. The circuit arrangement can improve reliability by providing alternate power sources that can be tapped when system equipment fails or is removed from service for maintenance, replacement or inspection. Often, however, provision of alternate power sources increases initial cost of the electrical system. This measure may also increase circuit complexity to an undesirable extent.

It is good practice to make circuitry simple and to use the minimum number of adequate high-quality pieces of equipment. Complex circuitry can be a source of mistakes that can cause disastrous system failures or accidents when equipment rarely used has to be operated in emergencies. Mistakes occur because personnel are not likely to be familiar with such equipment when emergencies occur infrequently.

Circuit Arrangements for Commercial Buildings

Basic circuit arrangements used for commercial buildings are similar to those used for industrial buildings. Different approaches to system design may have to be selected for commercial buildings, however, because of the high value of space in such buildings, limited floor-to-ceiling height and limited choice of supply voltage. Many commercial buildings are supplied at utilization voltage by the electric utility company.

Figure 13.25a shows an example of electrical distribution in a multistory building. The 480Y/277-volt main switchboard is placed in the basement. Three-wire feeders supply 480 volts to air-conditioning and elevator motors and to pumps. A busway riser extends vertically from the switchboard to serve each story of the building. In each story, in an electric closet (see Fig. 13.25b), a lighting panelboard and a 480-120-volt transformer receive power from the

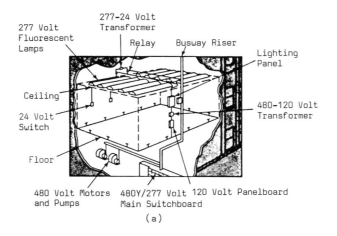

(a)

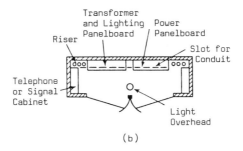

(b)

Fig. 13.25. Typical systems for a large commercial building. (a) Busway riser supplies power to panelboard in each story. (b) Plan of electrical closet on each building floor.

busway riser. As indicated in Fig. 13.25*a*, lighting is operated by a relay controlled by a remote 24-volt switch in a convenient location. Power for the control circuit is obtained from a 277-24-volt transformer.

One or more electric closets should be built on each floor of the building and located for the most economical distribution of power to utilization devices serving that floor.

Power-System Protection

Electrical systems should be designed to remove, with least system disturbance and minimum or no damage to equipment, any abnormal current that may arise in any part of the system. An abnormal current may develop gradually from overloading or suddenly from a short circuit. The National Electrical Code and local building codes contain requirements for capabilities and locations of protective devices.

Overload currents are caused usually by connection of too many load devices to a circuit or by overloading of a motor. Such currents may range up to six times normal full-load current. (Short-circuit currents generally are considerably larger.) Protection devices used for overloads usually are of the time-delay type, to prevent interruption of the circuit under transient conditions, such as starting of a motor or energizing of a transformer.

Short circuits are caused by connection of two unloaded ungrounded conductors or of an ungrounded conductor and a grounded conductor or a grounded building component, because of insulation failure or accident. The connection forms a completely new circuit in which the magnitude of the current is limited only by the circuit impedance. Because the impedance of such a circuit is usually very small, the short-circuit current is very large. It must be interrupted swiftly, before it can cause damage or injury. This is normally done with one or more of the overcurrent protective devices described in Sec. 13.5.

Only one combination of interrupting points, however, will produce the least system disturbance. Hence, detecting devices in the power system should be coordinated with each other to insure selectivity in circuit interruption so that minimum disturbance of the system results when an overcurrent actuates a protective device.

Coordination of short-circuit protection requires that the smallest possible section of the electrical system be isolated when a fault occurs. Coordination is achieved by setting the protective device closest to the fault to interrupt the circuit faster than the protective devices nearer to the power source. Protective devices should have progressively slower operating times the closer they are to the power source.

Voltage Spread and Regulation

Electric utilization equipment is designed for optimum performance at its rated voltage. Nevertheless, the equipment must be capable of satisfactory performance within a specified voltage spread.

Voltage spread is the difference between maximum and minimum voltages that may occur at a given voltage level. (Voltage spread does not apply to transient voltage changes, such as those due to faults, switching or motor starting.) Spread occurs because of voltage drop in system equipment and circuits as current fluctuates with variations in load on the system. Since maintenance of the specified voltage level is one of the main criteria of electrical-system performance, limitation of voltage spread is a major design objective.

The National Electrical Code recommends, for reasonable efficiency of operation, that the maximum total voltage drop over both feeders and branch circuits to the farthest outlet of power, heating and lighting loads not exceed 5% of the nominal system voltage. Also, conductors for feeders and for conductors for branch circuits should be sized to prevent a voltage drop exceeding 3% at the farthest outlet.

If voltage spread exceeds acceptable limits or is not suitable for a critical load, any of four basic ways may be used to reduce the spread:

1. Lengths of conductors carrying the lowest voltage should be shortened and lengths of

conductors carrying higher voltages should be lengthened.
2. System impedance should be reduced.
3. Voltage regulators should be installed on buses and feeders to compensate for voltage drop.
4. Capacitors should be installed to improve the power factor, because smaller currents flow as power factor is increased.

References

National Electric Code, National Fire Protection Association, Quincy, MA, 1984.

J. McPartland, editor, *National Electric Code Handbook*, McGraw-Hill, New York, 1983.

B. Stein et al., *Mechanical and Electrical Equipment for Buildings*, 7th ed., Wiley, New York, 1986.

D. Leach, *Basic Electric Circuits*, 3rd ed., Wiley, New York, 1984.

Words and Terms

Circuit arrangement	Substation
Demand factor	Voltage level
Diversity factor	Voltage regulation
Load factor	Voltage spread

Significant Relations, Functions and Issues

Power capability of systems: demand factor, diversity factor, load factor.
Considerations for selection of voltage level.
Circuit arrangements: basis for selection.
System protection: overload, short circuit, isolation of faults, protective devices.
Voltage control: Conductor sizes and lengths, voltage regulation, power-factor correction.

13.7. COMMUNICATION SYSTEMS

In this section, communication systems are considered to embrace electrical circuits used to convey information or to control other circuits. These systems are low-voltage systems, distinct from light and power systems and covered by different requirements in local building codes and the National Electrical Code. The latter, however, treats communication systems in a much narrower sense and promulgates separate requirements for them and for sound-recording, remote-control, signaling, power-limited, fire-

protective signaling, and radio and television circuits and equipment.

A remote-control system is any circuit that controls any other circuit through a relay or an equivalent device.

A signaling circuit is any electrical circuit that energizes signaling equipment.

The National Electrical Code applies the term *communication circuits* only to telephone, telegraph (except radio), district messenger, outside wiring for fire and burglar alarms, and similar central-station systems and to telephone systems not connected to a central-station system but using similar types of equipment, methods of installation and maintenance.

NEC requirements for sound-recording and similar equipment apply to equipment and wiring for sound-recording and reproduction, centralized distribution of sound, public-address systems, speech-input systems and electronic organs.

Remote-Control, Signaling and Power-Limited Circuits

Remote-control, signaling and power-limited circuits are divided into three classes. For Classes 2 and 3, current is limited to low values, specified in the National Electrical Code, by fuses or circuit breakers and by supply through transformers that allow only very small currents on short circuits. Current may be limited in some cases to 0.005 amp or less. Class 1 includes all remote-control and signaling circuits that do not meet the special current limitations of Classes 2 and 3.

Class 1 circuits are often merely extensions of light and power circuits. Permitted to carry up to 600 volts, Class 1 circuits generally are subject to the same installation requirements as light and power circuits. Power-supply and Class 1 conductors may be placed in the same enclosure, cable or raceway when they are connected to the same equipment. For Class 1 circuits, however, wires as small as No. 18 may be used, whereas for light circuits No. 14 or No. 12 is the usual minimum size. Also, the small conductors may have special types of insulation. In addition, in some cases, the general rules for overcurrent protection are modified and exceptions are made to the general requirements for system grounding.

Circuit wiring used for magnetically operated motor controllers is a common example of Class 1 remote-control circuitry. Wires as small as No. 18 may be used for control conductors, with overcurrent protection not exceeding 20 amp.

Conductors used with a remote-control switch for remote control of a feeder or branch circuit are usually in Class 1 circuits. In such circuits, wires as small as No. 18 may be used, if overcurrent protection does not exceed 20 amp.

Some signaling systems operate at 115 volts with 20-amp overcurrent protection. Examples are clock, programming, bank alarm and factory paging systems. Other signaling systems, such as nurses' call systems in hospitals, may operate at about 24 volts. All these circuits may contain wires as small as No. 18.

Classes 2 and 3 are used for circuits that need not comply with the general requirements for light and power systems, because they carry such small currents. Maximum permissible circuit voltage is 150 volts. Conductors of Class 2 and 3 circuits may not be placed, in general, in any enclosure, raceway, cable, compartment, outlet box or similar fitting with conductors of light, power or Class 1 circuits. In shafts, Class 2 and 3 conductors should be separated by at least 2 in. from other types of conductors that are not enclosed in noncombustible tubing, raceway or other protective sheathing.

Class 2 circuits differ from Class 3 in the degree of protection offered against potential hazards. Class 2 circuits are intended to offer protection against both fire and shock, whereas Class 3 circuits are concerned primarily with protection against fire.

Current limitations for Class 2 and 3 circuits may be met with use of batteries or current-limiting transformers. The latter have such a high secondary impedance that they cannot supply a current exceeding a specified maximum, even when a short circuit occurs. Transformers generally approved for use on doorbell circuits are suitable for Class 2 systems.

Most small bell, buzzer and annunciator systems have Class 2 circuits, operated at up to about 20 volts and 5 amp. Small intercommunication systems with the talking system supplied power by a battery and the ringing circuit delivered ac power by a transformer also may be Class 2.

Remote-control circuits to safety control equipment are required to be Class 1 if failure of the equipment to operate introduces a direct fire or life hazard. Room thermostats, water-temperature regulating devices and similar controls used in conjuction with electrically controlled household heating and air-conditioning, however, may be used with Class 2 circuits.

Alarm Systems

Signaling circuits are often used in buildings to warn of the existence of emergency conditions, such as fire or smoke or the presence of an intruder. Burglar alarm systems are often installed in conjunction with fire alarm systems because of the savings in cost resulting from dual use of some of the equipment. The two systems differ primarily in the types of detection devices used.

An alarm system consists basically of detection devices; circuits for carrying signals to a control panel when any of the devices are actuated; a control panel for interpreting the signals, selecting circuits through which the signals are to be forwarded and transmitting the signals to those circuits; and signaling circuits for carrying the signals to warning devices and other equipment for initiating protective measures. These final circuits may do any or all of the following: energize audible devices, such as gongs or bells; alert key personnel; shut off oil and gas flow; control fans; notify police or fire departments or a central station; and provide electrical supervision so that a trouble signal will indicate occurrence of a fault in the wiring circuit that would prevent proper alarm operation.

Fire-protective signaling circuits are classified as non-power limited or power limited. Non-power-limited circuits may carry up to 600 volts and are subject to the installation rules for light and power circuits, except that wires as small as No. 18 may be used if overcurrent protection does not exceed 20 amp. Power-limited circuits are restricted to low currents specified in the National Electrical Code. Wires in such circuits may be as small as No. 22.

Telephone Systems

The local telephone company usually designs and installs the telephone system for a building. Building designers should consult with telephone company representatives to insure that adequate telephone service will be provided, that sufficient space will be provided within the building at suitable locations for telephone equipment and that appropriate, accessible paths will be available for telephone wiring. Building contractors should cooperate with telephone company installers and coordinate the construction schedule so that telephone equipment and wiring will be installed during the proper construction stages, without delaying telephone installation, without interference with other construction trades and with a minimum of cutting and patching of building components for passage of telephone wiring.

Service-entrance telephone wiring normally follows the route of electrical power service conductors, but separate service entrances must be provided for telephone and power. Usually, the service-entrance telephone wiring is brought to a terminal in the basement. In small buildings, the terminal normally is a small piece of equipment. In large buildings, the equipment may be so large that a special terminal room and power supply must be provided.

Telephone conductors inside a building should be separated at least 2 in. from conductors of light, power and Class 1 circuits. Also, in general, telephone conductors, like other communication conductors, should not be placed in any raceway, compartment, outlet box, junction box or similar fittings with conductors of light, power and Class 1 circuits. Telephone circuits preferably should be installed during construction stages that will permit concealment of cables in walls, floors, ceilings and closets without the necessity of cutting holes in those components for the purpose. The conductors may be terminated at empty outlet boxes for connection to telephones.

For large multistory buildings, planning for telephone installations during the early design stages is especially important. Relatively large amounts of space, often in critical locations, are required for telephone equipment and wiring. Provision should be made for telephone service not only to predictable locations for telephones but also to possible future locations and for future expansion of service.

Typical facilities required for a large multistory office building are shown schematically in Fig. 13.26. A riser system brings cables from the main terminal room in the basement to the various stories. The riser shaft often serves riser closets on each floor, stacked vertically one above the other. Sleeves in the floors permit passage of the vertical cables. When complex switching equipment is needed to provide building occupants with private branch exchange (PBX) or central exchange (Centrex) services, a special equipment room should be provided for the purpose. Also, apparatus closets are needed to house telephone power equipment and terminating facilities. Such closets can serve telephones up to about 150 ft away. Finally, a cable distribution system is needed

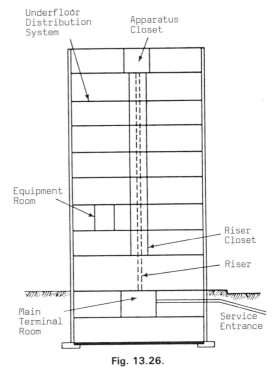

Fig. 13.26.

between the apparatus closets and telephone terminals. Like light and power wiring, telephone circuits in multistory office buildings may be placed conveniently in underfloor raceways.

Sound, Radio and Television Distribution

Planning for equipment and wiring for public address, intercommunication systems, paging, music and reception of radio and television in various parts of a building should start in the early design stages. As for signaling circuits, construction should be scheduled so that the necessary wiring can be concealed without the necessity of cutting holes in building components. The wiring can be terminated in empty outlet boxes, for later connection to appropriate terminal devices.

In general, the power-supply conductors from the light and power circuits to and between the equipment should be installed as required for light and power systems of the same voltage. Amplifier output circuits that carry audio-program signals of 70 volts or less and whose open-circuit voltage will not exceed 100 volts may be Class 2 or 3. Conductors, including antennas, should be kept away from light, power and Class 1 conductors. Coaxial cable may be employed for community television (CATV) systems to deliver low-energy power if the voltage does not exceed 60 volts and the current supply is from a transformer or other device with energy-limiting characteristics. The coaxial cable should contain a grounding conductor.

Data-Processing Systems

Planning for data-processing systems also should start in the early design stages. In some cases, equipment may occupy considerable space, require air conditioning and have extensive wiring.

Power-supply conductors to the equipment should be installed as required for light and power circuits of the same voltage. Data-processing equipment, however, may be connected by means of computer cable or flexible cord and an attachment plug. Separate units also may be interconnected with flexible cords

and cables. These may be exposed on the floor surface if they are protected against physical damage. Alternatively, they may be placed under a raised floor if they are protected by metallic sheathing and if ventilation in the underfloor area is used for the data-processing equipment and data-processing area only.

General requirements for data-processing equipment installation are given in National Fire Protection Association Standard No. 75, "Protection of Electronic Computer/Data-Processing Equipment."

References

NFPA, *National Electric Code,* National Fire Protection Association, Quincy, MA, 1984.

B. Stein et al., *Mechanical and Electrical Equipment for Buildings,* 7th ed., Wiley, New York, 1986.

Words and Terms

Alarm systems	Remote-control systems
Communication circuits	Signaling circuit
Low-voltage systems	Data-processing wiring

Significant Relations, Functions, and Issues

Features of communication systems: low-voltage, less concern for safety (as compared to power systems), use of small wires (easier installation and incorporation in construction), possible use of battery operation for independence from power system during power outages, less concern for protected installation, need for separation from power wiring.

13.8. SYSTEMS DESIGN APPROACH TO ELECTRICAL DISTRIBUTION

The purpose of electrical systems is to supply electrical power for operation of equipment installed in or on a building or on the building site. Most of this utilization equipment may be selected by personnel other than the electrical engineer; for example, HVAC engineers, mechanical engineers, lighting consultants and industrial or process engineers. Yet, the type, number, location and size of the loads imposed on the electrical system by the utilization equipment determine the basic characteristics required of the electrical systems.

Also, electrical systems are affected by decisions made by architects and structural engineers. These designers have to provide space and supports for electrical distribution equipment and wiring. For greatest economy in electrical distribution, the distribution equipment should be placed in locations, chosen by the electrical engineers, that provide the most efficient distribution, and the wiring should follow the shortest path from service entrance to the loads. However, for architectural reasons that take into account aesthetics and the value of space to the owner, distribution equipment is usually assigned to locations that decrease the efficiency of electrical distribution, and the wiring is required to be concealed in floors and walls and above ceilings. For structural reasons, wiring may have to be detoured around structural components. In addition, the electrical installation should not be permitted to contribute to fire and smoke hazard in the building nor to facilitate the spread of fire and smoke.

Because of the small initial cost of electrical systems compared with that of other building systems, the designs of those systems usually are given higher priority than the design of the electrical systems. Systems design, however, may be useful in coordinating the various building systems to produce an optimum overall design. For one thing, the electrical engineer should participate in all discussions of the building team that affect load type, number, location and size. Also, the electrical engineer should be consulted from the initial design stages on with regard to the space needs for electrical equipment and wiring and to the paths to be followed by wiring to the loads.

Data Collection and Problem Formulation

Basic information for design of electrical systems comes from examination of the owner's program of requirements, data from a utility company on the costs and characteristics of available electrical power at the building site and consultations with other members of the building team on the power requirements of equipment they specify. The information concerning loads usually will be vague in the initial design stages. Initial design of the electrical systems therefore has to be predicated on rough estimates and predictions. As design of the building and equipment to be installed in it and on the building site progresses and definitive decisions are made affecting power needs, design of the electrical system can be refined.

The members of the building team should collaborate in location of distribution-equipment rooms and provision of suitable environments in those rooms for the equipment and for maintenance and inspection personnel. Additional collaboration is desirable in determination of paths for the distribution wiring, to minimize cutting of building components during the electrical installation and to provide flexibility for changes in or extension of the electrical system. The results of the cooperative efforts should be displayed on architectural plans, elevations and cross sections and on schematic drawings of the distribution system and submitted to all members of the building team for review.

Further information for design of the electrical systems should be obtained from local building codes, national standards, requirements of the utility company that will supply the electrical power, and restrictions imposed by insurance companies. Also, essential information should be obtained from manufacturers of the electric equipment and devices that will be served by the electrical systems.

In later stages of design, drawings for all the building systems should be studied to insure compatibility of the electrical installation with the other systems and that all power needs will be met, including provision for estimated future needs.

The goal of electrical-system design may be stated as: to design a system that will provide, with complete safety and desired reliability, the necessary electrical power for functioning of a building at minimum life-cycle cost.

Objectives

The chief objective of electrical design should be to achieve the goal with complete safety. This requires adequate provision within the system primarily for prevention of injury to building occupants and secondarily for protection of property against damage. Cost of damage from an electrical system failure can far exceed the value of electrical components. A failure can cause a fire that destroys part or all of a building and endangers the lives of building occupants or that can deactivate equipment, with disastrous economic effects.

Another highly important objective is reliability of service. In some cases, an interruption of electrical service may be merely an inconvenience to building occupants. In other cases, a breakdown may cause only minor losses; for example, costs of spoiled food due to loss of refrigeration in a residence. In still other cases, for example, for an office building or an industrial plant, loss of power may be costly because of the loss of production and sometimes also because of high start-up costs after a shutdown of operations.

Another important objective is simplicity of operation of the system. Complicated systems may offer more opportunities for failures to occur and for mistakes to be made in emergencies.

Still another important objective is the aim of choosing from among all the possible electrical systems that can supply the needed power the one that will have the lowest life-cycle cost. While initial cost of the system is significant, operating costs may be even more important. For example, for an industrial plant, it usually is economical to spend several hundred dollars to eliminate 1 kw of loss. Higher initial costs are often amortized by annual savings from reliable and satisfactory operation of the system.

Another objective is energy conservation, which is closely related to life-cycle cost of the electrical system. Excessive consumption of electricity increases unnecessarily operation costs of the buildings. In addition, however, the wasted energy contributes to depletion of precious natural resources.

Other objectives of systems design of electrical systems include voltage quality, easy maintenance, and flexibility for change and expansion. Voltage quality is measured by stability (lack of major fluctuation in voltage level). Electrical devices often have limited tolerance for voltage fluctuation without suffering some disfunction or possibly even some permanent damage.

For some commercial and industrial occupancies, ease of access and general ability to accommodate change may be critical design goals. Modifications of building equipment involving connection changes and even rewiring without major disruption of the construction may be continuing needs. These situations call for cooperative decision making in the building design and well may affect some detailing of ceiling, wall, and floor construction, as well as the form of the general system and various components of the electrical system. Some form of modular system for circuits, conduits, and connections is likely to be indicated.

Constraints

Power supply may be a major constraint on design of electrical systems. When an adequate supply can be purchased at a reasonable cost, an electrical system usually can be designed to utilize that supply. When power cannot be purchased at an acceptable cost or is not available in adequate quantities, electricity will have to be generated at the building site or selection of a different site may be necessary.

Whether power should be purchased or generated should be determined by an economic study. The study should consider availability of large amounts of process steam, by-product heat, low-grade, inexpensive fuel or solar energy.

Loads are another major constraint on electrical-system design. Power requirements for lighting, motors, HVAC, process equipment, elevators and escalators are usually determined by building designers other than the electrical engineer.

Building codes impose additional constraints, generally to insure safety of electrical systems. The requirements affect voltages, currents, conductor sizes and other characteristics, insulation, overcurrent protection, housing of conductors, connections, transformers, switchgear, panelboards and many other components of electrical systems.

Other constraints often are imposed on design of electrical systems by architectural and structural considerations. A common constraint is the architectural requirement that wiring be concealed from view. This makes it necessary to install the wiring in walls and floors and above ceilings. Often, as a result, a preferred path for the wiring is obstructed by structural members or HVAC ducts.

Synthesis and Analysis

Because of the constraints and the tendency of owners to press for low initial costs, design of electrical systems usually cannot fully achieve all the design objectives. Efficiency, reliability, ease of operation and provisions for future growth may have to be compromised to keep initial costs down. Protection of human life, however, must remain a paramount design concern.

Important design decisions often have to be made in the early design stages when the electrical loads can only be roughly estimated. One of the first important decisions is likely to be whether power should be purchased or generated. If purchasable power is available in adequate quantities, a study should be made of the economics of the two alternatives.

Another important decision may be the proper location of a substation or service entrance relative to the load center of the building. The next important decision may be selection of a voltage for economical electrical distribution. This voltage depends on the voltage of the power supply, size of individual motors to be served and the area covered by the electrical system. Voltage for lighting usually will be different from that required for power. The proper voltage to use for electrical distribution should be determined from a cost analysis. This study should compare installed costs at different voltage levels and with various arrangements of equipment.

If power is purchased, voltage spread of the power supply should be investigated. If voltage variation may be excessive, the electrical system may require voltage regulators on feeders and buses. If critical utilization devices in the building will be voltage sensitive, added investment in them to make them less vulnerable to voltage fluctuations may be more economical than eliminating troublesome voltage fluctuations in the power supply. The decision as to which procedure to adopt should be based on economic and engineering factors.

Design of the electrical distribution system should aim at achieving the lowest life-cycle cost commensurate with safety requirements and the desired degree of reliability of service. For this purpose, circuits from load-center substations to the loads should be kept as short as possible. Economic studies should be made to determine the optimum number and size of load-center substations.

Design of the system and the cost studies should take into account provisions for changes in and expansion of the system in the future. These provisions may include use of load-center unit substations that may be expanded by addition of small units as required and use of plug-in busways, into which machine tools and other electrical devices may be connected by an attachment plug where needed.

Use of several load-center substations gives designers a choice of several different circuit arrangements. Selection of a circuit arrangement should be based on desired degree of reliability of service and cost comparisons. Circuit arrangement has a profound influence on system performance, cost and reliability. The wiring method should be selected and installed to meet requirements of the local building code or the National Electrical Code. Equipment and devices installed should have the approval of the Underwriters Laboratories and meet the standards of the National Electrical Manufacturers Association. Also, design of the system should incorporate flexible controls to insure proper operation of all devices, especially motors and lighting.

Maximum safety provisions should be incorporated in all parts of the system. These provisions should include:

1. Automatic switching devices with adequate short-circuit interrupting ability. They should be coordinated to insure selective removal of faulty system components.
2. High-quality electrical components
3. Equipment grounding
4. Electrical system grounding
5. Easy operation of the system
6. Metal-enclosed construction with inter-

locks to disconnect power when live parts become exposed by opening of doors or removal of panels

7. Installation in accordance with generally accepted standards

8. Good maintenance program

Design of the electrical system should also seek simple and low-cost maintenance. For one thing, the system should provide alternate pathways for power to loads, so when one circuit is disconnected for maintenance, the loads can continue to operate on power from the alternate circuit. Secondly, drawout circuit breakers should be used, to permit maintenance on them to be performed in a shop. Also, equipment that may require maintenance should be readily accessible and have adequate space around it for work to be performed on it or for easy removal and replacement. Light and auxiliary power should be available for performance of maintenance work on the equipment. In addition, switchboards, panelboards, circuit breakers and similar equipment should be located in clean, dry spaces where they will not be damaged by building operations.

Value Analysis and Appraisal

The benefits and costs of alternative electrical systems should be compared. Costs should be included for all parts of the systems from power source to utilization equipment, for comparison of systems on a uniform basis.

In these cost studies, value engineers should not place undue emphasis on lowest initial cost. Although this cost is important, an efficient, reliable system will cost less in the long run than a poor-quality system. When initial costs are compared, they should include the cost of installing distribution equipment as well as the cost of the equipment alone. Installation costs are a large percentage of the initial cost and may be larger than equipment costs. Because of the relatively high cost of installing equipment, it is often worthwhile to specify factory-assembled equipment, such as unit substations, to reduce field labor.

Initial cost of the distribution system sometimes can be reduced by use of equipment and wiring that will result in increased voltage drop in electrical distribution. The savings, however, may be more than offset by increased operating costs, because of higher energy losses in electrical transmission. In contrast, money spent on improving distribution efficiency will often prove to be a wise investment.

Studies of operating costs should take into account the different types of charges utility companies make for power. These charges cover energy consumption, penalty for low power factor and costs of supplying high peak demands.

Value analysis also should pay special attention to the utilization and distribution voltages selected for the system. These voltages generally affect the economics of distribution equipment more than any other single factor.

The power factor of the system should be checked when the characteristics of the loads become available. If the power factor will be low, the economics of improving it should be investigated.

In addition, the load factor should be checked, especially if the utility company will impose a demand charge for a low factor. If the load factor is low, consideration should be given to establishment of a demand-management program. This program should include demand control, an electromechanical procedure of load shedding, and scheduling for efficient operation of equipment and reduction of waste.

For the purpose, loads should be classified as primary or secondary. A primary load is one that cannot be shut down. A secondary load is one that can be shut down or delayed in starting with little or no economic loss and without causing discomfort to building occupants. Typcial secondary loads are lighting, electric heating and cooling units, water heaters and escalators. Such loads should be listed in order of priority, in accordance with the effect that shutdown or delay in operation will have on functioning of the building. With this information, decisions can be made on prevention of operation of some or all of the secondary loads at various times to reduce demand and on the order of restoration of the loads to service as demand drops.

Automatic demand controllers may be installed to keep peak demand below a specific threshold within a specific time interval, usually 15 or 30 min. A controller compares actual power consumption with an ideal, precomputed maximum, to minimize demand charges, while maintaining satisfactory facility operations. When it senses that the threshold level will be exceeded, the controller will shed loads in accordance with a predetermined schedule and similarly return the shutdown loads to service when the demand drops sufficiently.

GENERAL REFERENCES AND SOURCES FOR ADDITIONAL STUDY

These are books that deal comprehensively with several topics covered in this chapter. Topic-specific references relating to individual chapter sections are listed at the ends of the sections.

J. McPartland, editor, *National Electric Code Handbook,* McGraw-Hill, New York, 1983.

B. Stein et al., *Mechanical and Electrical Equipment for Buildings,* 7th ed., Wiley, New York, 1986.

D. Leach, *Basic Electric Circuits,* 3rd ed., Wiley, New York, 1984.

C. Kolstad, *Rapid Electrical Estimating and Pricing,* 4th ed., McGraw-Hill, New York, 1985.

EXERCISES

The following questions and problems are provided for review of the individual sections of the chapter.

Section 13.1

1. What happens when two particles with the same electrostatic charge are brought close together?
2. What happens when a north magnetic pole is brought close to a south pole?
3. What is a direct current?
4. A utility supplies electricity to a building at a nominal 220 volts. What voltage regulation should be specified if the supply voltage should not drop below 209 volts, to avoid damage to electric motors?
5. Four 1.5-volt batteries connected in series supply electricity to a 12-watt lamp.

(a) Sketch the electrical curcuit.
(b) How much current, amp, flows through the lamp?
(c) What is the resistance, ohms, of the lamp?
(d) What is the conductance, mhos, of the lamp?

6. How much power is expended:
(a) When a current of 10 amp flows through a 50-ohm resistance?
(b) When a voltage drop of 120 volts occurs across a 12-ohm resistance?

7. Four 150-watt lamps are connected in series and 120 volts are applied.
(a) Sketch the electrical circuit.
(b) How much current will the lamps draw?
(c) What is the resistance of each lamp?

8. Four 150-watt lamps are connected in parallel and 120 volts are applied.
(a) Sketch the electrical circuit.
(b) How much current will each lamp draw?
(c) How much current must the electric source supply?
(d) What is the equivalent resistance of the four lamps?

9. What is the difference between generators and motors?

Section 13.2

10. Why do utility companies transmit ac rather than dc?
11. What is the frequency of ac that is usually distributed in the U.S.A.?
12. What is the definition of a phase angle between a voltage and a current with the same frequency?
13. An alternating current that varies with time in the same way as a sine curve has a maximum magnitude of 20 amp. What is the effective current?
14. An ac circuit carries an effective current of 5 amp through a 24-ohm resistance.
(a) What is the effective voltage?
(b) How much power is expended?
15. A coil has an inductive reactance of 5 ohms. What is its inductive susceptance?

16. What is the phase relationship between current and voltage in:
    (a) A circuit with only pure inductance?
    (b) A circuit with only pure reactance?
17. A capacitor has a capacitive susceptance of 0.25 mhos. What is its capacitive reactance?
18. An ac circuit carries an effective current of 20 amp through a 6-ohm impedance.
    (a) What is the effective voltage?
    (b) If the phase angle between current and voltage is 60°, what is the power factor and how much power is being expended?
    (c) What is the admittance of the curcuit?
19. An ac circuit applies 225 volts at a frequency of 60 Hz to a motor with a resistance of 12 ohms in series with an inductive reactance of 9 ohms.
    (a) What is the impedance in this circuit?
    (b) What is the current in the circuit?
    (c) What is the circuit power factor?
    (d) How much power is being supplied to the motor?
    (e) A capacitor with a capacitive reactance of 27 ohms is connected between the motor terminals (in parallel). What is the current in the capacitor?
    (f) What is the phase relationship between the capacitor current and the applied voltage?
    (g) What is the phase relationship between the capacitor current and the motor current? (Note that the phase relationship between the motor current and the applied voltage is the same as before the capacitor was installed.)
    (h) How much current is supplied to the motor and the capacitor? (What is the vector sum of the currents to the motor and the capacitor?)
    (i) What is the power factor for the motor with the capacitor?
20. Why do utility companies that supply electricity charge extra for low power factors?
21. Describe briefly two types of equipment for converting large amounts of ac power to dc.
22. In a three-phase system:
    (a) What is the phase difference, degrees, between voltages?
    (b) What is the ratio of the potential difference, volts, between phase wires to the potential difference, volts, between a phase wire and neutral?
23. What is meant by balanced loading on a multiphase system?

Section 13.3

24. What electric-equipment ratings must be known for determination of the load the equipment will apply to an electrical distribution system?
25. What is the minimum lighting load, kw, for which electrical distribution should be designed:
    (a) An office building with 20,000 sq ft of floor area?
    (b) An apartment building with 20,000 sq ft of floor area?
26. A mill produces 1,600 lb of aluminum in an 8-hr day. What is the electrical load, kw, that must be served for the purpose?

Section 13.4

27. What is meant by ampacity of a conductor?
28. What prevents electric current from jumping from a power conductor to nearby metal?
29. Why are raceways required?
30. What devices are used for overcurrent protection of electrical circuits?
31. What device is used for intentional opening and closing of circuits?
32. What is the purpose of service equipment?
33. What type of conductors distribute electricity to branch circuits?
34. What is the purpose of an electrical box?
35. What is the major difference between a grounded conductor and a grounding conductor?
36. Why should a grounding conductor be provided for electric tools with metal housing?
37. A No. 6 wire has an area of about 26,000 CM.
    (a) What is the approximate area of a No. 3 wire?

(b) If a No. 6 wire has a resistance of 0.410 ohms per 1,000 ft of length, what is the resistance of the same length of No. 3 wire?

(c) What is the diameter, in., of the No. 6 wire?

38. What color should the outer covering of a grounded conductor be?

39. What color should the outer covering of an insulated grounding conductor be?

40. What type of supports are required for open wiring and knob-and-tube work?

41. Describe Type AC metal-clad (BX) cable. How does it differ from Greenfield cable?

42. Describe nonmetallic-sheathed (Romex) cable?

43. Which type of service-entrance cable may be used underground?

44. Which type of cable incorporates a support cable?

45. Describe bus duct.

46. Where should electrical boxes be located?

47. With which types of wiring should nonmetallic boxes be used?

48. What is the principal advantage of:

(a) Flexible conduit over rigid conduit and surface raceways?

(b) A multioutlet assembly over conduit, tubing and surface raceways?

(c) Underfloor raceways over other raceways?

(d) Wireways over other raceways?

Section 13.5

49. What range of voltages is generally considered low voltage?

50. What is the purpose of a substation?

51. What is meant by a secondary circuit?

52. What type of transformer can be used to correct low voltage?

53. A voltage of 2,400 volts is impressed across the primary coil of a transformer. The primary coil has 100 turns and the secondary coil 10 turns around the transformer core. What is the approximate voltage across the secondary coil?

54. What type of enclosure is required for an oil-filled transformer installed inside a building?

55. What type of instrument is used to measure consumption of electrical energy in a building?

56. If a single-pole switch is installed in a two-wire, single-phase circuit, in which of the wires should the switch be inserted?

57. Which overcurrent protective device can be used instead of a switch?

58. What is the purpose of an isolating switch? When may it be operated?

59. Why is it desirable that an overcurrent protective device not automatically open a circuit when a short-duration, moderately abnormal current flows?

60. What are the three ratings of an overcurrent protective device that determine its selection for a specific application?

61. How does a fuse interrupt an overcurrent?

62. What are the two major advantages of a curcuit breaker over a fuse?

63. How is a relay used in overcurrent protection?

64. What are the purposes of switchgear?

65. What are the components of a substation?

66. What are the purposes of a panelboard?

67. If a panelboard is used as service equipment, what is the maximum number of switches or circuit breakers that must be operated to disconnect the supply conductors?

68. What are the main components of a motor controller?

69. What is the purpose of a motor control center?

70. Where should lightning arresters be placed?

Section 13.6

71. What are the most important factors in design of an electrical distribution system for a building?

72. In Fig. 13.24, Branch Circuits *A* and *B* have a 100% demand factor and a diversity factor of 1.25. If the load on Branch Circuit *A* is 2 kva and on Branch Circuit *B*, 6 kva, what is the maximum demand on Secondary Feeder 1?

73. What factors influence selection of distri-

bution voltages for industrial and commercial buildings?

74. What major system components, other than utilization equipment, are primarily responsible for the cost of an electrical system?

75. What steps can be taken in design of an electrical system to achieve a specific degree of service reliability?

76. What is the purpose of an electrical closet?

77. What can be done to reduce utility charges for a low load factor?

78. Sketch a single-line circuit diagram for the electrical distribution for one floor of the building in Fig. 13.25a, serving the equipment shown.

79. What can be done in design of an electrical system to insure reliability of service?

80. What limits the magnitude of a short-circuit current?

81. How are overcurrent protective devices coordinated to insure selectivity?

82. Why must voltage spread be kept small?

Section 13.7

83. What is the purpose of a remote-control circuit?

84. What mainly distinguishes Class 2 and 3 circuits from Class 1?

85. What is the role of a control panel in an alarm system?

86. What are the responsibilities of building designers in planning a telephone system?

Chapter 14

Vertical Circulation

Vertical circulation comprises major subsystems that provide a means in multistory buildings for movement of people and goods between floors. The subsystems have substantially the same objectives but achieve them in different ways.

The subsystems may be divided into two classes, I and II. Class I is intended for movement of both people and goods. It includes ramps, stairs, escalators and elevators. Class II is not permitted for movement of people. It includes dumbwaiters and vertical conveyors.

Class I subsystems may be subdivided into two subclasses, A and B. Class IA comprises subsystems that can be used by people under normal conditions and as a means of egress under emergency conditions, such as fire. This class includes ramps, stairs and escalators meeting conditions specified in local building codes or the National Fire Protection Association (NFPA) *Life Safety Code*. Class IB consists of Class I subsystems not acceptable as a means of egress.

Although Class IB is not an acceptable means of egress, these subsystems nevertheless may be used in emergencies for evacuation of building occupants. The life-safety rules just require that Class IA alone provide sufficient capacity for rapid, safe evacuation of the maximum probable building population (see Sec. 6.5).

Every multistory building must have at least two Class IA means of egress. (Powered vertical transportation is considered unreliable, because, in an emergency, the mechanical equipment may become inoperative.) Some buildings, such as hotels and hospitals, usually are required to have some Class IA discharge to outdoors. In any case, under normal conditions, Class IA may be used in conjunction with Class IB for movement of people and goods and with Class II for movement of goods. To take advantage of this in design of a building, design of the different vertical-circulation subsystems should be coordinated to provide an optimum vertical-circulation system for the building.

An optimum vertical-circulation system provides, at the lowest life-cycle cost, safe means of movement of people and goods between floors of a multistory building, with the required degree of convenience and speed and with acceptable human expenditure of energy for the purposes of the building.

In providing the required convenience and speed of vertical movement and in limiting human effort in such movement, the vertical-circulation system should contribute to optimization of the total building system. Accordingly, to meet owner objectives, for some buildings, an optimum system may incorporate steep ramps and stairs to cut costs, despite the large human effort required to mount them. In contrast, for some buildings, to meet owner objectives, an optimum system may provide elevators and escalators for vertical transportation under normal conditions, although their life-cycle costs are much greater than that for stairs, which must be provided anyway.

Systems design can be useful in optimizing design of vertical circulation for a building. But vertical circulation is also the key to successful functioning of the building design. In fact, for a multistory building, location of stairs, escalators and elevators often control the floor plan. Consequently, systems design can also be useful in selecting the location of vertical-circulation subsystems for an efficient floor plan and thus further contributing to optimization of the building.

14.1. RAMPS

A ramp is a sloping passageway for movement of people and goods. It is a necessity for movement of handicapped persons, especially those restricted to wheelchairs, between floors of multistory buildings or between the street and the building interior when elevators are not available. Nearly every building should provide

ramps for this purpose, where necessary. Ramps are also advantageous for movement of vehicles between building levels, for example, in garages and for rapid movement of large groups of people. Consequently, ramps are often used in stadiums, garages, transportation terminals and exposition buildings.

For safety reasons and for convenience of people walking, ramps are built with a relatively flat slope. They have been constructed with slopes up to 15%, but 10% (5.7°) is a preferred maximum, to make ascension and descension easier.

Because of the flat slope, ramps require much more space within buildings for vertical circulation than do stairs. For example, with the 10% maximum slope and a story height of 12 ft, a ramp connecting two floors would have to be 12/0.10 = 120 ft long (see Fig. 14.1a).

The space occupied by a ramp generally cannot be used for any other building purpose than

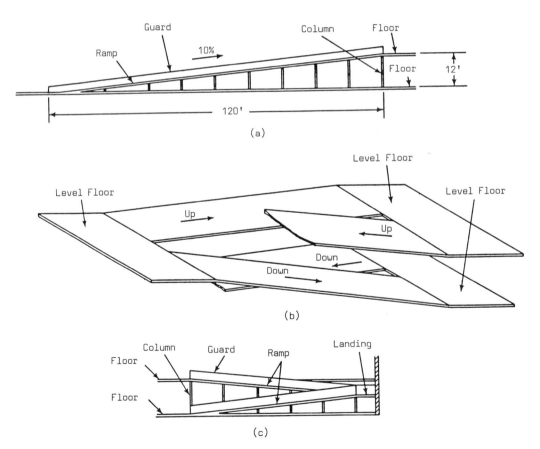

Fig. 14.1. Types of ramps. (a) Straight ramp. (b) Sloping garage floors used as ramps and for parking. (c) Zig-zag ramp.

vertical circulation. (To save space in garages, often every floor is designed to serve as a ramp. In such cases, each floor is split longitudinally, and each section thus formed is sloped gradually in opposite directions to landings at upper and lower levels, as shown in Fig. 14.1*b*). To reduce space requirements, however, ramps need not be straight over their entire length. They can be zigzagged (see Fig. 14.1*c*), curved or spiraled.

Inside Ramps

A ramp is acceptable as a means of egress when it conforms to the general requirements for emergency exits in the local building code or the NFPA *Life Safety Code* (see Sec. 6.5). Such ramps are classified as Class A or Class B.

The main difference between the two classes is that Class B may be as narrow as 30 in., whereas Class A must be at least 44 in. wide. (A ramp 44 in. wide can accommodate two adults abreast.) In addition, Class A is restricted by the *Life Safety Code* to slopes not exceeding 10% and Class B, to slopes up to $12\frac{1}{2}\%$.

No limit is established on the vertical distance between landings for Class A ramps, but this distance may not exceed 12 ft for Class B landings. The capacity, persons per 22-in. unit of exit width, may be taken as 100 in the downward direction and 60 in the upward direction for Class B ramps and as 100 for Class A ramps in either direction. Building codes usually make Class A ramps a required minimum only for places of assembly of more than 1,000 persons. For other occupancies, the choice between Class A and Class B depends on exit capacity needed.

Floors of stationary inside ramps are usually constructed in the same way as level floors. Floors of ramps, however, should have nonslip surfaces and should be solid; that is, without perforations. An exit ramp in a building more than three stories high or in any building of noncombustible or fire-resistant construction also should be of noncombustible construction.

Slope of a ramp should not vary between landings. If changes in direction of travel are required, they should be made only at landings, which should be level.

When a ramp is used as an exit, it should be protected by separation from other parts of the building in the same way as for other exits. Guards, vertical protective barriers, should be placed along the edges of ramps and along the edges of floor openings over ramps, to prevent falls over the open edges; also, handrails should be installed along both edges of Class B ramps.

There should be no enclosed usable space, such as closets, under ramps in an exit enclosure. Furthermore, open spaces under such ramps should not be used for any purpose, but other enclosed ramps are permitted to be located underneath.

Outside Ramps

Ramps permanently installed on the outside of a building may be used as a means of egress if they meet the life-safety requirements for inside ramps. Outside ramps, however, also should be constructed to avoid any handicap to their use by persons with a fear of high places. For protection of such persons, the NFPA *Life Safety Code* requires that, for ramps more than three stories high, guards along ramp edges should be at least 4 ft high.

Where ramps may be exposed to snow or ice, accumulations of these should be prevented.

Powered Ramps

A moving walk is a type of passenger-carrying powered device on which passengers stand or walk. The passenger-carrying moving surface is continuous and stays parallel to its direction of motion. Such powered walks provide horizontal, or nearly horizontal, transportation to make movement of people easy, convenient and rapid. Usually, moving walks convey passengers at speeds up to 180 fpm.

A powered ramp is an inclined moving walk. Powered ramps may be accepted as means of egress if they meet the requirements of stationary ramps, and moving walks may be accepted if they comply with requirements for exits (see Sec. 6.5). In both cases, however, the moving device should be incapable of operation in the direction opposite to normal exit travel.

Moving walks and powered ramps are especially useful for transporting people carrying large parcels or luggage. Hence, these devices sometimes are installed at airport terminal buildings.

Both moving walks and powered ramps should be designed, manufactured and installed in accordance with the requirements of the American National Standard Safety Code for Elevators, Dumbwaiters, Escalators and Moving Walks, ANSI A17.1, published by the American Society of Mechanical Engineers.

Moving walks and powered ramps consist basically of a treadway moved by a driving machine; a handrail on each side of the treadway that moves with the treadway; supports and guides for the treadway; balustrades, or guards, that enclose the treadway on each side and support the handrails; brakes, and threshold plates at the entrance to or exit from the treadway. The threshold plates are installed to provide smooth passage between treadway and landing. The plates are provided with a comb, or teeth, that mesh with and are set into grooves in the treadway in the direction of travel. The objective is to provide firm footing and to prevent anything from becoming trapped between the moving treadway and the stationary landing.

Treadways are usually one of the following types:

1. Belt type—a power-driven continuous belt treadway.
2. Pallet type—a series of connected, power-driven pallets. (A pallet is a short rigid platform, which, when joined to other pallets, forms an articulated treadway.)
3. Belt pallet type—a series of connected, power-driven pallets to which a continuous belt is fastened.
4. Edge-supported belt type—a belt supported near its edges by rollers in sequence.
5. Roller-bed type—a treadway supported throughout its width by rollers in sequence.
6. Slider-bed type—a treadway that slides on a supporting surface.

Powered ramps are supported on steel trusses, in much the same way as escalators. In fact, powered ramps resemble escalators in construc-tion and are subject to many of the safety regulations for escalators (see Sec. 14.3).

The driving machine for a moving walk or ramp may be connected to the main drive shaft by toothed gearing, a coupling or a chain. Movement of the treadway and handrails can be stopped by an electrically released, mechanically applied brake, located either on the driving machine or on the main drive shaft. The brake should be capable of stopping movement automatically when a power failure occurs, when the treadway or a handrail breaks or when a safety device is actuated. Safety devices required include starting, emergency stopping, and maintenance stopping switches and a speed governor that will prevent the walk speed from exceeding 40% more than the maximum design treadway speed.

The balustrade on each side of the treadway should be at least 30 in. high, measured perpendicular to the treadway. Motion of the handrails on the balustrade should be synchronized with that of the treadway. Hand or finger guards should be provided where the handrail enters the balustrade. At each end of a moving walk or ramp, the handrails should extend at normal height at least 12 in. beyond the end of the exposed treadway, to ease entry and exit of passengers from or onto a level landing.

Passenger capacity of a moving walk or ramp depends both on treadway width and speed. Information on capacity should be obtained from the manufacturer.

ANSI A17.1 limits the treadway width to a minimum of 16 in. and to a maximum that depends on the speed and the slope of the treadway. Standard widths are 24, 32, and 40 in. Maximum slope permitted is $15°$.

ANSI A17.1 also places restrictions on maximum treadway speed. These depend both on the maximum treadway slope at points of entrance and exit and on the maximum treadway slope at any other point on the treadway. More stringent speed limitations are applied to entry and exit slopes. Hence, powered ramps are usually designed to permit level entry and exit. For this condition, maximum speed may be 180 fpm for slopes up to $8°$ and 140 fpm for slopes between 8 and $15°$. Higher speeds may be used, however, to permit passage of passengers to or from other moving treadways.

14.2. STAIRS

Stairs, or stairways, like ramps, are sloping passageways for movement of people; however, stairs are built with much steeper slopes than stationary ramps, which rarely have a pitch angle exceeding 8°. The pitch angle of interior stairs often ranges between 30 and 35° and of exterior stairs, between 20 and 30°. To provide comfortable, safe footing for people on such steep slopes, stairs cannot have a uniformly sloping surface like ramps do. Instead stairs consist of a sequence of short, level platforms, called *steps*, placed at uniform intervals one above the other (see Fig. 14.2).

The top surface of a step is called the *tread*. The vertical surface between treads is called a *riser*.

Because of their steep slope, stairs occupy much less space in a building than do ramps. Hence, stairs usually are selected as the basic means for vertical circulation in multistory buildings. Stairs, however, have the disadvantages of being a hindrance to passage of wheeled vehicles, requiring considerable expenditure of human power for upward travel, posing tripping hazards in the downward direction and being an inconvenience or a barrier to vertical movement of handicapped persons.

Exit Capacity

The number of stairways required in a building is usually controlled by the local building code. This control is achieved in several ways. For some buildings, the code may require at least two exits from each floor. Also, the code may set a limit on the maximum horizontal distance from any point on a floor to a stairway, or on the maximum floor area contributing building occupants to a stairway in an emergency. In addition, the code may limit the maximum exit capacity of a stairway. For example, maximum capacity, persons per 22-in. unit of width, may be set at 15 for institutional buildings, such as hospitals and nursing homes; 30 for residential buildings; 45 for storage buildings; 60 for mercantile, business, educational and industrial buildings, theaters and restaurants; 80 for churches, concert halls and museums, and 320 for stadiums and amusement structures.

Thus, the capacity of a stairway, which may determine the number of stairways required in a building, depends on the number of units of width, or the total clear width. Clear width is the width free of obstructions except handrails and stringers, which are permitted to project up to $3\frac{1}{2}$ in. and $1\frac{1}{2}$ in., respectively, into the passageway on each side of it. To illustrate, suppose each floor of an office building above the lobby may be expected to be occupied by up to 300 persons. If each stairway in the building is to be two units wide, capacity of the stairs is $2 \times 60 = 120$ persons. The number of stairways required then is at least $300/120 = 2.5$. Hence, three stairways should be provided.

Types of Stairs

Stairs commonly used are straight, but sometimes, circular, curved, spiral or a combination stairway is used.

Straight stairs extend in only one direction in a *flight*, the series of steps between two floors or a floor and a landing. Such stairs may be arranged in a *straight run* between floors (see Fig. 14.3a) or in a series of flights, with one or more landings between floors, without change in direction (see Fig. 14.3b).

Straight stairs usually change direction only at floors or landings. When the stairs reverse direction, they are called *parallel stairs* (see Fig. 14.3c). They provide a more compact arrangement than straight runs. When the change in direction is at some other angle, for instance, 90° as in Fig. 14.3d, the stairs are called *angle*

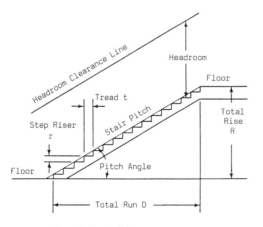

Fig. 14.2. Stairway parameters.

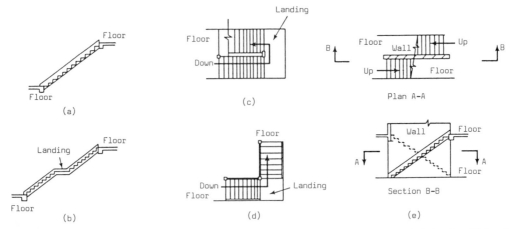

Fig. 14.3. Types of stairs. (*a*) Straight run. (*b*) Straight stairs with two flights. (*c*) Parallel stairs. (*d*) Angle stairs. (*e*) Scissors stairs.

stairs. Such stairs are often selected to utilize a corner or shorten a straight run.

A pair of straight-run stairs may be arranged in a pattern called *scissor stairs* (see Fig. 14.3*e*). In this case, the stairs extend in opposite directions to each other on each side of a fire-resistant wall.

Circular stairs are actually spiral stairs. Both appear to be circular when viewed from above. The term *spiral stairs*, however, is applied when the radius of the circle is small, usually less than 8 ft, and the inner portion of the treads are supported on a central post. *Curved stairs*, in contrast, when viewed from above appear to follow a curve with two or more centers of curvature, for example, an ellipse.

Winders are generally used as steps in sharply curved stairs. These are steps with tapered treads, with the inner end of the tread more than 3 in. less in width than the minimum that would be permitted for a straight-run exit stairs. Winders are not allowed in stairs that serve as a required means of egress in emergencies. Curving stairs may be used as exits, however, if, at a distance 18 in. from the narrower end of each step, the relationship of treads and risers is within permitted limits for straight stairs, and if no tread is more than 3 in. narrower or wider at any point than the width established 18 in. from the narrow end.

Dimensions of Straight Stairs

The total run D of a flight of stairs is the distance between the first and last risers (see Fig. 14.2). The total rise R of the flight is the vertical distance between floors or a floor and a landing. Every flight of stairs contains one less tread than the number of risers. Hence, the pitch of a flight of straight stairs is given by

$$p = \frac{R - r}{D} = \frac{r}{t} \qquad (14.1)$$

where

r = height of a riser
t = tread width

The pitch angle (see Fig. 14.2) is the angle whose tangent is p.

Because space used for a stairwell is unusable for the main functions of a building, stairs usually are set on a steep slope to conserve space for purposes other than vertical circulation. The NFPA *Life Safety Code* accepts as a means of egress stairs with slopes up to 41.5°. Such slopes, however, require expenditure of considerable human power to climb the stairs and pose a tripping hazard in descension.

For comfort, or low power expenditure, stairs should either have low risers and deep treads or shallow treads with deep risers. Local building codes generally express this relationship in a formula that establishes a range of acceptable tread widths and riser heights. Some common formulas are the following:

1. The product of riser height and tread width must be between 70 and 75.
2. Riser height plus tread width must equal 17 to 17.5.
3. The sum of the tread width and twice the riser height must fall between 24 and 25.5.

In addition, the building code may set a maximum for the riser height and a minimum for the tread width.

Experiments have indicated, however, that for fairly low human power expenditure at normal stair climbing speeds and a low rate of missteps, the slope of stairs should lie between 20 and 27°. To achieve this, stairs should have risers between 4 and 7 in. high with treads between 11 and 14 in. wide. If treads have to be 10 in. or less in width, they should have a nosing, or projection beyond the riser of about 1 in. to increase step width.

Monumental stairs, which are wide and have a low pitch, are sometimes used within or on the outside of buildings for aesthetic reasons.

Exit-Stair Details

Stairs may be designed in the same way as floors. Supports for stairs, however, usually are at the floors or landings between which they span.

In buildings more than 3 stories high or of fire-resistant construction, stairs, except for handrails, should be built of noncombustible materials. Treads and landings should be solid and have nonslip surfaces. Maximum rise of a single flight of stairs between floors or landings should not exceed 8 ft in institutional buildings or places of assembly, such as theaters, or 12 ft in other buildings. Landings should be at least 44 in. long in the direction of travel.

Stair and landing width should be at least 44 in., but the local building code may permit 36-in. width where the occupancy of all floors served by the stairs is less than 50 persons. There should be no decrease in width of stairways and landings in the direction of travel.

Headroom, or vertical clearance (see Fig. 14.2), should be at least 6 ft 8 in.

There should be no enclosed usable space, such as closets, under stairs in an exit enclosure. Furthermore, the open space under such stairs should not be used for any purpose, although stairs underneath are permitted.

A flight of stairs should contain at least three risers. At doors that open immediately on a stairs, a landing with at least the width of the stairs should be provided.

When stairs are to be used as an exit, protection by separation from other parts of the building should be provided in the same way as for other exits. Some means should be provided, such as a skylight or a vent, at the top of a stairwell to let escape any heat or smoke that may enter during a building fire.

Edges of stairs and edges of floor or landing openings around stairs should be guarded to prevent falls over the open edges. Handrails should be installed along both edges of the stairs. For stairways, other than outside monumental stairs, that are more than 88 in. wide, intermediate handrails should be placed to divide the stairways into portions not more than 88 in. wide.

Guards should be at least 42 in. high, measured vertically from the tread 1 in. back from the leading edge, for interior stairs and 48 in. high for outside stairs more than three stories high. Handrails should be between 30 and 34 in. high. They should provide a clearance of at least $1\frac{1}{2}$ in. from the wall to which they are fastened. No guards are required for inside stairs that reverse direction at landings, if the horizontal distance between successive flights does not exceed 12 in.

Guards may consist of a wall, longitudinal rails spaced to leave 10 in. or less of open space, vertical balusters spaced up to 6-in. apart, wire mesh, expanded metal or ornamental grilles.

Smokeproof Towers

A smokeproof tower is a continuous, fire-resistant enclosure protecting a stairway from fire or smoke that may develop in the building. Travel between the building and the tower must be over balconies open to the outside (see Sec. 6.5).

Stairs, enclosure walls, vestibules, balconies and other components of smokeproof towers are required to be built of noncombustible materials. The stairs must meet the same requirements as inside stairs that may serve as a means of egress in emergencies.

Wood Stairs

In buildings up to three stories high that are not required to be of fire-resistant construction, stairs may be built of wood. Construction of such stairs generally follows traditional practices, which experience has shown yields safe, economical stairs.

Treads are generally made of an abrasive-resistant wood, such as oak. Risers may either be omitted or also made of oak or a softer wood, such as pine. Both treads and risers usually are finished with a coating of paint or varnish and sometimes are carpeted.

Treads and risers are fastened to and sup-

ported on one or more longitudinal stringers, called *carriages* (see Fig. 14.4a). The upper edge of each carriage is cut out to receive the risers and treads, and the ends are shaped to frame into the supports for the carriage (see Fig. 14.4b). The end of the carriage at the upper floor or landing is usually supported on a transverse beam, called a *header*. The lower end may be supported on a floor or another header. The edges of the upper floor around the stairs, except at the landing, generally are supported on beams, which are called *trimmers* when they are parallel to the stairs.

For aesthetic reasons, outer edges of stairs are enclosed with finished stringers. A stringer placed along the wall is called a wall stringer (see Fig. 14.4d). A stringer shaped to fit between treads and risers is called an open stringer (see Fig. 14.4c and d). Treads project over such a stringer. A stringer that completely covers ends of risers and treads and is cut out to receive them is called a housed stringer.

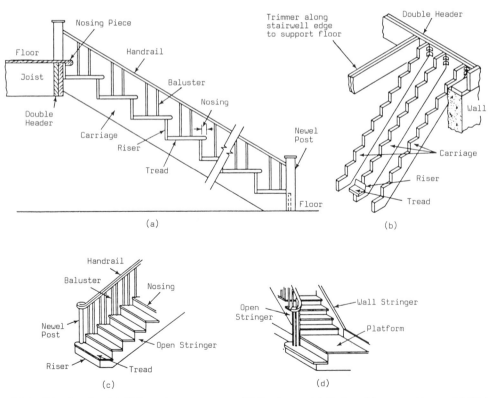

Fig. 14.4. Wood stairs. (*a*) Main stair components. (*b*) Carriages. (*c*) Stairs with open stringer. (*d*) Stairs with open and wall stringers.

Steel Stairs

For buildings in which noncombustible stairs are required, one of the materials that can be used to economic advantage is steel. Risers and treads may be made of cold-formed steel (see Fig. 14.5*a*) or floor plate (see Fig. 14.5*b*).

The cold-formed-steel type is often pressed into the shape of risers and subtreads, or pans, into which one of several types of tread may be placed. The treads may be made of stone, concrete, composition or metal, usually with a nonslip top surface.

Treads and risers of steel stairs generally are supported at their ends on carrier angles (see Fig. 14.5). These, in turn, are fastened to inclined stringers that span between headers at floors or landings.

Concrete Stairs

For buildings for which noncombustible stairs are required, another material that can be used to economic advantage is concrete. It may be reinforced or prestressed, cast-in-place or precast.

Concrete stairs are generally designed as a beam that spans between headers at floors or landings (see Fig. 14.6). The treads should have metal nosings to protect the edges and increase resistance to slipping.

14.3. ESCALATORS

Escalators are powered stairs, a sequence of continuously moving steps that transport passengers, standing or walking, between two floors of a multistory building. Except for the shape of the moving surface, escalators resemble pallet-type powered ramps. Many of the safety regulations noted in Sec. 14.1 for powered ramps apply also to powered stairs. In general, escalators may be accepted as a required means of egress in emergencies if they meet the requirements of exit stairs and if they are incapable of operation in the direction opposite to normal exit travel.

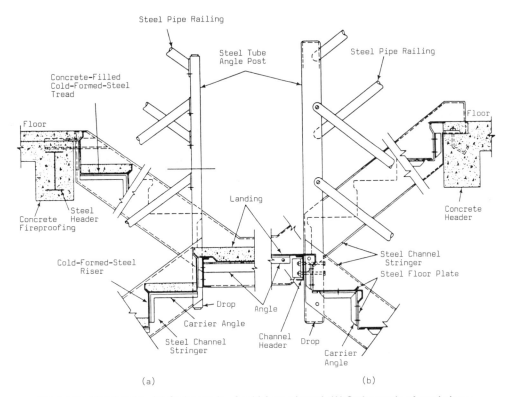

Fig. 14.5. Metal stairs. (*a*) Stairs made of cold-formed steel. (*b*) Stairs made of steel plates.

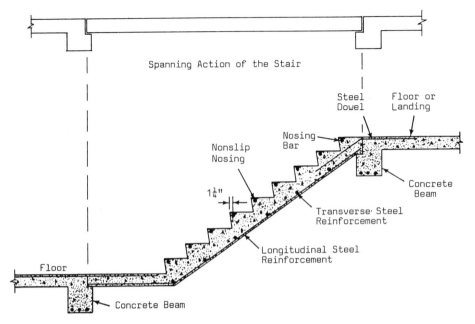

Fig. 14.6. Reinforced concrete stairs.

Escalators are useful for moving large numbers of people between floors of a multistory building, with little or no waiting time for transportation. The steps move continuously. Speed is usually 90 or 120 fpm, as required for peak traffic periods.

The capacity of powered stairs depends both on the speed and width of steps. Standard widths are 32 and 48 in. between handrails. A 32-in. escalator operating at 90 fpm can transport between 5,000 and 8,000 persons per hr. A 48-in. escalator operating at 120 fpm can transport as many as 10,000 persons per hr.

Escalators are installed at much steeper slopes than powered ramps. The pitch angle of powered stairs is standardized at 30°. Risers are usually 8-in. high and treads 16-in. wide. Because of the steep slope, escalators occupy less space than do ramps. Escalators, however, require more space than elevators and provide slower transportation in high-rise buildings.

Escalator Construction

Escalator steps are formed of pairs of pallets, hinged together in a continuous loop (see Fig. 14.7). One member of each pair serves as a tread that stays level while it is visible to passengers. The other member of each pair serves as a riser while traveling along the incline. This pallet, however, flattens to the horizontal at the top and bottom of the stairs, so that the pair of pallets provide a level platform to permit passengers to walk safely onto or off the escalator.

To prevent passengers from falling over the sides of the powered stairs, a solid balustrade is provided on each side of the steps (see Fig. 14.7). Each balustrade carries on its top a moving handrail about 30 in. above the nose of each step. The handrail speed should be synchronized with that of the steps. Each handrail should extend at normal height at least 12 in. beyond the end of the exposed pallets, to ease entry and exit of passengers from or onto a level landing.

The top of the balustrade outside a handrail is called a deckboard. Unless the intersection of the deckboard and a ceiling or soffit is at least 24 in. from the center of the handrail, a solid guard should be provided in the intersection. The vertical face of the guard should project at least 14 in. horizontally from the apex of the angle, and the exposed edge should be rounded to prevent injury to escalator passengers.

Main supports for an escalator usually are a

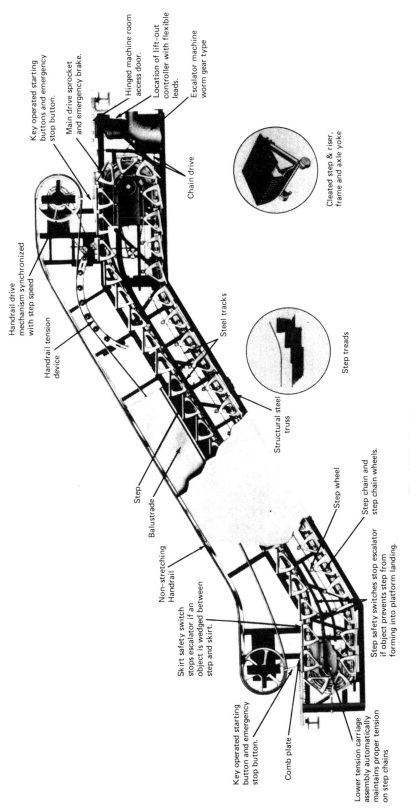

Key operated starting buttons and emergency stop button.

Hinged machine room access door.

Location of lift-out controller with flexible leads.

Main drive sprocket and emergency brake.

Escalator machine worm gear type

Chain drive

Handrail drive mechanism synchronized with step speed

Cleated step & riser, frame and axle yoke

Handrail tension device

Steel tracks

Step treads

Step.

Balustrade

Structural steel truss

Step wheel

Non-stretching Handrail

Step chain and step chain wheels.

Skirt safety switch stops escalator if an object is wedged between step and skirt.

Key operated starting button and emergency stop button.

Step safety switches stop escalator if object prevents step from forming into platform landing.

Comb plate

Lower tension carriage assembly automatically maintains proper tension on step chains

Fig. 14.7. Escalator components.

pair of inclined trusses, which are hidden from view by the balustrades. For rises up to about 20 ft, the trusses span between headers at upper and lower floors. For larger rises, a column may have to be provided to support the trusses at an intermediate point. Each truss is brought to the building in three sections, which are assembled at the escalator location.

The steps are carried on resilient rollers. These ride on a set of tracks attached to the trusses.

The steps are grooved in the direction of travel. The grooves mesh with combs, or teeth, on a threshold plate at the entrance to and exit from the escalator. The objective is to provide firm footing and to prevent anything from becoming trapped between the moving treadway and the stationary landing.

The driving machine for an escalator may be connected to the main drive shaft by toothed gearing, a coupling or a chain. Movement of the stairs and handrail can be stopped by an electrically released, mechanically applied brake, located either on the driving machine or on the main drive shaft. The brake should be capable of stopping movement automatically when a power failure occurs, the stairs become jammed, a pallet, chain or handrail breaks, or a safety device is actuated. Safety devices required include starting, emergency stopping, and maintenance stopping switches and a speed governor that will prevent the step speed from exceeding 40% more than the maximum design speed of the escalator.

Location and Arrangement of Escalators

The locations of escalators should be carefully planned to make maximum use of their capabilities. They should be placed where traffic between floors of a building is heaviest and where convenient for passengers. Their location should be obvious to persons approaching them.

In department stores, escalators usually carry 75 to 90% of the traffic between floors. Elevators transport the remainder. To take advantage of the heavy traffic carried by escalators, they should lead to strategic sales areas.

Escalators generally are installed in pairs, one escalator for transporting passengers upward

and the other for carrying traffic downward. In each story, an escalator pair may be crisscrossed or set parallel to each other (see Fig. 14.8).

The parallel type has a more impressive appearance and tends to draw passengers to it. This arrangement is often used for escalators in groups of three or four in transportation terminals. During peak traffic, most of the group can be operated in the direction of the heaviest traffic.

Crisscrossed escalators, however, are generally preferred, because they are more compact, thus requiring less space and reducing walking distance between stairs at landings to a minimum. Either arrangement may be used with the escalators separated by any desired distance. Separation eases the mixing of persons entering at each floor with passengers continuing their trip.

Sufficient space should be provided at each landing for such mixing, especially for movement of passengers exiting to allow time for ad-

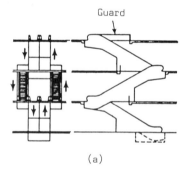

(a)

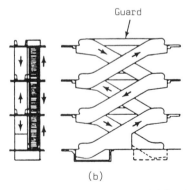

(b)

Fig. 14.8. Escalator arrangements. (a) Parallel. (b) Crisscross.

justment to the stationary landing. Also, space should be available for passengers waiting to board the powered stairs. Consequently, at each end of an escalator, a level landing should be provided with a width equal at least to the overall width of the powered stairs and a length in the direction of travel of at least 10 ft.

Fire Protection

When escalators qualify as exit escalators, 32-in. stairs may be given credit for one unit of width (step width is 24 in.) while 48-in. stairs may be given credit for two units of width. (Steps are 40 in. wide.)

Although an escalator may not be acceptable as a means of egress in emergencies, it nevertheless must have its floor openings enclosed or protected against fire as required for other openings in fire barriers. As an alternative, acceptable protection may be provided in buildings completely protected by a standard, supervised fire sprinkler system by any of the following:

Sprinkler-Vent Method. This is a combination of an automatic fire or smoke detection system, automatic air-exhaust system and an automatic water curtain.

Spray-Nozzle Method. This is a combination of an automatic fire or smoke detection system with a system of high-velocity water-spray nozzles.

Rolling-Shutter Method. This provides an automatic, self-closing, rolling shutter to enclose completely the top of each escalator.

Partial-Enclosure Method. This provides kiosks with self-closing fire doors to serve as a barrier to fire and smoke at each escalator.

The preceding methods are described in detail in the NFPA *Life Safety Code.*

Escalator trusses and machine spaces should be enclosed with fire-resistant materials. Step frames and treads should be noncombustible. Machine and control spaces should be adequately ventilated.

SECTIONS 14.1 TO 14.3

References

B. Stein et al., *Mechanical and Electrical Equipment for Buildings*, 7th ed., Wiley, New York, 1986.

C. Ramsey and H. Sleeper, *Architectural Graphic Standards*, 8th ed., Wiley, New York, 1988.

Life Safety Code, National Fire Protection Association, Quincy, MA, 1988.

Uniform Building Code, 1988 ed., International Conference of Building Officials, Whittier, CA.

The BOCA Basic National Building Code/1984, 9th ed., Building Officials Conference of America, Country Club Hills, IL.

Southern Standard Building Code, Southern Building Code Congress International, Birmingham, AL.

P. Hopf and J. Raeber, *Access for the Handicapped*, Van Nostrand Reinhold, New York, 1984.

Standard Safety Code for Elevators, Dumbwaiters, Escalators, and Moving Walks, ANSI A17.1, published by the American Society of Mechanical Engineers.

Words and Terms

Balustrade
Egress
Escalator
Flight
Header
Headroom
Landing
Ramp
Riser
Slope (of ramp, stair, escalator)
Smokeproof tower
Stairs: straight run, parallel, angle, scissors, circular
Tread
Winder

Significant Relations, Functions and Issues

Exit (egress) requirements of building codes.
Facilities for the handicapped.
Factors affecting choice of slope (pitch angle) of ramps, stairs, escalators, moving ramps.
Need and required details for smokeproof towers.

14.4. ELEVATORS

An elevator is a hoisting and lowering mechanism equipped with a car or platform that moves in guides in a substantially vertical direction and that transports passengers or goods, or both, between two or more floors of a building. A powered elevator employs energy other than gravitational or manual to move the car. The shaft in which the car travels is called a *hoistway.*

A *passenger elevator* is an elevator used primarily to carry persons other than the operator and persons necessary for loading and unloading.

A *hospital elevator* is a special type of elevator in which the cars are of suitable size and shape for transportation of patients in stretchers or standard hospital beds and of attendants accompanying them.

A *freight elevator* is an elevator primarily used for carrying freight and on which only the operator and the persons necessary for loading and unloading the freight are permitted to ride.

Elevators are useful for moving large numbers of people and heavy loads between floors of a multistory building. Elevators often are required by local building codes for vertical transportation in buildings more than four sto-

ries high, but they are generally not accepted as a required means of egress in emergencies. Although they are not relied on for this purpose, they nevertheless may be used as a means of escape by occupants in emergencies. Also, it is desirable that elevators be arranged for use by firemen in emergencies.

Two major types of elevators are in general use—electric and hydraulic. An *electric elevator* is a powered elevator to which energy is applied by an electric driving machine. This type is used exclusively in tall buildings and in most low buildings. An *hydraulic elevator* is a powered elevator to which energy is applied by a liquid under pressure in a cylinder equipped with a plunger, or piston. This type is used for low-rise freight service and may be used for low-rise passenger service where low initial cost is desired.

For an electric elevator, the car is raised or lowered along its guides by wire ropes controlled by the driving machine (see Figs. 14.9 and 14.10). For an hydraulic elevator, the car

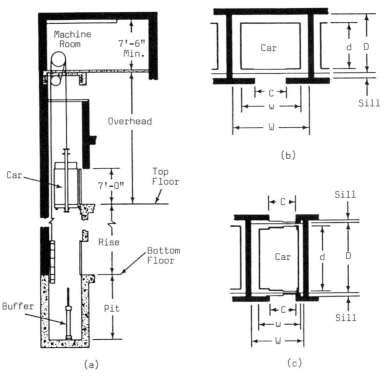

Fig. 14.9. Electric traction passenger elevator and hoistway. (*a*) Vertical section through hoistway. (*b*) Horizontal section through hoistway with one door at landing. (*c*) Horizontal section for two doors at landing.

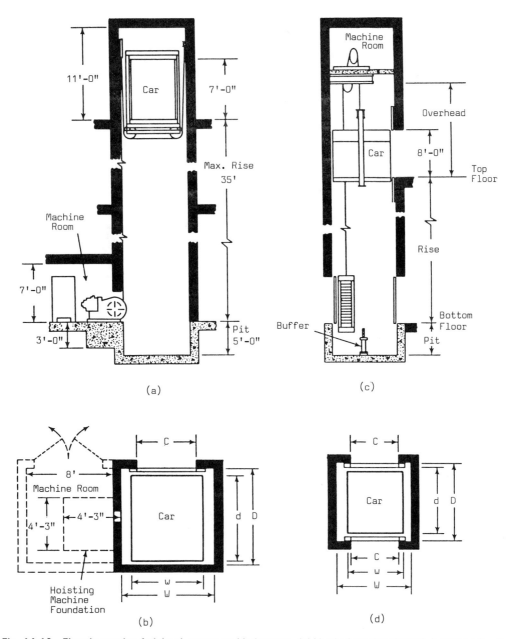

Fig. 14.10. Electric traction freight elevators and hoistways. (a) Vertical section through hoistway for elevator with basement machine room. (b) Horizontal section through hoistway of (a). (c) Vertical section through hoistway for elevator with overhead machine room. (d) Horizontal section through hoistway of (c).

is seated atop the plunger, which moves up and down under the control of hydraulic pressure (see Fig. 14.11).

The elevator manufacturer usually supplies and installs the elevators and required equipment, such as machinery, signal systems, cars, hoistway doors, ropes, guide rails and safety devices. The general contractor should guarantee the dimensions of the hoistway, its alignment and its freedom from encroachments on required elevator clearances. The owner's architect or engineer is responsible for design and construction of components needed for supporting the elevator installation.

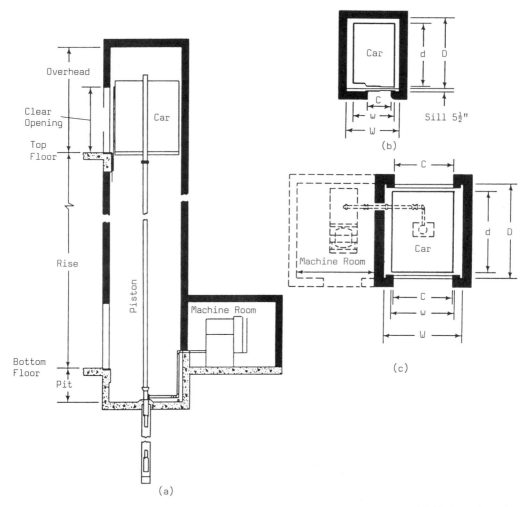

Fig. 14.11. Hydraulic elevator and hoistway. (*a*) Vertical section through hoistway. (*b*) Horizontal section through hoistway with one door at landing. (*c*) Horizontal section with two doors.

Hoistways

The hoistway in which an elevator travels extends from the bottom of a pit to the underside of the overhead machinery-space floor or grating, or to the underside of the roof if the hoistway does not penetrate the roof (see Figs. 14.9 to 14.11). Hoistways generally have openings through which elevator cars receive or discharge passengers at floors, balconies or platforms.

The elevator pit is that portion of a hoistway extending from the threshold level of the lowest landing floor to the floor at the bottom of the hoistway (see Figs. 14.9 to 14.11). The pit usually houses buffers or bumpers and other safety devices. A *buffer* is a device designed to stop a descending car or counterweight beyond its normal limit of travel, by storing or by absorbing and dissipating its kinetic energy. A *bumper* is a device, other than an oil or spring buffer, intended to halt a descending car or counterweight by absorption of the impact.

The *American National Standard Safety Code for Elevators, Dumbwaiters, Escalators and Moving Walks*, ANSI A17.1, published by the American Society of Mechanical Engineers, contains requirements for design and construction of hoistways as well as for elevators and elevator equipment. The code, for example, requires, in general, that hoistways be enclosed throughout their height with 2-hr-rated construction. All hoistway landing openings should be protected with $1\frac{1}{2}$-hr fire-rated door assemblies. Other hoistway-opening protective assemblies should have the same fire rating.

The code, however, does not require a fire-resistant enclosure for elevators that are entirely within one story or that pierce no solid floors while serving two or more open galleries, book stacks, etc., in such buildings as power-houses, libraries and open towers. Nevertheless, such hoistways should be fully enclosed. Enclosures and doors should be unperforated for a height of 6 ft above each floor or landing and above the treads of adjacent stairways. Open-work enclosures may be used above the 6-ft level. Made of a wire grille or expanded metal, such enclosures should not have spaces large enough to pass a 2-in. ball.

A hoistway for more than one elevator is called a *multiple hoistway*. Up to four elevators may operate in a multiple hoistway. If four or more elevators serve the same portion of a building, they should be installed in a minimum of two hoistways.

A metal or concrete floor should be provided at the top of each hoistway. (A floor, however, is not required below secondary and deflecting sheaves of traction-type driving machines located over the hoistway.) When the driving machine is placed over the hoistway, the floor should be set above or level with the top of the beams supporting the machine. If the machine is not over the hoistway, the floor should be placed below the overhead sheaves. If the cross-sectional area of the hoistway is 100 sq ft or less, the floor should cover the entire hoistway area. For larger areas, the floor should extend at least 2 ft beyond the general contour of the driving machine, sheaves or other equipment. Also, the floor should extend to the entrance to the machinery space at or above the level of the platform. If the floor does not cover the entire hoistway area, the exposed edges should be guarded with a toe board at least 4 in. high and a railing at least 42 in. high and conforming to the requirements of the *American National Standard Safety Code for Floor and Wall Openings, Railings and Toe Boards*, ANSI A12.1.

Hoistway Venting. Means for venting smoke and hot gases to the outdoors should be provided in hoistways of elevators serving four or more floors. (Automatic sprinklers responsive to smoke as well as heat may be used instead of vents in high-rise office and industrial buildings without overnight sleeping quarters.) Hoistway enclosure walls are not permitted to incorporate windows. Hence, vents must be installed. One permissible location for vents is in the hoistway walls directly below the floor at the top of the hoistway. These vents should exhaust directly to the outer air or through non-combustible ducts extending to the outside. Another permissible location for vents is in the wall or roof of the machinery penthouse, or overhead machinery space above the roof of the building.

Vent area should total 3.5% of the hoistway area but not less than 3 sq ft per elevator car. At least one-third of this vent area should be permanently open or automatically openable by a damper. The closed portions of the vent openings should consist of windows or skylights, with metal frames and sash, and glazed with plain glass not more than $\frac{1}{8}$-in. thick. (Thin glass is desirable to make breaking it easy in emergencies to increase the open vent area.) A wire-mesh or expanded-metal screen should be placed above each skylight to safeguard against accidental breakage of the glass. A similar screen should be placed under the skylight to prevent broken glass from falling into the hoistway.

Machinery Spaces. Enclosures of spaces containing machines, control equipment and sheaves should have construction equivalent to that used for the hoistway. The spaces should be ventilated to dissipate heat from the machinery.

If the elevator machine and control equipment are at the top of the hoistway, other machinery and equipment essential to building operation may be placed in the same room. The elevator equipment, however, should be separated from the others by a substantial metal grille at least 6 ft high and the entrance to the enclosure should be protected by a self-closing, self-locking door. If the elevator machine and control equipment are not located at the top of the hoistway, only equipment required for the operation of the elevator is permitted in the elevator machine space.

In general, at least 7-ft headroom above the floor should be provided in elevator machinery spaces, but the headroom may be only $3\frac{1}{2}$ ft in

Table 14.1. Sizes and Capacities of Electric Traction Passenger Elevators (Dimensions listed are indicated in Fig. 14.9.)

A. Gearless–Speeds in Excess of 300 Fpm

Building	Rated capacity		Platform		Hoistway		Doors:[a]
	Lb	Passengers	Width w	Depth d	Clear width W	Clear depth D	Clear opening C
Office	2,500	16	7'-0"	5'-0"	8'-4"	6'-4"	3'-6"
Office or hotel	3,000	20	7'-0"	5'-6"	8'-4"	6'-10"	3'-6"
Large office	3,500	23	7'-0"	6'-2"	8'-4"	7'-6"	3'-6"
Store	3,500	23	8'-0"	5'-6"	9'-5"	6'-10"	4'-6"
Large office	4,000	27	8'-0"	6'-2"	9'-4"	7'-6"	4'-0"

B. Gearless Hospital–Speeds in Excess of 75 Fpm

Rated capacity	Passengers	Platform		Hoistway		Doors:[a]
		Width w	Depth d	Clear width W	Clear depth D	Clear opening C
4,000 Lb						
Front opening	26	5'-8"	8'-8"	7'-8"	9'-1"	4'-0"
Front & rear opening	26	5'-8"	9'-1$\frac{1}{2}$"	7'-8"	9'-6"	4'-0"
5,000 Lb[b]						
Front opening	33	6'-0"	9'-6"	8'-0"	9'-11"	4'-0"
Front & rear opening	33	6'-0"	10'-0"	8'-0"	10'-4$\frac{1}{2}$"	4'-0"

C. Geared–Speeds up to 350 Fpm

Building	Rated capacity		Platform		Hoistway		Doors[a]		
	Lb	Passengers	Width w	Depth d	Clear width W	Clear depth D	Type	Sill	Clear opening C
Apartment	1,200	8	5'-0"	4'-0"	6'-4"	5'-3"	Single swing[c]	none	2'-8"
							Two speed	5$\frac{1}{2}$"	3'-0"
Apartment and small	2,000	13	6'-4"	4'-5"	7'-8"	5'-8"	Single swing[c]	none	2'-8"
							Single slide	4"	3'-0"
factory	2,500	16	7'-0"	5'-0"	8'-4"	6'-3"	Center opening	4"	3'-6"
Small office and factory	3,000	20	7'-0"	5'-6"	8'-4"	6'-9"	Center opening	4"	3'-6"
Department store	3,500	23	8'-0"	5'-6"	9'-5"	6'-10"	Center opening	4"	4'-6"

[a]Standard door height 7 ft. Rough sills are usually 4 in. for single-slide or center-opening doors and 5$\frac{1}{2}$ in. for two-speed doors.

[b]5,000-lb hospital elevators have two-speed, center-opening doors.

[c]Recommended only for elevators with speeds of 100 fpm or less.

spaces containing only overhead, secondary or deflecting sheaves, or $4\frac{1}{2}$ ft if such spaces contain also speed governors, signal machines or other equipment.

Car and Hoistway Dimensions. Tables 14.1 to 14.3 list typical car platform and hoistway dimensions for elevator cars with various rated load capacities. Table 14.1 gives details for electric passenger elevators, Table 14.2 for electric freight elevators and Table 14.3 for hydraulic elevators. (For speeds in excess of 700 fpm, an elevator manufacturer should be consulted for more accurate estimates of dimensions. Different manufacturers may have standard elevators available with sizes different from those

given in the tables for the capacities and speeds listed.) Dimensions referred to in the Tables 14.1 to 14.3 are indicated in Figs 14.9 to 14.11.

Horizontal Clearances. Clearance should be at least $\frac{3}{4}$ in. between the elevator car and the hoistway enclosure, except on the sides used for loading and unloading. Clearance between the car-platform sill and vertically sliding hoistway doors or the hoistway edge of any landing sill should be at least $\frac{1}{2}$ in. where side door guides are used and $\frac{3}{4}$ in. where corner guides are used. This clearance, however, should not exceed $1\frac{1}{2}$ in. The clearance between the loading side of the car platform and the hoistway

Table 14.2. Sizes and Capacities of Electric Traction Freight Elevators (Dimensions listed are indicated in Fig. 14.10.)

A. Light Duty—25 Fpm

| Capacity, lb | Platform | | Hoistway | | Doors:[a] |
	Width w	Depth d	Clear width W	Clear depth D	Clear opening C
1,500	5'-4"	6'-1"	6'-11"	6'-9"	4'-5"
2,000	6'-4"	7'-0"	7'-11"	7'-8"	5'-2"
2,500	6'-4"	8'-0"	7'-11"	8'-8"	5'-2"

B. General Purpose—Speeds 50–200 Fpm

| Capacity, lb | Platform | | Hoistway | | Doors:[c] |
	Width w	Depth d	Clear width W	Clear depth D[b]	Clear opening
3,000	5'-4"	7'-0"	7'-2"	7'-11"	5'-0"
2,500, 3,500, 4,000	6'-4"	8'-0"	8'-4"	8'-11"	6'-0"
4,000, 5,000, 6,000, 8,000	8'-4"	10'-0"	10'-4"	10'-11"	8'-0"
10,000	8'-4"	12'-0"	10'-6"	12'-11"	8'-0"

C. Heavy Duty—Speeds 50–200 Fpm

| Capacity, lb | Platform | | Hoistway | | Doors: |
	Width w	Depth d	Clear width W[e]	Clear depth D[b]	Clear opening
10,000	8'-4"	12'-0"	11'-4"	12'-11"	8'-0" × 8'-0"
12,000	10'-4"	14'-0"	13'-6"	14'-11"	10'-0" × 8'-0"
16,000	10'-4"	14'-0"	13'-10"	15'-3"	10'-0" × 10'-0"
18,000	10'-4"	16'-0"	13'-11"	17'-3"	10'-0" × 10'-0"
20,000	12'-0"	20'-0"	15'-9"	21'-3"	11'-8" × 10'-0"

[a] Standard door height is 7 ft.
[b] Hoistway, front to back, is large enough to allow for installation of a reverse or rear opening in the car.
[c] Standard door height is 8 ft.
[e] These dimensions include space for double column guide-rail supports.

Table 14.3. Sizes and Capacities of Hydraulic Elevators
(Dimensions listed are indicated in Fig. 14.11.)

A. Passenger

Capacity, lb	Passengers	Platform		Hoistway		Doors:[a] clear opening C	Pit	Overhead	Max. rise	Stops
		Width w	Depth d	Clear width W	Clear depth D					
1,500	10	5'-0"	4'-6"	6'-8"	4'-11"	2'-8"	4'-0"	10'-9"	29'-0"	3
2,000	13	6'-4"	4'-5"	7'-8"	4'-10"	3'-0"	4'-0"	11'-0"	41'-0"	5
2,500	16	7'-0"	5'-0"	8'-4"	5'-5"	3'-6"	4'-0"	11'-3"	42'-0"	5
3,000	20	7'-0"	5'-6"	8'-4"	5'-11"	3'-6"	4'-0"	11'-3"	42'-0"	5
3,500	23	7'-0"	6'-2"	8'-4"	6'-7"	3'-6"	4'-0"	11'-3"	42'-0"	5
4,000	26	8'-0"	6'-2"	9'-4"	6'-7"	4'-0"	4'-0"	11'-3"	42'-0"	5

B. Hospital

Capacity, lb	Passengers	Width w	Depth d	Clear width W	Clear depth D	clear opening C	Pit	Overhead	Max. rise	Stops
4,000	26	5'-8"	8'-8"	7'-3"	9'-1"	4'-0"	4'-0"	11'-3"	36'-0"	5

C. Freight

Capacity, lb	Platform		Hoistway		Doors: clear opening	Pit	Overhead
	Width w	Depth d	Clear width W	Clear depth D[b]			
2,500	5'-4"	7'-0"	6'-8"	7'-11"	4'-3" × 7'-6"	4'-3"	14'-0"
3,000	6'-4"	8'-0"	7'-8"	8'-11"	6'-0" × 7'-6"	4'-3"	14'-0"
3,500	6'-4"	8'-0"	7'-8"	8'-11"	6'-0" × 8'-0"	4'-6"	14'-3"
4,000	6'-4"	8'-0"	7'-8"	8'-11"	6'-0" × 8'-0"	4'-6"	14'-3"
5,000	8'-4"	10'-0"	9'-10"	10'-11"	8'-0" × 8'-0"	4'-6"	14'-3"
6,000	8'-4"	10'-0"	9'-10"	10'-11"	8'-0" × 8'-0"	4'-6"	14'-3"
8,000	8'-4"	12'-0"	9'-10"	12'-11"	8'-0" × . 8'-0"	4'-6"	14'-3"
10,000	8'-4"	12'-0"	11'-4"[c]	12'-11"	8'-0" × 8'-0"	4'-10"	14'-3"
12,000	8'-4"	14'-0"	11'-6"[c]	14'-11"	8'-0" × 8'-0"	5'-0"	14'-3"
16,000	8'-4"	16'-0"	11'-8"[c]	17'-3"	8'-0" × 10'-0"	5'-6"	16'-3"
18,000	10'-4"	16'-0"	13'-8"[c]	17'-3"	10'-0" × 10'-0"	5'-6"	16'-3"
20,000	12'-0"	20'-0"	15'-4"[c]	21'-3"	11'-8" × 10'-0"	5'-6"	16'-3"

[a]Standard door height 7 ft.
[b]Hoistway, front to back, is large enough to allow for installation of a reverse or rear opening in the car.
[c]These dimensions include space for double column guide-rail supports.

enclosure should at no place be more than 5 in. This clearance, however, may be increased to $7\frac{1}{2}$ in. if vertically sliding hoistway doors are used.

Between the car and its counterweight, the clearance should be at least 1 in. Between the counterweight and other items, the clearance should be a minimum of $\frac{3}{4}$ in. In multiple hoistways, the clearance between moving equipment should be at least 2 in.

Hoistway Doors. ANSI A17.1 lists types of doors that may be used at hoistway openings

at landings and gives requirements for their operation.

The code requires that horizontally sliding or swinging doors for automatic elevators be provided with door closers so that an open door closes automatically if the car leaves the landing zone. (A landing zone is the space 18 in. above and below a landing.) The code permits a horizontally sliding hoistway door to be open, if a car is at a landing, only when the car is being loaded or unloaded or when the door is under the control of an operator or an automatic-elevator dispatching system.

Hoistway doors may be equipped with door-locking devices, hoistway access switches and parking devices. A *locking device* secures a door in the closed position and prevents it from being opened from the landing side except under certain conditions, such as for repair, maintenance or emergency. An *access switch* is an electric switch at a landing that is used to permit operation of the car with the car door and hoistway door at the landing open, for access to the top of the car or to the pit. A *parking device* is equipment that permits opening or closing of a hoistway door from the landing side at any landing if the car is within the landing zone. At least one landing should be equipped with a parking device, unless hoistway doors automatically unlock when the car is in the landing zone.

Hoistway doors should also be equipped with hoistway-unit-system interlocks. These comprise a series of hoistway-door interlocks, hoistway-door electric contacts or hoistway-door combination mechanical locks and electric contacts that prevent operation of the elevator driving machine by the normal operating device unless all hoistway doors are closed and locked.

Hoistway doors should be so arranged that they may be opened by hand from the hoistway side when the car is within the interlock unlocking zone, except when the doors are locked out of service. (Means should not be provided for locking out of service the doors at the main-entrance landing or at the top or bottom terminal landing.) Automatic fire doors, controlled by heat, should not lock any hoistway door so that it cannot be opened manually from inside the hoistway. Also, such doors should not lock any exit leading from a hoistway door to the outdoors.

Car doors should also be equipped with electric contacts, to prevent operation of the driving machine by the normal operating device unless the car door is closed.

Car and hoistway doors may have vision panels of clear wired glass or laminated glass, to permit passengers in elevators to see if passengers are waiting to enter at landings. ANSI A17.1, however, requires all horizontally swinging hoistway doors and manually operated, self-closing, sliding hoistway doors for elevators with automatic or continuous-pressure operation to be provided with a vision panel. A vision panel, however, is not required at landings of automatic elevators provided with a device that indicates the location of the car in the hoistway (hall position indicator).

Guide Rails. Beams usually have to be provided at vertical intervals in the hoistway enclosure to support steel guide rails along which the elevator car and its counterweight run. The rails generally are tee-shape in cross section, with smooth guiding surfaces.

One rail is installed on each side of the hoistway to guide the car, which has upper and lower guide wheels attached to its supporting frame, or sling. A second pair of rails is placed along one wall of the enclosure for the counterweight, which also has upper and lower guide wheels attached to its frame.

Buffers and Bumpers. Buffers or bumpers should be installed at the bottom of hoistways to absorb the impact from cars that descend below their normal limit of travel. ANSI A17.1 requires buffers under all cars and counterweights in hoistways that are placed over accessible spaces. Buffers or solid bumpers should be used for passenger elevators with speeds up to 50 fpm and for freight elevators with speeds up to 75 fpm. For greater speeds, buffers are required. Spring buffers may be used for speeds up to 200 fpm, but for greater speeds, oil buffers or their equivalent are required.

Solid bumpers may be made of wood or other resilient material. Spring buffers utilize springs to bring a descending car to a gradual halt. Oil buffers resist a car impact with hydraulic pressure of oil applied against a plunger contacted by the car.

Electric Elevators

In addition to the hoistway components previously discussed, an electric-elevator installation requires a car, or cab, for transporting passengers or freight, wire ropes for raising and lowering the car and for other purposes, a driving machine, sheaves for controlling rope motion, control equipment, a counterweight and safety devices (see Figs. 14.12 and 14.13).

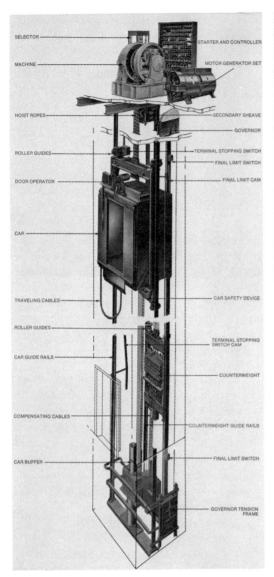

Fig. 14.12. Components of an installation of an electric traction (gearless) passenger elevator with overhead machine drive.

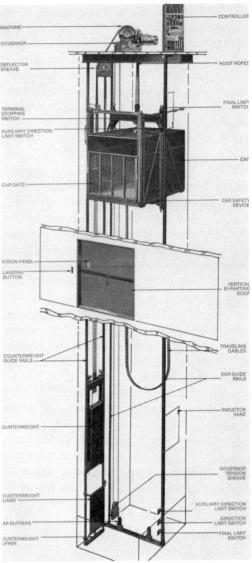

Fig. 14.13. Components of an installation of a freight elevator with geared machine and ac resistance control.

Elevator Cars. A car consists of a platform, on which passengers or freight are transported, and light metal walls and roof, with doors or gates on one or two sides. The platform and enclosure are supported on a structural steel frame.

Wire ropes that raise and lower the car are attached to the top of the frame or wind around sheaves connected to the frame. The guide wheels that roll along vertical guide rails in the

hoistway are attached to the sides of the frame. The frame also carries safety devices for preventing the car from attaining excessive speed, especially if the ropes should break and the car should fall.

The car doors may be horizontally or vertically sliding, and manually or power operated. They are equipped with safety devices to prevent them from opening while the car is in motion or outside the landing zone. Such devices also keep the car from moving while the doors are open. Additional devices reopen the doors when they start to close on a passenger or other object entering or leaving the car. Still other devices may be provided for other safety reasons, for instance, to keep the doors from closing and the elevator from starting when the car is overloaded.

The interior of the car should be illuminated with at least two lamps. Illumination provided at the landing edge of the car platform when the car and hoistway doors are open should be at least 5 ft-c for passenger elevators and 2.5 ft-c for freight elevators. An emergency lighting system also should be provided, to operate automatically within 10 seconds after failure of power for normal lighting. This system should be capable of providing for a minimum of 4 hr at least 0.2 ft-c at a level 4 ft above the car floor and about 1 ft in front of a car station.

The interior of the car should also be ventilated. A signal registering device, generally a set of pushbuttons, that control the stopping of the car at selected landings and an emergency stop switch should be installed in a convenient location for operation by passengers or an elevator operator. Alternatively, a car switch may be installed for an operator to control starting and stopping of the car. The car interior may also be equipped with a position indicator, an electrical device that shows, usually by lights, the position of the car in the hoistway.

The roof of the car should contain an emergency exit. Means should be provided for operating the car from its top during inspection, maintenance and repair.

Counterweights. The load placed on the driving machine by the car is counterbalanced by a counterweight. It consists of blocks of cast iron in a steel frame. It is connected to the elevator car by the hoisting ropes and moves up and down with the car. The counterweight is kept in a fixed vertical path by guide wheels on its frame that roll along vertical guide rails.

Use of the counterweight reduces power requirements for moving the car and maintains traction between the suspension ropes and the driving sheave. Weight of the counterweight usually is equal to the weight of the unloaded

car plus about 40% of the rated load capacity of the car.

Car and Counterweight Safeties. A safety is a mechanically operated device that is capable of stopping and supporting the weight of an elevator car and its rated load when the device is tripped by a car-speed governor. The governor should be set to actuate the safety when the car travels at more than 15% above its rated speed.

One or more safeties should be attached to the car frame, and one safety should be located within or below the bottom members of the frame (safety plank). The safeties, when tripped, should apply a squeeze to one pair of guide rails to retard and stop movement of the car. Safeties such as those used for cars should also be attached to the counterweight frame when a hoistway is located over an accessible space. The safeties may be released by upward motion of the car.

The governor has to be located where it will not be struck by the car or the counterweight if either should overtravel. The governor may be conveniently located in the machine space. The device may measure the car speed from the rotation of a sheave around which is wound a rope connected to the car and held under tension. When the car overspeeds, the governor actuates jaws that grip a rope connected through linkages to a safety. The gripping action on the rope releases a spring on the safety to apply to the guide rail a squeezing force that stops the car. Simultaneously, an electrical switch on the governor and one on the safety are actuated to remove power from the driving machine and apply a friction brake to the drive shaft.

Driving Machines and Controls. An electric driving machine consists of an electric motor, a brake, a drive shaft turned by the motor, a driving sheave or a winding drum, and gears, if used, between the drive shaft and the sheave or drum. The brake, actuated by springs or gravity and electrically released, applies friction to the drive shaft to retard or stop the movement of the elevator car. Movement of the hoisting ropes is controlled by the driving sheave or by the winding drum, whichever is used.

A **winding-drum machine** is a geared-drive machine with a grooved drum to which the hoisting ropes are fastened and on which they wind or unwind to move an elevator car. This type of machine may be used for freight elevators without counterweights, with speeds not exceeding 50 fpm and with a travel, or rise, of not more than 40 ft.

A **traction machine** has a motor directly connected mechanically to a driving sheave, with or without intermediate gears, and maintains and controls motion of the elevator car through friction between the hoisting ropes and the driving sheave. This sheave, also called a traction sheave, is a wheel with grooves in its metal rim for gripping the hoisting ropes. Traction machines are used for most electric elevators.

A **gearless traction machine** is a traction machine without intermediate gearing, which is not needed, because the traction, or driving, sheave and the brake drum are mounted directly on the motor shaft. Since there are no gears, the traction sheave must run at the same speed as the motor. As a result, for slow-speed elevators, the motor speed would be too slow for efficient operation. Hence, gearless traction machines usually are used only for elevators that operate at speeds of 350 fpm or more. Such elevators use dc motors.

Geared-traction machines are used for slower-speed elevators. The gears interposed between the motor and the driving sheave permit use of a high-speed ac or dc motor with low car speeds. For slow-speed freight elevators, ac motors usually are used.

The system governing starting, stopping, direction of motion, speed, and acceleration and deceleration of an elevator car is called *control*. Usually, either multivoltage, or variable-voltage, control or rheostatic control is used for electric elevators.

Multivoltage control is generally used with driving machines with dc motors. Control is achieved by varying the voltage impressed on the armature of the dc motor. In buildings supplied with ac power, the variable voltage usually is obtained from a motor-generator set, which converts ac to dc. This type of control is used because it combines smooth, accurate speed

regulation with efficient motor operation. It permits accurate car stops and rapid acceleration and deceleration, with low power consumption and little maintenance, but it costs more than rheostatic control.

Rheostatic control is accomplished by varying the resistance or the reactance of the driving-machine motor, to control its speed. This type of control is often used with an ac motor for low-rise, low-speed elevators that are used infrequently (less than five trips per hr).

Regardless of the type of control selected, provision must be made for stopping the elevator car level with landings. For the purpose, either of two types of leveling devices may be installed:

Inching, which permits an operator to move the car with continuous-pressure pushbuttons or levers to floor level from within a limited zone (about 9 in. above or below) with the car door open.

Automatic leveling, in which the driving motor is controlled to bring the car level with a landing. In the best types of elevator installation, a two-way, automatic maintaining leveling device is provided. This corrects the car level on both over-run and under-run at a floor and maintains the level during loading and unloading of the elevator. The device automatically compensates for changes in rope length with temperature and loading changes.

To prevent a car from traveling past the upper or lower terminal landings, stopping-device switches should be installed. *Normal terminal stopping devices* should be provided to slow down and stop the car automatically at or near the top and bottom terminal landings. Such devices should continue to function until the final terminal stopping device operates. *Final terminal stopping devices* should be installed to shut off automatically the electric power to the driving-machine motor and brake after the car has past a terminal landing; however, it should not function when the car has been halted by the normal terminal stopping device. Operation of the final terminal stopping device should prevent up or down movement of the car by normal car operating devices.

ANSI A17.1 lists the numerous electrical protective devices and acceptable emergency signaling devices required for elevators. The latter devices are especially required for elevators that may be operated at any time without a designated operator. Such devices include a bell, means for two-way conversation between each elevator and a readily accessible point outside the hoistway, or a telephone connected to a central telephone exchange.

Roping. Speed of the car, load on the driving machine and forces applied to building members by the elevator installation are considerably influenced by the arrangement of hoisting, or suspension, ropes and sheaves used.

ANSI A17.1 requires that cars be suspended on at least three ropes for traction-type machines and two ropes for winding-drum machines. When a counterweight is used, at least two ropes are required for it. All hoisting and counterweight ropes must be at least $\frac{1}{2}$-in. in diameter.

A *wire rope* consists of a group of steel strands, which, for elevator installations, are laid helically around a hemp core. A *strand* consists of an arrangement of steel wires laid

helically around a center wire to produce a symmetrical section.

A simple arrangement of hoisting ropes and sheaves is shown in Fig. 14.14a. The ropes attached to the top of the car frame pass over the driving sheave and are then guided by a deflector sheave to the side of the hoistway and downward to the top of the counterweight. Because the ropes pass over the sheave only once between the car and the counterweight, this rope arrangement is called *single wrap*. In addition the roping is referred to as 1:1, because the car speed equals the rope speed. The 1:1 single-wrap roping is suitable for high-speed passenger elevators.

To obtain sufficient traction, the driving sheave has wedge-shaped or undercut grooves that grip the ropes because of a wedging action between the sides of the grooves and the rope. The pinching tends to shorten rope life.

To improve traction with less wear of the ropes, *double-wrap roping* is often used for high-speed passenger elevators. The ropes in this case extend from the top of the car frame to the driving sheave, wind twice around an idler sheave and the driving sheave and are then deflected to the counterweight by the idler sheave (see Fig. 14.14b). The sheaves have U-shaped or round-seat grooves, which cause

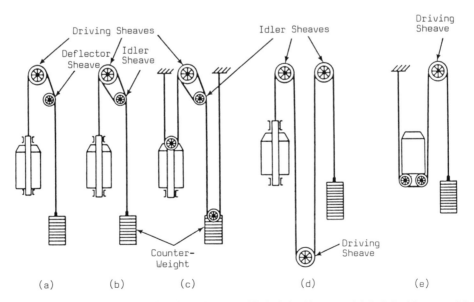

Fig. 14.14. Roping for elevators. (a) 1:1 single wrap. (b) 1:1 double wrap. (c) 2:1 double wrap. (d) 1:1 single wrap with driving machine in basement. (e) 2:1 for car with underslung frame.

less wear of the ropes. The 1:1 double-wrap roping applies twice the loading to the driving sheave for the same weight of car and counterweight than does the single-wrap roping and requires a heavier design.

For heavily loaded, low-speed freight elevators, 2:1 double-wrap roping may be used (see Fig. 14.14c). The rope speed, in this case, equals twice the car speed. Hence, a higher-speed, less-costly motor can be used than with 1:1 roping for a given car speed. For 2:1 roping, ends of the ropes are dead-ended at the top of the hoistway, at beams, instead of being attached to the car and the counterweight, as for 1:1 roping. The ropes pass around sheaves on the top of the car frame and the top of the counterweight. With this arrangement, the anchorages carry one-half the weight of car and counterweight. Hence, the load on the driving and idler sheaves is only about one-half that for 1:1 roping.

In Fig. 14.14a to c, the driving machine is shown at the top of the hoistway, a common location. Often, the machine space is provided in a penthouse. For some buildings, for architectural reasons, however, a penthouse may be undesirable. In such cases, the driving machine may be placed in a basement (see Fig. 14.14d). This arrangement, though, has numerous disadvantages. These include greater loads on overhead supports, tripling of rope length, more sheaves, higher fraction losses, greater rope wear and higher power consumption, with consequent greater operating costs.

For low-speed, low-rise elevators, the roping shown in Fig. 14.14e for a car with an underslung frame may be used. This is a form of 2:1 roping.

Operating Systems. The method of actuating elevator control is called elevator operation. Generally, the more complex and the more sophisticated the operating system is, the greater is the cost, but usually more things are accomplished automatically and handling of traffic is more efficient. Following are descriptions of several types of operation:

Car-Switch Operation. An operator controls movement and direction of travel with a manually operated car switch or continuous-pressure buttons in the car. Handles of lever-type operating devices for car-switch operation should return to the stop position and latch there automatically when the hand of the operator is removed, to insure that the operator controls car movement. In automatic car-switch floor-stop operation, the operator presses a button to initiate a stop at a landing, after which slowing and stopping of the elevator is accomplished automatically.

Preregister Operation. Signals to stop are registered in advance by passengers pushing buttons at landings or by an operator in the car. When the car approaches a stop point, the operator in the car is notified by a signal to initiate the stop, which is accomplished automatically.

Signal Operation. Predetermined stops are registered for one or more elevators serving a floor when either a button in the car is pressed or an up or a down button is pressed at that floor. Regardless of the direction of car travel or of the sequence in which buttons at various floors are pressed, the car stops automatically at the landings for which signals were received. When a landing is served by two or more elevators, the stops are made automatically by the first available car approaching the landing in the appropriate direction. The car, however, can be started only by pressing the car button for a specified floor stop.

Automatic Operation. The car starts when a button is pushed at a landing or in the car, or in response to an automatic starting mechanism. The car stops at landings, and car and hoistway doors open automatically. There are several types of automatic operation. In order of increasing sophistication, these are:

Single Automatic Operation. Stops are signaled by pressing a single button provided at a landing or a button in the car corresponding to a landing. After any button has been pressed, actuation of any other button has no effect on car operation until the stop signaled by the first button has been made.

Nonselective Collective Automatic Operation. Stops are signaled by pressing of buttons as for single automatic operation. All stops signaled, however, are made, regardless of the number of buttons that have been pressed or the sequence in which they were actuated. The car stops in the order in which landings are reached, regardless of its direction of travel. This has the disadvantage, however, that a passenger at an intermediate landing who wishes to descend may be served by an elevator traveling upward with passengers. This passenger then has the choice of traveling upward to the uppermost stop signaled and then descending or waiting at the landing for the downward trip of the elevator.

Selective Collective Automatic Operation. Stops are signaled by pressing one button in the car for each landing served or by pressing an up or down button at a landing. All stops called for by the car buttons are made in the same manner as for nonselective collective operation. All stops signaled by the buttons at landings (hall calls) are made in the order in which landings are reached in each direction of travel, regardless of the sequence in which signals are received. All up calls are answered when the car is traveling upward, and all down calls are answered when the car is traveling downward.

Group Automatic Operation. Selective collective operation may be extended to operation of groups of cars. A supervisory control system coordinates the operation, including automatic dispatching of the cars. Selected cars at designated dispatching points close their doors and proceed automatically on their trips in a regulated manner. A call is answered by the first car to pass in the proper direction. A timer signals the cars to leave a terminal at predetermined intervals. The automatic dispatching increases the number of passengers the elevators can carry in a given time.

Elevator Group Supervision. Supervisory control systems are widely used for groups of three or more operatorless passenger elevators. The systems are capable of adjusting to varying traffic conditions. They can control door action and car motion so that cars in the best location for responding to stop signals do so. Also, these systems can dispatch cars in a predetermined, regulated order from upper and lower terminals.

For an office building, the following operation may be used: When office employees report for work in the morning and traffic moves predominantly upward, the supervisory control sets itself for an up peak. For this setting, the control dispatches cars from the bottom terminal either at the end of a specific time interval or when they are loaded to 80% of capacity, as indicated by a load-weighting device in the car platform. As soon as a car answers the highest call registered, the control returns the car to the bottom terminal.

For a building with very heavy up traffic, the building may be divided into a high and low zone. An elevator then is assigned to serve each zone. The car acts as an express, bypassing floors outside its zone. This arrangement often reduces average round-trip time.

When employees leave at the end of the business day and traffic moves predominantly downward, the supervisory control sets itself for a down peak. Car parking at the bottom terminal is eliminated. Cars respond normally to up calls on the upward trip and start down as soon as the highest call registered has been answered. When the cars become fully loaded, they automatically bypass remaining stop calls. To improve round-trip time, the building may be divided into high and low zones and perhaps also one or more intermediate zones, with an elevator assigned to each zone.

Between peak periods, when traffic is moderate and generally a mixture of up and down, *multiple zoning* may be used, with the supervisory system automatically controlling the location of cars in their hoistways. The objectives are to minimize passenger waiting time and car travel. When traffic does not require a car to operate, the supervisory system parks it at a preselected location within its zone. Hence, when a call is registered within that zone, the parked car will be near the call. As traffic increases, for instance, at lunchtime, the supervisory system dispatches the parked cars at more rapid intervals. When necessary, the system shifts to down-peak or up-peak operation.

When traffic is very small, such as at night, weekends and holidays, cars are parked at the bottom terminal. Some or all of the motor-generator sets for the cars may be shut down. When a call is registered when all the motor-generator sets are off, the motor-generator set powering one car starts automatically. In off hours also, in buildings not equipped with a freight elevator, one or more of the elevators may be used as a service elevator, for transporting supplies, furniture and other goods to the various floors.

Several safety devices are incorporated in automatic elevators in addition to those normally installed in cars with operators. These devices include an automatic load weigher to prevent overcrowding, buttons in cars and at landings to stop the doors from closing and to hold them open, lights to indicate floor stops pressed, a system for communication with a supervisor outside the hoistway, means for preventing the doors from closing when a passenger or other object is standing in the door opening, and emergency power systems for use if the primary and supervisory systems should fail.

Hydraulic Elevators

For low-rise, low-speed elevators, cars may be seated on plungers and raised or lowered by hydraulic pressure, with lowering assisted by gravity. Oil serves as the pressure fluid. It is supplied through a motor-driven, positive-displacement pump, which is actuated by an electro-hydraulic system. The major parts of an electro-hydraulic elevator are illustrated in Figs. 14.11 and 14.15. Table 14.3 lists the sizes and capacities of typical hydraulic elevators.

To raise the car, the pump is started. Oil is pumped into the pressure cylinder to force the plunger up. When the car reaches the desired landing, the pump stops. The car is lowered when oil is released from the pressure cylinder, allowing the car to descend under the pull of gravity. The released oil is returned to a storage tank.

ANSI A17.1 contains requirements for design and installation of hydraulic elevators. Many of the requirements are the same as those required for safety of electric-elevator installations. Operation may also be the same as for electric elevators. Among the major differences in requirements are a prohibition on use of car safeties for hydraulic elevators; driving ma-

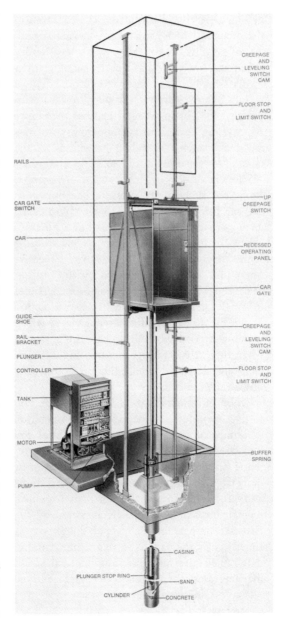

Fig. 14.15. Components for an installation of an hydraulic freight elevator.

chines must be the direct-plunger type; regulations for plunger, cylinder, tanks, safety devices, valves and supply piping and fittings. Also, each hydraulic elevator should be provided with an anticreep leveling device.

Although limited in rise and speed, hydraulic elevators have several advantages over electric elevators. Hydraulic installations are simpler.

They do not require wire ropes, sheaves or overhead equipment. They are suitable for applications where penthouse machine space is undesirable for architectural reasons. Electrohydraulic equipment may be placed in a basement. The hydraulic installation does not impose heavy overhead loads. Hence, lighter structural framing can be used. And because elevator speed is low, buffers need be only heavy springs that become fully compressed before the plunger reaches its lower limit of travel. The electric motor required for a hydraulic installation, however, may be larger and more costly than that required for an electric traction elevator of the same capacity.

Passenger Elevators

Selection of and planning for passenger elevators should take into consideration the type of building occupancy, anticipated traffic between floors, quality of service desired, life-cycle costs of the elevator installation, and spaces needed for hoistways, lobbies and elevator equipment. Also, the effects of these spaces on architectural floor plans and exterior elevations and on structural framing should be taken into account. Planning is complicated by the interdependency of these factors.

Space requirements for an elevator installation depend on the number of elevators to be installed, car platform dimensions, car shape, arrangement of hoistways, and shapes and sizes of lobbies. The number of elevators required depends on car capacity, volume of traffic and the interval between arrival of cars at any floor.

Interval is an important measure of the quality of elevator service. It is often taken as the average time between elevators leaving the main lobby. The shorter the interval, the better is the quality of elevator service, but the more expensive the elevator installation becomes because more, faster and larger elevators are needed to reduce an interval. Selection of an interval generally depends on the type of occupancy of a building. For example, an interval that may be considered satisfactory for an apartment building may be considered intolerable for an office building.

Interval is one of several factors determining travel time for passengers. Travel time is another measure of the quality of elevator service. The shorter the travel time between the arrival of a passenger at an elevator and arrival at the destination, the better is the quality of elevator service. Travel time preferably should be kept to less than one minute.

For a single elevator in a building, interval equals the round-trip time of the elevator. Round-trip time depends not only on the rated speed of an elevator but also on the probable number of stops a car has to make on its way up and down. In some cases, rated speed may have no influence on round-trip time because the short distance between stops does not allow sufficient time for the car to accelerate to full speed. Generally, however, round-trip time is composed of the time for a full-speed round-trip run without stops plus time for accelerating and decelerating per stop, time for leveling at each stop, time for opening and closing hoistway and car doors, time for passengers to move in and out of the car, reaction time of operator, lost time due to false stops, and parking time at upper and lower terminals.

Number of Elevators. If a floor will be served by n elevators, the average interval will be the round-trip time T divided by n. Similarly, the number of elevators n needed to maintain a desired interval is obtained by dividing T by the interval.

Selection of the number of elevators n is based also on the volume of traffic. Traffic flow for elevators is measured by the number of persons to be transported in 5-min. periods. The number of elevators required is calculated from the 5-min period with the maximum, or peak, traffic. For buildings under design, peak traffic generally can be estimated from studies of traffic flow in buildings of the same type located in the same or a similar neighborhood, or the peak flow may be predicted as a percentage of the probable building population, which in turn may be estimated from the total occupiable floor area.

The maximum traffic-handling capacity of an elevator car is the maximum number of passengers permitted in the car. This depends on the platform dimensions and rated capacity, lb, of the car. The handling capacity in a 5-min or 300-sec, period then is

$$HC = P \frac{300}{T} \qquad (14.3)$$

where

P = car capacity, passengers
T = round-trip, sec

The minimum number of elevators required then can be determined by dividing the peak 5-min traffic V by the 5-min handling capacity:

$$n = \frac{V}{HC} = \frac{VT}{300P} \qquad (14.4)$$

Thus, the number of elevators required is directly proportional to the round-trip time and inversely proportional to the car capacity.

Selection of the largest-capacity car available will reduce the number of elevators needed. Space needed for the elevator and initial cost, however, will be increased. Also, operating costs will increase, because the car often will travel unloaded or partly loaded. Hence, an optimum car is not likely to be the largest size available (see Table 14.1 and 14.3A).

Reduction of the round-trip time to reduce the number of elevators needed can be accomplished in several ways. For example, for tall buildings with long express-elevator runs, selection of faster cars will economically reduce round-trip time. The economics can be further improved by use of multiple zoning, discussed previously. Another way to decrease round-trip time is to reduce lost time at stops.

Opening and closing of doors, for instance, can contribute materially to lost time unless the doors are properly designed. A 42-in. opening is good, because two passengers can conveniently enter and leave a car abreast. A slightly wider door would be of little advantage. Department stores, hospitals and other structures served by passenger elevators of 4,000-lb capacity or more, however, require much bigger openings.

Power operation speeds door opening. Power-operated doors for passenger elevators are horizontally sliding. These doors may open at one side or at the center. To improve speed of opening, side-opening doors may consist of two leaves, which travel at different speeds. The leaf moving farthest travels at the greater speed. Similarly, to improve speed of opening, center-opening doors may be two speed. Center-opening doors are faster than either the single or two-speed side-opening types with the same width of opening. Transfer time also is less with center-opening doors, because passengers can move out as the doors start to open.

Lost time at stops and also round-trip time can be reduced by decreasing the time it takes an exiting passenger to move from the back of the car to the door. Thus, shape of car influences round-trip time and hence the number of elevators required. The narrower and deeper a car, the greater the time required for a passenger to leave the car is likely to be. It usually is economical to select a platform size that conforms to the standards of the National Elevator Industry, Inc. (see Table 14.1).

Diversified-Tenancy Office Buildings. The 5-min morning peak generally is the controlling factor in selection of elevators for volume of traffic. If the elevators can handle that peak well, they can also deal satisfactorily with the peaks at other times of the day. The 5-min morning peak may be about one-ninth of the building population. It may be as much as one-eighth of the population, however, if tenancy is not well diversified.

Intervals of up to 35 sec may be acceptable for high-rise elevators. In general, however, for busy, high-class office buildings in large cities, the interval should be 30 sec or less, whereas in small cities an interval of 30 sec or more may be satisfactory.

Car speeds used depend on building height:

4–10 stories	200–500 fpm
10–15 stories	600–700 fpm
15–20 stories	700–800 fpm
20–50 stories	900–1,200 fpm
over 50 stories	1,200–1,800 fpm

Elevators should be easily accessible from all building entrances. For maximum efficiency, the installation should be compact and centrally located. Not more than four elevators should be placed in a straight line. A good arrangement places elevators above the main floor in alcoves off the main corridor. Such a location eliminates interference between elevator and other traffic, makes possible narrow corridors, saves space in the upper floors, decreases passenger-transfer time and reduces walking distance to the individual elevators. Alcove width may be 10 to 12 ft. About 4 sq ft per person should be provided for passengers waiting for an elevator in a peak period.

For tall buildings, especially those with setbacks and towers, and for low buildings with large occupiable areas per floor, it is advisable to divide elevator groups into local and express banks. Such an arrangement should be considered when a building requires more than eight elevators. One or two of the elevators of the high-rise express bank should be provided with openings at all floors for weekend and holiday service. The division into local and express banks offers an additional advantage in that corridor space on the floors where there are no doors can be used for toilets, closets and stairs.

Single-Tenancy Office Buildings. Requirements and layouts for single-tenancy buildings, including government office buildings, are similar, in general, to those for diversified-tenancy buildings. In buildings occupied by one large organization, however, population density usually is large and peak traffic in the morning is very large, unless working hours are staggered. The 5-min peak may range from one-eighth to one-half the building population. Also, interfloor traffic is large and should be considered in design of the elevator system. So many elevators are usually required for handling traffic that the interval will be satisfactory.

Hospitals. Hospital elevators must handle two types of traffic: (1) medical staff and equipment and (2) transient traffic, such as patients and visitors. Peak traffic occurs when visitor traffic combines with regular hospital traffic. In large hospitals, pedestrian and vehicular traffic should be separated.

For vehicular traffic or a combination of vehicular and pedestrian traffic, platforms should be large enough to accommodate stretchers or standard hospital beds (see Table 14.1B and 14.3B). Depending on building height and car capacity, speeds range from 50 to 700 fpm. The cars should be self-leveling, for ease in loading and unloading vehicles. For pedestrian traffic alone, cars with wide shallow platforms, such as those used for office buildings, should be chosen.

Apartment Buildings. The 5-min evening peak generally is the controlling factor for selection of elevators for volume of traffic. This peak may equal 6 to 8% of the building population. One elevator rated at 2,000 lb and 200 fpm may be adequate for a six-story apartment building with 50 to 75 units. For high-rise buildings, two elevators with speeds of 400 to 500 fpm usually are sufficient. One of these elevators can be used for service work at times.

Department Stores. Elevators and escalators should be coordinated for vertical transportation in department stores. Elevators may be assumed to have to transport only 10 to 25% of the peak traffic. The required capacity of the whole vertical-transportation system may be calculated from the merchandising area above the first floor and the maximum density of shoppers on that area.

The density ratio, or ratio of peak hourly transportation capacity to the merchandising area, is about 1:20 for a busy department store. Thus, the required hourly handling capacity of the vertical-transportation system is numerically equal to 5% of the merchandising area, and elevators carry only 0.5%. Peak traffic usually occurs about midday. For supplementing escalators, a car with a capacity of 3,500 lb or more should be used (see Table 14.1C).

Freight Elevators

For low-rise, low-speed applications, electrohydraulic elevators may be advantageous for handling freight (see Table 14.3C). For buildings taller than about 50 ft, electric elevators usually will be more economical.

Note that ANSI A17.1 prohibits carrying of

passengers on freight elevators under normal conditions. In emergencies, however, elevators may transport a number of passengers equal to the rated load, lb, divided by 150.

Selection of a freight elevator should take into account the required travel, number of landings to be served, distance between landings and whether one or two openings are needed in the cars. Capacity and size of the car depend on the weight, size, type, number and method of handling of units to be transported by the elevator. Also, the probable cycle of operation and principal floors served during the peak of the cycle should be considered.

The hourly carrying capacity of a freight elevator is determined by the rated load, lb, of the elevator and the time required for a round trip. The round-trip time may be estimated as for passenger elevators, except that loading and unloading the car takes much more time for freight. This time may be predicted from studies of loading and unloading operations for a similar elevator in the same type of plant.

Control for freight elevators usually is multivoltage or ac rheostatic. The latter is preferred when the car is to be used for less than five trips per hour. The most useful and flexible type of operation is selective collective with an annunciator, an electrical device in the car that indicates visually the landings for which a stop signal has been registered. When operated without an attendant, the car automatically answers down calls when traveling downward and responds to up calls when traveling upward. An attendant, when present, can control car movement and can answer calls signaled by the annunciator by pressing the corresponding landing button in the car. For less-expensive-operation systems, single-automatic or car-switch may be used.

The standard hoistway door is vertically sliding, biparting at midheight. Doors should be power operated for busy freight elevators and those with openings wider than 8 ft. For elevators designed for motor-vehicle or industrial truck loading, car doors should be vertically sliding. For elevators used for general freight loading, however, horizontally sliding car and hoistway doors may be used.

Material-Handling Systems. For multistory warehouse or production facilities, automatic freight elevators can be integrated into material-handling systems. Infeed and outfeed, horizontal conveyors may be provided at each landing to deliver and remove loads, usually palletized, to and from a freight elevator. The elevator may be loaded, transported to another landing and unloaded—all automatically.

Automatic cart-lift elevators are useful in hospital, office and library buildings in freeing passenger and hospital elevators from disruptive cart traffic. Cars may have a capacity of 500 or 1,000 lb and 2-ft 4-in. × 4-ft or 2-ft 10-in. × 6-ft platforms, respectively. A cart to be delivered is simply placed in front of the hoistway door and dispatched by pushbutton. The cart then is automatically loaded onto the lift and unloaded at its destination.

References

B. Stein et al., *Mechanical and Electrical Equipment for Buildings*, 7th ed., Wiley, New York, 1986.

C. Ramsey and H. Sleeper, *Architectural Graphic Standards*, 8th ed., Wiley, New York, 1988.

American National Standard Safety Code for Elevators, Dumbwaiters, Escalators, and Moving Walks, ANSI A17.1, published by the American Society of Mechanical Engineers.

Words and Terms

Buffer
Bumper
Counterweight
Driving machine
Elevators: passenger, hospital, freight, electric, hydraulic
Hoistway
Interval
Roping
Terminal stopping devices

Significant Relations, Functions, and Issues

Character of elevator based on primary use: passenger, hospital, freight.
Hoistway features: multiple cars, stopping devices, venting, machinery location and space requirements, overhead clearance, general dimensional requirements.

Elevator system design features: type of operation (electric or hydraulic); type of basic service (passenger, etc.); speed, demand; number of cars and zoning; interval; hoistway and machinery requirements; operation systems; safety systems.

14.5. DUMBWAITERS

Dumbwaiters may be used in such multistory buildings as hospitals and restaurants to transport small items between levels. A dumbwaiter is a hoisting and lowering mechanism with a car of small load capacity and size. The car moves in guides in a substantially vertical direction and is used exclusively for carrying material. An undercounter dumbwaiter has its top terminal landing beneath a counter. The driving machine, usually winding drum but sometimes hydraulic for powered dumbwaiters, is placed at the bottom of the hoistway. For other types of dumbwaiters, the driving machine usually is electric traction for powered dumbwaiters and is located at the top of the hoistway. Dumbwaiters also may be hand operated.

ANSI A17.1 contains safety requirements for dumbwaiters. Often, powered dumbwaiters are constructed like elevators, but many of the safety requirements for elevators are waived for dumbwaiters. Standard height for a dumbwaiter is 3, $3\frac{1}{2}$ and 4 ft. Maximum platform area is 9 sq ft. Rated load capacity usually ranges from 20 to 500 lb. Speeds range from 45 to 150 fpm for powered dumbwaiters.

Dumbwaiters also are available with special equipment for automatic loading and unloading. They may also be built for floor-level loading suitable with cart-like conveyances.

14.6. PNEUMATIC TUBES AND VERTICAL CONVEYORS

Conveyors are often used in industrial buildings and warehouses for movement of materials. For horizontal movement of materials, conveyors usually are roller, belt or pallet type. They are useful when loads are uniform, materials move continuously, their path is fixed and speed may be constant and slow. Conveyors, however, also may be used for vertical transportation of materials. For rapid transport of small, light units, a vertical conveyor operates in a hoistway. Such installations are useful in high-rise office buildings and hospitals.

Pneumatic tubes also may be used for movement of small, light items horizontally, vertically, or both. The items usually are transported through the tubes in carriers slightly smaller in cross section than the tubes. The propelling force may be air pressure, suction or a combination of both.

In choosing between pneumatic tubes and vertical conveyors for movement of small items between floors of multistory buildings, the designer's first consideration should be the number of floors and stations to be served. Pneumatic tubes would generally be more efficient in a predominantly horizontal building. A vertical conveyor ususally would be more advantageous in a high-rise building. A conveyor also would be more efficient when the volume of items to be moved is large.

A typical vertical conveyor is similar to an escalator (see Sec. 14.3), except that the conveyor is enclosed in a fire-resistant shaft, like an elevator or a dumbwaiter. Carriers for trays are spaced at intervals along a continuous chain that is driven by an electric motor. The chain engages sprockets at the top and bottom of the shaft. The carriers may move up and down the shaft at speeds of about 70 fpm. Trays carrying materials to be transported are set on the carriers at stations on each floor.

Dispatching and delivery of trays may be automatic. When a tray is ready for dispatch at a station, an attendant sets a floor-selector dial or presses a button at the station. The tray is then automatically moved into the path of a continuously moving carrier on the chain. The tray moves upward and around the top sprocket and is automatically discharged on the downward trip at the selected floor. Short horizontal-conveyor sections at the discharge opening may lead from the vertical-conveyor station to a work area.

Doors at stations usually are vertically sliding. If a fire should occur, a fusible link, when actuated, will snap the doors down over the shaft openings, to seal off the conveyor shaft at every floor.

Vertical conveyors should be installed in a central location, preferably adjacent to other vertical shafts. A central location minimizes horizontal runs in collection and distribution of materials or correspondence in each story. Placement of a conveyor alongside a stairway is advantageous where it can provide access to the rear of the conveyor for maintenance and repair.

Mail Chutes

A mail chute is a vertical, unpressurized, rectangular tube for transporting of mail by gravity from upper floors of a multistory building to a mailbox in the main lobby. One or more chutes may be installed, with Post Office permission, in office buildings more than four stories high and in apartment buildings with more than 40 units. The chute is about 3 × 8 in. in cross section. It usually is made of 20-gage cold-formed steel, with a glass front, and supported by vertical steel angles. In each story, a slot is provided in the front of the chute for insertion of flat mail. The mail box generally is 20 in. wide, 10 in. deep and 3 ft high, with the bottom set 3 ft above the floor. The box must be located within 100 ft of the building entrance.

SECTIONS 14.5 AND 14.6

References

B. Stein et al., *Mechanical and Electrical Equipment for Buildings*, 7th ed., Wiley, New York, 1986.
C. Ramsey and H. Sleeper, *Architectural Graphic Standards*, 8th ed., Wiley, New York, 1988.
American National Standard Safety Code for Elevators, Dumbwaiters, Escalators, and Moving Walks, ANSI 17.1, published by the American Society of Mechanical Engineers.

Words and Terms

Dumbwaiter
Mail chute
Pneumatic tube
Vertical conveyor

14.7. SYSTEMS-DESIGN APPROACH TO VERTICAL CIRCULATION

Vertical circulation is an essential part of building design for any building with more than one level. Depending on the number of levels (stories) and the type of occupancy, the vertical movements of people, vehicles, and materials in the building may or may not require powered equipment to facilitate the movements. Very few single-family residences, for example, have powered equipment for vertical circulation; even ones with three or four interior levels. Most other multistory buildings, however, including multifamily housing and commercial and institutional occupancies, will have one or more major powered subsystems—powered ramps, powered stairs, elevators, dumbwaiters, for example.

When more than one subsystem is installed, the subsystems are kept physically independent of each other. Yet, they have a common purpose and often share the traffic flow. For architectural reasons, therefore, it is generally desirable that the subsystems be located close to each other. Furthermore, a compact arrangement often has a relatively low construction cost and minimizes the space required for vertical circulation, because two or more subsystems can be served by common lobbies and corridors. Also, a compact arrangement causes less physical disruption of usable floor area and usually is convenient for occupants and visitors. Consequently, the location of each vertical-circulation subsystem is often dependent on the location of the others.

Vertical circulation, in general, affects and is affected by other systems, principally architectural, structural, electrical and fire protection. Systems design, therefore, can be useful not only in coordinating the vertical-circulation subsystems but also the various building systems to produce an optimum overall design.

Data Collection and Problem Formulation

Basic information for design of vertical circulation comes from examination of the owner's program of requirements and decisions as to the number of stories in the building, story height and floor shape and dimensions. Additional information must be based on estimates of building population, distribution of the population on each floor, number of visitors to various destinations and characteristics and volume of materials to be transported. This information is useful in determining types, number, sizes and capacities of vertical-circulation subsystems needed.

The architect, mechanical engineer and vertical transportation consultants should collaborate in determination of space and other requirements for vertical circulation. Also, the architect and the structural engineer should collaborate on vertical circulation to minimize costs of structural framing and shaft enclosures. Structural costs often can be reduced by using enclosures required for safety and fire protection of stair, elevator and dumbwaiter shafts for structural purposes also. The enclosures can serve as load-bearing supports for floors and roofs and as shear walls for resisting lateral forces from wind and earthquakes. The results of the cooperative efforts should be displayed on architectural plans and elevations and submitted to all of the members of the building team for review.

Information for planning for powered installations should be obtained from prospective manufacturers and installers.

Further information for design of vertical circulation should be obtained from the local building code; *Life Safety Code*, NFPA No. 101, and *American National Standard Safety Code for Elevators, Dumbwaiters, Escalators and Moving Walks*, ANSI A17.1.

In the later stages of design, all drawings concerned with vertical-circulation subsystems should be studied to insure their compatibility with other subsystems.

The goal of vertical-circulation design may be stated as: to design, as a component of an optimum multistory building, a vertical-circulation system that is safe, reliable, convenient and of adequate capacity for movement, under normal and emergency conditions, of occupants and visitors from building entrances to all floors and from all floors to the exits from the building.

Objectives

The prime objectives of vertical-circulation design should be safety, reliability and adequate capacity for handling anticipated traffic. The design of systems to be used by people should minimize potentials for accidental falls, and powered systems should incorporate fail-safe devices. Material-handling systems also should be designed to prevent injury to people and property damage. In addition, systems to be used by people should be designed to be always available and usable under both normal and emergency conditions. Also, the systems should have adequate capacity, which often may be attained economically with powered equipment— powered ramps, escalators, and elevators for rapidly transporting people, and freight elevators, dumbwaiters and vertical conveyors for moving materials.

For convenience, vertical-circulation subsystems for people should be readily identified by occupants and visitors as means for moving from floor to floor. Access to and discharge from the subsystems from any point on any floor, under normal and emergency conditions, should be safe and easy. Often, an important objective is to require little human power for vertical movement, in which case use of powered equipment is essential.

Another important objective is the aim of choosing from among all the possible vertical-circulation systems that can achieve the desired goal the one that will have the lowest life-cycle cost. As for systems previously discussed, this cost includes the initial cost of the vertical-circulation subsystems and, for powered installations, operating and maintenance costs; however, for vertical circulation, a highly significant additional cost must be taken into account. This cost is that due to loss of usable floor area, because the space occupied by vertical-circulation subsystems cannot be used to serve the main purposes of the building. Thus, the objective should also aim at minimizing the space required for vertical circulation, including lobbies and access corridors.

For powered equipment, energy conservation should also be an objective. While the equipment may consume little energy under no-load operation, continuous movement unloaded is often a necessity and tends to waste energy. Elevators, which can be operated intermittently, may be more efficient than escalators and moving ramps. Elevators, when running, however, not only use electric power directly but also cause indirect energy losses through loss of conditioned air, because of stack effects created by the hoistway and infiltration around the cars. Vertical-circulation design therefore should keep energy requirements at a minimum.

Constraints

The most important constraints on vertical-circulation design are those imposed by type of occupancy, size and distribution of building population, number of stories, story heights and floor shape and dimensions. These combine to determine whether ramps alone, stairs alone, or combinations of these with powered equipment are required. Also, these constraints influence determination of the minimum number of units of each type required and the relative location of units.

Another important constraint is the requirement that vertical-circulation systems be available for use for safe evacuation of the whole building population in emergencies, such as fire, and for use in fire fighting. Thus, vertical circulation is part of the fire-protection system and must help achieve the goal of that system.

Building codes impose numerous constraints on vertical-circulation design in the interests of safety under normal and emergency conditions. They require elevators in buildings exceeding a specific height, usually four stories. But while codes permit use of elevators in emergencies for evacuation of the building population, they do not allow the capacity of elevators to be counted in computations of the capacity of the vertical-circulation system for meeting exit requirements. Also, codes usually require that one or more elevators be designed and equipped for fire-emergency use by fire fighters. Key operation should transfer automatic-elevator operation to manual and bring such elevators to the street floor for use by the fire fighters. These

elevators should be located so as to be readily accessible for the purpose.

Other constraints are imposed by the interaction of the vertical-circulation system with other systems, principally architectural, structural and electrical. Architectural constraints are generally the most significant, because of the effects of vertical circulation on floor plans and exterior elevations.

Synthesis and Analysis

Types of vertical-circulation subsystems that will be required for a multistory building can usually be selected early in the schematics phase of design. The number, approximate dimensions and locations of ramps, stairs, escalators and elevators can be tentatively chosen when the size and distribution of the building population can be predicted, the building height estimated and shape and dimensions of floors approximated.

Based on the early decisions affecting vertical circulation, the architect can develop schematic floor plans and elevations, which can be used by other members of the building team for development of their schematics. As changes are made in the architectural schematics, including final determination of the number of stories in the building and improved estimates of story heights, changes may be made in the layout and capacity of the vertical-circulation system.

Because of the significant influence of vertical circulation on floor plans, it is imperative that the vertical-circulation system selected by the start of design development be as close as possible to the one that will be installed in the building. Later changes may be very expensive in time and money. So, before completion of the schematics stage, building costs should be estimated and the effects of vertical circulation on them determined. Also, alternative systems should be studied. The system most likely to optimize the building design should be chosen.

In the preliminary design stage, when much more information becomes available on characteristics of the building and its occupancy, the layout of vertical circulation can be fixed. The individual subsystems and access to them at all floors can then be designed. The designs should be checked to verify that the goal, objectives and constraints have been met and that the vertical-circulation system will be compatible with other systems.

Value Analysis and Appraisal

Value analysis of vertical circulation should start in the schematics stage. Total costs of buildings with alternative vertical-circulation systems should be compared, rather than the costs of vertical-circulation subsystems alone, because of the effects of these subsystems on other building systems. In the design development phase, value analysis should be applied to each of the subsystems, and in the contract documents phase, to details.

For some types of low buildings, such as one- or two-family residences up to three stories high, the lowest-cost subsystem, stairs, may usually be chosen for an optimum building, but generally, such a choice will not yield an optimum structure for other types of buildings. Neither designers nor value analysts can always specify only stationary subsystems because they cost less than powered subsystems, for if such decisions are made, it is not likely that objectives and constraints can be met.

To illustrate, stationary stairs usually are the type of subsystem with lowest life-cycle cost, including the cost of lost usable space. Ramps, or elevators, or both, however, may have to be used in addition to stairs, when access to upper floors is required for handicapped persons. Also, building codes require elevators in tall buildings. Furthermore, where traffic is large and continuous, escalators or powered ramps may be necessary, despite operating costs and higher initial cost than that for stairs. Consequently, designers and value analysts have to pick the appropriate combination of subsystems for optimum results.

Initial building cost is considerably influenced by the size, location and arrangement of the selected subsystems. Total area of each floor of a multistory building, for example, is composed mainly of the sum of the areas required for vertical circulation and the usable floor area. Hence, the smaller the area for vertical circulation, the more usable floor area for a given perimeter, or the smaller the perimeter for a required usable floor area. Often, if more usable area can be obtained for each floor, the lower are the number of stories required for the building. Because of the lower height of the building, there will be savings in costs of interior and exterior walls, piping, wiring, stairs and elevators.

In tall buildings, floor area required for elevators may be a high percentage of the total floor area. For buildings with setbacks, this percentage increases in the upper stories with the decrease in floor area. Consequently, elevators should be divided into high and low banks, with only a few elevators in the high banks to serve the upper floors. Space above the discontinued hoistways then becomes usable. In very tall buildings, additional savings may be attained sometimes by starting high banks at upper floors, instead of in the main lobby, thus decreasing the weight of roping, size of driving machine and cost of elevator operation. Round-trip time may also be reduced; but this arrangement requires passengers to transfer from a low to a high bank at the upper terminal of the low bank. The transfer floor sometimes is referred to as a sky lobby.

When powered subsystems are to be installed, studies should be made to reduce power requirements, to decrease operating costs and conserve energy. For example, for elevators with multivoltage control, consideration should be given to use of solid-state rectifiers for supply of dc power to driving machines, to reduce power costs. Also, consideration should be given to use of computer control of elevator operation for greater efficiency in utilization of elevators.

Total cost of a building can often be reduced when part of any of its systems can also serve as part of any other system at little or no extra cost. Thus, there is an opportunity for cost reduction in utilization of the enclosures of stair, elevator and dumbwaiter shafts for structural purposes. Shaft walls are required usually to be fire resistant. Hence, they may be built of noncombustible materials, such as concrete or masonry. In such cases, the walls may not only be capable of supporting the stairs and elevator or dumbwaiter installation but also floors and roofs. Often, at little extra cost, the walls may be made strong enough to serve as shear walls, capable of resisting lateral forces on the building.

GENERAL REFERENCES AND SOURCES FOR ADDITIONAL STUDY

These are books that deal comprehensively with several topics covered in this chapter. Topic-specific references relating to individual chapter sections are listed at the ends of the sections.

B. Stein et al., *Mechanical and Electrical Equipment for Buildings*, 7th ed., Wiley, New York, 1986.

C. Ramsey and H. Sleeper, *Architectural Graphic Standards*, 8th ed., Wiley, New York, 1988.

American National Standard Safety Code for Elevators, Dumbwaiters, Escalators, and Moving Walks, ANSI A17.1, published by the American Society of Mechanical Engineers.

EXERCISES

The following questions and problems are provided for review of the individual sections of the chapter.

Section 14.1

1. Why are ramps necessary for most buildings?
2. For what type of multistory building are ramps essential for the functioning of the building?
3. What is the *desirable* maximum slope for stationary ramps? For which class of ramp is this slope set as a maximum by the *Life Safety Code*?
4. What is the safe maximum slope for powered ramps?
5. Which class of ramp is assigned greater downward exit capacity for a specific width?
6. Under what circumstances is a moving walk or a powered ramp acceptable as a means of emergency egress?
7. What means should be provided for a powered ramp to prevent excessive treadway speed?

Section 14.2

8. Why are stairs generally used instead of ramps?
9. What is the purpose of a tread?
10. Are risers essential in stair construction?
11. A multistory apartment building may be expected to house up to 48 occupants on each floor. If each stairway has a minimum clear width of 44 in., what is the least number of stairways that will provide adequate exit capacity in an emergency?
12. What is the advantage of parallel stairs over straight-run stairs?
13. What are the pitch and pitch angle of a stairway with 12-in. treads and 6-in. risers?
14. What is the maximum total rise that building codes permit for a flight of stairs between floors or landings?
15. Why is headroom important in stair design?
16. What is the purpose of carriages in wood-stair construction?
17. A two-story, single-family dwelling has a floor-to-floor height of 10 ft. Space available for a straight-run stairs between floors is 12 ft 10 in. long. The local building code sets a maximum of 8 in. for height of risers. Product of tread width and riser height must lie between 70 and 75.
    (a) How many steps (risers) are required?
    (b) What is the tread width of these steps?
    (c) What is the pitch angle of the stairs?

Section 14.3

18. What requirements must an exit escalator meet?
19. What are the main advantages of escalators over other vertical-circulation systems?
20. List the main components of an escalator.
21. What safety devices are required for control of escalators?

22. What are the advantages of crisscrossed escalators over parallel escalators?

Section 14.4

23. Under what conditions are elevators required by building codes?
24. Are elevators usually accepted by building codes as means of egress in emergencies?
25. How is the car of an electric elevator moved?
26. How is the car of an hydraulic elevator moved?
27. What is the purpose of a hoistway pit?
28. What is the maximum number of elevators permitted in a multiple hoistway?
29. What provisions should be made for removing smoke and hot gases from hoistways?
30. What is the purpose of a hoistway-unit-system interlock?
31. What is the purpose of electric contacts on a car door?
32. What is the purpose of a hall position indicator?
33. What keeps an elevator car in a fixed path?
34. What is the purpose of a counterweight?
35. What actuates a safety and what does a safety do?
36. What are the main components of a driving machine for an electric elevator?
37. What is the purpose of elevator control?
38. What devices are used to prevent a car from running past a terminal?

39. For 2 : 1 roping, what is the relationship between car speed and rope speed?
40. Which type of operation is more expensive: car switch or automatic?
41. How does nonselective collective automatic operation differ from single automatic operation?
42. What is the purpose of group automatic operation?
43. What are the advantages of zoning a building for elevator service?
44. What are the main components of a driving machine for an hydraulic elevator?
45. What factors that measure quality of elevator service should control elevator selection?
46. A passenger elevator makes a round trip in 90 sec on the average. If the desired average interval is 30 sec, how many elevators should be installed to maintain the interval?
47. An elevator in an office building has a rated capacity of 20 passengers. Peak 5-min traffic is 500 persons. Average round-trip time for a car is 90 sec. How many elevators should be installed to handle the peak traffic?

Sections 14.5 and 14.6

48. A tall hospital building has a continuous flow of small items between floors. Which would provide speedier delivery: a dumbwaiter or a vertical conveyor?

Chapter 15

Systems for Enclosing Buildings

Buildings are given an enclosure around their exterior so that the desired internal environment can be maintained while outside weather is excluded. Traditionally, the exterior enclosure consists of vertical walls and a roof, either of which may be punctured by such building components as doors, windows and ventilators, for access, light or ventilation. Other alternatives, though, are possible and have been used, such as a barrel arch or a dome, integrating walls and roof. Thus, a roof system and a wall system may be separate systems, physically independent of each other, or indistinguishable parts of a single enclosure system.

Roof and wall systems always affect or are affected by other systems, principally architectural, structural and HVAC. For example, walls, and sometimes also roofs, when visible from the ground outside a building, are major architectural features of the building. They determine whether or not the building exterior is aesthetically pleasing. Also, roofs and walls may merely be loads on the structural system or they may be part of that system, supporting or transmitting loads. For example, a roof may carry heat-collection panels for a solar heating system, an air-conditioning cooling tower or an elevator-machine penthouse. Similarly, a wall may support not only its own weight but also the weight of and loads on roofs and floors. In that case, the wall influences design of the foundation system, because the wall transmits its weight and loads to the foundations. In addition, roofs and walls interact with the HVAC system. The

heat flow through them often imposes a significant proportion of the total load on that system.

Because of the interaction of the various building systems, optimization of a roof or wall system, or both, does not necessarily yield an optimum building. Systems design, however, can be useful in achieving that goal.

15.1. ROOFS

A roof is the uppermost portion of a building constructed for the purposes of separating the building interior from the outdoors and excluding exterior environmental conditions from the interior.

While walls are vertical or nearly so, roofs may be horizontal or inclined at an angle considerably less than 90° with the horizontal. As a result, a major concern in the design of roofs is prevention of penetration of rain into the building interior and diversion of water to a plumbing drain, gutter or other means of disposal. (Although walls also have to be designed to prevent penetration of rain, their verticality aids shedding of rainwater.) In cold climates, another major concern in design of roofs is provision of a capability for supporting snow loads. This type of loading rarely is of concern in wall design, because in this case, too, verticality aids shedding of snow. For both roofs and walls, wind loads are an important design concern, but roofs usually also have to be designed to support some live loads, because people may have to walk on them.

In this last respect, roofs are loaded much like floors. In fact, the structural components of many flat, horizontal roofs resemble the structural components of floors. (Because of the economy of duplication in manufacturing components and of repetitive installation steps, the structural framing of floors of multistory buildings sometimes is repeated in the roof, although roofs, being more lightly loaded than the floors, could employ lighter framing.) For economy of structural framing, however, it is desirable, especially for multistory buildings, that the weight of roofs be kept as small as possible. For the load of the roof must be transmitted through columns, or walls, or both, all the way from the top of the building to the foundations at the bottom.

Roof Components

A roof basically requires only two components, a waterproof membrane, to exclude rain and wind, and supports for the membrane. To keep roof weight small, the membrane should be light and thin. Its thickness only need be sufficient to resist wind forces, installation forces and abrasion and to provide enough strength for the membrane to carry its own weight between supports. (Some roofs have been constructed with cable supports carrying only a glass-fiber fabric coated with a water-resistant plastic. With a $\frac{1}{8}$-in.-thick fabric, such a roof may weigh only 1 lb per sq ft.)

With a continuous support, or *deck*, under the waterproof membrane, less-expensive, lighter, thinner membranes can be used than those required when supports are spaced at intervals. When used with a deck, the membrane may serve merely as a covering. In such cases, the membrane is called *roofing*.

This type of construction, with roofing and a deck, is used for most buildings, because of its durability when properly constructed with good materials and of its ease of maintenance. The deck, whose main purpose is to provide continuous support for the roofing, may also be made light and thin. In that case, the deck must be provided with structural framing at appropriate spacing for support. In some buildings, relatively strong, thick decks are used with wide spacing of structural supports.

In addition to water resistance, strength, abrasion resistance and sometimes also fire resistance, one more characteristic is important for roofs. This desirable characteristic is low transmission of heat. To reduce the load on the HVAC system, roofs should be built of materials with a low heat-transmission coefficient and generally also should incorporate thermal insulation. (Materials with a high thermal capacity for storing heat may also be useful in reducing the HVAC load, but such materials generally are massive, and for some buildings, they may be uneconomic because of the increased cost of the structural system and foundations required for their support.) Also, when there are spaces between a roof and the ceiling of rooms below, those spaces should be ventilated to further reduce the HVAC load. By such treatments, passage of solar heat through the roof to the building interior can be considerably impeded, and also escape of heat from the interior to the outside through the roof in cold weather can be slowed.

Roof Shapes and Slopes

A simple shape for a roof is a single plane surface. If such a plane surface is placed horizontally, however, drainage of rainwater is more difficult than if the surface were placed on a steep incline. Also, more snow would pile up on the horizontal surface in cold climates. Consequently, a wide variety of shapes have come into use for roofs. Horizontal roofs, called *flat roofs,* are often used nevertheless (see Fig. 15.1a), but they are usually composed of surfaces with slight inclines to facilitate drainage. A pitch of at least $\frac{1}{8}$ in. per ft to a drain is desirable.

A *shed roof* (see Fig. 15.1b) is formed when the single plane surface is placed on an incline. The slope facilitates drainage of rainwater and shedding of snow. This type of roof, however, requires structural supports to be set along the incline, and walls must be of unequal height on two sides and have sloping tops on the other

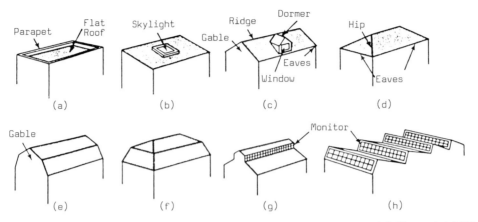

Fig. 15.1. Roof forms. (a) Flat. (b) Shed. (c) Pitched (gabled). (d) Hip. (e) Gambrel. (f) Mansard. (g) Ridge monitor. (h) Sawtooth monitor.

two sides. These effects, in some cases, may be economically or architecturally undesirable.

Shed and flat roofs may have glazed openings, called *skylights* (see Fig. 15.1b), for daylighting the building interior.

A *pitched roof* (see Fig. 15.1c) is formed by a combination of two inclined planes. The intersection of the planes at the peak is called the ridge. The bottom edge of each plane is called the eaves. Often, the slope and the distance between eaves and ridge is made the same on opposite sides of the ridge. The inverted-V-shaped end walls of the building are called gables.

This type of roof has some of the undesirable characteristics of shed roofs, but it has the advantage that if the roof pitch is sufficiently steep, an attic (space for storage or additional rooms) becomes available below the roof. Windows for attic rooms can be installed at the gables and in *dormers*, projecting through the roof (see Fig. 15.1c).

A *hipped roof* (see Fig. 15.1d) is formed by a combination of four inclined planes. As for a pitched roof, the hipped roof has a ridge but eliminates the gables. The tops of the walls all around the building are straight and may be at the same level. Also, eaves extend along the bottom edge of the roof all around the building. If the roof pitch is steep enough, the attic space may be usable for rooms. Dormers may be provided for windows. Hipped roofs require more complicated structural framing than pitched roofs. But the constant wall height produces some offsetting savings, and the appearance of the roof may be more architecturally desirable.

Gambrel roofs (see Fig. 15.1e) are also constructed of a combination of four inclined planes but they have gables, like pitched roofs. The two planes on each side of the ridge have different pitches, the lower plane being on a steeper slope. This type of roof is sometimes used to obtain larger attic space for use as rooms.

Mansard roofs (see Fig. 15.1f) are another means of obtaining more attic space for rooms. This type of roof resembles a hipped roof, but like gambrel roofs, mansard roofs combine a steep lower plane with a flatter upper plane. Figure 15.1f shows a pair of the upper planes intersecting at a ridge. In some mansard roofs, however, all of the upper planes may be replaced by a single flat plane.

Monitored roofs are sometimes used for industrial buildings to provide light and ventilation to the interior. A monitor is a row of windows placed vertically, or nearly so, above a roof (see Fig. 15.1g). Monitors may be installed on flat roofs, pitched roofs (see Fig. 15.1g) or sawtooth roofs (see Fig. 15.1h).

Roofs may also be constructed with curved surfaces. Such shapes are often used to provide large column-free areas in the building interior, because structural costs are lower than for other types of framing for such long spans. Single-curved surfaces, such as cylindrical, or barrel, arches, and double-curved surfaces, such as hyperbolic-paraboloid shells, are sometimes preferred because such sufaces can be generated by movement of a straight line. Hence, the roof deck can be formed with straight, linear components. In contrast, decks for such double-curved surfaces as domes are more difficult to form.

Roof Decks and Framing

When roofing is supported on a continuous deck, the characteristics and behavior of the deck determine to a significant extent the performance in service and durability of the roofing. Because the roofing usually is attached to the deck to prevent excessive shifting, the roofing, usually being thin, takes on the shape of the upper surface of the deck and moves as the deck moves under temperature changes and loads. Consequently, the deck and the framing supporting it must be designed and installed to enable roofing to perform in service as required. Roofing preferably should have an expected life, with little or no maintenance, of at least 10 years.

A roof deck and its framing should be capable of supporting their own weight and the weight of the roofing assembly, including insulation. They should also be capable of supporting transient loads, such as rain, snow, installation equipment and workmen, and wind loads. The deck, in addition, should be stiff enough to prevent deflections that would cause ponding of rainwater and to limit vibration. For the same reason, the framing should also be strong and stiff. Furthermore, the spacing of the framing should be commensurate with the strength, elastic characteristics and thickness of the deck.

The roof deck should be pitched sufficiently so that water will readily drain from the roofing, without collecting in the low areas of the deck between framing members. The roof should be completely dry within 48 hr after rain ceases. To prevent interference with drainage and to reduce abrasion when persons walk on the roof, the top surface of the deck should be smooth, free of humps, depressions and offset joints.

Roof failures often are caused by poorly designed or installed decks. Common causes of failures are faulty preparation of a deck to receive roofing and application of roofing to a deck that has been constructed with water, such as concrete or gypsum, before the deck has been properly cured and dried.

Types of Decks and Framing

The type of roof framing selected for a building generally depends on shape of roof, clear spans required under the roof, loads to be supported and structural characteristics of the deck.

When a deck is thin and flexible, closely spaced structural members are needed to support it. Such members then can usually be shallow and light, depending on the distances they must span. Whether such construction should be chosen depends on fire resistance required and local costs of materials and labor. Often, the familiarity of local contractors with the type of construction selected and the availability locally of required materials and labor have a considerable influence on the economics of roof construction.

When a deck is thick and stiff, structural members can be spaced far apart. In that case, the framing may be deep and heavy, depending on the distances the structural members must span. This type of framing, requiring fewer and stronger structural components, may be less costly than the lighter type, with more members. Before a decision is made, studies should be made of the relative costs of alternative appropriate roofing systems. Some of the more commonly used types of roof decks are the following:

Wood Decks. Plywood decks, or sheathing, is often used with light, wood structural members (see Fig. 8.62). Thickness of the plywood depends on the spacing of supporting members, which usually is 16 or 24 in. center-to-center (see Fig. 15.2). Wood decks also may be built of well-seasoned lumber. The pieces may be interconnected by tongue and groove, shiplapped joints or splines. Because of the moisture content of wood, it is important that a slip sheet be installed between the deck and roofing, to permit dimensional changes to occur in the wood, without damaging the roofing, as the moisture content changes.

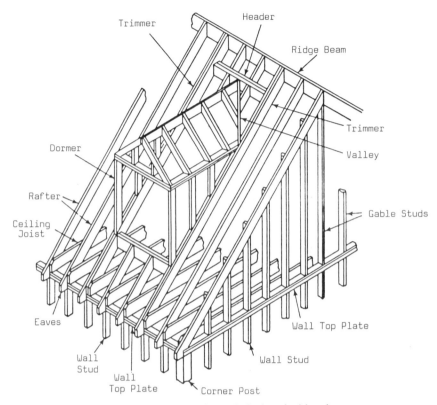

Fig. 15.2. Wood framing for a pitched roof with a dormer.

Steel Decks. Cold-formed steel is often used as a roof deck with wide spacing of structural steel members. Spacing of the framing depends on the gage of the steel deck and the depth of the deck ribs. This type of deck frequently is used for flat roofs. The roofing may be attached directly to the deck, to smooth-surface insulation board set on the deck or to a smooth-surface, rigid, cast-in-place insulating fill.

Concrete Decks. Roof decks to be supported on structural steel or concrete framing or on load-bearing walls are often made of concrete. The concrete may be reinforced or prestressed, the latter being preferred because of the smaller probability of objectionable cracks forming. The decks may be cast-in-place or precast in a fabricating plant, shipped to the site and erected with a crane. With precast construction, a light-weight fill or grout may be placed over the deck or between adjacent panels to provide a smooth top surface. Cast-in-place concrete decks usually are solid. Precast decks may be solid, hollow or shaped like a single tee, double tee or channel.

Gypsum-Concrete Decks. Decks similar to concrete decks, but usually lighter in weight, may be made of gypsum, wood chips or shavings, mineral aggregate and water and reinforced with steel bars or galvanized wire fabric. This type of deck, however, is not suitable for buildings in which the interior will have high humidity.

Insulation. To reduce the thermal load on the HVAC system, thermal insulation should be interposed between the building interior and the outdoors. In some cases, for example, when foamed concrete or concrete made with light-weight aggregates is used, the deck itself may provide sufficient insulation. Often, however, the insulation offered by the deck must be supplemented by at least one separate layer of thermal insulation.

The appropriate location for the insulation

depends on the roof construction and the types and uses of spaces directly below the roof. For example, if there is an unoccupied attic under a roof, the most advantageous location for insulation is in the floor of the attic or above the ceiling of the rooms below the attic. In hot weather, solar heating of the attic can be relieved by ventilation. In other cases, where there are occupied spaces below the roof, insulation may be placed above or below the deck.

With wood construction, insulation is usually nailed into position. When insulation is placed directly on top of steel or concrete decks, it may be attached with adhesive or a mopping of hot asphalt.

Ordinarily, insulation is placed beneath roofing. Sometimes, however, insulation is installed atop roofing, a type of construction called a ballasted system. Rigid insulation is used and bonded to the roofing with hot asphalt. The top surface of the insulation is protected by a layer of $\frac{3}{4}$-in. crushed stone, paving blocks or structural concrete. With this type of construction, the insulation not only resists passage of heat between the building interior and outdoors but also protects the roofing from harmful effects of thermal cycling, ultraviolet degradation, weathering and roof traffic. Exposed roofing is subject to such harmful effects, which decrease its durability.

Vapor Barriers. When the roof deck serves as insulation or a layer of insulation is placed below the roofing, precautions should be taken to prevent moisture from the building interior from condensing on the cold underside of the waterproof roofing. The condensate may saturate the insulation and reduce its effectiveness; or the moisture may drip back into the building and stain ceilings and wet floors. If the condensate is trapped in the roofing, the moisture will freeze in cold weather and rupture the roofing. To prevent such undesirable effects, a solid, unbroken vapor barrier should be placed on the underside (warm side) of the insulation.

When this is done, however, a different potential for damaging the roofing is created. The insulation becomes sandwiched between two vapor barriers, because roofing also is a vapor barrier. Air in the insulation cannot escape unless provision is made for it to do so. Trapped air will expand in the daytime when it is heated by the sun and contract at night on cooling. If the lower vapor barrier is restrained from moving with dimensional changes of the air, for example, by the deck, expansion of the air will cause bubbles under the roofing. These bubbles will collapse when the air cools. Repeated daily, the bubbling cycles will cause early failure of the roofing by fatigue. To prevent this, vent stacks or edge venting should be provided for the insulation to avoid buildup of air pressure.

Several different types of materials are available for use as vapor barriers under insulation set on top of roof decks.

Polyvinyl chloride (PVC) sheets are often used when noncombustible roofing is required. The sheets usually are attached to the deck with an adhesive or with mechanical fasteners that pass through the insulation.

Felts coated on opposite faces with bitumen are often used over light, insulating fills, such as concretes that have been foamed or made with lightweight aggregates and have a high moisture content. Vent stacks or edge venting should be provided to allow escape of moisture from such decks.

Kraft laminates, which are made of two layers of special heavy-duty kraft paper, may also be used as a vapor barrier with fire-retarding roofing. This type of vapor barrier also may be fastened to the deck with adhesives or mechanical fasteners through the insulation.

15.2 ROOFING

Choice of a roofing material and method depends on many factors. These include cost; desired appearance; purpose of the building; availability of roofers, materials and equipment; compatibility of materials to be installed; shape and slope of the roof and geographical location of the building.

For steep roofs, multiunit coverings generally are more advantageous than single-unit roofing. For flat roofs, single-unit roofing should be used. For curved roofs, single-unit, spray-on roofing, plastic panels and metal roofing are appropriate alternatives.

Before a roofing is selected, the compatibility

of potential materials with materials with which the roofing will be in contact or to which it will be attached should be investigated. If a chemcial reaction occurs between materials in contact, or coefficients of expansion differ, failure of the roofing may result.

Types of Roofing

Because bituminous materials are frequently used for roofing, a knowledge of the meaning of terms used in specifying them is necessary. Following are generally accepted definitions of bituminous materials:

Bitumen is a generic term used to indicate either asphalt or coal-tar pitch.

Asphalt is a by-product of the refining processes of petroleum oils.

Coal-tar pitch is a by-product of crude tars derived from coking of coal. The crude tars are distilled to produce coal-tar pitch.

Although coal-tar pitch and asphalt have a similar appearance, they have different chemical and physical characteristcs. Coal-tar pitch melts at a lower temperature and thus tends to be self sealing. But because of the low melting point, the upper surface of tar roofing must be restrained by a covering of gravel or slag from flowing when it is heated by the sun. Asphalt, which melts at a higher temperature, does not require such a covering, and because it does not flow as readily as tar under solar heating, it may be used on steeper slopes. Asphalt, though, is not as good a self sealer.

Roof coverings such as saturated felts, coated felts and prepared roofing, are made with either asphalt or tar as a saturant. They usually are delivered to the building site in rolls for installation.

Saturated felts are made of organic materials such as paper, wood fibers or a combination of these. The felts are designated by a number that indicates their approximate weight in pounds per square (100 sq ft). Felts that are both saturated and coated may be made of organic or inorganic materials and are surfaced with a mineral release agent to prevent adhesion of the felts while rolled. Prepared roofing comes with a surfacing material such as ceramic granules,

sand, mica, or talc added on the top face for decoration.

Roofing also may be made of materials without bitumen, for example, metal or plastics. Plastics may be supplied to the job in rolls containing sheets of polyethylene, PVC, Hypalon or rubber. These materials may be used also for flashing, expansion joints and vapor barriers. Plastics or bitumen also may be applied as a spray to form a roof covering.

Roofing generally may be separated into two main classes, single-unit or multiunit.

Single-unit roofing includes roof coverings consisting of one or more layers that become, after installation, a single entity. Such roofing may be used on flat or sloped roofs.

Multiunit roofing includes roof coverings consisting of many pieces that are installed individually, usually partly overlapped to prevent passage of water downward to the roof deck. Materials used are suitable for application on steep roofs (pitch of 4 in. or more on 12 in.).

Single-Unit Roofing

Materials and installation methods used for single-unit roofing generally permit its application on roofs that are flat or sloped. This type of roofing includes built-up, metal, synthetic and fluid-applied roofing.

Built-up Roofing. This usually consists of alternate layers, or plies, of saturated felt and hot bitumen applied by mopping. A top coating of bitumen in which slag or gravel is embedded is optional. The felts, organic or inorganic, may be nailed or bonded with adhesive or bitumen directly to the roof deck or to insulation. For good performance of the installation, specifications of the roofing manufacturer or of a roofers' association should be closely followed.

Three to five plies of felt usually are used for built-up roofing. For good performance, five plies should be used for wood decks, with the first two plies nailed to the decks. For concrete decks, equivalent results can be obtained with four plies of felts, with the first ply bonded to the decks with hot bitumen.

Cold-Process Built-up Roofing. Many different processes have been used for applying plies without hot bitumen. Some processes call for combinations of chemicals with bituminous products. For example, one cold-process method requires a heavy-duty glass-fiber fabric to be embedded in a high-viscosity, asphalt based emulsion. The emulsion is brushed or sprayed on a roof deck and a finish coat is applied to create a smooth top surface. In another process, polyurethane is combined with tar and sprayed on the deck.

Plies used with cold processes differ from those used with hot bitumen. Porous fabrics, often made with glass fibers, are used for cold processes, because the open weave allows evaporation of the solvent used to make the cementing material liquid. The cementing material may be an emulsion of bitumen and water, asphalt cut back with kerosene or tar cut back with toluene. A thick mastic-type emulsion is desirable with glass-fiber fabrics, because a heavy film is needed to hold the fibers in place. The cementing material may be applied by brush or spray to the plies.

Metal Roofing. Metals are often specified for roofing because of their durability and their ability to conform to a wide variety of roof shapes and slopes. Also, roof decks are not needed when the metals are formed into panels that are sufficiently strong and stiff for the spacing of structural framing.

Metals generally specified include steel, aluminum, copper and lead. Also used are alloys, such as terneplate, stainless steel, bronze and Monel metal, and multimetal systems, which bond two or more metals with heat and pressure to obtain properties not available in a single metal or alloy. Multimetal roofing has been made of galvanized steel, terne-coated stainless steel, copper-clad stainless steel, copper-clad stainless steel laminated to plywood, and brass-clad stainless steel, but other combinations may also be used.

Care must be taken in selection of metals and fasteners to avoid corrosion caused by contact between dissimilar metals. Also, contact between metal roofing and supports made of a different metal should be avoided. Hence, copper roofing should not be fastened to a wood deck with steel nails. Also, aluminum roofing should be separated from supporting steel rafters with felt strips.

In design and installation of metal roofing, provision should be made for dimensional changes in the roofing with variations in temperature. For example, for a 150°F rise in temperature, an 8-ft length of metal roofing may increase $\frac{1}{8}$ in. or more in length. If this movement were prevented, the metal would buckle and split at seams and joints. To allow for expansion and contraction, one of the following four types of construction is often used:

Corrugated-metal roofing (see Fig. 15.3a) with an undulated surface that absorbs thermal movements in one direction by bending. Connections that permit thermal movements in the perpendicular direction must be provided.

Standing-seam construction (see Fig. 15.3b), with narrow panels connected along their length by raised folded edges that permit expansion and contraction. Loose-locked seams are provided for longitudinal movement.

Batten-seam construction (see Fig. 15.3c), which is similar to standing-seam construction, except that a metal cap, or batten, is placed over the raised seams.

Flat-seam construction (see Fig. 15.3d), which connects small sheets with solder on all four edges. Because the sheets are small, the amount of expansion and contraction can be ignored.

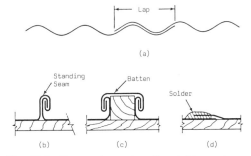

Fig. 15.3. Forms of metal roofing. (a) Lapped, corrugated sheets. (b) Standing-seam joint. (c) Batten seam. (d) Soldered flat seam.

Corrugated metal sheets can be connected directly to structural framing, despite the restraint to thermal movement offered by the framing. The undulations absorb such movement. But flat sheets should not be connected directly to the structural framing, because the metal must be free to move with temperature variations. For standing-seam and flat-seam construction, connections to structural framing can be made with small metal clips, locked into the folds of the sheets at seams. With batten-seam construction, connections can be similarly made with clips under the battens.

Where metal roofing sheets meet at a ridge, special cap sheets, shaped to the angle of intersection, should be placed over the intersection. The edges of the cap sheets should be folded inward to house outward folded edges of the intersecting sheets. Similarly, valley and hip sheets should be used at appropriate intersections, to cover the seams.

Rigid Vinyl Roofing. Polyvinyl chloride is used for roofing in the form of flat, corrugated or ribbed sheets. Any of several different methods for fastening the sheets may be used. Preferably, sheets should end at a structural support, to which they can be fastened. Corrugated sheets should be lapped one corrugation at joints. Ends of sheets should overlap at least 8 in.

Elastomeric Roofing. An elastomer is a synthetic polymer with rubberlike properties. Such materials, when water resistant, make good roofing for curved roofs. Elastomeric compounds suitable for roofing applications include the following:

Neoprene, a polymerized chloroprene, used as a liquid or a sheet and usually produced in dark colors. It does not support combustion. Adhesion is excellent. Resistance to weathering and thermal movement is good.

Hypalon, a chlorosulfonated polyethylene, used as a liquid or a sheet and produced in a variety of colors. It does not support combustion. Resistance to weathering and thermal movement is better than that of neoprene.

Silicone, a polymer with a structure of alternate silicone and oxygen atoms, having various organic groups linked to the silicon. Compounded with a filler and vulcanizing agent, the elastomer is applied as a liquid. Dirt retention is higher than for other elastomers.

Butyl rubber, a polymerized isobutylene containing isoprene or butadiene, used as a liquid. It is available only black.

Polyurethane rubber, a compound of diisocyanate and polyesters, with fire retardants added when fire resistance is required. Resistance to weathering and thermal movement is excellent. Fewer coats of this rubber are required for roofing than for other elastomers.

Elastomeric roofing may be classified as fluid applied or sheet applied.

Fluid-applied roofing, conforming readily to any roof shape, has better adhesion. On drying, the fluid forms a continuous membrane. It may be applied by roller, pressure-fed roller, spray gun, brush or squeegee in single or multiple coats. A typical application over a concrete deck might consist of a primer, two body coats of elastomer and two colored finishing coats of elastomer. The fluid cannot be relied on to fill voids.

Sheet-applied roofing may consist of a single ply, often with a reinforcing material laminated to the underside. Delivered to the job in rolls, the sheet is attached with an adhesive as it is unrolled on a deck. This type of roofing will withstand light foot traffic. The sheets should not, however, be used for steep roofs or roofs with intricate designs that require extensive cutting, fitting and seaming. A typical application might consist of a water-dispersed deck adhesive; the elastomeric ply; flashing tape, caps and seals for field joints, vents and roof edges; and a top coat of liquid synthetic rubber to cover the tapes and provide a seamless surface.

In all types of applications, manufacturers' specifications should be carefully followed. The roof deck for elastomeric roofing should be carefully selected and constructed. For example, structural concrete that has been properly finished and cured for at least 30 days provides an excellent deck for liquid-applied roofing. Plywood offers an excellent surface for neoprene and Hypalon. For foamed-in-place, closed-cell polyurethane insulation, elastomers are often the preferred roof covering. In contrast, lightweight concrete decks or fills should not be used with liquid-applied roofing. They entrap air and moisture, which may cause the elastomeric film to blister.

Multiunit Roofing

Multiunit roofing, which is suitable for use on steep slopes, includes roofing shingles and tiles. These are usually secured with nails to an inclined wood deck that is first covered with a waterproof membrane, such as asphalt-saturated felt. Application of the shingles or tiles starts at the eaves. For some types of multiunit roofing, a starting strip may be needed under the first course to slope it properly and prevent uplift by the wind. Subsequent courses are placed in succession up the slope, each course overlapping the one below in accordance with manufacturer's recommendations. Usually also, the units are laid with edges overlapping. To prevent penetration of water, joints between units should not be aligned. Specially shaped shingles and tiles or flashing should be used to fit and waterproof the roof at angle changes, such as ridges, hips and valleys.

Shingles and tiles are available in a wide variety of shapes, textures and colors. They may also be laid in a variety of patterns. Choice of roofing and method of laying the units usually depends basically on cost of the roofing and aesthetic effects desired by the architect and the owner.

Asphalt Shingles. The basic component of an asphalt shingle is an asphalt-saturated felt. It is usually given an asphalt coating on the underside for dimensional stability and as a barrier against moisture from the building interior. The felt is also given a thick top coating that is reinforced with mineral stabilizers and in which mineral granules are embedded for texture and color.

Asphalt shingles are available in weights ranging from 230 to 350 lb per square (100 sq ft). Cost of a specific type of shingle increases with weight. The lowest-weight asphalt shingle usually costs less than multiunit roofing made of other materials. Durability and especially resistance to wind uplift, however, is better for the heavier-weight shingles. To prevent wind damage, self-sealing shingles or interlocking shingles should be used on roofs with a slope less than 4 on 12. For steeper roofs, shingle tabs should be sealed to the underlying shingle with a dab of asphalt cement, for wind resistance.

Wood Shingles. Depending on the appearance desired, wood shingles may be obtained machine sawn or hand split. Either type is available with a fire-retardant treatment.

Roofing tiles. Clay tiles are made from special clays, such as shale, hardened by heat in a kiln. So treated, they develop the characteristics of stone. Based on shape, tiles may be classified as roll or flat. The shape of roll tiles may be semicircular, reverse-curve S, pan or cover.

Concrete tiles usually are made of a mix of portland cement, sand and water. While plastic, the mix is extruded on molds under pressure to shape the tiles. The exposed tile surface often is finished with a colored cementitious material. The tiles then are cured in a chamber in which temperature and humidity are controlled. Like clay tiles, concrete tiles may be roll or flat.

Roofing Slate. This type of multiunit roofing may be classified as standard commercial slate or textural, or random, slate. Random slate is delivered to the building site in a variety of sizes and thickness to be sorted by the slaters. Standard commercial slate, in contrast, is graded for size and thickness at the quarry. Hence, standard slate, as delivered to the site, costs more than random slate, but the cost of material plus installation is less. Either type of slate usually is supplied with a smooth and a rough face. The slate should be applied on the roof with the smooth face down.

Special Roof Waterproofing

Where roofs intersect other ·systems, such as walls and vent pipes, special precautions should be taken to provide a barrier between the building interior environment and the outdoors at the intersection. It is especially important to prevent water from penetrating at the junction.

Because materials used in the roof system and those used in the intersecting systems usually are different and have different coefficients of thermal expansion, waterproofing membranes placed to cover the junction and secured to the different materials should be elastic. Otherwise, differential movements will tear the membranes or pull them away and leave a gap through which water can penetrate. Waterproofing membranes consequently usually are made of bituminous felts, plastics or noncorrosive metals, such as copper, aluminum and stainless steel.

Flashing. Waterproofing membranes called flashing usually are installed where a roof system intersects another system at a sharp angle. Generally, two types of flashing are used at each intersection. This type of construction is illustrated in Fig. 15.4 for copper flashing at

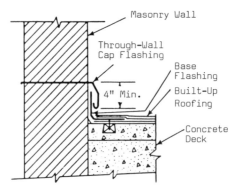

Fig. 15.4. Flashing at the intersection of a roof and a unit-masonry parapet.

the junction of built-up roofing on a flat concrete deck with a vertical masonry parapet.

In Fig. 15.4, base flashing is shown inserted between the plies of built-up roofing and extended upward along the face of the wall. Cap flashing, embedded in the mortar line of the wall, is turned down to cover the top of the base flashing and to overlap it at least 4 in. In this case, the cap flashing extends through the wall to prevent moisture that might seep through the masonry from leaking into the building interior.

Although Fig. 15.4 shows metal flashing,

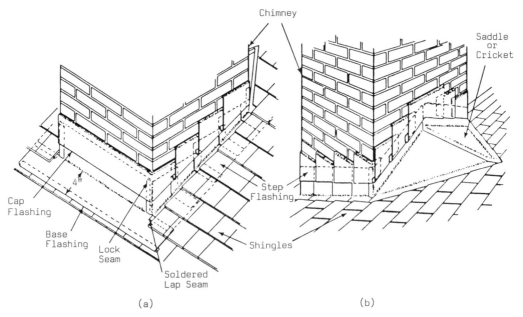

(a)   (b)

Fig. 15.5. Flashing at the intersection of a sloped roof and a brick chimney. (a) Cap and base flashing on the low side and step flashing along the slope. (b) Saddle or cricket on the high side.

other materials, such as bituminous felts or plastics, also may be used for flashing.

Where a steep roof intersects a vertical wall, step flashing may be used as indicated in Fig. 15.5. Short pieces of flashing metal, bent to the proper angle, are placed, with an overlap, along the intersection. One flange of the base flashing is placed under the shingles and the other flange against the wall. Similarly, one flange of the cap flashing is embedded in the mortar line of the masonry wall and the other flange is extended downward over the base flashing.

For the low edge of a chimney (see Fig. 15.5*a*) flashing may be placed in much the same way as for a flat roof. At the high end, a saddle, or cricket, may be used (see Fig. 15.5*b*), to divert rainwater past the sides of the chimney.

Expansion Joints. To permit dimensional changes to occur in a roof as temperatures vary, without buildup in stresses or strains, expansion joints should be provided between large sections of roof and between roof and walls. These joints usually are gaps filled with an elastic material, such as mastic or plastics.

At walls, such joints may be protected with flashing. Between abutting roof sections, a copper waterstop may be placed between the separated sections to prevent leakage. The waterstop is a bent sheet, which can absorb the thermal movements by bending.

SECTIONS 15.1 AND 15.2

References

Sweet's Architectural File, Division 7: Roofing, McGraw-Hill, New York, latest edition.
Also see general references for this chapter.

Words and Terms

Cricket
Deck
Dormer
Eave
Expansion joint
Flashing
Monitor
Roof form: flat, shed, pitched (gabled), hipped, gambrel, mansard, monitor, sawtooth

Roofing
Roofing type: single unit, multiunit
Saddle
Skylight
Vapor barrier

Significant Relations, Functions and Issues

Roof form related to structure, drainage, and roofing materials.
Roof angle (pitch, slope) related to roofing materials.
Enhancement of roof construction related to enclosure functions (as an exterior surface of the building): thermal insulation, acoustic transmission, vapor barrier, appearance as an architectural design concern.

15.3. EXTERIOR WALLS

An exterior wall extends from the roof to the foundation walls to separate the interior building environment from the outdoor environment. For economical construction, exterior walls are usually constructed vertically and along straight lines. Nevertheless, for architectural purposes, some walls are built curved and some, although built along straight lines, are inclined, inward or outward, away from the vertical.

The prime purpose of an exterior wall is to shelter the building interior from wind, rain, snow and extreme temperatures. An exterior wall, however, also must be capable of providing privacy to the occupants, if desired, and of excluding from the building dirt, noise and unwanted visitors. Preferably, in addition, an exterior wall should be fire resistant and have high resistance to heat flow. Furthermore, a wall plays an important role in the aesthetic treatment of the building interior and exterior.

To perform these functions, an exterior wall should be impermeable to water and have desirable acoustic and thermal insulating properties. Walls should also be strong and stiff. They must be able to support their own weight and other vertical loads that may be applied to them. Also, they must be capable of withstanding wind and accidental impact loads perpendicular to their exposed faces. Furthermore, surfaces exposed to the weather should be very durable and yet attractive in appearance.

Maintenance, including painting, repair and replacement of a building exterior, can be very expensive, especially for tall buildings. Hence, building owners often find it more economical to pay initially for a more-expensive exterior facing with a long life than to have installed a less-expensive material that will not last as long or will require much maintenance.

For economy, therefore, an exterior wall often is built with an exterior facing that is durable and attractive and with separate, less costly backup and interior facing. Both interior and exterior facings can be selected to meet decorative, sanitation, durability, fire resistance, impermeability and maintenance requirements. The backup can be chosen to meet strength, stiffness and thermal- and acoustic-insulation requirements.

Section 2.4 lists the ways in which walls are usually built: unit masonry, panel or framed walls, or combinations of these. The components are usually assembled, piece by piece, on the site. Often, however, walls can be assembled by mass-production methods, in a factory, into large panels that are joined on the site, with a lower total cost; however, care should be taken that savings through use of mass production and better quality control in a factory are not offset by higher handling, storage, delivery and financing costs.

Section 2.4 also indicates that walls can be classified as load-bearing or curtain walls. Description of some commonly used methods of constructing both classes follow.

Both classes of walls should possess the properties previously described. In addition, provision should be made for dimensional changes with temperature variations. To meet this requirement for long walls, the walls should be divided vertically into separate sections, with an expansion joint between them. The joint is a gap sufficiently wide to permit anticipated thermal movements and filled with an elastic material, such as mastic or plastic. Usually, a copper waterstop, a bent sheet that absorbs thermal movements by bending, is incorporated between the wall sections at each expansion joint, to prevent penetration of water through the joint. Spacing of joints generally should not exceed 200 ft for steel or concrete walls or 400 ft for stone or brick bearing walls.

Load-bearing Walls

Bearing walls are capable of supporting their own weight and other vertical loads imposed, such as those from roof and floors. Such walls often are of wood-frame, brick, concrete-block, stone, reinforced-concrete or combination construction.

Bearing walls may be classified as empirically designed or as engineered walls. Empirically designed walls are built in accordance with rules given in the local building code. These rules limit the height and length of wall between lateral bracing, specify maximum allowable bearing stresses and set minimums for thickness of wall for various heights or for sizes of structural members. Engineered walls, in contrast, are designed, in accordance with accepted design theories, for the specific material.

Wood-Framed Walls. A type of framing called *platform*, or *western*, *framing*, is frequently used for low wood buildings, such as houses. It is easy to erect, because one story of a multistory building can be built at a time over the foundations or lower stories.

Exterior walls and interior partitions are constructed one-story high. For one- and two-story houses, wall framing consists of 2 × 4-in. wood studs, spaced 16 or 24 in. center-to-center (see Fig. 15.6). The studs are seated on and nailed to a 2 × 4-in. sole plate. One or two 2 × 4-in. top plates are nailed across the tops of the studs. Construction above the wall rests on these plates. At each level of the building, the sole plate is supported on the floor.

A sill plate is anchored to the top of the foundation wall and carries a header that extends around the periphery of the building (see Fig. 15.6). First-floor joists and girders, of the same depth as the header, frame into it and rest on the sill. The first-floor subflooring is laid on and nailed to the joists. It serves as a working platform for erection of walls and partitions and for other first-story construction.

For the second story, a header is seated on the top plates of the walls below. The second-floor joists and girders frame into the header and also rest on the top plates. Then, the

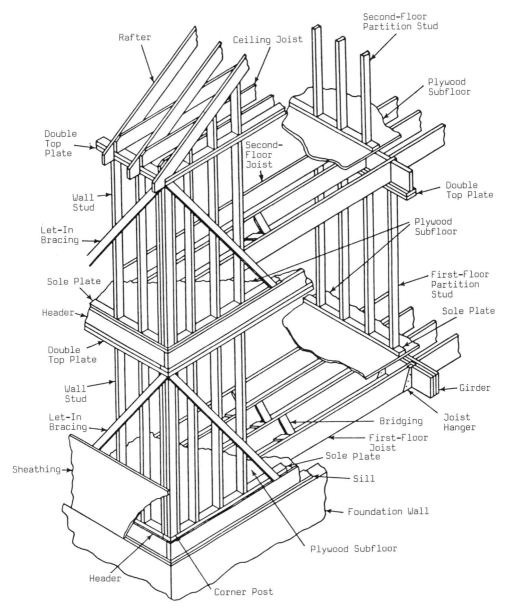

Fig. 15.6. Wood platform framing for a two-story building.

second-floor subfloor, walls and partitions are
erected over the joists and girders as for the
first story. Ceiling joists and rafters may be
seated directly on the top plates of the top-
story walls.

Insulation, with a vapor barrier on the interior (warm side) face, usually is installed between the studs. Sheathing is nailed to the outer
face of the studs and covered with building paper or other waterproofing membrane. An exterior facing is then nailed over the membrane.
For the purpose, aluminum siding, clapboard,
brick, stone, wood shingles or other facing materials may be used. Figure 15.7a shows how
a brick facing may be applied to a wood-framed
wall. The interior of the wall may be finished
with plaster, gypsumboard, wood panels or
other types of panelling, fastened to the studs.

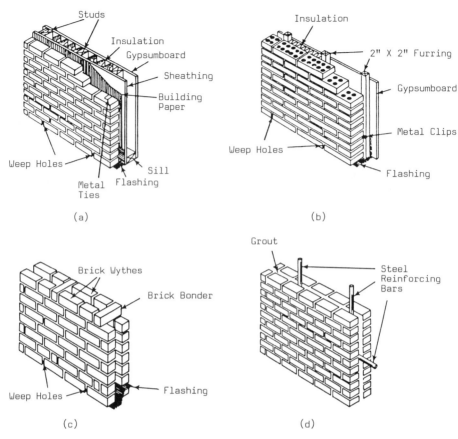

Fig. 15.7. Applications of brick in exterior walls. (*a*) Brick veneer facing with a wood-framed wall. (*b*) Brick wall with furring to form an insulating air space. (*c*) 8-in. brick wall formed with two wythes tied with brick bonders. (*d*) Reinforced wall, with reinforcing bars set in a grouted cavity between the two wythes.

Balloon framing is an alternative form of construction used for exterior walls of two-story buildings. A distinguishing characteristic of this type of wall is framing with two-story-high studs. The joists at all levels are nailed to the sides of the studs. Also, first-floor joists are seated on a sill plate anchored to the top of the foundation walls, and second-floor joists rest on a ribband, a horizontal beam nailed into a cutout in the inside faces of the studs, to form a ledge.

Unit-Masonry Walls. Bearing walls of unit masonry are usually built of brick, concrete block or stone. The units come in sizes and weights small enough for a mason to handle without additional help. To reduce weight and make it easy to grip them, brick and block are made with holes, or hollows. A unit generally weighs less than 25 lb.

A *brick* is a rectangular masonry unit, at least 75% solid, made from burned clay, shale or a mixture of these materials. Bricks usually are specified by their nominal dimensions. These differ from actual dimensions by the width of the mortar joint. For example, with a $\frac{1}{2}$-in. mortar joint, a brick nominally 8 in. long is actually $7\frac{1}{2}$ in. long. Bricks usually used are 4 or 6 in. thick, $2\frac{2}{3}$ or 4 in. high, and 8 or 12 in. long. Figure 15.7*b* illustrates use of 6-in.-thick bricks in construction of a one-story wall for a house.

A *concrete block* is a machine-formed rectangular masonry unit composed of portland cement, aggregate and water. Like bricks, concrete blocks are specified by their nominal dimensions. Blocks usually used are 8 × 8 in. by 16 in. long, with two large interior core holes and a half-size hollow at each end.

Stone may be used as rubble; that is, in irregular shapes and sizes, or cut into desired shapes. Stone cut into rectangular units larger in size than bricks is called *random ashlar*. Rubble may be classified as rough, or ordinary; random, or coursed. *Rough rubble* is composed of nonshaped field stones laid without regularity of coursing, but well bonded to form a wall. *Random rubble* is composed of roughly shaped stones, well bonded and brought at irregular vertical intervals to discontinuous but approximately level beds or courses. *Coursed rubble* is composed of roughly shaped stones that are fitted approximately on level beds, well bonded and brought at vertical intervals to continuous level beds or courses.

Brick and concrete block usually are laid on level beds, with mortar joints of constant thickness.

When walls of greater thickness than that of a masonry unit are required, they may be built up of two or more vertical layers, or wythes, of units. A solid wall consists of wythes bonded with mortar. A *bonder*, or *header* unit, may be laid flat across the wall, with end surface exposed, to bond the wythes more securely (see Fig. 15.7c). For better insulation and to prevent leakage, a gap may be provided between the wythes, which then may be tied together at vertical intervals with steel ties. Such walls are called *hollow*, or *cavity*, *walls*. For better insulation, the space between inner and outer wythes of a cavity wall may be filled with an insulating material.

Weep holes should be provided at the base of all masonry walls at intervals not exceeding 24 in., to permit rainwater that may penetrate the wall and flow down the back to drain to the outside. The weep holes may be easily formed in the mortar joints between the masonry units, by placing the mortar around sash cord or $\frac{3}{8}$-in.-diameter rubber tubing, which is withdrawn after the mortar has set.

To prevent a wall from buckling or overturning under eccentric loading or lateral forces, the wall should be braced horizontally and vertically. Floors, beams and girders, roof, cross

walls, partitions, buttresses and pilasters may serve this purpose. A *buttress* is a masonry column built as an integral part of a wall and decreasing in thickness from base to top, but never thinner than the wall. A *pilaster* is a column of masonry built as part of a wall, but thicker than the wall, and of uniform thickness throughout its height.

Mortar for unit masonry should meet the requirements of ASTM specifications C270 and C476. These define various types of mortar with a wide range of compressive strengths. Based on compressive strength, each type of mortar is used for a specific purpose (see Sec. 8.5).

Mortars are a mixture of portland cement and lime or masonry cement, sand and water. Often, *hydrated lime* or *lime putty* is substituted for masonry cement to obtain greater workability. Both hydrated lime and lime putty are prepared from *quicklime*.

Quicklime, made by heating limestone, consists essentially of calcium and magnesium oxides, plus impurities such as silica, iron, and aluminum oxides. When it is mixed with two or three times its weight in water, it slakes. In this process, the calcium oxides react with the water to form calcium hydroxide. The reaction develops sufficient heat to bring the mixture to a boil. Lime putty is formed when the mixture cools. This material is a semifluid, which may be shoveled or carried in a hod. *Hydrated limes* are a factory-made powder, produced by adding a small amount of water to quicklime.

Mortar mixes commonly used for unit-masonry walls are as follows:

For brick: 1 part cement, 1 part lime, 6 parts sand.

For concrete block: 1 part cement, 1 part lime putty, 5 to 6 parts sand.

For rubble: 1 part cement, 1 to 2 parts lime hydrate or putty, 5 to 7 parts sand.

Bagged, prepared mortar mixes also are available from material suppliers. For large construction projects, however, the services of a testing laboratory should be utilized to determine the most suitable mortar mix with available materials for specific applications. Plasticity and compressive strength are important characteristics that should be investigated. Higher-strength mortars, however, are not preferable to lower-strength mortars, where building codes or normal practice permit lower strength for the particular application. The smaller the compressive strength, the smaller is the modulus of elasticity and hence the smaller is the stress produced for a specific dimensional change.

Mortar should completely fill all joints between unit masonry in a wall. To accomplish this, a bed of mortar should be placed first on units already in position, before a new unit is laid. A heavy buttering of mortar should be applied to one end of the unit. Then, it should be pressed into the bed joint and pushed against an adjacent unit so that mortar is squeezed out of the joints on both sides of the unit. Excess mortar should be removed with a trowel. The joint may be tooled, preferably into a concave curve, to improve resistance to penetration of rainwater.

For empirically designed masonry walls, wall thickness is determined by building-code limitations. Either the ratio of unsupported height to nominal wall thickness or of length of wall to nominal wall thickness may be required to be 20 or less for solid walls or 18 or less for cavity walls or walls built of hollow masonry units. (A masonry unit is considered hollow when its net cross-sectional area in any plane parallel to its bearing surfaces is less than 75% of its gross cross-sectional area measured in the same plane.) In addition, the building code may require a masonry bearing wall to be at least 12 in. thick for the uppermost 35 ft of its height. Thickness then should be increased 4 in. for each successive 35 ft or fraction of this height measured downward from the top of the wall. Rubble stone walls should be at least 4 in. thicker but not less than 16 in. thick.

Building codes, however, usually permit many exceptions to these minimum thicknesses, except for rubble stone walls. For example, the top-story bearing wall of a masonry building not more than 35 ft high may be only 8 in. thick, if the wall is not more than 12 ft high and not subjected to lateral thrust from the roof. Similarly, in dwellings up to three stories

or 35 ft high, walls may also be 8 in. thick. For one-story houses, walls may be only 6 in. thick, if the height is 9 ft or less or if the height to the peak of a gable does not exceed 15 ft (see Fig. 15.7b).

In cavity or masonry bonded hollow walls, interior and exterior wythes should be at least 4 in. thick. The cavity should be between 2 and 3 in. wide. A 10-in.-thick cavity wall may not be permitted to exceed 25 ft in height, and a 35-ft height limitation may be placed on thicker walls.

Engineered Walls. Brick and concrete walls suitably reinforced with steel bars can usually be built thinner than empirically designed walls, if calculations made in accordance with accepted design theories indicate that such walls would be safe. Bars should be embedded horizontally and vertically in the walls to satisfy stress requirements and to control cracking, which may be caused by shrinkage as mortar or concrete dries or by dimensional changes as temperatures vary. Figure 15.7d illustrates placement of reinforcing in a brick wall.

Enclosure of a building with load-bearing panels of precast concrete is illustrated in Fig. 15.8. (The building is shown two stories high, but the technique is applicable to taller buildings.) In this case, the wall panels are two stories high. They support precast-concrete roof panels that span the width of the building. (Floor or roof panels, however, may be supported instead at one end on interior loadbearing walls or partitions or on structural framing.) The wall and roof panels may be steel reinforced or prestressed. They may be mass produced in a factory away from the building site, delivered by truck to the site, then speedily erected with a crane.

Curtain Walls

Supported on structural framing that carries floor and roof loads, exterior walls may function much like a curtain, separating the interior building environment from the outside environment. They need carry no loads other than their own weight and wind loads. Hence, cur-

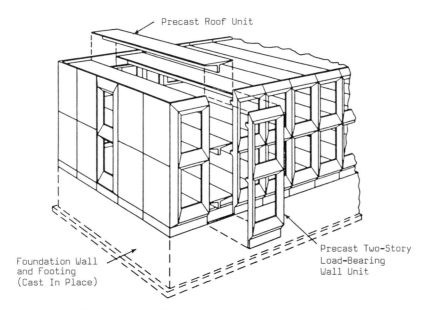

Fig. 15.8. Bearing-wall construction with precast concrete panels.

tain walls generally are much lighter and thinner than load-bearing walls. With curtain walls, building foundations therefore are subjected to much lighter loading and can be built much smaller and lighter than those required with load-bearing walls.

Thickness of a curtain wall need be no greater than that required for the wall to serve its principal functions. Many industrial buildings are enclosed only with cold-formed metal or concrete siding (see Fig. 15.9). For buildings close to others or with other types of occupancies, however, fire resistance may be important. For such buildings, fire resistance requirements in local building codes often determine the thickness and type of materials that may be used in curtain walls.

Also, it is advisable, for economy and energy conservation, that a curtain wall have good thermal insulation. In addition, the wall should have good acoustical properties, to exclude noise, and as for all types of walls, curtain walls should have a durable exterior surface and require little maintenance.

To meet these requirements, multilayered wall construction, incorporating thermal insulation, is usually used. The layers may be assembled on the site or may be factory prefabricated into panels for on-site assembly.

Siding. Solid curtain walls in sheet form are called siding. Made of cold-formed steel, aluminum, plastics or concrete, siding may be supplied flat or corrugated. For a specific thickness, flat sheets are not as stiff in the direction perpendicular to their plane as are corrugated panels. Corrugated siding also has the advantage of being able to absorb thermal movements by bending across the corrugations.

For low industrial buildings, siding may be supplied long enough to extend from the foundations to the roof (see Fig. 15.9). In that case, the panels may be supported at the foundations and at a spandrel beam at the roof. Also, horizontal secondary framing, called *girts*, may be placed between columns for intermediate support of the siding.

Details should be planned to insure that the siding will shed rainwater. Corrugated sheets may be overlapped for the purpose. Flat sheets may be installed in frames, like those used for window glazing, or edges may be flanged to interlock, or splices may be covered with battens. In all cases, allowances should be made at connections between panels and at supports for expansion and contraction with temperature variations. Manufacturers' recommendations should be observed in installation of the siding.

Fig. 15.9. Striated, precast-concrete panel being erected for a curtain wall. (Courtesy High Concrete Structures, Inc., of New Jersey)

Metal and Glass Facings. In multilayered wall construction, siding or metal and glass panels may serve as the exterior facing, backed up by thermal insulation, fire-resistant material and an interior finish. The glass may be supported in a light frame, as are windows. Metal panels also may be placed in a light frame. The frames may be supported at top and bottom at floors or spandrel beams or may be attached to *mullions*, secondary vertical structural members. The panels may be small and light enough for one worker to carry or prefabricated two or three stories high and erected with a crane.

After the siding or the metal and glass panels have been installed, insulation and an interior finish may be applied on the inside. If a unit-masonry backup wall is required for added fire resistance and noise control, it may be built

either before or after installation of the exterior facing. Details should be planned so that water that penetrates the facing will be drained to the outside.

Sandwich Panels. For more economical multilayered construction, curtain walls may be factory prefabricated into panels much larger than unit masonry. These panels usually contain an insulation core sandwiched between a thin lightweight facing and a backing. They may be supported in the same ways as metal and glass facings.

Sandwich panels may be classified as custom, commercial or industrial. *Custom panels* are those designed for a specific project, generally multistory buildings. *Commercial panels* are built-up of parts standardized by manufacturers. *Industrial panels* are an assembly of stock-size, ribbed, fluted or otherwise preformed sheets, standard sash and insulation. The panels of all classes may be designed for installation of glazing after the panels have been erected. (Factory-installed glazing is subjected to the risk of breakage during handling or erection, or even after the panels have been positioned.)

Sandwich panels may also be classified in accordance with the methods used for installation as stick systems (see Fig. 15.10a and b), mullion-and-panel systems (see Fig. 15.10c and d) or panel systems (see Fig. 15.10e and f).

For the *stick method*, the exterior wall is assembled piece by piece (see Fig. 15.10a). Because this system requires more parts and field joints than the other methods, it is not used as often. Figure 15.10b shows the completed wall assembly.

For the *mullion-and-panel method*, mullions, vertical supporting members, are erected first. Then, the sandwich panels are placed between them (see Fig. 15.10c). Figure 15.10b shows the completed wall assembly. Often, a cap is added to cover the vertical joints between the panels and mullions.

For the *panel method*, the exterior wall is composed only of sandwich panels (see Fig. 15.10d). The panels are anchored to the building frame and to each other. A completed wall assembly is shown in Fig. 15.10b. Panels usually are one-story high, but they may often be

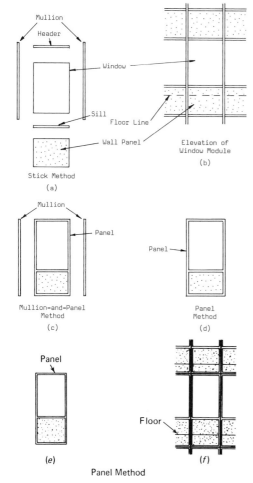

Fig. 15.10. Methods of constructing curtain walls.

two-stories high, with some savings in construction costs and time and with a reduction in field joints. The panel system requires fewer pieces and fewer field joints than the other methods.

Wall thermal movements, when mullions are used, may be accommodated by designing the mullions for the purpose. When mullions are not used, connections between panels often are made at deep flanges, which readily bend as the wall expands and contracts.

Lintels and Wall Flashing

The weight of and loads on curtain walls usually are transmitted through connections to the skeleton framing at the top and bottom of each story. One way of transferring the loads is to

anchor curtain walls to the floors, which then transmit the loads to the structural framing. Another way is to transfer the loads directly to the framing. An example of this is shown in Fig. 15.11b. There, an exterior masonry facing is carried on a steel shelf angle, or lintel, attached to a steel spandrel beam, which spans between a pair of exterior columns. Shims and slotted holes for the bolted connection permit horizontal and vertical adjustments in the position of the shelf angle.

At openings, such as those for doors and windows, provision must be made for supporting the wall above. The most common method for doing this is to install a horizontal beam, or lintel, over the opening and support it on the walls at the sides of the opening (see Fig. 15.11a). Many types of materials and beams may be used for the purpose, some of which are illustrated in Figs. 15.11 and 15.12.

For example, in the case of the lintel and beam that support a curtain wall (see Fig. 15.11b), a window or door opening can be provided below the lintel and placed anywhere along the length of the lintel. Figure 15.11c illustrates the use of a steel beam with a continuous steel plate welded to its bottom to span an opening in a thick bearing wall. Figure 15.11d shows how three steel angles can serve as a lintel, and Figure 15.11e demonstrates the use of an angle and a pair of steel channels as lintels in a thick wall.

For reinforced concrete bearing walls, the portion of the wall above an opening can be reinforced to act as a lintel. In addition, vertical reinforcing should be placed parallel to the sides of the opening, and diagonal bars should be laid close to the corners of the opening to prevent, or at least control cracking.

For masonry bearing walls, stone or cast-in-place or precast concrete lintels are often set above openings. For cast-in-place lintels, a convenient construction method is to use a precast-concrete U-shape as a form, to eliminate the need for shoring. Concrete is placed in the form after reinforcing steel has been positioned. For masonry cavity walls, the cavity may be widened, by use of half-thickness bricks at inner and outer wall faces, and then filled with reinforced concrete (see Fig. 15.11f).

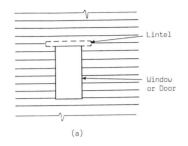

(a)

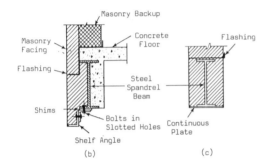

(b)   (c)

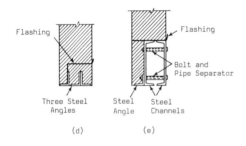

(d)   (e)

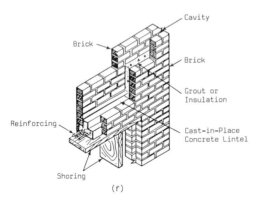

(f)

Fig. 15.11. Lintels and flashing in exterior walls. (a) Elevation of a bearing wall showing lintel over a window or door. (b) Lintel attached to a steel spandrel beam carries a curtain-wall facing. (c) Steel spandrel beam with plate on bottom flange supports a curtain wall. (d) Lintel composed of three angles frames an opening in a thick bearing wall. (e) Lintel formed with two steel channels and an angle. (f) Cast-in-place concrete lintel formed in a masonry cavity wall.

In wood-framed walls, the opening is framed with top and bottom headers and trimmers on the sides (see Fig. 15.12a). Short studs, called cripples, are placed between the headers and top and bottom plates for attachment of sheathing and inner facing. When the exterior facing is masonry, a steel-angle lintel may be placed over the opening to support the wall above it (see Fig. 15.12b). When the exterior facing is a thin siding or shingles, a lintel is not needed, because the facing is nailed to the sheathing, which, in turn, is attached to the studs (see Fig. 15.12c). Typical construction at the bottom of a window opening in a woodframe wall is shown in Fig. 15.12d and e.

Flashing is desirable above and below wall openings to prevent rainwater that may seep through the exterior facing from leaking into the building interior through joints at the openings. Figures 15.11 and 15.12 show some examples of how flashing may be installed.

15.4. SINGLE-ENCLOSURE SYSTEMS

An alternative to enclosure of a building with walls and roofs is a single-enclosure system. Such a system consists of curved surfaces, or of interconnecting planes approximating such surfaces, so that it is difficult or impossible to establish a boundary between roof and wall sections. The system rises from the foundations around the perimeter of the building and slopes inward to one or more peaks or plateaus in the interior. Structural systems that may be used for single enclosures are described in Secs. 8.11 to 8.14.

Single enclosures for a building may be thin-shell domes, barrel arches with curved ends, air-supported membranes, lightly framed single- and double-curved structures, or tentlike coverings hung on cables.

The shapes of single-enclosure systems generally permit use of very efficient structural systems, such as membranes and thin shells. As a result, less material is required than for wall-roof systems, and the materials may be thin and lightweight. Also, structural framing, if used, may be light. Hence, foundations may be much smaller than for wall-roof systems.

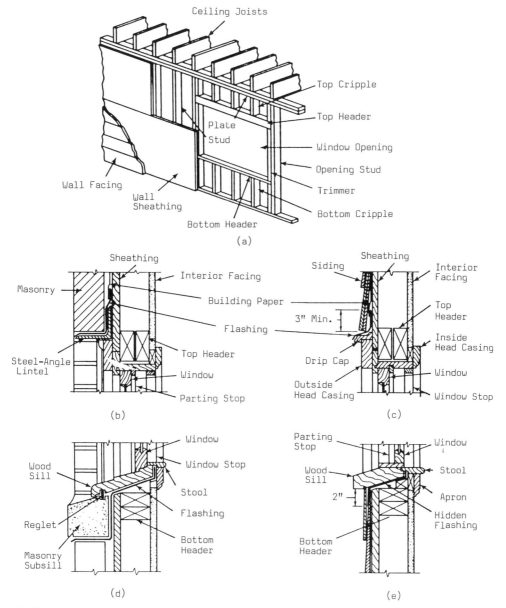

Fig. 15.12. Construction at windows in a wood-framed wall. (*a*) Headers and trimmers frame the opening. (*b*) Detail at head for a brick facing on exterior. (*c*) Detail at head for wood siding. (*d*) Detail at sill for a brick exterior facing. (*e*) Detail at sill for wood siding.

Such light construction is especially suitable for buildings that must be demountable for transfer to other sites.

Initial and life-cycle costs, however, may not necessarily be less than for wall-roof systems. For a given floor area, for example, a multistory building with walls and roof may cost less to build or can be erected on a smaller site. Also, operating costs may be greater for a single enclosure, because it has a larger volume to heat, cool or ventilate. If the enclosure is to be air supported, energy must be expended to maintain the interior pressure. In addition, the shape of a single enclosure may not always be aesthetically desirable.

Consequently, selection of a single-enclosure

system depends on the size and shape of the available site, purpose of the building, architectural objectives, demountability requirements, length of time occupants will be in the building, and comparative initial and life-cycle costs. Single enclosures have been used for a wide variety of building types. These include office and industrial buildings, warehouses, greenhouses, maintenance buildings, exhibitions and expositions, markets, radomes, concert halls, theaters, and athletic buildings, such as gymnasiums, field houses, natatoriums and tennis courts. Single enclosures have also been used to shelter construction operations for conventional buildings in winter, operations that would otherwise have to be halted because of inclement weather.

SECTIONS 15.3 AND 15.4

References

H. Sands, *Wall Systems: Analysis by Detail*, McGraw-Hill, New York, 1986.

Sweet's Architectural File, Divisions 7 (sealants, weatherproofing) and 9 (finishes), McGraw Hill, New York, (issued annually).

Building Code Requirements for Reinforced Concrete, ACI 318-83, American Concrete Institute, Detroit, 1983.

Building Code Requirements for Engineered Brick Masonry, Brick Institute of America, Washington, D.C.

J. Amrhein, *Masonry Design Manual*, 3rd ed., Masonry Institute of America, Los Angeles, 1979.

Also see general references for this chapter.

Words and Terms

Balloon framing
Platform framing
Curtain wall
Flashing
Lintel
Mullion
Random ashlar
Sandwich panel
Siding
Single-enclosure system
Stone
Unit masonry: brick, concrete block, stone
Weep holes

Significant Relations, Functions and Issues

Form and construction of exterior walls related to: building structure; nature of wall (structural or nonstructural); window type, shape and dimensions; need for enhancement (insulation, etc.)

15.5. WINDOWS

Windows are openings, glazed with a transparent or translucent material, in a building exterior. The glazing usually is transparent glass or plastic for admission of daylight and to provide occupants with a view of the outdoors, or, for retail stores, to give passersby a view of display items inside. Translucent glazing is used when a view is not desired but diffused daylighting is wanted.

Windows seldom are essential to the functioning of buildings. For some buildings, such as theaters and warehouses, windows may interfere with efficient functioning. They also have the disadvantage of decreasing the thermal insulation, fire resistance and noise-reduction capability of walls and roofs, unless special construction is used.

Building codes, however, may require a minimum amount of window area, for instance, 10% of the floor area of each room, in certain types of buildings. In addition, windows may be desirable for providing views, admitting daylight, ventilating the building interior and offering a means of escape from the interior in emergencies. Also, windows are an important feature of the aesthetic treatment of the building exterior. As a result, windows are usually installed in all buildings where they will not interfere with the functioning of the buildings.

Some of the disadvantages of windows can be reduced with proper construction or by installation of accessories. For example, window shades, blinds or curtains can be installed to cut off excessive sunlight. For the same purpose, on sun exposures, windows may be recessed, horizontal overhangs or vertical fins may be provided on the exterior, or trees may be planted to shade the glazing. Also, to reduce noise and heat gain and heat loss, windows may be double or triple glazed; that is, provided with two or three parallel panes of glass or plastic separated by a sealed air space. Alternatively, an additional window, called a *storm window*, may be installed, with an air space between it and the main window. Tinted or coating glazing, to absorb or reflect solar rays, are additional means of controlling heat gain. On the other hand, proper orientation of windows with respect to the sun can utilize solar heat for warming the building interior in cold weather. In addition, use of weatherstripping around windows and of windows with integral frame and trim will reduce air infiltration, thus decreasing the load on the HVAC system.

Weatherstripping is a barrier provided against water and air leakage through joints between separable building components; for instance, between windows and window frames, or between doors and door frames. It usually consists of thin strips of interlocking metal or resilient material.

Types of Windows

Numerous types of windows and in a wide variety of sizes are available. Some of the many types of windows are shown in Fig. 15.13.

The double-hung window (see Fig. 15.13a) consists of sash that slides vertically. Usually, movement of the sash is made easy by use of counterweights or springs that balance the weight of the sash. A horizontally sliding window is shown in Fig. 15.13b. The casement window in Fig. 15.13c has sash hinged at the sides. It offers the advantage over the preceding types of making the whole opening available for ventilation. In contrast, the picture window in Fig. 15.13d is fixed and provides no ventilation.

The vertically pivoted window in Fig. 15.13e makes it easier to clean both sides of the window from the building interior. The horizontally pivoted window in Fig. 15.13f permits ventilation in wet weather, because the inclined sash tends to exclude rain. A similar advantage, but with larger openable area for ventilation, is provided by an architectural projected window (see Fig. 15.13g), awning window (see Fig. 15.13h) and a jalousie (see Fig. 15.13i). These may be opened and closed mechanically.

Symbols used on architectural drawings to represent the action of windows are illustrated in Fig. 15.14. The symbols in Fig. 15.14a to

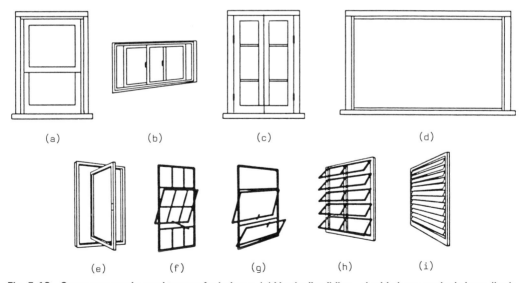

(a)   (b)   (c)   (d)

(e)   (f)   (g)   (h)   (i)

Fig. 5.13. **Some commonly used types of windows. (a) Vertically sliding—double hung or single hung (both sash slide or only bottom slides). (b) Horizontally sliding (also both or one). (c) Casement. (d) Fixed (nonoperating) (e) Vertically pivoted. (f) Horizontally pivoted. (g) Architecturally projected—upper as awning, lower as hopper. (h) Multiple awning. (i) Jalousie—unframed glass strips held at sides.**

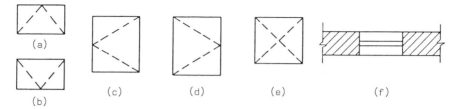

Fig. 5.14. Symbols for windows. In elevation, as viewed from the outside: (a) Outward projecting (awning). (b) Inward projecting (hopper). (c) Left-hand swing casement, hinged on left. (d) Right-hand swing casement, hinged on right. (e) Pivoted. In plan: (f) Double hung.

e are used in elevations. They represent windows that open outward (see Fig. 15.14a); windows that open inward (see Fig. 15.14b); windows with a left-hand swing (see Fig. 15.14c) or right-hand swing (see Fig. 15.14d), depending on whether the hinge is on the left or the right side when viewed from the outside, and horizontally or vertically pivoted windows (see Fig. 15.14e). For a fixed window, no dashed lines would be shown. The symbol used in plan views for double-hung windows is indicated in Fig. 15.14f.

Sash and window frames are often made of aluminum, painted wood or painted steel. Bronze, stainless steel or galvanized steel are sometimes used. Metal windows may be re-

quired for the purpose of meeting fire-resistance requirements of building codes.

Window Components

To support a window, a frame is usually provided around the perimeter of the window opening. Figure 15.15a shows the main parts of a wood frame. The upper and side parts of the frame are called *jambs*. The head jamb is the horizontal member forming the top of the frame. The side jambs form the vertical sides of the frame. The horizontal member forming the bottom of the frame is called a *sill*. An extension of the sill on the inner side of the window is known as a *stool* (see Fig. 15.12d and

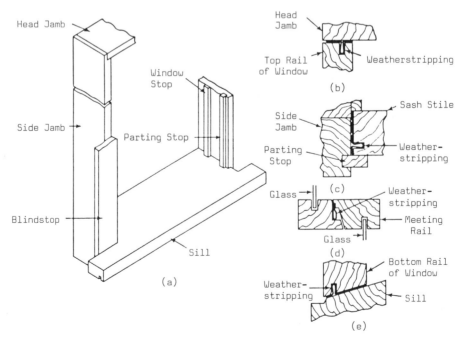

Fig. 15.15. Window-frame and sash details. (a) Main components of a wood frame for a double-hung window. (b) Detail at head (top). (c) Detail at jamb (side). (d) Detail at sill (bottom).

e). A projection below the sill is called a subsill (see Fig. 15.12*d*). For thick walls, a jamb liner may be needed. This is a thin strip of wood, either surfaced on four sides or tongued on one edge, which, when applied to the inside edge of a jamb, increases its width. *Casing*, molding of various widths and thicknesses, may be used to trim the window opening (see Fig. 15.12*c*).

Sash consists of *stiles*, which are vertical members, and *rails*, which are cross or horizontal members, made into a frame for holding glazing. The sash may also have *bars*, vertical or horizontal members that extend the full height or width of the glazed opening. Short, light bars are called *muntins*.

For double-hung windows, which consist of a pair of vertically sliding sash, the frame is provided with additional components, called *stops*, to form parallel grooves in which the sash slide along the side jambs. The inside strip is called the window stop, the center strip is a parting stop and the outer strip is a blind stop (see Fig. 15.15*a*). To accommodate the thickness of the parting stop, the bottom rail of the upper sash and the top rail of the lower sash are special rails, called *check*, or *meeting*, *rails*. They are made sufficiently thicker than the rest of the sash to fill the opening between the top and bottom sash. The meeting rails are usually beveled and rabbeted (see Fig. 15.15*d*).

If sash weights are used to counterbalance the weight of the double-hung sash, the weights and the sash are connected with sash cords that run over pulleys. The weights and pulleys are hidden from view in a box along a side jamb or in a hollow mullion between double-hung windows that are close together. The side jamb in which a pulley is fixed and along which the sash slides is called a *pulley stile*.

Dimensional Coordination. Window sizes, in general, are standardized. Sizes have been established to coordinate with wall materials, to reduce the amount of cutting and fitting of components in the field. Basis for the standard sizes is a 4-in. three-dimensional module. Design of exterior walls, therefore, should be based on this module and should provide standard openings for standard windows and doors. Standard sizes of many wall components are available based on the 4-in. module. An 8-in. brick, for instance, may be $7\frac{1}{2}$ in. long so that, with a $\frac{1}{2}$-in. mortar joint, the brick occupies two 4-in. modules.

Weatherstripping. Details showing application of weatherstripping to a wood double-hung window are presented in Fig. 15.15*b* to *e*. The top rail of the upper sash (see Fig. 15.15*b*) is grooved to receive weatherstripping attached to the head jamb. Similarly, the sash stiles are grooved to slide along weatherstripping on the side jamb (see Fig. 15.15*c*), and the bottom rail of the lower sash is grooved to close on weatherstripping on the sill (see Fig. 15.15*e*). Also, weatherstripping is incorporated in the rabbeted joint between the meeting rails (see Fig. 15.15*d*).

Glazing Supports. Prefabricated panels often are made with windows built in as an integral part of them. They may, however, be supplied unglazed, to avoid breakage of the glazing during building construction. Fig. 15.16*a* to *c* shows details of a metal window incorporated in a sandwich panel with concrete interior and exterior facings.

Glazing in many windows is held in place in the sash with putty, glazing compound, rubber or plastic strips and metal or wood moldings. Sometimes, for metal sash, metal clips are used for this purpose, and for wood sash, glazing points. Figure 15.16*d* shows glass bedded in putty and secured with glazing points driven into the wood sash. Such seating is desirable because it furnishes a smooth, resilient bearing surface for the glazing, prevents rattling and eliminates voids in which moisture can collect.

Putty is a mixture of about 10% raw linseed oil and 90% ground calcium carbonate. *Glazing compounds* often consist of about 15% mixed raw and bodied oils, with oil-absorbing pigments, such as fibrous silicates, and ground calcium carbonate or marble dust. Reactive pigments, such as lead carbonate, may be added to promote flexible skin formation.

Glazing beads (see Fig. 15.16*e*) may be used to cover exterior glazing compound and improve appearance. *Continuous glazing angles* (Fig. 15.16*f*) or similar supports are required

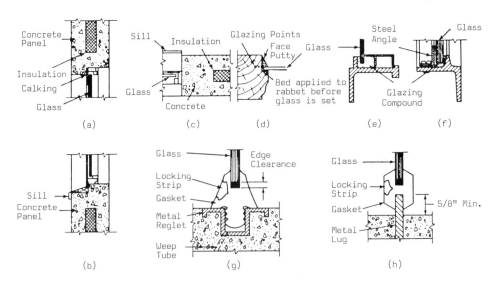

Fig. 15.16. Window details. For a metal window in a precast concrete panel: (*a*) Head. (*b*) Sill. (*c*) Jamb. For wood sash: (*d*) Bedding of glass in putty. For concrete or masonry walls: (*e*) Continuous bead glazing. (*f*) Continuous angle glazing. (*g*) Structural gasket with snap-in locking strip, fitted into groove for support. (*h*) H-shaped structural gasket locked to a projecting metal lug.

by the National Board of Fire Underwriters for so-called *labeled* windows used in fire-resistant construction.

As an alternative to sash, fixed windows may be held in place with structural gaskets made of preformed and cured elastomeric (rubber or rubber-like) material (see Fig. 15.16*g* and *h*). Extruded in a single strip, the gaskets are molded into the shape of the window perimeter and held by compression against the window frame and the glazing. The compressive force is provided when a locking strip of harder elastomer is squeezed into one side of the gasket. Figure 15.16*g* shows the installation of a reglet gasket, which fits into a recess in the window frame. Figure 15.16*h* illustrates use of an H gasket, which fits over a metal lug, or fin, projecting from the frame.

Types of Glazing

A wide variety of types and grades of glazing are available in appropriate thickness ranges. For large panes, thickness is generally determined by wind loads and by the loads to which the glazing may be subjected in being handled before and during installation. Some of the more common types of glazing are described in the following:

Clear window glass is the type used most often. It generally costs less than other types. Quality grades A and B, defined in Federal Government Standard DD-G-451*c*, are the grades most frequently used, A for buildings where appearance is important, and B for industrial buildings, basements and low-cost housing. Clear window glass also is classified in accordance with thickness. Single-strength glass, about $\frac{3}{32}$-in. thick, and double-strength glass, about $\frac{1}{8}$-in. thick, often are used for panes up to about 7 sq ft in area. Heavy sheet, up to about $\frac{7}{32}$-in. thick, may be used for larger panes.

Plate glass or **float glass**, however, is generally preferred for large panes, because the glasses are of superior quality and do not distort vision through them. They cost more than clear window glass. Polished plate or float glass is often used for double- or triple-glazed windows, showcases and picture windows.

Several types of glazing are available for use where special precautions are necessary to prevent breakage or where it is desirable to reduce the potential for injury when glazing is broken. *Tempered glass,* for example, made by a reheating and cooling process, is about five times as strong as annealed glass of the same thickness. *Wired glass* incorporates steel wire for reinforcement. It may also be specified where fire-retarding properties are desired. *Plastics*, such as acrylic or polycarbonate, provide much higher impact resistance than annealed or tempered glass. Also, *laminated glass*, an assembly of two or more layers of glass held together by two or more coatings of a transparent plastic, has high strength, and when fractured, shatters into small splinters with the glass adhering to the plastic, thus reducing the hazard of flying glass. Some types of laminated glass also provide heat-absorption, glare-reduction and sound-isolation properties.

Other types of glazing with heat-absorption and glare-reduction properties also are available. *Heat-absorbing glass* is made especially for the purpose. Glass may also be tinted or coated to achieve the same objective. In addition, *transparent mirror glass* offers not only heat absorption but also privacy. It acts like a mirror on one side but is transparent to a viewer on the opposite side. Installed on a building facade with the reflective side outward, the glass reflects much of the sunlight that strikes it.

Double glazing and even triple glazing is used to improve the thermal transmission character of windows, primarily in cold-weather climates where heat loss through single-thickness glazing is a major factor. Glazing units of this type may be produced in one of three forms. Small units may be produced by fusing two sheets of glass together at their edges, incorporating a very small air layer between them—typically about 1/4 in or so. Larger units, or heavier sheets of glass may be joined by using a special metal edge frame, permanently bonded to the glass. To create a larger air space—sometimes to allow the incorporation of a light-control device within the air space—it may be possible to simply build a conventional frame and permanently install two layers of glazing.

Use of special glazing (especially with tinted or reflective coated glazing) is a combined architectural-design and HVAC-design decision, HVAC considerations being predominant in very cold or very hot climates. Choice of the glazing must be made in conjunction with other aspects of the window design, including considerations for weatherstripping, size and number of windows, use of shading devices and sun orientation. HVAC problems usually cannot be effectively solved simply by choice of the correct glazing.

Structural Considerations

Windows generally constitute penetrations or voids in the structure of the exterior walls. In addition to losing a portion of its structural mass, a wall must usually provide support for a window. The effects of these considerations on the wall construction must be studied both in terms of the needs for the windows and required structural functions of the walls.

The need for lintels and headers to span across the top of the structural opening in the wall is discussed in Sec. 15.3. The form of such spanning members depends on the materials and details of the wall construction; several examples are shown in Fig. 15.11 and 15.12. A critical consideration is sizing the width of the opening, which determines the span for the lintels and headers, very large beam sections being required for large window openings. In some cases, large expanses of windows may be divided, into smaller units, with some vertical framing members of the window being utilized as structural columns to reduce the span of the lintels or headers over the windows.

Support for windows is often largely a matter of the development of attachments that are appropriate to the details of the wall construction and the form and materials of the window frames. Most windows are not exceptionally heavy, so support for gravity loads is not a major concern. However, where wind loads are high on the exterior walls, a major structural concern for both the window and the supporting wall structure may derive from the wind pressure on the window surface area. Thicknesses of window glazing and the sizes of window framing elements (mullions, etc.) may be determined by wind pressures. For the wall, provision must be made for the window support and for any necessary reinforcement of the wall structure to compensate for the void created by the opening for the window. Finally, the attachments of the window frame to the wall structure must be capable of transferring the total wind force on the window.

In order to provide for some adjustment in

accurately positioning of the windows, the opening in the wall construction—called the *rough opening*—is made slightly larger than the outer dimensions of the window framing. This requires some consideration in the development of support and attachment details, but is routinely accommodated by the use of shims and wedges of a type appropriate to the construction materials of the wall and window frame. Specific sizes for rough openings are often specified by manufacturers for their products.

Where exterior walls have major structural functions as bearing or shear walls, the planning of window openings must be carefully developed. This calls for close cooperation between the structural and architectural designers to assure the satisfaction of concerns for structural performance, aesthetics of the building exterior and the various practical requirements for the windows in terms of light, vision and ventilation.

15.6 Doors in Exterior Walls

Doors in exterior walls present many of the same load and support conditions as those discussed in Sec. 15.5. Wall openings must be structurally developed, weatherproofing and thermal transmission must be dealt with and support for the doors must be accommodated to permit the particular operation (swinging, sliding or folding) of the doors.

When doors contain glazing, their design becomes even more similar to that for windows. In storefronts, doors may be virtually all glass, truly constituting windows as well as functional doors.

Doors designed primarily for pedestrian entrance and exit functions must usually satisfy building-code requirements for fire safety, involving considerations of a number of issues that may include the following:

1. Width of the door opening usually a minimum of 3 ft for a single door and 5 ft for a double door (two swinging doors in a single opening).
2. Number of doors required to provide for the exiting (egress) of the building occupants. (See discussion in Sec. 6.5.)
3. Location of the doors for safe exiting: Separated if multiple doors are required and accessible to the building occupants, for example.
4. Direction of swing in the direction of the exiting occupants.
5. Operation allowing freedom and ease of unlatching by panic-stricken occupants (generating the need for so-called *panic hardware* in some cases).
6. Fire resistance of the door corresponds to any required fire resistance of the wall in which it is placed.

Exterior doors must ordinarily be provided with various features that relate to security or the prevention of undesirable entrance. Locking devices are a prime need. Special situations may require alarms, one-way vision, special lighting, voice communication or signaling devices.

If a door must accommodate the entrance of handicapped persons (in wheel chairs), there are special requirements for the door opening width, latching hardware and any details at the floor level, such as high sills, recessed mats or any feature that prevents ease of rolling of a wheel chair. Approach to such doors—from both the inside and outside—must also accommodate the wheel chair user; which may prompt concern for ramps and floor surfacing materials.

The most common door operation involves the use of hinges on one vertical edge of the door and opening and closing by horizontal swinging. Complete operation and control of such a door may involve the following items:

1. A stop, usually built into the door frame, that defines the closed position of the door.
2. A latch that secures the door in the closed position.
3. A lock to prevent opening of the door at determined times.
4. Swing-restraining devices that prevent opening beyond a certain arc of swing.

5. A hold-open device that holds the door in an opened position.

6. An automatic opener and/or closer device.

For special situations there may be many additional features or details required for a particular door. Most building construction drawings contain a door schedule, consisting of a table showing all of the features for each type of door in the building. Each type of door is given a code symbol which is placed near the door on the building plans.

Doors that accommodate vehicular traffic must be designed for the size of the vehicles. These may be modest, as for a typical single-car garage, or immense, as for the hangar doors of aircraft maintenance hangars. Operation of such doors may be by swinging in the manner of pedestrian doors, but due to their size, it is often more feasible to use horizontal sliding or folding or vertical sliding, rolling or folding as the form of movement.

SECTIONS 15.5 AND 15.6

References

Sweet's Architectural File, Division 7 (doors, windows, sealants, weatherproofing), McGraw-Hill, New York (updated yearly).

D. Watson, *Construction Materials and Practices*, 3rd ed., McGraw-Hill, New York, 1986.

Also see general references for this chapter.

Words and Terms

Glazing
Header
Jamb
Lintel
Mullion
Muntin
Operation (of door or window)
Panic hardware
Rough opening
Sash
Sill
Stiles
Weatherstripping

Significant Relations, Functions and Issues

Window type related to form, function or method of operation: fixed, double and single hung, horizontally sliding, casement, pivoting, awning, jalousie, storm.

Window components: jamb, sill, stool, casing, sash, stops, glazing, operating hardware, weatherstripping.

Choice of glazing for appearance, wind pressure and thermal effects.

Effects of door and window openings on wall structure.

Dimensional considerations: rough opening, modular (standard) window and door sizes; door widths required for codes.

15.7. SYSTEMS-DESIGN APPROACH TO BUILDING ENCLOSURE

A building may be enclosed with wall and roof systems or with a single-enclosure system in which the roof is indistinguishable from the wall. Roof and wall systems may be completely independent systems. When load-bearing walls are used, however, the roof depends on the walls for support.

Because roof, walls and single-enclosure systems determine the exterior appearance of the building, they are important components of the architectural system. They also, however, are important components of the structural system if they are to be constructed as load-bearing. In nonload-bearing construction, roof, wall and single-enclosure systems act as loads on and transmitters of loads to the structural system. Design of all these systems, therefore, are interrelated.

Rainwater runoff from roofs must generally be collected and disposed of in a sewerage system. Gutters, drains and drain pipes are needed for these purposes. Hence, roofs affect design of plumbing systems. In turn, plumbing systems may affect roof design, because the roof may be required to support storage tanks for water supply.

Roofs and walls also affect and are affected by the HVAC system. The most important influence exerted by HVAC is the requirement that roofs and walls be designed to impose as small a load as practicable on the HVAC system.

Because of the interrelationship between the enclosure systems and other building systems, systems design can be useful in coordinating design of the various systems to produce an optimum overall design.

Data Collection and Problem Formulation

Basic information for design of the building enclosure comes from examination of the owner's program of requirements and from floor plans. This information determines the shape of the building exterior in plan and also often in elevation. Decisions as to the number of stories in the building and story height provide additional information for wall design.

The architect should work closely with the owner or the owner's representatives to ascertain preferences affecting aesthetic decisions, such as choice of wall configurations, roof shapes, ornamentation, texture, types and sizes of windows and doors, and exterior colors. The architect should also collaborate with the structural engineer, especially in view of the influence on building design of the choice between load-bearing and nonload-bearing construction.

Further information for design of the building enclosure should be obtained from the building code and zoning ordinances. Fire-resistance requirements in the building code have a strong influence on enclosure design. Zoning requirements limiting building heights may affect decisions on building size and shape.

The results of the cooperative efforts of the building team should be displayed on architectural and structural plans, elevations and cross sections and further described in outline specifications. These documents should be submitted to all of the members of the building team for review. In the later stages of design, all drawings concerned with building enclosure, windows and exterior doors should be studied to insure compatibility with other systems.

The goal of design of the building enclosure may be stated as: to design, as components of an optimum building, wall and roof systems (or a single-enclosure system) that will separate the interior environment from the outdoor environment and exclude unwanted visitors, yet allow authorized persons to move to and from the building interior.

Objectives

The objectives of enclosure design may be classified as prime or optional. Prime objectives are those likely to be adopted for nearly every building. Optional objectives are those that depend on the functions of the building and the preferences of the owner.

Prime Objectives. A major objective of a building enclosure is to exclude rain, snow, wind and outside dirt from the building interior. It is also important that air leakage into and from the interior be controlled, to prevent drafts that would make occupants uncomfortable and to avoid excessive load on the HVAC system.

Another prime objective of a building enclosure is to prevent entry of intruders. Nevertheless, means for attaining this objective must be compatible with those for attaining the essential objective of permitting movement to and from the building interior of authorized persons and goods.

Optional Objectives. In many cases, the optional objectives are as important as the prime objectives. For example, the owner may choose as a major objective low initial cost. In that case, durability, low maintenance, aesthetics and other objectives may be given low priority or no consideration. The building enclosure would have to be the lowest-cost type that would meet the prime objectives.

On the other hand, the owner may adopt low initial cost as one objective but also require the building enclosure to be durable. In addition, little or no maintenance over a long period of time (20 or more years) may be chosen as an objective. In this case, design would involve trade-offs, because usually durable and maintenance-free construction costs more than less durable, troublesome construction.

Another important objective may be a good-looking enclosure, whether viewed from outside or inside the building. A related objective may be that the enclosure have eye-catching or unusual features, which may be of value to the owner for promotional purposes.

Another related objective is the installation and aesthetic treatment of windows. These have a strong influence on the appearance of the enclosure. Windows, however, can also meet objectives that may be established for daylighting and ventilating the interior, for providing a view of the outdoors from inside or of the interior from outside and for emergency escape routes.

The owner, though, may also choose as an objective privacy in all or part of the interior; that is, the enclosure may be required to conceal activities within the building. On the other hand, the owner may require opaque walls against which materials may be stored or furniture placed. In either case, windows would have to be selected and located to meet those objectives, or even eliminated.

Another important objective may be high resistance to passage of heat, for energy conservation and to keep the load on the HVAC system small. Similarly, an objective may also be to exclude outside noise or to contain sounds produced within the building.

Constraints

Significant constraints may be imposed on enclosure design by requirements in the local building code. These requirements may establish minimum thicknesses or maximum heights for some types of walls. The requirements may also influence thermal characteristics of the enclosure by establishing limits on energy consumption for HVAC. In addition, the code requirements may restrict choice of materials for walls and roofs and positioning of windows, to provide fire resistance and prevent spread of fire through window openings.

Other significant restraints may be imposed by zoning ordinances. Zoning restrictions on height or total floor area may influence the shape of the enclosure. Also, wall thickness may be limited by restrictions on the area of the lot that the building may occupy and the need to maximize the usable floor area in the building. With a limited perimeter, the thicker the exterior walls, the less the floor area.

Additional constraints on enclosure design may be imposed by the foundation and structural systems. These systems may require that weight of the enclosure be kept small. The smaller the weight of walls and roof, the smaller the load on the structural system and foundations and the lower their costs. Where soils are weak, a light enclosure is essential.

Synthesis and Analysis

In the earliest stages of design, schematic drawings consist essentially of floor plans and proposed elevations. Hence, selection of major components of the building enclosure, including building height and shape and fenestration (window characteristics), are among the earliest decisions made by the owner and the architect. As decisions are made as to whether the enclosure will be load-bearing or nonload-bearing, whether the building will be one story or multistory and whether the enclosure will consist of wall and roof systems or a single-enclosure system, the information is used by the members of the building team to prepare revised or new schematic drawings.

In the design development stage, enclosure dimensions are accurately established. Materials to be used for the enclosure are selected. Locations and sizes of windows and exterior doors are set. With this information, the structural engineer can prepare preliminary drawings for the foundations supporting the enclosure and for the enclosure framing. Also, the mechanical or HVAC engineers can make a preliminary estimate of the load on the HVAC system.

In the final design stages, details of the enclosure are prepared. As the information becomes available, the structural engineer can prepare details of the framing supporting the enclosure. Also, HVAC and plumbing details can be completed.

Value Analysis and Appraisal

Depending on the owner's objectives, value analysis may require only a straightforward cost comparison of alternative types of enclosure; however, value analysis can become very difficult if the owner's objectives require evaluation of such enclosure characteristics as beautiful appearance, durability and clear, unobstructed view of the outdoors. What is the money value of such features? Decisions may be determined not by equivalence but by whether a less costly component or method of construction will be acceptable to the owner.

Furthermore, changes made in the enclosure design to reduce costs may necessitate changes in other systems that will increase the cost of foundations, structural framing, HVAC and electric lighting. Also, cost-reducing changes may increase repair and maintenance costs of the enclosure. Therefore, value analysis should be based on the life-cycle cost of the whole building, not on the cost of enclosure components or of the total enclosure.

Construction costs, including the money value of construction time, should be an important consideration in appraising enclosure alternatives. For example, a load-bearing exterior wall may appear to be an economical choice for a multistory building, but if skeleton framing is selected for the interior of the building and that framing can be erected much faster than the exterior wall, construction cost for the whole building will be increased because of the choice of wall. The framing for the floors cannot be placed until the more slowly constructed wall can be built high enough to support them. As a result, the framing crew is able to work only intermittently. The inefficient use of manpower increases construction time and cost.

As another example, prefabricated wall panels may cost more than materials assembled on the site, but the panels can be erected faster. Hence, the total cost of the wall in place may be less than the cost when the materials are assembled on the site. When the panels become large and heavy enough that their installation requires mechanical aids, such as cranes, often the larger the panels, the lower the construction cost.

GENERAL REFERENCES AND SOURCES FOR ADDITIONAL STUDY

These are books that deal comprehensively with several topics covered in this chapter. Topic-specific references relating to individual chapter sections are listed at the ends of the sections.

F. Merritt, *Building Design and Construction Handbook*, 4th ed., McGraw-Hill, New York, 1982.

C. Ramsey and H. Sleeper, *Architectural Graphic Standards*, 8th ed., Wiley, New York, 1988.

D. Watson, *Construction Materials and Practices*, 3rd ed., McGraw-Hill, New York, 1986.

E. Allen, *Fundamentals of Building Construction: Materials and Methods*, Wiley, New York, 1985.

A. Dietz, *Dwelling House Construction*, 4th ed., MIT Press, Cambridge, MA, 1974.

H. Olin et al., *Construction Principles, Materials, and Methods*, 5th ed., The Institute of Financial Education, Chicago, 1983.

H. Sands, *Wall Systems: Analysis by Detail*, McGraw-Hill, New York, 1986.

Sweet's Architectural File, McGraw-Hill, New York (issued annually).

EXERCISES

The following questions and problems are provided for the review of the individual sections of the chapter.

Section 15.1

1. What are the purposes of a roof?
2. List the essential physical characteristics of a roof.
3. What is roofing?
4. Where should thermal insulation be placed for a roof over an unoccupied attic?
5. What are eaves?
6. What is a dormer?
7. What are the advantages of pitched and hipped roofs over shed roofs?
8. What are the advantages of gambrel and mansard roofs over pitched and hipped roofs?
9. What is a monitor?
10. Why are the characteristics of a roof deck important to performance of roofing?

Section 15.2.

11. How does a ballasted roof differ from an ordinary built-up roof?
12. On which side of roof insulation should a vapor barrier be placed?
13. What is the difference in the physical characteristics of tar and asphalt that is important for roofing?

14. Describe application of built-up roofing on a concrete deck.
15. Why are shingles or tiles overlapped when laid on a roof deck?
16. What is the purpose of flashing on a roof?
17. What is the purpose of expansion joints in a roof?
18. What precautions must be taken in installation of metal roofing that is not necessary with roofing made with other materials?

Sections 15.3. and 15.4

19. What physical characteristics are desirable for an exterior wall?
20. What are the usual ways in which an exterior wall is constructed?
21. What are the advantages of prefabricated walls?
22. What does a load-bearing wall do that a curtain wall cannot?
23. Why are expansion joints used in exterior walls?
24. How does balloon framing differ primarily from platform framing?
25. Why are brick and concrete block made with hollow cores?
26. What determines the difference in dimensions between actual and nominal dimensions of a brick or concrete block?
27. What is random ashlar?
28. Describe a masonry cavity wall.
29. What is the purpose of weep holes and where are they placed?
30. What are the main components of mortar for unit masonry?
31. What are the advantages of corrugated siding over flat siding?

32. What is a sandwich panel?
33. What are the advantages of the panel method of installing a curtain wall over the other methods?
34. What is the purpose of lintels?
35. Why is flashing used in exterior walls?

Section 15.5

36. Give five reasons why windows are used?
37. When should translucent glazing be used?
38. What is the purpose of weatherstripping?
39. Where should weatherstripping be placed?
40. How do casement windows differ from vertically pivoted windows?
41. When must metal windows be used?
42. Where are jambs used in windows?
43. Where are stiles used in windows?
44. What is the purpose of dimensional coordination of building materials?
45. What kind of support is desirable for glazing in sash?
46. Why might plate glass be used instead of clear window glass, which costs less?

Section 15.6

47. Describe the ways in which accommodation of doors in exterior walls is similar to the problem of accommodation of windows.
48. Describe the ways in which the planning and details of exterior doors is affected by building-code requirements for fire.
49. What features of exterior doors are affected by requirements to facilitate entry and exit by handicapped persons?

Chapter 16

Systems for Interior Construction

The interior of a building is usually divided into many spaces that are enclosed so that they are separated from each other. There are numerous reasons for this.

One very important reason is that often the land area a building is permitted to occupy is less than the floor area required for the activities to be performed in the building. To obtain sufficient floor area in such cases, it is necessary to build a multistory building; that is, to divide the building into stories by placing one floor above the other. The objective is to make the sum of the areas of the floors at least equal to the required area. As a result, in multistory buildings, an upper floor not only serves as a deck for performance of activities but also separates the activities being carried out on one level from those being performed at a lower level. The top story of the building is capped with a roof.

Usually, a roof or a floor is supported on beams. Also, ducts, pipes and electric wiring may be placed below the floor or roof. The beams, ducts and other items often are not a pretty sight. For aesthetic reasons, therefore, a ceiling may be placed below them for concealment. The ceiling is generally supported by the beams or the floor or roof. Because of this relationship of a ceiling to the floor above, a floor and associated ceiling may be considered as a subsystem of the interior-enclosure system.

The space in each story of a building is usually divided into several spaces, or rooms, principally for reasons of privacy. Examples of such rooms include bathrooms, bedrooms, offices and conference rooms. Some spaces, however, may be formed for aesthetic reasons. The objectives may be to conceal certain activities, such as cooking in kitchens; or to cover stored materials, such as storage in closets; or to shelter equipment, such as that required for HVAC, elevators and electrical transformers, switchboards and panelboards.

Some spaces must be created so as to meet building-code requirements. These requirements call for use of fire walls to prevent spread of fire (see Sec. 6.3). Other spaces may be enclosed for noise, heat, light or odor control. Still other spaces, such as corridors, stairwells and elevator shafts, may be formed to accommodate horizontal and vertical circulation.

All the required spaces in each story may be produced by appropriate use of walls and partitions. (For the purposes of this chapter, interior load-bearing walls are referred to as walls and interior nonload-bearing walls are called partitions.)

Openings must be provided in the walls and partitions to permit passage of people and goods between the spaces of each story. Doors are usually installed in these openings, to control access to the spaces, prevent spread of fire or preserve the capabilities of the walls and partitions for cutting off noise, heat, light or odors.

Thus, a system for enclosing an interior space may consist of a floor at the bottom of the space, a ceiling or the underside of a floor or

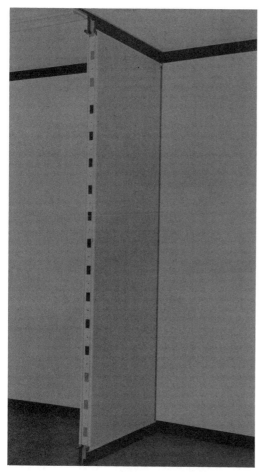

Fig. 16.1. Prefabricated modular partition unit, supported at floor and ceiling. (Courtesy Johns-Manville Corp.)

roof at the top, and walls or partitions, with one or more doors, along the perimeter.

Each subsystem of this system may usually be designed independently of the other subsystems and other building systems. While there is some interaction between the various systems, it is not usually of major import in optimization of a building. Consequently, in design of an optimum building, it is generally sufficiently accurate to optimize each of the interior-enclosure subsystems separately.

16.1. INTERIOR WALLS AND PARTITIONS

The space dividers in each story of a building may be vertical bearing walls or partitions. Bearing walls may be designed and built in much the same way as the load-bearing exterior walls described in Sec. 15.2. For exterior walls, however, the outside facing must be weather resistant, but the inside facing need not be. For interior walls, neither facing need be weather resistant. Cost and appearance are usually the major factors in selection of facing materials for interiors, but fire resistance, durability and low maintenance must also be considered, depending on the type of occupancy and the characteristics of the interior environment of the building.

Partitions may extend the full distance from floor to ceiling (see Fig. 16.1) or only part of this distance. Low partitions are often used in large offices and in some areas of schools to provide a limited degree of privacy and noise control.

Partitions may be permanently fixed in place, folding or sliding, or temporary. The last type is installed so that it may be easily moved to another location when desired.

Because the main function of partitions is to separate space, any of a wide range of materials and types of construction may be used (see Sec. 2.5). Partitions may be opaque, translucent or transparent. They may be louvered or hollow or solid. Also, they may serve several purposes. For example, a cabinet or closet may act as a space divider. As another example, a partition may be housing for piping and electrical conduit. This type of partition is often used in hospitals and laboratories.

The type of construction selected for an interior wall or partition should take into account requirements for movability, fire resistance, noise control and light transmission.

Bearing walls should not be used if it is likely that a space divider will have to be moved in the future. Changes in position of a bearing wall will be very expensive, because of the cost of replacement of the support capacity being removed. Folding, sliding or temporary partitions, in contrast, are intended to be movable. For this purpose, they should be of light construction.

The type of construction required for a wall or partition may be dictated by building-code requirements for fire resistance. Codes may specify that the types of facings used for space dividers in certain locations, such as along exit passageways, limit or prevent spread of flame when a fire occurs. Codes also may require dividers between hazardous and other occupancies, for instance, between a garage and a residence, to have a specific minimum fire rating. In addition, codes may limit the amount of floor area in any story to a specific maximum enclosed within fire walls (see Sec. 6.3). Partitions or walls serving as fire walls must have the minimum fire rating specified in the code.

The type of construction selected for a wall or partition may also depend on acoustic requirements. Thin construction may vibrate like a sounding board when struck by sound waves. Such construction should be avoided if sound control is desired. In some cases, massive or double walls may have to be installed where severe restrictions on sound transmission have to be imposed. In some cases also, sound-absorbing materials may have to be applied to the face of a wall or partition to provide desired acoustical treatment for sound control in an enclosed space.

Materials and type of construction of partitions may, in addition, depend on light-transmission requirements. If transparency or translucency is desired, a partition may be partly or completely constructed of glass, glass block or plastics.

Structural Requirements

Bearing walls should be capable of supporting not only their own weight but also loads from walls, floors and roofs above. The walls, in turn, should rest on sufficiently strong supports that will not deflect excessively.

Folding partitions may be supported only at the top. Sliding partitions may be supported only at the bottom. In both cases, guides may be used at top and bottom to control movement. Folding and sliding partitions are, in a sense, large doors, and they should be framed and supported in much the same way as large doors (see Sec. 16.3).

Fixed and temporary partitions should always be supported at the bottom. Story-high partitions should also be supported at the top to prevent them from falling over. The bottom

support, which may be a floor or beam, should be capable of carrying the weight of the partition and all vertical loads imposed on it without excessive deflection. Also, partitions should be constructed to be stable laterally between lateral supports.

Some of the finishes commonly used for interior walls and partitions are described in Sec. 16.5.

References

Sweet's Architectural File, Division 10, McGraw-Hill, New York (updated annually).
Also see general references for this chapter.

Words and Terms

Fire wall
Partition
Space divider

Significant Relations, Functions and Issues

Character and form of partitioning: fixed or movable; opaque, translucent or transparent; fire-resistance rating; partial height or full height (floor-to-ceiling); integration with floor and ceiling systems.

Functional requirements for partitions; demountability or movability; support for doors, shelves, etc.; incorporation of wiring.

Requirements for acoustic performance; desired level of privacy of individual spaces.

Use of furniture and equipment for space division.

16.2. ORDINARY DOORS

At least one opening must be provided in an enclosure of a usable interior space in a building to provide access between that space and outdoor spaces or other interior spaces. Usually, a door is installed in the opening to control access and sound and light transmission and to prevent spread of fire and smoke between spaces. A door is, in effect, a part of a partition or wall that can be moved to permit or block passage of persons or goods, or both, past the space divider.

Openings and doors should be sized to provide access between spaces under both normal and emergency conditions. Doors should be constructed so that traffic will flow smoothly through the openings. For this purpose, building codes usually place maximum and minimum limits on door sizes. Typical limits are as follows:

A single leaf in an exit door should be at least 28 in. wide, but not more than 48 in. wide. (Exit doors should be the swinging type, installed to swing in the direction of travel to the exit.)

Minimum opening width of door should be at least:

36 in. for single corridor or exit doors

32 in. for each door of a pair of corridor or exit doors with central mullion

48 in. for a pair of doors with no central mullion

32 in. for doors to all occupiable and habitable rooms

44 in. for doors to rooms used by bedridden patients and for single doors used by patients in such buildings as hospitals, sanitariums and nursing homes

32 in. for toilet-room doors

When the door is open, jambs, stops and door thickness that project into the nominal opening width should not decrease it by more than 3 in. for each 22-in. unit of opening width. The floor on both sides of an exit door should be level, preferably for a distance of at least 4 ft.

Also, for exit and corridor doors, nominal opening height should be at least 6 ft 8 in. Jambs, stops, sills and closures, however, usually are permitted to project into the opening so long as the clear opening is not reduced to less than 6 ft 6 in. in height.

Table 6.5 in Sec. 6.5 lists the maximum capacity, persons per unit of width, that a building code may permit for an exit door, depending on the type of occupancy. Table 6.3 gives the maximum size of opening that may be permitted in a fire wall.

Most doors are available from door manufacturers in standard sizes or can be assembled from standard stock parts. When they do not have to be engineered for a specific installation, such doors are considered ordinary doors. Other types of doors are considered special-purpose doors. Ordinary doors are discussed in the rest of this section, whereas special-purpose doors are discussed in Sec. 16.3.

Ordinary doors may be classified as interior doors or entrance doors. The latter are installed in exterior walls and must satisfy more requirements than interior doors because entrance doors must exclude weather and have heavier traffic flows (see Sec. 15.6 and p. 627).

Some ordinary doors must meet fire-resistance requirements. Such doors may be classified as fire or smokestop doors.

Door Materials

Doors usually are made of wood, glass, aluminum, steel or bronze and often with cores to improve fire or heat resistance or to limit sound transmission. Doors may be solid or hollow. They may be completely opaque or they may be built wholly or partly of plastics or tempered glass for light transmission. Vision panels in doors are often referred to as *lights*.

Components of doors are named after similar parts in windows. The vertical sides of a door are called *stiles*, and the horizontal members are called *rails* (see Fig. 16.2a).

As for windows, doors also require a frame, and the frame parts have names like those in window frames. The sides of a frame are called *jambs*; the top, a *header* and the bottom a *sill* (see Fig. 16.2). Projections on the frame against which a door closes are called *stops*. The stop at the lock side of a swinging door is called a *strike*. Wood members that frame an opening for a door are often referred to as *bucks*. Lintels or headers are required over doors to support walls above the doors.

Wood Doors. Doors made of wood generally cost less than doors made of other materials. An inexpensive door, for example, can be built simply by nailing boards together. When the boards are vertical and held together by nails to a few horizontal boards, the door is called a *batten door*. Panels set in a frame or with flush construction, however, usually offer better appearance and greater durability, although at higher cost.

Paneled doors are an assembly of solid wood or plywood panels supported by stiles and rails

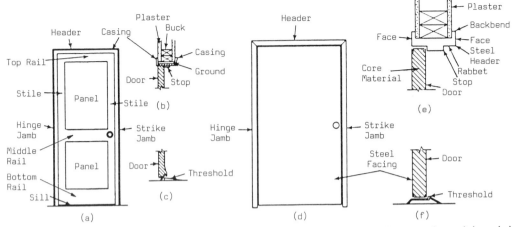

Fig. 16.2. Components of doors. (*a*) Parts of a wood door and frame. (*b*) Detail at top of wood door. (*c*) Detail at sill of wood entrance door. (*d*) Parts of a steel flush door and frame. (*e*) Detail at top of steel door. (*f*) Detail at sill of steel entrance door.

(see Fig. 16.2*a*). (Some of the panels may be replaced by tempered glass or plastics.) Joints between panels and their supports permit dimensional changes with variations in atmospheric moisture content.

A paneled door is called solid when rails and stiles are made of a single piece of wood. When doors are made of hardwood or better-quality woods, the doors usually are veneered, with rails and stiles consisting of cores of softwood sandwiched between the veneer facings.

Flush doors may also be solid or veneered. The veneered type may have a core of softwood between flat faces of hardwood veneer. Or the door may be of the hollow-core type, with plywood surfaces on a supporting grid with edges of solid wood boards.

Metal Doors. These may be constructed in one of the following ways:

1. Cast as a single unit or as a separate frame and panel pieces. Relatively expensive, such doors are used principally for monumental structures.
2. Metal frame (see Fig. 16.2*d*) covered with cold-formed steel. Such hollow doors may be of flush or panel design.
3. Wood or other type of insulating core clad with cold-formed metal. This type of door may also be flush or panel design. *Kala-*

mein doors, used as fire doors, have metal-covered wood stiles and rails and insulated panels with cold-formed steel facings. They are available as the swinging type only. Flush doors may have polyurethane or polystyrene foam, a honeycomb core or other insulation sandwiched between the metal facings and acting also as a lightweight support for the facings.

Fire and Smokestop Doors

Openings in walls and partitions that building codes require to have a minimum fire-resistance rating are also required to have protection with a corresponding rating. Doors for such openings are usually referred to as fire doors. Typical requirements are listed in Table 16.1.

Fire ratings, hr, of doors are obtained by testing complete assemblies, including frames and hardware. Consequently, only assemblies that have been labeled or listed by Underwriter's Laboratories, Inc. (UL) or approved by Factory Mutual Research Corp. (FM) should be specified for fire doors. Some products carry a self-certified label, provided by the manufacturer, but such labels are not accepted by the National Fire Protection Association and may not be approved by some building-code officials.

All fire doors should be the swinging type. They should be self closing or else they should close automatically when a fire occurs, and they should also be self latching so that they remain closed.

Exit doors for places of assembly for more than 100 persons should be equipped with panic hardware. This is a horizontal bar capable of releasing the door latch when pressure of 15 lb or less is applied to the bar in the direction of exit.

For some critical locations, building codes require smokestop doors to bar passage of smoke. These doors need not be fire rated. They may be metal, metal covered or $1\frac{3}{4}$ in. of solid wood ($1\frac{3}{8}$ in. in building less than three-stories high), with 600-sq-in. or larger clear, wire-glass panels in each door. Such doors should close openings completely. They should have only enough clearances for proper operation.

Combustible materials, such as flammable carpeting, should not be allowed to pass under a fire door.

Entrance Doors

Doors in exterior walls should be constructed to exclude rain and snow, to withstand solar heat, to limit air leakage and to operate when the ground outside is covered with snow. Ordinarily, however, entrance doors, even when fully closed, may not be capable of completely excluding water or preventing air leakage, be-cause of clearances provided. These clearances are necessary to allow easy operation and thermal expansion and contraction of doors. Measures can be taken, though, to reduce vulnerability of doors to water and air penetration.

It is helpful, for example, to shelter exterior doors by setting them back from the building face or by placing a canopy, marquee or balcony over them. Such shelters will also be helpful in reducing collection of snow and ice at door thresholds. Also, weatherstripping around doors decreases penetration of water and air. As for windows, flashing should be incorporated in walls above entrance doors.

An entrance vestibule on the inner side of an entrance door is advantageous for several reasons. For one thing, it serves as a weather barrier, because an interior door leading to the vestibule acts as a second line of defense against wind, rain and snow. In addition, a vestibule is desirable to combat the stack effect in tall buildings. The stack effect is the difference in air pressure between the building interior and the outdoors that exists at entrances because of air movements through stairwells and shafts. When a building is heated in cold weather, warm air rises in the vertical passageways. In warm weather when the building is cooled, cool air flows down the passageways. This stack effect may create large air flows through entrance doors, making door operation

Table 16.1. Typical Fire Ratings Required for Doors[a]

Door Use	Rating, hr
Doors in fire barriers with 3- or 4-hr rating	3[b]
Doors in fire barriers with 2- or $1\frac{1}{2}$-hr rating	$1\frac{1}{2}$
Doors in fire barriers with 1-hr rating	$\frac{3}{4}$
Exit doors	$1\frac{1}{2}$[c]
Doors to stairs and exit passageways	$\frac{3}{4}$
Doors in corridor walls with 1-hr rating	$\frac{3}{4}$
Other corridor doors	0[d]

[a]Based on New York City Building Code. All fire doors are required to be self-closing, swinging type and normally kept closed.
[b]Some codes require two $1\frac{1}{2}$-hr doors, with one door installed on each face of a fire barrier.
[c]No rating is required for street-floor exit doors with a separation from other construction on the outside of at least 15 ft, or for exit doors of one- or two-family houses.
[d]Although no rating is required, such doors should be noncombustible or $1\frac{3}{4}$in. solid-core wood doors. Some building codes do not require these doors to be self closing in such buildings as hospitals, sanitariums and nursing homes.

difficult and increasing the load on the HVAC system.

A vestibule with interior and exterior doors can serve as a barrier to such air flows. When traffic flow is so large, however, that interior and exterior vestibule doors may be open simultaneously, the effectiveness of the vestibule in preventing air flow is decreased. The effectiveness of the vestibule can be improved in such cases by venting it to the outdoors and providing compensating heating or cooling in the vestibule.

The stack effect can also be reduced by use of revolving doors at entrances. The outer edges of such doors maintain continuous contact with their enclosures during rotation, barring passage of air.

When swinging dorrs are used and the stack effect makes operation difficult, the doors may be hung on balanced pivots or may be equipped with automatic devices to make operation easier. An alternative may be replacement of the swinging doors with automatic, horizontal sliding doors.

Swinging Doors

Doors that open or close by rotation about hinges along one vertical edge are called swinging doors. Single-acting doors can swing 90° or more in only one direction. Double-acting doors can swing 90° or more in each of two directions.

For specification of hardware for a door, the direction of swing, or hand of the door, relative to the key, or locking, side of the door must be given. As indicated in Fig. 16.3a, a left-hand swing door is hinged at the left jamb. A right-hand swing door is hinged at the right jamb (see Fig. 16.3b). In both cases, the door opens inward, away from the holder of the key. Reverse swing doors (see Fig. 16.3c and d) open toward the key holder. The double-swing frame in Fig. 16.3e incorporates a left-hand reverse and a right-hand reverse door, with an astragal, or molding, on the edge of one door to serve as a strike for the other door.

The part of a door frame to which a door is hinged is called a hinge jamb. The opposite jamb is called the strike jamb (see Fig. 16.2). The header and hinge jamb are equipped with a stop, and the strike jamb, with a strike against which the door closes, to stop drafts and prevent passage of light (see Fig. 16.2b and e).

Entrance doors are provided at the base with a sill (see Fig. 16.2e and f), which covers the junction between outside construction and the finished floor on the inside of the door. The top of the sill should be sloped to drain water away from the door. The sill may have a raised section so that water dripping from the door will fall on the slope. The raised section may be integral with the sill or a separate threshold. Weatherstripping may be attached to the bottom and along other edges of the door or on stops and strike.

Swinging doors are available in thicknesses of $1\frac{3}{8}$ and $1\frac{3}{4}$ in., with widths of 30, 36 and 42 in., and opening heights of 6 ft 8 in. and 7 ft. They may be obtained in a package including

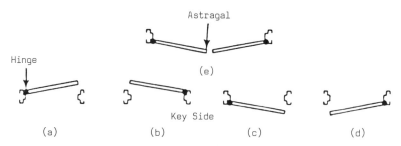

Fig. 16.3. Hand (swing) of doors - plan views. Reference orientation is with respect to a person approaching from the bottom side of the figure as viewed in plan here. (a) Left-hand swing. (b) Right-hand swing. (c) Left-hand reverse swing (toward approaching user). (d) Right-hand reverse swing. (e) Double-swing, reverse, with no center post.

a single door or a pair of doors, door frame and all required hardware. Nonstandard sizes are available on special order.

Hinges. The hinge about which a swinging door rotates consists of two metal plates, or leaves, with a vertical barrel, or knuckle, along one edge into which a pin is placed to join the leaves (see Fig. 16.4a to d). In a strap hinge, both leaves are elongated (see Fig. 16.4a). In a tee hinge, one leaf is elongated and the other is a butt, or short leaf (see Fig. 16.4b). Both types of hinge are suitable for mounting on the surface of a door.

To improve the appearance of a door, however it is often desirable to mount hinges on the edge of the door and the face of the jamb, so that the leaves are concealed when the door is closed. In such cases, both leaves are shortened into butts, with length less than the thickness of the door. Figure 16.4c shows a left-hand butt and Fig. 16.4d a right-hand butt.

Some butts, however, are not handed and may be mounted on either a right-hand or a left-hand door. For such butts, the number of bearing units actually supporting the vertical load of the door equals only one-half the bearing units available. Hence, with a two-bearing butt, only one bearing carries the vertical load, and with a four-bearing butt, only two bearings carry the load. The number and strength of hinges and bearings required for a door depends on its weight, frequency of use and need for maintaining continued floating, silent operation.

Hinges are often placed with the top of the upper hinge 5 in. below the rabbet of the head jamb and the bottom of the lower hinge 10 in. above the finished floor line. If a third hinge is required, it is placed midway between the two other hinges.

The pin is an important element of the hinge and should be selected to meet service requirements. Pins may be either fast or loose. A fast, or tight, pin is permanently set in the barrel of

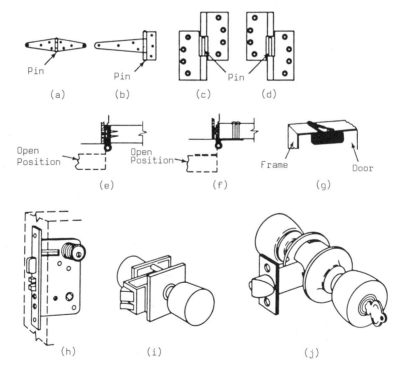

Fig. 16.4. Door hardware. (*a*) Strap hinge. (*b*) Tee hinge. (*c*) Left-hand loose-joint hinge. (*d*) Right-hand loose-joint hinge. (*e*) Full-mortise butt hinge for wood door with steel jamb. (*f*) Half-surface hinge for steel door with steel jamb. (*g*) Door closer. (*h*) Mortise lock. (*i*) Unit lock. (*j*) Cylindrical lock.

the hinge at the time of manufacture. Such pins are often used in hospital- or asylum-type hinges. Ordinary loose pins can be pulled out of the hinge barrel at any time so that the leaves of the hinge can be separated and the door lifted out of the frame. This type of hinge pin, however, has the disadvantage of working upward and out of the barrel with repeated door operation. One alternative to overcome this disadvantage is a nonrising, or self-retaining, loose pin. Another alternative is a nonremovable loose pin, which is desirable for security reasons for entrance doors or other doors that open outward and have the barrel of the hinge on the outside of the door. This pin is secured by a setscrew that is inaccessible when the door is closed. Consequently, when the door is closed, it is impossible to remove the pin, without cutting or breaking it, and lift the door out of the frame.

Hinges are fastened to doors and jambs with screws. Butts are usually mortised into the jamb. When the leaves are mortised into both the jamb and the edge of the door, a full-mortise hinge (see Fig. 16.4e) is required. When one leaf is mortised into the jamb and one leaf is surface applied to the door, a half-surface hinge (see Fig. 16.4f) should be specified.

Exterior doors and interior doors exposed to dampness should have hinges of nonferrous metal. Butts for other interior doors may be of ferrous metal, preferably hardened cold-rolled steel.

Door-Closing Devices. Many types of devices are available for closing a door automatically after it has been released by the person opening it. Most types are a combination of a spring, which is tensioned when the door is opened, and a pneumatic- or oil-cushioned piston, which dampens the closing action (see Fig. 16.4g). Adjustment screws are provided to control the rate of closing.

To close a door automatically in case of fire, a fusible-link closer may be installed. It normally holds a door in the open position, but it shuts the door when fire-heated air reaches the doorway and melts the link.

Locks and Latches. A latch is a device that prevents a closed door from being opened until the device is retracted. It is usually mounted on the door and secures it by projecting into a receptacle, or lock strike, on a strike jamb, head jamb or floor. A lock is a device that prevents retraction of a latch.

Locks or latches that are attached on the surface of a door are called rim locks or latches. Those mortised into the edge of the door are called mortise locks or latches (see Fig. 16.4h). A latch may be operated by a knob, lever or key.

A latch controlled by a lock is often referred to as a locking bolt. When the bolt is rectangular in shape, it does not slide into position automatically when the door is closed. It must be projected with a thumb turn or a key. This type of bolt is called a dead bolt, and the lock, a dead lock. A combination of a dead bolt and a latch is often referred to simply as a lock.

A latch bolt is a beveled locking bolt that automatically slides into position when the door is closed. (Specifications for such bolts should indicate the hand of the doors on which they are to be mounted.) Latch bolts should be equipped with a dead latch. This is a small plunger that is held depressed when the door is closed to prevent the latch bolt from being retracted by a shim, card or similar device inserted between the latch and door frame.

Unit locks (see Fig. 16.4i) are complete assemblies of latches, locks and knobs, which control latch movement. Eliminating most of the adjustments that would otherwise be necessary during installation, unit locks need only to be inserted into a standard notch in a door.

Bored-in locks also are complete assemblies but they are installed by boring standard-size holes in a door. They may be a tubular-lock or cylindrical-lock set, depending on how the holes have to be bored to accommodate them. Tubular locks have a tubular case that is inserted horizontally perpendicular to the edge of the door. A locking cylinder is placed in a small hole perpendicular to the tubular case. Cylindrical locks (see Fig. 16.4j) have a cylindrical case that is inserted in a large hole perpendicular to the face of the door and containing the main body of the lock. The latch bolt is placed in a small hole bored perpendicular to the door edge.

Exit devices, such as panic hardware, are a special type of lock required by building codes for certain exit doors in public buildings. They must be openable from the inside by merely pressing down on a bar that extends at least two-thirds the width of the door about 3 ft above the floor. These devices should be labeled for safe egress by a nationally recognized independent testing laboratory. For fire doors, fire exit hardware, labeled to indicate that it has been investigated for fire, should be specified. All these devices may be constructed to prevent the door from being opened from the outside except with a key.

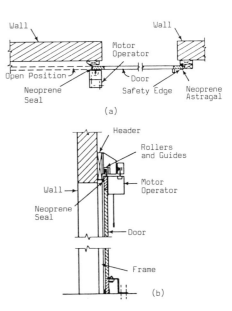

Fig. 16.5. Horizontal sliding door. (*a*) Horizontal section through opening. (*b*) Vertical section through opening.

Horizontally Sliding Doors

For a large opening, a swinging door would require too much space for opening. One alternative is a horizontally sliding door. A single-leaf door may be used when there is ample room on one side of the opening (see Fig. 16.5). A double-leaf door, parting at the center and moving to both sides, may be used when there is room for a door only half the width of the opening or when faster opening is required. When space on the sides is tight, a folding, or accordion-type, door may be used.

A horizontally sliding door may move on rollers on tracks at the top (see Fig. 16.5) or at the bottom. In the latter case, guides are required at the top of the door for stability. When a door is suspended from tracks at the top, bottom guides are optional, depending on the height of the opening and rigidity required of the door.

Vertically Sliding Doors

Another alternative to a swinging door for a large opening is a vertically sliding door. It can be used where there is insufficient space along the sides of the opening for a horizontally sliding door but ample headroom. A vertically sliding door may rise straight up (see Fig. 16.6), may rise up and swing in, roll up, or pivot outward to form a canopy. Some doors may be furnished in two sections, one moving upward and the other dropping downward. All types are usually counterweighted.

Vertically sliding doors should be installed so as to exclude weather from the building interior. For this purpose, the top of the door may be recessed into the bottom of the wall above the door. Or the top part of the door may be extended slightly above the inside of the wall. Similarly, the sides of the door may be recessed into the jambs or may lap the inside of the wall and be held firmly against it. Also, the surface of the inside floor should be a little higher than the outside grade to prevent entry of rainwater and snow.

Revolving Doors

Revolving doors are suitable for use in entrances that carry a continuous traffic flow without very high peaks. In such applications, they offer the advantage of restricting interchange of inside and outside air. They are generally required by building codes to be used in combination with swinging doors as emergency exits,

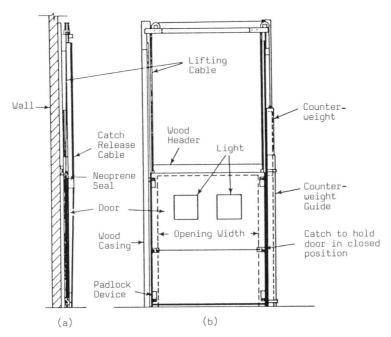

Fig. 16.6. Vertically sliding door. (a) Vertical section through opening. (b) Elevation of door.

because of the inability of revolving doors to pass large numbers of persons in a short time.

A revolving door consists of four perpendicular leaves that rotate about a vertical axis in a cylindrical enclosure. The enclosure usually is at least 6 ft 6 in. in diameter and the opening to it is usually between 4 and 5 ft wide.

As for other doors, the capacity of a revolving door for passing people in an emergency depends on the number of units of width. Some codes assign one unit of width to a revolving door, other codes only one-half unit. In addition, revolving doors may not be permitted to account for more than one-half the required exit capacity at any location; the rest must be supplied by swinging doors.

Rotational speed of a revolving door should not exceed 15 rpm. Each leaf should be provided with at least one push bar. Also, leaves should be glazed with transparent plastic or tempered glass. In addition, some codes require the doors to be collapsible.

16.3. SPECIAL-PURPOSE DOORS

Doors have to be specially designed for openings larger than those for which ordinary doors are available. Large doors may be required for airplane hangars, garages and craneway openings in walls and for subdividing gymnasiums and auditoriums. Not only must the doors be designed for the openings but also the structure around the openings must be designed to accommodate the doors. Usually, the doors are large and heavy. As a result, the structure must be designed to carry the heavy weights of the doors and operating mechanisms. Also, space must be provided for the doors when open and for the operating mechanisms.

Types of doors used for special purposes generally are much the same as ordinary doors except for size and weight. Some other types, however, sometimes are used, for example, horizontal-hinge doors for craneway entrances and plug doors for radiation shielding.

Horizontally Sliding Doors

Because of their weight, horizontally sliding, special-purpose doors are usually supported on bearing-type wheels and ride rails set in the floor. For stability, the doors are also equipped with top rollers that operate in overhead guides. The doors may be telescoping or folding.

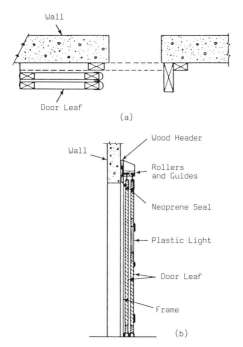

Fig. 16.7. Horizontally sliding, two-leaf door. (*a*) Horizontal section through opening. (*b*) Vertical section through opening.

Telescoping doors often are used at large entrances, for example, for airplane hangars. Built of wood, steel or a combination of these materials, the doors are composed of several parallel leaves, each of which moves in a separate track (see Fig. 16.7). Open doors may have the leaves stacked alongside a wall or in a pocket in the wall. The doors may be opened by moving all the leaves in one direction or they may

be center parting, with leaves moving in opposite directions. Often, the doors are operated by motors that drive an endless chain attached to the top of one leaf. The other leaves are moved by a series of interconnecting cables attached to the powered leaf. Generally, the arrangement is such that all leaves arrive at the open or closed position simultaneously.

Folding doors are used at entrances with very wide openings, such as those for hangars, or for dividing large rooms, such as gymnasiums, auditoriums and cafeterias, into smaller rooms. The doors consist of a series of leaves connected by hinges along their vertical edges. The leaves unfold when the doors close (see Fig. 16.8*a*). When the doors open, the leaves rotate about the hinges and roll into a tight assembly at either or both sides of openings. Figure 16.8 shows details of a manually operated wood folding door.

For a center-parting, motor-operated door, motors are usually located in mullions at the center of the opening. Cables connect the mullions to the sides of the opening. When a closed door is to be opened, the cables draw the mullions to the sides, sweeping the leaves along.

The main advantage of folding doors over telescoping doors for large openings is that folding doors require only one top and one bottom track or guide channel, regardless of the width of openings. Each leaf of a telescoping door, in contrast, requires its own guide channels. Hence, folding doors require less material for rails, guide channels and supporting members.

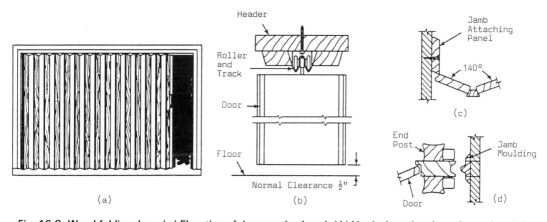

Fig. 16.8. Wood folding door. (*a*) Elevation of door nearly closed. (*b*) Vertical section through opening. (*c*) Detail at fixed end of door. (*c*) Detail at strike jamb.

Vertically Sliding Doors

Special-purpose vertically sliding doors are similar to ordinary doors that operate in the same way, except for size and weight. This type of door is suitable for high openings, when space is available above and below them into which door leaves can be moved. Usually counterweighted, the doors may be operated manually or electrically.

Figure 16.9 shows a telescoping vertically sliding door. The leaves decrease in thickness from the top down and are hollow (see Fig. 16.9a). When the door is open, the leaves nest within each other in a compact assembly (see Fig. 16.9b). Hung on cables attached to a reel in a header box, the leaves are released from the assembly one at a time as the cables unwind (see Fig. 16.9c).

Swinging Doors

Special-purpose swinging doors are used in a much wider variety of forms than ordinary swinging doors. Often, special-purpose doors are double swing, with hinges on both jambs; twofold, with a vertical hinge on one jamb and another vertical hinge between two leaves, so that the leaves fold when the door opens; or fourfold, which is center parting and each half is twofold. Such doors may be used instead of sliding doors when there is insufficient space around openings for storing open sliding doors.

Top-Hinged Doors

Doors with horizontal hinges at the top are often used for craneway entrances to buildings. This type of door has been employed for openings more than 100-ft wide and from 4 to 18-ft high.

To increase the depth available for openings, horizontally sliding doors sometimes are placed under a top-hinged door. In such cases, the top guides for the sliding doors are placed along the bottom of the upper door. Hence, when the doors are closed, the sliding doors must be opened before the top-hinged door. Closing must proceed in the reverse order.

SECTIONS 16.2 AND 16.3

References

Sweet's Architectural File, Division 8, McGraw-Hill, New York (updated annually).
Also see general references for this chapter.

Words and Terms

Door-closing device
Entrance door
Fire door
Hinge
Latch
Lock
Smokestop door
Threshold

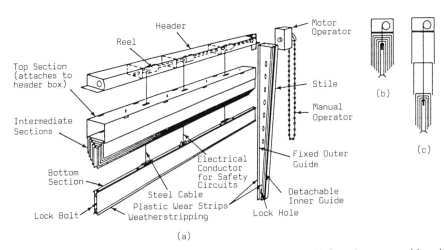

Fig. 16.9. Vertically sliding, telescoping door. (*a*) Door components. (*b*) Door in open position. (*c*) Door partly open.

Significant Relations, Functions and Issues

Door dimensions, materials, operation and hardware related to: appearance, code requirements, usage (traffic, users).

Door operation: swinging (horizontally), sliding, revolving, folding, telescoping, top hinged (swinging vertically).

16.4. FLOOR–CEILING AND ROOF–CEILING SYSTEMS

In general, a building requires basically a floor, or horizontal deck on which activities can be performed to serve the purposes of the building, and a roof to shelter the floor from rain and snow. Such simple construction, however, usually cannot meet buildings owners' objectives or satisfy constraints, such as building-code requirements for fire resistance. Consequently, a floor or roof in most buildings is only one component in a major subsystem, which is called in this chapter a floor–ceiling, or roof–ceiling system.

Floor–ceiling and roof–ceiling systems differ principally in the characteristics and construction of the top surface. As discussed in Sec. 15.1, the main objective of a roof is to separate the indoor environment from the outdoors. The main objective of a floor is to provide a surface for performance of necessary activities in the building interior. The top surfaces of floors and roofs, therefore, are constructed to meet these different objectives. Below the top surfaces, however, the systems may be similar. This section deals mainly with the portions of floor-ceiling and roof–ceiling systems below the top surfaces. Because roofs are covered in Sec. 15.1, the following discussions emphasize floor–ceiling systems, but the principles apply equally well to roof–ceiling systems.

A floor–ceiling system may consist of one or more components in addition to the floor, ceiling and means of supporting the ceiling from the floor. Depending on the objectives and constraints, the system may also incorporate a floor covering, beams, fire protection, and a plenum, or space between floor and ceiling (see Fig. 16.10).

Floor Coverings

Because of the traffic to which a floor is subjected, it is usually desirable to cover the deck with a nonstructural material to resist the wear. The main objective of the floor covering may be long-term durability; or the objective may be low initial cost, with the aim of replacing the covering when wear damages it or a different appearance for the covering is desired. Also, in selection of a floor covering, aesthetics may be an important consideration. To meet these objectives, a thin, lightweight material may be adequate. Materials used for floor coverings are discussed in Sec. 16.5.

Floors or Subfloors

Traditionally, wood floor coverings have been called floors. The decks on which such coverings are laid are known as subfloors (see Fig. 16.10*e* and *f*). This terminology is generally used for the floor construction in one- and two-family dwellings, even when the floor covering is some material other than wood. For other types of buildings, the top material is referred to as a floor covering, or flooring, and it is considered laid on a floor. The purpose of the floor is to provide a continuous support for the floor covering and loads on it, because it is usually too thin to be self supporting.

Floors have to be strong enough to support their own weight, the weight of the floor covering and live loads, including weight of partitions, plus impact. Also, the floors should be sufficiently stiff that loads do not cause excessive deflection or vibration. Floors also may have to support the weight of ceilings, ducts, pipes and wiring that may be suspended from them. In addition, for some occupancies, floors may be required to be incombustible.

Floors normally are part of the structural system. Consequently, they are built of structural materials, such as portland-cement concrete, gypsum concrete, wood, plywood, cold-formed steel or combinations of these materials. Concrete may be reinforced or prestressed with steel, and cast in place or precast. Often, it is made with lightweight aggregates to reduce the weight of the floor. When concrete is used in

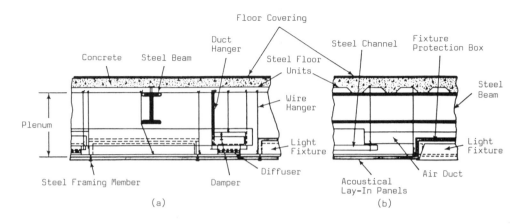

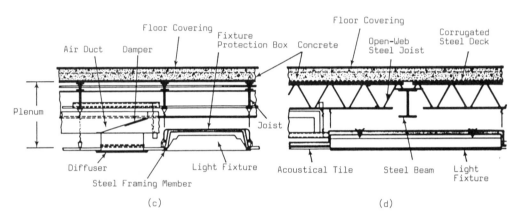

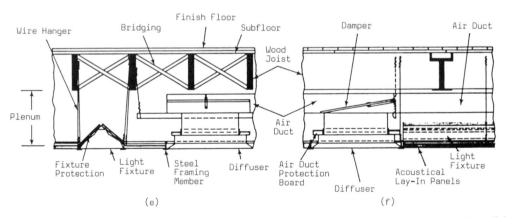

Fig. 16.10. Vertical sections through floor–ceiling systems. (*a*) and (*b*) Sections perpendicular and parallel, respectively, to structural steel beams supporting a concrete floor cast on steel deck. (*c*) and (*d*) Sections perpendicular and parallel, respectively, to open-web steel joists supporting a concrete floor cast on a corrugated steel deck. (*e*) and (*f*) Sections perpendicular and parallel, respectively, to wood joists supporting a wood subfloor.

combination with a cold-formed-steel deck, the concrete may be nonstructural, merely a fill for housing electrical conduit and pipes. In some cases, the underside of a concrete floor may serve as a finished ceiling, often only painted.

In construction of a cast-in-place concrete floor, a form must be provided to support the concrete until it has hardened. Forms and their supports that serve only during construction are part of the cost of construction. An alternative is to leave the forms in place after use, as part of the floor subsystem. Figure 16.10*a* and *b* show a concrete floor cast on steel decking, which is capable of supporting the concrete before it has hardened while spanning between widely spaced beams. Figure 16.10*c* and *d* show a concrete floor cast on a thin, corrugated steel deck, which is required to span only 2 ft between open-web joists.

In buildings with wood framing, the subfloor often is plywood (see Fig. 16.10*e* and *f*). The thickness of the plywood depends on the live loads to be carried and on the joist spacing, which usually is 16 or 24 in.

Floor Supports

In flat-plate and flat-slab concrete construction, floors are thick, strong and stiff enough to span, without additional support, between columns or between beams along column lines. For other types of construction, widely spaced beams or trusses or closely spaced joists usually are used to support thin, lightweight floors. Selection of the type of floor and supports should be based on cost studies of alternative designs.

When concrete or combinations of concrete and steel floors are used, composite construction is often economical. In this type of construction, the floors assist the beams, trusses or joists in carrying loads.

In buildings that are required by building codes to be fire resistant, the structural members must be protected against fire. (Concrete members are inherently fire resistant, but steel tendons and reinforcing in them must be protected by an adequate depth of concrete cover.) A concrete floor of sufficient thickness may be used to protect the tops of structural members. The rest of the members may be protected by

encasement in concrete, which is very heavy, or by a coating of insulating material, which may be lightweight. Alternatively, the protection may be provided by a continuous, fire-resistant ceiling, with openings of limited area for lighting fixtures and air supply and exhaust. This is an economical alternative when a ceiling is to be installed for other reasons.

Plenums

The space between the floor and ceiling may be used to house structural members, pipes, ducts and electrical wiring. The plenum may also be used without ducts to distribute air for ventilation or air conditioning.

In Fig. 16.10, the ceilings are shown at a level far enough below the beams and joists to provide space for air-conditioning ducts. The depth of the floor–ceiling system, however, can be reduced, with considerable savings in construction costs, by placing the ceiling along the underside of the structural members. For this purpose, the ducts would have to be run parallel to the beams or joists. With very deep beams, open-web joists or trusses, ducts may be run perpendicular to the structural members by cutting holes in the deep beams or by making the ducts small enough to pass between the openings in the joists and trusses.

Ceilings

A ceiling may be the underside of a floor or a separate construction, usually installed at or below the bottom of floor structural members, to conceal them. In either case, the ceiling is normally supported by the floor or the floor structural members. In Fig. 16.10, the ceilings are shown suspended on steel wire hangers.

Because ceilings cover entire spaces, aesthetics is usually an important consideration in choices of ceiling materials, shapes and textures. The bottom surface may be smooth, textured or sculptured; plain white or colored; opaque, mirrored or luminous (translucent to provide uniform illumination); and completely flat, waffle shaped, arched, vaulted or interrupted to expose structural members. The choice of the numerous alternatives possible is generally made by an architect or an interior decorator, with the approval of the owner.

A ceiling is also part of the lighting system. Light reflectance of the bottom surface is an important consideration in illumination of interior spaces. Also, lighting fixtures may be recessed in, mounted on or suspended from ceilings. In some cases, the whole ceiling of a room may be made luminous. Figures 16.10 and 16.11 show ceilings with recessed fixtures. Arrangements of the lighting fixtures, as may be seen in Figs. 16.11a and b, play an important role in the aesthetic design of the ceiling.

Another important consideration in selection of a ceiling design is fire resistance. For one thing, building codes may require the ceiling to have a low flame-spread rating. (Flame-spread rating is a measure of the speed at which a flame will travel along an exposed surface and is a guide to the relative flammability of a material.) In addition, building codes may specify a minimum fire rating for the ceiling. Also, when a ceiling serves as a protective membrane for the construction above, building codes may specify a minimum fire rating for the whole floor–ceiling or roof–ceiling system. In such cases, fire tests are made for specific assemblies and a rating is assigned to each assembly tested. (Time–temperature designs for fire-resistant constructions are listed in Underwriters' Laboratories Fire-Resistance Index.) These fire-resistance requirements limit the choice of ceiling materials. Note also in Fig. 16.10b and c the installation of a box over the light fixture for fire protection, because only limited opening areas are permitted in ceilings serving as a protective membrane.

Often, also, acoustical considerations are important in ceiling design. The ceiling, for example, may be shaped, sculptured or textured for acoustical purposes. Also, it may be treated to achieve desired sound absorption. Furthermore, the ceiling may be required to be in a high sound-transmission class (STC), to limit sound transmission. If sound is transmitted readily through a ceiling, it may travel unhindered along the plenum and pass from room to room, with annoying effects on occupants.

In addition, ceilings play an important role in air-conditioning. They often contain diffusers,

(a)

(b)

Fig. 16.11. Modular ceiling systems. (a) Acoustic tile with matching light and HVAC registers. (b) Integrated system with fluorescent fixtures in panels and HVAC linear diffusers between panels. (Courtesy Johns-Manville Corp.)

registers or grilles for supplying or exhausting air. In some cases, ceilings may even be part of the HVAC system. Figure 16.11b, for example, shows a ceiling with a linear diffuser. Air is discharged into the room through slots in the supports of the acoustical ceiling panels. Sometimes, when the plenum is used for air distribution, part or all of the ceiling may be perforated to discharge air uniformly into a room.

For ceilings in the top story of a building, it may be advantageous to select ceiling materials that provide good thermal insulation. Such construction is economical because it reduces the amount of separate insulation that must be added for the roof.

In selection of ceiling materials, maintenance should also be considered. In general, it is desirable that the ceilings be stain resistant. But for some environments, it may also be necessary that the ceiling be resistant to moisture, steam or chemicals. Materials that are easily cleaned are advantageous for ceilings. Also, paintable surfaces are often preferred. In addition, when maintenance is being considered, the possible need for access to the plenum, to reach pipes, ducts or wiring for repair or replacement, should be taken into account. Use of easily removed ceiling panels makes such repair or replacement easier and less costly.

For buildings with a tight construction budget, cost of ceiling materials and their installation may be an overriding concern. In such cases, a single material that meets several objectives and all the constraints and can be installed easily and quickly is highly desirable, but it may not be feasible to meet all objectives and the constraints. Ceiling materials are discussed in Sec. 16.5.

Access Flooring

For office buildings and other occupancies where extensive wiring is required for individual work stations, conduit and raceway systems have traditionally been built into the floor system. Various forms for these systems are discussed in Sec. 13.4. These systems typically provide for modification to accommodate changing needs for location of work stations or for the alteration or addition of equipment.

Where extensive use of computers is anticipated, provisions for frequent and possibly large-scale modification of wiring for work sta-

tions are even more important. For the purpose, an elevated floor system, also called access flooring, may be installed. This consists of a second floor structure supported a short distance above the structural floor, creating a space that can be accessed by removal of small modular units of the elevated floor and that can be occupied by electrical wiring (see Fig. 16.12).

Some early advocates of the system visualized its use as a substitute for suspended ceiling space as a location for wiring, HVAC ducts and other items for service. Although some installations have been made without a complete suspended ceiling, some items—such as sprinkler piping, light fixtures and HVAC registers—must stay at the ceiling level, making installation of both an access floor and a suspended ceiling advantageous.

One possible drawback to the use of the access flooring system in combination with a suspended ceiling is the increase in the floor-to-floor distance for multistory buildings. This disadvantage may be offset by benefits deriving from the use of the access system, but cost–benefit analysis for the whole building should be done to provide a basis for system choices and to pinpoint areas that may be dealt with to improve the design of the subsystems.

References

Sweet's Architectural File, McGraw-Hill, New York (updated annually):
Div. 9—suspended ceilings,
Div. 10—access flooring,
Div. 15—HVAC systems,
Div. 16—electrical and lighting,
Also, for structural floors:
Div. 3—concrete,
Div. 5—metal (steel deck)
Also see general references for this chapter.

Words and Terms

Access flooring
Flooring
Plenum
Subfloor
Suspended ceiling

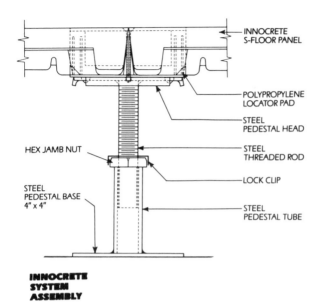

INNOCRETE
S-FLOOR PANEL

POLYPROPYLENE
LOCATOR PAD

STEEL
PEDESTAL HEAD

HEX JAMB NUT

STEEL
THREADED ROD

LOCK CLIP

STEEL
PEDESTAL BASE
4" x 4"

STEEL
PEDESTAL TUBE

**INNOCRETE
SYSTEM
ASSEMBLY**

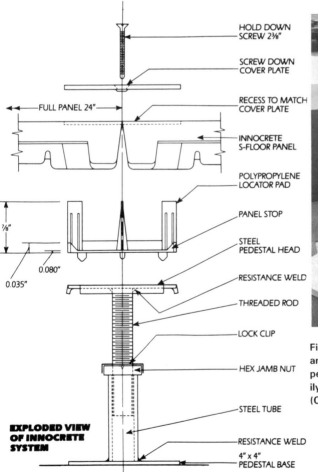

HOLD DOWN
SCREW 2⅜"

SCREW DOWN
COVER PLATE

RECESS TO MATCH
COVER PLATE

FULL PANEL 24"

INNOCRETE
S-FLOOR PANEL

POLYPROPYLENE
LOCATOR PAD

PANEL STOP

⅞"

STEEL
PEDESTAL HEAD

0.080"

RESISTANCE WELD

0.035"

THREADED ROD

LOCK CLIP

HEX JAMB NUT

STEEL TUBE

**EXPLODED VIEW
OF INNOCRETE
SYSTEM**

RESISTANCE WELD

4" x 4"
PEDESTAL BASE

Fig. 16.12. Components and installation views for an access flooring system. Vertically adjustable pedestals support modular deck units that are readily removable for access to wiring or other services. (Courtesy Innocrete Systems, Inc.)

Significant Relations, Functions and Issues

Floor–ceiling and roof–ceiling systems designed to accommodate building service systems.

Modular ceiling systems with integrated ceiling construction, HVAC units, and lighting units.

Use of concealed, enclosed spaces in suspended ceilings and access flooring; plenum returns for HVAC systems; concealed but accessible wiring and equipment.

16.5. INTERIOR FINISHES

Interior finishes provide the surfaces of the systems for enclosing interior spaces and that are visible to occupants within those spaces. The finish may be a major component of a system, such as the exposed stone or brick of a wall, acoustical tile, or glazing of a window; or the finish may be a surfacing component, such as paint, wallpaper, plastic sheeting, wood paneling or floor covering.

Selection of a finish is often made by an architect or interior decorator, with the approval of the owner. The choice is normally governed by considerations of aesthetics and life-cycle costs. But in addition, the finish used for any system enclosing an interior space should meet the objectives of and constraints on all the building systems of which the enclosure is a component. (Sections 16.1 and 16.4 discuss objectives and constraints for partitions, interior walls, doors and floor–ceiling and roof–ceiling systems.) It is especially important that building-code requirements for flame-spread and incombustibility be satisfied. Also, designers should bear in mind that walls, floors, doors and windows, like ceilings, interact with or are components of lighting and acoustical systems and should meet requirements for light reflectance and sound absorption.

Floor Coverings

Specifications have been written to set minimum standards for quality of the various flooring materials. Principal sources for such specifications are ASTM; Federal Specifications, obtainable from the Standardization Division of the General Services Administration, Washington, D.C. 20025; Commercial Standards, obtainable from the Office of Commodity Standards, National Institute for Standards and Technology; and trade associations for specific materials, such as Portland Cement Association, Resilient Tile Institute, Tile Council of America, and Wood Flooring Institute. Flooring materials should be installed in accordance with the recommendations of the manufacturers of the specific materials.

Floor coverings may be classified as resilient or rigid. Methods of installation for each type is different.

Resilient Flooring. This type of floor covering is usually thin, generally $\frac{1}{4}$ in. or less in thickness. It therefore must be laid on a smooth, rigid floor. The floor covering is usually secured to the floor with an adhesive, such as an asphalt cutback or emulsion, latex, linoleum paste or rubber cement, as recommended by the flooring manufacturer. Hence, the floor surface should also be dry and clean when the flooring is installed.

When a concrete floor is used, protection of the floor covering against moisture is an important consideration. The installation should not start until the concrete has dried. For concrete at or below grade, in particular, at least 30 days' drying time should elapse after placement of concrete before flooring is installed. Preferably, such floors should be cast on a moisture barrier, to prevent moisture in the ground from being absorbed by the concrete.

Also, when a concrete floor is used, the possible effects on the flooring adhesive of treatments, such as curing compounds or parting agents used to insure separation from formwork, which may have been applied to the concrete surface, should be investigated. If deleterious treatments have been used or if the concrete surface has come in contact with oils, kerosene or waxes, the surface should be cleaned in accordance with recommendations of the adhesive manufacturer.

When wood subfloors are used and they are nailed to structural members, the nails should be driven flush with the underfloor top surface without denting it around the nail head. Also, the nails should be ringed or barbed to prevent them from working upward and causing bulges in the floor covering. With thin, resilient floor coverings, not only do imperfections in the floor cause unsightly waves in the flooring but they also cause uneven wear of the flooring, shortening its useful life.

Among the most frequently used resilient materials are carpets, asphalt tiles, cork tiles, vinyl flooring, backed vinyl, rubber and linoleum. Cork, vinyl, rubber flooring or linoleum should not be installed on concrete floors at or below grade, because of the risk of damage from moisture. Otherwise, when properly maintained, all these materials perform well, unless subject to heavy traffic such as that in industrial buildings and warehouses.

Rigid Flooring. More rigid floorings, such as ceramic tiles, terrazzo, concrete or wood, are more suitable for heavy traffic. But such floor coverings may also be selected for aesthetic reasons.

Ceramic tiles are usually set in a portland-cement mortar bed or cemented with an adhesive. Three main types of tiles are available: mosaic, paver and quarry. Mosaic tiles are relatively small. They are assembled in a factory in desired patterns, usually with the top face temporarily glued to a sheet of paper, to simplify installation. Paver tiles are similar to mosaic tiles but larger. They usually range in size from 3×3 to 6×6 in. Quarry tiles are much denser than the other types. Resistant to freezing, abrasion and moisture, quarry tiles may be used outdoors.

Terrazzo is a decorative wearing surface much like concrete. It is made with two parts of marble chips to one part of portland cement. The finished surface therefore has the appearance of marble particles set in a cement matrix. A terrazzo topping may be cast integral with a concrete floor, bonded to a hardened concrete floor, or seated on a sand cushion, to prevent injury from settlement, expansion, contraction or vibration. The topping may also be precast. Terrazzo toppings may also be constructed with a matrix of rubber latex, epoxy or polyester.

Concrete wearing surfaces are especially suitable for floors subjected to heavy traffic. Set on a concrete slab, they may be cast integral with the structural slab or placed after the slab has hardened. Integral toppings usually are $\frac{1}{2}$-in. thick, whereas separate toppings may be about 1-in. thick.

Wood floorings may be made of hardwoods or softwoods. They may be applied as strips or blocks. Blocks are normally used for floors subjected to heavy traffic. Wood should not be used on concrete below grade.

Hardwoods most often used include maple, beech, birch, oak and pecan. As strip flooring, these woods are available in thicknesses of $\frac{11}{32}$, $\frac{15}{32}$ and $\frac{25}{32}$ in. and widths ranging from $1\frac{1}{2}$ to $3\frac{1}{4}$ in.

Softwoods frequently used include yellow pine, Douglas fir and western hemlock. These woods are less resistant to wear and indentation than the hardwoods. As strip flooring, softwoods are available $\frac{25}{32}$-in. thick and in widths from $2\frac{3}{8}$ to $5\frac{3}{16}$ in.

Wood blocks for flooring may be solid or laminated. Solid blocks are composed of two or more units of strip-wood flooring, which are fastened together with metal splines or other devices. A laminated block is made of plywood with three or more plies glued together. Both types of blocks are usually supplied square in plan, with tongues and grooves either on opposite or adjacent sides.

Solid blocks may be attached to a subfloor with nails or an asphalt adhesive. The blocks normally are $\frac{25}{32}$ or $\frac{1}{2}$-in. thick. Laminated blocks are usually installed with an asphalt or rubber-base adhesive. Block thickness generally is in the range of $\frac{1}{2}$ to $\frac{13}{16}$ in.

Gypsum Finishes

Several forms of gypsum products are widely used for interior finishes of buildings. The main ingredient of these products is commonly known as *plaster of paris*. It is a fine powder, usually white, produced by heating gypsum rock, a crystallized calcium sulfate, to about 130°C, driving off about three-quarters of the water of crystallization. When mixed with water, plaster of paris sets rapidly into a solid mass and gains strength on drying. When exposed to water again, however, it disintegrates. An admixture, or dope, may be added to plaster of paris to accelerate or retard its set.

Plaster of paris is used as a molding plaster or as a gaging plaster. *Molding plasters* are used to form ornamental objects. *Gaging plasters* are used with finishing hydrated lime to form smooth, white-coat finishes on plaster walls and ceilings. Manufacturers also use unretarded-set plaster of paris to produce gypsum block, gypsum tile and gypsumboard.

Hardwall plaster, often used for a base coat in plastering, is made of a retarded-set plaster of paris mixed with sisal. Different effects are obtained by combining hardwall plaster with various aggregates, such as wood fiber, sand, perlite, vermiculite and pumice. The lightweight aggregates produce lightweight plasters with high fire resistance. Plasters made with porous aggregates, such as pumice, are sometimes used in acoustical treatments.

Keene's cement is produced by heating gypsum rock to a high temperature to drive off all the water and by adding alum as a set accelerator. The resulting product is harder and more durable than gaging plaster and much more resistant to water, but Keene's cement costs more. Hence, it is usually used only for special finishes, such as for bathrooms.

Gypsum plasters provide a hard finish, which can be smooth or textured; it can readily be painted; and it is suitable for use where a fire-resistant finish is desired.

Gypsumboard consists of a core of set gypsum faced with paper or other fibrous material. It is used in dry-type construction, instead of plaster, as an interior finish.

Gypsum lath is similar in content to gypsumboard but it is constructed and treated for application of plaster coats to its surface. The lath may be perforated or may incorporate mechanical key devices to hold the plaster.

Gypsum sheathing board also is similar to gypsumboard except that the sheathing is faced with a water-repellent paper. Some sheathing boards also have a water-repellent core.

Gypsum tile or block are sometimes used for fire protection of columns and for building nonload-bearing fire walls or walls around elevator shafts. The blocks are made from set gypsum. They are 12 in. high, 30 in. long and range in thickness from $1\frac{1}{2}$ to 6 in. The blocks $1\frac{1}{2}$ or 2 in. thick are solid. They are commonly used for furring, for example, to build a fire-protection layer around a column, which is then plastered. Thicker blocks have holes through them to reduce their weight.

Gypsum plank is a precast product sometimes used as a roof deck. Reinforced with wire fabric, the plank usually is provided with tongue-and-groove metal edges and ends to form tight joints.

The preceding descriptions of gypsum products indicate that gypsum finishes may be partly or completely prepared in the field (wet-type construction, because of addition of water on the site) or prefabricated (dry-type construction). Often, prefabricated products are selected because of the speed of construction and lower initial cost, but wet-type construction may be chosen to obtain greater durability, lower maintenance and special decorative effects.

Regardless of whether wet or dry construction is used, application of the finish should take into account the physical characteristics of the materials. Gypsum or portland-cement products have low resistance to strains induced by structural movements. They also are subject to dimensional changes because of variations in temperature or humidity. Hence, proper provisions should be made in installation of these products for these conditions. Otherwise, the finishes may crack or chip or fasteners may pop out.

To prevent these occurrences, gypsum and cement products should be installed so that movement is not restrained. For example, at intersections with obstructions, such as those between walls and ceilings, the surfacing materials should be isolated by control joints, floating angles or other means.

Control joints should be placed not more than 30 ft apart in long walls or partitions, nor more than 50 ft apart in large ceilings. Such joints interrupt the continuity of the construction, to permit small movements. The joints may be conveniently located along lighting fixtures, heating vents or air-conditioning diffusers, where continuity would ordinarily be broken to accommodate the devices.

A floating angle is a means of providing for movement at the intersection of a ceiling and a partition. Figure 16.13 illustrates a floating an-

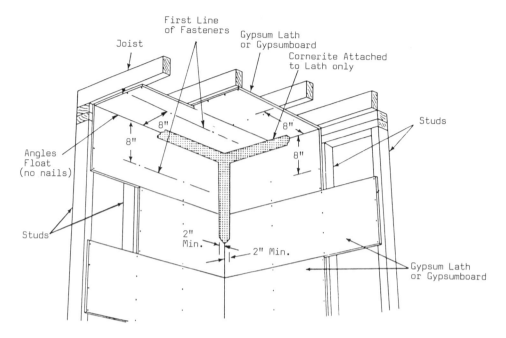

Fig. 16.13. Floating angle construction at intersection of ceiling and partitions.

gle constructed with gypsum lath, ready to receive a base coat of plaster (wet construction). Construction with gypsumboard is similar.

Construction of the floating angle at the intersection starts with installation of gypsum lath or gypsumboard in the ceiling. Away from a reentrant angle, the boards would be attached to joists with nails about $\frac{1}{2}$ in. away from the ends of the boards, but at the ceiling-partition intersection, the first line of nails in the ceiling should be placed at least 8 in. from the wall, for greater flexibility in absorbing movements. In the rest of the ceiling, normal nail spacing should be used.

Then, gypsum lath or gypsumboard should be secured to the partition framing. The topmost wallboards should make firm contact with the ceiling, to support the ceiling boards. Again, to provide flexibility in absorbing movements, the first line of nails in the topmost wallboards should be placed at least 8 in. from the intersection.

Wet-Type Plaster Construction

A plaster finish prepared on the building site consists of a base and one or more thin coats of portland-cement or gypsum plaster. (Portland-cement plaster is often referred to as *stucco*.) The base consists of a material—gypsum lath, metal lath or masonry—with a surface to which plaster will bond. Special finish-coat plasters, such as veneer plasters, may be applied as a single coat on a gypsum lath made for this type of application. Other plasters generally require two or more coats on the base.

The outermost coat is called the finish coat. The underneath coats are called base coats. In a three-coat plaster installation (see Fig. 16.14b), the coat applied to the base is called scratch coat, because its outer surface is roughened with a toothed tool to form a good bonding surface for the intermediate, or brown, coat. In a two-coat installation (see Fig. 16.14a), the coat applied to the base is a combined scratch and brown coat.

Bases. The most commonly used bases for plaster are gypsum lath, metal lath and masonry. The laths generally are fastened to and supported by studs or joists spaced 16 or 24 in. center-to-center. Nails, screws, clips or staples may be used as fasteners with wood framing, whereas metal lath may be tied with steel wire

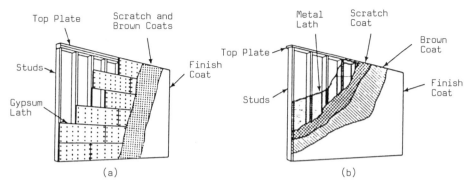

Fig. 16.14. Plaster wall construction. (a) Two-coat system. (b) Three-coat system.

to metal studs or joists. Masonry provides a continuous surface for bonding plaster.

Gypsum lath is usually supplied as boards 16 × 48 × $\frac{3}{8}$ or 16 × 48 × $\frac{1}{2}$ or 16 × 96 × $\frac{3}{8}$ in. The $\frac{1}{2}$-in.-thick lath should be used with 24-in. stud or joist spacing. When applied to studs, gypsum lath should be placed with long direction horizontal, and vertical joints should be staggered. When applied to ceilings, the lath should be laid with long sides spanning across joists. The ends of every board should rest on or be fastened to framing, or else headers or nailing blocks should be installed to support the lath ends. Ends and edges of each board should be in contact with adjoining boards, but if spaces more than $\frac{1}{2}$-in. wide are necessary, the plaster should be reinforced at the gaps with self-furring metal lath stapled or tied with wire to the gypsum lath.

Reentrant corners at intersections, except those required to be unrestrained, should be reinforced. Cornerite, a bent metal strip produced for this purpose, is usually used for such reinforcement. It should be stapled or tied with wire to the gypsum lath. Similarly, a corner bead, a strip of metal with flanges and a nosing at the junction of the flanges, should be used to reinforce exterior angles at intersections. Casing beads should be installed around wall openings and at intersections of plaster with other finishes and of lath with lathless construction, to separate the different materials.

Metal lath is fabricated to hold plaster by mechanical bond. Basic types frequently used are expanded-metal, punched-sheet metal and paper-backed, welded-wire lath. All provide

openings into which plaster keys when it hardens. The lath metal usually is copper-bearing steel with a protective coat of paint or galvanized steel.

Expanded metal lath is made by slitting sheet steel and expanding it to form a mesh. Types available include diamond mesh, flat rib and high rib. The deeper the ribs, the more rigid is the lath. Hence, the ribbed lath is suitable for long spans or heavy weights, but the more flexible lath is preferable for contour plastering.

Welded-wire lath is made with wire forming 2 × 2-in. or smaller meshes with a paper backing. The wires, 16 gage or thicker, should be stiffened continuously parallel to the long dimension of the lath at intervals not exceeding 6 in. The paper backing, forming a base to which plaster can adhere while hardening around the wire, should be attached to the wires so as to allow full embedment of most of the wires in at least $\frac{1}{8}$ in. thickness of plaster.

Application of metal lath should start with the ceiling. Flexible sheets may be carried 6 in. down on walls and partitions. With rigid lath, sides and ends of the lath may be butted into reentrant corners between ceilings and partitions, which may then be reinforced with cornerite.

Masonry bases for plaster may be provided by gypsum tile, brick, clay tile, concrete block or precast or cast-in-place concrete. Plaster, however, should not be applied directly to exterior masonry walls. There is a possibility that moisture will seep through such walls and damage the plaster. To prevent such damage, the plaster should be furred at least 1 in. away from the masonry. The furring may be wood strips, protected against moisture, that are fastened to the masonry and to which lath can be attached.

Gypsum partition tile is produced with scored faces to improve bond with plaster. Bricks or clay tiles provide a good base for plaster if they are not smooth-surfaced or nonporous. Aged concrete block may have plaster directly applied when used for walls or partitions, but for ceilings, a bonding agent or a special bonding plaster should be applied to obtain reliable bond. A bonding agent or special bonding plaster should also be applied initially to precast or cast-in-place concrete with a smooth, dense surface. For thick coats of plaster (more than $\frac{3}{8}$ in. for ceilings or $\frac{5}{8}$ in. for walls), metal lath should be fastened to the concrete to serve as a base for the plaster.

Preparation of Plaster. Plasters are usually made by mixing with water portland cement or gypsum, lime and an aggregate, such as sand, vermiculite or perlite. For application to metal lath, hair or sisal fiber may be added to a scratch-coat plaster to limit the amount of plaster that passes through the openings in the lath to what is needed for good bond. (The fibers, however, do not increase the strength of the plaster.) The ingredients may be mixed manually or mechanically. A mechanical mixer not only reduces the amount of labor required for mixing but also usually provides better dispersion of the ingredients.

Plaster should be protected against freezing until it has dried. When outdoor temperatures are less than 55°F, a temperature of at least 55°F should be maintained in interior spaces for plastering. This temperature should be sustained for at least one week before application of the plaster and one week after the plaster has dried. In hot, dry weather, measures should be taken to prevent water from evaporating before the plaster has set. After the plaster sets, ventilation should be provided to allow moisture from the plaster to evaporate.

Dry-Type Construction

Wet-type construction of interior finishes may slow construction of a building because of the necessity of waiting for plaster to dry before decoration, installation of lighting fixtures or HVAC grilles and similar finishing jobs can proceed. Instead, construction can be speeded and often, also, costs can be cut by use of dry-type finishes. These include prefabricated partitions and site-assembled rigid or semirigid boards or panels, which are fastened directly to masonry, furring or structural framing.

With board or panel finishes, various types of joints may be used, depending on the visual effects desired. The boards may abut each other and the joints may then be concealed, or battens, moldings or beads may be applied to cover and accentuate divisions between panels.

If the boards are thinner than plaster finishes, framing members should be furred out to the thickness that would be used with plaster, to permit use of stock doors and windows.

Dry-type site-assembled finishes often used include plywood, acoustical tile, fiber boards, pulp boards and gypsumboards.

Types of Gypsumboards. These are available in numerous forms: wallboard, backing board, coreboard, fire-resistant gypsumboard, water-resistant gypsumboard, sheathing and formboard, for use as formwork for concrete. For interior finishes, gypsumboards usually are available in 4-ft widths and lengths of 8 to 12 ft.

Gypsum wallboard is used for the surface layer of interior walls, partitions and ceilings. It normally comes with gray liner paper on the back and a special paper covering, usually cream colored, on the facing side and edges, for application of paint, wallpaper or other decoration. Aluminum foil may be applied to the backing to produce an insulating wallboard. Gypsum wallboard also is available predecorated, so that no added finish is required after installation. Board thickness ranges from $\frac{1}{4}$ to $\frac{5}{8}$ in.

Gypsumboard may be applied to framing as a single ply (see Fig. 16.15*a*) or combined in multi-ply systems (see Fig. 16.15*b*). Multi-ply systems are stronger and provide better sound control and greater fire resistance.

Backing board is used as a base layer in multi-ply construction. This type of board has

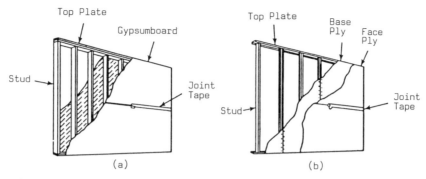

Fig. 16.15. Gypsumboard construction (drywall) with studs. (*a*) Single-ply system. (*b*) Two-ply system.

gray liner paper on front and back faces. It is also available as an insulating board, with aluminum foil on the back face.

Gypsum coreboard is used as a base in multi-ply construction without structural framing. This type of construction, by eliminating studs from walls, makes possible much thinner walls.

Type X gypsumboard is used in fire-rated assemblies. It may be a wallboard, backing board or coreboard made more fire resistant than normal gypsumboards by addition of glass fibers or other reinforcement.

Water-resistant gypsum backing board comes with a water-resistant core and water-repellent face paper. It may be used as a base for wall tile in bathrooms and other areas subject to wetting.

Fire-block gypsumboard is used under combustible roofs to protect rafters and other internal construction from roof fires.

Gypsum sheathing is manufactured for use on the exterior of frame walls as fire protection and bracing. It must, however, be protected from the weather by an exterior water-resistant facing.

Gypsum formboard is made to serve as a permanent form for the casting of gypsum-concrete roofdecks.

Installation of Gypsumboards. When temperatures outdoors are below 55°F, the building interior should be maintained at a temperature of at least 55°F for at least 24 hr before installation of gypsumboards and held at that temperature until a permanent heating system is in operation.

For single-ply construction (see Fig. 16.15*a*),

gypsumboard should be applied in the same manner as recommended for gypsum lath. Wallboard should be applied to studs with the long dimension horizontal, and ceiling panels, with the long dimension spanning structural framing. Joints at board ends should be staggered. Ceiling panels should be applied first, then the wallboards. The boards may be secured to framing with adhesives or special nails or screws.

Special treatment materials are available for concealing joints between gypsumboards. Joint tape, for example, is a strip of strong paper reinforcement with feathered edges, for embedment in a joint compound. The tape usually is about 2 in. wide and $\frac{1}{16}$ in. thick. Joint compound is available as a joint, or taping, compound or as a topping, or finishing, compound. The taping compound is applied as an initial coat for filling depressions at joints and fastener heads and for adhering joint tape. The finishing compound is used to conceal the tape and for final smoothing and leveling at joints and fasteners.

Before decoration is applied to gypsumboard, the surface should be sealed or primed. A primer is needed as a base for paint and to conceal color and surface variations. An emulsion sealer should be used to block the surface pores of the gypsumboard and reduce suction and temperature differences between paper and joint compound. When such a sealer is used under wallpaper, the paper can be removed later without damaging the gypsumboard surface and will leave a base suitable for redecorating.

In multi-ply gypsumboard construction, face layers may be glued, nailed or screwed to the base layer. Backing board is often used for the base layer because of lower cost, but wallboard or sound-deadening board may be used instead in some applications. When adhesives are used between face and base plies, some fasteners are usually used in addition, to insure bond. The base layer may be attached to structural framing with special nails, screws, staples or clips.

Face and base plies may be installed with long dimensions in perpendicular directions (see Fig. 16.16b). Joints in the two layers should be offset at least 10 in. For walls or partitions, face ply may be applied horizontally or vertically. There are fewer joints with horizontal application. Vertical application may be preferred, however, for predecorated wallboard that is to have joints covered with battens for aesthetic effect.

Ceramic Tiles

For decorative and sanitary purposes, ceramic tile is an excellent, durable finish, with high fire resistance. It has a burned-clay body on which a decorative glaze is usually imposed. If a glaze is used, color of the tile depends on that of the glaze. Usually flat, tiles vary in size from about $\frac{1}{2}$ in. to 6 in. or more. They are applied to a solid backing, such as masonry or plaster, by embedment in a mortar or an adhesive. Joints between tiles are usually grouted.

References

Sweet's Architectural File, Division 9 (finishes) McGraw-Hill, New York (updated annually).
Also see general references for this chapter.

Words and Terms

Control joint
Dry-type construction
Flooring
Wet-type construction

Significant Relations, Functions and Issues

Criteria for interior finishes: cost, appearance, wear and maintenance, flame-spread, combustibility, sound and light reflectivity.

Components of construction required to achieve finishes of various kinds for walls, floors and ceilings.

16.6. SYSTEMS-DESIGN APPROACH TO INTERIOR SYSTEMS

In general, the application of systems design to the development of interior systems is similar to that for enclosure systems. Therefore, the discussion in Sec. 15.7 may be reasonably extended to include the design of interior systems. There are, however, some particular issues that are quite significant for interior systems, while they are either of less importance or totally lacking from the concern for the building enclosure. The material in this section is limited to discussion of these particular issues.

Of critical concern for enclosure systems are the thermal, acoustic and light-transmission character of the enclosing construction. Of these the only major concern for interior systems is that of acoustic transmission. Obtaining privacy and preventing noise transmission between adjacent offices, apartments or school classrooms, for example, is a major design factor for interior wall or floor construction. As discussed in Chapter 12, design for the control of sound and vibrations is best solved by considerations of the choice of materials, components and details of the basic building-construction systems. If these can be developed to maximize acoustic benefits, without compromising economy or other performance requirements, the overall cost–benefit aspects of the building design will likely be optimal. If sound and vibration problems are dealt with only as corrections of the fully developed design—largely after the fact of all the other design decisions—a good total building system design is unlikely.

Of major concern for elements of the interior construction is the choice of materials and details that generally prevent or retard spread of fires. Details of wall and floor construction and choice of surface finishes are typically subject to critical limitations due to building-code requirements. Considerations for safe egress will provide major input for minimum design standards for doors, corridors and stairs. Except for exit doors, these issues are generally of less concern for the building enclosure systems; an exception being the case for closely situated buildings in densely builtup urban areas.

Floor–ceiling and roof–ceiling systems present a major opportunity for integrated design of the building construction and the various service systems for HVAC, lighting, electric power, communication and fire control. For commercial interiors particularly, this area of building design is one with a relatively long history of integrated system development. Changing needs and evolving new technology present possibilities for new solutions, but this area of building construction is already stocked with products and systems for multipurpose, integrated utilization.

It is also important to select interior finishes that not only retard the spread of flames from fires in interior spaces but also provide acceptable light-reflectance characteristics. In addition, the finishes must satisfy the aesthetic requirements of the owner.

There is, of course, no clear divorce between the systems for building enclosure and those for interior development. Interior walls, floors and ceilings meet the building envelope at some points. Exterior doors and windows must relate to locations of interior partitioning. Utilization of natural light and ventilation must be coordinated with the artificial lighting and mechanical ventilation systems. In the end, the building is a whole, single system and its totality is more important than its parts.

GENERAL REFERENCES AND SOURCES FOR ADDITIONAL STUDY

These are books that deal comprehensively with several topics covered in this chapter. Topic-specific references relating to individual chapter sections are listed at the ends of the sections.

F. Merritt, *Building Design and Construction Handbook*, 4th ed., McGraw-Hill, New York, 1982.

C. Ramsey and H. Sleeper, *Architectural Graphic Standards*, 8th ed., Wiley, New York, 1988.

D. Watson, *Construction Materials and Practices*, 3rd ed., McGraw-Hill, New York, 1986.

H. Olin et al., *Construction Principles, Materials, and Methods*, 5th ed., The Institute of Financial Education, Chicago, 1983.

Sweet's Architectural File, McGraw-Hill, New York (issued annually).

For building code requirements, use local code or the latest edition of one of the following model codes:

Uniform Building Code, International Conference of Building Officials, Whittier, CA, 1988 (new edition every three years).

The BOCA Basic National Building Code/1984, 9th ed., Building Officials and Code Administrators International, Country Club Hills, IL.

Southern Standard Building Code, Southern Building Code Congress International, Birmingham, AL.

EXERCISES

The following questions and problems are provided for the review of the individual sections of the chapter.
Section 16.1

1. What is the major difference in requirements for exterior and interior walls?
2. In addition to acting as a space divider, what other purposes may a partition or interior wall serve?
3. Explain why bearing walls are generally not economical as space dividers in a department store, except at permanent installations, such as stairs and elevators.

Sections 16.2 and 16.3

4. What are the purposes of doors?
5. What is the minimum width of opening permitted for an exit door?
6. What is the minimum height permitted for an exit door?
7. What are the major differences in requirements for entrance and interior doors?
8. Where are stiles used in doors?
9. For what purpose are Kalamein doors used?
10. Which type of door operation is the only one permitted for fire doors?
11. What are the advantages of vestibules at building entrances?
12. What types of doors may be used where wind makes operation of swinging doors difficult?

13. How does a butt differ from a strap hinge and for what purpose?
14. What is the purpose of a pin in a hinge?
15. How is a half-surface hinge installed?
16. What is the purpose of a fusible-link closer?
17. What is the purpose of a latch?
18. What is the advantage of a latch bolt over a dead bolt?
19. What are the advantages of unit and bored-in locks?
20. How may a closed door equipped with panic hardware be opened from inside a building?
21. Which types of doors are suitable for large openings?

Section 16.4

22. Name the usual components of a floor–ceiling system.
23. What is the main advantage of a floor covering?
24. What are the major purposes of a floor?
25. Of what building systems are beams a component?
26. How can a plenum between a floor and a ceiling be utilized?
27. Is a separate ceiling necessary?
28. Of what building systems may ceilings be a component?

Section 16.5

29. Discuss the interaction of interior finishes with the lighting system.
30. Discuss the interaction of interior finishes with the acoustical system.
31. What are the requirements for floors under resilient flooring?
32. What is terrazzo?
33. Which flooring materials should not be used on concrete basement floors?
34. What is the main ingredient of gypsum building products?
35. What are the advantages of Keene's cement?
36. What are the advantages of dry-type finishes over wet-type?
37. What is the purpose of a floating angle?
38. What are the components of a wet-type plaster finish?
39. What are the components of plaster?
40. What is the purpose of grounds?
41. For what bases is three-coat plaster required?
42. What are the components of gypsum wallboard?
43. What is the purpose of gypsum coreboard?
44. What treatments are needed for gypsumboard (not supplied predecorated) before application of paint or wallpaper?

Chapter 17

Building Systems

Preceding chapters describe the various major subsystems that usually comprise a complete building. The chapters also explain the various design methods applicable to the individual subsystems. One of the goals of this chapter is a discussion of the coordination and integration of the subsystems into the overall design of the building.

In addition, this chapter presents a number of case studies of recent building-design projects. Some are presented to show applications of value engineering and systems-design techniques that led to some design innovations or improvements. Others are presented primarily to describe situations in which design would have benefited greatly from the ability to apply such methods—whether they actually were or not.

It is possible to get the impression that systems-design procedures are highly sophisticated and somewhat mysterious, and are really only well suited to major, highly complex projects. In truth, systems-design methods can be applied in a variety of ways, at a range of levels of sophistication, to projects of different nature and complexity. Of course, the more complex and seemingly unmanageable the project, the more systems design is needed. But the basic systems approach can be beneficial and justified for just about any design project. For projects of modest scope, the benefits may be modest, but nontheless real.

17.1. MISHAPS AND CORRECTIVE MEASURES

Development of a building project is both predictable and unique. Uniqueness comes largely from the specifics of time and place and the particular persons and organizations that are involved in the project. Thus, every building is literally unique, and the unfolding of events in the design and construction processes is truly a one-time occurrence. The quality of the end product—the constructed building—as a measure of the success of the participants and the soundness of design judgments in general must be viewed against the background of the events that occurred as the work progressed. It is always easier to play the big game over on Monday, while watching the recordings of last Saturday's events.

Most of the events that occur during the progress of a building project are predictable as to nature and timing, and in many cases are subject to reasonable anticipation or control. Successful projects are seldom accidents; rather they come about because of the ability of designers and managers to anticipate events and the actions necessary to deal with them, to control well what can be controlled, and to provide for eventualities that are unpredictable with some degree of flexibility and versatility.

Given the number of groups and persons ordinarily involved in design and construction of a building, the length of time for the whole process and the interdependence of many actions and events, the potential for chaos is quite real. Although some mistakes, unforeseen conditions and controversies cannot be predicted, some must certainly be expected and provisions made to cushion progress of the work from mishaps. Some common potential mishaps are the following:

Mistakes—in judgment, computations, estimates, drawings, specifications, verbal instructions, filing, for example.

Accidents—files destroyed, design-document originals lost, computer data erased, construction-site accidents.

Unanticipated changes in the rules—new building codes, new products replacing old ones, new tax laws, research showing old tried-and-true methods to be wrong.

Fires, floods, earthquakes, windstorms, sabotage and armed conflict. Strikes, crime crack downs, bankruptcies, mergers, takeovers, swindles, thefts, etc.

Unforeseen site conditions—unusable subsurface rock, unreliable soil, high water table, undiscovered utilities tunnels—conditions that may make planned foundation work or site grading impractical or uneconomic.

Large shifts in the national or international economic or political conditions.

The objective should be not to be caught totally unprepared for these events and for the unstablizing effects they may have on the progress of the work. At the least, some corrective measures should be possible, preferably provided for beforehand and taken up automatically when a mishap occurs.

Persons highly organized in a systems-design operation may be especially vulnerable to unstablizing effects. The linear progression and high efficiency level of such an operation may be easily derailed. Persons whose normal operation is a day-to-day, fly-by-wire type of program may work inefficiently, but are naturally suited to dealing with things as they come up—good or bad, anticipated or not.

For the tightly programmed operation, it is essential to have recovery, reassessment and adjustment built into the operation as much as possible. This requires the incorporation of a lot of "what if?" steps and a large storage of corrective or alternative procedures.

Well-planned management programs deliberately provide for many strategic checkpoints in any progressive operation. The building-design process, as illustrated in Fig. 1.5, clearly provides for major checkpoints at the end of the several phases of the design work. This general procedure of deliberate pauses for approval and evaluation should continue through any production process. Once the parties involved are conditioned for these checkpoints, the procedure may be easier to invoke in the event of emergencies due to unanticipated events.

17.2 DESIGN OF A BUILDING SYSTEM

A building system is composed of numerous subsystems, as described in preceding chapters. The subsystems must perform tasks assigned by the building designers to insure that the building system meets the goal set for it by the building owner. The tasks are defined by establishing for each subsystem a goal and listing objectives and constraints that it must meet. The objectives introduce design variables that the owner and designers can control. The constraints impose conditions that are beyond the control of the owner and designers.

Performance of the total of the tasks of all the subsystems should be equivalent to performance of the tasks required of the building system. Hence, the complete set of objectives and constraints for all the subsystems also forms the objectives and constraints of the building system. To meet its goal, the building system therefore must satisfy all the objectives and constraints of its subsystems. For an optimum building, the system should do neither more nor less than its goal, objectives and constraints require.

For some buildings, the objectives and constraints may be familiar to the designers because of repeated occurrence on previous design projects and may also be small in number, or at least can be reduced to a few high-priority items. In that case, the designers may be able to design an optimum building directly from a well organized list of objectives and constraints.

(While it is desirable to record objectives and constraints as a checklist for design, it is not necessary to list every detail. Many of the constraints, for example, are contained in building codes, zoning ordinances and generally accepted design standards and codes of practice and are customarily satisfied by designers without the necessity of listing them or referring to those documents for every design task. Hence, only unusual and major constraints must be included in a checklist.)

A one-story gymnasium or field house may be one example of a type of building that can be designed directly from a complete listing of objectives and constraints. Such a building might be built with a single-enclosure system (thin-shell construction, for instance) on linear spread footings. The floor, placed on leveled ground, might be a concrete slab with appropriate floor coverings in various sections of the building. A few partitions would be erected for privacy in offices, dressing rooms, showers and toilets and for protection of storage items. Entrance and interior doors would be installed in suitable locations. Similarly, other subsystems, such as lighting, heating, ventilation, grandstands, lockers and athletic equipment would be provided to satisfy remaining objectives and constraints. A study of a relatively few alternatives would then enable the designers to optimize the design speedily.

Most buildings, however, are too complex and have too many objectives and constraints to permit such a simple design procedure. In such cases, the only practical method is separate design of each subsystem, or in some cases, of each subsubsystem, and integration of the components to form an optimum system.

It is conceivable that a design could be optimized by completion of design of optimum subsystems in succession. As each subsystem design is finished, the preceding completed subsystems would then be redesigned to optimize the design of the combined subsystems. In general, however, this procedure would be time consuming and inefficient.

For practical reasons, therefore, design of the various subsystems and subsubsystems should be executed simultaneously. Sometimes, it may even be essential that building construction commence before completion of design. As a result, construction of some parts of a building and design of other parts will proceed concurrently.

To insure integration of the subsubsystems and subsystems and successful functioning of the whole building system, and to reach an optimum design speedily, expert design management is required. The activities of the members of the building team should be well organized, supervised and coordinated. The members should communicate frequently with each other so that each member is kept informed of the designs of others, as work proceeds or revisions are made. The team leader should insure that the system goal, objectives and constraints are being satisfied and should guide design efforts toward optimization of the total building system. Frequent estimates of initial and life-cycle costs and of the completion date of construction should be made to ascertain that neither costs nor time will violate objectives or constraints. Also, value engineers should review the designs as work progresses to see if better alternatives should be adopted.

In the schematic stage of design, the generation, rapid analysis and quick appraisal of alternative systems is highly important to speedy achievement of an optimum building design. Toward this end, designers and value engineers should cooperate in suggesting better ways to satisfy objectives and constraints. The team should investigate possibilities of eliminating or combining subsystems, to reduce the number of components required, to obtain better performance, higher quality or lower costs. (Use of one component to serve several purposes; for instance, use of a hollow floor that also serves as a ceiling, fire barrier, sound absorber, air-conditioning ducts and conduit for pipes and electric wiring, can often replace numerous components, with a savings in initial and life-cycle costs.) The team should also consider designs that can be prefabricated to yield higher-quality products, speedier construction and lower costs.

At the end of the schematics stage, choice of the subsystems, or at least of the major subsystems, should be final. (Later changes in choices of subsystems might cause abandonment of nearly all completed design work and require several members of the building team to start design again at the beginning, with considerable expense and lost time.)

Cooperation of designers and value engineers is equally important in later design stages. In those stages, however, the emphasis of value analysis should be shifted to smaller building components than the major subsystems. In the design-development stage, designers and value engineers should investigate elimination or combination of subsubsystems, or even smaller components. Also, the team should continue to investigate possibilities of prefabrication. In the final design stages, the team should concentrate on improvement and cost reduction of details and small components incorporated in them. In those cases where small components are repeatedly used throughout a building and represent a substantial life-cycle cost, however, value analysis should be performed during an early design stage, because changes made as a result of the analysis might cause substantial revisions of subsystem or subsubsystem designs.

It is especially important that value analysis be applied to construction specifications and contracts. These may present many opportunities for design and construction improvements, cost reduction and energy conservation.

After each cost estimate has been made and the necessity for cost reduction has been established, value analysis should determine which items have a significant effect on costs. These items should be given special attention in cost-reduction investigations.

In studies aimed at improvement of performance of building components and at energy conservation, members of the building team experienced in building operation and maintenance should be consulted, to obtain criticisms and recommendations. Similarly, in studies of cost effectiveness of building components, availability of materials and equipment, and potential construction methods, the construction experts on the building team should be consulted.

Until recently, the areas of building design in which systems-design techniques were most utilized were those linked to the use of particular commercial products and systems, or those related to special concerns of current major interest. Typically, the concentration of time and effort and the availability of funding for research is greater in such areas. In recent times, prime concerns of this kind have supported major development of products and extraordinary expenditures of design efforts on building design projects relating to concerns such as energy use, daylighting, hazard abatement, barrier-free environments and so-called smart buildings (responding automatically to changing conditions). Areas of design of less public or commercial concern have been slower to utilize the full potential of developing technology and knowledge for sophisticated design methodologies.

This uneven development is really only a matter of the typical nature of the progress of science and technology. As public interests shift (from winning a war to more mundane concerns, for example), as the directions of product-development-oriented research by industries swings back-and-forth and—quite simply—as time marches on, less-developed areas of effort tend to catch up. Thus, the disproportionate concentrations of money, design talents and time and energy that were directed toward getting man on the moon resulted in some major advancements in science and technology and the eventual redirection of highly trained people into areas such as city planning, medical science, information handling, management techniques and general systems-design methodologies.

Good design does not necessarily come about because of slick methods or highly organized work processes. With the latest CAD (computer-aided design) programs and systems methods at their disposal, incompetent designers will most likely still produce poor designs. Nevertheless, as building projects steadily become more complex and public demand for quality in the constructed project becomes more insistent, successful management of the design process becomes a necessity for survival as a business and profession.

Building design in general—in all areas—is steadily gaining the potential for routine application of advanced design methodologies. As education for engineering, construction, architecture, landscape design and interior design improves, including computer applications, skilled manpower for design and construction becomes more available. Coupled with this is the increasing pace of development of software and hardware for design-office utilization and the simple accumulation of design-application experience. As stated previously, this is not necessarily going to result automatically in better design, unless production capability falls in the hands of talented persons. But for the competent, creative designer, computer assistance is certain to increase to some degree the likelihood of competent treatment of complex design situations with greater ease.

17.3. CASE-STUDY ONE: McMASTER HEALTH SCIENCES CENTER

Location: Hamilton, Ontario, Canada.
Architects: Craig Zeidler Strong, Toronto (later reorganized as The Zeidler Partnership).
Design period: Late 60s to early 70s.

The design goal for this project was to provide, in one or more buildings for regional use, facilities for health services, related research and health education, at lowest life-cycle cost. Future growth and change at minimal cost and with least disturbance of current activities was a principal design requirement.

Two concepts, given widespread attention by designers in the 1960s, were used as major factors in the design of this project. The first involves the use of space often produced within a building but not occupied for the routine functioning of the building users. This is the space inside the construction components—most notably between the ceilings and the floor or roof above. This space, called interstitial space or plenum space, is often occupied by various components of the building's service subsystems: ducts, piping, wiring, recessed lighting and some equipment. An idea that emerged in the 1960s was to make this space deep enough for a person to stand up and walk in, so that maintenance and alterations of the service elements could be routinely and repeatedly accomplished with minimum disturbance

of the spaces above or below that were being served by the items in the interstitial space.

The second design concept involves the development of space as an assembly of modular units, which can be added to, subtracted from, or reassigned as space needs change or new functions are required. For the Health Sciences Center at McMaster University, the modular unit consisted of both a basic plan module and a three-dimensional component utilizing interstitial space in a multistory form. The success of the design stimulated the use of these concepts in other buildings—particularly large health facilities.

In conventional construction interstitial floor–ceiling or roof–ceiling space is usually developed on a minimal basis, in order to reduce the total building volume, exterior wall surfaces and vertical runs of stairs, elevators, ducts, piping and wiring. This puts constraints on horizontal-span framing and limits the possibilities for clear-span, column-free areas in the occupied spaces. The system generally developed to form walking-height interstitial space and also to develop maximum column-free occupied spaces employs longspan trusses. The top chords of the trusses form the basic framing for the floor or roof above and the bottom chords serve as framing for both the ceiling below and the floor of the interstitial space.

Hospitals have special requirements that make interstitial spaces advantageous and possibly economical. For example, changes in health-care services with time require that provision be made in building design for future changes in space use. Hospital interiors, though, must be supplied with numerous mechanical-electrical services. If alterations cannot be easily made to meet changing future needs, changes in the services will be very costly. In addition, maintenance of the services is an important concern not only because of cost but also because long shutdowns for repairs may be intolerable in a hospital. Walk-in interstitial spaces offer hospitals the advantage of easy access for repair, maintenance and alterations, with consequent lower costs for these operations.

In a multistory hospital, interstitial spaces are alternated with functional spaces, those used for the prime purposes of the building (see Fig.17.1). Generally, from the horizontal services in an interstitial space, HVAC ducts serve the functional space below, plumbing the functional space above, and electrical cables both the functional space above and the one below. Vertical shafts are provided in appropriate locations to house the risers that feed the horizontal runs. The shafts usually also contain electrical and telephone equipment rooms.

Although interstitial spaces have not been widely used in multistory buildings, modular planning, in contrast, is frequently used. In office buildings, for example, each of the various stories of a building is used in about the same way. Different tenants, however, usually desire to divide the floor area they occupy in different ways. Hence, flexibility in changing sizes of spaces is desirable, but alterations of mechanical-electrical services may not often be necessary. Modular planning is a useful tool in providing the desired flexibility.

A planning module usually extends from the floor to the ceiling and may have any desired floor area. The planning module is a complete unit; it has its own HVAC and lighting and telephone services. Consequently, it may be enclosed by placement of partitions along its borders without impairment of function. The space assigned for any use in any story of the building therefore, can be assembled by addition of as many modules as needed to supply the required floor area. If, in the future, more or less space is needed for that use, modules may accordingly be added or subtracted.

The concept can be expanded to include a group of planning modules that form a megamodule, or very large module. A megamodule is also a complete unit, with its own mechanical-electrical services. A building may be constructed as an assembly of identical or nearly identical megamodules. Future expansion may be accomplished by addition of more megamodules.

Parts of a building can be classified as either permanent or nonpermanent. Permanent parts are those that are not changed when alterations are made for changes in space use. These elements usually are the structural framing, stairs, elevators, mechanical-electrical risers and such

components as partitions and curtain walls that, when removed, can be reused in a different location. Nonpermanent building parts are elements that serve only special functions and usually cannot be reused when alterations are made. To keep cost of alterations low, the cost of nonpermanent elements in a building should be a low percentage of the total capital investment in the building.

This objective is achieved with the combination of interstitial spaces and modular planning. Major components, such as the structural framing and main mechanical and electrical elements, need not be changed when space use changes. Generally, only minor alterations in architectural components and secondary mechanical and electrical branches should be necessary. Consequently, compared with alteration costs for conventional multistory buildings, cost of alterations to meet changing space-use needs should be small for a building with modular planning and interstitial spaces.

This was the prime reason for use of modular planning and interstitial spaces for the Mc-Master Health Sciences Center. About 40 acres of floor area had to be provided for the use of 49 independent departments, which could share some facilities. More than 5,000 persons could be expected to use the space daily. The hospital was to provide about 400 beds. The outpatient department was expected to accommodate about 400,000 visits per year. Research needed about 400,000 sq ft of floor area. Teaching facilities were required for about 900 students. Also, a parking garage was to be planned for 1,000 automobiles.

The buildings were to be provided with lighting, acoustical treatment, telephone, television, electrical services and all-year-round air conditioning appropriate for the various space uses. Services required included plumbing, domestic hot water, cold water, water for fire protection, steam, deionized water, compressed air, vacuum, heating gas, oxygen and nitrous oxide. For the plumbing, sanitary wastes had to be kept separate from laboratory and isotope wastes. For HVAC, special provisions had to be made for exhausting fumes from laboratories.

As the need for the facilities was urgent, it was desired to begin construction as early as

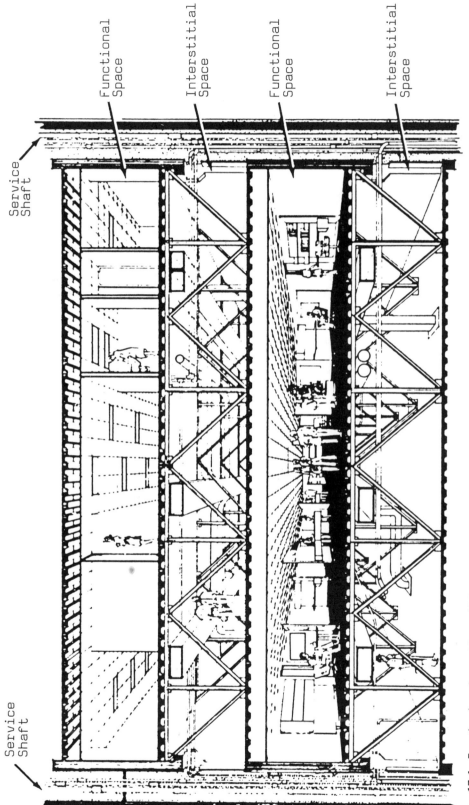

Fig. 17.1. Functional stories and interstitial mechanical-electrical spaces alternate vertically in St. Joseph Hospital, Tucson, AZ. (Courtesy of Willam Wilde and Associates, architects.)

657

possible. It was also desired that the design permit future changes in space use, to keep pace with progress in medicine, medical research, and hospital management. Although the required functions to be housed were discrete (inpatient services, outpatient services, administration, research, and teaching), it was desired to have them physically integrated in order to avoid isolation of involvements and promote an orientation to the central purpose of health care. While this might have worked best in a high-rise building, it was desired to keep the structure low to harmonize with the rest of the McMaster campus.

Life-cycle costs were to be minimized. These included not only initial construction cost, maintenance costs, and operational costs, but also the cost of future alterations necessitated by changes in space use.

Because the interstitial spaces had to permit unobstructed runs of mechanical-electrical services, vertical fire barriers could not be erected in the spaces to prevent spread of fire. Nevertheless, local authorities required 2-hr fire protection for individual structural members and 2-hr resistance to spread of fire from room to room through the plenum. (Some building codes also require fire sprinklers in plenums not firestopped.)

In addition, the Canadian Department of Labour required that a flush walk-on surface be provided in the interstitial spaces. This requirement could be met only by a deck at or above the level of the truss bottom chords. A ceiling below those chords could not serve the double function of a walk-on deck and a fire barrier, at what would have been a considerable savings in cost.

A team approach to design was adopted from the beginning for the Health Sciences Center. Also, from the beginning, the users (doctors, nurses, administrators, researchers, etc.) were included as a source for information on the functional requirements of the Center. An effective system for transfer of a vast amount of information from the users to the other members of the building team had to be provided.

For this purpose, planning was divided into three stages:

1. Conceptual and basic programming.

2. Basic design and department-area definition.

3. Final department planning.

Stages 1 and 2, which involved department-planning teams, did not require development of as much information as Stage 3. Hence, during Stages 1 and 2, use was made of a simple organization that was controlled by the team leader and the vice president of the Health Center. User groups with special assignments fed information directly into the planning process. Stage 3 commenced with the start of design development, when more detailed information was needed. User groups then conferred directly with design teams dealing with specialties. The team leader and Center vice president, although relieved of the necessity of transmitting huge amounts of information between user groups and the designers, nevertheless were able to monitor the information flow from reports received.

In Stage 3, the project architect and seven architect area planners maintained contact with the user groups. Also, in establishment of mechanical and electrical requirements, the engineering members of the building team became closely involved with the users.

In the early planning stage, a basic decision was made that permitted an early start on construction of the Center. This decision was to keep the mechanical and electrical systems separate from the user areas. Construction then could proceed before the users' needs were determined in detail. Consequently, when the architects' schematics were approved, establishing the building module and the basic size and relationships of the departments, foundation design commenced. When it was completed, excavation and subsurface contracts were let. While the substructure was being bid and built, the structural steel for the building framing was designed and a contract awarded for its fabrication and erection. Also, design of the mechanical and electrical systems proceeded. Finally, design of the department areas was completed.

The owner engaged a construction manager to handle, coordinate, expedite and generally supervise the subcontract work. This manager was also part of the building team. He was available for consultation during the design stages.

The early start on construction is estimated to have saved about two years compared with the time required for the conventional method in which construction starts after completion of design.

Two decisions were made to achieve physical integration of the facilities for health care, research and education without disruption of individual department functions. One decision was to place all facilities in only one building. The second decision was to make that structure a low one.

The Health Sciences Center consequently was designed with four functional levels, with provision for addition of a fifth level in the future. The parking garage was placed below the first level. A mechanical-electrical interstitial space was located above each functional story.

The lowest functional level accommodates all aspects of materials handling and associated departments. It also houses some teaching and research facilities. It may be entered from the side of the building facing the campus.

The main entrance to the hospital is at the second level. This level has facilities for administration of emergency treatment, admission of inpatients, family practice, accounting, administration offices and records, nursing school, operating suite, radiology, intensive care and several other health-related departments.

On the third and fourth levels, hospital inpatient facilities are located on the south, research on the north, with related outpatient and clinical services between.

Floor Layouts. Figure 17.2 shows a plan of the fourth functional level. Surgical inpatient units A occupy the southeast corner of the floor. They partly encircle surgical outpatient clinics B and are contiguous with offices of the specialized surgical teams. To the north are the main areas for surgical research J. The department of neurosciences H links the inpatient, outpatient and research areas. Classrooms and conference rooms used by the medical staff, researchers and students are distributed throughout this eastern half of the floor. Facilities for obstetrics and gynecology are similarly placed in close relationship in the western half of the fourth level.

Each functional level provides nearly 350,000 sq ft of floor area. To simplify and guide traffic flow to destination points throughout this large area, a "Ring Street," or main corridor, envelops the central portion of the level and links all departments (see Fig. 17.2). The Ring Street is given distinguishing characteristics so that it can readily be differentiated from corridors connected to it. Elevators are located at intersections of perpendicular branches of the Ring Street. Maximum distance from any part of the main corridor to an elevator is 140 ft.

Examination of Fig. 17.2 reveals that the building is an assembly of megamodules 94 ft 6 in. $\times$ 115 ft 6 in. in plan. The building is planned for future expansion by addition of such megamodules. Figure 17.3 shows a typical megamodule, with part of the floor at the second functional level removed to expose the structural framing and interstitial mechanical-electrical space below. The megamodule, in turn, is an assembly of $10\frac{1}{2}$-ft-square planning modules.

Floor–Ceiling System. Each functional floor is supported on a grid of perpendicular steel trusses. The trusses lie along the borders of the planning modules and hence are spaced $10\frac{1}{2}$ ft apart. The trusses are $8\frac{1}{2}$ ft deep to provide sufficient height for the walk-in interstitial spaces, as illustrated in Fig. 17.3.

Supports for the truss grid consists of a steel-framed tower at the four corners of the megamodule (see Fig. 17.3). The towers are hollow and may house stairs or mechanical-electrical risers. As indicated in Fig. 17.3, the trusses span $73\frac{1}{2}$ ft between the towers.

The functional floors consist of cold-formed steel decking, which spans the $10\frac{1}{2}$ ft between the top chords of the trusses, and a lightweight concrete fill. The ribs of the decking are used for horizontal distribution of electrical wiring. Because the concrete slab is nonstructural, openings can be made in it where needed for access to the electrical services.

To form a ceiling in the functional spaces, cold-formed metal channels were hung 30 in. apart from the bottom chords of the trusses. These channels may house lighting, HVAC

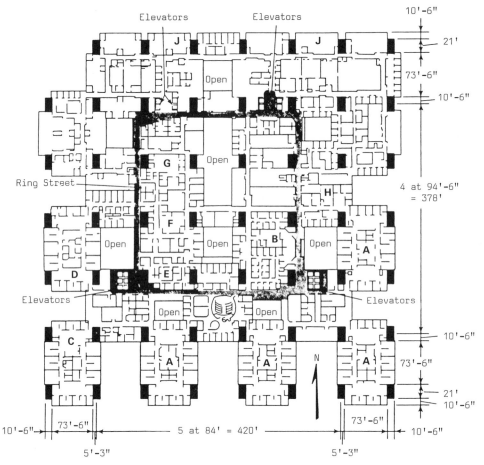

Fig. 17.2. Plan of the fourth functional level of the McMaster Health Sciences Center. As noted on plan: A. surgical inpatient unit. B. surgical outpatient unit. C. gynecology inpatient unit. D. obstetrics inpatient unit. E. obstetrics-gynecology outpatient unit. F. delivery unit. G. neonatal unit. H. Neurosciences. J. research laboratories.

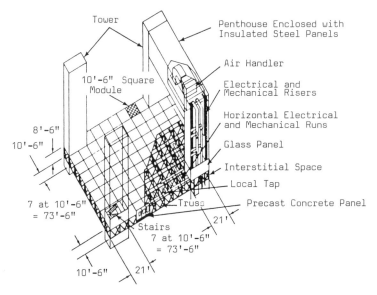

Fig. 17.3. Megamodule used for the McMaster Health Sciences Center is an assembly of 10.5-ft-square plan units.

grilles, acoustical tiles or other items to serve a module. Acoustical tiles span the 30 in. between the channels. For fire protection, a gypsum-board membrane was laid on top of the channels. Above the membrane, supported on the truss bottom chords, is cold-formed steel decking that provides a flush walk-on surface in the interstitial spaces. Removable panels were placed along the lines of the channels to allow access to them from above.

Movable Facades. Curtain walls were selected for vertical enclosure of the building to permit changes in the exterior without altering the structural framing. The walls between towers are formed with precast-concrete panels with only four different sizes. One size provides a window opening that occupies almost the whole panel. Two sizes contain window openings about half as large. The fourth size is used for blank panels. This variety of panels made it possible to arrange window panels to suit interior layouts and permit future changes in those layouts or expansion of the building. The towers are glass enclosed, and penthouses are clad with insulated, galvanized-steel panels.

HVAC System. During design of the building, studies indicated that greater economy and flexibility in providing HVAC could be achieved with a combination of 80,000-cfm air-handling units than with a large central system. Accordingly, factory-built units of this capacity were specified. They were placed in a penthouse high enough above the roof to allow enough space for a possible future fifth functional level (see Fig. 17.3). They supply air to risers in the megamodule towers.

Air is distributed through a medium-pressure, single-duct, terminal reheat system. This system is adaptable to changing loads, is flexible for future modifications, can accommodate many zones and requires less ductwork than a double-duct system. Junctions of branch ducts and diffusers were designed to permit all ductwork changes to be made in interstitial spaces. Only HVAC grilles need be changed from the story below.

Electrical Power. The electrical system was also designed to provide flexibility in operation

and making changes. The power distribution system is primary selective, served by two feeders, with automatic transfer. Local substations are located in penthouses above the roof. Also, diesel engines for generating emergency power are placed in the penthouses.

Because of the large electrical loads in the building and the lengthy feeders required, a 347/600-volt, 3-phase, 4-wire system was chosen for secondary distribution to all major mechanical loads and fluorescent lighting. Use of the high voltages cuts costs of feeders and equipment and permits large areas to be served from a single substation. Convenience outlets, incandescent lighting, small motors and most of the laboratory and instructional equipment are served by a 120/208-volt, 3-phase, 4-wire distribution system.

Layered Services. The various mechanical and electrical services are assigned to horizontal layers in the interstitial spaces. This layout reduces interference between the services. Also, during construction of the building, the separation of services permits the various building trades to work simultaneously in the interstitial spaces, thus reducing construction time.

Analysis. Although the building is low and thus has large floor areas at each level to meet the space requirements of the users, it covers less than ten acres of the 13-acre initial site. Compared with conventional construction with a service base and nursing tower above, the low building offers the advantages of lower cost, better functional integration, less vertical traffic and efficient material handling. Operational costs are lower because of fewer elevators, shorter elevator runs, and fewer windows with consequent smaller heat loss.

The design of the building satisfies the goal, objectives and constraints. It can be readily expanded by removal and reuse of curtain-wall panels and addition of megamodules. Each megamodule provides a floor area of more than 8,000 sq ft unobstructed by columns. These large column-free areas eliminate the need for structural alterations when space use changes in the future. Concentration of all mechanical and electrical services in interstitial spaces and shafts offers flexibility in meeting future space needs at low cost and reduces maintenance and repair costs. This flexibility is further improved by stratification of the services in the interstitial spaces, because access is easy and space is available for extra services that might be required in the future.

SECTIONS 17.1 THROUGH 17.3

References

1. *Energy Management Case Studies*, Edison Electric Institute, 1111 19th Street NW, Washington, D.C. 20036.
2. *Engineering News-Record*, May 27, 1976.
3. E.H. Zeidler, *Healing the Hospital*, The Zeidler Partnership, Toronto.

17.4. CASE-STUDY TWO: XEROX INTERNATIONAL CENTER FOR TRAINING AND MANAGEMENT DEVELOPMENT

Location: Near Leesburg, Virginia, and Dulles Airport; 40 miles from Washington, D.C.
Architects: The Kling Partnership, Philadelphia.
Design Period: Late 60s to early 70s.

This facility—opened in 1974—was built to replace ten regional facilities and to centralize the many training and orientation programs of this large, internationally operating corporation. The objective was to provide a pleasant residential/educational/communal setting for trainees that could handle simultaneously many intensive two-day to two-week programs in a university-like environment.

Although the purpose for the Xerox Center was quite different, there were many similarities in design criteria to those for the McMaster Health Center, described in Sec. 17.3. These included:

1. Need for an early start of construction to permit early occupancy of the facility.
2. A design period in which energy conservation was emerging as a major concern; relating mostly to passive considerations of the building and operating efficiency of the HVAC and lighting systems.

3. Desire for a modular building planning unit with some ease of modification to accommodate changes in functions—primarily in the sizes of groups and the types of instruction—from technical, hands-on laboratory work to management seminars.
4. A plan that would permit future growth of the facility with minimum disturbance of current operations.
5. Arrangement of spaces and organization of pedestrian traffic to promote a communal atmosphere and group interactions—both formal and casual.

The site for the Center was required to be near an airport, because students might be sent to the center from any place in the world. For proximity to the educational, cultural and public-affairs resources of a metropolitan area and convenient delivery of food and other supplies, the site had to be near a city. Yet, it was desirable that the site be far enough away to avoid disturbing traffic noises and other urban distractions. Also, the site was required to be large. The aim was not only to accommodate the large buildings anticipated initially and to provide space for future expansion but also to insure tranquil surroundings conducive to study and recreation.

Xerox wanted the Center to provide more than just classrooms and laboratories. It was also to provide room and board, opportunities for exchanges of information between all students, health care and recreation. On completion of the first construction stage, the center was to accommodate 1,000 students at a time and a staff of 500. The students might reside at the Center from a few days to several weeks. Subsequent construction stages would expand the Center's long-range capability for service as a training center for all levels of the corporate organization. In general, classes would be small, ten students or less.

Sleeping quarters for the students were not to be luxurious, but sleeping, conference and learning spaces were required to be comfortable. All buildings were to be well lighted and air conditioned all year round.

The architects were engaged in time to assist in final site selection. After that, for nearly 2 months, a team of six architects and a Xerox task group worked together to develop the basic program and schematics.

Xerox specified that opening of the first classes was to occur within 24 months after selection of the architects. Hence, a contractor was engaged at an early stage so that construction could start as soon as site and foundation design drawings became available. Generally, construction followed only about a step behind building design.

In its initial stage, the Xerox International Center for Training and Management Development contained three buildings on a hilltop of the 40-acre site. Two of the buildings house living–learning facilities. The third building is a gymnasium–auditorium, with a large swimming pool.

The two living-learning buildings are large (1,000,000 sq ft of floor area) and have setbacks at each story to follow the slopes along the hill. In each building, residential and educational facilities are contiguous.

Each building is an assembly of pairs of megamodules. One megamodule of each pair contains residential facilities, whereas the second megamodule houses classrooms and laboratories (see Fig. 17.4). One building consists of two such megamodule pairs. The other building comprises three pairs and administration facilities. In each building, the residential megamodules are interconnected in the form of a crescent, partly encircling the classroom megamodules (see Fig. 17.4*a*).

Residential Megamodules. Each of the five residential megamodules contains private rooms for 204 persons. The megamodule is six stories high, each story set back from the story below to follow the hill slope (see Fig. 17.4*b*). The private rooms are placed along the outside wall, which is serrated to provide each room with a window offering a view of the wooded site. The rooms are arranged in groups of six, each group sharing a common lounge at the same level.

The private rooms and their lounges ring a high-bay commons area, which is at ground level (see Fig. 17.4). The commons connects the residential and classroom megamodules and contains facilities shared by all the students.

These facilities include dining room, cocktail lounge, game lounge, health services, service and registration desk and faculty offices. A pedestrian street above the commons and along the inner wall of the megamodule connects it with the other megamodules in the building and provides access to the classroom megamodule. A utility street at ground level serves the classroom megamodule.

Classroom Megamodule. Three stories high, each of the five classroom megamodules contains classrooms and laboratories on two floors and a service floor at ground level. The classroom areas are an assembly of square modules, each with a floor area of 4,000 sq ft. Within this area, classrooms or laboratories of various sizes can be formed at any time by shifting of partitions.

Engineering. The residential megamodule has reinforced concrete structural framing and architectural concrete exterior walls. It also has clay-tile pitched roofs. The classroom megamodule has structural steel framing and ribbed-faced, concrete-block walls.

Artificial illumination in the living spaces is provided by incandescent fixtures. Low-brightness lighting is supplied in the commons by 250-watt quartz floodlamps. And in the learning areas, fluorescent lamps provide 130 ft-c of illumination at working level.

Each building is heated or cooled by a closed-loop, water-to-air, electric heat-pump system (see Fig. 17.5). The heat pumps range in size from $\frac{3}{4}$ to 20 tons. The smallest of these are the cabinet type, with integral thermostats, and are placed under the windows of the private rooms. Ducted heat pumps are installed above ceilings or in equipment closets to serve larger spaces, such as lounges and classrooms.

In each system, the heat pumps are connected to a closed loop of circulating water, the temperature of which is kept between 70 and 92°F. When cooling a space, the heat pumps reject heat extracted from the space into the circulating water. To heat the space, they extract heat from the water. Supplementary heating, when required, is provided by two 1,500-kw electric boilers in the larger of the living-learning build-

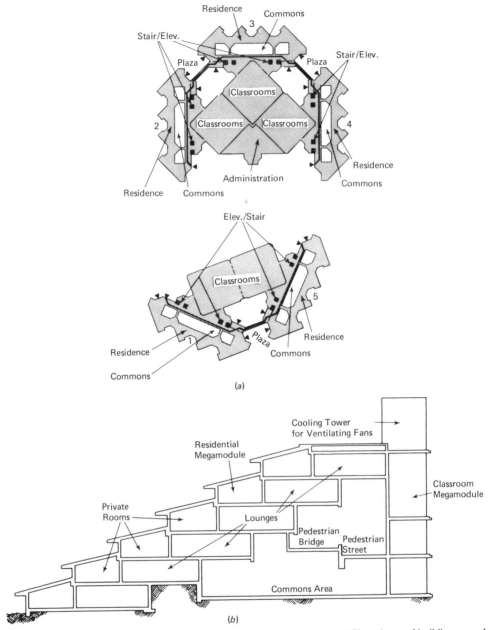

Fig. 17.4. Arrangement and layout of megamodules in the Xerox Center. (*a*) Plan views of building complex and the residential–classroom units. (*b*) Vertical section through a residential module.

ings and by two 1,020-kw boilers in the smaller building. The system, however, usually operates with a net rejection of heat via roof-top evaporative cooling towers.

Analysis. The initial stage of the Center provides facilities for comfortably housing and efficiently training 1,000 students at a time, as

required. The interior design of the buildings, coupled with the wooded site and opportunities for recreation, create a reflective environment that is conducive to study. It also permits employees of various specialties to meet as equals to pursue the goals of individual and corporate development.

The private rooms are essentially efficiency

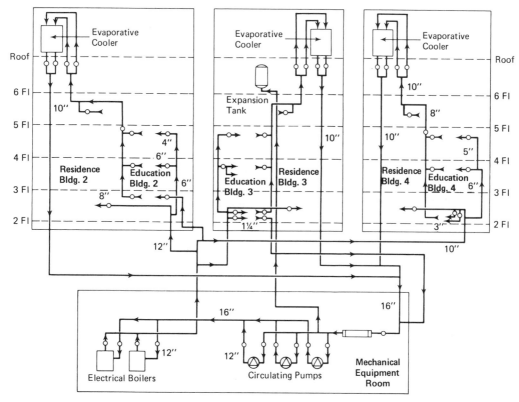

Fig. 17.5. Closed-loop heat-pump system.

units, with bed, built-in desk, chair, closet and shared bathrooms. As a result, students are encouraged to venture into the lounges or commons, where they will meet other students.

The octagonal classrooms promote learning in the round, involving participation of all the students in a class in classroom activities. Outside the classrooms, the junctions of the octagonal shapes creates alcoves in the corridors where groups can gather for discussions during study breaks.

The design of the classroom megamodule provides flexibility for meeting future teaching needs. As these needs change, the spaces can be rearranged or changed in size without the need for structural alterations. The Center can be expanded in the future by addition of megamodules, as needed.

The HVAC system achieves energy conservation through recovery of heat that would otherwise be wasted. Heat transmitted by the heat pumps to circulating water in one space may be recovered to heat another space. For

example, the learning areas with a heavy lighting and equipment load require cooling even in cold weather. Heat extracted from those spaces by the heat pumps in cooling them becomes available for heating the commons or residential spaces.

Water-to-air heat pumps such as those used in the Xerox Center usually have a higher coefficient of performance than air-to-air heat pumps. (This coefficient is the ratio of heating output by a heat pump to the electrical energy supplied to the heat pump.) For the Xerox Center, the coefficient may be as large as 5.

Appraisal

The HVAC installation in the Xerox Center illustrates an important principle: Energy is conserved and operating costs are reduced by recovering and utilizing heat that would otherwise be wasted. For the Xerox Center, the engineers specified heat pumps and a loop of circulating water in each building for the heat-recovery system. While this particular method may not be suitable for other buildings, other heat-recovery methods are available and may prove advantageous for specific design conditions.

Reference

Energy Management Case Studies, Edison Electric Institute, 1111 19th Street NW, Washington, D.C. 20036.

17.5 CASE-STUDY THREE: SUBURBAN OFFICE BUILDING FOR AT & T

Location: Basking Ridge, New Jersey (near New York City).
Architects: The Kling Partnership, Philadelphia.
Design Period: Early 1970s.

The design goal for this project was to provide an optimum building with office space for about 3,000 employees of administrative departments of the American Telephone & Telegraph Company for which space was not available at the national headquarters in New York City.

Numerous objectives and constraints had to be satisfied to meet this goal. Only the few pertinent to the purpose of this section are presented in the following.

AT & T required an office building with about 1,300,000 sq ft of floor area and 1,400,000 sq ft of parking area for automobiles. Because the offices were to be provided for administrative departments, it was desirable that the building be located close to the national headquarters in New York City. The company wanted to provide its employees with excellent working conditions in pleasant surroundings. Hence, the building should have good lighting, all-year-round air conditioning, dining facilities and lounges. Other facilities needed included conference rooms, computer rooms, TV studio, library and storage.

AT & T also wanted to maintain good public relations in whatever community the building would be located. Consequently, the building had to be acceptable to the community in both purpose and appearance and harmonize with its surroundings.

The HVAC system was to be designed to save energy and keep operating costs low. The method chosen employs the heat-recovery principle used for the Xerox Center, which is discussed in Sec. 17.4, but in a different way. Heat recovery in this building is accomplished with a central system, whereas in the Xerox Center numerous heat pumps are used for the purpose.

The architects assisted AT & T in selecting a site. Available parcels of land in the New York City area were rated on the basis of criteria that included cost, soil conditions, topography, flood plains, and transportation facilities. The site selected comprises 150 acres of rolling meadowland, crossed by a stream.

The community wanted the economic benefits of the office building. Yet, the local people did not want the tranquil, rural environment spoiled. Consequently, to maintain good public relations, the new building had to be designed to harmonize with the rural surroundings. Although more floor area had to be provided than that contained in many record-height skyscrapers, a tall building was ruled out, because it would be too obtrusive and unacceptable by the community.

Instead, the new building ranges from two to four stories in height, at most only 48 ft above grade. To provide 2,700,000 sq ft of floor area in such a low building, the architect had to spread it over a large ground area. There are, however, no huge windowless office spaces. The offices are placed in narrow wings with continuous bands of windows in the exterior walls of each story. The wings extend from a central core and branch into additional wings that are wrapped around large courtyards.

Parking for 3,000 automobiles is provided indoors, in two levels under the offices and surrounding terraces. The parking areas are kept out of sight because the enormous paved surfaces would have been out of character with the rural setting. Large openings in the courtyards and elsewhere admit light and air to the garage.

The facades and roof of the office building are designed to avoid a long, monotonous appearance for the wings. For this purpose, the length of the exterior walls is fragmented by large projections or indentations. The projections have mansard roofs and the connections between them have pitched roofs, in both cases with long overhangs. The roofs are covered with brown terra-cotta tile. As a result, the building harmonizes with the residential character of the community in which it is located.

The central core houses a two-story-high reception lobby, lounges, barber and beauty shops, library, TV studio, dining facilities and computer rooms. On the third level of the core are five large conference rooms.

Windows throughout the office areas are double glazed with tinted, heat-absorbing glass. Window frames are bronze anodized aluminum. Curtain walls between the bands of windows are precast-concrete panels. Framing for the building is structural steel.

The offices were planned with a 5-ft-square module. The ceiling of each module was factory assembled with acoustical tile, a fluorescent fixture, an HVAC diffuser and perimeter slots for return air.

Heat-Recovery System. Air conditioning for each module is provided by a variable-volume system. For this purpose, the office spaces are divided into interior zones and zones along the exterior walls (peripheral zones). Supply air is distributed to the modules in ducts. Return air is transmitted from the perimeter slots of each module through the plenums between ceilings and floors.

The temperature in each office is controlled by regulation of the flow of cool air into the space. Regulation is accomplished by adjustment of dampers above the module diffusers.

Cooling usually is required in the interior zones, even in cold weather. The HVAC system is designed to recover the heat ejected from those zones and use it to warm the peripheral zones.

Figure 17.6 shows diagrammatically how this heat recovery is accomplished with a central air-conditioning system. It employs four centrifugal chillers with a total capacity of 4,800 tons, located in the basement. Two of the chillers are used only for cooling. The other two are capable of supplying hot and chilled water simultaneously. When heating is required, they operate as heat pumps. They recover heat from

interior zones, which require cooling, and transfer the heat to convectors along the exterior walls. For the purpose, chillers Nos. 3 and 4 transmit the refrigerant to double-bundled condensers.

The condenser is called double bundled because the shell houses two separate coils, or bundles, of pipes in which water circulates. The water in each bundle extracts heat from the refrigerant in cooling it. The water from one bundle is pumped to evaporative cooling towers, where the water is cooled and then recirculated to the condensers. The water from the second bundle is piped into the heating circuit.

When cooling of the interior zones produces an excess of heat, some of the hot water is diverted into four storage tanks and the rest is mixed with cool water from the tanks. When the tanks are full, they are disconnected from the circuit. In warm weather, the condenser water is pumped to the evaporative cooling towers for cooling. In intermediate weather, cooling requirements may be reduced by use of outdoor air. In very cold weather, supplementary heating is supplied by a 6,000-kw electric boiler. In this boiler, water is flashed into steam by an arc struck between two electrodes, which operate at 12,470 volts, the primary voltage of the utility feeders. (The high voltage eliminates a step-down transformer and its associated initial, operating and maintenance costs.) The steam from the boiler is fed to a hot-water heater.

Computer Control. HVAC, lighting and security systems for the building are monitored and controlled by a high-speed electronic computer. The device optimizes operations of the systems based on data received from numerous points throughout the building.

Appraisal

The AT & T building successfully adapts for a central HVAC system the heat-recovery principle presented in Fig. 17.6 and at the end of Sec. 17.4. Use of double-bundled condensers enables chillers to be used simultaneously for heating and cooling. The heat rejected in cooling interior zones and peripheral zones that re-

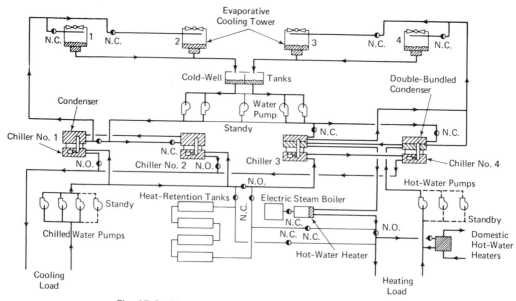

Fig. 17.6. Heat-recovery system used for AT & T building.

quire cooling is recovered by the water in one bundle of the condensers and used to warm zones that call for heat. This system may be adapted for use in other types of buildings.

17.6 CASE-STUDY FOUR: A GLASS-ENCLOSED OFFICE TOWER

Location: Southfield, Michigan.
Architects and Engineers: Smith, Hinchman, and Grylls, Detroit.
Design Period: Mid-1970s.

Tall buildings have very large exterior-wall areas compared with the roof area. If a good deal of the wall areas are occupied by windows, the heat gains and losses through the walls when the building is air conditioned can consume large amounts of energy and make HVAC expensive. Energy losses and HVAC costs could become even larger when the exterior of a building is all glass.

The purpose of this section is to illustrate what was done in the design of one glass-enclosed building to conserve energy and hold down HVAC costs. The design goal was to provide an optimum building with about 150,000 sq ft of office area for American Motors Corp. and an additional 250,000 sq ft of rental office area. Numerous objectives and constraints had to be satisfied to meet this goal. Only the few pertinent to the purpose of this section are presented in the following.

The office building was to share a 33-acre site with a two-level shopping mall. American Motors wanted the building to have an attractive appearance for promotional value for the company, to stand as a symbol of the beauty and quality it puts into its products. The company also required that the building should provide comfortable, pleasant conditions for its employees and tenants.

The HVAC system required for this purpose should be designed for energy conservation and low operating costs.

To meet space requirements and promotional objectives, the architects designed a 25-story sheer tower. Rising 338 ft above grade without setbacks, the building is the most prominent structure for miles around. Headquarters for American Motors Corp. occupy the upper eight stories. The remainder of the office space is leased to tenants.

In keeping with the promotional objectives and also for aesthetic reasons, the tower exterior is all glass, except for the narrow aluminum framing that holds the glass in place. With the sky and clouds reflected from the glass, the effect of the facade is dramatic. In addition, the large windows daylight the interior and offer the occupants pleasant views of the outdoors.

The windows in each story are story high. To prevent such large windows from imposing an excessive heat load on the HVAC system, several precautions were taken. For one thing, the windows were made unopenable. Hence, air leakage is small. Also, the indoor environment is not subject to unbalanced air distribution that would result when windows are opened or closed at random and therefore is easier to control. (Another advantage of the fixed sash is that it facilitates cleaning of the exterior with automatic equipment.) In addition, the windows have insulating glazing. They are double glazed and consist of two $\frac{1}{4}$-in. panes separated by a $\frac{1}{2}$-in. air gap. Also, the outer pane is transparent mirror glass. As a result, the pane reflects solar rays, while occupants of the building can see through the glass.

The spandrel wall between the windows of each story is made of opaque mirror glass. It is backed by compressed mineral-fiber insulation. The wall has an overall heat-transmission coefficient U of 0.2.

To insure that the window wall would have adequate wind resistance, a model of the building was tested in a wind tunnel to determine the pressures that might be produced by the strongest wind that might occur locally in the next 100 years (95 mph). Performance tests, in addition, were made on a prototype section of wall. The tests indicated that it would be able to withstand pressures up to those that would be produced by a 180-mph wind, which is nearly twice as fast as the 100-year wind.

Another step the designers took to reduce the heat load imposed by the glass exterior on the HVAC system was to make the building square in plan. Of all the rectangular shapes, a square has the smallest perimeter for a specific enclosed area. For the sake of appearance, however, the corners of the building were beveled.

The center of the building is occupied for its full height by a concrete-enclosed core. It houses such services as elevators, stairs, toilets and utilities. The core walls also serve as load-bearing walls and as shear walls that resist the wind pressures on the building. They support steel beams that are supported at their outer ends on steel columns along the exterior walls. As a result, office spaces are unobstructed by columns, and partitions can be placed wherever desired along boundaries of modules.

The offices are planned with a 5-ft-square module. Each module provides fluorescent lighting and HVAC, electrical and telephone services.

Heat Recovery System. Heating and cooling needs of the building are met through the use of three 600-ton centrifugal chillers in the basement. Capable of supplying hot and cold water simultaneously, they operate as heat pumps when heating is required. They transmit the heat ejected in cooling interior building spaces to the building exterior where heat is needed. Supplementary heat, when necessary, is supplied by two 2,000-kw electric boilers.

For HVAC, the building is divided into interior and peripheral zones. Each zone is served by a separate and different single-duct system.

Air is supplied all year round at 55°F to the interior zones, which require cooling even in winter. Two reheat coils in the ducts in each story temper the air (adjust the air temperature) to provide comfort conditions. For this purpose, the heat pumps or the boilers supply hot water to the reheat coils.

Heating and cooling of the peripheral zones are provided mainly by a two-pipe fan-coil system. It supplies hot or cold water, as needed, to fan-coil terminals set at 10-ft intervals at the base of the windows. For ventilation, supply air is distributed to the peripheral zones through ceiling diffusers. The temperature of this air is varied in accordance with outdoor temperatures by heating and cooling coils in the main air-handling units.

The HVAC system is monitored and controlled by a high-speed electronic computer. It optimizes operation of the system based on data received from numerous points throughout the building. In particular, it can activate certain economizer features incorporated in the HVAC system.

One such feature enables the HVAC system to handle low cooling loads without operating the compressors. This is achieved by the inclusion of refrigerant spray pumps in the refrigerant lines of two of the chillers. These refrigerant pumps and pipes allow the gas to bypass the tight clearances of the compressor rotors when they are not rotating. When the pumps operate, liquid refrigerant is drawn from the chillers and returned as a spray over the chiller coils in much the same way as if the compressors were operating (see Fig. 17.7).

This feature is useful when there is only a light load on the compressors and the outdoor wet-bulb temperatures are 5 to 10°F lower than the 55°F chilled-water design temperature. With only a small load, the evaporative cooling tower can cool condenser water to within a few degrees of the outdoor wet-bulb temperature. This cool water, transmitted to the condenser, liquefies the refrigerant, which then flows by gravity to the spray pumps. With the compressors not operating, hot refrigerant gas from the chillers bypasses the compressors and flows directly to the condensers, to complete the cycle, while the refrigerant spray pumps circulate liquid refrigerant to and from the chillers. With this arrangement, 25% of the rated cooling capacity of the chillers can be attained without operation of the compressors, thus reducing operating costs.

Appraisal

Glass facades offer the benefits of an attractive building exterior and excellent views of the outdoors from the building interior. But HVAC costs are high when large windows are used. These costs, however, can be reduced by installation of fixed windows glazed with insulating glass and by insulating the spandrel walls above and below the windows.

In northern climates, where peripheral zones of buildings require heat most of the year, HVAC costs can be further reduced by recovering heat from interior zones and using it to warm the peripheral zones. In southern climates, it is advisable to reduce the solar heat load on windows by shading them with overhangs and vertical fins. In all climates, installation of blinds is desirable to enable occupants to reduce the penetration of the solar rays yet admit some daylight.

17.7. CASE-STUDY FIVE: AN OFFICE BUILDING ON A TIGHT SITE

Location: Boston, Massachusetts.
Architects: The Architect's Collaborative.

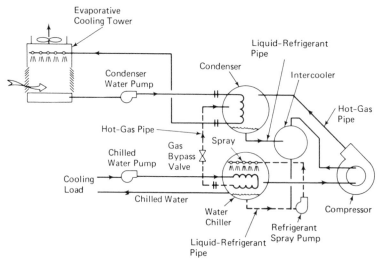

Fig. 17.7. System for cooling without operation of compressors.

Structural Engineer: LeMessurier Associates/ SCI.

Mechanical Engineer: Consentini Associates.

The purpose of this section is to show how a 17-story office building was designed for a site congested with underground utilities and hemmed in by rapid-transit tunnels. The designers also had to cope with relatively small money and energy budgets.

The building is the Fiduciary Trust Building in Boston, Mass. The building team included the owners, a joint venture of The Beacon Companies and Carpenter & Company and the general contractor, Beacon Construction Co.

The design goal was to provide, for a specific site in downtown Boston, a building with about 200,000 sq ft of rental office area and with low life-cycle costs. Numerous objectives and constraints had to be satisfied to meet this goal. Only the few pertinent to the purpose of this section are presented in the following.

The owners wanted to provide quality office space that would attract tenants to the new building in a renewal area of downtown Boston. Nevertheless, the owners also specified that the building should have low life-cycle costs. In addition, the owners established a tight target of 28 kwhr per sq ft of floor area per year as maximum energy use for all purposes.

Investigation of the site revealed conditions that would impose severe constraints on building design. The site is crisscrossed underground by utility lines, some nearly 200 years old. Probers discovered old brick water and sewer conduits and numerous service lines for gas, steam, telegraph and electrical power. Site investigations also disclosed that a rapid-transit tunnel ran along one side of the property and that a utility tunnel for the subway skirted a second side. The building foundations consequently had to be constructed to avoid the underground maze.

The designers also had to contend with constraints imposed by recently adopted fire-safety rules in the Boston building code. These include requirements for installation of fire sprinklers and smoke detectors that act with control circuits for HVAC and elevators.

Foundations for the building consist of 14 concrete caissons. This type of construction was chosen because the caissons could be placed to avoid the underground utilities. (Piles could have been driven, but the hammering and consequent vibration of the soil might have damaged the piping or the tunnels.) Concrete for the caissons was cast in steel-lined holes that were drilled with augers. Some of the caissons are $8\frac{1}{2}$ ft in diameter and penetrate to a maximum depth of 50 ft, where they are belled to a diameter of $14\frac{1}{2}$ ft.

To avoid the utilities, however, the caissons had to be placed several feet in from the boundaries of the lot. In fact, eight of the large caissons had to be set in a relatively small area around the middle of the parcel. They support a high percentage of the building load, including the service core, which houses elevators and stairs. Six smaller caissons in an outer ring help carry a ground-level lobby and a basement (see Fig. 17.8).

The large caissons in the middle of the lot support structural steel columns 56 ft high. The columns extend through the lobby to the first office floor, which is five stories above the street. Because the columns, like the caissons on which they are seated, carry most of the building load, they are very large structural members. Box-shaped in cross section, they were fabricated by welding together rolled sections and steel plates up to 8 in. thick.

Tower on a Pedestal. For better utilization of the lot, the building projects well beyond its

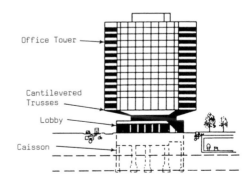

Fig. 17.8. Vertical section through Fiduciary Trust Building.

supports in the middle of the parcel (see Fig. 17.8). To support the projections as well as the center of the building, cantilever trusses are seated on the core columns. The trusses have to cantilever as much as 40 ft out from the columns and have a total length of up to 158 ft. To provide adequate stiffness, the trusses were made 30 ft deep. For additional support for the building, smaller trusses extend perpendicular to the cantilever trusses along column lines. The first office floor is seated on the top chords of the trusses. The floors above and the roof are supported by structural steel framing, which is carried by the trusses.

The locations of the core columns and caissons and the irregular shape of the lot were major factors in the determination of the shape of the building. In plan, it is an irregular hexagon, having no sides of the same length. To meet the office-area requirements of the owners within the hexagon, twelve stories of office space are provided in a tower without setbacks. With the lobby at ground level and the office tower commencing five stories above ground level both extending well beyond the central core, the building when viewed from street level resembles an enormous vase with pedestal base (see Fig. 17.8).

Energy Conservation. The exterior of the office tower is interrupted at regular intervals by continuous horizontal bands of windows. To conserve energy, the windows are double glazed with heat-absorbing, tinted glass. About 20% of the exterior consists of insulation-backed opaque glass and about 45% of the exterior, of precast-concrete panels.

Also to conserve energy, the office areas are illuminated with 35-watt, 4-ft-long fluorescent lamps, instead of the conventional 40-watt lamps of the same length. The rated light output of the 35-watt lamps, however, is less than that of the 40-watt lamps. To compensate for this, the 35-watt lamps are kept cooler than usual, as a result of which their light output is increased 10% above the rated capacity. Cooling of the lamps is accomplished by passing over them return air from the air-conditioned offices. This arrangement has the additional benefit of reducing fan energy requirements,

because heat is removed from the lighting fixtures before it reaches the offices.

For air conditioning, the building is divided into interior and peripheral zones. A single-duct system serves the interior zones, which require year-round cooling during working hours. A four-pipe fan-coil system handles the heating and cooling loads of the peripheral zones, with some ducted air provided for ventilation.

Chillers and other mechanical equipment are located on the fourth floor of the building. To keep duct runs short and fan power requirements small, however, an individually controlled air handler is placed on each office floor. Use of many individual units instead of a single large air handler on the mechanical floor has the additional advantage of providing more uniform air distribution. Also, use of individual units makes it possible to reduce or cut off conditioned air to unoccupied floors while conditioned air is supplied to offices that are occupied after normal working hours or on weekends.

Two chillers were installed to meet cooling needs and to recover heat from interior zones for heating the peripheral zones in cold weather. One chiller with a capacity of 300 tons supplies only chilled water. The second chiller with a 275-ton capacity can supply both chilled and hot water simultaneously. For this purpose, it is equipped with a double-bundled condenser (see Sec. 17.5). In cold weather, this chiller operates as a heat pump, transmitting heat ejected in cooling interior zones to the peripheral fan-coil terminals.

Additionally, to conserve energy, the HVAC system recirculates 85% of the conditioned air. Yet, ventilation air supplied is considerably above recommended minimums.

Fire Safety. Normal automatic operation of the HVAC system is changed, however, when a smoke detector is activated. Controls immediately cut off the flow of conditioned air to the story in which the activated detector is located, to deprive the fire of oxygen. They also close return-air dampers to prevent smoke from traveling through the ducts to other parts of the building. In addition, the controls start up blowers to exhaust smoke from the building, especially from stairwells.

Activation of the smoke detector transmits an alarm to the city fire department. Also, elevator hoistway doors for the story in which the detector is located are locked, to prevent heat and smoke from entering an elevator car if it should stop at that floor. In addition, building management is alerted to issue evacuation instructions.

Appraisal

Many of the building's features were determined by the shape of the lot and subterranean conditions. Use of caissons located well within lot boundaries led to cantilever construction, for which light weight was desirable, and the pedestal shape of the building. To keep weight small, steel skeleton framing and curtain walls were advantageous.

Design of other features of the building was governed by energy conservation requirements. These led to selection of lamps with lower-than-usual wattage and an HVAC system with heat-recovery capabilities.

17.8. CASE-STUDY SIX: OFFICE BUILDING FOR PRUDENTIAL INSURANCE COMPANY

Location: Thousand Oaks,California, 35 miles west of central Los Angeles.
Architects: Albert C. Martin, Los Angeles.
Design Period: Mid to late 1970s.

This large single-tenant office building was designed to be used as a major center for Prudential's west coast operations, including insurance, real estate and other activities. These activities had been housed in various locations in the Los Angeles area, including a high-rise building in the west San Fernando Valley.

The site was in a new town development, for which Prudential had been a principal financial backer and retained a substantial investment in developed and undeveloped properties. For the benefit of the community, as well as protection of its own investments, the company did not want to make an overwhelming presence by building the necessary high-rise building (16 to 20 stories) that would be required for a single

structure. They therefore decided to build a low-rise building and to further reduce its apparent size by use of ground contours and plantings, so that it would have a very modest overall presence, (see Fig. 17.9).

In some ways, the design goals for the general planning of this building were similar to those for the AT & T building (Sec. 17.5). However, while the AT & T building had a very large, semi-rural site, this building was on a relatively small site in a highly visible location in the community. Thus the views of the building from a number of points had to be considered. In addition, the local zoning (which Prudential had helped to legislate) required minimal changes in existing land conditions and a strict control of the preservation of any existing oak trees—the numerous, hundreds-of-years-old craggy remnants from which the community derived its name.

What the building team produced is a gentle giant of a building, occupying a ravine that was enhanced by earth berms and additional planting to mask the long horizontal spread of the building. The external visible impression is primarily of a modest size, three-story building, visible only in selected, partial glimpses that do not permit a sense of the total amount of the structure. Use of brown granite, bronze-tinted glass, a horizontal strip form, and a sloped-back exterior surface further assist in blending the building into the landscape (see Fig. 17.10).

Because of the timing of the design work, major design criteria included concerns for energy conservatrion and use of daylight for interior lighting. In the southwestern, coastal climate of the Los Angeles area, these factors combine for a complex pro and con relationship with the sun—making it a source for useful light (particularly for daytime office occupancy) and a hostile source of solar heat gain for a major portion of the year.

The design period also included the development of major interest in a shift from dependency on equipment and an emphasis on use of the natural attributes (passive resistances) of the building to help achieve interior environmental control. This added major concerns for the form, detail, sun exposure and window loca-

(a)

(b)

(c)

Fig. 17.9. Views of Prudential Building (now General Telephone offices). (a) From adjacent freeway, showing the only clear, generally nonobstructed view. (b) and (c) From nearby adjacent streets. (d) From nearby, unoccupied hilltop—the only way to see the extent of the building from a nearby location. (e) View in (a) with dashed-line profile indicating probable size of an equivalent high-rise building with comparable floor area.

(d)

(e)

Fig. 17.9. (*Continued*)

tions to all of the other concerns for appearance and occupant needs.

Because the large, low-rise building was the equivalent of a 20-story building laid on its side, a major problem was that of horizontal traffic of people and materials. In a high-rise building, no person is ever very far from an elevator; thus going to see someone or to deliver something to another floor is quite efficient. For the horizontally extensive building this is another matter; the more so with a single tenant. Major studies were made of plan layouts and occupant groupings in the interest of reducing the need for people to move great distances in the normal course of their work day. Components of this effort included use of communication by computer terminals and use of material-moving facilities that ranged from so-

phisticated, high-tech robots to low-tech roller-skated messengers. Proposed building plans were tested for man-hour conservation using various typical work scenarios for major office operations.

The ability to deal with all of these interactive design issues relating to the building form and exterior was made more feasible because the design firm was both well equipped and generally experienced in the use of computers for design support. All designers in the office—including architects and engineers—were equiped with computer terminals and were trained in their use. Design-development work was shared almost instantly by inputs to the central data-storage system. Implications of design changes by any individual could be evaluated in terms of current general design devel-

Fig. 17.10. Detail view of Prudential building, showing general form of building exterior. Spandrels are covered with panels of dark-brown granite and glazing is brown-tinted glass, further achieving visual harmony with the landscape.

opment in all areas of the work. Value-based testing of alternates could be done as fast as they could be described. While this coordination of design work could be done without the computer (and largely was until recently), it is unlikely that it would have proceeded as quickly or smoothly, have involved as few people, or allowed the serious study of as wide a range of alternate possibilities in all the areas of design.

Shortly after Prudential took occupancy of the building, a decision was made by the company to centralize administration in the main headquarters in the midwest and to reduce the extent of administrative work in the regional offices. (It is possible that some of the studies done to show the feasibility of work in the long, horizontal building may have fed this decision— showing that computer communications could allow for effective operations with a major reduction of dependency on person-to-person meetings.) The building was then sold to General Telephone, who was able to make use of the facility with only minor alterations.

This was a case where a major amount of design study was funded by the client because of a special desire to create the unobtrusive (but not necessarily unnoticeable or unassuming) building. That the designers were able to achieve this, while also satisfying other complex design goals, is a tribute to well-organized design work.

17.9. CASE-STUDY SEVEN: ROWES WHARF HARBOR REDEVELOPMENT PROJECT

Location: Boston, MA, downtown harbor-front site.
Architects and Engineers: Skidmore, Owings, & Merrill, Chicago.
Design Period: Early to mid 1980s.

The design for this project was the winning solution for a competition sponsored by the Boston Redevelopment Authority (BRA) in 1982. The program for the project called for a mixed-use development (commercial, residential, retail and maritime), emphasis on a major link between the harbor and the adjacent Boston financial district, facilitation of a ferry landing in operation, use of architectural forms and materials to reflect the character of older buildings in the area, and building-height-zoning stepping down to the water's edge. A view of the harbor was also required. This necessitated splitting the site at ground level.

As completed, the building complex contains:

1. The ferry terminal with access through the south edge of the site.
2. A 50-slip marina along the water's edge promenade.
3. A 700-car, 4-story, underground parking garage.
4. A public observatory.
5. A 230-room hotel.
6. 100 condominium apartments.
7. 320,000 sq ft of office space.
8. 24,000 sq ft of retail space and restaurants.
9. A 15,000-sq-ft health club.

Incorporation of all of this on the 5-acre site, squeezed by the harbor on one side, a major elevated highway on the opposite side and existing buildings on the other two sides, presented a monumental design problem in terms of management of pedestrian and vehicular traffic. A major problem was the control of ground-level traffic. However, the separation of access and the control of internal traffic were also of concern—recognizing the needs of the offices, stores, restaurants, hotel, health club, apartments and parking.

Design schemes for the site and the individual buildings, as well as for interior space distribution between mixed uses in the buildings, were tested against a number of scenarios that incorporated the usage by the various occu-

pants. Time-of-day traffic peaks for the various occupants were superimposed to pinpoint bottlenecks and point out issues of security, elevator use, exit pathways, and ease of circulation. All this, plus the special requirements of the BRA and the zoning controls for the harbor-edge site, had to be added to typical concerns for subsystem integration in any large building.

A continuing dialogue concerns the advantages of an architectural firm that uses outside consultants for its non-architectural work (mainly engineering) versus the complete (full-service) design firm that has a full staff of in-house consultants. Skidmore, Owings, & Merrill's (SOM's) Chicago office has always been the latter type, with substantial in-house personnel in all the main branches of design: structures, electrical, mechanical, civil and interiors. For systems-design operations that incorporate the inputs of all of the design participants, the full-service firm has the advantage of a fully integrated design team, deliberatley organized for such cooperative efforts on a routine operational basis. The SOM office, being organized especially to do this complex design task, was prepared for the task.

At street level the building complex presents four separate, closely spaced structures, two of which are extended into the harbor on finger wharves. (See ground-level plan in Fig. 17-11.) The two portions of the complex parallel to the water's edge are separated by a tall archway that provides both access and view through the center of the site. Above the archway, these two buildings join, with the dome-roofed observatory capping the center of the building above the archway. The building ends, then continue up as short towers, with spaces for mechanical equipment on their tops. (See section in Fig. 17–11.)

In addition to the multiple design constraints, there was a special condition related to the harbor-edge site. The desire for a five-story structure below grade to house the health club and the parking garage, produced a major problem regarding hydrostatic uplift on the subgrade structure, especially at high tide. The solution for this involved the use of an unusual construction process called up-and-down construc-

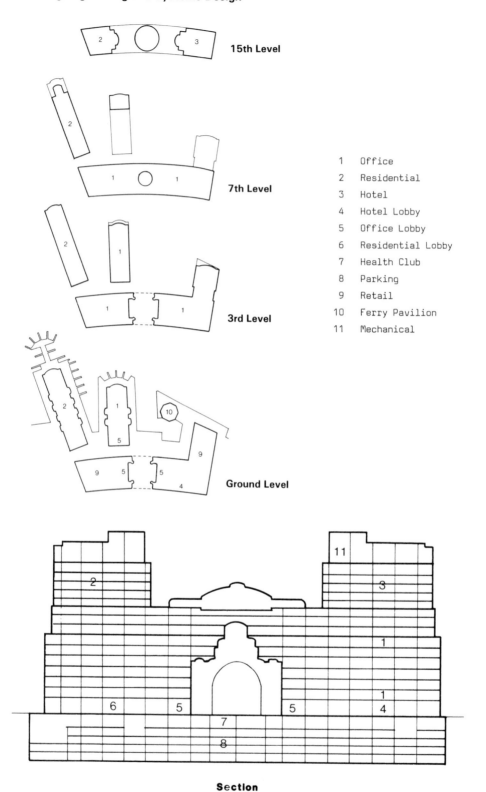

1	Office
2	Residential
3	Hotel
4	Hotel Lobby
5	Office Lobby
6	Residential Lobby
7	Health Club
8	Parking
9	Retail
10	Ferry Pavilion
11	Mechanical

15th Level

7th Level

3rd Level

Ground Level

Section

Fig. 17.11. Rowes Wharf Harbor Redevelopment, Boston. General form of the building complex.

tion whereby the subgrade excavation and construction proceeds simultaneously with the erection of the building superstructure. Thus the accumulation of the dead weight of the superstructure is used to counteract the buoyancy effect on the subgrade structure. The geotechnical and foundation-construction problems for this work were handled with the help of consultations from Haley & Adrich, Geotechnical Engineers, and the contractor for the subgrade work, Perini Corp. However, the up-and-down process also calls for more than the usual co-

ordination between the designs of the subgrade and superstructure portions of the building. Thus yet another coordination input was given to the building designers and another construction scheduling problem was given to the builders.

Reference

"Harboring Tradition," in the *Architectural Record*, March, 1988, pp. 87–93.

Index